THE CORRESPONDENCE OF
CHARLES DARWIN

Editors

FREDERICK BURKHARDT DUNCAN M. PORTER

JANET BROWNE MARSHA RICHMOND

Production Director

PETER SAUNDERS†

Assistant Editors

SARAH BENTON CHARLOTTE BOWMAN

HEIDI BRADSHAW PERRY O'DONOVAN

Research Associates

ANNE SCHLABACH BURKHARDT

NORA CARROLL STEVENSON

Office Manager

HEDY FRANKS

This edition of the Correspondence of Charles Darwin is sponsored by the American Council of Learned Societies. Its preparation is made possible by the co-operation of the Cambridge University Library and the American Philosophical Society.

Support for editing has been received from the National Endowment for the Humanities, the National Science Foundation, the Andrew W. Mellon Foundation, the Alfred P. Sloan Foundation, the Royal Society of London, the British Academy, the Pilgrim Trust, and the Stifterverband für die Deutsche Wissenschaft. The National Endowment's grants (Nos. RE-23166-75-513, RE-27067-77-1359, RE-00082-80-1628, RE-20166-82, RE-20480-85, RE-20764-89, and RE-20913-91) were from its Program for Editions; the National Science Foundation's funding of the work was under grants Nos. SOC-75-15840, SES-7912492, SES-8517189, and DIR-9020874. Any opinions, findings, conclusions, or recommendations expressed in this publication are those of the editors and do not necessarily reflect the views of the grantors.

Charles Darwin. Photograph by Maull & Fox, *c*.1857.
(DAR 225: 175. By permission of the Syndics of Cambridge University Library.)

In Memoriam

PETER SAUNDERS

(1966–1992)

THE CORRESPONDENCE OF
CHARLES DARWIN

VOLUME 8 1860

CAMBRIDGE
UNIVERSITY PRESS

Published by the Press Syndicate of the University of Cambridge
The Pitt Building, Trumpington Street, Cambridge CB2 1RP
40 West 20th Street, New York, NY10011, USA
10 Stamford Road, Oakleigh, Melbourne 3166, Australia

First published 1993

Printed in Great Britain at the University Press, Cambridge

British Library cataloguing in publication data

Darwin, Charles *1809–1882*
The correspondence of Charles Darwin.
Vol. 8: 1860
1. Organisms. Evolution. Darwin, Charles, 1809–1882
Correspondence, diaries, etc.
I. Title II. Burkhardt, Frederick H. *1912–*
575′.0092′4 QH31.D2

Library of Congress cataloguing in publication data

Darwin, Charles, 1809–1882.
The correspondence of Charles Darwin.
1. Darwin, Charles, 1809–1882.
2. Naturalists—England—Correspondence.
I. Burkhardt, Frederick, 1912–
II. Title.
QH31, D2A33 1985 575′.0092′4[B] 84–45347

ISBN 0 521 44241 9 hardback

CONTENTS

ILLUSTRATIONS

LIST OF LETTERS

The following list is in the order of the entries in the *Calendar of the correspondence of Charles Darwin*. It includes all those letters that are listed in the *Calendar* within the date range covered by this volume of the *Correspondence*. Some letters have been redated since the publication of the *Calendar*, so this list is necessary to enable users of the *Calendar* to locate such letters in the *Correspondence*. Alongside the *Calendar* numbers are the current dates ascribed to every item. In the list, a date printed in italic type indicates that the item has been omitted from the volume entirely.

Letters included in this volume that were not available for inclusion in the *Calendar* appear at the end of the list. These letters will be listed in a forthcoming 'Supplement' to a new edition of the *Calendar* (Cambridge University Press, in preparation). The numbers given to the new letters are those assigned in the *Calendar* Supplement: they correspond to the chronological ordering of the original listing with the addition of an alphabetical marker.

1571. 17 June [1860]
2196. [1860?]
2197. [1860?]
2389. [January? 1860]
2391. [after 5 June 1860]
2394. [after 28 April 1860?]
2395. [April 1860]
2468. 5 June [1860]
2563. [17 January 1860]
2564. 1 December [1860]
2609. 27 December [1860]
2616. 30 [January 1860]
2617. 31 December [1860]
2618. *[1856–7]*. *See* Supplement to vol. 7.
2619. Second CD note of 2980.
2620. [12 November 1860]
2621. [November? 1860]
2622. [after 7 April 1860]
2623. [15–16 October 1860]
2624. [1860?]
2625. *Wide date range.*
2626. *Wide date range.*
2627. [December? 1860]
2628. *[c. 28 February 1861]*
2629. *Wide date range.*
2630. 9 [July 1860]
2631. [10 January 1860]
2632. [25 January 1860]
2633. 1 January [1860]
2634. 2 January 1860
2635. 3 January [1860]
2636. [3? January 1860]
2637. 4 [January 1860]
2638. 5 January 1860
2639. [5–11 January 1860]
2640. 6 January 1860
2641. 6 January [1860]
2642. 6 January 1860
2643. [October 1860 or after]
2644. 7 January [1860]
2645. 7 January [1860]
2646. 9 January [1860]
2647. 10 January [1860]
2648. *10 January [1859]*
2649. 11 January [1860]
2650. 14 January [1860]
2651. 14 [January 1860]
2652. 15 January [1860]

INTRODUCTION

On 7 January 1860, John Murray published the second edition of Darwin's *Origin of species*, printing off another 3000 copies to satisfy the demands of an audience that surprised both the publisher and the author. One week later Darwin was stunned to learn that the book was on sale even in railway stations (letter to Charles Lyell, 14 January [1860]). By May, with the work continuing to sell well in England and with editions out in the United States and in Germany, he expressed to Asa Gray his astonishment at the widespread interest it had aroused: 'No doubt the public has been *shamefully* imposed on! for they bought the book, thinking that it would be nice easy reading.' (letter to Asa Gray, 22 May [1860]).

But it was the opinion of scientific men that was Darwin's main concern. He eagerly scrutinised each new review and was heartened to find that many of the early notices were favourable. He later learned that they came from some of those whose support he most wanted: Thomas Henry Huxley, William Benjamin Carpenter, and Joseph Dalton Hooker. Others were not quite as supportive but were nonetheless appreciated for their honest critiques of his views. 'One cannot expect fairness in a Reviewer', Darwin commented to Hooker after reading an early notice that gave 'good and well deserved raps' on his discussion of the geological record; but this criticism, he told Hooker, did not at all concern his main argument (letter to J. D. Hooker, 3 January [1860]).

Darwin's magnanimous attitude soon faded, however, when 'the stones began to fly'. His 'dearly beloved' theory suffered a series of attacks, the most vicious of which came from Richard Owen in the April issue of the *Edinburgh Review*. Indeed, after reading not only critical but 'unfair' reviews that misrepresented his ideas, Darwin began to feel that without the early favourable notices his theory would have been '**utterly** smashed' (letter to T. H. Huxley, 3 July [1860]). (A chronological list of all the reviews mentioned in the volume is given in Appendix VII.)

The difficulties that members of the scientific community found in *Origin* in the twelve months after its publication make an impressive list. Adam Sedgwick, not surprisingly, attacked the book on a number of fronts. But it was his methodological criticism in the accusation that Darwin had 'deserted the inductive track, the only track that leads to physical truth' (Sedgwick 1860) that most wounded Darwin. Having spent years gathering the evidential underpinning for his theory, he had hoped to deflect such criticism. 'I can perfectly understand Sedgwick or any one saying that nat. selection does not explain large classes of facts; but that is very different from saying that I depart from right principles of scientific investigation.—' (letter to J. S. Henslow, 8 May [1860]).

Above all else Darwin prided himself on having developed a theory that explained several classes of facts—those of geological succession, geographical distribution, classification, homology, and embryology—which were inexplicable by the theory of creation. Asa Gray's statement in his March review that natural selection was a hypothesis, not a theory, therefore also displeased Darwin. Comparing natural selection to the undulatory theory of light or to the theory of gravity, he retorted: 'It seems to me that an hypothesis is *developed* into a theory solely by explaining an ample lot of facts.' (letter to Asa Gray, 18 February [1860]). To those who objected that his theory could not be a *vera causa,* he similarly stated that 'it seems to me fair in Philosophy to invent *any* hypothesis & if it explains many phenomena it comes in time to be admitted as real.' (letter to C. J. F. Bunbury, 9 February [1860]). This helps to explain why Darwin was delighted by the defence of his scientific method by the young Cambridge fellow Henry Fawcett in the December issue of *Macmillan's Magazine.* Fawcett asserted that Darwin's theory accorded well with John Stuart Mill's exposition of the deductive method of scientific investigation, consisting of direct induction, ratiocination, and then verification.

Specific difficulties were raised against the theory on the basis of existing scientific evidence. Several correspondents, such as his cousin Hensleigh Wedgwood and Heinrich Georg Bronn, expressed their concerns about the question of the origin of life itself, which the theory did not address. Darwin chose to treat this as an entirely separate problem from the origin of species, believing that current knowledge could not illuminate this 'mystery'. Charles Lyell worried, among other things, about the multitude of still living simple forms. Darwin readily admitted that his failure to discuss this point was a 'most serious omission' in his book and explained how natural selection did not necessarily lead to progression (letter to Charles Lyell, 18 [and 19 February 1860]). To this and Lyell's many other queries he responded carefully and patiently, knowing that Lyell was earnestly attempting to understand natural selection and incorporate it into his method of reasoning about global change. Darwin also knew that Lyell was a powerful potential ally. Indeed, the letters between Darwin and Lyell are some of the most fascinating in the volume.

George Henry Kendrick Thwaites questioned Darwin about how natural selection could explain the production of beauty, pointing to organisms such as the desmids, whose intricate patterned forms seemed to offer no obvious selective advantage to the creatures themselves. The evolutionist Hewett Cottrell Watson argued for the concept of 'convergence' of species to join that of divergence. Andrew Murray challenged the explanation of the origin and distribution of blind cave animals. Darwin attempted to answer each of these issues in his letters; several were considered in future editions of *Origin.* William Henry Harvey wondered, in addition to questioning gradual versus saltatory species change, how natural selection could ever alter and improve various 'simple' protozoans that were seemingly identical to one another. Harvey's letters reveal aspects of Darwin's theory that gave contemporary naturalists the greatest conceptual difficulty, and theological discomfort. After several long letters were exchanged, Darwin finally decided that Harvey and other working naturalists simply did not yet understand the concept of natural selection.

Even Huxley, an avowed supporter, proved a formidable critic. Huxley extolled the analogy between artificial selection among domestic varieties and natural selection in a lecture before the Royal Institution. Yet he also noted the difficulty that varieties are not sterile when crossed, whereas sterility had long been recognised by naturalists as a criterion of specific difference. He concluded the lecture by asserting that until naturalists could find 'derivatives from a common stock, whose offspring should be infertile, *inter se*,' Darwin's theory would remain unproven (T. H. Huxley 1860a).

Darwin had long reflected on the origin of sterility between incipient species but knew that much work was needed before the problem could be conclusively resolved. He also recognised that, owing to experimental difficulties, advances were more likely to come from studies of crossing among plant species and varieties than from animal breeding. With Lyell also questioning how interbreeding among animal groups could give rise to new species, Darwin found Huxley's lecture irritating and ultimately considered it more a failure than a success (see letter to J. D. Hooker, 14 February [1860]).

As Darwin himself well recognised and fully anticipated in *Origin*, several major criticisms arose from evidence (or the lack thereof) in the geological record. Several critics pointed out that the earliest geological formations already contained relatively advanced forms of life. Many singled out Darwin's own discussion of the absence of transitional or intermediate forms between related groups and the vast time required for one or more primordial forms to have evolved into the multitude of the earth's present inhabitants. Darwin agreed, for example, with Alfred Russel Wallace's assessment that the imperfection of the geological record was one of the weakest parts of his theory. Yet he nonetheless noted that in fact 'there are almost more Geological converts than of pursuers of other branches of natural science.' As for why this should be so, he confided to Wallace: 'I think geologists are more converted than simple naturalists because more accustomed to reasoning.' (letter to A. R. Wallace, 18 May 1860).

Darwin began to tabulate (and categorise) his various followers as early as March. Dividing up his supporters on the basis of their professional affiliations, he counted among this number four geologists, four zoologists or palaeontologists, two physiologists, and five botanists (see letter to J. D. Hooker, 3 March [1860]). Others, like François Jules Pictet de la Rive, he judged would eventually be converted, for he found them somewhat 'staggered' by his theory—and once staggered, he believed, it was only a matter of time before a person would become an adherent.

Certainly Darwin was disappointed by the small number of physiologists who initially supported his theory. Even Carpenter, whom he included as a proponent in this group, offered only partial support, for he hesitated at going so far as to derive all vertebrates from a single progenitor. The major stumbling block for most anatomists and physiologists was the difficulty of conceiving of the selection of chance variations being able to produce such a marvellously perfected structure as the eye. As Darwin admitted to Lyell, Gray, and others, imagining how selection could account for highly adapted organs had sometimes given even him a 'cold shudder'. Yet it was more trifling structures, ones for which it was difficult to see clear selective advantages, that caused him greater discomfort. As he readily admitted to Gray: 'The sight of a feather in a peacock's tail, whenever I gaze at it, makes me sick!' (letter to Asa Gray, 3 April [1860]).

By the end of 1860, Darwin was disheartened that so few of his reviewers had noticed what he considered to be 'the strongest single class of facts in favour of change of form', namely those of embryology (letter to Asa Gray, 10 September [1860]). Only his theory, he believed, could explain the basis for the correlation of the facts of embryology, homology, morphology, and classification and thereby account, as the creationist view could not, for anomalous structures such as rudimentary organs and the general resemblance of embryos of the same class. But it was precisely the physiologists, steeped in a heavily developmental conception of nature, who had more difficulty reconciling the seemingly directed process of embryogenesis with a theory of change based upon the selection of chance variations. Few could imagine how such a process could actually alter the developmental patterns of species. Darwin took some comfort from news that the doyen of embryology, Karl Ernst von Baer, had expressed support for species change based on his study of the geographical distribution of species (see letter from T. H. Huxley, 6 August 1860). But Baer in fact eventually opposed Darwin's theory on the grounds that it failed to account for the purposiveness of embryonic development.

Among the formidable array of problems confronting Darwin's theory, perhaps most telling of all was that arising from the implications it had for human ancestry. Certainly this was a major difficulty standing in the way of Lyell's acceptance of the theory, as Darwin well knew. Even though he had not discussed this problem explicitly in *Origin*—only one sentence, he told Lyell, showed that he believed 'man is in same predicament with other animals' (letter to Charles Lyell, 10 January [1860])—he and others were well aware that the continuity of humankind with the animal kingdom was an obvious logical consequence of his theory. The question of man's 'descent from apes' was explicitly raised in February in Thomas Vernon Wollaston's review in the *Annals and Magazine of Natural History*. The question would not disappear with Darwin's dismissal of the issue as having more to do with theology than with science. It emerged with force and in full public view in June at the Oxford meeting of the British Association for the Advancement of Science.

Darwin first learned that he and his book had become 'topics of the day' at the meeting in a letter from Hooker written from Oxford. Hooker's letter, one of the few known eyewitness accounts, well captures the excitement of the legendary episode. After avoiding earlier sessions, including the Thursday meeting at which Huxley and Owen 'had a furious battle over Darwins absent body', Hooker attended the fabled Saturday session of Section D. He told Darwin how 'between 700 & 1000 people' crowded into the as yet unfurnished library of the new Oxford Museum to hear

Samuel Wilberforce, the bishop of Oxford, reply to John William Draper's paper giving a Darwinian view of the development of Western civilisation. Wilberforce, Hooker recounted, responded by shouting 'for half an hour', ridiculing Darwin 'badly & Huxley savagely'. Huxley rose in response and 'answered admirably', but 'could not throw his voice over so large an assembly, nor command the audience'. With his blood boiling and his heart pounding, Hooker threw down the gauntlet and became a 'referee on Natural Selection'. His performance surprised many: he 'smashed' Wilberforce 'amid rounds of aplause' and the meeting adjourned, leaving Darwin 'master of the field after 4 hours battle' (letter from J. D. Hooker, 2 July 1860). Other correspondents informed Darwin about further, less dramatic incidents, including John Lubbock's retort to Wilberforce on the embryological evidence supporting Darwin's views and Robert FitzRoy's denunciations in both the Saturday meeting and in the geography section. Darwin, who had been kept away by illness, was glad to have avoided such public confrontation. 'I would as soon have died as tried to answer the Bishop in such an assembly,' he told Huxley; nonetheless he believed that 'this row is best thing for subject.—' (letter to T. H. Huxley, 3 July [1860]). Further details of the meeting, taken from the contemporary report in the *Athenæum*, are given in Appendix VI.

Wilberforce's review of *Origin*, published in the *Quarterly Review*, appeared shortly after the Oxford meeting. In its arguments Darwin clearly recognised the clever touch of his now arch-foe Owen. It specifically addressed the question of man, among other difficult scientific problems, and set the tone for future theological controversy over evolution theory.

The response to *Origin* by theologians, however, was not entirely negative, as the approval of the scientifically literate clergymen Baden Powell and Charles Kingsley attested. Moreover, theological opinion has to be gauged as much in terms of the response to the publication of the theological reform tract *Essays and reviews* in January 1860 as to that of *Origin* itself. This volume of seven essays, whose authors (six of whom were clergymen) were liberal critics of prevailing Church doctrine, was more inspired by the German movement of historico-literary criticism of biblical texts than it was by new developments in science. But the correspondence shows that the two separate attacks on orthodoxy were related. Sedgwick, for example, stated publicly at a meeting of the Cambridge Philosophical Society in May that 'his chief attacks were directed against Powell's late Essay' in *Essays and reviews* rather than against Darwin's book per se. Prodded by Henslow's defence of the integrity of Darwin's scientific motivation, Sedgwick admitted that he was distressed that Powell had, in his opinion, 'accepted all Darwin had suggested, & applied these suggestions (as if the whole were already proved) to his own views.—' (letter from J. S. Henslow to J. D. Hooker, 10 May 1860). What worried Darwin most about such attacks was what the effect might be 'of so many heavy guns fired by great men', believing that they might retard the theory's gaining any new converts or even cause earlier proponents (like Lyell) to retract their support altogether (letters to Charles Lyell, 1 June [1860] and 11 August [1860]).

As the months passed by, Darwin read each review with less trepidation, commenting on each, and the private communications he received, in letters to his closest confidants Hooker, Lyell, and Gray. Initially he found it curious 'how differently different opposers view the subject' (letter to Charles Lyell, 15 February [1860]); later he became 'fairly sick' of the notices as he increasingly realised that one of the reasons for the paucity of novel objections to his views was that very few reviewers actually understood his theory. Somewhat exasperated after reading William Hopkins's hostile critique of his geological argument, he wrote to Lyell on 6 June [1860]:

> I am beginning to despair of ever making the majority understand my notions. Even Hopkins does not thoroughly . . . I must be a very bad explainer.

This was not, however, his opinion of Asa Gray, who Darwin thought understood his argument perhaps better than anyone else. Having been impressed by Gray's review in the *American Journal of Science and Arts*, Darwin was elated by his series of three (unsigned) articles published in the *Atlantic Monthly*. Although intended to counter the powerful opposition to *Origin* in America by Louis Agassiz and his followers, Gray's essays, Darwin believed, could also serve a purpose in Britain. He immediately wrote to Gray on 10 September after studying the first published piece:

> I said in a former letter that you were a Lawyer; but I made a gross mistake, I am sure that you are a poet. No by Jove I will tell you what you are, a hybrid, a complex cross of Lawyer, Poet, Naturalist, & Theologian!— Was there ever such a monster seen before?

Gray's essays attempted to show that the operation of natural selection was compatible with continued belief in Design in nature. Even though unable to believe in design himself, Darwin recognised that Gray's argument could be a powerful antidote to the theological attacks on his book that were rife in England. He immediately set about arranging for the publication of Gray's articles in Britain: the first article appeared in the *Annals and Magazine of Natural History*, and all three pieces were published together under Gray's name in a pamphlet (Gray 1861) with the publication costs shared equally by Darwin and Gray.

Darwin was not, however, entirely preoccupied in 1860 with his critics and the reception of *Origin*. This work was, after all, only an 'abstract' of his theory. His critics merely reinforced his agenda: their remarks led to detailed arguments in his subsequent works. Only two days after the second edition was issued, on 9 January 1860, he turned to preparing the first part of his planned three-volume work intended to buttress the compressed arguments of *Origin*. Many of the letters of 1860 pertain to his collection of further facts in support of his postulate of a vast amount of variation among plant and animal species on which natural selection could operate. He wrote to various correspondents in the hope of finding more cases of striping in dray and cart horses, of inheritance in fowls, of the intercrossing among sweet-pea plants, of variation in the nests of bees and wasps, and of a myriad of other phenomena that provided empirical support for his arguments. He recommenced writing his chapter on pigeons (interrupted in 1858 by the receipt of Wallace's manuscript and the subsequent drafting of *Origin*), and he wrote up anew a chapter discussing the origin of various breeds of dogs. By June it was clear that his materials on variation alone would fill one volume exclusively; in the event, they ran to two. The work drained him, yet he persevered, strongly believing in the importance of the task. It would provide evidence in support of his theory that would help to 'stagger' the new generation of naturalists, who Darwin felt were crucial to the ultimate triumph of his theory. 'I can pretty plainly see', he commented to Huxley on 2 December, 'that if my view is ever to be generally adopted, it will be by young men growing up & replacing the old workers, & these young ones finding that they can group facts & search out new lines of investigation better on the notion of descent, than on that of creation.—' The new book on which he laboured was intended to provide just that mass of documentary evidence by which this process could best be advanced.

But *Variation* was not the only 'evolutionary tract' on which Darwin worked in the months after publishing *Origin*. Turning to pursue a number of topics about which he had long been curious, he began a serious study of several different botanical problems. For more than twenty years, Darwin had observed the role of insects in the fertilisation of plants. In the spring and summer of 1860, he began to investigate the problem in detail, focusing on specific aspects of the structure of the flower parts of various species. He drew up a list of those species in which the pistil appeared to be bent toward the nectar-secreting surface of the flower. 'Why I care about it,' he confided in Hooker, 'is that it shows that visits of insects are so important, that these visits have led to changed structure.' (letter to J. D. Hooker, 27 April [1860]). Tracing the complicated contrivances by which insects, in getting at the nectar, inadvertently become transporters of pollen (and hence agents effecting cross-pollination between different plants) necessitated meticulous microscopic examination of the fructifying structures of many plant species. Orchids in particular came to his attention as being 'beautifully adapted' to insect visitation. The project engrossed Darwin, as he examined one genus after another. 'I am intensely interested on subject,' he told Hooker several months later, 'just as at a game of chess.' (letter to J. D. Hooker, 19 [July 1860]). With the work halted by the coming of winter, Darwin carried over the investigation to the next year and published the results of the orchid study in 1862.

Another botanical project was a study of primroses and cowslips, which Darwin had discussed in *Origin* in regard to whether the oxlip, intermediate in form between the two, could be considered as a hybrid offspring and hence an example of an incipient species. But it was the evolution of sex, not speciation, that captured his interest, just as it had in his earlier study of the cirripedes. A chance observation of two different forms of the flower parts of cowslips soon evolved into an extensive piece of botanical research. Keeping Hooker closely informed on the progress and results of his study, he wrote on 7 May [1860]:

> I have this morning been looking at my experimental Cowslips & find some plants have all flowers with long stamens & short pistils which I will call "male plants"—others with short stamens & long pistils, which I will call "female plants" . . . I cannot help suspecting that the cowslip is in fact dioicous—but it may turn out all blunder, but anyhow I will mark with sticks the so-called male & female plants & watch their seeding. It would be fine case of gradation between an hermaphrodite & unisexual condition.—

He presented the results of his study in a paper of 1862 and in *Cross and self fertilisation* (1876).

Serendipity led to the commencement of what was to become a fifteen-year-long investigation. During visits to Hartfield in July and Eastbourne in September and October, he amused himself by observing how the common sundew of the Sussex heaths, *Drosera rotundifolia*, was able to capture insects. A casual observation soon turned into a passionate investigation of one very curious point, soon revealed to Hooker: 'The leaves are first rate chemists & can distinguish even an incredibly small quantity of any nitrogenised substance from non=nitrogenised substances.' (letter to J. D. Hooker, 31 [August 1860]). Relying in part on his youthful experiences in chemical experimentation (see *Correspondence* vol. 1), and the tenaciousness exhibited in all his undertakings in natural history, he tested the sensitivity of various insectivorous plants to a large variety of substances, observing their digestive processes at the cellular level. Describing her husband's current enthusiasm, Emma Darwin wrote to Mary Lyell: 'At present he is treating Drosera just like a living creature, and I suppose he hopes to end in proving it to be an animal.' (*Emma Darwin* 2: 177). As was so frequently the case with his research topics, Darwin was able to solicit the assistance of a number of other specialists who also became deeply engrossed in his experimental regime. In this instance, he drafted the services not only of Hooker but also of a newly appointed Kew botanist, Daniel Oliver; his old friend and neighbour Edward Cresy; and Cresy's acquaintances August Wilhelm von Hofmann, the renowned German organic chemist, and Alfred Swaine Taylor, an authority on poisons and tests for chemical sensitivity. After finding 'one suspicious case' in his experimental record, however, Darwin abandoned his original intention to publish a short paper on the subject, fearing that his estimate of the astonishing sensitivity of the leaves of these plants to minute quantities of nitrogenous substances would scarcely be believed without further supporting evidence (letter to Edward Cresy, 12 December [1860]). This work was not published until 1875, when *Insectivorous plants* appeared.

These studies offered naturalists different perspectives of the marvellous adaptive mechanisms of plants; moreover, they showed how the design of such structures could be interpreted on the basis of the theory of natural selection. As the letters between Darwin, Hooker, and Oliver indicate, the novelty of this approach to botanical research was striking even to experts in that field who were favourably disposed to his views. In future years, his botanical work served as models for investigation in natural history, graphically illustrating the conceptual and methodological power of the theory. This is not to say that Darwin himself was conscious of attempting to initiate an evolutionary 'research programme'. Indeed, he was rather apologetic about having spent so much time on the *Drosera* study in particular, admitting to Lyell 'how much better fun observing is than writing.—' (letter to Charles Lyell, 12 September [1860]). Despite the fact that he cared 'more about Drosera than the origin of all the species in the world', he told Oliver he was going to put away his mass of notes, 'for I am convinced that I ought to work on Variation & not amuse myself with interludes.—' (letters to Charles Lyell, 24 November [1860], and to Daniel Oliver, 20 October [1860]).

In the event, however, Darwin's work on variation had to be interrupted for other reasons. His health was still a source of continued anxiety and annoyance, disturbing his work and causing him to miss the British Association meeting. But it was the health of his daughter Henrietta that caused the family the greatest concern. Having suffered intermittent bouts of illness for the past few years, Etty fell seriously ill late in April 1860 with what was eventually diagnosed as a form of typhus fever. With the memory of their older daughter Anne's fatal illness never far from their minds, Charles and Emma did whatever they could to promote Etty's recovery, even if it meant interrupting his precious studies.

His work was also halted abruptly late in November when Murray again called for a new edition of *Origin*. This gave Darwin the opportunity to make a number of corrections and additions 'in the hopes of making my many rather stupid reviewers at least understand what is meant.— I hope & think I shall improve the Book considerably.—' (letter to John Murray, 5 December [1860]). Although he took Lyell's advice not to name any names when responding to his reviewers, he nonetheless added a considerable amount of new material in support of his various arguments, as well as including in the third English edition the historical preface (acknowledging naturalists who had propounded views about species change) previously printed in the German and American editions.

In light of the criticism his book had received over the past year, Darwin was determined to do what he could to make his doctrines more generally understood, realising that the process of acceptance would be a long and gradual one. 'I never expected to convert people under 20 year,' he told his friend John Innes, 'though firmly convinced *now* that I am in the main right.— For a week hardly passes without my hearing of some good judge coming some little way with me.' (letter to John Innes, 28 December [1860]).

ACKNOWLEDGMENTS

The editors are grateful to Mr George Pember Darwin for permission to publish the Darwin letters and manuscripts. They also thank the Syndics of the Cambridge University Library and other owners of the manuscript letters who have generously made them available.

The work for this edition has been supported by grants from the National Endowment for the Humanities (NEH) and the National Science Foundation (NSF). The Alfred P. Sloan Foundation and the Andrew W. Mellon Foundation provided grants to match NEH funding, and the Mellon Foundation in 1981 and 1984 awarded grants to Cambridge University that have made it possible to put the entire Darwin correspondence into machine-readable form. Research and editorial work have also been supported by timely grants from the Royal Society of London, the British Academy, and the Pilgrim Trust. The Royal Society has helped meet the costs of publication. The Stifterverband für die Deutsche Wissenschaft has provided funds to edit Darwin's correspondence with German naturalists.

The Cambridge University Library and the American Philosophical Society have generously made working space and many services available to the editorial staff.

Since the project began in 1975, we have been fortunate in benefiting from the interest, experience, and practical help of many people in many places, and the editors hope that they have adequately expressed their thanks to them individually as the work proceeded. There are some, however, who have helped over so long a period that it would be ungracious not to thank them personally in the work that has profited so much from their co-operation.

Over the years, Kathy Fuller, formerly assistant director of the Division of Research Grants of NEH, and Ronald J. Overmann, program director for History and Philosophy of Science of NSF, have given far more time and attention to the project than their formal duties required. The editors also appreciate the interest and encouragement they received from John E. Sawyer, former president of the Andrew W. Mellon Foundation, and his colleague, James Morris. The editors would like to acknowledge with gratitude the assistance rendered to the project by the late Shepard Stone and by Frank Rhodes.

Without the expert help of John L. Dawson of the Literary and Linguistic Computing Centre of Cambridge University, the computerisation of the correspondence would not have been possible, for the work on both the *Calendar* and the *Correspondence* required the solution of many novel technical problems.

The computer of the Cambridge University Library, as well as that of the University, was used for initial work on this edition and for proof production.

W. D. S. Motherwell, former head of automation at the Library, arranged for the installation of terminals in the project's office in the Manuscripts Department and solved the operational problems that arose. More recently, Paul Ayris and Chris Sendall have assisted us in accessing the library's laser printer.

This book was type-set at the Oxford University Computing Service on a Monotype Prism Imagesetter. We are grateful to Stephen Miller and to the Oxford operations staff for advice and assistance with the Prism, and for enabling the type-setting of the volume to be carried out.

The late Sir Hedley Atkins and Philip Titheradge, formerly curator of the Darwin Museum at Down House, Downe, Kent, welcomed the editors on numerous visits and responded most generously to frequent requests for information and for material from the collections at Down. The new curator, Solene Morris of the British Museum (Natural History), has enabled our close relationship with Down to continue. Professor Richard Darwin Keynes kindly made available Emma Darwin's diaries and other Darwin family material in his possession. The help of Peter J. Gautrey, who until 1989 directed the Darwin Archive at the Cambridge University Library, has been of special importance. His great knowledge of Darwiniana has always been readily shared with the editors, as with countless other Darwin scholars.

Libraries all over the world have given help that was literally indispensable by making available photocopies of Darwin correspondence and other manuscripts in their collections. The institutions and individuals that furnished copies of letters for this volume can be found in the List of provenances. The editors are extremely indebted to them all. We are also grateful to the many people who have transmitted information regarding the whereabouts of particular letters and, in many cases, have generously provided copies of such letters for the project's use.

The editors make daily use of the incomparable facilities of the Cambridge University Library and have benefited greatly from its services and the help and expertise of its staff, particularly the staff of the Manuscripts Department. We are especially grateful to the University Librarian, Frederick W. Ratcliffe, and the Underlibrarian of Manuscripts, Patrick Zutshi, for their generous support. Other members of the library's staff who have frequently responded to the editors' requests are: David Hall, Elisabeth Leedham-Green, Jayne Ringrose, William Noblett, Gerry Bye, Godfrey Waller, Vicki Newman, Roger Fairclough, Cynthia Webster, and Christine Fenn. Paul Emmines, Natasha Charles, and Michelle Gower, all of the Manuscripts reading room, have patiently dealt with the editors' often complex requirements.

At the American Philosophical Society Library, a splendid collection of Darwiniana and works in the history of science has been continuously available to the editorial staff since the inception of the project. Whitfield J. Bell Jr, secretary of the society until 1983, serves on the United States Advisory Committee for the project and has done his utmost to further its work. The editors have also benefited from the co-operation of Edward Carter II, Roy C. Goodman, Carl F. Miller, Elizabeth Carroll-Horrocks, and Bertram Dodelin, all of the APS Library.

Douglas W. Bryant, formerly of the Harvard University Library; Rodney Dennis, Jennie Rathbun, and Susan Halpert of the Houghton Library; Constance Carter of the Science Division of the Library of Congress; Lothian Lynas of the New York Botanical Garden Library; Judith Warnement and Jean Cargill of the Gray Herbarium of Harvard; and the staff members of the New York Public Library have all been exceptionally helpful in providing material from their great collections.

In Britain, the editors have often received assistance from Mary Sampson, archivist of the Royal Society; Rex E. R. Banks and John Thackray of the British Museum (Natural History) Library; Gina Douglas, librarian of the Linnean Society of London; and Sylvia FitzGerald and Leonora Thompson of the Library of the Royal Botanic Gardens, Kew. We would also like to thank Anthony M. Carr, Miss Williams, and Mrs Ion of the Local Studies Library, Shrewsbury; James B. Lawson of Shrewsbury School Library; Richard G. Williams, Mrs Felton, and Anne Barrett (college archivist) of Imperial College of Science, Technology and Medicine; Christine Fyfe, archivist of Keele University; and Dr and Mrs Courtney of Christ's College, Cambridge. Virginia Murray of John Murray (publishers) Ltd has graciously provided the editors with copies of the Darwin correspondence in the Murray archive and with related information. Leonard G. Wilson kindly provided the editors with photocopies of manuscript material in the possession of the family of Sir Charles Lyell.

Among the others who have advised and assisted the editors in their work are Julius Held, Garry J. Tee, Hugh Torrens, Thomas Junker, Antonello La Vergata, and Leonard Wilson. The late Fredson Bowers and G. Thomas Tanselle were frequently called upon for advice on editorial method. We are also pleased to acknowledge the invaluable support of Desmond King-Hele and Geoffrey Lloyd of the British Advisory Committee.

Continued thanks are due to William Montgomery, former associate editor with this project, for his assistance in editing the letters in the collection of the American Philosophical Society. We are also grateful for the many years of service and scholarly expertise given to the project by Janet Browne, who left the staff during the preparation of this volume. We wish Janet continued success in her future endeavours. We also gratefully acknowledge the valuable assistance of Stefanie Cookson.

It is with utmost sadness that we record the death, on 30 January 1992, of our dear colleague Peter Saunders. Only through Peter's untiring persistence in tackling the daunting challenge of developing a new system of computerised typesetting was this volume prepared for publication. His skills, intelligence, and wit are greatly missed by us all.

LIST OF PROVENANCES

The following list gives the locations of the original versions of the letters printed in this volume. The editors are grateful to all the institutions and individuals listed for allowing access to the letters in their care.

Academy of Natural Sciences of Philadelphia, PA, USA
American Philosophical Society, Philadelphia, PA, USA
Bath Royal Literary and Scientific Institute, Bath, England
Beinecke Library, Yale University, New Haven, CT, USA
Bibliothèque Publique et Universitaire de Genève, Geneva, Switzerland
British Library, The Manuscript Collections, Great Russell Street, London, England
British Museum (Natural History). *See* The Natural History Museum
Brown University Library, Providence, RI, USA
Buckinghamshire Record Office, County Hall, Aylesbury, Bucks., England
Cambridge University Library, Manuscripts department, Cambridge, England
Christ's College Library, Cambridge, England
Cleveland Health Sciences Library, Cleveland, OH, USA
Cottage Gardener (publication)
CUL. *See* Cambridge University Library
Dana Biomedical Library, Dartmouth College, Hanover, NH, USA
DAR. *See* Cambridge University Library
Darwin Pamphlet Collection–CUL. *See* Cambridge University Library
Down House, Downe, Kent, England
Dunedin Public Library, Dunedin, New Zealand
Entomologist's Weekly Intelligencer (publication)
Fitzwilliam Museum Library, Cambridge, England
Gardeners' Chronicle and Agricultural Gazette (publication)
H. R. Glennie (private collection)
Gray Herbarium of Harvard University, Cambridge, MA, USA
R. A. Hooker (private collection)
Imperial College of Science, Technology and Medicine, South Kensington, London, England
Quentin Keynes (private collection)
Kinnordy MS. *See* Lady Lyell
Library Board of Western Australia, State Archives of Western Australia, Perth, Western Australia

Lincolnshire Archives Office, The Castle, Lincoln, England
Linnean Society of London, Burlington House, London, England
Lady Lyell (private collection)
Medical Center Library, Duke University, Durham, NC, USA
Moray District Record Office, Tolbooth, Forres, Scotland
John Murray (publishers), London, England
The Natural History Museum, South Kensington, London, England
New York Botanical Garden Library, Bronx, New York, NY, USA
New York Public Library, New York, NY, USA
Oxford University Museum, Oxford, England
Phillips, Son & Neale, Inc. (dealers)
Pierpont Morgan Library, New York, NY, USA
Praeger 1935 (publication)
Proceedings of the Royal Society of London (publication)
R. D. Pyrah (private collection)
Quaritch (dealers)
Hans Rhyn collection, Berne, Switzerland
Paul C. Richards (dealer)
Shrewsbury School Library, The Schools, Shrewsbury, Shropshire, England
Smithsonian Institution, Washington, D.C., USA
Kenneth Spencer Research Library, University of Kansas Libraries, Lawrence, KS, USA
University College London, London, England
University of London Library, Senate House, London, England
University of Virginia Library, Charlottesville, VA, USA
Uppsala University Library, Uppsala, Sweden
Wellcome Institute Library, London, England
West Suffolk Record Office, Bury St Edmunds, Suffolk, England
Jeffries Wyman Jr (private collection)
Yale University Library, New Haven, CT, USA

A NOTE ON EDITORIAL POLICY

The first and chief objective of this edition is to provide complete and authoritative texts of Darwin's correspondence. In so far as it is possible, the letters have been dated, arranged in chronological order, and the recipients or senders identified. Darwin seldom wrote the full date on his letters and, unless the addressee was well known to him, usually wrote only 'Dear Sir' or 'Dear Madam'. After the adoption of adhesive postage stamps in the 1840s, the separate covers that came into use with them were usually not preserved, and thus the dates and the names of many recipients of Darwin's letters have had to be derived from other evidence. The notes made by Francis Darwin on letters sent to him for his editions of his father's correspondence have been helpful, as have matching letters in the correspondence, but many dates and recipients have had to be deduced from the subject-matter or references in the letters themselves. These tasks, together with the deciphering of the handwriting of Darwin and others, have presented the most troublesome problems.

Whenever possible, transcriptions have been made from manuscripts. If the manuscript was inaccessible but a photocopy or other facsimile version was available, that version has been used as the source. In many cases, the editors have had recourse to Francis Darwin's large collection of copies of letters, compiled in the 1880s. Other copies, published letters, or drafts have been transcribed when they provide texts that are otherwise unavailable.

The method of transcription employed in this edition is adapted from that described by Fredson Bowers in 'Transcription of manuscripts: the record of variants', *Studies in Bibliography* 29 (1976): 212–64. This system is based on accepted principles of modern textual editing and has been widely adopted in literary editions.

The case for using the principles and techniques of this form of textual editing for historical and non-literary documents, both in manuscript and print, has been forcefully argued by G. Thomas Tanselle in 'The editing of historical documents', *Studies in Bibliography* 31 (1978): 1–56. The editors of the *Correspondence* followed Dr Tanselle in his conclusion that a 'scholarly edition of letters or journals should not contain a text which had editorially been corrected, made consistent, or otherwise smoothed out' (p. 48), but they have not wholly subscribed to the statement made earlier in the article in which he says, 'In the case of notebooks, diaries, letters and the like, whatever state they are in constitutes their finished form, and the question of whether the writer "intended" something else is irrelevant' (p. 47). The editors have preserved the spelling, punctuation, and grammar of the original, but they have found it impossible to set aside entirely the question of authorial intent. One obvious reason is that in reading Darwin's writing, there must necessarily be reliance upon both context and intent. Even when Darwin's general intent is clear, there are cases in which alternative readings are, or may be, possible, and therefore the transcription decided upon must to some extent be conjectural. In this work, when the editors are uncertain of their transcription they have enclosed the doubtful text in italic square brackets.

A major editorial decision was to adopt the so-called 'clear-text' method of transcription, which so far as possible keeps the text free of brackets recording deletions, insertions, and other alterations in the places at which they occur. Darwin's changes are, however, recorded in the back matter of the volume as alteration notes keyed to the printed text by paragraph and line number, under 'Manuscript alterations and comments'. All lines above the first paragraph of the letter (i.e., date, address, or salutation) are referred to as paragraph '0'. Separate paragraph numbers are used for subscriptions and postscripts. These notes enable the reader who wishes to do so to reconstruct the manuscript versions of Darwin's letters while furnishing printed versions that are uninterrupted by editorial interpolations. The Manuscript alterations and comments record all alterations made by Darwin in his letters and any editorial amendments made in transcription. For copies and drafts of Darwin letters included in the correspondence no attempt has been made to record systematically all alterations to the text, but ambiguous passages in copies and major revisions of drafts are noted. The editors believe it would be impracticable to attempt to go further without reliable information about the texts of the original or final versions of the letters involved. Letters to Darwin have been transcribed without recording any of the writers' alterations unless they reflect significant changes in substance or impede the sense; in such cases footnotes will bring them to the reader's attention.

Misspellings have been preserved, even when it is clear that they were unintentional as, for instance, 'lawer' for 'lawyer'. Such errors often indicate excitement or haste and may exhibit, over a series of letters, a habit of carelessness in writing to a particular correspondent or about a particular subject.

Capital letters have also been transcribed as they occur except in certain cases, such as 'm', 'k', and 'c', which are frequently written somewhat larger than others as initial letters of words. In these cases the normal practice of the writers has been followed.

In some instances that are not misspellings in a strict sense, editorial corrections have been made. In his early manuscripts and letters Darwin consistently wrote 'bl' so that it looks like 'lb' as in 'albe' for 'able', 'talbe' for 'table'. Because the form of the letters is so consistent in different words, the editors consider that this is most unlikely to be a misspelling but must be explained simply as a peculiarity of Darwin's handwriting. Consequently, the affected words have been transcribed as normally spelled and no record of any alteration is given in the textual apparatus. Elsewhere, though, there are misformed letters that the editors have recorded because they do, or could, affect the meaning of the word in which they appear. The main example is the occasional inadvertent crossing of 'l'. When the editors are satisfied that the intended letter was 'l' and not 't', as, for example, in 'stippers' or 'istand', then 'l' has been transcribed, but the actual form of the word in the manuscript has been given in the alteration notes.

If copies of letters provide the only source, the editors have frequently retained corrections made to the text when it is clear that they were based upon comparison with the original. Francis Darwin's corrections of misreadings by copyists have usually been followed; corrections to the text that appear to be editorial alterations have not.

Editorial interpolations in the text are in square brackets. Italic square brackets enclose conjectured readings and descriptions of illegible passages. To avoid confusion, in the few instances in which Darwin himself used square brackets, they have been altered by the editors to parentheses with the change recorded in the back matter. In letters to Darwin, square brackets have been changed to parentheses silently.

Material that is irrecoverable because the manuscript has been torn or damaged is indicated by angle brackets; any text supplied within them is obviously the responsibility of the editors. Occasionally, the editors are able to supply missing sections of text by reference to transcripts or photocopies of manuscript material made before the damage occurred.

Words and passages that have been underlined for emphasis are printed in italics in accordance with conventional practice. When the author of a letter has indicated greater emphasis by underlining a word or passage two or more times, then the greater emphasis is indicated by printing the text in bold type.

Paragraphs are often not clearly indicated in the letters. Darwin and others sometimes marked a change of subject by leaving a somewhat larger space than usual between sentences; sometimes Darwin employed a longer dash. In these cases, and in very long stretches of text when the subject is clearly changed, a new paragraph is started by the editors without notice. The start of letters, valedictions, and postscripts are also treated as new paragraphs regardless of whether they appear as such in the manuscript. Special manuscript devices delimiting sections or paragraphs, e.g., blank space left between sections of text and lines drawn across the page, are treated as normal paragraph indicators and are not specially marked or recorded unless their omission leaves the text unclear.

Additions to a letter that run over into the margins, or are continued at its head or foot, are transcribed at the point in the text at which the editors believe they were intended to be read. The placement of such an addition is only recorded in a footnote if it seems to the editors to have some significance or if the position at which it should be transcribed is unclear. Enclosures are transcribed following the letter.

Occasionally, punctuation marking the end of a clause or sentence is not present in the manuscript, but the author has made his or her intention clear by allowing, for example, extra space or a line break to function as punctuation. In such cases, the editors have inserted an extra space following the sentence or clause to set it off from the following text.

Some Darwin letters and an occasional letter to Darwin are known only from entries in the catalogues of book and manuscript dealers or mentions in other published sources. Whatever information these sources provide about the content of such letters has been reproduced without change.

For every Darwin letter, the text that is available to the editors is always given in full. Some other items, however, are not printed in their entirety. Some memorandums and other documents that are not letters but are relevant to the correspondence have been summarised, as have those letters to Darwin that the editors consider can be presented adequately in shortened form. In a few cases the editors have included letters that are not to or from Darwin if they relate significantly to the published correspondence. Letters that can only be given a wide date range, in some instances spanning several decades, will be printed together in one of the concluding volumes.

The format in which the transcriptions are printed in the *Correspondence* is as follows:

1. *Order of letters*. The letters are arranged in chronological sequence. A letter that can be dated only approximately is placed at the earliest date on which the editors believe it could have been written. The basis of a date supplied by the editors is given in a footnote unless it is derived from a postmark, watermark, or endorsement as recorded in the physical description of the letter (see section 4, below). Letters with the same date, or with a range of dates commencing with that date, are printed in the alphabetical order of their senders or recipients unless their contents dictate a clear alternative order. Letters dated only to a year or a range of years precede letters that are dated to a particular month or range of months, and these, in turn, precede those that are dated to a particular day or to a range of dates commencing with a particular day.

2. *Headline*. This gives the name of the sender or recipient of the letter and its date. The date is given in a standard form, but those elements not taken directly from the letter text are supplied in square brackets.

3. *The letter text*. The transcribed text follows as closely as possible the layout of the source, although no attempt is made to produce a type-facsimile of the manuscript, that is, word-spacing and line-division in the running text are not adhered to. Similarly, the typography of printed sources is not replicated. Dates and addresses given by authors are transcribed as they appear, except that if both the date and address are at the head of the letter they are always printed on separate lines with the address first, regardless of the exact manuscript order. If no address is given on a letter by Darwin, the editors have supplied one, when able to do so, in square brackets at the head of the letter. Similarly, if Darwin was writing from an address different from the one given on the letter, his actual location is given in brackets. Addresses on printed stationery are transcribed in italics. Addresses, dates, and valedictions have been run into single lines to save space, but the positions of line-breaks in the original are marked by vertical bars.

4. *Physical description*. All letters are complete, signed, autograph letters unless otherwise indicated. If a letter was written by an amanuensis, or exists only as a draft or a copy, or is incomplete, or is in some other way unusual, then the editors provide the information needed to complete the description. Postmarks, endorsements, and watermarks are recorded only when they are evidence for the date or address of the letter.

5. *Source*. The final line provides the provenance of the text. Some sources are given in abbreviated form (e.g., DAR 140: 18) but are listed in full in the List of provenances unless the source is a published work. Letters in private collections are also indicated. References to published works are given in author–date form, with full titles and publication details supplied in the Bibliography at the end of the volume.

6. *Darwin's annotations*. Darwin frequently made notes in the margins of the letters he received, scored significant passages, and crossed through details that were of no further interest to him. These annotations are transcribed or described following the letter text together with details of their location on the manuscript and the writing medium. They are keyed to the letter text by paragraph and line numbers. Most notes are short, but occasionally they run from a paragraph to several pages, sometimes written on separate sheets appended to the letter or otherwise genetically related to it. Extended notes relating directly to a letter and physically associated with it are transcribed whenever practicable following the annotations.

Quotations from Darwin manuscripts in footnotes and elsewhere, and the text of his annotations and notes on letters, are transcribed in 'descriptive' style. In this method the alterations in the text are recorded in brackets at the places in which they occur. For example:

'See Daubeny ['vol. 1' *del*] for *descriptions of volcanoes in [*interl*] S.A.' *ink*

means that Darwin originally wrote in ink 'See Daubeny vol. 1 for S.A.' and then deleted 'vol. 1' and inserted 'descriptions of volcanoes in' after 'for'. The asterisk before 'descriptions' marks the beginning of the interlined phrase, which ends at the bracket. The asterisk is used when the alteration applies to more than one preceding word. The final text can be read simply by skipping the material in brackets. Descriptive style is also used in the Manuscript alterations and comments.

Several letters include hand-drawn illustrations or diagrams. These have been reproduced and positioned in the text as accurately as possible. Captions have been transcribed using the same criteria applied to the letter text, as described above. In some cases, it has been necessary either to reduce the size of the diagram or to enhance the outline for clarity; sometimes they have been repositioned for type-setting purposes. Footnotes record any further pertinent information.

Volumes of the *Correspondence* are published in chronological order. Each volume is self-contained, having its own index, bibliography, and biographical notes. A chronology covering the period of each volume is supplied, and Appendixes give additional material as appropriate to aid understanding of the correspondence. A comprehensive index is planned for the final volume. References are supplied for all persons and subjects mentioned, even though some repetition of material in earlier volumes is involved.

If the name of a person mentioned in a letter is incomplete or incorrectly spelled, the full correct form is given in a footnote. Brief biographies of every identifiable person mentioned in a volume and dates of each correspondent's letters to and from Darwin are given in the Biographical register and index to correspondents.

From time to time, supplements will be published containing letters that have come to light or have been redated since the relevant volumes of the *Correspondence* have appeared. The first such supplement was included in Volume 7, and covered the years from 1821 to 1857.

The editors use the abbreviation 'CD' for Charles Darwin throughout the notes. A list of all abbreviations used by the editors in this volume is given on p. xl. For references to Darwin's books, articles, and collections of his letters, short titles are used (e.g., *Descent*, *Collected papers*, *LL*). Short titles are also used for some standard reference works (e.g., *Alum. Cantab.*, *Wellesley index*, *DNB*). For all other works, author–date references are used. The full titles of all the books referred to are given in the Bibliography; the short titles and author–date references are listed together in alphabetical order.

The Wedgwood and Darwin

Robert Waring Darwin = Susannah Wedgwood
1766–1848 — 1765–1817

- Marianne 1798–1858 = Henry Parker 1788–1856
  - Robert b.1825
  - Henry 1827–92
  - Francis 1829–71
  - Charles b.1831
  - Mary Susan 1836–93
- Caroline Sarah 1800–88 = Josiah III (Jos) 1795–1880
  - Sophy Marianne 1838–9
  - Katherine Elizabeth Sophy (Sophy) 1842–1911
  - Margaret Susan 1843–1937
  - Lucy Caroline 1846–1919
- Susan Elizabeth 1803–66
- Erasmus Alvey 1804–81
- Charles Robert 1809–82 = Emma 1808–96
  - William Erasmus 1839–1914
  - Anne Elizabeth 1841–51
  - Mary Eleanor Sept.–Oct. 1842
  - Henrietta Emma (Etty) 1843–1927
  - George Howard 1845–1912
  - Elizabeth 1847–1926
  - Francis (Frank) 1848–1925
  - Leonard 1850–1943
  - Horace 1851–1928
  - Charles Waring 1856–8
- Emily Catherine (Catherine) 1810–66

Families up to 1860

Josiah Wedgwood II 1769–1843 = Elizabeth (Bessy) Allen 1764–1846

- Sarah Elizabeth (Elizabeth) 1793–1880
- Mary Anne 1796–8
- Charlotte 1797–1862 = Charles Langton 1801–86
  - Edmund 1841–75
- Henry Allen (Harry) 1799–1885 = Jessie Wedgwood 1804–72
  - Louisa Frances 1834–1903
  - Caroline b. 1836
  - John Darwin 1840–70
  - Anne Jane 1841–77
  - Arthur 1843–1900
  - Rowland 1847–1921
- Frances Mosley d. 1874 = Francis 1800–88
  - Godfrey 1833–1905
  - Amy 1835–1910
  - Cicely Mary 1837–1917
  - Clement 1840–89
  - Lawrence 1844–1913
  - nstance Rose 1846–1903
  - Mabel Frances 1852–1930
- Hensleigh 1803–91 = Frances (Fanny) Mackintosh 1800–89
  - Frances Julia (Snow) 1833–1913
  - James Mackintosh (Bro) 1834–64
  - Ernest Hensleigh 1838–98
  - Katherine Euphemia (Effie) 1839–1931
  - Alfred Allen 1842–92
  - Hope Elizabeth 1844–1934
- Frances (Fanny) 1806–32

ABBREVIATIONS

A	autograph, i.e., in the hand of the sender
L	letter
mem	memorandum
S	signed by the sender
AL	autograph letter
ALS	autograph letter signed
LS	letter in hand of amanuensis, signed by sender
LS(A)	letter in hand of amanuensis with additions by author
CD	Charles Darwin
CUL	Cambridge University Library
DAR	Darwin Archive, Cambridge University Library
del	deleted
illeg	illegible
interl	interlined
underl	underlined

TRANSCRIPTION CONVENTIONS

[some text]	'some text' is an editorial insertion
*[*some text*]*	'some text' is the conjectured reading of an ambiguous word or passage
[some text]	'some text' is a description of a word or passage that cannot be transcribed, e.g., '*3 words illeg* '
⟨ ⟩	word(s) destroyed
⟨some text⟩	'some text' is a suggested reading for a destroyed word or passage
⟨*some text*⟩	'some text' is a description of a destroyed word or passage, e.g., '*3 lines excised* '

THE CORRESPONDENCE OF CHARLES DARWIN
1860

From Bernard Philip Brent [1860?][1]

you that ducks when they sleep on the water take out their feet shake them and place them under their feathers.[2] ⟨*section missing*⟩

I trust you will not take it amiss in me for thus freely expressing my views, and again thanking you for the entertain- ⟨*section missing*⟩

AL incomplete
DAR 205.2 (Letters)

CD ANNOTATIONS
1.1 on the water . . . shake them] *double scored pencil*
2.1 I trust . . . entertain- 2.2] *crossed pencil*
Top of page: '18'[3] *brown crayon*

[1] The date is based on the suggestion in the text that Brent may recently have read CD's discussion of the possible relation of the behaviour of ducks to the dispersal of plants and animals in *Origin*, pp. 383–8. See n. 2, below.

[2] Brent was a noted pigeon-fancier and breeder of domestic poultry. He had assisted CD in his investigations of pigeons and fowls since 1856 (see *Correspondence* vols. 6 and 7). Brent's remarks suggest that he was commenting on CD's view that young molluscs or floating plants could become attached to ducks' feet while they were swimming and thus be transported elsewhere. CD had been actively investigating means of geographical dispersal since 1855 (see *Correspondence* vols. 5 and 6). The results of his study are summarised in *Origin*, pp. 383–410, where an experimental trial on young molluscs clinging to ducks' feet is mentioned: 'I suspended a duck's feet, which might represent those of a bird sleeping in a natural pond, in an aquarium' (*Origin*, p. 385). The experiments, carried out in 1857 and 1859, are recorded in CD's Experimental book, pp. 22–3 (DAR 157a).

[3] The number of CD's portfolio of notes on the means of dispersal of plants and animals.

From Robert Scot Skirving [1860?][1]

Last year I had 1,200 killed in one week on one field.[2] A rather distinguished naturalist in Scotland insists they are not birds of passage, yet I know I have shot them with acorns in their crop that never grew here, & the keeper of the May Island light in the Firth of Forth says he has seen them arrive there in a state of exhaustion.[3] The question of these migrations has been discussed in Agricultural

Clubs. ⟨In the past⟩ members [killed last winter] there ⟨*section missing*⟩ whether one was ⟨*section missing*⟩ why not millions?

I have only now to ask pardon for this intrusion—unwarranted on my part, as, in begging for information as to the clover plant, I presume not to think I can in any way return more than thanks.

I am | Your Obt & faithful Serv | R Scot Skirving

Incomplete
DAR 205.2 (Letters)

CD ANNOTATIONS
1.1 Last year . . . in Scotland 1.2] *crossed ink*
1.2 birds of passage,] 'Wood Pigeon' *added red crayon*
1.2 yet I know . . . exhaustion. 1.5] *square brackets added brown crayon*; 'see Back'[4] *added red crayon*
1.7 why not millions? . . . Skirving 3.1] *crossed ink*

[1] It has not been possible to ascertain the exact date of this and the following letter. The period between 1858 and 1860 was one in which CD sought information from a wide variety of correspondents on the possible means of dispersal of seeds by birds. Skirving lived in Edinburgh and was the author of several tracts on the labour question in Scotland. He had apparently written to CD with an inquiry about one of CD's references to clover plants. Although it is possible that the inquiry related to CD's letter to the *Gardeners' Chronicle*, [13 November 1858] (*Correspondence* vol. 7), Skirving's second letter to CD indicates that a post-*Origin* date is more likely. CD discussed the relation between bees and clover in *Origin*, pp. 73–4. See also following letter.

[2] Skirving refers to wood-pigeons. See CD's annotations and following letter.

[3] The Isle of May lighthouse is situated at the entrance to the Firth of Forth, on the east coast of Scotland. CD discussed the transport of seeds in the crops of migratory and other birds in *Origin*, pp. 361–2.

[4] CD's annotation refers to the fact that a subsequent note from Skirving, supplying information pertinent to this letter, was attached to the letter (see following letter). Both letters were cut into sections, presumably to retain only those portions that were of interest to CD.

From R. S. Skirving [1860?][1]

Then as to the sitting on trees. An Arab village in these parts is a conglomeration of tall dome shaped mud hovels each much in the form of a thimble. The top part of each is a dove cot. I see no reason why the pigeons did not light on the tops of these houses, but it was certainly not their custom to do so—They seemed always to light upon the few trees (other than Palms) which ⟨ ⟩ has about the ⟨ ⟩ to excess, ⟨*section missing*⟩

With regard to the acorn in the crop of the wood pigeon shot here, I cannot state the fact in the same positive manner I can as to the habits of the Egyptian birds.[2] It is long since the incident, & I was indeed a boy at the time. Still I have no doubt in my mind that the acorn was larger & longer than any that grow in Scotland.

Not long ago I saw what seemed similar acorns growing on dwarf oaks in the Island of Corfu. The bird from which I took the acorn, I saw shot out of an immense flock, which seemed to be making a steady flight from the German Ocean, & I conjectured they were arriving from some part of the continent. ⟨*section missing*⟩

It may never ⟨ ⟩ be all mortal, in *death* he may ⟨ ⟩ lately for the first time. Of course you might ask is a dead foetus then immortal. I would ask, was it ever *a being*? However, once admit that man has at any period a soul, & it would follow he was utterly different from all the animals that lived *before* he was introduced into the world. If so, a new & special act of creative power has been exhibited, & if one act why not any number?[3]

Yours faithfully | R Scot Skirving

Incomplete
DAR 205.2 (Letters)

CD ANNOTATIONS
1.2 dome shaped mud hovels] 'of a single story' *added pencil*
4.1 It may never . . . Skirving 5.1] *crossed pencil*

[1] For the basis of the date see the preceding letter, n. 1.
[2] The letter was glued to the bottom of the preceding letter. The reference to an acorn in the crop of a wood pigeon is presumably in response to a query from CD about the preceding letter.
[3] The reference to a 'special act of creative power' suggests Skirving was writing to CD after the publication of *Origin*.

From Hensleigh Wedgwood [January? 1860][1]

not seen it myself and cannot get a very distinct account of the extent of the complaint.

I feel the origin of things wholly beyond the scope of our imagination and should by no means be surprised if there never was any beginning The difficulties of that hypothesis do not seem to me to be greater than the converse. But to my mind the problem is not in the least degree smoothed by supposing that life began with a single cell. If you suppose a cell started in a mineral world with only such a tendency to variation in the offspring as we are acquainted with I cannot think that an infinity of ages would have made any material advance in organisation.[2]

Is it a fact that we have any geological evidence of a gradual progress in organisation? I think that when the Vestiges were in discussion the opinion of scientific people was that there was not.[3] As far as I see the only positive evidence we have is of degradation, not of advance as in the loss of sight in cave animals moles &c. loss of limbs by snakes and other rudimentary cases. If all vertebrate animals have descended from a common ancestor I should think it must have been of the complete four limbed type, otherwise we might expect to find the fundamental vertebra with utterly discordant modifications. Because we recognise the same senses in the lower animals and in vertebrates does not seem to me to tend towards evidence of their original identity. They both have to live in the same world and to avail themselves of the same powers of nature and one would expect à priori that there should be resemblance in the organs with which they are furnished for that purpose.

I am rather ashamed of sending my crude speculations to you who have thought so much on these subjects but it is well for you to see the light in which they strike all sorts of people

Many thanks for your criticism on my dictionary.[4] It agreed for the most part all too well with my own conviction at the time of writing. Video meliora but I cannot reach them. If I could illustrate and fill out as I ought to do my book on the Understanding would be recognised as having attained what Locke aimed at.[5]

Adieu | H W.

Incomplete
DAR 48: 83–5

CD ANNOTATIONS
1.1 not seen . . . the converse. 2.3] *crossed pencil*
4.1 I am rather . . . H W. 6.1] *crossed pencil*

[1] The date is suggested by the reference to Wedgwood's *Dictionary of English etymology*, the first volume of which was published in December 1859 (*Publishers' Circular*, 15 December 1859, p. 692).

[2] Wedgwood's point relates to CD's concluding remarks in *Origin* that he was led by analogy to believe 'that all animals and plants have descended from some one prototype.' (*Origin*, p. 484). For CD's thoughts on this point, see the letter to Charles Lyell, 18 [and 19 February 1860].

[3] [Chambers] 1844. One of the most prominent critics of this work was Adam Sedgwick, who attempted to deny progression, and hence transmutation, by stressing the discontinuity of the geological record. See Yeo 1984, pp. 19–20.

[4] Wedgwood 1859–65. There is a copy of the second edition (1872) in the Darwin Library–CUL. CD's comments have not been found, but they may have concerned the points made in Wedgwood's introduction concerning the origin of language. Wedgwood believed that language had arisen gradually through the elaborated imitation of natural sounds. CD had formerly corresponded with Wedgwood about the ancestry of related words in different languages as an example of transitional forms (*Correspondence* vol. 6, letter from Hensleigh Wedgwood, [before 29 September 1857]). See also *Natural selection*, p. 384, and *Origin*, pp. 422–3.

[5] Wedgwood's book on the development of the understanding (Wedgwood 1848) discussed John Locke's famous attack on the existence of 'innate ideas' (Locke 1690). There is a copy of Wedgwood 1848 in the Darwin Library–CUL.

To Thomas Henry Huxley 1 January [1860][1]

Down Bromley Kent
Jan 1st

My dear Huxley

I write one line merely to thank you for your pleasant note & to say that I will keep your secret.[2] I will shake my head as mysteriously as Ld Burleigh.—[3] Several persons have asked me who wrote that "most remarkable article" in the Times. As a cat may look at a king, so I have said that I strongly suspected you.— Mrs Hooker (do not repeat this) was so sharp that the **first** sentence revealed to her the authorship.[4] The Lubbocks Father & Son (God save the mark) thought it was

Owen's!!![5] You may rely on it that it has made deep impression, & I am heartily glad that the subject & I owe you this further obligation. But for God's sake take care of your health: remember that the Brain takes years to rest whilst the muscles take only hours.— There is poor Dana to whom I used to preach by letter writes to me that my prophecies are come true:[6] he is in Florence quite done up, can read nothing & write nothing & cannot talk for half-an-hour

I noticed the "naughty sentence about Owen, though my wife saw its bearing first.—[7]

Farewell you best & worst of men.—Most sincerely Yours | C. Darwin

That sentence about the Bird & the Fish dinners charmed us.—[8]

Lyell wrote me style like yours.—[9]

Have you seen the slashing article of Dec. 26th in Daily News against my stealing from my "Master", the Author of Vestiges.—[10]

Imperial College of Science, Technology and Medicine Archives (Huxley 5: 94)

[1] The date is given by the reference to Huxley's anonymous review of *Origin* in *The Times*, 26 December 1859, p. 8 ([T. H. Huxley] 1859a).

[2] Huxley had written the review of *Origin* as a favour to Samuel Lucas of *The Times*, who had been assigned to review the work but felt unqualified to do so. See *Correspondence* vol. 7, letter to T. H. Huxley, 28 December [1859], n. 4, and *LL* 2: 255.

[3] A character in Richard Brinsley Sheridan's *The Critic*, whose brief role consists solely of silently shaking his head (act I, scene I).

[4] Frances Harriet Hooker, the wife of Joseph Dalton Hooker, was a close family friend of both the Darwins and the Huxleys. CD's allusion is to the fact that the first two paragraphs of the review were actually written by Lucas (see *LL* 2: 255).

[5] John William Lubbock and his son John Lubbock were CD's neighbours. Richard Owen had expressed his willingness to consider CD's 'heterodox' views on species change in a letter thanking CD for a presentation copy of *Origin*. He had also given CD further reason to believe he was favourably disposed to the theory when the two met in London early in December 1859. See *Correspondence* vol. 7, letter from Richard Owen, 12 November 1859, and letter to Charles Lyell, [10 December 1859].

[6] James Dwight Dana, with whom CD had long corresponded, had suffered a mental collapse in 1859 and travelled to Italy to convalesce. See *Correspondence* vol. 7, letters to Charles Lyell, 29 [December 1859], and to J. D. Dana, 30 December [1859]. CD had previously warned Dana that overwork might damage his health (*Correspondence* vol. 7, Supplement, letter to J. D. Dana, 15 June [1851]).

[7] The sentence reads: 'no one has asserted the incompetence of the doctrine of final causes, in its application to physiology and anatomy, more strongly than our own eminent anatomist, Professor Owen' ([T. H. Huxley] 1859a). Huxley had for several years been conducting a public, verbal attack on Owen's scientific methodology. See *Correspondence* vols. 6 and 7.

[8] Huxley discussed Jean Baptiste Pierre Antoine de Monet de Lamarck's explanation of the development of the long neck of the heron. Lamarck believed it resulted from the long-continued efforts of short-necked birds to catch fish without getting themselves wet. Huxley wryly commented: 'The bird, in our example, would surely have renounced fish dinners long before it had produced the least effect on leg or neck' ([T. H. Huxley] 1859a).

[9] The letter from Charles Lyell has not been found.

[10] CD refers to the *Daily News*, 26 December 1859, p. 2. He was particularly anxious to distance his theory from the discredited views expressed in *Vestiges of the natural history of creation* ([Chambers] 1844). See *Correspondence* vol. 3, letters to J. D. Hooker, [7 January 1845] and [10 September 1845].

From William Whewell 2 January 1860

Trin. Lodge
Jan. 2. 1860

My dear Mr Darwin

I have to thank you for a copy of your book on the 'Origin of Species'.[1] You will easily believe that it has interested me very much, and probably you will not be surprized to be told that I cannot, yet at least, become a convert to your doctrines. But there is so much of thought and of fact in what you have written that it is not to be contradicted without careful selection of the ground and manner of the dissent, which I have not now time for. I must therefore content myself with thanking you for your kindness[2]

believe me | Yours very truly | W Whewell

C. Darwin Esqe

DAR 98 (ser. 2): 19

[1] CD had sent Whewell a copy of the first edition of *Origin* (see Appendix III).

[2] CD had known Whewell since he was an undergraduate at Cambridge University. See *Autobiography*, p. 66, and *Correspondence* vol. 2.

To Joseph Dalton Hooker 3 January [1860]

Down Bromley Kent
Jan. 3d.—

My dear Hooker

I have finished your Essay.[1] As probably you would like to hear my opinion though a non-Botanist, I will give it without any exaggeration. To my judgment it is by far the grandest & most interesting Essay on subjects of the nature discussed I have ever read. You know how I admired your former essays,[2] but this seems to me far grander. I like all the part after p. xxvi better than the first part; probably because newer to me.[3] I daresay you will demur to this, for I think every author likes the most speculative parts of his own productions. How superior your essay is to the famous one of Brown (here will be sneer first from you).—[4]

You have made all your conclusions so admirably clear that it would be no use at all to be a Botanist (sneer nor 2.). By Jove it would do harm to affix any idea to the long names of outlandish orders,—one can look at your conclusions with the philosophic abstraction with which a mathematician looks at his $a \times x + \sqrt{3^2}$ &c &c I hardly know which parts have interested me most; for over & over again I exclaimed this beats all. The general comparison of Flora of Australia with rest of world strikes me (as before) as extremely original, good & suggestive of many reflexions

p. xxxiii Orders next most abundant in species in S. Africa, & yet several of these orders being most abundant in S. **E** corner seems to weaken your remarks on affinity of S.**W** corner & S. Africa, but I suppose I cannot weigh affinity of the different orders.—By the way I call that affinity between S.W. corner & S. Africa, a nasty ugly inexplicable fact.—[5]

The invading Indian Flora **very** interesting; but I think the fact you mention towards close of Essay that the Indian vegetation in contradistinction to Malayan vegetation is found on low & level parts of Malay Is.[d] *greatly* lessens the difficulty which at first (p. L) seemed so great.[6] There is nothing like one's own hobby Horse, I suspect it is same case as of Glacial migration & of naturalised productions of productions of greater area conquering those of lesser: of course the Indian forms would have greater difficulty in seizing on the cool parts of Australia.—[7] I demur to your remarks (p. L) as not "conceiving anything in soil climate or vegetation of India" which could stop the introduction of Australian plants.— Towards close of Essay (p. civ) you have admirable remarks on our profound ignorance of cause of possible naturalisation or introduction; I would answer p. L. by a later page. viz p. civ.—[8]

Your contrast of SW & SE. corners is one of the most wonderful cases I ever heard of (do you not rather too much pass over Eyere's intermediate deserts?);[9] you show the case with wonderful force.

Your discussion on mixed invaders of S.E. corner (& of N. Zealand) is as curious & intricate a problem as of the races of man in Britain. Your remark on mixed invading Flora keeping down or destroying an original Flora which was richer in number of species, strikes me as *eminently new & important.*[10]

I am not sure whether to me the discussion on N. Zealand Flora is not even more instructive. I cannot too much admire both. But it will require long time to suck in all the facts. Your case of the largest Australian orders having none or very few species in N. Zealand is truly marvellous. Anyhow you have now *demonstrated* (together with no mammals in N. Zealand) (bitter sneer no.[r] 3.) that N. Zealand has never been continuously or even nearly continuously united by land to Australia!!

At p. Lxxxix is only sentence (on this subject) in whole Essay at which I am much inclined to quarrel, viz that no theory of transoceanic migration can explain &c &c.—[11] Now I maintain against all the world that no man knows anything about power of transoceanic power of migration. You do not know whether or not the absent orders have seeds which are killed by sea-water like almost all *Leguminosæ* & like another order which I forget. Birds do not migrate from Australia to N. Zealand, & therefore flotation *seems* only possible means; but yet I maintain that we do not know enough to argue on question; especially as we do not know the main fact whether seeds of Australian orders are killed by sea-water.

The discussion on European Genera profoundly interesting; but here alone I earnestly beg for more information viz to know which of these genera are absent in Tropics of world, ie confined to Temperate regions.—I excessively wish to know, *on the notion of Glacial migration*, how much modification has taken place in Australia: I had better explain when we meet, & get you to go over & mark the list.—[12] (N.B with respect to Glacial migration my remark is simple, & in the spirit of Agassiz, it

must be true, for Nature does not lie.)[13] What capital lists you give of temperate forms in S. Africa & Fuegia!

The list of Naturalised plants extremely interesting; but why at the end in the name of all that is good & bad do you not sum up & comment on your facts? Come, I will have a sneer at you in return for the many which you will have launched at this letter. Should you have remarked on the number of plants naturalised in Australia & U. States *under extremely different climates*, as showing that climate is so important; & that the considerable sprinkling of plants from India, N. America & S. Africa as showing that the frequent introduction of seed is so important? With respect to "abundance of unoccupied ground in Australia" do you believe that European plants introduced by man now grow on spots in Australia, which were absolutely bare? But I am an impudent dog. One must defend one's own puny theories, against such cruel men as you.

I daresay this letter will appear very conceited, but one must form an opinion on what one reads with attention, & in simple truth I cannot find words strong enough to express my admiration of your Essay.

My dear old friend | Yours affec[ly] | C. Darwin

I received half-an-hour ago your note.[14] Sir J. Richardson might have added to his exploded fallacies the Quinarian System, in which he formerly believed.[15] I can quite sympathise in the old not being ever staggered. To stagger any one quite satisfies me.—

I differ about Saturday R. One cannot expect fairness in a Reviewer, so I do not complain of all other arguments besides the G. Record being omitted. Some of the remarks about the lapse of years are very good, & the Reviewer gives me some good & well deserved raps,—confound it I am sorry to confess the truth. But it does not at all concern main argument.—[16]

That was a nice notice in G. Chronicle. I hope & imagine that Lindley is almost a convert.—[17] Do not forget to tell me if Bentham gets at all more staggered.[18]

With respect to Tropical plants during Glacial period, I throw in your teeth your own fact at base of Himalaya on possibility of the coexistence of at least forms of Tropics & Temperate regions. I can give parallel case for animals in Mexico.— Oh my dearly beloved puny child how cruel men are to you.—[19]

I am very glad you approve of Geographical Chapters.

P.S. | Lenny has got the Measles & it is sure to run like wild-fire through the house, as it has been *extraordinarily* prevalent in village. If your Boy Willy has *not* had measles, I fear it will not be safe for you to bring him here.—

DAR 115.1: 1

[1] The 'essay' was the introduction to Hooker's *Flora Tasmaniæ* (Hooker 1855–60). It was also issued separately in December 1859 (Hooker 1859). There are copies of both works in the Darwin Library –CUL; the text of Hooker 1859 was extensively annotated by CD. CD and Hooker discussed the work at length during its composition (see *Correspondence* vol. 7).

[2] CD refers to the introductory essay of the *Flora Indica* (Hooker and Thomson 1855) and a similar introduction to Hooker's *Flora Novæ-Zelandiæ* (Hooker 1853–5). CD's annotated copies of these works

are in the Darwin Library–CUL. In both essays Hooker discussed his views on the species question and on other leading problems in botany. See *Correspondence* vols. 5 and 6.

[3] CD had read the first part of Hooker's essay in proof-sheets (see *Correspondence* vol. 7). The rest of the essay discusses the plants of Australia in detail (Hooker 1859, pp. xxvii–cxxviii).

[4] CD refers to Robert Brown's essay on the Australian flora (Brown 1814), which Hooker praised in his essay (Hooker 1859, p. i).

[5] Hooker listed the 'orders' ('families' in modern terminology) of plants that were highly characteristic of the Australian flora, but not entirely confined to it, and recorded the geographical area in which they were next most abundant. In seven orders out of the total of eight, this area was South Africa. CD marked the passage in his copy and wrote 'Curious' beside it in pencil (Hooker 1859, p. xxxiii). Hooker also listed the regions of Australia in which these orders were represented by most species. Four of the six listed had most species in the south-west; two had most in the south-east (Hooker 1859, p. xxxiv).

[6] Hooker listed the Indian species found in Australia (Hooker 1859, pp. xlii–l). He believed that to account for the presence of Indian plants in Australia by proposing transoceanic migration failed to explain the lack of a reciprocal Australian contribution to the Indian flora (Hooker 1859, p. l). CD believed that Hooker's remark about Indian plants being found in Malaysia, which is nearer to Australia than India, was evidence for such a northward migration (Hooker 1859, p. civ). CD explained the absence of Australian plants in India by arguing that natural selection would favour species that were products of a larger land-area and had therefore been subjected to more intense competition.

[7] CD wrote a note to this effect in his copy of Hooker 1859, p. liv: 'Indian Tropical plants formed in big area & fitted for Tropics & not for temperate parts have invaded & almost exterminated Australian Flora of Tropics.—'

[8] In his essay (Hooker 1859, p. l), Hooker remarked how curious it was that Australian tropical species could grow in India if planted there, yet they had not become established by natural means. On p. civ, Hooker remarked that the future of the Australian flora depended upon its power to compete with immigrant species from the north, and he observed that the more widely distributed plant forms of the northern hemisphere tend to spread and the more peculiar southern forms to die out. CD's 'answer' was presumably that Australian forms were less successful in the 'struggle for existence' when in competition with Indian species.

[9] In noting the remarkable difference between the flora of south-west Australia compared with that of the south-east, Hooker suggested that the difference was all the more difficult to explain because of the continuous land connection (Hooker 1859, p. li). Next to this passage, CD wrote in his copy: 'Eyre's desert between'. The desert, which CD believed acted as a barrier, was named after the Australian explorer Edward John Eyre.

[10] Hooker 1859, p. liv. CD marked the passage in his copy of the work and noted: 'One intermediary dominant form might well destroy several closely allied representative species.'

[11] The sentence was marked by CD in his copy. In the margin, he wrote: 'You do *not* know effect of Salt-Water on the seeds.—'

[12] CD commented in his copy (Hooker 1859, p. xcv):

> I wish I knew which of these *genera* are **not** found in lowland Tropics & include species representative of those f[d] in Northern Europe *or N. Asia [*added*] for on Glacial view these have been modified since Glacial period. (next page 38 *identical* species are given) It has always been my greatest fear that there has been so much modification since Glacial, that it w[d] upset view.— Some few genera may formerly have been mundane & Tropical & not now so.—

The list referred to in the letter is tipped into Hooker 1859 following p. xcvi. It is headed 'Notes for Hooker' and was later dated 'Ap 8th 1860'. Hooker wrote a few comments on the list.

[13] An allusion to a remark attributed to Louis Agassiz that had become a joke between CD, Hooker, and Asa Gray (see *Correspondence* vol. 6, letters to Asa Gray, 1 January [1857] and [after 15 March 1857]).

[14] Hooker's letter has not been found.

[15] CD refers to John Richardson, the Arctic explorer and zoologist, who had apparently discussed *Origin* with Hooker. Richardson had collaborated with William Swainson, a leading advocate of the quinarian system of classification, in describing the birds collected during John Franklin's second polar expedition of 1824.

[16] An anonymous review of *Origin* appeared in the *Saturday Review of Politics, Literature, Science, and Art*, 24 December 1859, pp. 775–6. The reviewer singled out the geological arguments that CD put forward in the exposition of his theory, in particular the vast amount of time required for the gradual transition from one species to another. He criticised CD's calculations concerning the time required for the deposition of various formations and particularly for the denudation of the Weald (*Origin*, pp. 285–7), stating that 'in this delicate part of his case he has committed some manifest errors, and allowed himself assumptions to which he would scarcely have resorted had his need been less urgent.' (p. 776).

[17] CD believed that the review of *Origin* in the *Gardeners' Chronicle and Agricultural Gazette*, 31 December 1859, pp. 1051–2, had been written by the editor, John Lindley. It was printed in the opening columns of the *Gardeners' Chronicle* that were normally reserved for the editor's comments. In fact, Hooker was the author (see letter to J. D. Hooker, 14 [January 1860]). The reviewer stated (p. 1052):

> In fine, whatever may be thought of Mr. Darwin's ultimate conclusions, it cannot be denied that it would be difficult in the whole range of the literature of science to find a book so exclusively devoted to the development of theoretical inquiries, which at the same time is throughout so full of conscientious care, so fair in argument, and so considerate in tone.

[18] George Bentham, who worked closely with Hooker at the Royal Botanic Gardens, Kew, was one of the foremost botanists of the period. CD was anxious to know his opinion of *Origin*, particularly since Bentham had given great thought to the species question and had previously maintained the fixity of forms. See *Correspondence* vol. 7, letter to J. D. Hooker, 14 December [1859], and letter from J. D. Hooker, [20 December 1859].

[19] CD refers to his belief in a former cold period during which northern plants migrated south through the tropics, a view set out in detail in *Origin*, pp. 365–82. Hooker discussed the possibility of European species having migrated through the tropics to Australia in Hooker 1859, pp. cii–cv, concluding that such an occurrence was unlikely. CD's comment about the flora at the foot of the Himalayas alludes to a phenomenon described in Hooker 1854a, p. 109, and cited by CD in *Origin*, p. 378.

From Hewett Cottrell Watson [3? January 1860][1]

Notes on Darwin's 'Origin of Species[2]

1. Admitted the probability of a gradual 'variation' from parent to progeny, from ancestors to descendants, to an unlimited extent, if traced through a vast series of successions.[3]

2. Also, that among the countless numbers of organic beings produced, a process of 'natural selection' will in the long run determine which kinds shall die out, and which shall continue to produce descendants, themselves in turn more or less changing still.

3. Also, that a constant divergence from the lineal types is thus effected; because variation is necessarily divergence, and because no extremely antient type appears to have been exactly continued to the present epoch.

4. Such divergence nepotally onwards necessarily implies an equal convergence

ancestrally backwards; which must therefore be admitted, although unproved except for slight variations and brief periods.

5. Constant convergence backwards must lead at length to one single prototype;—which itself must have been originated in some mode different from the mode in which all its successors have been produced; because that prototype could have had no organic parent.

6. Constant convergence onwards from that prototype must have led successively to types,—

1----10----100----1,000----100,000----1,000,000.

We may assume the million to represent the number of specific or quasi-specific types now existing on the Earth.

7. Are we now at or near the end of the divergent variations? It would seem not. No grounds are shown for supposing that variation from parent to progeny has ceased, or that it is slower or less complete than heretofore. And the theory of the 'Origin' proceeds on an assumption that present facts evidence a continuance of variation now as in eras past.

8. Therefore this figure of 1,000,000 may and will, for aught we can see to the contrary, go on indefinitely increasing until the Earth has its present million increased to millions

2----3----4----5----6----7----8----9----10----

&c. &c. &c.

9. Eventually, in this course, there will be as many millions of species as of beings or pairs of beings; every one or pair still tending to vary. If it be argued that 'natural selection' will prevent this vast increase of specific types, at any one time in existence, the answer is, why did it not arrest them at 10,000 or 100,000?—Why now, rather than then?—Or, why at 2 millions rather than at 4 millions?—&c. &c.

10. Thus, whether we go ancestrally backwards, to arrive at one prototype, not produced like all the rest that have succeeded it;—or go nepotally diverging onwards to a termination in countless million of species,—in either course we arrive at something so totally different from what is now observed, so utterly incapable of proof, so inconceivable as a reality, that it must seem in itself an improbability amounting to an absurdity almost.

11. Ought we not to seek carefully among present facts or events for some counter-action in nature, adequate to save us from being forced upon this almost reductio ad absurdum in either direction, either backwards or forwards?—I think that the counter-action can be seen and specified.

12. Would it not be better to reason up to no beginning, and to no end, than thus to make out a seeming absurdity, by arriving at results totally different from, if not absolutely inconsistent with, the present course of natural events, as seen in our own small span of time?[4]

13. It would appear that the difficulty could be met and neutralized by the

hypothesis or inference that individuals *converge* into orders, genera, species, *as well as diverge* into species, genera, orders, through nepotal descent.

14. No doubt that hypothesis (rejected in the 'Origin') is far more difficult to support by evidence. But various facts in nature seem explicable only on this assumption. And it tends to remove other difficulties which might be urged against the theory of the 'Origin', and which nepotal divergence by itself seems unequal to explain;—for example, the intricately complex manner in which characters are crossed and combined in species; the effect of which would likely be to throw the same species into different groups, not subordinate to each other, if classification were attempted on the hypothesis, that any character in common between two species must have been derived from some common ancestor.

A mem
DAR 47: 135–8

CD ANNOTATIONS

0.1 Notes . . . Species] 'by H. C. Watson' *added ink*

3.1 Also . . . epoch. 3.3] 'I see this is common opinion & was mine but I have come to doubt. To my mind at the time, the cause of divergence was an important step.' *added ink over pencil*

4.1 Such . . . periods. 4.3] 'I do not see why variations should not fluctuate backwards & forwards, without much Divergence.' *added ink*

8.1 Therefore . . . &c. 8.4] 'I *think* I can show that this is not so.—' *added pencil, del pencil*

9.1 Eventually . . . &c. &c. 9.6] 'C. of Good Hope.' *added pencil, del pencil*

12.1 Would . . . ancestor. 14.9] '*Ch 6*'[5] *added brown crayon*; 'I daresay this has taken place to small extent; it is when "*adaptive*" characters affect closely related forms.—' *added ink*

CD note: [6]

Species which are rare (& all w^{d} become rare if infinitely many species) will be very liable to extermination. Moreover I fully believe that rare species, (from fewness of individuals & therefore less chance of *profitable* variations) will vary less quickly,— *(possibly close interbreeding would come into play with very rare species: mem. decrease in size in Auroch, Red Deer &c.) [*interl*] so that the great increase of number of species would act as a sort of regulator to further increase. Moreover Physical conditions are not infinitely different in kind, & as far as adaptations to physical conditions are concerned the number of species would soon come to a limit. But I fully, most fully admit, that the organic [*interl*] relations will tend, with increase of number of species, to go on getting more & more complex, & thus there will be tendency to increase in number of new species. Then will come in checks above alluded to. We do not know the full number anywhere arrived at.—mem. naturalised plants at C. of Good Hope.—Also **dominant** invading species will destroy many representative species: Mem. Hooker on Australia.—'[7] *ink and pencil, del pencil*

[1] The date is suggested by CD's remark in his letter to Charles Lyell, 4 [January 1860], that he had received a long letter from Watson.

[2] The accompanying letter from Watson, alluded to in the letter to Charles Lyell, 4 [January 1860], has not been found.

[3] Watson responded enthusiastically to *Origin* (see *Correspondence* vol. 7, letter from H. C. Watson, 21 November [1859]). In 1845, Watson had publicly announced his own belief in the transmutation of species (Watson 1845), and he again discussed the subject in the final volume of Watson 1847–59. CD had read this volume on publication (see *Correspondence* vol. 7, letter to J. D. Hooker, [26 May 1859]).

[4] This and the following two paragraphs were marked 'Ch 6' by CD (see CD annotations). He refers to chapter 6 of *Origin*, 'Difficulties on theory' (*Origin*, pp. 171–206). CD later responded to Watson's notion of a convergence of species in the third edition of *Origin*, using the arguments set out in the annotations to this letter (*Origin* 3d ed., pp. 141–3; Peckham ed. 1959, pp. 267–70). He concluded that there must be limits to the 'tendency to an indefinite augmentation of specific forms' (*Origin* 3d ed., p. 143). See also letter to H. C. Watson, [5–12 January 1860].

[5] See n. 4, above.

[6] The note was begun at the top of paragraph 9 and continued on the verso of Watson's letter. See also letter to H. C. Watson, [5–12 January 1860].

[7] CD had just finished reading Hooker 1859, in which Joseph Dalton Hooker discussed the invasion of Australia by Indian plant species (see preceding letter).

From Leonard Jenyns[1] 4 January 1860

Jan. 4. 1860.

My dear Darwin

I have read your interesting book with all carefulness as you enjoined,—have gleaned a great deal from it, & consider it one of the most valuable contributions to Nat. Hist Literature of the present day.[2] Perhaps you may like to know what I think of your particular theory, tho' you will doubtless have many opinions offered far more deserving to be weighed than mine. As you are aware, I am no stickler for the multitude of so-called species created by so many naturalists of late years, & I always thought the time was not distant—when, after the brain-splitting process had been carried to it's utmost length—some at least could see the necessity of retracing their steps, & again uniting a large number of the forms they had so carefully separated.[3]

But I frankly confess I did not look for any such large assemblages of species to be brought together in this way, as the descendants from one & the same stock, similar to what you have attempted in your volume. By this you will see that I embrace yr theory in part, but hardly to the full extent to which you carry it.[4] Still I allow you have made out a very strong case, and I will not pretend to say what future researches in the same direction may not ultimately establish.

I can quite fall in with the view that those fossil animals which so closely resemble their living representatives at the present day, are in fact the progenitors of these last;—such indeed has been my opinion for many years, tho' a contrary one, I know had been adopted by many of our first Geologists & Naturalists. I can also well believe that whole families have had a common parentage at some remote period of the past,—& that the same *may* have been the case, reasoning analogically (tho' this is not always safe in Nat. Hist. speculations),—with groups of even higher denomination. But I cannot think that *all* the difficulties which stand in the way of so extensive a generalization have been entirely got over—

It seems to me that if "all organic beings that have ever lived on the earth, had descended from some one primordial form", as you seem to think possible,—we should find (either among the fossil or living species) the same connecting links, &

"fine gradations" between the highest groups, that we *do* find among the lower, & which renders classification & definition so difficult.[5] Thus birds as a class may be shown to be all more & less closely connected, with only here & there a saltus which we cannot hedge over: the same perhaps of all vertebrate animals—if we had all the lost & unknown living forms at hand to fill up the gaps;—but when we come to another quite distinct type of organization—the annulose, for instance, how little of intermediate between this & the vertebrate, or between either of these & the molluscous, &c. Still more what an hiatus between vegetables & animals, tho' just at their respective origins, or where the vital powers seems least active, they are scarcely separable by any distinguishing characters. But why should this never be the case afterwards—& not merely at the beginning of things—if they originally sprung from the same primordial germ?

One great difficulty to my mind in the way of your theory is the fact of the existence of Man. I was beginning to think you had entirely passed over this question, till almost in the last page I find you saying that 'light will be thrown on the origin of man & his history'. By this I suppose is meant that he is to be considered a modified & no doubt *greatly* improved orang! I doubt if this will find acceptance with the generality of readers— I am not one of those in the habit of mixing up questions of science & scripture,[6] but I can hardly see what sense or meaning is to be attached to *Gen*: 2.7. & yet more to vv. 21. 22, of the same chapter, giving an account of the creation of *wo*man,—if the human species at least has not been created independently of other animals, but merely come into the world by ordinary descent from previously existing races—whatever those races may be supposed to have been. Neither can I easily bring myself to the idea that man's reasoning faculties & above all his *moral sense*, c[d] ever have been obtained from irrational progenitors, by mere natural selection—acting however gradually & for whatever length of time that may be required. This seems to be doing away altogether with the Divine Image which forms the insurmountable distinction between man & brutes.

Incomplete
Copy
Kinnordy MS, Charles Lyell's journal V, pp. 95–103

[1] The letter has not been found. CD apparently sent it to Charles Lyell, who copied it into his journal with the heading: 'Letter from Rev[d]. Leonard Jenyns to C. Darwin, Jan. 4. 1860.' The text has been transcribed from Lyell's journal entry; it has also been published in Wilson ed. 1970, pp. 349–51.

[2] CD sent Jenyns a presentation copy of *Origin* (see Appendix III and *Correspondence* vol. 7, letter to Leonard Jenyns, 13 November [1859]).

[3] Jenyns had published a paper on the species question (Jenyns 1856). In 1858, he sent CD his notes on the variability of species (see *Correspondence* vol. 7, letters to Leonard Jenyns, 9 April [1858], 18 April [1858], and [28 April 1858]).

[4] Jenyns was one of the few naturalists to whom CD revealed his views on the transmutation of species before publication (see *Correspondence* vol. 3, letter to Leonard Jenyns, 12 October [1844]).

[5] Jenyns refers to *Origin*, p. 484.

[6] Jenyns was ordained in 1824 and had served as a vicar since 1828.

To Charles Lyell 4 [January 1860]

Down.
4th

My dear L.

Gardener Ch. returned safe.—[1] Thanks for note. I am beyond measure glad that you get more & more roused on subject of species; for, as I have always said, I am well convinced that your opinion & writing will do far more to convince the world than mine.—You will make **grand** discussion on Man.[2] You are very bold in this & I honour you.—

You saw I suppose Saturday Review: argument confined to Geology, but has given me some perfectly just & severe raps on knuckles.—[3] I have been like you quite surprised at want of originality in opposed arguments & in favour too. G. Jeffrey attacks me justly in letter about strictly littoral shells not being often embedded at least in Tertiary deposits.[4] I was in muddle for I was thinking of Secondary, yet Chthamalus applied to Tertiary.— Jeffrey write me, I think nonsense, about non-migration of sea-shells.—

Possibly you might like to see enclosed note from Whewell, merely as showing that he is not horrified with us.—[5] You can return it, whenever you have occasion to write, so as not to waste your time.

C.D. I think it likely I shall have to send long letter from H. C. Watson, not yet read by me.[6]

Endorsement: 'Jan[y] | 1860'
American Philosophical Society (190)

[1] The *Gardeners' Chronicle and Agricultural Gazette* published a review of *Origin* in the issue of 31 December 1859. CD lent Lyell his copy. CD believed that John Lindley had written the review, but he later learned from Lyell that Joseph Dalton Hooker was the author (see letters to Charles Lyell, 10 January [1860], and to J. D. Hooker, 14 [January 1860]).

[2] Lyell was investigating the antiquity of man with the intention of incorporating his results in a new edition of the *Elements of geology*. He subsequently decided that the material demanded a separate work, which he published in 1863 (C. Lyell 1863).

[3] *Saturday Review*, 24 December 1859, pp. 775–6. See also letter to J. D. Hooker, 3 January [1860].

[4] The letter from John Gwyn Jeffreys has not been found; for CD's reply, see *Correspondence* vol. 7, letter to J. G. Jeffreys, 29 December [1859].

[5] Letter from William Whewell, 2 January 1860.

[6] Letter from H. C. Watson, [3? January 1860].

From Asa Gray to J. D. Hooker[1] 5 January 1860

Cambridge, Mass.
Jan[y]. 5/1860

My Dear Hooker

Your last letter, which reached me just before Christmas—has got mislaid during the upturnings in my study which take place at that season, & has not yet been discovered. I should be very sorry to lose it, for there were in it some botanical mem. which I had not secured. It will turn up, I trust. I remember you wrote of dear Lady Hooker being laid up with a bad leg.[2] We hope it will not prove serious,—I fear it will be tedious.

The principal part of your letter was high laudation of Darwin's book.—Well, the book has reached me, and I finished its careful perusal 4 days ago!—And I freely say that your laudation is not out of place.

It is done in a *masterly manner*,—it might well have taken 20 years to produce it. It is crammed full of most interesting matter—thoroughly digested—well expressed—close, cogent—and taken as a system it makes out a better case than I had supposed possible.

D^r. Wyman is just reading it—is struck with its ability,—but I shall know more what he thinks of it presently.[3]

I gave a copy to a hard-headed friend of mine, of very impartial mind, familiar with physical science, thoughtful about general problems in Nat. Hist. but neither naturalist nor geologist. He is much impressed by it.

Agassiz—when I saw him last, had read but a part of it.[4] He says it is *poor*—*very poor*!! (entre nous). The fact he growls over it, like a well cudgelled dog,—is very much annoyed by it—to our great delight—and I do not wonder at it. To bring all *ideal* system within the domain of pure science, & give good physical or natural explanations of all his capital points, is as bad as to have Forbes take the glacier-materials A. had long muddled over, and give scientific explanation of all the phenomena.—[5]

Tell Darwin all this. I will write to him when I get a chance.[6] As I have promised he & you shall have fair play here,—& Dana is in Italy, *ill.*—[7] I must myself write a review of Darwin's book for *Sill. Journal*, (the more so that I suspect Agassiz means to come out upon it)—for the next (March) no.—[8] And I am now setting about it (When I ought to be every moment working Expl. Expedition Compositæ, which I know far more about).—[9] And really it is no easy job, as you may well imagine.

I doubt if I shall please you altogether. I know I shall not please Agassiz at all. I hear ⌈one more⌉ reprint is in press,—and the book will excite much attention here, and some controversy.[10]

Your last letter had the address to *Cambridge* so blotted that the Boston Post office kept it a day or so to decypher. Else I should have had it in time to acknowledge in my last to Sir William.[11] It answers the queries I asked, after the safety of the case I sent last summer, &c, &c.—

AL incomplete
DAR 98 (ser. 2): 20–1

CD ANNOTATIONS
9.1 Your . . . &c.— 9.4] *crossed ink*

[1] Hooker sent the first sheet of this letter to CD (see letter to J. D. Hooker, [22 January 1860]).

[2] Maria Hooker was J. D. Hooker's mother. She was suffering from 'inflammation in the cavity of the Tibia' (see *Correspondence* vol. 7, letter to J. D. Hooker, [12 December 1859] and n. 4).

[3] Jeffries Wyman, professor of anatomy at Harvard University, was a friend of Gray's (Dupree 1959, p. 149). According to Appel 1988, p. 84 n. 40, Wyman borrowed the work from Charles Eliot Norton, who later recalled: 'I had the first copy of the 'Origin of Species' that came to Boston, & after reading it, before it was reprinted, lent it to Jeffries Wyman.'

[4] CD had sent Louis Agassiz a presentation copy of *Origin* (see Appendix III and *Correspondence* vol. 7, letter to Louis Agassiz, 11 November [1859]).

[5] Gray refers to James David Forbes, who had investigated the structure and movement of glaciers at length. Controversy arose between Forbes and Agassiz over an alleged want of recognition on Forbes's part of Agassiz's previous work on the structure of glaciers.

[6] See letters from Asa Gray, [10 January 1860], [17 January 1860], and 23 January 1860.

[7] James Dwight Dana was convalescing in Italy. See letter to T. H. Huxley, 1 January [1860] and n. 6.

[8] [Gray] 1860a.

[9] Since 1848, Gray had been working on a description of the flowering plants collected on the United States Exploring Expedition, the first volume of which was published in 1854. The final volume did not appear until 1873.

[10] See letter from Asa Gray, [10 January 1860].

[11] William Jackson Hooker, the director of the Royal Botanic Gardens, Kew, and J. D. Hooker's father, was a close friend of Gray.

To H. C. Watson[1] [5–11 January 1860][2]

I remember putting the case to myself almost in your words,—Why does not every individual become converted into distinct species? The answer, which I framed, and the discussion on this curious point, which I have somewhere in M.S. is nearly to the following effect: Although the number of species supported over equal areas under apparently nearly similar conditions is now very different in different quarters of the world; and although we are far from knowing that the most prolific area is fully stocked with species, as perhaps may be inferred if not so from some European plants having become naturalized even at Cape of Good Hope, yet Geology shows us, at least within the whole immense Tertiary period, that the number of species of shells & probably of mammals has not increased. How is this? As far as mere physical conditions are concerned (i.e. heat, moisture, height, etc., etc.) it seems probable that a sufficient number of species would soon become sufficiently adapted to the most diversified conditions, but the relation of organic being to organic being, I fully admit is far more important, and as the principle of divergence goes on augmenting the number of species in any area, or in the world, the relation between being and being will become more and more complex, and there seems at first no limit to the amount of possible diversification of structure and therefore no limit to the number of species. But I think the following causes will tend to limit the number (1) the *amount* of life (not number of specific forms) supportable on any area cannot be illimitable, depending so largely as it does on physical conditions; and therefore where very many species are supported, each or nearly each will be few in individuals; and any species with scanty numbers will be liable to extermination from accidental fluctuations in seasons and number of enemies. The process of *exter*mination would in such case be rapid, whereas the process of production of new species would always be slow. Just imagine the extreme case of as many species as individuals in England, and the first severe winter or very dry summer would exterminate forever thousands on thousands of species. (2) I suspect that when any species becomes very rare, close interbreeding will tend to exterminate it; at least authors have thought that this comes into play in accounting for the deterioration of Aurochs in Lithuania and of Red Deer in Scotland, Bears in Norway, etc.[3] (3) As far as animals are concerned some species owe their origin to being adapted to prey on some one other being, but if this other one being be rare, it will not have any advantage to an animal to have been produced in close relation to it, whilst so poor in number, and therefore it will not have been produced. (4) When any species becomes few in number, the process of modification will be slower for the chance of favourable variations arising will be lessened. Therefore if we suppose all or most of the species in any area to be very numerous and consequently each poor in individuals, the proofs of modification and of giving birth to new forms will probably be retarded. (5) And this I am inclined to think is the most important element: a dominant species which has already beaten many competitors will tend to beat more; the relation of organism to organism being more important than relation to physical conditions; and Alph. de Candolle has shown that those species which spread widely tend generally to spread *very* widely,[4] and consequently they will tend to supplant several species in several areas and thus check the inordinate increase of specific forms throughout the world. Hooker has shown that in S.E. corner of Australia, where there are many invaders, (i.e. dominant forms) from different quarters of the world, the endemic Australian species have apparently been greatly reduced.[5]

How much weight to attribute to these several causes I do not pretend to assign; but conjointly I think they must limit the constant tendency to augmentation of specific forms. What do you think? I should like to hear.[6]

With respect to "convergence"[7] I daresay, it has occurred, but I should think on a very limited scale, (owing to strong principle of inheritance retaining resemblance to parents) and only in case of closely related forms.[8] In case of varieties and subvarieties of cultivated plants, I dare say (I have vague remembrance that I have heard such a case) that the offspring of one variety might easily so closely resemble the offspring of a distinct variety as to be classed as a subvariety of it, instead of its own parent. When thinking of analogical or adaptive characters, as of external resemblance of whale and fish, it occurred to me that the same cause acting on two closely related forms (i.e. those which closely resembled each other from inheritance from a common parent) might confound them together so closely that they would be (*falsely* in my opinion) classed in same group. But I should require pretty strong evidence before I should be willing to admit that this had often occurred.—[9]

C. Darwin

Incomplete
Copy
Kinnordy MS, Charles Lyell's journal V, pp. 77–87; DAR 47 (ser. 2): 136a (verso)

CD note:[10]
Jan 15 | Mr H C. W. in letter sticks up for convergence, but says too lengthy to give facts & reasons.— Says truly convergence would convert my diagram into net-work. My argument about same [*above del* '*f*'] species arising in different areas from 2 other species, is in fact applicable to the question; for it comes to this would species *belong to distinct genera [*interl*] in two countries become modified into species of same genus. I doubt.—

[1] Only the last paragraph of the letter has been found in the Darwin Archive (see n. 6, below). The remaining text has been transcribed from Charles Lyell's scientific journal (see also Wilson ed. 1970, pp. 344–6). In addition to CD's letter, Lyell also copied into his journal the text of the letter from H. C. Watson, [3? January 1860]. For CD's intention to send these two letters to Lyell for his comments, see the letter to T. H. Huxley, 11 January [1860].

[2] The letter seems to be a response to the letter from H. C. Watson, [3? January 1860], of which only a part—the notes on *Origin*—is extant; CD received Watson's letter on 4 January (see letter to Charles Lyell, 4 January [1860]). According to the letter to T. H. Huxley, 11 January [1860], CD had already sent his response to Watson. The note presumably refers to points raised by Watson in the part of his letter that is now missing.

[3] See CD's note transcribed following the letter from H. C. Watson, [3? January 1860].

[4] CD refers to A. de Candolle 1855. Alphonse de Candolle did not explicitly make this statement, but in the index of the book's main points that CD compiled for himself, there is the following entry: 'There is p 484. *great* distinction between local species & very widely extended species, i e when a species is once a spreader it spreads *widely*.'

[5] Hooker 1859, p. liii.

[6] The last page of the letter, beginning with 'like to hear.', has been transcribed from a draft in DAR 47 (ser. 2): 136a. It is in the hand of an amanuensis, although CD added several corrections and his signature.

[7] Watson discussed his idea of convergence at length in the letter from H. C. Watson, [3? January 1860].

[8] CD made an outline of his argument in annotations written on the letter from H. C. Watson, [3? January 1860]. CD responded to Watson's criticism in *Origin* 3d ed., pp. 141–3 (see also Peckham ed. 1959, pp. 267–70).

[9] CD discussed the convergence of species in *Origin* 3d ed., p. 141:

> I will only say that if two species of two closely allied genera produced a number of new and divergent species, I can believe that these new forms might sometimes approach each other so closely that they would for convenience sake be classed in the same new genus, and thus two genera would converge into one; but from the strength of the principle of inheritance, it seems hardly credible that the two groups of new species would not at least form two sections of the supposed new single genus.

[10] The note is in DAR 47 (ser. 2): 136a. It is bound with the letter from H. C. Watson, [3? January 1860].

To Thomas Bridges[1] 6 January 1860

Do the Fuegians or Patagonians, or both, nod their heads vertically to express assent, and shake their heads horizontally to express dissent?[2]

Do they blush? and at what sort of things? Is it chiefly or most commonly in relation to personal appearance, or in relation to women?

Do they express astonishment by widely open eyes, uplifted eyebrows and open mouth?

Do they evince anger or fear by same expression of countenance and actions as we do?

When out of spirits or dejected do they turn down the corners of the mouth?

Do they express contempt by the same gestures as we do, namely, by turning up nose and puffing out their breath or even by spitting?

Do they sneer, which is chiefly shown by turning up the corners of upper lip?

Do they frown when trying to understand anything or considering any difficulty?

Do they ever shrug their shoulders to show that they are incapable of doing or understanding anything?

Any information on the manner of expression of countenance of any emotion in savages would be curious, and I believe is a subject, which has been wholly overlooked.—[3]

The only satisfactory method to collect information is to make notes at the time.

What ideas of feminine beauty have the Fuegians? do they admire women with strong American cast of countenance, or such as at all approach Europeans in appearance?

Do the Fuegians take any pains in breeding or matching their dogs; or is all left to chance.— The habits of the Fuegian dogs would be worth observing.[4]

Are the young of the wild Pigs at the Falkland Islands striped lengthways on the back when first born?

What colour are the calves of the wild *White* cattle with red ears, in the Falkland Islands?[5]

Charles Darwin

Down Bromley Kent
Jan. 6th— 1860.

LS[6]
DAR 185: 72

[1] Bridges had lived on Keppel Island, in the West Falklands, since 1856. As a boy, he accompanied George Packenham Despard on an expedition to found an Anglican mission in Tierra del Fuego. Although Despard had the help of Jemmy Button, the Fuegian who had come to England with Robert FitzRoy and returned to his country on the *Beagle* voyage with CD, the attempt ended in disaster. Despard returned to England leaving Bridges in charge of mission work on Keppel Island. Bridges was ordained in 1869 during a visit to England. See Bridges 1933, pp. xiv–xv. The text of the letter has been taken from a copy made by an amanuensis and dated and signed by CD (see also n. 6, below).

[2] CD believed that human facial expressions, their relation to those of animals, and the expression of emotions in general could be explained by natural selection. As early as 1838, he began to record and make observations on expressions, noting the behaviour of animals as well as the development of his own children in this respect (see *Correspondence* vol. 4, Appendix III). See also the letter to

Charles Lyell, 10 January [1860], in which CD states that he had collected much information on the subject of expression. Notes on this subject dating from January 1860 are in DAR 189. The information requested from Bridges was published in *Expression* and his contribution acknowledged in the introduction (*Expression*, p. 22). See letter from Thomas Bridges, [after October 1860?].

[3] CD distributed a printed circular of questions similar to these in 1867. See Freeman and Gautrey 1972; Freeman 1977, pp. 120–2; and *Collected papers* 2: 136–7.

[4] CD cited information from Bridges about Fuegian dogs in *Variation* 1: 39 and 2: 207. See also letter from Thomas Bridges, [after October 1860?].

[5] In January 1860, CD began looking over his manuscript on the variation of domesticated animals and plants ('Journal'; Appendix II). He had received information in 1843 about the colour variation of Falkland cattle from his *Beagle* shipmate Bartholomew James Sulivan, who had surveyed the islands. Sulivan also described the Falkland wild pigs. See *Correspondence* vol. 2, letter from B. J. Sulivan, [10 May 1843].

[6] Both a draft of the memorandum and a copy in the hand of an amanuensis are in the Darwin Archive (DAR 185: 72, 73). The draft was marked by CD 'Expression of Savages' in brown crayon and includes the note: 'Mr Norman | write on both sides of Paper' (DAR 185: 73). CD refers to Ebenezer Norman, his copyist.

To William Benjamin Carpenter 6 January [1860][1]

Down Bromley Kent
Jan. 6th

My dear Carpenter

I have just read your excellent article in the National.[2] It will do great good; especially if it becomes known as your production. It seems to me to give an excellently clear account of M^r^. Wallace's & my views.[3] How capitally you turn the flanks of theological opposers by opposing to them such men as Bentham & the more philosophical of the Systematists![4] I thank you sincerely for the *extremely* honourable manner in which you mention me.— I sh^d^. have liked to have seen some criticisms or remarks on Embryology, on which subject you are so well instructed.[5] I do not think any candid person can read your article without being much impressed with it. The old doctrine of immutability of specifics will surely but slowly die away. It is a shame to give you trouble, but I sh^d^ be very much obliged if you could tell me where differently coloured eggs in individuals of the Cuckoo have been described & their laying in 27 kinds of nests.[6] Also do you know from your own observation that the lambs of sheep imported into W. Indies change colour. I have had detailed information about the loss of wool; but my accounts made the change slower than you describe.—[7]

With most cordial thanks & respect, Believe me | My dear Carpenter | Yours very sincerely | Ch. Darwin

Down House (MS 6: 4)

[1] The year is given by the reference to Carpenter's review of *Origin* (see n. 2, below).

[2] [Carpenter] 1860a. Carpenter told CD in December 1859 that he was going to review *Origin* (see *Correspondence* vol. 7, letter to W. B. Carpenter, 5 December [1859]).

[3] Although entitled 'Darwin on the origin of species', Carpenter's review also discussed Alfred Russel Wallace's contribution to Darwin and Wallace 1858 and Baden Powell's consideration of the species question (Powell 1855).

[4] Carpenter attempted to place the doctrine of transmutation in perspective by comparing it to the recent results of systematists like George Bentham, who had greatly reduced the number of reputed species by reassessing variations that earlier taxonomists regarded as an indication of specific differences. Carpenter went on to state ([Carpenter] 1860a, p. 193):

> We do not expect to see, even in our "most straitest" sectarian organs, any accusations brought against Mr. Bentham for impiety, because he affirms that three or four hundred of the reputed species of British plants are really descendants of others from which they have gradually diverged.

[5] Carpenter was one of the leading figures in comparative anatomy and physiology. He discussed in great detail the embryological doctrines of Karl Ernst von Baer in the fourth edition of his *Principles of comparative physiology* (Carpenter 1854). CD owned a copy of this work (Darwin Library–CUL) and studied it carefully in 1855 (*Correspondence* vol. 4, Appendix IV, 128: 12). In his *Autobiography*, p. 125, CD remarked how disappointed he had been that no reviewers commented on the embryological arguments put forward in *Origin*, pp. 439–50: 'Hardly any point gave me so much satisfaction when I was at work on the *Origin*, as the explanation of the wide difference in many classes between the embryo and the adult animal, and of the close resemblance of the embryos within the same class.' See also *Correspondence* vol. 7, letter to J. D. Hooker, 14 December [1859].

[6] Carpenter discussed the differently coloured eggs of cuckoos as an example of the natural specialisation of habit, stating ([Carpenter] 1860a, pp. 204–5):

> Recent observations on the Cuckoo have shown that this bird deposits its eggs in the nests of no fewer than twenty-eight different species, and that the cuckoo's egg almost invariably agrees so closely in colour with the eggs among which it is laid, as only to be distinguished from them by a practised eye.

[7] Carpenter gave as an illustration of acclimatisation the case of Leicestershire sheep introduced to the West Indies, which had 'their thick matted fleeces replaced in a year or two by short crisp hair' ([Carpenter] 1860a, p. 195).

From Joshua Toulmin Smith 6 January 1860

Highgate
6 Jan 1860

Dear Sir

By the same post with this, I have the pleasure of sending you a copy of my "Ventriculidæ"—of which I beg your acceptance.[1] When you have a little leisure, I think you will find in its contents some matters more than usually illustrative of your views on the development of species. Not to take up, now, too much of your time, I will note two points:—first, here is a group of fossils entirely distinguished, by a very remarkable structure, from any other. We find them running thro' all the Chalk formation—*but with a difference.* Indeed, when I opened your plate of graduating species,[2] I immediately remarked—"this is a table of the Ventriculidæ". There is both the change & branching off of many, & the retention of single forms, just as you have pictured. (Let me refer to pp. 41–43, 55, 73, 88).

Second, there is the difficulty of structure—a structure unapproached by any known structure, recent or fossil. This is a more difficult point than any of the *modifications* of parts. But no doubt you will see the way, from your vast stores of comparison, to illustrate it. I can thus far illustrate it:—that, for years, I have lost no opportunity of pointing out to visitors to my collection, the "*imperfection of the geological record*", as illustrated by the Chalk fossils, & the beds above & below them. I could enlarge much on this,—but will only add, that two flints, compared, are enough to satisfy the most sceptical,—while we have, between the chalk & the London clay, the distinct & unquestionable **marks** of a *lost record*—of which all the details are gone; & what is left is only enough to tell us how much is wanting. I compare it to Livy's History. We have some chapters tolerably complete. Of others we have the mere heads—but enough to know that they once existed.[3]

If it will be agreeable to you, at any future time, to see my collection, I shall have pleasure in showing it to you—& pointing out the illustrations to which I have alluded.

Apologising for the length at which I am troubling you, | I am Dear Sir | Yours faithfully | Toulmin Smith

C. Darwin Esq

Down House (MS 11: 26)

[1] Smith was the first president of the Geologists' Association, founded in 1859, and an expert on the Ventriculidae, fossilised siliceous sponges from the Cretaceous period. His papers on the Ventriculidae of the Chalk, printed in the *Annals and Magazine of Natural History*, were issued as a volume in 1848. A copy of this work (J. T. Smith 1848) is in the Darwin Library–Down.

[2] Smith refers to the diagram in *Origin*, inserted between pp. 116 and 117, illustrating the principle of divergence.

[3] Smith alludes to CD's remark in *Origin*, pp. 310–11, that the geological record is like a history of the world imperfectly kept, of which only a short chapter here and there has been preserved. Titus Livius's *History of Rome* originally had 142 books, of which only 35 survive complete, along with fragments of others. For CD's discussion of the fragmentary nature of the fossil record, see *Origin*, pp. 279–311.

To Asa Gray 7 January [1860][1]

Down Bromley Kent
Jan 7th

My dear Gray

I have just finished your Japan memoir & I must thank you for the extreme interest with which I have read it.[2] It seems to me a most curious case of distribution & how very well you argue & put the case from analogy on the high probability of single centres of creation. That great man Agassiz, when he comes to reason seems to me as great in taking a wrong view as he is great in observing & classifying.[3]

One of the points which has struck me as most remarkable & inexplicable in your memoir is the number of monotypic (or nearly so) genera amongst the representative forms of Japan & N. America. And how very singular the preponderance of identical & representative species in Eastern compared with Western America.—

I have no good map showing how wide the moderately low country is on the west side of the Rocky Mountains: nor of course do I know whether the whole of the low Western territory has been botanised; but it has occurred to me looking at such maps as I have, that the Eastern area must be larger than western which would account to certain small extent for preponderance on Eastern side of the representative species. Is there any truth in this suspicion? Your memoir sets one marvelling & reflecting.— I confess I am not able quite to understand your geology at p 447, 448; but you would probably not care to hear my difficulties. & therefore I will not trouble you with them.[4]

I was so grieved to get a letter from Dana at Florence, giving me a very poor (though improved) account of his health.[5]

Believe me, my dear Gray | Yours very sincerely | C. Darwin

What an admirable memoir on the distribution of Australian plants is that by Hooker![6]

Gray Herbarium of Harvard University (15)

[1] The year is given by the references to Gray 1859 and Hooker 1859.

[2] Gray had sent CD a copy of his description of the plants collected by Charles Wright in Japan (Gray 1859) (see *Correspondence* vol. 7, letter to Asa Gray, 24 December [1859]). Several months earlier CD had read proof-sheets of the concluding section of the paper, sent to him by Joseph Dalton Hooker (see *Correspondence* vol. 7, letter to J. D. Hooker, 11 May [1859]). There is a copy of the work in the Darwin Pamphlet Collection–CUL.

[3] In his paper, Gray discussed the views that had been put forward to explain the distribution of species. He contrasted the opinion of Louis Agassiz that species had originated in great numbers in the geographical area they now inhabit with the theory that each species had originated in a single place from which it dispersed. Gray gave his reasons for supporting the latter view and mentioned CD's contribution to this topic. See Gray 1859, pp. 445–8.

[4] CD felt that there was insufficient geological evidence for Gray to suggest that there had been a post-glacial warm period during which Asian and North American plants had mingled. CD described his objections in detail in letters to Gray and to Hooker in 1859. See *Correspondence* vol. 7, letters to J. D. Hooker, 11 May [1859] and 28 [December 1859], and to Asa Gray, 11 November [1859] and 24 December [1859]. Gray had based his geological arguments on information given to him by his friend James Dwight Dana (Gray 1859, p. 449).

[5] Dana was in Italy convalescing from a breakdown in his health. He had written to CD in December 1859 (see *Correspondence* vol. 7, letter to Charles Lyell, 29 [December 1859]). See also letter to T. H. Huxley, 1 January [1860].

[6] Hooker 1859. See letter to J. D. Hooker, 3 January [1860].

To Leonard Jenyns 7 January [1860]

Down Bromley Kent
Jan. 7th

My dear Jenyns

I am very much obliged for your letter.[1] It is of great use & interest to me to know what impression my Book produces on philosophical & instructed minds. I thank you for the kind things which you say; & you go with me much further than I expected.— You will think it presumptuous, but I am convinced, *if circumstances lead you to keep the subject in mind*, that you will go further. No one has yet cast doubt on my explanation of the subordination of group to group, on homologies, Embryology & Rudimentary organs; & if my explanation of these classes of facts be at all right, whole classes of organic beings must be included in one line of descent.— The imperfection of the geological Record is one of greatest difficulties (by the way, Lyell, who is convert, does not think that I have exaggerated imperfection). During earliest period the record would be most imperfect, & this seems to me sufficiently to account for our not finding intermediate forms between the classes in the same great Kingdoms.—

It was certainly rash in me putting in my belief of probability of all beings having descended from *one* primordial form; but as this seems yet to me probable, I am not willing to strike it out.—[2] Huxley alone supports me in this, & something could be said in its favour.—

With respect to man, I am very far from wishing to obtrude my belief; but I thought it dishonest to quite conceal my opinion.—[3] Of course it is open to everyone to believe that man appeared by separate miracle, though I do not myself see the necessity or probability.—

Pray accept my sincere thanks for your kind note. Your going some way with me gives me great confidence that I am not very wrong. For a very long time I halted half-way; but I do not believe that any enquiring mind will rest half-way. People will have to reject all or admit all,—by *all* I mean only the members of each great Kingdom.—

My dear Jenyns | Yours most sincerely | C. Darwin

Endorsement: '1860.'
Bath Royal Literary and Scientific Institution[4]

[1] Letter from Leonard Jenyns, 4 January 1860.

[2] See letter from Leonard Jenyns, 4 January 1860, in which Jenyns discussed CD's theory of the origin of organic life from one primordial form as given in *Origin*, p. 484. CD did not alter the passage in any substantive way in subsequent editions (Peckham ed. 1959, pp. 752–3).

[3] CD refers to his remark in *Origin*, p. 488: 'Light will be thrown on the origin of man and his history.' Although CD did not explicitly discuss the evolution of man in *Origin*, his concluding remarks imply that his views encompassed human evolution (*Origin*, pp. 484–90).

[4] The letter is bound into Jenyns's copy of the first edition of *Origin*, sent to him by CD in 1859 (see *Correspondence* vol. 7, letter to Leonard Jenyns, 13 November [1859]).

To T. H. Huxley 9 January [1860]

Down Bromley Kent
Jan 9th

My dear Huxley

If on 11th you have half-an-hour to spare you might like to see a very good show of Pigeons, & the enclosed card will admit you.—[1]

The History of Error is quite unimportant, but it is curious to observe how exactly & accurately my Grandfather (in Zoonomia Vol. I. p. 504 1794) gives Lamarcks theory.[2] I will quote one sentence Speaking of Birds Beaks, He says "All which seem to have been gradually produced during many generations by the perpetual endeavour of the creatures to supply the want of food, & to have been delivered to their posterity with constant improvement of them for the purposes required.—"[3] Lamarck published Phil. Zoolog in 1809. The Zoonomia was translated into many languages.—[4]

Ever yours | C. Darwin

Grote, the Historian, was much struck with the article in Times.—[5]

Endorsement: '1860'
Imperial College of Science, Technology and Medicine Archives (Huxley 5: 96)

[1] The annual show of the Philoperisteron Society was held on 11 January 1860 at the Freemason's Tavern, London (*Cottage Gardener*, 17 January 1860, p. 248). CD had been a member of the society since 1855 (see *Correspondence* vol. 5, letter to W. B. Tegetmeier, 31 August [1855]). Huxley intended to describe CD's pigeon work in a lecture on natural selection at the Royal Institution of Great Britain on 10 February 1860; to assist Huxley, CD had provided him with information on pigeon breeding in a previous letter. CD had also mentioned the expertise and quaint practices of pigeon fanciers and recommended that Huxley attend an exhibition (see *Correspondence* vol. 7, letter to T. H. Huxley, 27 November [1859]).

[2] CD refers to his grandfather Erasmus Darwin and Jean Baptiste Pierre Antoine de Monet de Lamarck. Huxley had ridiculed Lamarck's theory of transmutation in his review of *Origin* in *The Times*, 26 December 1859, p. 8. See letter to T. H. Huxley, 1 January [1860].

[3] E. Darwin 1794–6, 1: 504. CD's copy of the work, inherited from his father Robert Waring Darwin, is in the Cambridge University Library. CD marked the passage quoted in the letter and wrote next to it: 'Lamarck!!'. A slip pasted on the same page reads: '504 Lamarck concisely prestalled by my Grandfather'. CD inserted a note to this effect in the 'historical sketch' included in the American edition of *Origin*, p. vi. See Appendix IV.

[4] In spite of CD's inference, there is no evidence to suggest that Lamarck was aware of Erasmus Darwin's work when he composed *Philosophie zoologique* (Lamarck 1809).

[5] George Grote, the author of *A history of Greece* (1846–56), was a fellow of the Royal Society. CD met him when he lived in London (*Autobiography*, p. 111). He had read early volumes of Grote's history in 1853 and finished the sixth volume in December 1859 (*Correspondence* vol. 4, Appendix IV, 128: 5, 25).

From Asa Gray [10 January 1860][1]

Copy-right.[2] And it may be useful, since it is largely occupied with a defence of you against Agassiz—who has been helping the circulation of your book by denouncing it as *atheistical* in a public lecture![3] I suspect, also, he means to attack it in the Atlantic Monthly.[4] The book annoys him; and I suppose the contrast I run between his theories and yours will annoy him still more.[5]

Ever since I read your book I have been busy—absorbed—in a review of it—no easy job, I tell you.— It is only half written now, and is too long, & the printer will want the MSS. this week. So I must hurry up the close.[6]

Lose no time, but send me over a lot of new matter.[7]

Wyman—the best of judges—& no convert, but much struck with it;—says your book is "thundering able",—"a thoroughly scientific & philosophical work".[8] ⟨*section missing*⟩

There are more very strong points made than I can mention here,—but 2 or 3 great gaps in the evidence—some of them you frankly admit,—the others I suppose you do not feel to be as important as I do— I will mention them hereafter— No time now.

Ever dear Darwin | Yours sincerely | Asa Gray

Incomplete
DAR 98 (ser. 2): 26a

[1] Dated by the relationship to the letter to Asa Gray, 28 January [1860], in which CD refers to 'your letter to me of the 10th of Jany.' The manuscript was, however, marked 'Jan 5 1860' by Francis Darwin, perhaps on the basis of the letter from Asa Gray to J. D. Hooker, 5 January 1860.

[2] The missing portion of this letter may have included the suggestion that Gray's review ([Gray] 1860a) be included as a preface to the revised American edition of *Origin* in part to secure CD's copyright in the United States. See n. 7, below, and letter to Asa Gray, 28 January [1860].

[3] Details concerning the public lecture given by Louis Agassiz, in which he discussed *Origin*, have not been traced. Francis Darwin suggested that it was given before the Mercantile Library Association (*LL* 2: 269 n.). Gray subsequently sent CD passages from it. For a transcript of the extract, see the letter to Asa Gray, 28 January [1860], n. 10. At a meeting of the American Academy of Arts and Sciences on 25 January 1860, Agassiz and others 'discussed several points in natural history and geology in reference to their bearing upon the origin and distribution of species' (*Proceedings of the American Academy of Arts and Sciences* 4 (1857–60): 362). Agassiz also denounced CD's views at a meeting of the Boston Natural History Society on 15 February 1860 (*Proceedings of the Boston Natural History Society* 7 (1860): 231–5). For the debate between Gray and Agassiz over CD's theory, see Dupree 1959, chaps. 14 and 15.

[4] Agassiz in fact discussed *Origin* in the July issue of the *American Journal of Arts and Sciences* (Agassiz 1860). Gray wrote a three-part article on *Origin* for the *Atlantic Monthly* ([Gray] 1860b).

[5] Gray alludes to the preface he suggested writing for the American edition of *Origin*. See n. 7, below.

[6] Gray refers to his review of *Origin* published in the March number of the *American Journal of Science and Arts* ([Gray] 1860a). Gray had received his copy of *Origin* around Christmas 1859 (see Dupree 1959, p. 267–8, and letter from Asa Gray to J. D. Hooker, 5 January 1860).

[7] Gray had asked CD about the possibility of an American edition of *Origin* in a letter (now missing) that CD received late in December 1859 (see *Correspondence* vol. 7, letter from J. D. Hooker, [20 December 1859], and letter to Asa Gray, 21 December [1859]). CD consequently arranged for the unbound sheets of the second edition of *Origin*, compiled in November and December 1859, to be forwarded to Gray by John Murray, the English publisher (see *Correspondence* vol. 7, letter to John Murray, 22 December [1859]), and Gray began negotiations with Ticknor and Fields, a Boston publishing firm. Gray probably asked CD in the missing part of the letter to supply as soon as possible any corrections that he would like to make for an authorised version of *Origin*. See also letter from Asa Gray, 23 January 1860, and letter to John Murray, [25 January 1860].

[8] Jeffries Wyman, a colleague of Gray's at Harvard University, taught courses on comparative anatomy and physiology. For his importance alongside Gray and Louis Agassiz in the development of natural history in the United States, see Appel 1988. See also letter from Asa Gray to J. D. Hooker, 5 January 1860, and letter from Asa Gray, 23 January [1860].

To Charles Lyell 10 January [1860]

Down Bromley Kent
Jan. 10th

My dear Lyell

I will run through your letter.—Parthenogenesis (p. **9**6) is nothing & I know not why I inserted it in list.[1] It is perfectly true that I owe nearly all the corrections to you & several verbal ones to you & others: I am heartily glad you approve of them.[2] As yet only two things have annoyed me; those confounded millions of years[3] (not that I think it is probably wrong) & my not having by inadvertence mentioned Wallace towards close of Book in summary,—not that anyone has noticed this to me.— I have now put in Wallace's name at 484 in conspicuous place.—[4]

I cannot refer you to tables of mortality of children &c &c. I have notes somewhere, but I have not the **least** idea where to hunt, & my notes would now be old. I shall be truly glad to read carefully any M.S. on man & give my opinion. You used to caution me to be cautious about man, I suspect I shall have to return the caution a hundred-fold! Yours will no doubt be a grand discussion; but it will horrify the world at first more than my whole volume; although by the sentence (p. 489 new Edit) I show that I believe man is in same predicament with other animals.—[5] It is in fact impossible to doubt it.— I have thought only vaguely on man. With respect to the Races, one of my best chances of truth has broken down from impossibility of getting facts.—[6] I have one good speculative line, but a man must have entire credence in N. Selection before he will even listen to it.—[7] Psychologically I have done scarcely anything. Unless indeed expression of countenance can be included, & on that subject I have collected a good many facts & speculated: but I do not suppose I shall ever publish; but it is an uncommonly curious subject.—[8] By the way I sent off a lot of questions the day before yesterday to Tierra del Fuego on expression![9] I *suspect* (for I have never read it) that "Spencer's Psychology" has a bearing on Psychology, as *we* should look at it.—[10] By all means read Preface in about 20 pages, of Hensleigh Wedgwoods new Dictionary on **first** origin of Language:[11] Erasmus could lend it.[12]

I agree about Carpenter,—a very good article, but with not much original.[13]

I am very sorry that Lindley did not write in Gardeners' Chronicle.—[14]

Andrew Murray (the Entomologist & dabbler in Botany) has criticised in Address to Botanical Soc. of Edinburgh the notice in Linn. Journal,[15] & "has disposed of" the whole theory by an ingenious difficulty, which I was very stupid not to have thought of; for I express surprise at more & analogous cases not being known. The difficulty is that amongst the blind insects of the caves in distant parts of world, there are some of the same genus, & yet the genus is not found out of caves, or living in the free world.[16] I have little doubt that like the fish Amblyopsis & like Proteus *in Europe*, these insects are "wrecks of ancient life" or "living fossils", saved from competition & extermination.[17] But that formerly seeing insects of the same genus roamed over the whole area in which the caves are included.

Farewell | Yours affect^y | C. Darwin

Our ancestor was an animal which breathed water, had a swim-bladder, a great swimming tail, an imperfect skull & undoubtedly was an hermaphrodite! Here is a pleasant genealogy for mankind.—[18]

Postmark: JA 11 60
American Philosophical Society (191)

[1] CD apparently sent Lyell a list of the alterations he had made in the second edition of *Origin.* For a transcript of the list, which was also sent to Asa Gray, see the enclosure to the letter to Asa Gray, 28 January [1860]. Lyell's letter in response has not been found. CD added a reference to 'the curious and not well-understood cases of parthenogenesis' to the second edition of *Origin*, p. 96, in his discussion of the intercrossing of individuals. See also Peckham ed. 1959, p. 185.

[2] For the numerous corrections that Lyell suggested for the second edition of *Origin*, see *Correspondence* vol. 7.

[3] In the discussion of the great length of time required to wear down geological strata (*Origin*, pp. 282–7), CD estimated that the process of denudation of the Weald had probably taken three hundred million years. This estimate was challenged by a reviewer of *Origin* in the *Saturday Review*, 24 December 1859, pp. 775–6. CD had accepted this criticism as justified (see letters to J. D. Hooker, 3 January [1860], to Charles Lyell, 4 [January 1860], and to Asa Gray, 28 January [1860], n. 22).

[4] Alfred Russel Wallace's name was added to the concluding passage of the second edition of *Origin.* The sentence reads: 'When the views advanced by me in this volume, and by Mr. Wallace in the Linnean Journal, or when analogous views on the origin of species are generally admitted, we can dimly foresee that there will be a considerable revolution in natural history.' (*Origin* 2d ed., pp. 484–5; Peckham ed. 1959, p. 754).

[5] CD refers to Lyell's intention to publish an account of the recent work on the antiquity of man. The page reference he gives is to his famous remark: 'Light will be thrown on the origin of man and his history.' CD did not expand this comment in the second edition of *Origin* (see Peckham ed. 1959, pp. 757–8).

[6] CD may be referring to his inquiries concerning the relation between skin and hair colour (and hence ethnic group) and resistance to disease in humans. His attempts to gather material on this subject had not been very successful. See *Correspondence* vol. 6, letters from Thomas Hutton, 8 March 1856, and from W. F. Daniell, 8 October – 7 November 1856; and vol. 7, letter to Asa Gray, 18 November [1858].

[7] CD probably alludes to his idea of sexual selection, expanded at length in *Descent.*

[8] CD had long collected notes on the expression of the emotions. His earliest comments are in his notebooks (*Notebooks*). See also CD's observations on the expressions and behaviour of his children (*Correspondence* vol. 4, Appendix III). He published the results of his research in *Expression.*

[9] Letter to Thomas Bridges, 6 January 1860.

[10] Spencer 1855, which Herbert Spencer presented to CD in 1856 (see *Correspondence* vol. 6, letter to Herbert Spencer, 11 March [1856]). CD's copy of this work is in the Darwin Library–CUL.

[11] Wedgwood 1859–65. Hensleigh Wedgwood's account of the origin of language ran counter to contemporary opinion: he believed that words first emerged as an elaborated imitation of natural sounds. CD's copy of a later edition of the work is in the Darwin Library–CUL.

[12] Erasmus Alvey Darwin, CD's brother.

[13] William Benjamin Carpenter reviewed *Origin* in the *National Review*, 10 (1860): 188–214.

[14] CD initially believed that the favourable review in the *Gardeners' Chronicle*, 31 December 1859, pp. 1051–2, had been written by John Lindley, the editor. The author, however, was Joseph Dalton Hooker. See letter to J. D. Hooker, 14 [January 1860].

[15] Andrew Murray's presidential address of 10 November 1859 was reported in the *Edinburgh New Philosophical Journal* n.s. 11 (1860): 141–51. He discussed Darwin and Wallace 1858 and also *Origin*, whose arguments he had first learned of 'from a friend who had seen the proof-sheets' (*ibid.*, p. 148).

[16] Murray believed that he had found a flaw in CD's theory. He pointed out, as CD states, that eyeless insects of the same genera were found in widely separated caves (sometimes even on different continents) without the genus being represented outside the caves. Murray believed this showed that migration to the interior of caves was out of the question and that particular forms were created for particular conditions (*Edinburgh New Philosophical Journal* n.s. 11 (1860): 149). See also letters to Andrew Murray, 28 April [1860] and 28 [April 1860].

[17] See *Origin*, p. 139.

[18] Lyell made annotations relating to the letter on the cover: 'C. Darwin Mortality of children— Man originally an hermaphrodite— Blind *genus* of insect with wide range. Man & Spencer's Psychology—'.

To T. H. Huxley 11 January [1860]

Down Bromley Kent
Jan. 11th

My dear Huxley

I fully agree that the difficulty is great, & might be made much of by a mere advocate.[1] Will you oblige me by reading again slowly from p. 267–272.—[2] I may add to what is there said, that it seems to me quite hopeless to attempt to explain why varieties are not sterile; until we know precise cause of sterility in species.— Reflect for a moment on how small & on what very peculiar causes the unequal *reciprocity* of fertility in same two species must depend.—Reflect on the curious case of species *more* fertile with foreign pollen than their own. Reflect on many cases which could be given, & shall be given in my larger book (independently of hybridity) of very slight changes of conditions causing one species to be quite sterile & not affecting a closely allied species.—[3] How profoundly ignorant we are on this intimate relation between conditions of life & impaired fertility in pure species.—

The only point which I might add to my short discussion on this subject, is that I think it probable that the want of adaptation to *uniform* conditions of life in our domestic varieties has played an important part in preventing their acquiring sterility when crossed. For this want of uniformity & changes in the conditions of life seems only cause of the elimination of sterility (When crossed) under domestication. This elimination though admitted by many authors rests on very slight evidence, yet I think is very probably true,—as may be inferred from the case of dogs.— Under nature it seems improbable that the differences in the reproductive constitution, on which the sterility of any two species when crossed depends, can be acquired directly by natural selection; for it is of no advantage to the species. Such differences in reproductive constitution must stand in correlation with some other differences; but how impossible to conjecture what these are!

Reflect on case of the vars. of Verbascum which differ in no other respects whatever besides the fluctuating element of colour of flower, & yet it is impossible to resist Gärtner's evidence, that this difference in the colour does affect the mutual fertility of the varieties.[4] The whole case seems to me far too mysterious to rest valid attack on the theory of modification of species, though, as you say, it offers excellent ground for mere advocate.—

I am surprised considering how ignorant we are on very many points, that more weak parts in my Book have not as yet been pointed out to me. No doubt many will be. H. C. Watson founds objection in M.S. on there being no limit to infinite diversification of species: I have answered this I think satisfactorily, & have sent attack & answer to Lyell & Hooker.[5] If this seems to you good objection, I would send papers to you.—

Andrew Murray "disposes of" the whole theory by ingenious difficulty from distribution of blind cave insects; but it can, I think, be fairly answered.[6]

My dear Huxley | Yours most sincerely | C. Darwin

Endorsement: '1860'
Imperial College of Science, Technology and Medicine Archives (Huxley 5: 98)

[1] Huxley was preparing a lecture on CD's theory to be delivered as one of the Friday evening lectures at the Royal Institution on 10 February 1860. In his lecture, Huxley stated his general approval of CD's work, but he nonetheless said that it 'falls short of being a satisfactory theory', primarily because 'there is as yet no *proof* that, by selection, modifications having the physiological character of species (*i.e.*, whose offspring are incapable of propagation, *inter se*) have ever been produced from a common stock.' (T. H. Huxley 1860a, p. 198). A note in CD's hand in DAR 205.7 (2): 149 discusses Huxley's point:

> Jan 11. 1860. Huxley objects to domestic vars. not being sterile. I answer in addition to my Book. Too ignorant of causes of sterility in species.— *Reflect on what slight causes in conditions render pure species sterile [*interl*] Reflect on unequal reciprocity, on Hippeastrum case—on Verbascum case. Probably want of *adaptation to [*interl*] uniform [*altered from* 'uniformity'] conditions in domestic productions important; for this seems to cause elimination of sterility. The Verbascum case shows how ignorant we are on what differences [*interl*] the difference in reproductive condition are correlated: it is probably correlation.—

The 'addition to my Book' is a reference to *Variation*, in which both *Hippeastrum* and *Verbascum* are discussed (*Variation* 2: 138–9, 105–7, 136–7).

[2] These pages of *Origin* discuss the 'Fertility of varieties when crossed, and of their mongrel offspring'. CD contended that the sterility of interspecific hybrids, when contrasted with the fertility of crosses between different varieties, does not imply an essential distinction between species and varieties. He argued that the sterility of interspecific hybrids is not a special endowment but is gradually acquired through divergent modifications in the reproductive systems of the forms that are crossed (*Origin*, p. 272).

[3] CD intended to publish a comprehensive work on natural selection giving all his references and sources and a more considered view of the subject. His plan was never realised beyond the publication of *Variation*, which was the first part of the planned 'larger book'. In *Variation* 2, chap. 18, the beneficial and detrimental effects of changed conditions of life on fertility are discussed in detail.

[4] Gärtner 1849, pp. 215–16. CD's annotated copy is in the Darwin Library–CUL. These pages were heavily marked by CD. Karl Friedrich von Gärtner's case of the incipient sterility of different varieties of *Verbascum* according to the colour of the flowers had interested CD greatly and was cited in *Origin*, pp. 270–1. See also *Correspondence* vol. 5, letter to M. J. Berkeley, 7 April [1855], and letter from C. C. Babington, [*c.* June 1855].

[5] Letter from H. C. Watson, [3? January 1860], and letter to H. C. Watson, [5–11 January 1860]. The letters to Charles Lyell and to Joseph Dalton Hooker have not been located.

[6] See preceding letter and nn. 15 and 16.

To *Gardeners' Chronicle* [14–19 January 1860][1]

I hope that some of your readers will respond to Mr. Westwood's wish, and give any information which they may possess on the permanence of cross-bred plants and animals.[2] Will Mr. Westwood be so good as to give a reference to any account of the variability of the Swedish Turnip?[3] I did not even know that it was reputed to be a cross-bred production. I am aware that this is supposed to be the case with some Turnips; but I have searched in vain for any authentic history of their origin. No one, I believe, doubts that cross-bred productions tend to revert in various degrees to either parent for many generations; some say for a dozen, others for a score or even more generations. But cannot breeders adduce some cases of crossed breeds of sheep and pigs (such as the Shropshire or Oxford sheep, or Lord Harborough's pigs) which are now true?[4] With respect to the Cottagers' Kale, I was so much surprised at the accounts of its trueness that I procured seed from the raisers; but in my soil the plants were far from presenting a uniform appearance.[5] In addition to the tendency to reversion to either parent form, it is almost universally asserted that cross-bred productions are highly variable, and often display characters not observed in either parent. I do not wish to dispute this common belief, but I suspect it would puzzle any one to adduce satisfactory cases; and certainly Gärtner has advanced a mass of evidence on the opposite side.[6] I am not at all surprised at Mr. Westwood demurring to the belief that occasionally crossing the strain is advantageous or necessary with productions in a state of nature. The subject is only just alluded to in my volume on the "Origin of Species."[7] I do not pretend that I can prove the truth of the doctrine; but I feel sure that many important facts and arguments can be adduced in its favour. The ill effects of close inter-breeding between the nearest relations, especially if exposed to the same conditions of life, would be, I believe, the same under Nature as under domestication,—namely, some degree of sterility and weakness of constitution. Variability arises from quite independent causes, and is to a certain extent counteracted in its early stages by the free crossing of the individuals of the same species. Mr. Westwood misunderstands me if he supposes that it is my opinion that the Ibis, for instance, keeps true to its kind "by occasional crosses with individuals of the same species which have not sprung from the same grandfather or great-grandfather."[8] I only believe that if individuals of the Ibis did vary, such crosses would tend to keep the species true; and further, if the young from a single pair increased so slowly that they all continued to inhabit the same small district, and if brothers and sisters often united during successive generations, then that the Ibis would rapidly deteriorate in fertility and constitution. Mr. Westwood advances the hive-bee, as probably a case of constant intercrossing. Andrew Knight, however, who specially attended to this point, has published his belief (Whether founded on sufficient evidence I will not pretend to say) that the queen-bee commonly unites with a drone from another community.[9]

Charles Darwin, Down, Bromley, Kent.[10]

Gardeners' Chronicle and Agricultural Gazette, 21 January 1860, p. 49

[1] The letter responds to a note published in the *Gardeners' Chronicle and Agricultural Gazette* (see n. 2, below) and was itself published in the following number of the journal. It is reprinted in *Collected papers* 2: 31–2.

[2] John Obadiah Westwood published a notice in the *Gardeners' Chronicle and Agricultural Gazette*, 14 January 1860, p. 26, entitled 'Doubtful permanence of cross-breeds'. He stated his belief that plants of hybrid or mongrel origin would ultimately revert to their original type and suggested that this was 'at once contradictory to the theory of progressive development'. In his closing remarks, Westwood stated that he hoped to induce a discussion on 'these and other questions raised by Mr. C. Darwin in his very remarkable volume lately published'.

[3] Westwood referred to the search under way to find a substitute for the Swedish turnip, which was rapidly deteriorating despite the efforts of agriculturalists to keep it true-breeding. The report was printed in the *Gardeners' Chronicle and Agricultural Gazette*, 10 December 1859, p. 997.

[4] CD discussed crossed breeds of sheep that remain true in *Variation* 2: 95–6. The pigs are the breed known as the Improved Essex, bred by Robert Sherard, Earl of Harborough (H. D. Richardson 1847, p. 37).

[5] Cottagers' kale is produced by crossing common kale with Brussels sprouts and then recrossing the progeny with purple broccoli. It was reported to breed true (*Gardeners' Chronicle and Agricultural Gazette*, 10 February 1858, p. 128).

[6] Gärtner 1849. CD studied this work closely when compiling the manuscript of his species book (see *Correspondence* vols. 5 and 6). His notes on the book are in DAR 116. CD discussed Karl Friedrich von Gärtner's assertion that plants of hybrid origin were not variable in *Variation* 2: 265.

[7] *Origin*, pp. 96–7.

[8] Westwood had asserted: 'To say that wild animals or plants which keep true to their kind—as, for instance, the Egyptian Ibis (Which is still identical with the individuals found in the mummy pits)—are only enabled to do so by occasional crosses with individuals of the same species which have not sprung from the same grandfather or great-grandfather is fallacious.' (*Gardeners' Chronicle and Agricultural Gazette*, 14 January 1860, p. 26). He further maintained that even if only two Ibises were left, they would breed true for the next 4000 years without needing to cross-breed with a dissimilar strain.

[9] Knight 1828, p. 319.

[10] Westwood's reply to CD's letter appeared in the *Gardeners' Chronicle and Agricultural Gazette*, 11 February 1860, p. 122. He emphasised the point that 'progressive development' could not take place while there was such a pronounced tendency for the reversion of varieties and mongrels back to the original type.

To J. D. Hooker 14 [January 1860]

Down.—
14th

My dear Hooker

I must just thank you for your long & interesting letter. I am very sorry to hear about Lady Hooker: it must be a great anxiety to you. An operation is so horrid a thing.[1]

Do not forget, when free from care & if you ever have time how heartily glad we sh^{d} be to see you here.

I heard from Lyell this morning & he tells me a piece of news.— You are a good-for-nothing man: here you are slaving yourself to death with hardly a minute to spare & you must write a review on my Book! I thought it a very good one, & was so much struck with it, that I sent it to Lyell.[2] But I assumed as a matter of course that it was Lindleys:[3] now that I know it is yours I have reread it, & my kind & good friend it has warmed my heart with all the honourable & noble things you say of me & it. I was a good deal surprised at Lindley hitting on some of the remarks, but I never dreamed of you. I admired it chiefly as so well adapted to tell on the readers of the G. Chronicle; but now I admire it in another spirit.

Farewell with hearty thanks. What a lot we shall have to talk about if we ever meet.— There are several points in your Essay which I wish to discuss.[4] (I have just lent it to J. Lubbock) & some point in Asa Grays.[5]

Farewell your affect. | C. Darwin

I shall be very glad to hear that Lady Hooker has got over her dreadful trial.

We have three children in bed with measles.[6]

Lyell is going at Man with an audacity that frightens me: it is a good joke he used always to caution me to slip over man.—

Endorsement: '/Jany 60'
DAR 115.1: 36

[1] Hooker's mother, Maria Hooker, was suffering from a leg ailment.

[2] See letter to Charles Lyell, 4 [January 1860].

[3] CD refers to Hooker's anonymous review of *Origin* published in the *Gardeners' Chronicle and Agricultural Gazette*, 31 December 1859, pp. 1051–2. John Lindley was the editor of the *Gardeners' Chronicle*.

[4] Hooker 1859.

[5] Gray 1859. CD had recently received a copy of this work, an analysis of the Japanese flora (see *Correspondence* vol. 7, letter to Asa Gray, 24 December [1859]). His annotated copy is in the Darwin Pamphlet Collection–CUL.

[6] According to Emma Darwin's diary, Elizabeth, Francis, and Leonard Darwin contracted measles in January 1860. See also letter to J. D. Hooker, 3 January [1860].

To Charles Lyell 14 January [1860]

Down Bromley Kent
Jan. 14th

My dear Lyell

Your gossip about Reviewer has interested & amused me much.— I have reread Gardeners Ch. with extreme pleasure now that I know Hooker is author.—[1]

Jeffreys' letter is really not worth sending:[2] he says nothing about migration but only refers me to his paper on Testacea of Piedmont in Annals & Mag. of Nat. History for Feb. 1856.[3] He refers, to another paper on British shells, *but gives no sort of reference*, & I have looked over all his papers, which I have, & can find nothing. Perhaps I spoke too strongly about his non-migration views;; I remember thinking him far too narrow & decided in his opposition to Forbes. for casting my eye now over his paper I find my memory seems to˛have exaggerated his non-migration doctrine.[4] I hardly understand what he objects to, in my Book; unless it be that I follow Forbes on Glacial migration—[5] I am convinced from littoral shells of Galapagos Arch. that such shells must have some great power of migration; for if you require continuous coasts for the migration of littoral shells, you would have to unite these islands, within the period of living shells *both* to America & the far distant Pacific islands, which, I presume would stagger even the warmest followers of Forbes.[6] By the way it strikes me that one of grandest points in Hooker's Essay is the (in my eyes, but I presume by no means in his eyes) *demonstration* that N. Zealand has not been continuously, or even nearly continuously, united with Australia within recent times.[7]

I shall be much interested in reading your man discussion, & will give my opinion carefully whatever that may be worth; but I have so long looked at you as the type of cautious scientific judgment (to my mind one of the highest & most useful qualities) that I suspect my opinion will be superfluous.— It makes me laugh to think what a joke it will be, if I have to caution you, after your cautions on same subject to me!.—

I will order Owen's Book:[8] I am very glad to hear Huxley's opinion on his classification of man; without having due knowledge it seemed to me from very first absurd.[9] All classifications founded on single character, I believe have failed.

I agree with Huxley that it is a difficulty about no ascertained varieties known to have been raised by man, being sterile together: varieties of same kind, I believe, not rarely prefer pairing together; I have fact on this head. But I think Huxley had not considered my discussion. (p. 268 new Edit.) The sterility of the **vars** of Verbascum is a most wonderful case, & that of Tobaccos. But the subject of sterility is so profoundly mysterious, as I could show, & as anyone will see who will read even my short chapter, that it would be truly wonderful if no difficulties were presented; & as you say the subject would be no subject for discussion.[10]

With respect to Geoffroy St. Hilaire, I have been glancing over his Life by Isidore & his Principes;[11] & it seems to me that he was a rather doubtful maintainer of change of species.— Isidore writes to me that he himself is a firm maintainer of such views: he says he has sent me a publication of his to show this, but it has not arrived.[12]

With respect to cave insects the same *seeing* genera do range over, as you well know, Europe & N. America, & the difficulty applies chiefly in this case of America & Europe. *Do you know of any miocene or pliocene fossil insect in* **N. America**; your hint would be **very valuable**, if I could show that some **extinct** genera were common to both countries?—

Goodnight— | Yours affect | C. Darwin

What a grand, immense benefit you conferred on me by getting Murray to publish my Book.—[13] I never till to day realised that it was getting widely distributed; for in a letter from a lady today to Emma, she says she heard a man enquiring for it at *Railway Station*!!! at Waterloo Bridge; & the Bookseller said that he had none till new Edit. was out.— The Bookseller said he had not read it but had heard it was a very remarkable book!!!

Can you give me any suggestion how to get a German Translator. As for France I suppose there will be no chance: Madame Belloc finds it too difficult.—[14] (Since this written by odd chance I have had letter from Frenchman who wishes much, to translate.)[15]

Postmark: JA 15 60
American Philosophical Society (192)

[1] See preceding letter.

[2] John Gwyn Jeffreys's letter has not been found, but its contents can be inferred from CD's reply (see *Correspondence* vol. 7, letter to J. G. Jeffreys, 29 December [1859]).

[3] Jeffreys 1856. There is an annotated offprint in the Darwin Pamphlet Collection–CUL.

[4] In his paper, Jeffreys criticised Edward Forbes's division of the Mediterranean fauna into 'elements', each of which originated in different geographical areas. CD marked the passages discussing this point in his copy of Jeffreys's paper (Darwin Pamphlet Collection–CUL). However, Jeffreys went on to state that all he contended was 'that the areas of geographical distribution, as proposed by the late Professor Forbes and others, are much too restricted to existing circumstances, and that they ought rather to be referred to a prior state of things.' (Jeffreys 1856, p. 171).

[5] On the question of plant migration during the glacial period, Forbes hypothesised that as the ice retreated, alpine plants continued to survive on mountain tops in the temperate zones (E. Forbes 1846).

[6] CD had debated the land-bridge theory of geographical distribution and other means of dispersal with both Lyell and Joseph Dalton Hooker in 1846 and again in 1856 (see *Correspondence* vols. 3 and 6). CD particularly alludes to the heated correspondence he had with Lyell about Forbes's resort to hypothetical former continents in order to explain patterns of geographical distribution (*Correspondence* vol. 6, letter to Charles Lyell, 25 June [1856]).

[7] Hooker 1859, p. lxxxvii. In his copy (Darwin Library–CUL), CD noted next to this passage: 'very striking, looks to me conclusive that never continuous land'.

[8] R. Owen 1859a. See letter to Williams and Norgate, 16 January 1860.

[9] Thomas Henry Huxley's comments on Richard Owen's classification of man have not been located, but he discussed the subject at length in T. H. Huxley 1861b. CD had made much the same remark about an earlier paper by Owen on classification (see *Correspondence* vol. 6, letters to J. D. Dana, 5 April [1857] and to J. D. Hooker, 5 July [1857]).

[10] See letter to T. H. Huxley, 11 January [1860]. CD discussed the varieties of *Verbascum* and the sterility of tobacco hybrids in *Variation* 2: 105–7, 136–7, 108–9.

[11] CD was composing a historical sketch of evolutionary theories for future editions of *Origin*, in which he mentioned the work of Étienne Geoffroy Saint-Hilaire. He refers to Isidore Geoffroy Saint-Hilaire's biography of his father (I. Geoffroy Saint-Hilaire 1847) and Étienne Geoffroy Saint-Hilaire's *Principes de philosophie zoologique* (E. Geoffroy Saint-Hilaire 1830). CD recorded having read the biography in 1855 (*Correspondence* vol. 4, Appendix IV, 128: 14).

[12] The letter has not been found, but see the letter to Isidore Geoffroy Saint-Hilaire, 28 January [1860]. The reference is to I. Geoffroy Saint-Hilaire 1851, a copy of which is in the Darwin Pamphlet Collection–CUL.

[13] For the circumstances involving John Murray's publication of *Origin*, see *Correspondence* vol. 7, letters to Charles Lyell, 28 March [1859] and 30 March [1859], and to John Murray, 31 March [1859].

[14] Louise Swanton Belloc had contacted CD in November 1859 offering to translate *Origin* into French. See *Correspondence* vol. 7, letter to John Murray, 14 November [1859].

[15] See following letter.

To Edward Cresy 15 January [1860][1]

Down Bromley Kent
Jan. 15th

My dear Sir,

If you kindly would, you could *perhaps* do me a considerable service.— A M. Talandier, French Master at Royal Military School at Sandhurst writes to me that he wishes to translate the "Origin" into French.—[2] He refers me as witnesses of his capability to Louis Blance[3] of 13 George St. Portman Sqe & to M. Alexandre **H**erzen,[4] of Park House, Percy Cross *[illeg]*[5]

Now I heard Mrs Cresy say she knew L. Blanc. Could you enquire for me whether he thinks M. Talandier is capable of translating my Book, ie a scientific Book with a good deal of reasoning in it.— I do not think it would be very easy to translate. Can you aid me & let me have a line.—

In Haste to catch Post—

Yours sincerely | C. Darwin

I hope you received copy of my Book.—

Private collection

[1] The year is given by CD's reference to a French translation of *Origin*.

[2] Pierre Théodore Alfred Talandier was one of four masters teaching French at the Royal Military College, Sandhurst, in Berkshire. CD had previously called upon Cresy to help him obtain information from French sources (see *Correspondence* vol. 4). For CD's previous negotiations concerning Louise Swanton Belloc's proposed translation, see *Correspondence* vol. 7, letters to John Murray, 14 November [1859], 2 December [1859], and 4 December [1859].

[3] Louis Blanc was a French socialist living in exile in London.

[4] Alexander Ivanovich Herzen was a Russian socialist and revolutionary currently living in England.

[5] CD apparently could not read the writing. He seems to have written 'Froohham' over which he then placed a large question-mark, partially obscuring some letters.

To Williams and Norgate 16 January [1860][1]

Down Bromley Kent
Jan 16th

Dear Sir

Please to send to 57 Queen Anne St Cavendish Sqre.— Sir Emerson Tennants work on Ceylon.—[2] and Owen's Classification & Distribution of Mammalia: Parker: 1859.—[3]

I have received your account for the Parcels which you so kindly sent for me & I will pay it, when I get the account for above books.—[4]

Yours very faithfully | C. Darwin

American Philosophical Society (B/D25.291)

[1] The recipient and the year are provided by an entry in CD's Account book (see n. 4, below).

[2] Tennent 1859.

[3] R. Owen 1859a, published by John William Parker and Son. The work is not in the Darwin Library.

[4] CD recorded a payment of £3 11*s.* 4*d.* to the London booksellers Williams and Norgate on 29 January 1860 (CD's Account book (Down House MS)). CD engaged this firm in November 1859 to send presentation copies of *Origin* abroad (see *Correspondence* vol. 7, letters to John Murray, 15 October [1859], and to J. D. Hooker, [23 October 1859]).

From Asa Gray [17 January 1860][1]

⟨reprint⟩, your new book,—promise to allow you copy right equivalent— will not wait for electrotypes of the cuts.—[2]

But they (& I) want you to send a note or two—to go at end, and a *preface*—a few words—to identify it as *your* ed. & to secure the market against any other reprint. Send at once any *corrections* you are making for your 2^{d} ed.[3]

I noted, with pleasure, your son's success at Cambridge.[4]

At my house, a neighbor mentioned a case of man, he knew, born legless—two of whose children were equally so. I am trying to get authentic evidence about it. ⟨*section missing*⟩

Do you know of dogs which—without ever having been taught, lick their paws and then wash their faces like a cat? A black- & tan we have does so regularly, I know not where he picked up the habit.— he must have inherited it.

Ever Yours, | A. Gray

Incomplete
DAR 165: 102

CD ANNOTATIONS

1.1 ⟨reprint⟩ . . . about it. 4.3] *crossed pencil*

5.1 Do you . . . regularly, 5.2] *double scored blue crayon*

End of letter: '(Instinct—)' | 'Animaux fossile de *[*Pi Vermi*]*' *pencil*; 'Index | London'[5] *ink, crossed ink*; 'Extraordinary kind— *another set of [*interl*] new sheets—& scored sheets— return latter. A few additions for final note Preface consult Lyell—'[6] *ink, crossed ink*

[1] Dated on the assumption that this is the letter referred to by CD as 'your letter of the 17th' in his letter to Asa Gray, 1 February [1860].

[2] See letter from Asa Gray, [10 January 1860], in which Gray discusses his attempt to secure an authorised American edition of *Origin*. His remarks may relate to his negotiations with the Boston firm of Ticknor and Fields. But see letter from Asa Gray, 23 January 1860, in which Gray relates how the news that a New York publishing house had already printed an edition has changed his plans. However, CD did eventually supply a preface, several additional notes, and a supplement to D. Appleton & Co.'s revised and augmented edition of *Origin*. See Appendix IV.

[3] Gray was about to receive proof-sheets of the second English edition of *Origin*, sent to him by John Murray in December 1859 (see *Correspondence* vol. 7, letter to John Murray, 22 December [1859]). He was not yet aware that the second edition had been issued on 7 January 1860 ('Journal'; Appendix II).

[4] William Erasmus Darwin had won a scholarship at Christ's College, Cambridge (*Cambridge University Calendar* 1860).

[5] The annotations are possibly an aide-mémoire for books that CD wished to acquire. See letter to Williams and Norgate, 29 [January 1860].
[6] See letter to Asa Gray, 1 February [1860].

To Baden Powell 18 January [1860]

Down Bromley Kent
Jan. 18th

My dear Sir

I am much pleased by your approbation of my book, as everyone must admit that you are a master in philosophical logic;[1] I am the more pleased at this, as one eminent scientific man writes to me that I have violated the whole spirit of inductive philosophy.[2]

My health was so poor, whilst I wrote the Book, that I was unwilling to add in the least to my labour; therefore I attempted no history of the subject; nor do I think that I was bound to do so. I just alluded indeed to the Vestiges & I am now heartily sorry I did so.[3] No educated person, not even the most ignorant, could suppose that I meant to arrogate to myself the origination of the doctrine that species had not been independently created. The only novelty in my work is the attempt to explain *how* species become modified, & to a certain extent how the theory of descent explains certain large classes of facts; & in these respects I received no assistance from my predecessors. To the best of my belief I have acknowledged with pleasure all the chief facts & generalisations which I have borrowed. If I have taken anything from you, I assure you it has been unconsciously; but I will reread your Essay.[4] Had I alluded to those authors who have maintained, with more or less ability, that species have not been separately created, I should have felt myself bound to have given some account of all; namely, passing over the ancients, Buffon (?) Lamarck (by the way his erroneous views were curiously anticipated by my Grandfather), Geoffry St. Hilaire & especially his son Isidore; Naudin; Keyserling; an American (name this minute forgotten); the Vestiges of Creation; I believe some Germans. Herbert Spencer; & yourself.—[5]

The task would have been not a little difficult, & belongs rather to the Historian of Science than to me. I ought also to have alluded to chief maintainers of opposite doctrines.— I had intended in my larger book to have attempted some such history; but my own catalogue frightens me. I will, however, consult some scientific friends & be guided by their advice.

Permit me to add that I read your Philosophy of Creation with great interest: it struck me as excellently & vigorously argued & written with a clearness, which I remember excited my warmest admiration.[6] I most fully agree that your work must have had a great effect with philosophical minds in removing prejudices on the subject; in a higher degree but in nearly the same manner as the Vestiges has had with a less highly endowed class of readers. I have had to make by letter the same acknowledgement to the Author (as I believe) of the Vestiges.[7] By the above remarks

I do not by any means intend to say that your work has not entirely converted many readers, & induced them to give up the doctrine of creation; in simple truth I do not at all know how the case stands.— I sh^{d} not have *presumed* to have made these remarks on your work, had not your letter induced me.

Believe me, with sincere respect, My dear Sir | Yours sincerely | Charles Darwin

Endorsement: '60'
Quentin Keynes

[1] Baden Powell was Savilian professor of geometry at Oxford University. He was a frequent contributor to scientific periodicals on mathematical topics, but was most famous for his works on doctrinal and philosophical questions. His letter to CD has not been found, but its substance can be inferred from a favourable notice of *Origin* that Powell inserted in Powell 1860 at the proof stage. He considered *Origin* to be a 'masterly volume' (Powell 1860, p. 139). Powell died in June 1860.

[2] CD refers to Adam Sedgwick. See *Correspondence* vol. 7, letter from Adam Sedgwick, 24 November 1859.

[3] CD refers to *Vestiges of the natural history of creation* ([Chambers] 1844). He mentioned the work briefly in the introduction to *Origin*, pp. 3–4, and had been accused of borrowing from its content (see letter to T. H. Huxley, 1 January [1860]).

[4] Powell had enthusiastically adopted the doctrine of the uniformity of nature in the 1830s and discussed the origin of species in several works, most notably in Powell 1855, in which he treated *Vestiges* ([Chambers] 1844) sympathetically. He rejected the idea of supernatural intervention and endorsed a history of living beings based on the fulfilment of a divine plan by natural means. See Corsi 1988.

[5] CD refers to Georges Louis Leclerc, Comte de Buffon; Jean Baptiste Pierre Antoine de Monet de Lamarck; Étienne and Isidore Geoffroy Saint-Hilaire; Charles Victor Naudin; Alexandr Andreevich Keyserling; and [Chambers] 1844. CD prepared a 'historical' essay for the revised American edition of *Origin*, expanded for subsequent editions, in which he mentioned all these authors (see Appendix IV). The American to whom he alludes was Samuel Steman Haldeman (see following letter).

[6] Powell 1855. CD recorded reading this work in January 1856 (*Correspondence* vol. 4, Appendix IV, 128: 14). His notes on it are in DAR 71: 43–50.

[7] The author of *Vestiges* was believed by many to be Robert Chambers, but his authorship was not publicly acknowledged until after his death (see A. Desmond 1982, pp. 210–11 n. 28, and Secord 1989). CD guessed that Chambers was the author in 1847 (see *Correspondence* vol. 4, letter to J. D. Hooker, [18 April 1847]). CD's letter to Chambers has not been found.

To Baden Powell 18 January [1860]

Down Bromley Kent
Jan 18th

My dear Sir

Thinking over my letter addressed to Athenæum Club to you this morning,[1] as far as I can remember it, it has just occurred to me that you might misunderstand one passage; & though I do not suppose that you would care much for my opinion, I sh^{d} be very sorry that anyone should suppose that I ranked your Essay & the Vestiges in the same class. I coupled them merely in relation to both having produced a good effect on the public mind;—the Vestiges probably on a greater number but on a very inferior class.—[2]

The more I think of the whole subject the more difficult I feel it would be to give a fair account of the several authors who have maintained on various grounds the modification of species.— I beg pardon for troubling you with this second note & remain

My dear Sir | Yours sincerely | Charles Darwin

Haldeman is name of American Author forgotten this morning.[3]

I have just bethought me of a Preface which I wrote to my larger work, before I broke down & was persuaded to write the now published Abstract.[4] In this Preface I find following passage, which on my honour I had as completely forgotten as if I had never written it. "The "Philosophy of Creation" has lately been treated in an admirable manner by the Revd. Baden Powell in his Essay &c &c 1855. Nothing can be more striking than the manner in which he shows that the introduction of new species is "a regular not a casual phenomenon",[5] or as Sir John Herschel expresses it "a natural in contradistinction to a miraculous process"."[6]

Endorsement: '60'
Quentin Keynes

[1] See preceding letter.
[2] For a survey of the responses to [Chambers] 1844, see Secord 1989.
[3] Charles Lyell reminded CD about Samuel Steman Haldeman's paper (Haldeman 1843–4) in June 1859 (see *Correspondence* vol. 7, letter to Charles Lyell, 21 June [1859]). CD included Haldeman among those who had previously discussed the species question in the 'historical sketch' prepared as a preface for the American edition of *Origin*. See Appendix IV.
[4] CD refers to the manuscript on species that he wrote between 1856 and 1858 (*Natural selection*), from which *Origin* was largely compiled. The preface has not been preserved.
[5] The quotation is taken from Powell 1855, p. 359.
[6] John Frederick William Herschel's remark was printed in Babbage 1837, p. 204. CD included this statement in his historical sketch (see Appendix IV).

To Edward Cresy 20 January [1860]

Down, Bromley, Kent
Jan: 20th.

My dear Sir

You have been extraordinarily kind in taking so much trouble for me.[1] Your account this morning has been so satisfactory, that I have written to M. Talandier to agree to his proposal.[2] It is an immense advantage to any scientific work and subject to be translated into French, as it can then be read in any country. I am a **very** poor French scholar, though I read it with fluency, and it would no doubt be an immense advantage if you were to read over the proofs; but then I know how hard you are worked, and I have scruples about accepting your extremely kind offer. Indeed it is too much; but perhaps you would not object to look at any passage which seemed to me doubtful.—

With my hearty thanks to you and Mrs. Cresy | Believe me, Yours very sincerely | Charles Darwin

I am very much pleased at what you say about my Book in your first note.—You have praised exactly what pleases me most to be praised.

Copy
DAR 143

[1] See letter to Edward Cresy, 15 January [1860].
[2] CD's letter to Pierre Théodore Alfred Talandier has not been found. The name 'Talandier' was added to the copy of the letter in pencil.

To William Bernhard Tegetmeier 20 January [1860][1]

Down Bromley Kent
Jan. 20th

My dear Sir

I have been looking over the Poultry Book again & have found so very many remarks, which have interested me greatly, that I am anxious to know whether more than No XI have been published, which is all that I have got.[2] Will you kindly take the trouble to inform me.—

You make like to hear about the mongrel poultry, on which you gave me much all-important assistance.[3] The result has now become interesting. The chickens from Spanish Cock & White Silk fowl were jet black in the Down & in first plumage, but late in Autumn to my astonishment red feather after red very slowly appeared *in the Cocks* & now one of the Cocks is *nearly* as splendid as the wild Gallus Bankiva. Another young cock from game Hen was *quite* white, but now has very much red about it, ie the hackles red. Are not these curious facts?[4]

How goes on the Apiarian Soc.?[5] And I hope that you yourself & family are well. As soon as I finished my Book (& I hope you received your copy safely) was finished, I went to Ilkley Wells for 10 weeks to recruit;[6] but it did not do me much good, & presume I shall remain a wretched valetudinarian to the day of my death.

I received a note, when at Ilkley; but as it did not particularly require an answer, & as I was then extra unwell, I did not write.—

Yours sincerely | C. Darwin

Endorsement: '**1860**'
New York Botanical Garden Library (Charles Finney Cox collection)

[1] The year is given by the reference to CD's visit to Ilkley (see n. 6, below).
[2] Tegetmeier ed. 1856–7. The work was issued in parts, and publication ceased with number 11. CD's annotated copy is in the Darwin Library–CUL.
[3] CD first asked Tegetmeier 'whether Fowls when crossed throw odd & unexpected colours like Pigeons do.—' in 1857 (see *Correspondence* vol. 6, letters to W. B. Tegetmeier, 12 [May 1857] and 18 May [1857]). In 1859, CD acquired several different breeds of fowl from Tegetmeier and initiated a cross-breeding experiment to ascertain whether the mongrel offspring resemble the colouring of the ancestral breed *Gallus bankiva*. The results are recorded in CD's Experimental book, pp. 41 and 49 (DAR 157a).
[4] CD described these cases in *Variation* 1: 240–2.

[5] Tegetmeier was a founding member and secretary of the Apiarian Society (E. W. Richardson 1916). He had a keen interest in bee studies and assisted CD with his researches in 1858 and 1859 on the construction of bees' cells (see *Correspondence* vol. 7).

[6] CD visited a hydropathic establishment in Ilkley, Yorkshire, from 2 October to 9 December 1859 (see *Correspondence* vol. 7, 'Journal'; Appendix II).

To T. H. Huxley 21 [January 1860][1]

Down Bromley Kent
21st

My dear Huxley

I have told Murray to send you copy of 2[d] Edit of my Book.[2] I ought to have thought of this before, as you have been beyond all or nearly all the warmest & most important supporter. I did not think of it, simply from the corrections being so few (of which I send list)[3] & now I really hardly know whether you will care to have copy; but you can give it away, if you do not care.—

I long to have a little talk with you. I had firmly resolved to come up & dine with you all at Athenæum, but my accursed health made it impossible.[4] I intend coming up on Tuesday evening & will call early on Wednesday at Museum for chance of seeing you; & sh[d] I fail on Wednesday in being able to come or in your not being there I will call on Thursday.—[5]

Could you let me have on Wednesday at Museum Pigeon M.S.—[6] I am beginning to think of, & arrange my fuller work;[7] & the subject is like an enchanted circle; I cannot tell how or where to begin.—

By strange chance, since sending you the Drawings, I have had specimen & have now prepared the skull of the Bagadotten (of which I send Plate out of German Book) & the extraordinarily curved beak is not exaggerated.—[8]

I cannot think it possible that you can wish to keep, but I do not want M.S. on Hybrids.

In Haste | My dear Huxley | Most truly yours | C. Darwin

I have never received from Ray Soc. your Volume on Hydrozoa:[9] I must enquire what cause is.

Imperial College of Science, Technology and Medicine Archives (Huxley 5: 102)

[1] The date is given by the reference to CD's trip to London (see n. 5, below).

[2] The second edition of *Origin* was published by John Murray on 7 January 1860 ('Journal'; Appendix II).

[3] For CD's list of corrections, see the enclosure with the letter to Asa Gray, 1 February [1860].

[4] Huxley became a member of the Athenaeum Club in 1858. CD had been a member since 1838. The dinner referred to took place on 19 January, and those attending included Huxley, Joseph Dalton Hooker, and certain unidentified 'Naturalists'. See letter to J. D. Hooker, [22 January 1860].

[5] On 27 January 1860, CD recorded the expenses of a trip to London in his Account book (Down House MS). Emma Darwin's diary indicates that CD was away from home from 24 (Tuesday) to 27 (Friday) January. Huxley had an office in the Museum of Practical Geology.

[6] CD had lent Huxley his manuscript describing the results of his study of domestic pigeons to assist Huxley in the preparation of his Royal Institution lecture on 10 February 1860 (T. H. Huxley 1860a). See *Correspondence* vol. 7, letters to T. H. Huxley, 16 December [1859] and 24 December [1859].

[7] CD had begun to prepare the more comprehensive work on natural selection in which he planned to give further examples and citations not included in *Origin*. The first part of this project, *Variation*, appeared in 1868. The second and third parts were ultimately abandoned.

[8] The Bagadotten-Tauben is a breed of pigeon in which the beak curves downwards 'in a highly remarkable manner' (*Variation* 1: 141, 163). The German book to which he refers is Riedel 1824. There is an annotated copy of the work in the Darwin Library–CUL.

[9] T. H. Huxley 1859b. A copy of the work, published by the Ray Society, is in the Darwin Library–Down.

To Jean Louis Armand de Quatrefages de Bréau 21 January [1860][1]

Down Bromley Kent
Jan. 21st

My dear Sir

I am sorry to hear that you have been ill, & that you should have had the trouble of writing to me. I thank you cordially for all your great kindness. The gentleman who wishes to translate my Book is M. Talandier, Professor of French at the Royal Military College at Sandhurst. He was a procureur-general, & I imagine is an exile; but I hear on very good authority that he writes a remarkably good style, & is a very clever man.[2] Should he not succeed in getting a Publisher, or for any reason change his mind, I will in that case venture to apply to you again. But I have no reason to doubt his success.

I have heard this morning from N. America that my Book is exciting considerable attention there amongst the naturalists. But I further hear that Agassiz is very savage at it.—[3]

With my most sincere thanks for all your kindness, I remain, with much respect. My dear Sir | Yours very sincerely | C. Darwin

P.S. How I should like to know whether Milne Edwards has read the copy which I sent him & whether he thinks I have made a pretty good case on our side of the question. There is no naturalist in the world for whose opinion, I have so profound a respect. Of course I am not so silly as to expect to change his opinion.—[4]

I did not understand in your former note whether your Lectures on Anthropology are published; but I shall soon know for I have sent to my Bookseller to procure them, if they are published.—[5]

American Philosophical Society (193)

[1] Dated by the reference to a French translation of *Origin* (see n. 2, below).

[2] Pierre Théodore Alfred Talandier had written to CD offering to translate *Origin* into French (see letters to Edward Cresy, 15 January [1860] and 20 January [1860]). CD had previously asked Quatrefages de Bréau whether he knew of anyone who might be willing to prepare a French translation (see *Correspondence* vol. 7, letter to J. L. A. de Quatrefages de Bréau, 5 December [1859]). Talandier's only published work was as co-author of a radical pamphlet, *Lettre au Parlement et à la presse* (London 1858).

[3] Letter from Asa Gray, [10 January 1860]. See also following letter and letter to Asa Gray, 28 January [1860].

[4] CD corresponded with Henri Milne-Edwards about crustacean embryology and development when he was composing his monograph on Cirripedia, and his work in this area had had a great impact on CD (see *Correspondence* vol. 4 and Appendix II). As a token of his esteem, CD had dedicated *Living Cirripedia* (1854) to him. So far as is known, Milne-Edwards never wrote to CD about *Origin*. In 1867, he discussed the importance of CD's theory for the development of general zoology and was generally positive, although he criticised aspects of the work, particularly the attempt to explain all species change on the basis of natural selection. See Milne-Edwards 1867, pp. 420–32.

[5] Quatrefages de Bréau's letter of 19 November 1859 has not been found, but see *Correspondence* vol. 7, letter to J. L. A. Quatrefages de Bréau, 5 December [1859]. Quatrefages de Bréau was appointed professor of the natural history of man at the Muséum d'histoire naturelle in Paris in 1855. His lectures on anthropology were published in Quatrefages de Bréau 1867–8.

To J. D. Hooker [22 January 1860][1]

Down. Bromley Kent
Sunday

My dear Hooker

What a splendid, magnificent letter from Asa Gray![2] I should out of pure vanity rather like to keep first sheet, so do not throw it away.—[3] It is rich about Agassiz.

I mean to come to London on Tuesday evening for the vain purpose of consulting a new Doctor for my stomach;[4] & for the Club on Thursday, where, if I can possibly, attend, I very much hope to see you.[5] What a time it is since I have seen you my dear old friend, & such kind & generous sympathy you have shown me.—

God Bless you | Yours affect | C. Darwin

I tried to come up last Thursday to dine at Athenæum with you Naturalists, but failed.

I hope Lady Hooker goes on favourably.—[6]

Is there a cool-greenhouse Goodenia, so that I could get a plant & examine the process of impregnation?[7]

A capital clear article in todays Gardener Ch. I suppose by you.—[8]

Endorsement: 'Jany 60'
DAR 115.1: 37

[1] Dated by the reference to CD's intention to attend a meeting of the Philosophical Club of the Royal Society (see n. 5, below) and to the review of *Origin* in the *Gardeners' Chronicle* (see n. 8, below).

[2] See letter from Asa Gray to J. D. Hooker, 5 January 1860.

[3] The first sheet of the letter is in DAR 98 (ser. 2): 20–1.

[4] There is an entry dated 24 January 1860 in CD's Account book (Down House MS) that records a payment to 'Mr Headland & Physic'. Edward Headland was the leading general physician in London; his address in Portland Place is recorded in CD's Address book (Down House MS).

[5] CD refers to a meeting of the Philosophical Club of the Royal Society held on 26 January 1860, at which he was present (Philosophical Club minutes, Royal Society).

[6] See letter to J. D. Hooker, 14 [January 1860].

[7] CD feared that *Goodenia* presented a strong case against his doctrine that all organic beings occasionally cross. Auguste de Saint-Hilaire maintained that in *Goodenia* the pollen is 'enclosed in a cup surrounding

the stigma & is then hermetically sealed; so that here a cross would appear physically impossible.' (*Natural selection*, pp. 63–4, 72).

[8] 'Natural selection', the continuation of Hooker's anonymous review of *Origin*, appeared in the *Gardeners' Chronicle and Agricultural Gazette*, 21 January 1860, pp. 45–6.

From Asa Gray 23 January 1860

Cambridge.
Jany. 23, 1860

My Dear Darwin

You have my hurried letter telling you of the arrival of the remainder of the sheets of the reprint, and of the stir I had made for a reprint in Boston.[1] Well, all looked pretty well, when, lo, we found that a 2d New York publishing house had announced a reprint also! I wrote, then to both N.Y. publishers, asking them to give way to the *author* & his reprint of a revised edition. I got answer from the Harpers' that they withdrew,—from the Appletons that they had got the book *out*—(and the next day I saw a copy); but that "if the work should have any considerable sale we certainly shall be disposed to pay the author reasonably & liberally".[2]

The Appletons being thus out with their reprint, the Boston house declined to go on.— So I wrote to the Appletons.—taking them at their word—offering to aid their reprint,—to give them the use of the alterations in the London reprint, as soon as I find out what they are &c And I sent them the 1ry leaf, and asked them to insert, in their future issue the additional motto, from Butler—which tells just right.[3]

So there the matter stands. If you furnish any new matter in advance of London 3d edition, I will make them pay for it.—[4] I may get something for you—all got is clear gain—but it will not be very much, I suppose.

Such little notices in the papers here as have yet appeared are quite handsome and considerate.

I hope next week to get printed sheets of my review from New Haven and send to you,[5] and will ask you to pass them on to Dr. Hooker.

To fulfil your request I ought to tell you what I "think the weakest, & what the best parts of your book. But this is not easy, nor to be done in a word or two. The *best part*, I think, is the *whole*, i.e. its *plan* & *treatment*,—the vast amount of facts and acute inferences handled as if you had a perfect mastery of them. I do not think 20 years too much time to produce such a book in.

Style clear & good, but now & then wants revision for little matters. (p. 97, self-fertilises *itself*—&c)

Then your candor is worth everything to your cause. It is refreshing to find a person with a new theory who frankly confesses that he finds difficulties—insurmountable, at least for the present. I know some people who never have any difficulties, to speak of.

The moment I understood your premises, I felt sure you had a real foundation, to build on. Well, if one admits your premises, I do not see how he is to stop short of your conclusions, as a probable hypothesis, at least.

It naturally happens that my review of your book does not exhibit any thing like the full force of the impression the book has made upon me. Under the circumstances I suppose I do your theory more good here, by bespeaking for it a fair and favorable consideration, and by standing non-committal as to its full conclusions, than I should if I announced myself a convert,—nor could I say the latter, with truth.

Well, what seems to me the weakest point in the book is the attempt to account for the formation of organs,—the making of eyes, &c by natural selection. Some of this reads quite Lamarckian.

The Chapter on *Hybridism* is not a *weak*, but a *strong* chapter. You have done wonders, there. But still you have not accounted as you may be held to account, for divergence up to a certain extent producing increased fertility of the crosses—but carried one short, almost imperceptible step more, giving rise to sterility, or reversing the tendency Very likely you are on the right track; but you have something to do yet in that department.

Enough for the present.

By the way, *Wyman*—the person best prepared to criticise your book of any one in America—for all matters of embryology, zoology, anatomy &c—and as cautious as possible—could give you many interesting facts & hints.[6] He promised me he would write to you; but he has gone to Florida for winter vacation—to be absent till March.

He says it is all gammon about Agassiz, not knowing to what class an *unticketed embryo* belonged.[7] Agassiz denies that he ever said so, only that he could not tell whether bird or reptile. Pray send me a reference to the authority of the statement on p. 439, line 13, &c. I dare say you have it complete.[8] A. often says *too much* and the reporters take it down.

Wyman can give you plenty of cases of barred or striped horses, &c—and other interesting facts picked up on your own ground in S. America, &c—

Hooker was quite right in letting me know frankly your opinion & Lyell's about warmer period after glacial; and you quite wrong in wishing he had not.[9]

I relied wholly on Dana for all this,[10]—and it still seems to me that the facts look that way moderately It is hard to believe that elephants, ever so woolly could live in our present Arctic regions—or the *Mastodon* in New England and New York with such winters as we have. I see well enough that the difference could not have been very great,—but, in spite of the musk-ox—which is a stumper—I should expect something of a case to be made out. But this is all out of my line. I do not pretend to have any right to speak or think about it. Dana, unfortunately is ill—quite broken down & in Italy. Next spring when I hope he will be well again, and in England, you must see him; and I shall write him that he must give you good grounds for his opinion, or give it up.[11] He is a capital fellow, generally very sober-minded, a good investigator—but as a reasoner apt to be seduced by fallacious analogies.—

I know from you that the Forbesian doctrine about alpine plants &c, &c originated with you, and have always put your name forward in relation to it.[12]

I am not insensible to your compliments—the very high compliment which you pay me in valuing my opinion. You evidently think more of it than I do—tho' from the way I write you, and especially Hooker, this might not be inferred from the reading of my letters.

I am free to say that I never learned so much from one book as I have from yours. There remain 1000 things I long to say about it.—

Ever Yours | Asa Gray.

DAR 98 (ser. 2): 22–5

CD ANNOTATIONS

0.1 My Dear . . . Dr Hooker. 5.2] *crossed pencil*
2.3 give them the use . . . are &c 2.4] *cross added, brown crayon*
7.1 (p. 97, . . . &c) 7.2] *cross added, brown crayon*
10.2 Under . . . consideration, 10.4] *cross added, brown crayon*
10.5 nor could I . . . truth. 10.6] *cross added, brown crayon*
11.1 Well, . . . organs,— 11.2] *cross added, pencil; circled pencil*
12.2 But still . . . divergence 12.3] *cross added, pencil; circled pencil*
12.5 Very likely . . . department. 12.6] *cross added, brown crayon*
14.3 He promised . . . you;14.4] *cross added, brown crayon*
15.3 Pray send . . . &c. 15.4] *cross added, brown crayon*
16.1 striped horses,] *cross added, brown crayon*
18.7 Dana, . . . ill—] *cross added, brown crayon*

[1] Letter from Asa Gray, [10 January 1860]. Gray refers to the proof-sheets of the second edition of *Origin* sent to him by John Murray. See also letter from Asa Gray, [17 January 1860].

[2] Gray refers to the New York publishing firms Harper & Brothers and D. Appleton & Co. His quotation is taken from a letter from D. Appleton & Co. to Asa Gray, dated 19 January [1860], in the Gray Herbarium Archives, Harvard University. The letter reads:

> We have stereotyped & printed the 'Origin of Species' and could publish it almost any day. We dont know what may be the changes in the 2^d^ ed^n^. or if they could appear at the end— If the work sh^d^ have any considerable sale we certainly shall be disposed to pay the author reasonably & liberally

[3] A quotation from Joseph Butler's *Analogy of revealed religion* (1736) was added to the frontispiece of the second English edition. It reads:

> The only distinct meaning of the word 'natural' is *stated, fixed*, or *settled*; since what is natural as much requires and presupposes an intelligent agent to render it so, *i.e.*, to effect it continually or at stated times, as what is supernatural or miraculous does to effect it for once.

[4] The third edition of *Origin* was published in April 1861 (Freeman 1977, p. 78). Appleton distributed four separate printings of *Origin* in 1860. Although the text of the first three printings is identical to that of the first English edition, the second printing included the quotation from Butler (Freeman 1977, p. 83). The fourth printing, which was issued in May 1860 (see 'Journal'; Appendix II), is described on the title page as 'a new edition, revised and augmented by the author'. It differs textually from the first and second English editions of *Origin* by including the 'historical sketch' as a preface, a section added to the end of chapter four responding to certain criticisms of CD's theory, and a supplement containing further 'additions and alterations . . . received too late to be incorporated in their proper places'. See letter to Asa Gray, 1 February [1860], and Appendix IV. This edition is hereafter referred to as

the revised American edition (*Origin* US ed.). For the financial arrangements that Gray negotiated, see Dupree 1959, pp. 270–2.

[5] Gray's review of *Origin* ([Gray] 1860a) was published in the *American Journal of Science and Arts*, whose editorial office was in New Haven, Connecticut.

[6] Jeffries Wyman, Hersey professor of anatomy at Harvard, wrote to CD later in the year (letter from Jeffries Wyman, [*c*. 15] September 1860). He had recently returned from a scientific expedition to South America.

[7] In *Origin*, p. 439, CD mentioned Louis Agassiz in support of his statement that the embryos of distinct animals of the same class are often strikingly similar: 'a better proof of this cannot be given, than a circumstance mentioned by Agassiz, namely, that having forgotten to ticket the embryo of some vertebrate animal, he cannot now tell whether it be that of a mammal, bird, or reptile.'

[8] In the revised American edition and the third English edition of *Origin*, CD replaced the statement attributed to Agassiz with a quotation to the same effect from Karl Ernst von Baer (*Origin* US ed., pp. 381–2; *Origin* 3d ed., p. 471; Peckham ed. 1959, pp. 685–6). See also letter to Asa Gray, [15? February 1860].

[9] CD was critical of the geological argument Gray had advanced in Gray 1859 to explain the distribution of northern plant species, and he had mentioned this to Joseph Dalton Hooker. See letter to Asa Gray, 7 January [1860] and n. 4.

[10] James Dwight Dana believed that there was evidence in North America of a post-glacial warm period, which would account for Gray's postulated second commingling of the Arctic and northern flora of Asia and America (Gray 1859, pp. 447–8).

[11] Gray eventually relinquished the idea of a second mingling (Dupree 1959, pp. 251–2).

[12] Edward Forbes was the first to publish the theory that alpine plants, following the worldwide cold period, would have remained on mountain tops in temperate regions (E. Forbes 1846). CD had earlier developed the same view as an explanation for certain patterns of geographical distribution. In a letter to Asa Gray, 11 August [1858] (*Correspondence* vol. 7), CD told Gray that he had 'written out' such a theory some four years before Forbes published. On this basis, Gray credited CD as being the originator of the theory in Gray 1859, p. 446 (*ibid.*, letter to Asa Gray, 24 December [1859]).

To Richard Henry Meade[1] 23 January 1860

Down, Bromley, Kent.
January 23, 1860.

Dear Sir,

I hope that you will excuse the liberty I take in writing to you and requesting a favour. In the 'Annals of Nat. Hist.,' vol. 15, p. 39, you remark "The variations of form in the maxillæ are of no value among the Phalangida in affording generic or specific characters as with the true spiders."[2] Am I to understand from the latter part of sentence that with the individuals of the same undoubted species the maxillæ vary in form? Is not this a very surprising fact? Would you have the great kindness, if the fact be so, to give me some details on the amount and kind of variations and in what species. And further would you permit me to quote any such facts on your authority?[3]

With many apologies for troubling you, I beg to remain, Dear Sir, | Yours very faithfully, Charles Darwin.

Proceedings of the Royal Society of London ser. B 93 (1922), p. xi[4]

[1] In the published version of the letter, the recipient was incorrectly identified as Louis Compton Miall (see n. 4, below). The author of the paper to which CD refers, however, was the entomologist Richard Henry Meade.

[2] Meade 1855, p. 396. The Phalangida, or Opiliones, is an order of arachnids commonly known as daddy-long-legs or harvestmen.

[3] There is no record of a reply from Meade or of any further use of his paper by CD.

[4] The letter was published in an obituary notice of the entomologist and palaeontologist Louis Compton Miall and is followed by the statement: 'One wonders what answer was sent, and whether Darwin was aware that his correspondent was a boy of eighteen.' (*Proceedings of the Royal Society of London* ser. B 93(1922), p. xii).

To John Murray 23 [January 1860][1]

Down Bromley Kent
23[d]

My dear Sir

As you consented to M. Belloc's wish to translate my Book,[2] I concluded that I need not trouble you before agreeing (after making *careful* enquiry with most satisfactory results) to permit "M. Talandier, Professor of French Royal Military College, Sandhurst" to translate my Book.—

Please send him copy of 2[d] Edit. by post to above address.

I hope I have not acted wrongly in agreeing without consulting you; but I did so merely to save you trouble.— I shall see you, I hope, on Wednesday or Thursday[3]

My dear Sir | Yours very sincerely | Charles Darwin

John Murray (publishers)

[1] Dated by the reference to CD's trip to London (see n. 3, below).

[2] See *Correspondence* vol. 7, letter to John Murray, 14 November [1859]. Louise Swanton Belloc had found *Origin* too scientific to translate easily (see letter to Charles Lyell, 14 January [1860]).

[3] CD went to London on 24 January 1860 and returned to Down on 27 January (Emma Darwin's diary). See also following letter.

To John Murray [25 January 1860][1]

57. Queen St. Cavendish Sqe
Wednesday night

My dear Sir

Will you be so kind as to inform me in note addressed to Down how soon you intend going to press with my Journal; as I think I had better look over it, & see whether progress of science makes any corrections necessary.[2]

I shall be very glad to escape labour of correcting press, & if M[ess] Clowes print it (as I hope) they print so correctly that they may safely be trusted.[3] I, also, hope that they can correct index. I can do so little work per diem, that I grudge every day taken from my larger work.—

It would be **great advantage** to have outline map of S. part of S. America, such as was published by Colburn in his 1st Edition of my Journal.[4]

My dear Sir | In Haste | Yours very sincerely | C. Darwin

Let me hear how many copies of Journal sold altogether.

P.S. I have just received letter from Prof. Asa Gray, who says the U. States in a few days "will be flooded" with my Book.[5] He says it was too late to make any arrangement; that the sheets ought to have been sent him of 1st Edit.[6] But he still thinks of republishing a long Review which he is writing as an Introduction to a separate Edition; & he believes this Introduction would give him copy-right— I feel sure that I may tell him, that if he likes to do anything of the sort, we will share any profit with him.[7] I presume it would be a mere trifle. He is as noble & good a man as ever lived,—but much overworked.—

It will be worth while to remember & send over sheets of my larger work.— I am amused by Asa Gray's account of the excitement my Book has made amonst Naturalists in U. States. Agassiz has denounced it in Newspaper, but yet in such terms that it is in fact a fine advertisement!—[8]

Yours most sincerely | C. Darwin

N.B | I have forgotten the object of my note to say that at date of A. Grays letter he had not received remainder of sheets of the 2d Edit: if not already sent, please send immediately.—

Endorsement: '1860'
John Murray (publishers)

1 The date is given by the reference to the publication of *Origin* in the United States and by CD's stay in London from Tuesday, 24 January, to Friday, 27 January 1860 (Emma Darwin's diary). The address is that of CD's brother Erasmus Alvey Darwin, with whom CD usually stayed when in London.

2 It seems likely that CD had met Murray earlier in the day, as arranged in the preceding letter, and discussed a new edition of CD's *Journal of researches*. Murray published the second edition in 1845 (see *Correspondence* vol. 3) and had issued several reprints since then, including a so-called 'new edition' in 1852 (see Freeman 1977, pp. 39–40).

3 William Clowes and Co. were the printers generally employed by John Murray. They printed *Journal of researches* 2d ed. and *Origin*.

4 The first edition of *Journal of researches* was published in 1839 by Henry Colburn. For CD's unsatisfactory relations with Colburn at that time and in 1845, when he began negotiations with Murray, see *Correspondence* vols. 2 and 3. Murray did not include a map in the second edition (see *Correspondence* vol. 3, letter to John Murray, [31 May 1845], and Freeman 1977, p. 39).

5 Letter from Asa Gray, [10 January 1860]. The part of the letter to which CD refers is now missing.

6 CD had asked Murray to send Gray the proof-sheets of the second edition of *Origin* in December 1859 (see *Correspondence* vol. 7, letter to John Murray, 22 December [1859]).

7 See letter to Asa Gray, 1 February [1860]. CD had not yet received the letter from Asa Gray, 23 January 1860.

8 See letter from Asa Gray, [10 January 1860] and n. 3.

To T. H. Huxley [26 January 1860][1]

[London]
Thursday

My dear Huxley

I have arranged all about the Pigeons.—[2] All the main kinds & some Dovecot & some pretty Toy Pigeons will be sent in open show Baskets.—

But please observe you must write 4 clear days before to "John Baily Esq^e^ Junr 113 Mount St. Grosvenor Sq^e^" & tell him where & at what hour the Pigeons must be sent.[3] And please remember I promised he sh^d^ have ticket for Lecture, though I suspect the old Father-Poulterer will seize on the Ticket.—

They would not hear of any payment, but I promised that the man who brought & attended the Pigeons should have a douceur of ^s^2 · ^d^6 or 5^s^.— Perhaps two men will be sent.—

Ever yours | C. Darwin

Do not forget 3 tickets for my Family as you promised.—

Please put down my Brother as Subscriber to Herbert Spencer Book—[4]

E. Darwin Esqr
57 Queen Anne St | (W.)

Imperial College of Science, Technology and Medicine Archives (Huxley 5: 119)

[1] Dated by the preparations for Huxley's forthcoming lecture, given on 10 February 1860 at the Royal Institution (T. H. Huxley 1860a). CD's reference to having made arrangements about the pigeons that Huxley wanted to exhibit suggests that he was in London. CD visited London from 24 to 27 January 1860 (Emma Darwin's diary).

[2] Huxley planned to exhibit several of the major breeds of the domestic pigeon at his Royal Institution lecture on Friday, 10 February 1860.

[3] John Baily, and his father, John Baily Sr, were dealers in fancy poultry and pigeons in London. They had provided CD with many birds for his study of the effects of artificial selection (see *Correspondence* vols. 5 and 6).

[4] Spencer 1860–2. The work, entitled *First principles*, formed the first volume of Herbert Spencer's five-part series entitled 'A system of philosophy'. It was issued in parts to subscribers only, at an annual cost of 10*s*. Huxley was a close friend of Spencer's; he later read the proof-sheets of this work (L. Huxley ed. 1900, 1: 212). Like Erasmus Alvey Darwin, CD also subscribed to the series (see letter to Herbert Spencer, 2 February [1860]). A copy of Spencer 1860–2 is in the Darwin Library–CUL.

From John Murray 27 January 1860

Jan 27 1860

I have just got in the amount of the Expence of producing the New Edition of 3000 Copies of your Work on Species and I find that after deducting the cost of production including advertising and binding, the Sale of that Number will leave a total profit of £950—two thirds of which will amount to *636. 13. 4*[1] for which Sum I am prepared to give you my notes of Hand at 6 & 9 Months from the day of publication according to the terms of our agreement signed J. Murray

Mem M[r] M. offers by Same Letter to give author anything that may be paid by the Americans for *early Sheets*[2]

Incomplete copy
John Murray (publishers)

[1] A correction of this sum has been written on the copy. It reads: '633/6/8'. For a transcript of John Murray's ledger listing the cost of and profits from the first and second editions of *Origin*, see Peckham ed. 1959, pp. 775–6.

[2] See letter to John Murray, [25 January 1860].

To Isidore Geoffroy Saint-Hilaire 28 January [1860]

Down Bromley Kent
Jan. 28

Sir

I beg pardon for troubling you again. The pamphlet on the origin or variation of species, which you kindly said you had sent me has never arrived & I fear is lost.[1] In your letter you say it was published in the Magasin de Zoologie; but your reference is defaced by a blot. It seems like 1850, but I can find nothing; would you be so good as to give me the volume year & page of the Magasin, for I am very anxious to see the paper.[2] To save you as much trouble as possible I enclose an envelope.—

With may apologies for troubling you & with much respect, I remain | Dear Sir | Your obliged servant | Charles Darwin

Perhaps your paper was printed in the Revue de Zoologies: I searched many volumes in vain.—

Postmark: 30 Jan 1860
Uppsala University (Waller autograph collection)

[1] Geoffroy Saint-Hilaire's letter has not been found. CD referred to it in the letter to Charles Lyell, 14 January [1860].

[2] CD refers to I. Geoffroy Saint-Hilaire 1851, published in the *Revue et Magasin de Zoologie Pure et Appliquée*. The article contains a section entitled 'Résumé des leçons sur la question de l'espèce', in which Geoffroy Saint-Hilaire stated his belief that species are subject to limited variability (I. Geoffroy Saint-Hilaire 1851, pp. 15–20). There is a lightly annotated copy of the work in the Darwin Pamphlet Collection –CUL. CD cited the article in the 'historical preface' added to the revised American edition of *Origin* (see Appendix IV), to the German translation (Bronn trans. 1860), and to the third English edition (1861).

To Asa Gray 28 January [1860][1]

Down Bromley Kent
Jan. 28th

My dear Gray

Hooker has forwarded to me your letter to him; & I cannot express how deeply it has gratified me.[2] To receive the approval of a man, whom one has long most sincerely respected, & whose judgment & knowledge are universally admitted, is the highest reward an author can possibly wish for; & I thank you heartily for your most kind expressions.—

I have been absent from home for a few days, & so could not earlier answer your letter to me of the 10th of Jan.y—[3] You have been extremely kind to take so much trouble & interest about the Edition. It has been a mistake of my publisher not thinking of sending over the sheets. I had entirely & utterly forgotten your offer of receiving the sheets as printed off.[4] But I must not blame my publisher; for had I remembered your most kind offer, I feel pretty sure I shd not have taken advantage of it; for I never dreamed of my Book being so successful with general readers: I believe I shd. have laughed at the idea of sending the sheets to America.—

After much consideration & on the strong advice of Lyell & others, I have resolved to leave the present book as it is (excepting correcting errors or here & there inserting short sentence) & to use all my strength *which is but little* to bring out the first part (forming a separate volume with index &c) of the three volumes which will make my bigger work; so that I am very unwilling to take up time in making corrections for an American Edition.— I enclose list of few corrections in the 2d Reprint, which you will have received by this time complete;[5] & I could send 4 or 5 corrections or additions of equally small importance, or rather of equal brevity.—[6] I, also, intend to write a *short* Preface with brief history of subject.—[7] These I will set about, as they must some day be done & I will send them you in a short time,—the few corrections first, & the preface afterwards, unless I hear that you have given up all idea of separate Edition. You will then be able to judge whether it is worth having new Edition with *your Review prefixed*.[8] Whatever be the nature of your Review, I assure you I should feel it a *great* honour to have my Book thus preceded.— In business matters it is always best to speak out plainly: as I should not, as we have managed it, get anything from an American publisher, if you do print an Edition & can get any profit, nothing should induce me to touch a penny of it.[9] My terms with Murray are that I receive $\frac{2}{3}$ of Profits, & he $\frac{1}{3}$;—[10] with respect to latter, I can of course say nothing.— But I expect when you come to consider the case, you will not think a new Edition worth thinking about; though an answer to Agassiz would be a **great** advantage to subject.—

Thank you much for telling me the magnificent compliment of Wyman & for sending me extract from Agassiz.[11] I cannot see the force of his argument; & if he wished to puff my book he could not have been more ingenious.— I am delighted to hear that H. D. Rogers, the Professor at Glasgow & so excellent a geologist goes very long way with my views.—[12]

Believe me, my dear Gray I feel all your most generous sympathy & assistance
Yours most truly | C. Darwin

I shall value much at any time your criticisms, either in your Review or by letter.

Gray Herbarium of Harvard University (43)

[1] Dated by the reference to an authorised American edition of *Origin*.

[2] See letter from Asa Gray to J. D. Hooker, 5 January 1860, and letter to J. D. Hooker, [22 January 1860] and n. 3.

[3] CD was in London from 24 to 27 January (Emma Darwin's diary).

[4] See *Correspondence* vol. 7, letter to Asa Gray, 4 April [1858].

[5] The enclosure has not been found. CD had asked his publisher to send Gray the unbound sheets of the second English edition of *Origin* late in 1859 (*Correspondence* vol. 7, letter to John Murray, 22 December [1859]). For some of the corrections CD made to the second edition of *Origin*, published on 7 January 1860, see *Correspondence* vol. 7.

[6] See the letters to Asa Gray, 1 February [1860] and [8 or 9 February 1860], in which CD enclosed additional changes for the American edition of *Origin*. The latter changes, however, arrived too late to be incorporated into the text and were included in a 'Supplement' to the revised fourth printing (*Origin* US ed., pp. 426–32). This edition also included other previously unpublished material: a long section responding to criticisms of natural selection by naturalists such as Hewett Cottrell Watson was added to the end of chapter four (*Origin* US ed., pp. 116*–21*). CD substantially reworked this discussion for the third English edition of *Origin*. See also Appendix IV.

[7] CD's historical preface was included in the revised American edition of *Origin*, pp. v–xi. This issue was advertised on the title page as 'a new edition, revised and augmented by the author'. See Appendix IV.

[8] Gray may have suggested this arrangement in the missing section of the letter from Asa Gray, [10 January 1860]. His review of *Origin* however, was not included in the revised and augmented American edition.

[9] Gray's letter of 23 January [1860], discussing details of the American negotations, did not reach CD until mid-February (see letter to Asa Gray, [15? February 1860]).

[10] For CD's arrangement with John Murray concerning the publication of *Origin*, see *Correspondence* vol. 7, letter from John Murray, 1 April 1859. See also letter to Asa Gray, 1 February [1860].

[11] CD refers to Jeffries Wyman and Louis Agassiz. The 'extract from Agassiz' may refer to a passage taken from a lecture delivered by Agassiz in January 1860 that was reported in a Boston newspaper. It is possible that CD sent the extract on to Charles Lyell, as he did other letters and material concerning *Origin*, for in Lyell's scientific journal for the period there is an entry headed 'Extract from lecture by Agassiz, Boston, January 1860' (Wilson ed. 1970, pp. 348–9) that reads:

> What has the whale in the arctic regions to do with the lion or the tiger in the tropical Indies? There is no possible connection between them; and yet they are built respectively according to one & the same idea. There is behind them & anterior to their existence, a thought. There is a design according to which they were built, which must have been conceived before they were called into existence; otherwise these things could not be related in this general manner. Whenever we study the general relations of animals, we study more than the affinities of beasts. We study the manner in which it has pleased the Creator to express his thoughts in living realities; and that is the value of that study for intellectual Man; for while he traced these thoughts as revealed in nature, he must be conscious that he feels, and attempts as far as it is possible for the limited mind of man to analyze the thoughts of the Creator, to approach if possible into the counsels that preceded the calling into existence of this world with its inhabitants, and there lies really the moral value of the study of nature; for it makes us acquainted with the Creator in a manner in which we cannot learn [of] Him otherwise. As the Author of Nature, we must study Him in the revelation of nature in that which is living before our eyes.

[12] Henry Darwin Rogers, an American-born geologist, was Regius professor of natural history at Glasgow University. On 23 December 1859, he wrote to his brother William Barton Rogers and mentioned his opinion of *Origin* (Rogers ed. 1896, 2: 17–18):

> The only matter of any interest is the appearance of Charles Darwin's book "On the Origin of Species by Means of Natural Selection." It is a suggestive book, full of ingenious arguments

in favour of the Lamarckian hypothesis . . . When you read it you will often say, I think, that in his geology Darwin outdoes Lyell himself in ignoring paroxysmal actions. This is its chief blemish with me.

To Williams and Norgate 29 [January 1860][1]

Down Bromley Kent
29th

Dear Sir

I enclose draft for your account with my thanks for all your very kind assistance.—[2]

Please procure for me Agassiz Index Generum—it is a one volume Book & distinct from the Nomenclator.—[3]

M. Quatrefage is going to send me a book of his through you,[4] when it arrives, please send it to 57 Queen Anne St Cavendish Sqe— Please send there, also, the Index.—[5]

Yours very faithfully | C. Darwin

Endorsement: 'Jan 30 1860'
American Philosophical Society (194)

[1] The recipient and date are inferred from an entry in CD's Account book (see n. 2, below) and from the endorsement.

[2] See letter to Williams and Norgate, 16 January [1860]. There is an entry in CD's Account book (Down House MS) dated 29 January 1860 that records a payment of £3 11*s.* 4*d.* to Williams and Norgate. CD engaged this firm to send presentation copies of *Origin* abroad.

[3] Agassiz 1848. The work is an alphabetical list of animal names with their first proposers, and forms the second part of the *Nomenclator zoologicus* (Agassiz 1842). CD's copy of Agassiz 1848 is in the Darwin Library–CUL.

[4] Probably Quatrefages de Bréau 1859. CD thanked Jean Louis Armand de Quatrefages de Bréau for presenting him with his work on the 'maladies' of silkworms in the letter to J. L. A. Quatrefages de Bréau, 30 March [1860]. Quatrefages de Bréau issued a second work with a similar title in 1860. Both books are in the Darwin Library–CUL and are inscribed by the author.

[5] In CD's Account book (Down House MS), there is a record of a payment on 10 March 1860 of £2 19*s.* 5*d.* made out to Williams and Norgate. The address CD gives is that of his brother, Erasmus Alvey Darwin.

To Charles Griffin & Co.[1] 29 January [1860][2]

Down Bromley Kent
Jan. 29th

Sir

Absence from home has prevented my returning your M.S. earlier.[3]

Part was so inaccurate & there was so important an omission, that I have had a new copy made. If too long you can easily strike your pen through any part.—

Sir | Your obent sert | Charles Darwin

I return your own M.S. that you may see alterations.—[4]

[Enclosure][5]

Darwin, Charles. M.A. Cantab. F.R.S an eminent naturalist, is distinguished both as an author and a man of Science. He accompanied Capt. FitzRoy in his voyage round the world in H.M.S Beagle during the years 1831 to 1836.—[6] His Journal first appeared in 1839 as part of the general narrative of the voyage, and was subsequently republished in a modified form under the title of Journal of Researches into the Natural History and Geology of the countries visited by the Beagle. In 1842 his work on the Structure and Distribution of Coral Reefs appeared, which was followed by two Volumes, namely, his Geological Observations on Volcanic Islands and in 1846 on South America. Since this period he has contributed several papers to the Geological Transactions and to other scientific periodicals.[7] His chief contribution to Zoology is the "Monograph on the Family Cirripedia" in two large volumes in which he points out many curious and interesting particulars in relation to the history and economy of the barnacles and sea-acorns, and furnishes a minute description of every known species of the Family. He has recently (November 1859) published a work entitled "The Origin of Species by means of Natural Selection or the preservation of favoured races in the struggle for life;" this volume, as stated in the introduction, gives only in a condensed form the result of more than twenty years study; and will hereafter be followed by more detailed treatise on the same subject.[8] Mr Darwin's writings exhibit close observation and untiring industry in collecting and arranging facts. As a geologist he stands in one of the highest ranks and was for some time Secretary to the Geological Society. As a Zoologist he is, though learned, very cautious in arriving at conclusions without possessing sufficient data;[9] while his style is so clear that it at once affords pleasure to the student and information to the professor.[10]

British Library (Add. 28509: 408)

[1] The recipient was the proprietor of the publishing firm Richard Griffin and Co. It was based in London, having moved from Glasgow in 1848.

[2] The date is inferred from the publication of [Griffin] ed. 1861.

[3] Charles Griffin intended to include an entry on CD in his forthcoming *Dictionary of contemporary biography* ([Griffin] ed. 1861). He had sent a provisional entry to CD (see n. 4, below).

[4] Griffin's manuscript, bearing CD's corrections, was returned with the letter. It is in the British Library (Add. 28509: 410). CD also enclosed a fair copy of the new text (see enclosure).

[5] The enclosure, which is a fair copy of the revised text, is bound with the letter and Griffin's original manuscript in the British Library (Add. 28509: 409). It is in the hand of an amanuensis, with minor corrections and alterations in CD's hand.

[6] In the published version, the first two sentences were altered to read: 'Darwin Charles. M.A. Cantab. F.R.S., an eminent naturalist and author. When a very young man he accompanied Captain Fitzroy in his voyage round the world, in H.M.S. "Beagle," during the years 1831–1836.' ([Griffin] ed. 1861, p. 122).

[7] The original wording of this section reads (British Library (Add. 28509: 410)):

> His "Journal of Researches into the Geology and Natural History of the Various Countries visited by H.M.S. Beagle" was published in 1839. In 1846 his "Geological Observations on South America" was published; and from time to time he contributed a great number of papers

to the Geological Society and to the "Magazine of Natural History". But his profoundest and most philosophic work is the "Monograph of the Family Cirripedia".

[8] The sentence about *Origin* was added by CD to Griffin's manuscript, which did not refer to the book.

[9] The original reads: 'unimpractable data'.

[10] The two final sentences were not included in the published version. Instead, the text reads ([Griffin] ed. 1861, pp. 122–3):

> Mr. Darwin, although he has adopted conclusions contested by other naturalists, has always been very cautious in arriving at results without sufficient data. He is a clear and elegant writer; and his works, independently of their scientific value, are written in a style well calculated to render them highly attractive.

To John Murray 29 January [1860][1]

Down Bromley Kent
Jan. 29th

My dear Sir

I had quite forgotten that my Journal was stereotyped.[2] I find that I could here & there improve style, but that is not of course worth while. At p. 378 I much wish to make a few corrections, which I shall be able to do in a few days, & I will then send the M.S. & you must decide whether it is possible to introduce them.[3]

Very many thanks for your note—I write in great Haste to catch our early Sunday Post

Yours | C. Darwin

John Murray (publishers)

[1] Dated by the reference to a new edition of *Journal of researches*.

[2] John Murray published the second edition of *Journal of researches* in 1845 using the stereotype technique, which involves preparing fixed metallic printing plates. The text for each page was cast from a mould and was therefore unchangeable.

[3] The main text of the edition was not changed by Murray. The two alterations CD wished to make in the 1860 edition were printed in a short 'Postscript' to the original preface (*Journal of researches* (1860), p. vii). The change relating to page 378 amends the discussion of Galápagos birds (see letter from P. L. Sclater, [3? February 1860]).

From Charles James Fox Bunbury 30 January 1860

Barton
January 30, 1860

My dear Darwin,

I will now begin as I told you I would, to write down my remarks on your book, tho' I dare say it may be some time before I finish them.[1] I must begin by saying, more explicitly than when I met you the other day, how much I admire your book.[2] The astonishing labour & patience with which you must have collected the materials during so many years; the wide range of reading as well as observation; the skill with which the facts are brought to bear on your argument; the extraordinary power of condensation by which so much matter is brought within the compass of a single volume; the candour with which you have tried at least to state the case fairly against yourself; as well as the extreme ingenuity of your theory; all these qualities undoubtedly render your book (Whether its theory be ultimately established or no) one of the most important that have ever appeared on Natural History.

It is moreover no small triumph to your cause, that you have already made converts of the greatest geologist & the greatest botanist of our day.[3] Nevertheless you will not be surprised at my saying, that I am not convinced.[4] In the first place, as I told you the other day, I cannot feel satisfied till I see your "Pièces Justificatives"—the body of evidence which you are to bring forward in your big book. Not that I have not perfect confidence in your accuracy & honesty, but the same facts

AL incomplete
DAR 98 (ser. 2): 26

CD ANNOTATIONS
Top of first page: 'C. J. F. Bunbury' *pencil*

[1] See *Correspondence* vol. 7, letter to C. J. F. Bunbury, 3 December [1859].
[2] CD was in London from 24 to 27 January 1860 (Emma Darwin's diary).
[3] Bunbury refers to Charles Lyell (his brother-in-law) and Joseph Dalton Hooker.
[4] Bunbury had been among the first to know about CD's views on the species question and had discussed them with him on several occasions. See particularly *Correspondence* vol. 3, letter from Charles Lyell, [after 2 August 1845]; and vol. 6, letter to C. J. F. Bunbury, 21 April [1856].

To John Murray 30 [January 1860][1]

Down Bromley Kent
30th

My dear Sir

I am really sorry to trouble you so often; but it has this moment occurred to me that it would be a far easier & cheaper plan to give me one (or at most two) pages of additional Preface, in which I could state two or three *important* errors, which further examination of species referred to in text has led me to detect.[2] What do you say to this? Is the present Preface stereotyped; if not I should rather like to leave out one sentence, but it is unimportant.

The shabbiest appearance in whole Book is Table of Contents in such small type,—not that I care about it.—[3]

My dear Sir | Yours very sincerely | C. Darwin

John Murray (publishers)

[1] Dated by the relationship to the letter to John Murray, 29 January [1860].
[2] The new issue of *Journal of researches* (1860) included a 'Postscript' to the original preface, in which CD corrected a few errors. See *Journal of researches* (1860), p. vii.
[3] The table of contents was reset in larger type for the new printing.

To J. D. Hooker 31 [January 1860]

Down Bromley Kent
31

My dear Hooker

I have resolved to publish a little sketch of the progress of opinion on the change of species—.[1] Will you or Mrs Hooker do me the favour to copy *one* sentence out of Naudins paper in Revue Horticole 1852 p. 103, namely that on his principle of Finalité.—[2] Can you let me have it soon, with those confounded dashes over the vowels put in carefully. Asa Gray, I believe, is going to get a 2^d^ Edit of my Book, & I want to send this little preface over to him soon.—

I did not think of necessity of having Naudins sentence on finality, otherwise I would have copied it.—

Yours affect | C. Darwin

I shall end by just alluding to your Australian Flora. Introduction— What was date of publication December 1859 or Jan. 1860?[3]

Please answer this.—

My Preface will also do for French Edition which I *believe* is agreed on.—[4]

Endorsement: 'Jany/60'
DAR 115.1: 38

[1] CD's 'historical sketch' was first published as a preface to the 'revised and augmented' American edition of *Origin* (Freeman 1977, p. 85). It was also included in the German translation (Bronn trans. 1860). See Appendix IV. CD expanded the text for the third English edition (1861).

[2] Late in 1859 Hooker had drawn CD's attention to the similarity between his theory and an essay by Charles Victor Naudin and sent CD the journal in which Naudin 1852 appeared. See *Correspondence* vol. 7, letters to J. D. Hooker, 21 [December 1859] and 23 [December], and to Charles Lyell, 22 [December 1859]. In the historical sketch included in *Origin* US ed., p. ix, CD referred to Naudin's belief 'that species are formed in an analogous manner as varieties are under cultivation' through man's selection. However, he further noted that Naudin believed in a 'puissance mystérieuse', a concept that CD interpreted as a 'principle of Finalité' guiding the transformation of species.

[3] CD refers to the introductory essay of Hooker's *Flora Tasmaniæ* , published as the concluding number of Hooker 1855–60. It was available in December 1859 (Wiltshear 1913) and was also issued as an independent volume, (Hooker 1859). CD gave the date as December 1859 in *Origin* US ed., p. xi.

[4] See letters to J. L. A. de Quatrefages de Bréau, 21 January [1860], and to John Murray, 23 [January 1860].

From George Robert Waterhouse [February 1860][1]

les sinus qui se trouvent ⟨sur le⟩ trajet de l'air, et, dès lors, les frontaux sont dans ce cas aussi bien que tous ceux qui entourent les fosses nasales; dès lors aussi, par l'écartement des deux lames de l'os, le gonflement des bosses frontales, indépendantes et séparées par un silon—"—[2]

I suppose you have Owen's British Fossil Mammalia—look to pp 88 & 89 &

you will find one or two remarks having reference to DeBlainvilles conclusions—[3] It is true what DeB. states—that the frontal sinuses & the facial part of the skull is particularly prone to vary ⟨*section missing*⟩

Geo R Waterhouse

It is not true, in my opinion, that all these Fossil Cave Bears are of one & the same species—there are certainly two very distinct species— The one with the enormous frontal sinuses (U. *spelæus*) varies *much* in the development of the part in question, but there are characters in the teeth & form of the lower jaw (especially in the condyle) of *U. priscus* which make it easy of separation—

Incomplete
DAR 47: 152

CD ANNOTATIONS
1.1 les sinus . . . to vary 2.4] *crossed pencil*
4.1 It is . . . separation— 4.5] 'Ch. 7'[4] *brown crayon, underl brown crayon*; 'Feb. 1860' *added pencil*
4.2 The one . . . in question, 4.4] *double scored pencil*

[1] The date is given by CD's annotation.
[2] The passage quoted by Waterhouse comes from Blainville 1841–55, 1: 56, and forms part of a larger discussion by Henri Marie Ducrotay de Blainville on variation in fossil bears. Blainville criticised the distinction that had been drawn between two alleged species of fossil bears, pointing out that variations occurred between young and old, male and female, and even between individuals of the same sex. He thought the shape and size of the frontal sinuses were not sufficiently constant to be used as taxonomic characters.
[3] Owen 1846b, p. 99. Richard Owen disputed Blainville's conclusion that there was no distinction between *Ursus spelaeus* and other species of fossil cave bears. CD's annotated copy of Owen 1846b is in the Darwin Library–CUL.
[4] CD's annotation refers to chapter 7 of his species book, 'Laws of variation: varieties & species compared' (see *Natural selection*). CD intended to revise and publish this larger work.

To Asa Gray 1 February [1860][1]

Down Bromley Kent
Feb 1

My dear Gray

I received yesterday your letter of the 17th— I am *heartily* glad you have so kindly determined to reprint; but I especially hope that your Review will appear in it.—[2]

The Edition might be entitled "Reprinted from corrected Second Edition with additional corrections".[3] I enclose today some few additions, all that I have. I will send in a fortnight a Preface giving a short History of opinion on origin of species.—[4]

Since writing last I have had a letter from Murray, the publisher; & he says "Whatever benefit may arise from the Reprint of your Book in U.S. you are fairly entitled to".— So he claims nothing. Perhaps you would like me to stand in Murray's position & take $\frac{1}{3}$ of the profits whatever they may be.— Nothing would pain me so much as to take all the profit, as I **wholly** & **absolutely** shall owe all to you; & if you will print your Review, it will be a joint publication.

Believe me, My dear Gray | Yours most truly & gratefully | C. Darwin

(The Article in Times by Huxley.—)[5]

If my Additions & Preface should arrive too late; ⟨they⟩ will not be labour in vain, as they will do for French Edition ⟨*several words missing*⟩ Edition.–

[Enclosure][6]

Mere verbal corrections not here noticed & the omission 2 or 3 paragraphs

New Edit	
17. 18.	Pallasian doctrine made clearer.[7]
49	names of Primula veris & elatior corrected.[8]
72	age of fir-tree corrected[9]
73	case of clover made stronger.[10]
96	parenthesis on parthenogenesis added[11]
165	mule of U. States striped[12]
214	sentence about dogs pointing added[13]
219–223	Slave-ants made clearer[14]
253	sentence about crossed pheasants added.[15]
286	Weald Denudation made milder.—[16]
303	fossil Birds instead of fossil Whales[17]
336	sentence added on advancement of organisation[18]
390, 391	about crossing keeping birds of Madeira & Bermuda unchanged[19]
452	paragraph on "nascent" organs added[20]
480	Kingsley's sentence on theological bearing added.—[21]

Gray Herbarium of Harvard University (44)

[1] Dated by the reference to the revised American edition of *Origin*.

[2] See letters from Asa Gray, [10 January 1860] and [17 January 1860], and letter to Asa Gray, 28 January [1860]. CD refers to the suggestion that Gray's review of *Origin* ([Gray] 1860a) might be printed as an introduction to an authorised American edition.

[3] The title page of the new edition issued in May 1860 by D. Appleton & Co. reads: 'A new edition, revised and augmented by the author.' Since Appleton had electrotyped *Origin* (see letter from Asa Gray, 23 January 1860, n. 2), CD's corrections and additions had to be fitted in with as little change to the plates as possible.

[4] See the enclosure and also the letter to Asa Gray, [8 or 9 February 1860].

[5] CD refers to [T. H. Huxley] 1859a. See letter to T. H. Huxley, 1 January [1860].

[6] The enclosure is pasted in the back of Gray's presentation copy of *Origin* (Gray Herbarium of Harvard

University). The page numbers refer to the second English edition. Although the enclosure may have been sent in the letter to Asa Gray, 28 January [1860], there CD states that he is enclosing only a list of the corrections made in *Origin* 2d ed. The corrections listed here were first included in the revised American edition of *Origin* and later incorporated into the third edition of *Origin*, published in March 1861. Hence, the American edition of *Origin* was 'in some respects even ahead of the current English edition' (Dupree 1959, p. 271).

7 The reference is to the section in the discussion of the origin of domesticated animals in which the views of Pyotr Simon Pallas are treated. According to the 'Pallasian doctrine', domestic animals arose from distinct wild species that were able, after a long period of domestication, to interbreed successfully (*Origin*, pp. 17–18; see also *Natural selection*, p. 440). Wishing to show that man had domesticated animals long before the time in which they were depicted on Egyptian monuments, CD cited Leonard Horner's recent study of the antiquity of man in Egypt (Horner 1858). After Horner's findings were thrown into doubt ([W. Smith] 1859), CD substituted in the revised American edition of *Origin*, p. 23, the following passage:

> After the recent discoveries of flint tools or celts in the superficial deposits of France and England, few geologists will doubt that man, in a sufficiently civilized state to have manufactured weapons, existed at a period extremely remote as measured by years; and we know that at the present day there is hardly any tribe so barbarian as not to have domesticated at least the dog.

The passage was slightly altered for the third edition of *Origin* (*Origin* 3d ed., p. 18).

8 *Origin* US ed., p. 51.

9 *Origin* US ed., p. 70.

10 *Origin* US ed., p. 71. The change was also made to *Origin* 3d ed., pp. 76–7.

11 *Origin* US ed., p. 90.

12 *Origin* US ed., pp. 148–9.

13 *Origin* US ed., p. 191.

14 *Origin* US ed., pp. 195–9.

15 *Origin* US ed., p. 224.

16 In *Origin* US ed., p. 252, the following note was added to the description of the denudation of the Weald:

> I have left the foregoing passages as they stand in the second edition, but I confess that an able and justly severe article, since published in the Saturday Review (Dec. 24th, 1859), shows that I have been rash. I have not sufficiently allowed for the softness of the strata underlying the chalk; the remarks made are more truly applicable to denuded areas composed of hard rocks. Nor have I allowed for the denudation going on on both sides of the ancient Weald-Bay; but the circumstance of the denudation having taken place within a protected bay would prolong the process. It has long been my habit to observe the shape and state of surface of the fragments at the bases of lofty retreating cliffs, and I can find no words too strong to express my conviction of the extreme slowness with which they are worn away and removed. I beg the reader to observe that I have expressly stated that we cannot know at what rate the sea wears away a line of cliff: I assumed the one inch per century in order to gain some crude idea of the lapse of years; but I always supposed that the reader would double or quadruple or increase in any proportion which seemed to him fair the probable rate of denudation per century. But I own that I have been rash and unguarded in the calculation.

This discussion was altered in the third English edition of *Origin*. See Peckham ed. 1959, pp. 483–4.

17 The passage referred to is actually in *Origin* 2d ed., p. 304, and in *Origin* US ed., p. 266.

18 *Origin* US ed., p. 294.

19 *Origin* US ed., pp. 340–1.

20 *Origin* US ed., p. 393.

21 The statement by Charles Kingsley was actually added to *Origin* 2d ed., p. 481, and included in *Origin* US ed., p. 417.

To Richard Kippist?[1] 1 February [1860][2]

Down Bromley Kent
Wednesday Feb: 1st

Dear Sir

The Carrier[3] shall call tomorrow (Thursday) *morning* & bring some Books & I shd be much obliged if you would give him (if you have) the **1st** &—*likewise* the **10th** Editions of the Vestiges of Creation[4] Also the **2d** Edit of Baden Powell's Unity of Worlds.[5]

No other Editions will be of any service.—[6]

Yours faithfully | C. Darwin

Phillips, Son & Neale, Inc. (dealers)

[1] CD had an arrangement with the Linnean Society of London enabling him to borrow books from the society's library (see *Correspondence* vol. 5, letter from J. D. Hooker, [before 7 March 1855]). Kippist was the librarian of the society.

[2] The year is given by the preparation of the historical preface for *Origin* (see n. 6, below).

[3] George Snow operated a parcels carrier service between London and Down.

[4] [Chambers] 1844 and 1853. CD owned a copy of the sixth edition of this work (1847). It is in the Darwin Library–CUL.

[5] The second edition of Baden Powell's *Essays on the spirit of the inductive philosophy* (Powell 1856) was much enlarged. See CD's two letters to Baden Powell, 18 January [1860].

[6] CD cited these works in the historical sketch added to the revised and augmented American edition of *Origin* and included in all subsequent editions.

To T. H. Huxley 2 [February 1860][1]

Down Bromley Kent
2d

My dear Huxley

I have had this morning a letter from old Bronn (who to my astonishment seems *slightly* staggered by Nat. Selection)[2] & he says a publisher in Stuttgart is willing to publish a Translation & that he Bronn will to certain extent superintend.[3] Have you written to Kölliker:[4] if not perhaps I had better close with proposal: what do you think?? if you have written, I must wait, & in this case will you kindly let me hear as soon as you hear from Kölliker.

My poor dear friend you will curse the day when you took up the "general agency" line; but really after this I will not give you any more trouble.

Yours most truly | C. Darwin

I forgot to say that if you care in least about Agassiz on Museum, you had better cut off & keep all that part & return the bit about my Book anytime.—[5]

Do not forget the 3 tickets for us for your lecture, & the ticket for Baily the Poulterer.—[6]

Old Bronn has published in the Year-Book for Mineralogy a notice of the Origin;[7] & says he has himself published elsewhere a foreboding of the theory!—[8]

Imperial College of Science, Technology and Medicine Archives (Huxley 5: 80)

[1] The date is given by the reference to Huxley's forthcoming lecture, presumably that given at the Royal Institution on 10 February 1860.

[2] The letter from Heinrich Georg Bronn has not been found (see n. 5, below). Bronn was professor of natural and technical sciences and director of the zoological collection at Heidelberg University. CD was well acquainted with his work, having read carefully volume two of Bronn's *Handbuch einer Geschichte der Natur* (Bronn 1841–9) in February 1846 (*Correspondence* vol. 4, Appendix IV, 119: 16a). Copies of the first and second volumes of this work are in the Darwin Library–CUL, the first volume being a reprint dated 1842. Volume two, which contains an extensive discussion of variation in organic life, has been heavily annotated by CD (see *Marginalia*). Further notes on the work are in DAR 91: 73.

[3] The publishing firm of E. Schweizerbart was founded in 1826 by Wilhelm Emanuel Schweizerbart in Stuttgart. It was noted for its publication of scientific works, particularly those connected with palaeontology, and published many of Bronn's works. In 1860 the owner of the firm was Christian Friedrich Schweizerbart.

[4] CD had asked Huxley to write to Rudolf Albert von Kölliker concerning a possible German translation of *Origin* (see *Correspondence* vol. 7, letter to T. H. Huxley, [9 December 1859]).

[5] CD apparently sent Bronn's letter to Huxley, but it has not been found in the Huxley archive. Bronn may have written about Louis Agassiz's recent visit to Heidelberg, during which time he purchased Bronn's substantial palaeontological collection for the new zoological museum of Harvard University (Marcou 1896, 2: 81). The Museum of Comparative Zoology had opened in November 1859 (Winsor 1991, p. 37 n.). Agassiz had studied in Heidelberg with Bronn in the late 1820s.

[6] See letter to T. H. Huxley, [26 January 1860].

[7] Bronn reviewed *Origin* in the *Neues Jahrbuch für Mineralogie, Geognosie, Geologie und Petrefakten-Kunde* (Bronn 1860a). This journal was co-edited by Bronn and Karl Cäsar von Leonhard and published by E. Schweizerbart. An offprint of the review, annotated by CD, is in the Darwin Pamphlet Collection –CUL.

[8] Bronn had been working for a number of years on an ambitious attempt to discover the general laws of development regulating the appearance of both organic and inorganic phenomena in nature. In Bronn 1858b, which had received the prize of the French Académie des Sciences in 1857, Bronn discussed the sequence of the appearance of animal and plant life in geological strata. In Bronn 1858a, he attempted to formulate laws governing the underlying plans revealed in both organic and inorganic forms.

To Herbert Spencer 2 February [1860][1]

Down Bromley Kent
Feb 2^d

My dear Sir

I know so few people, that I can really think of only one person, to whom it would be any good to send your gigantic programme.[2] This one is

Dr. Drysdale
care of Dr. Lane
Moor Park
Farnham
Surrey[3]

I asked Huxley to put my own name & that of my Brothers on the list for copies.—[4]

From your letter I infer that you have not received a copy of my Book, which I am very sorry for: I told Mr Murray to send you one, *amongst the first* distributed, in November: it was addressed, I am almost sure, to care of Mess Longman.[5] Will you enquire, if you think it worth while, & let me know if not there; & then I will write to Murray to see what has become of it—[6]

I was so much out of health when I was writing my Book, that I grudged every hour of labour, & therefore gave no sort of history of progress of opinion.—

I have now written a Preface for the foreign Editions & for any future English Edit (shd there be one) in which I give a very brief sketch, & have with much pleasure alluded to your excellent essay on Development in your general Essays.[7] Can you give me precise date of its publication in *Leader*, as I arrange my notices chronologically? I am sorry to say that I have never read your Psychology,[8] having no strength to spare but I have just looked at the latter part.— May I say in my Preface that you have treated Psychology on the principle "*of the necessary acquirement of each mental power & capacity by gradation*"? You will find that I use these words in inverted commas towards close of my volume (P. 489 of Reprint) & when I wrote them, I did not think of your work.—[9]

My dear Sir | Yours sincerely | C. Darwin

Kindly answer my two questions soon

University of London Library MS.791/47. (Courtesy of the Athenaeum.)

[1] The year is given by the reference to the first volume of Spencer 1860–2.

[2] CD refers to the subscription list for Spencer's ten-volume series entitled 'System of synthetic philosophy', the first work of which was *First principles* (Spencer 1860–2). CD's subscriber's copy of the book is in the Darwin Library–CUL.

[3] Edward Wickstead Lane was the proprietor of Moor Park hydropathic establishment, which CD had visited on several occasions since 1857 (see *Correspondence* vols. 6 and 7). Dr Drysdale may have been related to Lady Drysdale, Lane's mother-in-law.

[4] See letter to T. H. Huxley, [26 January 1860].

[5] The publishing firm of Longman, at that time headed by Thomas Longman, published the first volume of Spencer's *Essays* (Spencer 1858–63). Spencer's name appears on the presentation list CD drew up for *Origin* (see Appendix III).

[6] On 10 February 1860 Spencer wrote to a friend (Duncan ed. 1908, p. 98):

> I am just reading Darwin's book (a copy of which has been searching for me since November and has only just come to hand) and want to send him the 'Population' [Spencer 1852] to show how thoroughly his argument harmonizes with that I have used at the close of that essay.

See also letter from Herbert Spencer, 22 February 1860.

[7] Spencer sent a copy of the first volume of his *Essays* (Spencer 1858–63) to CD late in 1859 (*Correspondence* vol. 7, letter to Herbert Spencer, 25 November [1859]). CD's copy is in the Darwin Library–Down. Spencer's essay 'The development hypothesis', number 2 of 'The Hawthorne papers' originally published in *The Leader* in 1852, was reprinted in Spencer 1858–63, 1: 389–95. In this essay, Spencer advocated the view that species had been modified and that the modification could be attributed to a change of external conditions. CD cited the essay in his historical preface (*Origin* US ed., pp. viii–ix, and *Origin* 3d ed., p. xvii; see Appendix IV).

[8] Spencer 1855. CD's presentation copy is in the Darwin Library–CUL. See *Correspondence* vol. 6, letter to Herbert Spencer, 11 March [1856].
[9] Spencer apparently approved of CD's statement: it appears in the historical preface (*Origin* US ed., p. ix; see Appendix IV).

To J. S. Henslow 3 February [1860]

Down.
Feb 3^{d}

My dear Henslow

Many thanks for L. Jenyns very sensible letter.—[1]

We shall be truly delighted to see you, whenever you can come, after the 11th or any other time.—[2]

In Haste | Most truly yours | C. Darwin

Endorsement: '*1860*'
DAR 93: 62

[1] Leonard Jenyns was Henslow's brother-in-law. He had probably written to Henslow about *Origin*. For Jenyns's views about CD's work, see letter to Leonard Jenyns, 7 January [1860].
[2] Henslow visited the Darwins from 14 to 16 February 1860 (Emma Darwin's diary). See letter to J. S. Henslow, 9 February [1860].

To John Murray 3 February [1860][1]

Down Bromley Kent
Feb. 3^{d}

My dear Sir

I have not yet succeeded in getting the requisite information for the most important correction for my Journal:[2] I expect it daily; if it does not come in time, it cannot be helped: unless I can get this *one*, I will not trouble you with any. I will send it when it comes & then you must decide whether you can insert it for me.

Now I am sorry to say I must trouble you on another point, viz to send a copy immediately of the Origin (you must keep account against me for extra copies of this Reprint) to

M. Elisée Reches[3]
10 rue Bénard
Batignolles
Paris

M. Talandier says he is in treaty with Bailliere about Translations & Bailliere asks "si les droits d'auteur vous sont reserves ou s'ils appartiennent a M. Murray".[4] In answer I have said that I have full power to give to M. Talandier the right & profit from a Translation, & that you would if necessary take any necessary step to declare to M. Bailliere that this is so.— If you object to this, please send me a line by return of Post; if I do **not** hear I will understand that you will so declare to M. Bailliere.—

I hope I shall soon cease to cause trouble
Yours sincerely | C. Darwin

Endorsement: 'Feb 3/60'
John Murray (publishers)

[1] The year is given by the reference to a French translation of *Origin*.
[2] See following letter.
[3] Elisée Reches has not been identified.
[4] CD refers to J. B. Baillière et fils, a firm of publishers and booksellers with offices in Paris and London.

From Philip Lutley Sclater [3? February 1860][1]

Land-birds of Galapagoes[2]

1.	*Craxirex galapagoensis.*	I believe to be diff.t from anything on the main.
2.	*Strix punctatissima*	occurs in Ecuador. I have examined specimens collected by Fraser at Quito.[3]
3.	*Otus galapagoensis*	occurs in S. A. as I believe
4.	*Progne modesta*	I have never seen from the main.
5.	*Pyrocephalus nanus*	occurs in Ecuador (Fraser) at least I believe it the same[4]
6	— *dubius*	a *doubtful species*? probably = *nanus*.
7	*Myiobius magnirostris,*	probably peculiar to Galapagoes
8 9 10 }	Three *Mimi*	all peculiar but *M. parvulus* is hardly separable fr. *M. melanotis*
11.	'Wren' Journal p. 378.	I do not know what this is: not ment.d in Zool. of Beagle?[5]
12.	*Zenaida galapagoensis*	*probably* occurs on mainland—

13–25— *Geospizinæ* — I have every reason to believe this group of Finches quite peculiar to these islands and "Bow island" whence one was obt[d] during voyage of the 'Sulphur' see Zool. of Sulphur p. 42[6]

P. L. S.

DAR 205.3 (Letters)

CD ANNOTATIONS

2.1 *Strix punctatissima*] *cross added pencil*
3.1 *Otus galapagoensis*] *nought added pencil*
5.1 *Pyrocephalus nanus*] *cross added pencil*
7.1 *Myiobius*] '= Tyrannula' *added ink over pencil*
11.1 'Wren'] 'Sylvicola aureola' *added ink over pencil*
12.1 *Zenaida*] *nought added pencil*
Top of list: 'X certainly manl[d] | 0 probably mainland' *pencil, circled pencil*
Bottom of list: 'Feb 4th 1860 | L. P. Sclater'[7] *ink*; 'Copy of Journal'[8] *pencil*; '(19)'[9] *brown crayon*

[1] The date is based on CD's annotation and on his remark in the preceding letter that he had not yet received Sclater's information. Although the section of the preface to *Journal of researches* (1860) that includes Sclater's revisions is dated 1 February 1860, it seems that CD did not send the manuscript to John Murray until 4 February (see letter to John Murray, 4 February [1860]).

[2] In earlier editions of *Journal of researches*, CD stated that all but one of the twenty-six species of land birds found on the Galápagos Islands were peculiar to the islands (*Journal of researches*, p. 461; 2d ed., p. 378). CD repeated this information in *Origin*, pp. 390, 397–8. He may have asked Sclater, who was secretary of the Zoological Society in 1860, to check the validity of this statement after having heard from William Jardine that some of these species had since been found on the mainland (*Correspondence* vol. 7, letter from William Jardine, 20 December 1859, and letter to Charles Lyell, 22 [December 1859]). Sclater's list gives the additional localities in which several Galápagos birds had been found.

[3] Louis Fraser, a former curator of the Zoological Society museum, was collecting in South America. Sclater had described the birds Fraser collected near Quito at a meeting of the Zoological Society on 24 January 1860, commenting in the description of *Strix punctatissima* that it was 'Hitherto only known from the Galapagos.' (*Proceedings of the Zoological Society of London* pt 28 (1860): 82).

[4] At a later meeting of the society, Sclater listed *Pyrocephalus nanus* Gould as possibly the species obtained by Fraser in Ecuador (*Proceedings of the Zoological Society of London* pt 28 (1860): 282).

[5] Sclater refers to *Journal of researches* 2d ed., p. 378. In *Birds* (part 3 of *The zoology of the voyage of H. M. S. Beagle*), the wren is named *Sylvicola aureola* (see letter to P. L. Sclater, 4 February [1860]).

[6] The Geospizinae are the Galápagos finches, now popularly known as Darwin's finches. In 1843 the exploring expedition led by Edward Belcher reported the discovery of another species belonging to the Geospizinae, *Cactornis inornatus*, on Bow Island in the Low Archipelago (now known as Hao in the Tuamotu Islands, French Polynesia). See Belcher 1844, vol. 1, p. 42; see also *Correspondence* vol. 2, letters from R. B. Hinds, 21 April 1843, and 19 July [1843]. This finch has subsequently been attributed to Cocos Island, south-west of Costa Rica, instead of to Bow Island (Richmond 1902). CD had expressed doubts about the place of origin of the specimen in *Birds*, p. 105.

[7] The date given by CD is presumably that of receipt. See letter to P. L. Sclater, 4 February [1860]. CD mistakenly transposed Sclater's initials.

[8] See letter to John Murray, 4 February [1860].

[9] The number of CD's portfolio of notes relating to the geographical distribution of animals.

To Heinrich Georg Bronn 4 February [1860][1]

Down. | Bromley. Kent.
Feb 4.

Dear & much honoured Sir.

I thank you sincerely for your most kind letter,[2] I feared that you would much disapprove of the "Origin" & I sent it to you merely as a mark of my sincere respect. I shall read with much interest your work on the productions of Islands whenever I receive it[3] I thank you Cordially for the notice in the Neuen Jahrbuch für Mineralogie[4] & still more for speaking to Schweitzerbart about a Translation;[5] for I am most anxious that the great & intellectual German people should know something about my book

I have told my publisher to send immediately a copy of the *new edition* to Schweitzerbart, & I have written to Schweitzerbart that I give up all right to profit for myself; so that I hope a translation will appear—[6] I fear that the Book will be difficult to translate, & if you could advise Schweitzerbart about a *good* Translator it would be of very great service. Still more if you would run your eye over the more difficult parts of the Translation, but this is too great a favour to expect. I feel sure that it will be difficult to translate from being so much condensed

Again I thank you for your noble & generous sympathy & I remain with entire respect.

Yours truly obliged. | C. Darwin

The new edition has some few corrections & I will send in M.S some additional corrections & a short Historical Preface to Schweitzerbart

How interesting you could make the work by *editing* (I do not mean translating) the work & appending notes of *refutation* or confirmation[7] The Book has sold so very largely in England that an Editor would, I think, make profit by the Translation.

Copy
DAR 143

[1] Dated by the reference to a German translation of *Origin*.

[2] Bronn's letter has not been found, but see the letter to T. H. Huxley, 2 [February 1860].

[3] Bronn 1860b. There is an annotated copy in the Darwin Pamphlet Collection–CUL.

[4] Bronn 1860a.

[5] The reference is to Christian Friedrich Schweizerbart, owner of the Stuttgart publishing firm E. Schweizerbart.

[6] See letter to John Murray, 4 February [1860]. CD's letter to E. Schweizerbart has not been found.

[7] CD told Huxley that Bronn 'seems *slightly* staggered by Nat. Selection' (letter to T. H. Huxley, 2 [February 1860]). Bronn himself took on the task of providing the first German translation of *Origin*, to which, as CD suggested, he appended his own remarks (see letter to H. G. Bronn, 14 July [1860]).

To T. H. Huxley 4 February [1860][1]

Down.—
Feb. 4th

My dear Huxley

On principle of Bird in Hand, I will write to Bronn & agree; so will you write a line to Kölliker.[2] God forgive me for being so troublesome. Do not hate me. Hurrah for the Bishop![3]

Ever yours | C. Darwin

I have had endless bother about French Translation, between two stools, which makes me gladder to close with any one for German Translation.—

Imperial College of Science, Technology and Medicine Archives (Huxley 5: 105)

[1] Dated by the relationship to the preceding letter.

[2] See preceding letter and letter to T. H. Huxley, 2 [February 1860].

[3] The reference has not been traced.

To John Murray 4 February [1860][1]

Down Bromley Kent
Feb. 4th

My dear Sir

I send M.S.[2] which will print on two pages: it might with advantage be printed in smaller type than rest of volume: it will make p. vii & viii of Preface, if you print new Table of Contents, which Table looks so poor.—

I **must** see proof of these additions on account of proper names;[3] but proof shall be returned same day. I do not at all care for corrections given to p. vi of present Preface: it is not worth printing new page.— I should like one copy of this Reprint just to see it & to give to Mr Sclater who has aided me.—[4] If you give it me, please send it to 57 Queen Anne St. W.—[5]

Please to send *immediately* a copy of the "Origin" (for German Translator) to

Herr Schweitzerbart[6]
Buch=Verleger
Stuttgart
Wurtemburg

Can it be sent by Post without outrageous Postage? Please send it in quickest & best manner.

My dear Sir | Yours very sincerely | C. Darwin

Prof. Asa Gray is going to bring out an Edition of the Origin; so that I believe there will be two Editions in U. States.—[7]

John Murray (publishers)

[1] Dated by the reference to a new printing of *Journal of researches*.

[2] CD refers to the manuscript of the corrections for the new edition of *Journal of researches* (see letters to John Murray, [25 January 1860] and 30 [January 1860]). The corrections were added in a 'Postscript', p. vii, dated 1 February 1860.

[3] See letter from P. L. Sclater, [3? February 1860]. In the postscript to the preface of *Journal of researches* (1860), p. vii, CD stated that *Strix punctatissima*, *Pyrocephalus nanus*, *Otus galapagoensis*, and *Zenaida galapagoensis*, previously believed to be confined to the Galápagos Archipelago, probably also occurred on the mainland.

[4] See letter from P. L. Sclater, [3? February 1860], and following letter.

[5] The address is that of CD's brother, Erasmus Alvey Darwin, with whom he usually stayed when in London.

[6] Christian Friedrich Schweizerbart. See letter to H. G. Bronn, 4 February [1860].

[7] The first printing of *Origin* in America, published without CD's authorisation, was issued in mid-January by the publishing firm D. Appleton & Co (see letter from Asa Gray, 23 January 1860). CD sent corrections to Gray that were included in a 'revised and augmented' edition published by D. Appleton & Co. in May 1860 (see letter to Asa Gray, 1 February [1860]).

To P. L. Sclater 4 February [1860][1]

Down Bromley Kent
Feb 4th

My dear Sir

I thank you extremely for your list of the Galapagos Birds, & I trust that it will be in time for a note at end of my Journal, & will likewise do to correct the Origin of Species.—[2]

The wren is Sylvicola aureola figured in Zoology of Beagle;[3] I have assumed that it confined to the Archipelago. I am surprised that the Zenaida "probably occurs on Mainland".—[4]

Do you remember my asking you to be so very kind as to keep in mind & make memorandum for me of any cases of abnormal parts in Birds being variable. Will you kindly run your eye through enclosed list[5] (& after having looked through it, tear it up) & then you will see the sort of cases, which interest me, & very grateful I shall be for any assistance on this head.—

With my best thanks for your kind assistance | Believe me | Yours sincerely | C. Darwin

American Philosophical Society (195)

[1] Dated by the publication in 1860 of a new issue of *Journal of researches* that included Sclater's information.

[2] See letter from P. L. Sclater, [3? February 1860]. CD cited Sclater's information concerning the Galápagos birds in *Journal of researches* (1860), p. vii. The first edition of *Origin* (p. 389) refers to twenty-five of the twenty-six land birds being ranked as distinct species. CD corrected the relevant passage in *Origin* 3d ed., p. 429:

> There are twenty-six land-birds, and twenty-one, or, perhaps, twenty-three, of these are ranked as distinct species, and are supposed to have been created here; yet the close affinity of most of these birds to American species in every character, in their habits, gestures, and tones of voice, was manifest.

[3] See the letter from P. L. Sclater, [3? February 1860] for Sclater's query about this species. CD refers to *Birds*, p. 86.

[4] *Zenaida galapagoensis* was described in *Birds*, p. 115, as a bird that was confined to the Galápagos Archipelago. In *Journal of researches* (1860), p. vii, CD instead included it among the Galápagos birds that were also found on the mainland. Recent authorities state that *Z. galapagoensis* is endemic to the islands (Harris 1974, p. 119).

[5] The list has not been found. The variability of abnormally developed structures had particularly engaged CD in 1857 while he was writing the chapter on 'Laws of variation: varieties and species

compared' for his species book (see *Natural selection*, pp. 309–11, and *Correspondence* vol. 6, letter to J. D. Hooker, 16 [May 1857]).

To John Murray 5 February [1860][1]

Down Bromley Kent
Feb. 5th

My dear Sir

Very many thanks about Bailliere & extract (highly flattering) from German Newspaper.—[2]

If you have not already taken, will you please to take proper steps in Germany for reserving to me power of Translation.—

Shall you have new title-page for my Journal? if so would it not be splendid puff to put in 7th Thousandth, or if Colburn's Edition of 1500 might be added 8th thousandth[3]

Yours very truly | C. Darwin

How many copies shall you print? Do not think of answering this on purpose.

John Murray (publishers)

[1] Dated by the reference to a German edition of *Origin*.
[2] See letter to John Murray, 3 February [1860]. The extract from a German newspaper has not been found.
[3] The title page of the 1860 edition of *Journal of researches* reads 'Tenth Thousand'.

To J. D. Hooker 8 February [1860]

Down Bromley Kent
Feb—8th

My dear Hooker

The extract from Naudin, which I owe to Mrs Hooker's kindness, was amply & more than amply sufficient.—[1] Pray thank her also for her nice little note received yesterday which has pleased me much.— I am very sorry to hear about the price of your Introduction; it seems very high, but I suspect that it contains much more matter or words than anyone would at first think. Your presentation copies will cost you a little fortune.— It is really a horrid shame to think that after all your immense labour that you should be heavily fined in hard cash.—[2]

By an *odd* chance, on my honour it was only night before last, I was thinking to myself how you could work up your published materials into one volume. It seemed to me very difficult; but I thought you might make a new Essay, & beginning by a sketch of the general great affinities & classification of the great orders, & then pass on to their geographical Distrb. & Geological Succession. And add a chapter on effect on Landscape.— But I clearly foresaw that you would object to the noble art of compiling; but for a general subject surely compiling is necessary. You know how I admire a Compiler! Such a Book would be very difficult; & style would come much into play; & yet I believe you could, if you would, produce a grand Book of the kind.—

You did not answer my query about date of publication of Essay, but I have put down December 1859, & that ends my little Historical sketch.—[3]

Farewell my dear Hooker | Ever your affect | C. Darwin

I walked home with Grove from last Royal Soc. meeting[4] & he harangued me to that extent I was half-dead, & he did not at *all* clearly see what he was talking about. I had thought you rather unjust about Grove; I humbly axe your pardon.—

From a talk a year or two ago, I am sure he then saw *nothing* about means of modification, & he knew *nothing* of difficulties—

Endorsement: '/60'
DAR 115.1: 39

[1] See letter to J. D. Hooker, 31 [January 1860]. In the preface added to the revised American edition of *Origin*, CD cited Charles Victor Naudin as having 'expressly stated his belief that species are formed in an analogous manner as varieties are under cultivation; and the latter process he attributes to man's power of selection.' (*Origin* US ed., p. ix; see Appendix IV).

[2] Hooker had told CD of his intention to buy 100 copies of Hooker 1859 for distribution (see *Correspondence* vol. 7, letter from J. D. Hooker, [8–10 April 1859]). The publisher of the work, Lovell Augustus Reeve, had apparently charged Hooker a high price for each copy. For CD's offer to pay Hooker for the copies he had sent to CD, see the letter to J. D. Hooker, 14 February [1860].

[3] On p. xi of the preface added to the revised and augmented American edition of *Origin* (see Appendix IV), CD cited Hooker 1859 as follows:

> In December, 1859, Dr. Hooker published his Introduction to the Tasmanian Flora: in the first part of this admirable essay he admits the truth of the descent and modification of species; and supports this doctrine by many original and valuable observations.'

[4] William Robert Grove was an active participant in Royal Society affairs. CD was in London to attend the meeting of the Royal Society on 26 January 1860 ('Journal'; Appendix II).

To Asa Gray [8 or 9 February 1860][1]

Down Bromley Kent

My dear Gray

I now send my short Historical Preface & one page more of corrections.[2] With respect to Agassiz, I have shamefully blundered but Von Baer is fully as good an authority; some would say better authority.[3] Please forward enclosed to Agassiz & read if you think worth while. I wrote all latter part of volume from memory, & it is a blessing more blunders have not yet been detected: but the state of my health left me no choice: I felt on the point of quite breaking down.— If the M.S. now (& before) sent is of no use, no harm is done.—[4] The pages all refer to the new Edition or Reprint—

I have now received your letter of the 23^d^ of Jan. by Hooker, in which you tell me all the trouble (& you so much overworked) which you have taken about an American Edition. I fear there is no chance of your Review now appearing at the head, which would have *greatly* pleased me. If there is another Reprint in America, will you send the English 2^d Edit (or Reprint) on account of several *verbal* corrections & likewise the M.S corrections; I would send you another copy, or make the Publishers give you one. If you notice any errors like "self-fertilise itself" pray correct.—[5] Again & again, I thank you heartily for all your generous kindness to me.—

Now I will just run through some points in your letter. What you say about my Book gratifies me most deeply & I wish I could feel all was deserved by me.

I quite think a Review from a man, who is not an entire convert, if fair & moderately favourable, is in all respects the best kind of Review. About weak points I agree. The eye to this day gives me a cold shudder, but when I think of the fine known gradations, my reason tells me I ought to conquer the cold shudder.[6] I agree about sterility, & especially about fertility of the strongest marked varieties, & in the M.S already sent to you, I have confessed more plainly the difficulty. But a vast number of facts show how mysteriously & easily the reproductive system is affected.

Pray kindly remember & tell Prof. Wyman, how **very grateful** I sh^d. be for any hints, information or criticisms: I have the highest respect for his opinion. I am so sorry about Danas health; I have already asked him to pay me a visit.—[7]

Farewell, you have laid me under a load of obligations,—not that I feel it a load. It is the highest possible gratification to me to think that you have found my Book worth reading & reflexions from you & these others I put down in my own mind as the judges whose opinions I should value most of all.—

My dear Gray | Yours most sincerely | C. Darwin

PS. I feel pretty sure from my own *[experience]*, that if y⟨ou⟩ ⟨ ⟩ by your studies to keep the subject of Origin of Species before your mind, that you will go further & further in your belief.— It took me long years & I assure you I am astonished at the impression my Book has made on many minds— I fear 20 years ago I sh^d not have been half as candid & open to conviction.—

Gray Herbarium of Harvard University (11)

[1] Dated by the reference to the historical preface for the revised American edition of *Origin*, which in the preceding letter CD says he has just finished. CD also states that he has received Gray's letter of 23 January 1860, to which he also alludes in the following letter.

[2] CD's historical sketch appeared as a preface to the revised and augmented American edition of *Origin*. It was inserted before the table of contents, presumably so as not to affect the pagination of the electrotyped plates, and is dated 'Feb. 1860.' The 'one page more of corrections' that CD sent has not been identified. Since the historical sketch was included, they too were probably incorporated into the text. Subsequent corrections that CD sent (letter to Asa Gray, 8 March [1860]), had to be listed separately in a supplement (*Origin* US ed., pp. 426–32). For a transcription of the historical sketch, textual alterations, and the supplement, see Appendix IV. See also letter to Asa Gray, 1 February [1860].

[3] See letter from Asa Gray, 23 January 1860. In the authorised American edition of *Origin*, the reference to Louis Agassiz's being unable to identify whether an unticketed vertebrate embryo was that of a mammal, reptile, or bird was replaced by a similar statement attributed to Karl Ernst von Baer (*Origin* US ed., pp. 381–2). CD made the same correction to the third English edition of *Origin*. Baer's anecdote was related in T. H. Huxley trans. 1853, pp. 186–7.

[4] See the enclosure with the letter to Asa Gray, 1 February [1860].
[5] See letter from Asa Gray, 23 January 1860. The sentence was corrected in *Origin* US ed., p. 91.
[6] See *Origin*, pp. 186–7.
[7] Jeffries Wyman and James Dwight Dana. See letter from Asa Gray, 23 January 1860.

To C. J. F. Bunbury 9 February [1860][1]

Down Bromley Kent
Feb. 9th

My dear Bunbury

I am *very* much obliged to you for writing to me at such length; for I value your opinion much. Ever since our sojourn at Capel Kurig[2] I have formed (if you will not think it impertinent in me to say so) the highest opinion of your knowledge, acuteness & judgment. Therefore I thank you much for your praise of my Book. I am sorry, but not *in the least* surprised that you do not go a little further with me.[3] I changed so slowly myself that I am indeed surprised at anyone becoming a convert.—

I will just run through some of the salient points in your letter: with respect to admitting representative species to be descendants of one species & not extending the same view to whole classes, the point seems to me to turn whether you think the explanation which I offer of the classification of all Beings, the principles of Homology & of Embryology, & Rudimentary organs are satisfactorily explained.[4] The explanation seems to me to be nearly satisfactory; & this being so, I can to myself allow of no difficulties in imagining what were the steps of transition to weigh against the understanding such large classes of facts.— It is curious how differently different men view the same subject; Asa Gray thinks the Hybrid Chr. one of strongest in Book![5] But he says (& I quite agree) more wants making out why varieties are not sterile when crossed. But this seems hopeless till we can say on what difference two close species are to certain extent fertile, & another two close species absolutely sterile.

With respect to Nat. Selection not being a "vera causa"; it seems to me fair in Philosophy to invent *any* hypothesis & if it explains many phenomena it comes in time to be admitted as real.[6] In your sense the undulatory theory of the *hypothetical* ether (the undulations themselves being not recognised) is not a vera causa in accounting for all the phenomena of Light. Natural selection seems to me in so far in itself not be quite hypothetical, in as much if there be variability & a struggle for life, I cannot see how it can fail to come into play to some extent. You speak of "Degeneracy",[7] I wish you could give me any instance of a race, *kept under the same conditions.* & in a moderately large body together degenerating or modifying: this is always assumed as certain; I do not dispute it, but I wish much for some evidence.

The extinction of *intermediate* forms seems to me explained in Ch. IV & more especially in the Xth Chapt.: if my Principle of Divergence of Character be admitted this almost necessarily follows.

I quite agree with what you say on effect of admission of theory on Systematic work; (see p. 485), not but what I was haunted with endeavouring to guess what cirripedes would be ranked as species by other naturalists.[8] I think the importance of theory bears on opening up new fields of enquiry & in giving a rational, instead of theological explanation of many known facts I am much pleased to hear that you intend reading the Book again. If not too troublesome I shall be most grateful for any other remarks. Your letter is as clear as daylight, & mine, I fear, like a London foggy day.—

Again I thank you heartily for all your kind trouble & believe me | Dear Bunbury | Yours very sincerely | Charles Darwin

As you give up representative species, I suspect & hope that you will go further; but this will depend wholly whether you think the large classes of facts in Ch. XIII are explained by Theory.— See remark (at p 127) where I say "Whether nat. Select. has really thus acted in nature in modifying &c &c must be judged of by the balance of evidence given in the following Chapters"

West Suffolk Record Office

[1] Dated by the relationship to the letter from C. J. F. Bunbury, 30 January 1860.

[2] CD and Bunbury met in the Welsh village of Capel Curig in 1842 during CD's expedition to examine the erratic boulders of Caernarvonshire (see *Correspondence* vol. 2, letter to W. H. Fitton, [*c.* 28 June 1842]). The occasion is described in Bunbury ed. 1891–3, *Early life* 1: 367 n.

[3] See letter from C. J. F. Bunbury, 30 January 1860.

[4] The portion of the letter from C. J. F. Bunbury, 30 January 1860, in which he addressed this point is now missing. In the terminology of the time, 'representative' species were species of similar structure living in different locales; they were said to take the place of, or 'represent', each other. Orthodox naturalists believed that such species had been designed with a similar form but created separately for different geographical areas.

[5] See letter from Asa Gray, 23 January 1860.

[6] CD's view of the concept of a 'vera causa' was primarily derived from John Frederick William Herschel's discussion in Herschel 1831, a text that he had read while at Cambridge University (see *Correspondence* vol. 1, letter to W. D. Fox, [15 February 1831], and Ruse 1975). CD's copy of the work is in the Darwin Library–CUL. Herschel used the notion to define what can legitimately be proposed as the underlying factors on which a phenomenon depends (Herschel 1831, p. 197):

> They must be *veræ causæ* , in short, which we can not only show to exist and to act, but the laws of whose action we can derive independently, by direct induction, from experiments purposely instituted; or at least make such suppositions respecting them as shall not be contrary to our experience, and which will remain to be verified by the coincidence of the conclusions we shall deduce from them, with facts.

[7] It was a long-held belief among many naturalists, systematised through the writings of Georges Louis Leclerc, Comte de Buffon, that some modification or variation in species had occurred through the degeneration or 'falling off' of an original created type. This concept was especially invoked to explain the production of new varieties of domesticated plants and animals by man. See Roger 1989, pp. 392–401.

[8] CD refers to the passage in *Origin*, p. 485, in which he stated that some 'forms now generally acknowledged to be merely varieties may hereafter be thought worthy of specific names, as with the primrose and cowslip'. For CD's attempts to balance the requirements of taxonomic nomenclature

with his views on the genealogical relationships between Cirripedia, see *Correspondence* vol. 4, especially his letters to Hugh Edwin Strickland, and Appendix II.

To J. S. Henslow 9 February [1860]

Down Bromley Kent.
Feb. 9th

My dear Henslow

We shall be *delighted* to see you on 14th.[1] Trains leave London Bridge (N. Kent Station) for Bromley (**not** Shortland Station) at 1° 30′. 3.30 4.40. 5.40. If you can tell me by what Train you will come, I will endeavour to send carriage to meet you; if it shd not be there, you must take Fly at Station—6 miles here.—

In Haste | Yours affect | C. Darwin

Endorsement: '1860'
DAR 93: 63–4

[1] Entries in Emma Darwin's diary record that Henslow visited Down from 14 to 16 February 1860.

From P. L. Sclater 11 February 1860

Zoological Society of London, | 11, Hanover Square, | London, W.
Feb. 11th. *1860*

My dear Sir,

I send you an extract from the minutes of one of our scientific meetings relating to the subject I was speaking of today—[1] I find nothing was said about it in the printed volume of Proceedings—

I find I never answered your letter of the 4th. and now have only to say that *Sylvicola aureola may* be a distinct species, but it is a close ally of *S. æstiva* of *N.* and *S.* America and perhaps only a "*climatic variety*" or as the Germans say "*climatische Abart*"—[2]

The next time I go to the B.M. I will examine *Otus galapagoensis Zenaida galapagoensis* and *S. aureola.*[3]

Very faithfully Your's | P. L Sclater

[Enclosure]

Zool. Soc. May 11, 1858.
(Extract from minutes)

Mr. Gould made some observations on the Indian Phasianidæ imported last year and now laying in the Gardens of the Society.[4] Drawings of the Eggs of the Impeyan Pheasant the Cheer, the Purple Pheasant and two species of Kaleege were exhibited to the meeting.

With reference to the eggs of the Indian *Phasianidæ* Mr. Sclater remarked that though the Egg of *Gallophasis albo-cristata* and *G. melanota* were easily recognizable as of distinct species—as indeed were the birds themselves—yet it was well known in India that in the region where these two species inosculate, a transitional variety is found passing from one to the other. This was not so surprizing in a gallinaceous bird, but the same thing occurred in two instances in birds of the Passerine order and was very remarkable. No one would deny the specific distinctness of *Coracias bengalensis* of the Indian Peninsula from *C. affinis* of Assam, or of *Colaptes auratus* of the Eastern United States of America from *Colaptes mexicanus* of California and Mexico yet in the country where these species respectively inosculate intermediate varieties are found.[5]

DAR 205.3 (Letters), DAR 205.7 (2): 143

CD ANNOTATION
Top of first page: '19'[6] *brown crayon, circled brown crayon*

[1] See the enclosure. Sclater was secretary of the Zoological Society of London, and he frequently contributed notices and reports at the society's meetings.
[2] See letter to P. L. Sclater, 4 February [1860]. In the postscript to the preface of the new issue of *Journal of researches* (1860), p. vii, CD stated: 'Mr. Sclater thinks that one or two of these endemic [Galápagos] forms should be ranked rather as varieties than species, which always seemed to me probable.'
[3] See letter from P. L. Sclater, [3? February 1860], and letter to P. L. Sclater, 4 February [1860].
[4] The reference is to the ornithologist John Gould.
[5] Inosculate: 'To pass into; . . . to blend.' (*OED*).
[6] The number of CD's portfolio of notes on the geographical distribution of animals.

To Charles Lyell 12 [February 1860]

Down Bromley Kent
12th

My dear Lyell

I send Bronn's letter with translation of latter half made for *me* by my nephew.[1]

I send also letters by Asa Gray & Bunbury for the *chance* of your liking to see them.[2] I had not intended to have sent them, as it seems so boastful to send them,—not that I have a shred of modesty in me.— I have answered Bunbury that in his sense the undulatory theory of Light is very far from a vera causa.[3] Bunbury's letter seems to me uncommonly well written. There is one sentence in A. Gray's (if you read it) which you might misunderstand; he put my name before Forbes on Glacial distribution; & I told him in answer that I had written out the notion 3 or 4 years before Forbes, but that I had no sort of claim to notice on this head, as he published first, & that in the Origin I shd. of course take no notice of this.—[4]

In few days I will send you 1st part of Asa Grays excellent Review[5] & notice by Bronn in Jahr-buch fur Mineralogie.—[6]

It is good job that I have heard today that Bronn will *superintend* the German Translation.—[7] All these letters &c may be left at Erasmus' marked *not* to be forwarded, as I shall be up in fortnight.—[8]

I think it was a great pity that Huxley wasted so much time in Lecture on preliminary remarks:[9] he hardly gave idea of my notions; but his Lecture seemed to me very fine & very bold. I have remonstrated (& he agrees) against impression that he would leave that sterility was universal & infallible criterion of species.—[10]

You will, I am sure, make a grand discussion on Man. I am so glad to hear that you & Lady Lyell will come here.—[11] Pray fix your own time & if it does not suit us we w^{d}. say so. We could then discuss man well

Ever yours | C. Darwin

How much I owe to you & Hooker; I do not suppose I sh^{d}. hardly ever have published, had it not been for you

Endorsement: 'Febr 1860'
American Philosophical Society (196)

[1] The letter from Heinrich Georg Bronn has not been located, but see the letters to T. H. Huxley, 2 [February 1860], and to H. G. Bronn, 4 February [1860]. CD possibly refers to his nephew Henry Parker, who was a fellow of Oriel College, Oxford.

[2] Letters from Asa Gray, 23 January 1860, and from C. J. F. Bunbury, 30 January 1860.

[3] See letter to C. J. F. Bunbury, 9 February [1860].

[4] CD refers to Edward Forbes. See letter from Asa Gray, 23 January 1860, and *Correspondence* vol. 7, letters to Asa Gray, 11 August [1858] and 11 November [1859].

[5] [Gray] 1860a. Gray's review of *Origin* was published in the March number of the *American Journal of Science and Arts*. There is a copy of the review in the Darwin Pamphlet Collection–CUL.

[6] Bronn 1860a.

[7] The letter has not been found, but see letter to H. G. Bronn, 14 February [1860].

[8] Emma Darwin's diary records that CD went to London on 27 February 1860. He stayed at the house of his brother Erasmus Alvey Darwin.

[9] Thomas Henry Huxley delivered a Friday evening lecture at the Royal Institution on 10 February 1860, in which he addressed CD's work on species (T. H. Huxley 1860a). CD attended the lecture (see letters to T. H. Huxley, [26 January 1860] and 2 [February 1860]).

[10] The printed version of Huxley's lecture contains a passage that directly relates to CD's comment (T. H. Huxley 1860a, p. 195):

> Leaving open the question whether the physiological distinction just mentioned is, or is not, a universal character of species, it is indubitable that it obtains between many species, and therefore has to be accounted for by any theory of their origin.

[11] The Lyells visited Down from 9 to 12 March 1860 (Emma Darwin's diary).

From Charles Lyell [13–14 February 1860][1]

vast that a variety of phases may have occurred in climate— Heer & Gaudin with their leaves & plants in modern travertines will I expect clear up the point in time.[2]

Auguste Bravard has sent me in a lithographed letter an astounding list of new mammalia found by him in Buenos Ayres, Bahia Blanca & Paranà—basin of Plata—Pampas &c—[3] He says (& he is generally right) he has found 56 species, raising the number from 10 to 66 Three new species of Scelidotherium, one Megalonix Two new Mylodons five new Glyptodons, about as many of Dasypus

Five new pachyderms of new genera— Several new felidæ, canidæ New ruminants Camelotherium & Cervus besides rodents &c—

All the above post- or newer-pliocene but there are also some upper Miocene [whale], & what is surely a very grand discovery among the Eocene mammalia Anoplotherium americanum Brav—& Paleotherium Paranense Brav—both from Paranà. Is not the occurrence of Paris Gypsum Upper Eocene European forms quite new?[4] Then he has with the above an Eocene Palæophis which I think is the genus of Owen's Sheppey Sea-serpent.—[5] On referring to Morris' catalogue[6] I see it also ascends to middle Eocene in Hants—two species— But Bravard is not such good authority for reptiles. How much more significant such facts are if one can once feel sure of transmutation—& hereditary descent.

I have had a good half hour's argument with the Bp of Oxford, Wilberforce, he maintaining that your book "was the most unphilosophical he had ever read". Of course he seized on the most vulnerable points with no small skill, as if "all having come from one egg" was the Keystone of your theory!— Bell did the same in argument—[7] Some day I will tell you all about the fight— If he reviews you I think I shall recognize his hand.[8] Have you seen the attack in Annals & Mag. Nat. Hist. Febr^y 1860 (v. 5. N°. 26)

Surely it must be Lowe or Wollaston—some one who knows Madeira p. 141. probably the Rev^d—for tho' T. V. W. is very orthodox I do not think he would experience such a "cold shudder"—[9] It is written by one who personally has a liking for you I should think—on the whole ably done— But if it be Low he has only put one side of the (Madeira?) land-shell question for out of his *sub*-species of certain polymorphic species in each island I c^d. raise a good argument to meet the equally striking persistency of the majority of species from a remote post- or newer-pliocene epoch—I speak of course of the land shells— It is strange when each rock & island has an insular variety of certain species that others remain so entirely unchanged from the fossil period— Still on the whole the Madeira evidence cuts as much or more your way as the opposite.[10]

Have you read Wallace in the last Linn. Soc^y Proceedings.[11]

If Madeira was joined to the main land & Porto Santo I cannot understand how the land shells of the two can be so very dissimilar & also so peculiar & unlike the continental species. What a deal Wallace gives one to think of— He gains with some by stating his *antitype* theory because he makes that a law at any rate & so the anti-

AL incomplete
DAR 205.3 (Letters), DAR 205.9 (Letters)

CD ANNOTATIONS

1.1 vast . . . time. 1.2] *crossed pencil*

3.2 what is surely . . . descent. 3.9] 'Mem. Nebraska Grand showing same law on Land as sea.' *added ink*

4.1 I have . . ably done—5.4 *crossed pencil*

4.5 If he . . . hand. 4.6] *cross added pencil, circled pencil*

4.7 Febr^{y} . . . 26)] *underl pencil*
5.3 "cold shudder"] *cross added pencil, circled pencil*
5.4 But if . . . opposite. 5.11] '19'[12] *brown crayon, circled brown crayon*
6.1 Have . . . anti- 7.5] *crossed pencil*
Top of first page: '21'[13] *brown crayon*

1 Dated by the reference to the reviews of *Origin* (see nn. 8 and 9, below) and by the relationship to the letter to Charles Lyell, 15 and 16 [February 1860]).

2 Oswald Heer and Charles Théophile Gaudin were Swiss palaeobotanists. Both were currently investigating the Tertiary flora of Europe.

3 Pierre Joseph Auguste Bravard emigrated to Argentina from France in 1852 and became director of the museum in Paraná .

4 The gypsum quarries around Paris provided some of the fossils on which Georges Cuvier based his pioneering reconstructions of extinct mammals, including the genera *Anoplotherium* and *Palaeotherium* described in Cuvier 1804–8.

5 In R. Owen 1839, Richard Owen described the fossilised remains of what he called an 'extinct species of serpent', found in the London Clay deposits on the island of Sheppey.

6 Morris 1854. John Morris's work was the standard source of information on British fossils.

7 Thomas Bell was president of the Linnean Society of London.

8 Samuel Wilberforce, the bishop of Oxford, reviewed *Origin* anonymously in the *Quarterly Review* 108 (1860): 225–64.

9 Both Richard Thomas Lowe and Thomas Vernon Wollaston had collected extensively on the island of Madeira. Wollaston was the anonymous author of the review of *Origin* in the *Annals and Magazine of Natural History* ([Wollaston] 1860).

10 Both CD and Lyell had corresponded with Lowe and Wollaston about the distinct species of land molluscs found on islands in the Madeira group. See *Correspondence* vols. 5 and 6, and Wilson ed. 1970.

11 Wallace 1860. Wallace had sent CD the paper and asked him to communicate it to the Linnean Society. CD told Wallace he thought it was "*admirable*". See *Correspondence* vol. 7, letter to A. R. Wallace, 9 August 1859.

12 The number of CD's portfolio of notes on the geographical distribution of animals. The letter was cut in two and filed separately.

13 The number of CD's portfolio of notes on palaeontology and extinction.

To H. G. Bronn 14 February [1860][1]

Down. Bromley Kent.
Feb. 14th

My dear & much honoured Sir

I thank you cordially for your extreme kindness in superintending the Translation I have mentioned this to some eminent scientific men, & they all agree that you have done a noble & generous service— If I am proved quite wrong yet I comfort myself in thinking that my book may do some good; as truth can only be known by rising victorious from every attack. I thank you also much for the Review & for the kind manner in which you speak of me.—[2] I send with this letter some corrections & additions to M. Schweitzerbart, & a short historical Preface.[3] I am not much acquainted with German Authors as I read German very slowly. Therefore I do not know whether any Germans have advocated similar views with mine; if they have would you do me the favour to insert a footnote to the Preface?[4] M Schweitzerbart has now the reprint ready for a Translator to begin—

Several scientific men have thought the term "Natural Selection" good, because its meaning is *not* obvious & each man could not put on it his own interpretation, & because it at once connects variation under domestication & nature Is there any analogous term used by German Breeders of Animals? "Adelung" ennobling would or perhaps be too metaphorical. It is folly in me but I cannot help doubting whether "Wahl der Lebens-weise" expresses my notion— It leaves the impression on my mind of the Lamarckian doctrine (which I reject) of habits of life being all important— Man has altered & thus improved the English Race Horse by *selecting* successive fleeter individuals; & I believe owing to the struggle for existence that similar *slight* variations in a Wild Horse *if* advantageous to *it* would be *selected* or *preserved* by nature; Hence Natural Selection— But I apologise for troubling you with these remarks on the importance of choosing good German terms for "Natural Selection".[5]

With my heartfelt thanks & with sincere respect—I remain Dear Sir | Yours very sincerely—| Charles Darwin—

I am very much obliged for your "Stuffengang &c" which I am now reading—[6] I wish I knew what was the authority for a Batrachian in the New Hebrides.—[7]

Copy
DAR 143

[1] Dated by the reference to the German translation of *Origin*.

[2] Bronn 1860a. There is a copy of the review in the Darwin Pamphlet Collection–CUL.

[3] Christian Friedrich Schweizerbart, owner of the publishing firm E. Schweizerbart of Stuttgart, was the publisher of the German edition of *Origin* (see letter to H. G. Bronn, 4 February [1860]). The historical preface that CD had composed for the authorised American edition of *Origin* (see Appendix IV) was also published in the German edition of 1860. It was subsequently expanded and included in the third English edition.

[4] Bronn did not follow CD's suggestion, although he did include a footnote to the discussion of Étienne Geoffroy St Hilaire's views on species (Bronn trans. 1860, p. 2 n.):

> Bekanntlich kam er in der Akademie mehrmals zu heftigen Auftritten noch mit Cuvier, welcher die Beständigkeit der Species gegen ihn vertheidigte. D. Übers. [It is well known that on several occasions in the academy he had violent clashes with Cuvier, who defended against him the view of the constancy of species. The translator.]

In the historical sketch included in *Origin* 3d ed., CD cited the German naturalists Hermann Schaaffhausen, Karl Ernst von Baer, and Karl Friedrich Burdach as having expressed views on species change. In a note (p. xviii), he added to these Franz Unger, Joseph Wilhelm Eduard d'Alton, and Lorenz Oken on the basis of references given in Bronn 1858b.

[5] In his review of *Origin*, Bronn had referred to natural selection as 'Wahl der Lebens-Weise' (Bronn 1860a, p. 112). The term Bronn used in his translation was 'natürliche Züchtung'.

[6] Bronn 1860b. There are two copies in the Darwin Pamphlet Collection–CUL, one of which is annotated.

[7] Bronn 1860b, p. 13 n. According to CD's views on the colonisation of oceanic islands, it would be very difficult for frogs or their eggs, which are damaged by salt-water, to reach isolated areas such as the New Hebrides.

To J. D. Hooker 14 February [1860][1]

Down Bromley Kent
Feb. 14

My dear Hooker

I succeeded in persuading myself for 24 hours that Huxley's lecture was a success. Parts were eloquent & good & all *very* bold, & I heard strangers say "what a good lecture". I told Huxley so; but I demurred much to time wasted in introductory remarks; especially to his making it appear that sterility was a clear & manifest distinction of species,[2] & to his not having even alluded to the more important part of subject. He said that he had **much** more written out, but time failed.[3] After conversation with others & more reflection I must confess that as an Exposition of the doctrine the Lecture seems to me an entire failure.— I thank God I did not think so, when I saw Huxley; for he spoke so kindly & magnificently of me, that I could hardly have endured to say what I now think.

He gave no just idea of *natural* selection. I have always looked at this doctrine of Nat. Selection as an hypothesis, which if it explained several large classes of facts would deserve to be ranked as a theory deserving acceptance; & this of course is my own opinion.— But, as Huxley has *never* alluded to my explanation of Classification, Morphology Embryology &c, I thought he was thoroughily dissatisfied with all this part of my Book. But to my joy I find it is not so & that he agrees with my manner of looking at subject; only that he rates higher than I do the necessity of Natural Selection being shown to be a vera causa always in action.— He tells me he is writing long Review in Westminster.—[4]

It was really provoking how he wasted time over the idea of a species as exemplified in Horse[5] & over Sir J. Hall's old experiment on marble.—[6] Murchison was very civil to me over my Book after Lecture, in which he was disappointed.—[7]

I have quite made up my mind to savage onslaughts; but with Lyell, you, & Huxley I feel confident we are right & in **long** *run* shall prevail. I do not think Asa Gray has quite done you justice in beginning of Review of me.[8] The Review seemed to me **very** good, but I read it only hastily.

I am very glad to hear that you are thinking of a general work—do not be in a hurry.[9] Do give a resume on propagation of Crytogamic plants: it would interest general reader & pocket your pride & compile! It will be a precious difficult Book to write.—

When we meet I will pay 14^{s} with thanks.—What a poor sneaking fellow L. Reeve must be.—[10]

Adios | I am tired. | Your affect | C. Darwin

Henslow comes here tonight for 2 or 3 days. He is going to Anniversary of Geolog. Soc.— By the way I fancy I shall be cut up by Phillips.[11]

DAR 115.1: 40

[1] Dated by the reference to Thomas Henry Huxley's lecture on CD's theory at the Royal Institution, which took place on 10 February 1860.

[2] See also letter to Charles Lyell, 12 [February 1860].

[3] Huxley's outline for the lecture is in the Imperial College Archives (Huxley 41: 10). The printed version of the address was only an abstract (T. H. Huxley 1860a).

[4] [T. H. Huxley] 1860b.

[5] T. H. Huxley 1860a, pp. 195–6.

[6] The geologist James Hall heated limestone until it turned into marble in order to demonstrate the igneous origin of such rocks. Huxley's remarks on this experiment do not appear in the published abstract of the lecture.

[7] Roderick Impey Murchison, director-general of the Geological Survey of Great Britain, did not accept CD's theories. Writing to the geologist Robert Harkness, Murchison stated that he denied all CD's inductions and remained 'as firm a believer as ever that a monkey and a man are distinct species, and not connected by any links,—*i.e.* are distinct creations.' See Geikie ed. 1875, 2: 322.

[8] An opening note in Asa Gray's review of *Origin* in the *American Journal of Science and Arts* stated that the concluding part of the review of Hooker 1859 was being deferred 'for want of room'. Gray further stated: 'Fully to understand the foregoing Essay of Dr. Hooker, it should be read in the light of Mr. Darwin's book. The Essay is a trial of the Theory' ([Gray] 1860a, p. 153).

[9] See letter to J. D. Hooker, 8 February [1860].

[10] CD refers to the copies of Hooker's introductory essay of the *Flora Tasmaniæ* (Hooker 1859), presented to him at the end of 1859. The publisher Lovell Augustus Reeve had set a high price for the volume (see letter to J. D. Hooker, 8 February [1860]) and his conditions for publication meant that Hooker had to pay for any presentation copies (see *Correspondence* vol. 7, letter to J. D. Hooker, [8–11 April 1859], n. 3). An entry in CD's Account book (Down House MS), dated 6 April 1860, records payment of 14*s.* to Hooker for "copies of my Pamphlet." CD's copy of the separately printed essay (Hooker 1859) and of the larger work (Hooker 1855–60) are in the Darwin Library–CUL.

[11] John Phillips, president of the Geological Society, was to give his presidential address at the anniversary meeting of the society on 17 February 1860. Contrary to CD's expectation, Phillips discussed *Origin* in a temperate tone (*Quarterly Journal of the Geological Society of London* 16 (1860), Proceedings, p. xxxvi). CD expressed his belief that Phillips was 'dead against' his views in the letter to John Phillips, 14 November [1860].

To P. L. Sclater 14 February [1860]

Down Bromley Kent
Feb. 14th

My dear Sir

Very sincere thanks for your kind information, which is very valuable to me.— I asked G. R. Gray about the Otus & Zenaida—but whether he will give me satisfactory information I know not.—[1] I had intended asking you whether the Strix punctatissima & Pyrocephalus nanus differed in any *degree* whatever, in size & duskiness from the same species on the Mainland.[2] This would be very interesting to me; but I fear I am too troublesome.—

With sincere thanks | Yours very faithfully | Charles Darwin

Endorsement: 'Feb. 1860'
American Philosophical Society (197)

[1] See letter from P. L. Sclater, 11 February 1860. George Robert Gray, an assistant in the zoological department of the British Museum, had worked on *Birds* after John Gould's departure for Australia. Gray described species of owls (*Otus*) and doves (*Zenaida*) for the volume (see *Correspondence* vol. 2). CD's

letter to Gray has not been found. CD seems to have pre-empted Sclater's offer to obtain information regarding these species at the British Museum (see letter from P. L. Sclater, 11 February 1860).

[2] In *Birds*, pp. 34–5 and 45–6, *Strix punctatissima* and *Pyrocephalus nanus* were stated to be species that were confined to the Galápagos Islands. Sclater had recently advised CD that they also occurred in Ecuador (see letter from P. L. Sclater, [3? February 1860]).

From George Henry Kendrick Thwaites [14 February 1860][1]

in some simple cells—*Desmidieæ*, *Diatomaceæ* &c?[2]

How could mere natural selection produce such exquisite beauty & symmetry as we see in every organized structure?

It is not easy to understand how, with a similar capability of modification in all, some cells should remain in their originally simple condition whilst others should, within the same period, have become modified by mere natural selection into the most highly organized & complicated structures.

Something might be said, it seems to me, in favour of a general degradation or degeneration taking place under certain conditions, just as a single organ, complicated in structure in some species, is found rudimentary only in allied ones.

You would be extremely interested to see how the *Lantana aculeata*, an introduction from the Mauritius about 35–40 years ago, is spreading all over the uncultivated parts of the country which are not under forest— Some of the indigenous plants would indeed have a hard struggle for existence if it were not for the interposition of man's labour. M. Humbert,[3] a very intelligent Swiss naturalist who is out here collecting for the Geneva Museum of which he is the Curator, has been surprised to find how very many of our native birds feed upon the fruit of this Lantana, their crops being often quite filled with it.

I very much regret to hear that your health has not been very good. I wish I could persuade you that a visit to Ceylon would benefit it & that I could enjoy the pleasure of your society for a few months as my guest here.

Believe me always, | Your's very sincerely | G. H. K. Thwaites

Incomplete
DAR 205.4 (Letters)

CD ANNOTATIONS

1.1 in some . . . allied ones. 4.3] *crossed pencil*
5.1 You would] '20'[4] *brown crayon, circled pencil*
5.3 Some of . . . man's labour. 5.5] *cross added brown crayon*
6.1 I very . . . Thwaites 7.1] *crossed pencil*
Bottom of last page: '14 Feb 1860' *pencil*

[1] Dated by CD's annotation.

[2] Thwaites, superintendent of the botanic gardens at Peradeniya, Ceylon, was an expert on the anatomy and reproduction of the Diatomaceae. CD had met Thwaites in 1847 and subsequently asked him to forward specimens of Ceylonese pigeons and poultry to him (see *Correspondence* vol. 5, letter to G. H. K. Thwaites, 10 December 1855).

[3] Alois Humbert was a Swiss naturalist and palaeontologist.
[4] The number of CD's portfolio of notes on the geographical distribution of plants.

To Charles Lyell 15 and 16 [February 1860]

Down Bromley Kent
15th

My dear Lyell

It is *very* goodnatured in you to write & tell me so many things which are very interesting to me. Bravard's discoveries seem to me magnificent, & especially interesting is the fact of Palæotherium Paranense, taken with (I think) the Nebraska Palæotherium.[1] Bravard has sent me two Spanish pamphlets (which I find to my surprise I can hardly translate) in which he has strange geological doctrine, of whole enormous Pampean deposit being a *subaerial* deposit.—[2] He disputes the coembedment of the Bahia Blanca fossils with recent shells; but I am by no means convinced. It seems to me impossible that a whole skeleton, (even to knee-cap) could be washed out of one formation & embedded in another & that other formation a turbulent one with largish pebbles & cross layers.—

I am perfectly convinced (having read this morning) that the Review in Annals is by Wollaston:[3] no one else in the world would have used so many parentheses. I have written to him & told him that the "pestilent" fellow thanks him for his kind manner of speaking about him.[4] I have also told him that he w^{d} be pleased to hear that the B. of Oxford says it is the most unphilosophical work he has ever read.—[5] The review seems to me clever & only misinterprets me in a few places. Like all hostile men he passes over the explanation given of Classification, Morphology, Embryology & Rudimentary organs &c.— —

I read Wallace's paper in M.S & thought it admirably good: he does not know that he has been anticipated about *depth* of intervening sea determining distribution.[6] The expression "coincidence" in time & space between new & old species is unfortunate, as he believes, as we do, that new species are very *slowly* formed.—[7] The most curious point in Paper seems to me that about the African character of the Celebes productions;[8] but I sh^{d} require further confirmation: I believe the aberrant Anoa, or so-called Antelope is really a small Buffalo.—

That is a *very very* interesting fact of the Loess Man belonging to peculiar of man;[9] do get that well worked out.—[10] Remember what you told me of fossil monkey; very man-like in middle Tertiaries. I will send (when I get from Hooker) Asa Gray's capital review.—[11] Henslow is staying here:[12] I have had some talk with him: he is in much same state as Bunbury[13] & will go a very little way with us, but brings up no real argument against going further. He, also, shudders at the Eye![14] It is really curious (& perhaps is an argument in our favour) how differently different opposers view the subject.— Henslow used to rest his opposition on imperfection of Geolog: Record, but he now thinks nothing of this, & says I have got well out of it; I wish[15] I could quite agree with him. Baden Powell says he never read anything so conclusive as my statement about the Eye!![16] A stranger writes to me about "sexual selection" & regrets that I boggle about such a trifle as the brush of Hair on Male Turkey. And so on.—

As L. Jenyns has a really philosophical mind, & as you say you like to see everything, I send an old letter of his.[17] In a later letter to Henslow which I have seen, he is more candid than any opposer I have heard of; for he says though he *cannot* go as far as I do, yet he can give no good reason why he should not.—[18] It is funny how each man draws his own imaginary line at which to halt.— It reminds me so vividly what I was told about you, when I first commenced geology, to believe a *little* but on no account to believe all.—[19]

Ever yours affect | C. Darwin

I hope you will succeed in finding out what the *great* Celts were used for, it bears on state of civilisation of the old natives.—[20]

Henslow means this spring to visit the Celt-Beds in France[21]

Many thanks for Bunbury letter received this morning & for your note.—[22] I doubt whether I use term Natural Selection more as a Person, than writers use Attraction of Gravity as governing the movement of Planets &c. but I suppose I could have avoided the ambiguity.

C. D.

16th—

Endorsement: 'Feb^{y} | 1860'
American Philosophical Society (198), Kinnordy MS

[1] See letter from Charles Lyell, [13–14 February 1860].

[2] Bravard 1857 and 1858. Both works are in the Darwin Pamphlet Collection–CUL; they were signed and presented by the author. In the papers, Pierre Joseph Auguste Bravard criticised CD's interpretation of the formation of the estuarial deposits near Bahia Blanca, Argentina, that contained the remains of extinct mammals and recent shells (*South America*, p. 101).

[3] Thomas Vernon Wollaston's unsigned review of *Origin* appeared in *Annals and Magazine of Natural History* 5 (1860): 132–43. See letter from Charles Lyell, [13–14 February 1860].

[4] Wollaston's review referred to CD's theory as a 'pestilent abstraction like dust cast into our eyes to obscure the workings of an Intelligent First Cause of all' ([Wollaston] 1860, p. 138).

[5] Samuel Wilberforce, bishop of Oxford, had made this remark to Lyell. See letter from Charles Lyell, [13–14 February 1860].

[6] Wallace 1860 (see letter from Charles Lyell, [13–14 February 1860]). CD refers to George Windsor Earl's work on the depth of the straits in the Malay Archipelago, to which CD had drawn Alfred Russel Wallace's attention when he first read Wallace 1860 in manuscript (see *Correspondence* vol. 7, letter to A. R. Wallace, 9 August 1859). Earl's findings had been published in Earl 1853. Wallace discussed the relationship between animal distribution patterns and the depth and extent of the stretches of ocean that separated geographical regions in Wallace 1860, pp. 179–82.

[7] The term 'coincidence' appears to have been removed from the published version of Wallace 1860.

[8] Wallace described the fauna of Celebes as being composed of species almost entirely confined to the island. They 'find their allies nowhere nearer than in tropical Africa' (Wallace 1860, pp. 176–8).

[9] CD omitted a word when he began a new page of the letter.

[10] Lyell was at work on a study of early human remains. Late in 1859, he had visited the Loess formations in Belgium and the valley of the Rhine, where fossil human bones had been found together with

Hensleigh Wedgwood
(Courtesy of the late Hensleigh Cecil Wedgwood.)

Heinrich Georg Bronn
(Courtesy of the Archive of Ruprecht-Karls-Universität, Heidelberg.)

those of extinct animals. Lyell discussed these findings in his work on the antiquity of man (C. Lyell 1863).

11 CD received proof-sheets of Asa Gray's review of *Origin* ([Gray] 1860a) on or around 4 February. After reading the review, CD sent it on to Joseph Dalton Hooker. See letter to Asa Gray, 18 February [1860].

12 John Stevens Henslow visited Down from 14 to 16 February 1860 (Emma Darwin's diary).

13 For Charles James Fox Bunbury's reaction to *Origin* and CD's rejoinder, see the letter from C. J. F. Bunbury, 30 January 1860, and the letter to C. J. F. Bunbury, 9 February [1860].

14 CD had used this expression in his letter to Asa Gray, [8 or 9 February 1860], in response to Gray's remark that the explanation of the formation of 'organs,—the making of eyes, &c' seemed the weakest point in his book (see letter from Asa Gray, 23 January 1860). CD had sent Gray's letter to Lyell (see letter to Charles Lyell, 12 [February 1860]).

15 The manuscript from this point up to the signature is now held separately from the rest of the letter in the collection of Lady Lyell, Kinnordy, Scotland.

16 *Origin*, p. 186. Baden Powell's letter to CD has not been found, but see the two letters to Baden Powell, 18 January [1860].

17 Letter from Leonard Jenyns, 4 January 1860. Lyell copied this letter into his scientific journal (Wilson ed. 1970, pp. 349–51).

18 The letter from Jenyns to Henslow has not been located, but CD referred to it in his letter to J. S. Henslow, 3 February [1860].

19 This probably alludes to the time, just before the *Beagle* sailed, when Henslow advised CD to read Lyell's *Principles* (C. Lyell 1830–3), 'but on no account to accept the views therein advocated.' (*Autobiography*, p. 101).

20 This and the following paragraph were written in pencil on a separate sheet of paper and enclosed with the letter. 'Celts' are large, shaped flints. Lyell was investigating the possible origin of the flints as part of his study of the antiquity of man (see *Correspondence* vol. 7, letters to Charles Lyell, 2 September [1859] and 20 September [1859]).

21 Henslow was not convinced that the celts were of the same age as the fossil remains with which they were found; he reported his doubts in a series of letters printed in the *Athenæum* in 1859 and 1860. In the autumn of 1860 he visited the gravel pits in Abbeville and Amiens, where various shaped flints had been found, and returned 'impressed with the conviction, that the facts I have witnessed do not *of necessity* support the hypothesis of a pre-historic antiquity for these works of man.' (*Athenæum*, 20 October 1860, p. 516).

22 This paragraph was also written in pencil on a second slip of paper and enclosed with the letter. CD refers to the letter from C. J. F. Bunbury, 30 January 1860, which he had sent to Lyell (letter to Charles Lyell, 12 [February 1860]). The note from Lyell has not been found.

To Neil Arnott 16 February [1860?][1]

Down. | Bromley Kent.
Feb. 16th.[2]

My dear Sir.

I am much obliged for your kind note & pamphlet—[3] The latter seems to me very interesting, & I agree with all that you say about the course of development of the intellect of man; especially about the dreadful waste of time, on the dead languages, German would be difficult enough if learning a language be good exercise for the mind, & how incomparably more useful.—[4] I really have no suggestions or criticisms worth giving it would indeed be presumptuous to think that I had, for I have never thought much about Human Progression or on the all important subject of Education— I can, however, say from my own personal experience with my five Boys[5] that it is surprising how very early in life they take *vivid* interest in & understand something of Natural Philosophy— If you enlarge the pamphlet into a Book, it might be very valuable; especially if it were any how possible to make it, when enlarged interesting to the general reader.

You put the Malthusian *great* truth of the "Struggle for existence" very forcibly—[6]

To give one infinitely little criticism; I demur to your saying—p. 17. that animals are governed only by selfish motives.—[7] look at the maternal instincts & still more at the social instincts. How unselfish is a Dog! At p. 18, ought not conscience as well as "the far seeing reasoning nature" to be introduced?[8] To me it seems as clear that we have a conscience as that the lower animals have a social instinct: indeed I believe they are nearly the same— But these are mere trifles.

With my best thanks & with my hopes that you may produce a larger work on the same subject—

I remain. | My dear Sir. | Yours very faithfully | Charles Darwin.

Copy
DAR 143

[1] The year is suggested by CD's reference to Arnott's enlarging into a book the pamphlet he sent CD. This was apparently the book he published in 1861 (Arnott 1861).

[2] Although the copyist wrote '18th', this was altered to '16th?', presumably by Francis Darwin.

[3] Arnott's pamphlet has not been identified. It was probably a preliminary version of his book on human progress (Arnott 1861); the passages CD cites are included in this work.

[4] Arnott was a prominent educational theorist who advocated the revision of school and university curricula to take more account of science and less of classics. See Arnott 1861, pp. 87–92.

[5] The copyist wrote 'four Boys'. This was altered, presumably by Francis Darwin, to read 'five'.

[6] Arnott 1861, pp. 60–3.

[7] 'The lower animal is shortsightedly or almost blindly selfish' (Arnott 1861, p. 51).

[8] 'The mere animal nature would yield to the impulses of the moment; the far-seeing reasoning nature calculates remote consequences, and can plan very complex arrangements to bring about desirable ends.' (Arnott 1861, p. 52).

To Smith, Elder, and Company?[1] 17 February [1860]

Down Bromley Kent
Feb. 17th

Dear Sir

I am infinitely obliged to you for so kindly offering to ship the Box with ear-trumpet & pay the charges for me thereon.— I will give as little trouble as possible.— I *have* paid the maker (W. Harrowin)[2] & told him to put Trumpet in a Box & address it to

Mr Syms Covington[3]
Post-Office
Pambala
Twofold Bay
Australia

To be left at your house.—

I thank you sincerely for aiding me in this little act of charity for the poor man writes to me that he cannot get Ear-trumpet which will suit him in Australia.—[4]

I will write to him & say that the Box will be sent by the next P. & O. Steamer after Feb. 20th, for the Box will not be in time for 20th—

Dear Sir | Yours faithfully & much obliged | Charles Darwin

Endorsement: '1860 | March 2'
Pierpont Morgan Library (Gordon Ray collection)

[1] The identity of the recipient is suggested by a previous occasion on which CD sent an ear-trumpet to Syms Covington. The publishing firm of Smith, Elder, and Co. sent CD's present in one of their regular shipments to their agent in New South Wales (see *Correspondence* vol. 2, letter to Syms Covington, 7 October 1843).

[2] The maker has not been identified. There is an entry dated 3 February 1860 in CD's Account book (Down House MS) that reads: 'Ear-trumpet. 13^{s} . 6^{d}'.

[3] Covington was CD's servant and assistant during the *Beagle* voyage and for some years afterwards. In 1842 he had emigrated to Australia (see *Correspondence* vol. 2).

[4] Covington's letter has not been found. In a previous letter, he asked CD to send him a book on curing deafness. See *Correspondence* vol. 7, letter to Syms Covington, 16 January [1859].

To Asa Gray 18 February [1860][1]

Down Bromley Kent
Feb. 18th

My dear Gray

I received about a week ago two sheets of your Review; read them, & sent them to Hooker; they are now returned & reread with care, & tomorrow I send them to Lyell.—[2]

Your Review seems to me *admirable*; by far the best which I have read.— I thank you from my heart both for myself, but far more for subject-sake. How curious your contrast between the views of Agassiz & such as mine is very curious & instructive.[3] By the way if Agassiz writes anything on subject, I hope you will tell me.— I am charmed with your metaphor of the streamlet never running against the force of gravitation.[4]

Your distinction between an hypothesis & theory seems to me very ingenious; but I do not think it is ever followed.—[5] Everyone now speaks of the undulatory *theory* of light; yet the ether is itself hypothetical & the undulations are inferred only from explaining the phenomena of light.— Even in the *theory* of gravitation, is the attractive power in any way known, except by explaining the fall of the apple & the movements of the Planets? It seems to me that an hypothesis is *developed* into a theory solely by explaining an ample lot of facts.

Again & again I thank you for your generous aid in discussing a view, about which you very properly hold yourself unbiassed.—

My dear Gray | Yours most sincerely | C. Darwin

There is Review in last Annals & Mag. of N.H. on my Book—rather bitter: I feel moral certainty it is by my good friend the Entomologist Wollaston: I have not derived any new idea from this Review.—[6] Several Clergymen go far with me.—Rev. L. Jenyns, a very good naturalist: Henslow will go a very little way with me & is not shocked at me; he has just been visiting me.

Gray Herbarium of Harvard University (22)

[1] Dated by the reference to CD's having received a copy of [Gray] 1860a.

[2] CD's annotated copy of Gray's review ([Gray] 1860a) is in the Darwin Pamphlet Collection–CUL.

[3] Gray's review took the form of an extended comparison between CD's views on the origin of species and those held by Louis Agassiz.

[4] Writing of artificial selection, Gray stated that 'man leads the course of variation as he leads a streamlet,—apparently at will, but never against the force of gravitation,—to a long distance from its source, and makes it more subservient to his use or fancy.' ([Gray] 1860a, p. 166). The phrase is marked in CD's copy (Darwin Pamphlet Collection–CUL).

[5] Gray distinguished two ways in which CD's theory might be substantiated ([Gray] 1860a, p. 162):

> Either, 1, to assign real and adequate causes, the natural or necessary result of which must be to produce the present diversity of species and their actual relations; or 2, to show the general conformity of the whole body of facts to such assumption, and also to adduce instances explicable by it and inexplicable by the received view,—so perhaps winning our assent to the doctrine, even though the cause of the assumed variation remain as occult as that of the transformation of tadpoles into frogs . . . The first line of proof . . . would establish derivation as a true physical theory; the second, as a sufficient hypothesis.

Gray made this distinction prior to discussing the nature of CD's theory and its strengths and weaknesses.

[6] [Wollaston] 1860. See letter to Charles Lyell, 15 and 16 [February 1860].

To Charles Lyell 18 [and 19 February 1860]

Down Bromley Kent
18th

My dear Lyell

I send by this Post, Asa Gray which seems to me very good, with the stamp of originality on it.[1] Also Bronn's out Jahrbuch fur Mineralogie.[2] The united intellect of my family has vainly tried to make it out.— I never tried such confoundedly hard German: nor does it seem worth the labour.— He sticks to Priestley's green matter & seems to think that till it can be shown how life arises, it is no good showing how the forms of life arise.[3] This seems to me about as logical (comparing very great things with little) as to say it was no use in Newton showing laws of attraction of gravity & consequent movements of the Planets, because he could not show what the attration of Gravity is.—

The expression "wahl der lebensweise" makes me doubt whether B. understands what I mean by natural selection,—as I have told him.—[4] He says (*if I understand him*) that you ought to be on same side with me.—[5]

Ever yours affecly | C. Darwin

P.S. | Sunday afternoon | I have kept back this to thank you for your letter with much news received this morning.[6] My conscience is uneasy at the time you waste in amusing & interesting me.— I was very curious to hear about Phillips.—[7] The Review in Annals *is*, as I was convinced, by Wollaston; for I have had very cordial letter from him this morning.[8]

I send by this Post an attack in G. Chroni. by Harvey (a first-rate Botanist as you probably know).[9] It seems to me rather strange: he assumes the permanence of monsters, whereas monsters are generally sterile & not often inheritable. But grant his case it comes that I have been too cautious in not admitting great & sudden variations. Here again comes in the mischief of my *Abstract*: in fuller M.S. I have discussed parallel case of a normal fish like a monstrous Gold-fish: I end my discussion by *doubting*, because all cases of monstrosities which resemble normal structures, which I could find were not in allied groups.—[10]

Trees like Aspicarpa with flowers of two kinds, (in the Origin) led me also to speculate on same subject;[11] but I could find only one doubtfully analogous case of species having flowers like the degraded or monstrous flowers.— Harvey does not see that if only a few (as he supposes) of the seedlings inherited his monstrosity natural selection would be necessary to select & preserve them.— You had better return G. Chron. &c &c to my Brother's.—

The case of Begonia in itself is very curious: I am tempted to answer the notice, but I will refrain for there w[d] be no end to answers.

With respect to your objection of multitude of still living simple forms, I have not discussed it anywhere in the Origin, though I have often thought it over.—What you say about "progress being only occasional & retrogression not uncommon" I agree to; only that in animal kingdom I greatly doubt about retrogression being *common*.— I have always put it to myself what advantage can we see in an infusory animal, or an intestinal worm—or coral-polyps—or earth-worm being highly developed? If no advantage, they would not become highly developed,—not but what all these animals have very complex structures (except infusoria) & they may well be higher than the animals which occupied similar places in the economy of nature before the Silurian Epoch.— There is a blind snake, with appearance & in some respects (Perhaps falsely appearing so) habits of worm, but this blind snake does not tend, as far as we can see, to replace & drive out worms. I think I must in any future Edit. discuss a few more such points—& will introduce this & H. C. Watson's objection about infinite number of species—& the general rise in organisation.[12] But there is a directly opposite objection to yours, which is very difficult to answer, viz how at first start of life, when there were only simplest organisms, how did any complication of organisation profit them? I can only answer that we have not facts enough to guide any speculation on the subject.—

With respect to Lepidosiren Ganoid Fishes, perhaps Ornithorhynchus I suspect, as stated in Origin, that they have been preserved from inhabiting F. Water & isolated parts of world, in which there has been less competition & less rapid progress in Nat. Selection, owing to the fewness of individuals which can inhabit small areas. &c &c—& where there are few individuals, variation almost must be slower.—[13]

There are several allusions to this notion in the Origin—as under Amblyopsis the blind cave fish. & under Heer about Madeira plants resembling fossil & extinct plants of Europe.—[14]

With hearty thanks for all your kindness—

Yours affect[y] | C. Darwin

Postmark: FE 20 60
American Philosophical Society (199)

[1] [Gray] 1860a. See preceding letter.

[2] Bronn 1860a.

[3] The reference is to Heinrich Georg Bronn's assertion that CD's theory of evolution was incomplete until it could account for the initial creation of organic matter from inorganic elements. He referred to Joseph Priestley's report of having isolated in a sealed glass jar 'green matter' that was neither animal nor vegetable in nature 'but a thing *sui generis*' (Priestley 1779–86, 1: 342). Bronn also discussed the recent debate in France over Félix Archimède Pouchet's claim to have produced unicellular organisms through spontaneous generation (Pouchet 1859). This passage is scored in CD's copy of Bronn 1860a, p. 114, which is in the Darwin Pamphlet Collection–CUL. For a translation of Bronn 1860a, see Hull 1973, pp. 120–4.

[4] See letter to H. G. Bronn, 14 February [1860].

[5] The passage to which CD refers follows a discussion of the evidence for metamorphic processes in some of the oldest geological formations, which bears on the question of the existence of life before the Silurian period. The passage contains an element of ambiguity (Bronn 1860a, p. 115):

> Ja Lyell nimmt bekanntlich einen endlosen Wechsel-Prozess dieser Art an; daher wir kürzlich nicht ohne einige Überraschung fanden, dass er die Darwin'sche Schrift denjenigen Geologen entgegenhält, welche an eine progressive Entwickelung der organischen Welt glauben.

[Indeed Lyell assumes, as is well known, a continual process of change of this sort [metamorphic processes]; therefore it is not without some surprise that we have recently found that he contrasts Darwin's writing with those geologists who believe in a progressive development of the organic world.] Bronn believed that CD's theory necessarily entailed a tendency toward progression of organisms, a principle that Lyell vehemently rejected (Bronn 1860a, p. 115). See also Hull 1973, p. 123.

[6] Lyell's letter has not been found.

[7] John Phillips discussed *Origin* in his presidential address to the Geological Society of London. See letter to J. D. Hooker, 14 February [1860] and n. 10.

[8] The letter from Thomas Vernon Wollaston has not been found. CD refers to [Wollaston] 1860.

[9] A letter from William Henry Harvey, professor of botany at Trinity College, Dublin, was published in the *Gardeners' Chronicle and Agricultural Gazette*, 18 February 1860, pp. 145–6. Harvey gave the example of *Begonia frigida*, in which it appeared that a new species had originated through the abnormal development of the existing form. If this proved to be the case, he stated, CD's 'theory would receive a serious damage' (*ibid.*, p. 145). Lyell included in his scientific journal a brief memorandum about Harvey's article (Wilson ed. 1970, p. 353).

[10] CD discussed his belief that monstrosities could not give rise to new specific forms in his unpublished 'big species book' (*Natural selection*, pp. 318–21).

[11] *Origin*, p. 417. Lyell recorded CD's statement in his scientific journal (Wilson ed. 1970, p. 353).

[12] Hewett Cottrell Watson discussed this objection in his letter of [3? January 1860]. CD had sent the letter to Lyell and Joseph Dalton Hooker (see letters to Charles Lyell, 4 [January 1860], and to T. H. Huxley, 11 January [1860]). CD discussed these and other objections to his theory in the third

edition of *Origin*, in a new section entitled 'On the degree to which organisation tends to advance.—' (*Origin* 3d ed., pp. 133–43).

[13] *Origin*, p. 130.

[14] *Origin*, pp. 139, 107.

From François Jules Pictet de la Rive[1] 19 February 1860

Geneva
19. Feby 1860

Your book makes science young clear, elevated, but no facts to prove that principle that slight modification multiplied by any factor of time no matter how long could reach the character of families &c or c^{d} produce profound modific^{s} in organization.—[2]

Incomplete
Copy
Kinnordy MS, Charles Lyell's journal V, p. 110

[1] Pictet de la Rive's letter has not been found. The text has been transcribed from Charles Lyell's scientific journal. It is also printed in Wilson ed. 1970, p. 354. The entry in Lyell's journal is headed: 'Pictet to C. Darwin, Geneva 19 Feby 1860'.

[2] CD sent Pictet de la Rive a presentation copy of *Origin* (see Appendix III). CD had been impressed by the discussion of the species question in Pictet de la Rive 1853–7 and thought that Pictet might be swayed by his theory. See *Correspondence* vol. 7, letters to F. J. Pictet de la Rive, 11 November [1859], and 24 December [1859].

From Asa Gray 20 February 1860

Feb. 20, 1860

My Dear Darwin,

I got this to-day.[1] I send Appletons, now, the sheets of ed. 2, & your additions appended in their places.[2] I promised the Historical Preface next week,[3] and I put it in their hands—trusting to their *promise* of 5 percent and to their honor for more if they are not molested by reprinters, which we shall keep off.

The offer of check for £50—(which I might send to Mrs. Darwin for pin-money, since you scorn it) tempts me,—but I think it wiser to wait and hope for more.[4]

I mailed you a clean copy of my Review. I have sent it to Agassiz. He is childishly apt to be offended at any opposition, but I have, as you see been very careful to avoid all cause of personal offence.[5]

Ever, Dear Darwin, | Yours cordially | Asa Gray

[Enclosure]

New York Feb 17/60

Dear Sir

Yr favor of 15th is at hand. We cant say what we can do respecting the notes & additions till we see them but we shall be anxious to make our edition conform to any future English Ed^{n}—[6]

You are under a mistake in supposing that new matter unpublished in England secures a copyright in this country if written by Mr Darwin—on the contrary if the entire work had never been published in England and first appeared here no copyright would hold in this country as no one can hold a copyright here for what he has written unless he be a citizen of the U.S—

We proposed to ourselves to pay 5% on retail price as suggested in yr letter as there is no reason why a work without any legal rights, sh^d pay the same as one that is secured by law—We desire to act liberally altho' we printed the work after it had reached this country some little time; not having rec'd even early sheets which is usual when any payments are made.

We regret very much there is no protection to the foreign author, think it a monstrous shame, but we are obliged to take things as they actually exist—We are quite willing if it w^d be agreeable to Mr Darwin to send him a check for 50£ stg. and very likely that will be as much as he could receive by the sales.

We remain | Very respectfully | D. Appleton & Co

Gray Herbarium of Harvard University (37)

[1] See enclosure.

[2] See letter to Asa Gray, 1 February [1860] and enclosure.

[3] Gray apparently had not yet received CD's letter of [8 or 9 February 1860], in which CD had enclosed the historical preface and a few more corrections for the new American edition of *Origin*.

[4] See letter to Asa Gray, 28 January [1860].

[5] Gray's review of *Origin* ([Gray] 1860a) contrasted CD's theory of species change with Louis Agassiz's views on the multiple creation of species.

[6] The 'revised and augmented' American edition of *Origin* included many corrections and additions that were later incorporated into the third English edition, published in 1861.

To J. D. Hooker [20 February 1860][1]

Down
Monday

My dear Hooker

If you come to Club on Thursday do pray come in good time & agree to sit together & I will say so, if anyone asks me to sit with them, & do you say same.[2]

I have just read the very curious case by Harvey on the Begonia.[3] I suppose of course a wild plant? Considering how often monsters are sterile (as with peloric flowers) & likewise non-inherited, Harveys supposition seems rather bold; but he overlooks that as he omits in all probability not that all the offspring would be similarly characterised, so Natural Selection would have to come in to preserve the new form.—

Here is the evil of an abstract; in my fuller M.S. I have discussed a very analogous case viz of a normal fish like an extremely monstrous Gold Fish. Also I have discussed such cases as the Aspicarpa with two kinds of flowers, & have speculated on the degraded flowers alone being preserved, so that a new genus might thus be suddenly formed.[4] If my memory is right I have such a case amongst the Campanulaceæ. I remember asking Bentham whether any Leguminosæ existed having degraded flowers alone like those borne on certain Leguminosæ together with perfect flowers.—[5] I write all this to ask you to keep this subject in mind.[6]

As the "Origin" now stands Harvey's is a good hit against my talking so much of insensibly fine gradations; & certainly it has astonished me that I sh[d] be pelted with the fact that I had not allowed abrupt & great enough variations under nature.[7] It would take a good deal more evidence to make me admit that forms have often changed by saltum.—

Have you seen Wollaston's attack in Annals?[8] The stones are beginning to fly. But theology has more to do with these two attacks than science.

I much enjoyed the day Henslow spent here.—[9]

Yours affect[y]—| C. Darwin

Endorsement: 'Feby/60'
DAR 115.1: 41

[1] Dated by the endorsement and the reference to William Henry Harvey's article in the *Gardeners' Chronicle and Agricultural Gazette* (see n. 3, below). The Monday following the publication of Harvey's letter was 20 February.

[2] Because of illness, CD was unable to attend the meeting of the Philosophical Club of the Royal Society on 23 February 1860 (see letter to J. D. Hooker, [23 February 1860]).

[3] Harvey's letter was printed in the *Gardeners' Chronicle and Agricultural Gazette*, 18 February 1860, pp. 145–6.

[4] *Natural selection*, pp. 318–21.

[5] See *Correspondence* vol. 6, letters to George Bentham, 30 November [1856] and 3 December [1856].

[6] Hooker published a response to Harvey's letter in the *Gardeners' Chronicle and Agricultural Gazette*, 25 February 1860, pp. 170–1. Hooker pointed out that Harvey was wrong in believing that the particular *Begonia* in question illustrated the abrupt formation of a new species; rather it was 'a good instance of how slow and partial such a change is at the commencement'. He further suggested that if the modified form were subjected to natural selection, 'there are prodigious odds against its ultimate success' (*ibid.*, p. 171).

[7] See letter to Charles Lyell, 18 [and 19 February 1860].

[8] [Wollaston] 1860.

[9] John Stevens Henslow visited Down from 14 to 16 February 1860 (Emma Darwin's diary).

From Andrew Crombie Ramsay[1] 21 February 1860

21. Feb. 1860

I had for years been using to myself the terms mutability of species, and transmutation of species, for want of better words to express the feeling, (amounting to a conviction) that in Time species passed insensibly into each other instead of being produced by separate acts of creation. But the idea implied by your principle of Natural Selection taken in connection with the struggle for existence is very different indeed from the vague gropings towards the light in which myself and others with various degrees of hesitation indulged.[2]

My faith such as it was, was founded on the following considerations. The disputes among Naturalists as to what constitutes a species the wranglings as to whether such & such were distinct species or varieties, the acknowledgment of permanent varieties, the existence of specific & still more of generic centres, of representative species in space, & in time in a geological sense, & the links (especially the rudimentary organs) that bind genera together, especially in the higher Mammalia, when living & fossil forms are considered together. The succession of small miracles required to produce certain species in a formation just a very little different from those in the preceding formation went sadly against my mental stomach & I never could reconcile myself to the idea of a Creator making a number of small experiments as it were of feeling his way in the work of Creation— And when asked where are the perfect gradations, that show the passage of fossil species, I have always maintained that the record as shown in the succession of formations was too imperfect for us to find this & that it is not till an aberrant branch, multiplied to a vast extent, that we were likely to find fossilized any of its representatives at all.

Incomplete
Copy
Kinnordy MS, Charles Lyell's journal V, pp. 112–16

[1] The letter has not been found. The text has been transcribed from Charles Lyell's scientific journal. It has also been published in Wilson ed. 1970, pp. 355–6. The entry in Lyell's journal is headed: 'Extract of Letter from A. C. Ramsay to C. Darwin, 21. Feb. 1860'.

[2] CD sent Ramsay a presentation copy of *Origin* (see Appendix III). Thomas Henry Huxley had mentioned to CD that Ramsay was favourably inclined towards the book. See *Correspondence* vol. 7, letter to T. H. Huxley, 27 November [1859].

From Herbert Spencer[1] 22 February 1860

22^d Feby. 1860

You have wrought a considerable modific^n in the views I held— While having the same general conception of the relation of species genera, orders &c as gradually arising by differentiation & divergence like the branches of a tree & while regarding these cumulative modifications as wholly due to the influence of surrounding circums^s. I was under the erroneous impression that the sole cause was adaptation to changing conditions of existence brought about by habit, using the phrase conditions of existence in its widest sense as including climate, food, & contact with other organisms (for general statement of this view see Essay pp. 41. 45.)[2] But you have convinced me that throughout a great proportion of cases, direct adaptation does not explain the facts, but that they are explained only by adaptation through Natural Selection—

Many (&c) must have been struck with the fact that among all races of organisms the tendency was for the best individ^s. only to survive & that so the goodness of the race was preserved

I have in Essay on Population &c remarked this as a cause of improvement among mankind—[3]

But I & every one overlooked the selection of "spontaneous" variations without which I think you have clearly shown that many of the phenomena are insoluble.

You have shown that the doctrine furnishes explanations to phenomena otherwise inexplicable. However the argument may as yet fall short of direct demonstration yet the indirect demonst^{n}. is to me conclusive. I take it to be incredible that so many different kinds of evidence sh^{d} coincide in supporting a doctrine that was untrue—

Incomplete
Copy
Kinnordy MS, Charles Lyell's journal V, pp. 107–9

[1] The letter has not been found. The text has been transcribed from Charles Lyell's scientific journal; it is also printed in Wilson ed. 1970, pp. 353–4. The entry in Lyell's journal is headed: 'Herbert Spencer to C. Darwin, 22^{d} Feby. 1860 | H. S.'s own doctrine of "evolution" put on so satisfactory a basis.'

[2] Spencer refers to his article 'The development hypothesis', reprinted in Spencer 1858–63, 1: 389–95. For CD's opinion of Spencer's work, see letter to Herbert Spencer, 2 February [1860], and *Correspondence* vol. 7, letter to Herbert Spencer, 25 November [1858].

[3] [Spencer] 1852. A copy of this paper was apparently enclosed with the letter (see letter to Herbert Spencer, 23 [February 1860]).

To J. D. Hooker [23 February 1860][1]

[Down]
Thursday

My dear Hooker

As usual when I set my heart on anything I cannot do it—a bad cold has made my stomach intolerably bad.— It is the more provoking as I must, if I can, be in London tomorrow to see my Doctor— I am so very sorry that you will have gone early to club. for my sake.—[2]

I enclose A. Gray's letter as it is partly Botanical; please return it in empty cover.—[3] I send, also, the remainder of Gray's Review.— Please forward it to Dr Boott[4] as requested & ask him to send it to Lyell.—

Yours ever | C. D.

Endorsement: 'Feb/60'
DAR 115.1: 42

[1] Dated by the endorsement and by the relationship to the letter to J. D. Hooker, [20 February 1860].

[2] CD had hoped to attend a meeting of the Philosophical Club of the Royal Society on 23 February 1860 (letter to J. D. Hooker, [20 February 1860]).

[3] Asa Gray's letter has not been found.

[4] Francis Boott, the American-born physician and botanist, was a close friend of Hooker's and Gray's. The reference is to [Gray] 1860a.

From James Lamont [23 February 1860][1]

to breed more and more to the color that afforded them most protection.

I may mention with deference that I do not think you have *made the most of* (Page 49) the similarity between our British red grouse and the Norwegian species: I have shot many thousands of British grouse and many hundreds of Norwegian Ripas, and may therefore be permitted to express my opinion viz. that not the smallest doubt exists on my mind as to their being merely "strongly marked varieties" of the same species—the difference in plumage being caused by the necessity for the British grouse to assimilate his plumage to the color of brown heath, whereas the Norwegian *Dal-ripa* must more resemble grey rocks, birch-bushes (and snow in the winter) in order to protect his life from the numerous hawks, gulls, and ravens frequenting that country: these birds exactly resemble one-another in their size, their voice, their flight, and their general habits: and I have a few times killed a British grouse *so white* as exactly to resemble *an accidentaly rather dark* Dal-Ripa.

I suppose no one will pretend that there is any specific difference between the British ptarmigan and the "fjëld-ripa" or hill-ptarmigan of Norway? or between the Blackcocks of either country? Only, because the "fjëld-ripa" frequents the high, grey, rocky mountains as does the Scottish Ptarmigan; and the Blackcocks of both countries inhabit the wooded valleys—there is no necessity for any alteration of plumage in them.

I entertain no doubt that if a number of Dal-Ripas were taken over and liberated on a well-preserved Scottish moor, they would in the course of a few generations come to resemble the indigenous red grouse, or vice versa.

I may mention a curious instance of the hereditary propensities of birds which is notorious in this district of Argyleshire: it is well-known that for ages the grouse on the extensive moors of Ayrshire and Renfrewshire have always been very "wild"—far more so than on this side of the Firth of Clyde: now some 20 years ago a neighbour of mine, thinking to improve the breed, introduced a few brace of Lowland grouse into this district, and in a very few years a *marked increase of wildness* took place among the grouse for many miles around, and throughout the district of Cowal I fancy they are now as wild as in the Lowland Counties, whereas in Cantyre on the opposite shore of Loch-Fyne they are still comparatively tame.

Your remarks about an "over-stock of "game" causing epidemics[2] will be fully borne out by any observant game-preserver or any intelligent gamekeeper.

I have no doubt that the grouse-disease so prevalent in Scotland for 13 or 14 years back was caused by nothing else than high preservation (aided by 2 or 3 wonderfully favorable breeding seasons consecutively) having got up a stock of grouse which the soil was unable to supply with a sufficiency of *something* essential to their existence: I have always suspected that this something was *insects*: it clearly was not for want of *heather* as has been nonsensically written, because in 1846–47 when it commenced there was heather enough to have kept all the grouse in Cowal on this estate, and yet they were dying in hundreds. It may be worthy of observation that during that winter of '46–'47 the grouse *in this district first commenced* the practice of coming down to the stubble-fields and low grounds as if in quest of something wh. they could not find in sufficient quantities on the hills: they have ever since more or less continued that habit, but previous to '46 it was unknown in this district.

Your illustration of the Magpie (Page 212) as showing the hereditary tendency of birds to become tame or wild, is a very happy one, as no bird is so much persecuted in Britain, whereas in Norway the people have a superstition that it is "unlucky" to touch them and they are consequently so audaciously tame as to sit on the houses.

It is a common remark all over Britain that "game is becoming much wilder": this seems to me very easy of explanation when we consider that formerly the greatest danger to game-birds was *from hawks* and their greatest security from them lay in *cowering* as much as possible, but now from the gradual extermination of hawks &c and from the frequency of shooters and goodness of guns, their principal danger is from the latter, and they are no doubt gradually finding out from generation to generation that their best mode of security is to fly off whenever they see a man.

I have passed the last two summers in hunting in Spitzbergen and the great ice-fields lying around it:[3] I shot 100 Reindeer there, and I am almost certain that many of these Deer had never seen a human being in their lives before, because there were some so ridiculously tame as to come up—even to windward—*to look*

AL incomplete
DAR 47: 150–1

CD ANNOTATIONS
1.1 to breed . . . protection.] *crossed ink*
2.1 I may] 'Mr James Lamont of Knockdow Argylshire' *added ink*
2.2 I have shot] *cross added brown crayon*
3.1 I suppose . . . vice versa. 4.3] *crossed ink*
5.1 I may] 'James Lamont Esqre of Knockdow Argylshire.— | (shows what a cross will do.—' *added ink*
5.1 I may . . . *wildness* 5.6] 'Bears on crossing keeping Birds uniform.' *added pencil*
5.5 a few brace] *double underl ink*; 'C. D' *added ink*
6.1 Your remarks . . . in hundreds. 7.8] *crossed ink*
7.8 It may be] 'Ch X'[4] *added brown crayon*
7.11 they have . . . district. 7.12] *cross added brown crayon*
8.1 Your illustration . . . houses. 8.4] 'So in Normandy.' *added pencil*
8.3 superstition . . . houses. 8.4] *cross added brown crayon*
10.1 I have . . . *to look* 10.4] *crossed ink*
Top of first page: 'Ch IV'[5] *brown crayon*; 'Feb. 23— 1860' *ink*

[1] The identity of the correspondent and the date are given by CD's annotations on the letter. The letter was at some stage divided into two and presumably stored in two different portfolios of notes (see CD's annotations and nn. 4 and 5, below).

[2] *Origin*, p. 70.

[3] Lamont related his observations on Spitzbergen at a meeting of the Geological Society of London on 28 March 1860 (*Quarterly Journal of the Geological Society of London* 16 (1860): 428–44). A note by CD discussing the paper is in DAR 205.2: 189.

[4] The annotation refers to chapter 10 of *Natural selection*, on the 'Mental powers and instincts of animals'.

[5] The annotation refers to chapter 4 of *Natural selection*, on 'Variation under nature'.

To Charles Lyell 23 February [1860][1]

Down Bromley Kent
Feb.—23^{d}

My dear Lyell

That is a splendid answer of the father of Judge Crompton.[2] How curious that the Judge sh^{d} have hit on exactly same points as yourself.— It shows me what a capital Layer you would have made,—how many unjust acts you would have made appear just! But how much grander a field has science been than the law,—though the latter might have made you Lord Kinnordy.[3]

I will, if there be another Edition, enlarge on gradation in eye & on all forms coming from one prototype, so as to try & make both less glaringly improbable.—

Hooker is going to answer Harvey (ie if Lindley will admit)[4] & I am very glad of it;—you can see answer, when (as I am delighted to think) you will be here. Aspicarpa is a most interesting case, like the difference, on which I have enlarged a little, of difference in most important characters in the outer & inner florets of Compositous & Umbelliferous plants.— I think these facts are most important as showing how easily what naturalists call very important characters may be modified by correlation of growth.[5] But I doubt whether they throw light on abrupt origin of new forms. At least I have tried long & hard with respect to such cases as Aspicarpa; & I could find only one apparent case in the Campanulaceæ.— With respect to animals, besides the case of monstrous Gold-fish with analogous fish in state of nature alluded to, I have wondrous case of monstrous eels, (examined by Agassiz) & apparently produced by darkness, but I cannot satisfy myself on case;[6] nor does it appear certain that they breed.— On the whole I still feel excessively doubtful whether such abrupt changes have more than very rarely taken place.—

With respect to Bronn's objection that it cannot be shown how life arises, & likewise to certain extent Asa Gray's remark that natural selection is not a vera causa,—I was much interested by finding accidentally in Brewster's life of Newton, that Leibnitz objected to the law of gravity, because Newton could not show what gravity itself is.[7] As it has chanced I have used in letters this very same argument, little knowing that anyone had really thus objected to Law of Gravity.— Newton answers by saying that it is philosophy to make out the movements of a clock, though you do not know why the weight descends to ground.— Leibnitz further objected that the Law of Gravity was opposed to natural Religion!—[8] Is this not curious? I really think I shall use these facts for some introductory remarks for my bigger book.—

I cannot conceive what Bronn can mean by his quotation about you:[9] I do not remember even mentioning your name in my two brief notes to him.—

I have seen something about the Infusorial experiments in Paris: Quatrefage objected to their accuracy.—[10] Some old experiments were several years ago tried in Germany with *astonishing* precautions (air all passed through sulphuric acid & caustic potash) & infusoria *never* appeared.—[11]

You ask, (I see) why we do not have monstrosities in higher animals; but when they live they are almost always sterile (even giants & dwarfs are *generally* sterile) & we do not know that Harvey's monster would have bred.—[12] There is, I believe, only one case on record even of a peloric flower being fertile & I cannot remember whether this reproduced itself.—

To recur to the Eye, I really think it w.[d] have been dishonest not to have faced the difficulty; & worse (as Talleyrand would have said) it would have been impolitic I think;[13] for it would have been thrown in my teeth,—as H. Holland threw the bones of the ear;[14] till Huxley shut him up by showing what a fine gradation occurred amongst living creatures.

Thank you much for your most pleasant letter | Yours affet | C. Darwin

I send a letter by Herbert Spencer which you can read or not as you think fit.— He puts to my mind, the philosophy of the argument better than almost anyone at close of letter.—[15]

I could make nothing of Dana's idealistic notions about species;[16] but then as Wollaston says I have not metaphysical Head. By the way I have thrown at Wollastons Head, a paper by Alex. Jordan who demonstrates metaphysically that **all** our cultivated races are god-created species.—[17]

Wollaston misrepresents accidentally to wonderful extent some passages in my Book: he reviewed without relooking at certain passages.[18]

Endorsement: '1860'
American Philosophical Society (200)

[1] The year is confirmed by CD's reference to reading Brewster 1855. See n. 7, below.

[2] Peter Crompton, a medical practitioner in Derby, was the father of the eminent judge, Charles John Crompton. The reference has not been traced.

[3] CD misspelled 'Lawyer'. Lyell trained as a barrister but never practised law. Kinnordy was the Lyells' family estate near Kirriemuir, Scotland.

[4] See letter to Charles Lyell, 18 [and 19 February 1860]. John Lindley was the editor of the *Gardeners' Chronicle*, in which William Henry Harvey had recently published a letter about monstrosities (*Gardeners' Chronicle and Agricultural Gazette*, 18 February 1860, pp. 145–6). Joseph Dalton Hooker answered Harvey's points in the issue of 25 February 1860, pp. 170–1.

[5] CD discussed correlation of growth in *Origin*, p. 143:

> I mean by this expression that the whole organisation is so tied together during its growth and development, that when slight variations in any one part occur, and are accumulated through natural selection, other parts become modified.

[6] CD described the case of eels possessing huge eyes taken from deep wells in *Natural selection*, p. 297. Louis Agassiz, who examined one of the specimens, believed that it was identical to the common eel.

[7] Probably Brewster 1855. CD recorded having read 'Sir D. Brewsters life of Newton' on 23 February 1860 (see *Correspondence* vol. 4, Appendix IV, 128: 27). A second edition of this greatly augmented version of David Brewster's biography of Isaac Newton was published in November 1860 (*Publisher's Circular* (1860), p. 579). Brewster referred to Gottfried Wilhelm Leibniz's attack on Newton's theory of gravity in Brewster 1855, 2: 282–3.

[8] Brewster 1855, 2: 284–5.

[9] See letter to Charles Lyell, 18 [and 19 February 1860] and n. 5.

[10] CD refers to the experiments on spontaneous generation and fermentation conducted by Félix Archimède Pouchet in 1858 and published at length in Pouchet 1859; this work brought Pouchet

into fierce controversy with Louis Pasteur. Pouchet's work was criticised by Jean Louis Armand de Quatrefages de Bréau in Quatrefages de Bréau 1858. Heinrich Georg Bronn referred to the failure of these attempts to create organic life as depriving CD's theory of an important means of empirical verification (Bronn 1860a, p. 114).

[11] Probably a reference to the experiments of Theodor Ambrose Hubert Schwann (see Farley 1974, p. 50).

[12] See n. 4, above.

[13] The expression 'worse than a crime, a blunder' was commonly attributed to the French statesman Charles Maurice de Talleyrand-Périgord.

[14] See *Correspondence* vol. 7, letter from Henry Holland, 10 December [1859].

[15] Letter from Herbert Spencer, 22 February 1860.

[16] Dana 1857. There is a copy in the Darwin Pamphlet Collection–CUL. For CD's poor opinion of the paper, see *Correspondence* vol. 6, letter to J. D. Hooker, 25 December [1857]. Lyell may have mentioned the paper in connection with Asa Gray's reference in his review of *Origin* to James Dwight Dana's comparison of the fixity of animal species to that of chemical elements ([Gray] 1860a, p. 154).

[17] CD's letter to Thomas Vernon Wollaston has not been found, but see the letter to Charles Lyell, 15 and 16 [February 1860]. CD had discussed Alexis Jordan's work in a letter to the *Gardeners' Chronicle* (*Correspondence* vol. 5, letter to *Gardeners' Chronicle*, [before 29 December 1855]). He refers to Jordan 1852, a copy of which is in the Darwin Library–CUL.

[18] [Wollaston] 1860.

To F. J. Pictet de la Rive 23 February [1860][1]

Down Bromley Kent
Feb. 23^{d}

Dear Sir

I must trouble you with a few lines to thank you most truly for your very kind note. What you say about my Book, pleases me *extremely*, & I am far from surprised that you go with me a very short way. I remember how slowly I changed my own opinion; & even supposing for the moment that my views were in the main right, I do not think anyone could at once undergo so great a revolution in opinion.— I thank you cordially for the notice which you intend to publish, & for so kindly offering to send me a copy.[2] This will be invaluable, as showing me what parts you think weakest; & it will largely spread the knowledge of my book.—

Do you ever see the American Phil. Journal? there will appear in the next number an excellent Review by that admirable Botanist Asa Gray.[3] He informs me that your illustrious countryman, Agassiz, is very bitter against my Book, as I fully expected would be the case.— Prof. Bronn of Heidelberg although very much of course opposed to my doctrine, with noble liberality of sentiment is going to superintend the work of a Translator into German; so that my Book will be pretty widely known, & consequently what is true will soon be known from what is false in it.—[4] You will think me very presumptuous, but as your studies naturally lead you to reflect much on Geological Succession, Geograph. Distribution, Classification, Homology & Embryology, I expect & hope you will be led ultimately to go a little way further with me: as these facts receive some sort of explanation on the theory of descent; whereas they are inexplicable on the theory of creation.

Pray believe me, dear Sir | with sincere respect & cordial thanks for your kindness | Your faithful servant | Charles Darwin

Bibliothèque Publique et Universitaire de Genève (MS. fr. 1651, f. 8–9)

[1] Dated by the relationship to the letter from F. J. Pictet de la Rive, 19 February 1860.
[2] Pictet de la Rive reviewed *Origin* in *Bibliothèque Universelle* 7 (1860): 233–55. CD's annotated reprint of the article is in the Darwin Pamphlet Collection–CUL.
[3] CD refers to [Gray] 1860a, which was published in the *American Journal of Science and Arts.*
[4] See letters to H. G. Bronn, 4 February [1860] and 14 February [1860].

To A. C. Ramsay 23 February [1860][1]

Down Bromley Kent
Feb.y 23d

Dear Ramsay

I must write a line to say how sincerely rejoiced I am to hear that your health is nearly reestablished. I remember Sir H. Holland telling me that a gradual & slow recovery of health was always better than a sudden jump.[2] I am *extremely* much pleased to hear that you like my Book. I look at every geological believer of the mutation of species as a most important gain.— There will be a long & stiff battle before such doctrines are generally admitted. The progress of subject will now depend far more on such men as yourself than on anything I can do.—

You put, as it seems to me, the general arguments against the eternal immutability of species with great clearness.—

With very sincere thanks for your note. | Believe me | Yours very truly | C. Darwin

I hope to be in London in a day or two & will call at Museum for chance of seeing you.—[3]

Down House (MS 9: 2)

[1] Dated by the relationship to the letter from A. C. Ramsay, 21 February 1860.
[2] Henry Holland, a distant relative of CD's, was a prominent London physician.
[3] The Museum of Practical Geology, located in Jermyn Street, London, was run by the Geological Survey of Great Britain. Ramsay was the local director for Great Britain and a lecturer at the Government School of Mines. Emma Darwin's diary records that CD went to London on 27 February 1860.

To Herbert Spencer 23 [February 1860][1]

Down Bromley Kent
23d

My dear Sir

I write one line to thank you much for your note. Of my numerous (Private) critics, you are almost the only one who has put the philosophy of the argument, as it seems to me, in a fair way.— Namely, as an hypothesis (with some innate probability as it seems to me) which explains several groups of facts.—

You put the case of selection in your Pamphlet on population in a very striking & clear manner.[2] You do not say whether I am to return it; I have not yet had time to read it, for my very small power of work of any kind is much overtaxed.— If you require the pamphlet back, kindly send me *one* line: if I do **not** hear I shall understand that I am to keep it.

My dear Sir | Yours sincerely | C. Darwin

University of London Library MS. 791/51. (Courtesy of the Athenaeum.)

[1] Dated by the relationship to the letter from Herbert Spencer, 22 February 1860.
[2] [Spencer] 1852, pp. 496–501.

To Asa Gray 24 February [1860][1]

Down Bromley Kent
Feb 24th

My dear Gray

I received yesterday your kind letter & the remaining sheets of the Review, which I forwarded to Hooker, to go to Dr Boott & then to Lyell.—[2]

This remainder of the Review most fully confirms my opinion that it is by far the most able which has appeared, & you will have done the subject infinite service. And now I most sincerely hope that it will cost you no more labour.— Lyell quite agreed with me that the two first sheets were the best thing which has appeared on the subject.—

Thank you much for your letter, which as it contains such curious botanical facts, I have sent to Hooker. Why should you call one of the best essays ever written (at least in my opinion) on Geograph. Distribution of Plants "unfortunate". I really hope, as a punishment to myself, that I may be found egregiously wrong about the warmer period subsequent to the Glacial Epoch.—[3] Certainly very much more has to be made out about the latest geological periods.—

I am very glad to think (if I understand rightly, but I must reread your letter) that the greater resemblance of N.E. America than of N.W. America to Asia in its plants, is not quite so great an anomaly as it at first appears.—

I have been interested by your theological remarks in the Review, but I must reconsider them. It has always seemed to me that for an Omnipotent & omniscient Creator to foresee is the same as to preordain; but then when I come to think over this I get into an uncomfortable puzzle *something* analogous with "necessity & Free-will" or the "Origin of evil", or other such subject quite beyond the scope of the human intellect. I was interested the other day in reading the Life of Newton by Brewster to find that Leibnitz actually attacked the Law of Gravity as "subversive of all Natural Religion"!![4] He further attacked Newton for having used gravity "*an occult quality*" to explain the motions of the Planets.— Newton answered that it is philosophy to explain movement of wheels of clock, though the cause of descent of philosophy to explain movement of wheels of clock, though the cause of descent of the weight could not be explained. This seems to me rather to bear on what you say of Nat. selection not being proved as a vera causa.—[5]

I cannot but think you lay rather too much stress on new organs appearing in the animal kindgom, at least, it is most difficult to show any number of new organs.— Milne Edwards has tried with very little success.[6] One sh[d] never forget such transitions or at least such gradations, as a swim-bladder into a Lung.— So even with the Eye, as numerous fine gradations can be shown to exist, the perfecting this wondrous organ by Nat. Selection I must look at as a difficulty to our imagination & not to our reason.—

In the M.S. additions sent you for any future edition, there is one blundering sentence, herewith corrected. If I knew there w[d] be a second American Edition I would send a few more additions & corrections

With most sincere thanks for all your extreme kindness—Believe me | Yours most sincerely | Charles Darwin

Gray Herbarium of Harvard University (23)

[1] Dated by the reference to [Gray] 1860a.

[2] Gray's letter to CD has not been found. See letter to J. D. Hooker, [23 February 1860].

[3] CD refers to a paper on the flora of Japan (Gray 1859), in which Gray proposed that there had been a post-glacial, as well as a pre-glacial, warm period during which the plants of northern America and northern Asia intermingled. For CD's comments on the postulated post-glacial warm period, see *Correspondence* vol. 7, letters to J. D. Hooker, 11 May [1859], and to Asa Gray, 24 December [1859].

[4] Brewster 1855, 2: 284–5. See letter to Charles Lyell, 23 February [1860].

[5] See letter to Asa Gray, 18 February [1860] and n. 5.

[6] CD followed Henri Milne-Edwards in believing that organs were generally transformed through performing increasingly specialised functions rather than formed *de novo*. See Milne-Edwards 1851 and *Natural selection*, p. 354.

To W. B. Tegetmeier 24 [February? 1860][1]

Down Bromley Kent
24th

My dear Sir

I write a line to thank you sincerely for your two notes.— Very curious your facts about the crossing the Game Fowls. Good Heavens what a hopelessly difficult subject is that of inheritance!

The crosses from Spanish & Silk Hen are *all* black-combed, & as you say *not* silky feathers.—[2]

I crossed the Barbs which you had (ie plain-headed) the parents of which got 2[d] prize at C. Palace, *& likewise* the turn-comded[3] Barbs with the Fan-tails, & recrossed their offspring, but I cannot positively say, from which I procured the blue-bird.[4] The turn-commed were from Sir J. Sebrights. & it is simple nonsense to say they were not pure.[5] It was a notion which M[r] Brent got owing to a German author (of whom I told him) saying this cross was sterile.[6] *The* **one** *German Book is a very poor one & no particulars are given.*— When M[r] Brent gets a notion into his head nothing gets it out.—

In Haste | Yours very sincerely | C. Darwin

P.S I have just remembered positively that I reared *several* mongrels from the plain-headed Barbs & Fantails. I have *several* other cases of blueish birds reared from other crosses of other kinds not given in my abstract.—[7]

New York Botanical Garden Library (Charles Finney Cox collection)

[1] Dated by the relationship to the letter to W. B. Tegetmeier, 20 January [1860], and by CD's remark that he had received two notes from Tegetmeier in the interval.

[2] See letter to W. B. Tegetmeier, 20 January [1860], in which CD described the results of experiments in which he cross-bred fowls to see whether the progeny would show the coloration or other characteristics of the ancestral breed.

[3] CD misspelled 'turn-combed' here and elsewhere in the letter.

[4] Tegetmeier helped CD to procure many different breeds of pigeon for his study of variation and artificial selection during the years 1855 to 1858 (see *Correspondence* vols. 5, 6, and 7). Tegetmeier also introduced CD to some of the leading breeders and exhibitors at poultry and pigeon shows and at pigeon-fanciers' clubs. CD refers to the Crystal Palace poultry show, the largest in the London area, which he first attended in 1857 (see *Correspondence* vol. 6, letter to John Lubbock, 11 August [1857]). CD began crossing pigeons in August 1856, in part to seek evidence for reversion. The records of his crosses are in DAR 205.7: 166–89, and in the Catalogue of Down specimens (Down House MS). The barb–fantail crosses are described in *Variation* 1: 200.

[5] John Saunders Sebright was a noted breeder of domestic animals.

[6] Bernard Philip Brent was a pigeon-fancier whom CD acknowledged as one of his chief sources of information on the subject (*Variation* 1: 132 n. 2). The book referred to is Pistor 1831 (see *Correspondence* vol. 7, Supplement, letter to B. P. Brent, [after August 1856]). CD's annotated copy is in the Darwin Library–CUL. The work is cited in *Variation* 1: 192 n., where CD stated that, contrary to the assertion of C. M. Wilhelm Pistor, these cross-bred pigeons were not sterile.

[7] CD discussed his crosses of domestic pigeons in *Origin*, pp. 20–9.

To H. G. Bronn [*c.* 25 February 1860][1]

[Down]

Shoulder-pad. is the thick, almost horny skin on the shoulder of the wild Boar, which is said to serve as defence to tusks of other boars: I have not examined this myself.—[2]

Cross & intercross.—There is no difference: I ought not to have used both terms: an English friend has complained of this mistake of mine.—[3]

Structure anything visible to eye in appearance of parts

Constitution. rate of pulse—period of gestation—length of life—liability to certain diseases—degree of fertility &c &c &c.

Habit—whether diurnal or nocturnal—what kind of food an animal prefers—Habit may, also, be applied to instincts; but it is rather a loose word, often used in English. It differs little from instinct; but would never be applied to the more wonderful instincts. We might say that the owl had nocturnal *habits*; but *instinct* impelled it to build its nest.— The optic nerve being fitted to see in the dark we should call a *constitutional* difference—the great size of eyes a structural difference.—

I am astonished & much pleased that you think the Translation will be ready by May.—[4] Most truly do I thank you for the great honour & benefit which you have conferred on me.—

Honoured Sir, Yours very sincerely | C. Darwin

Incomplete
American Philosophical Society (340)

[1] The recipient and the date are inferred from CD's remark in the following letter that he has had a second letter from Bronn containing queries relating to Bronn's translation of *Origin* into German. Bronn's letter, to which CD's is a reply, has not been found.

[2] The term 'shoulder-pad' was used in *Origin*, p. 88.

[3] Both terms occur frequently and interchangeably in *Origin*. CD did nothing to correct the usage in later editions.

[4] The letter to H. G. Bronn, 10 April [1860], indicates that CD received copies of the first part of the German translation of *Origin* in April. According to the *Börsenblatt für den Deutschen Buchhandel und die mit ihm verwandten Geschäftszweige* 27 (1860), pp. 683, 874, 1166, the three parts of Bronn trans. 1860 were issued on 4 April, 2 May, and 11 June 1860.

To Charles Lyell 25 February [1860][1]

Down Bromley Kent
Feb 25th

My dear Lyell

I am very glad to hear about Owen; not but what I fully expect many & bitter sneers from him.—

I am glad to hear that you used the same line of argument as Spencer's to the Bishop: it seems to me much the safest & truest.—[2]

I cannot help wondering at your zeal about my Book. I declare to Heaven you seem to care as much about my Book, as I do myself.—You have no right to be so eminently unselfish! I have taken off my spit a letter of Ramsays; as every geologist convert, I think *very* important.[3] By the way I saw some time ago letter from H. D. Rogers to Huxley in which he goes very far with us.—[4] I send also first page of 2d letter from Bronn; showing that he is thinking more about the Book. It seems he is going to translate it himself. The rest of letter was only doubt about meaning of unusual English terms.—[5]

I had letter today from Sir W. Jardine,—strongly opposed to my views—but his attack on my ornithological accuracy has frittered into absolutely nothing.—[6] He says Andrew Murray has just read Paper to Royal Socy. versus my views.,—I presume brought in somehow incidentally.—[7]

I do not know whether the degraded flowers of Aspicarpa fruit, but in some other cases the degraded flowers are *more fertile* than the perfect.—[8]

I can easily believe in your criticisms on part of H. Spencer's work: I have just read his Essay on population, in which he discusses life & publishes such dreadful hypothetical rubbish on the nature of reproduction.[9]

My dear Lyell | Yours ever gratefully | C. Darwin

American Philosophical Society (201)

[1] Dated by the reference to Murray 1860a (see n. 7, below).

[2] CD refers to Richard Owen, Herbert Spencer, and Samuel Wilberforce, the bishop of Oxford. See letter from Charles Lyell, [13–14 February 1860], and letter to Charles Lyell, 15 and 16 [February 1860].

[3] CD refers to his practice of filing letters, or portions of letters, on metal spikes ('spits') until needed. According to Francis Darwin, CD periodically burnt the accumulated letters. 'This process, carried on for years, destroyed nearly all letters received before 1862.' (*LL* 1: V). The letter from Andrew Crombie Ramsay was probably the one dated 21 February 1860.

[4] Henry Darwin Rogers, the American geologist, was professor of natural history at Glasgow University. A change in university regulations meant that Rogers was obliged to teach zoology, and he received coaching from Thomas Henry Huxley during the summer of 1858 (Gregory 1916, p. 12). In a letter to his brother William Barton Rogers, H. D. Rogers called *Origin* 'a suggestive book, full of ingenious arguments in favour of the Lamarckian hypothesis.' However, he disagreed with CD's emphasis on slow and gradual geological change. See Rogers ed. 1896, 2: 17–18.

[5] The letter from Heinrich Georg Bronn has not been found, but see the preceding letter for CD's reply.

[6] The letter from William Jardine has not been found. For Jardine's earlier criticisms of *Origin*, see *Correspondence* vol. 7, letter from William Jardine, 20 December 1859.

[7] Murray 1860a. Andrew Murray read the paper at a meeting of the Royal Society of Edinburgh on 20 February 1860. See also letters to Andrew Murray, 28 April [1860], and 28 [April 1860].

[8] See letter to Charles Lyell, 23 February [1860].

[9] See letter to Herbert Spencer, 23 [February 1860].

To J. D. Hooker 26 [February 1860]

Down Bromley Kent
26th

My dear Hooker

Your answer to Harvey seems to me *admirably* good.[1] You would have made a gigantic fortune as a Barrister.— What an omission of Harveys about the graduated state of the flowers! But what strikes me most, is, that surely I ought to know my own Book best, yet by Jove you have brought forward ever so many arguments, which I did not think of! Your reference to classification (viz I presume to such cases as Aspicarpa) *excellent*; for the monstrous Begonia no doubt in all details w^{d} be Begonia.—[2] I did not think of this; nor of the *retrograde* step from Separated sexes to an hermaphrodite state; not of the lessened fertility of the monster.—[3] Proh pudor to me.—

The world would say what a Lawyer has been lost in a *mere* Botanist! Farewell my dear master in my own subject.—

Your affect | C. Darwin

I am so heartily pleased to see that you approve of the Ch on Classification.—

I wonder what Harvey will say.[4] But no one, hardly, I think, is able at first to see when he is beaten in an argument.— I shall want some time to hear a little more about the Crassulaceæ.[5]

I wonder when we shall ever meet.—

Endorsement: 'Feby | 1860'
DAR 115.1: 43

[1] The reference is to Hooker's article in the *Gardeners' Chronicle and Agricultural Gazette*, 25 February 1860, pp. 170–1, entitled 'The monstrous Begonia frigida at Kew, in relation to Mr. Darwin's theory of natural selection'. It was a response to a previous piece written by William Henry Harvey. See letter to Charles Lyell, 18 [and 19 February 1860].

[2] After explaining the origin of the monstrous begonia, Hooker stated (*Gardeners' Chronicle and Agricultural Gazette*, 25 February 1860, p. 171):

> Instead of this being a case which (according to Dr. Harvey) 'was not contemplated by Mr. Darwin's hypothesis,' it is one of a class which he had specially in view; it is a beautiful illustration of the truth and wisdom of his chapter on classification.

[3] Harvey had argued that the change from single-sexed to hermaphrodite flowers was so great that the monstrous begonia should be considered an entirely new species. Hooker countered this argument by stating that, according to CD's views, the switch would lead to self-fertilisation and a consequent loss of vigour in the progeny (*ibid.*).

[4] For Harvey's response to *Origin*, see letters from W. H. Harvey, 24 August 1860 and 8 October 1860.

[5] In the concluding section of his article in the *Gardeners' Chronicle*, Hooker stated that the begonia was not unusual in presenting abnormalities like those under discussion: the genus *Sempervivum* also offered 'another most curious instance of the conversion of stamens into carpels' (*ibid.*). *Sempervivum* belongs to the family Crassulaceae.

To J. D. Hooker [26 February or 4 March 1860][1]

[Down]
Sunday night.

My dear Hooker

I forgot in this morning's note to say that I have tried in vain to get any Goodenia from large nursery here.— If you have any plant which you can **perfectly** spare & would send it as by enclosed address.— But it does not of course really at all signify; as it half mere amusement——[2]

Do get some of your men (is Mr Oliver to whom you alluded one of them?)[3] to cross with distinct individual *2 or 3 times in morning* a bunch of flowers of some healthy Australian mimosa. perhaps also Eucalyptus I see in Lecoq's work he speaks of sterility of the group under culture; it may *possibly* be from want of crossing; but not probably.[4]

Remind your man that the stigma *may* be ready after or before pollen of own flower is shed.—

Ever yours | C. Darwin

Endorsement: 'March 1860'
DAR 115.1: 44

[1] Dated by the relationship to the preceding letter, also written on a Sunday. However, the reference to 'Mr Oliver' links the letter with that written to Hooker on Saturday, 3 March [1860]. The *Goodenia* plants referred to in the letter arrived at Down House on 9 March (see letter to J. D. Hooker, 12 March [1860]).

[2] See letter to J. D. Hooker, [22 January 1860]. Plants of the genus *Goodenia* have flowers with a cup immediately under the stigma into which the pollen drops before the flower opens. CD wished to investigate the mechanism by which insects might bring about pollination and any adaptations exhibited by the flowers that might facilitate this process. However, Hooker could only supply plants of the closely related genus *Leschenaultia*. See letters to J. D. Hooker, 12 March [1860] and 18 April [1860].

[3] Daniel Oliver was an assistant in the herbarium of the Royal Botanic Gardens, Kew.

[4] CD refers to Lecoq 1845, pp. 103–4. CD's annotated copy of this work is in the Darwin Library–CUL.

From Joseph Beete Jukes 27 February 1860[1]

(Febry 27. 1860)

I beleive in the perfect indefiniteness & frequently the vast length of the interval between the formation of any one bed & that which rests upon it

I beleive any amount of beds may be intercalated almost any where not only between any two formations, but anywhere in any formatn. & that the ordinary ideas of Geologists as to contemporaneity will have to be greatly expanded—

I am therefore prepared to treat with very little respect the argument agt. you derived from missing links.[2]

When you have anything approaching to a continuous series of beds you have the links, moreover, when you get species ranging through a great series of beds they do vary. Calymene Blumenbachii for instance is found in both Upper & Lower Silurian, but Salter says that those found in Lower Silurian vary so much from the Wenlock ones that he has doubted their being the same species & was inclined to give them another name.[3] So Baillie & I have arrived at the conviction that Spirifer Verneuilli. S. disjunctus & S. strictus are mere varieties of one species, that found in the Lower beds having been christened Verneuilli.[4] So Terebratula hastata has about 30 synonyms & it is very likely that some of the varieties are of greatly difft. ages.[5]

Your theory accounts for just so much of progressive improvement as we see has taken place i.e. an improvement & a multiplication & a complexity in the whole but not necessarily absolute in every step or every instance, & relative but not always absolute—

Incomplete
Copy
Kinnordy MS, Charles Lyell's journal V, pp. 125–7

[1] The date as given in Charles Lyell's journal, into which the letter was copied. See also Wilson ed. 1970, p. 361.

[2] Jukes refers to arguments against transmutation that pointed to the absence of transitional forms in the geological record.

[3] John William Salter, palaeontologist to the Geological Survey of Great Britain, was an expert on the fossils of older rocks like the Silurian (Secord 1986). He had worked with Jukes on the classification of the Devonian and Carboniferous formations of the south of Ireland in the early 1850s.

[4] William Hellier Baily worked with Jukes on the staff of the Geological Survey in Ireland. He was a specialist in the palaeontology of ancient formations.

[5] *Spirifer* and *Terebratula* are brachiopods. Because of their common occurrence in geological deposits beginning with the Lower Cambrian, they are frequently used for dating strata. CD had attempted to gain information on the variability of species of fossil shells in 1856. See *Correspondence* vol. 6, letter from S. P. Woodward, 2 May 1856, and letter to S. P. Woodward, 27 May 1856.

From Francis Boott 29 February 1860

24 Gower St. W.C
Feby 29, 1860

My Dear Mr Darwin

I return you Grays paper which I have twice read, & which I shall read again, as he has promised me a copy.[1] He does not grapple with your hypothesis, as Huxley calls it, nor do I see how he could, in the wish he had to oppose Agassiz. I do not see how anyone can grapple with it, for your Book is like the outward aspects of nature, so full, that the Author of nature alone can deal with it. I am a reader, anxious for light, which my mental optics can only accomodate to the degree of their structure, & I can only admire, consider, scratch my head & gape, as the varying impressions crowd upon me enlightening & confusing me. You have not had a more devout reader. I have owed to you untold benefits. I have always said, condemn me to a Robinson Crusoe Islet, & let me have the choice of the Books of the World for my companions, & I would claim for my portmanteau the 4 Gospels (excluding the Epistles) White's Selbourne—[2]your voyage in the Beagle Shakespeare, Byron, Burns & everything of Milton, & I would be content to work out in solitude my claims for salvation.

I greatly admire your Book, which I have read twice & which I shall read as I do the precepts of Christ & the parable of the prodigal son, till my eyes fail me, & I have to trust to my memory for my devotion to that which quickens my intellectual & moral sense. I owe you (a poor debtor) a thousand obligations. You have enlarged my mind, filled it with power, given an upward direction to my thoughts, & conciliated my respect for humanity. I have for long years sat a humble listener to your teaching—but untill lately you asked nothing of me but admiration for the most perspicuous interpreter of nature. Now you demand of me a faith "in the substance of things not known & the evidence of things not seen"[3] —which requires all those dispositions of mind which faith implies. So far, my faith has been individual & personal. I have a profound reverence for Abraham & Moses & Jesus Christ, & I have much of the same reverence for you, whom I look upon as the High priest of nature. But the Church would condemn me to the stake for my religious creed, & you would commiserate my difficulties in adopting all your inferences.

So little reliance have I however on my own discernment, that I read your book with all the spirit of faith & all the consciousness of self-weakness, so that my reverence is constantly at war with my doubts, & I feel as much debased on the one hand as you are exalted on the other. My own side of the balance has a natural tendency to sink & yours to rise. It is the fullest Book I ever read, the most candid—the most ingenious, & the most suggestive. From my humble state I have said again & again, as I have floundered often in difficulties "Quid æternis minorem conciliis animum fatigas?"—[4]

You have done me one essential service. You have put me at rest about the perplexing variations in Carex—a genus which at times I wish to God I had had nothing to do with for its mazes lead me in wanderings lost.[5] I can follow your effects of natural selection in it safely, but when you bring millions of years into play, & ask me to believe in a common ancestor to the Sedge-palm, Oak &c you are like the Church, which demands of me a faith I have no elements for. I cannot attain to the belief in the Divinity of Christ, his resurrection—the miracles or the Apostolic succession or the verbal inspiration—& if I am to be d— for it, it is as least a graceful recollection that the rose & violet, which you make my fellow mortals go with me. I do not envy you meeting our common mother Eve hereafter. You will have a rougher reception than Æneas had from Dido,[6] for tho' Decandolle had alluded to the struggle for existence in plants as well as animals,[7] You have above all shewn what Eve produced in the fall of our race.

What a mystery is nature, & the destiny of man! To one, like myself, approaching the psalmists age,[8] it is natural to look beyond the curtain that falls between the present & the future scene. While I see much in the instinctive religion which nature plants in the mind, I separate myself the more & more from the Church—& knowing nothing beyond my admiration for the beauties & the mysteries of this glorious world, I shall go with a conviction that the power which created & has sustained me will provide for me hereafter—& that is enough for me. Of this world & its marvels, nothing has so much impressed me as the human mind, & no one has more conciliated my homage than your own—

With the deepest respect | Yrs sincerely | F. Boott

C. Darwin Esq

DAR 98 (ser. 2): 27–8

[1] [Gray] 1860a, which CD had forwarded to Boott via Joseph Dalton Hooker (see letter to J. D. Hooker, [23 February 1860]).

[2] White 1789.

[3] Heb. 11:1.

[4] Horace *Odes* 2.11.12: 'Why, with planning for the future, weary your soul unequal to the task'.

[5] Boott was an expert on the taxonomy of the genus *Carex*, the sedges.

[6] Boott refers to Virgil *Aeniad* 6: 450–76, in which Dido coldly refuses to speak to her former lover, Æneas.

[7] A. P. de Candolle 1820, p. 384. CD's annotated copy of this work is in the Darwin Library–CUL. In

Origin, p. 62, CD referred to Augustin Pyramus de Candolle's discussion of the competition in nature of all living beings.

[8] Boott, aged sixty-eight, alludes to the expression 'threescore years and ten' (Psalms 90:10).

To Asa Gray 2 March [1860][1]

Down.
March 2^d

My dear Gray

I have had very short but sharpish touch of illness,—a slight touch of pleurisy, & am weak.—but write just to thank much for letter received today[2] (your two enclosed letters have been posted) & to say that in one week I hope & think I shall send you a few more rather important additions & expansions for the "Origin". The first & most considerable one is for p. 126. If you can anyhow manage to get this inserted I shall be very glad— It is of some pages length.— There are a few of the most objectionable parts later in volume expanded a little.[3] I am tired, so no more

C. Darwin

Gray Herbarium of Harvard University (24)

[1] Dated by the relationship to the following letter and to the letter to Asa Gray, 8 March [1860].

[2] Probably the letter from Asa Gray, 20 February 1860. CD replied to it in his next letter to Gray (letter to Asa Gray, 8 March [1860]).

[3] See letter to Asa Gray, 8 March [1860].

To J. D. Hooker 3 March [1860]

Down Bromley Kent
Saturday March 3^d

My dear Hooker

What a days work you had on that Thursday.[1] I was not able to go to London till Monday, & then I was a fool for going, for on Tuesday night I had an attack of fever (with a touch of pleurisy) which came on like a Lion, but went off as a lamb, but has shattered me a good bit.—[2]

I was much interested by your last note. That was a good sneer by the Bishop at Huxley.—[3] I think you expect too much in regard to change of opinion on the subject of species. One large class of men, more especially I suspect of naturalists, never will care about **any** general question, of which old Gray of Brit. Mus. may be taken as a type;[4] & secondly nearly all men, past a moderate age either in actual years or in mind, are, I am fully convinced physically incapable of looking at facts under a new point of view. Seriously I am astonished & rejoiced at progress which subject has made; look at enclosed memorandum.

Owen says my book will be forgotten in 10 years;[5] perhaps so, but with such a list, I feel convinced the subject will not. The outsiders, as you say, are strong.—

You say that you think that Bentham is touched, "but like a wise man holds his tongue".—[6] Perhaps you only mean that he cannot decide: otherwise I sh^{d} think such silence the reverse of magnanimity; for if others behaved the same way, how would opinion ever progress? It is a direliction of actual duty.—

I am so glad to hear about Thwaites.[7] I do not know who "Greene" & "Oliver" are.—[8] I have had an astounding letter from Dr Boott: it might be turned into ridicule against him & me, so will not send it to anyone.— He writes in a noble spirit of love of truth.—[9]

I wonder what Lindley thinks;—probably too busy to read or think on the question.—[10]

I am vexed about Bentham's reticence; for it would have been of real value to know what part appeared weakest to a man of his powers of observation.—

I sh^{d} like to have a copy of Huxley's lecture.—[11]

I am very very sorry for Huxley about his Lecture: he seems to me vexed. I suppose H. C. Watson was there for he noticed in a letter to me one surprising lapsus linguæ.—[12]

Farewell | My dear Hooker | Yours affect | C. Darwin

I am very glad to hear that Lady Hooker improves gradually though slowly.—

Is not Harvey in class of men who do not at all care for generalities?[13] I remember your saying you could not get him to write on Distribution.— I have found his works very unfruitful in every respect.—

[Enclosure]

Geologists[14]	Zoologists & Palæontologists[15]	Physiologists[16]	Botanists[17]
Lyell	Huxley	Carpenter	Hooker
Ramsay	J. Lubbock	Sir H. Holland to large extent	H. C. Watson
Jukes	L. Jenyns (to large extent)		Asa Gray to some extent
H. D. Rogers	Searles Wood		Dr Boott (to large extent)
			Thwaites

Endorsement: '1860'
DAR 115.1: 45

[1] CD had intended to attend a meeting of the Philosophical Club of the Royal Society on Thursday, 23 February (see letter to J. D. Hooker, [23 February 1860]).

[2] Emma Darwin's diary indicates that CD went to London on 27 February 1860 and 'came home unwell' on the following day. On 29 February she recorded 'his attack of the chest'.

[3] Perhaps a reference to Samuel Wilberforce, bishop of Oxford. See letter from Charles Lyell, [13–14 February 1860].

[4] John Edward Gray was keeper of the zoological department of the British Museum. CD had mentioned Gray's criticisms of *Origin* in the letter to J. D. Hooker, 14 December [1859] (*Correspondence* vol. 7).

[5] The occasion on which Richard Owen made this comment has not been traced.

[6] Hooker told CD in December 1859 that George Bentham was 'much agitated' by *Origin* and that he did not know 'how it will go' (see *Correspondence* vol. 7, letter from J. D. Hooker, [20 December 1859]).

[7] Hooker had received a letter from George Henry Kendrick Thwaites, the director of the botanic garden in Peradeniya, Ceylon. Thwaites had also written to CD praising *Origin* (see letter from G. H. K. Thwaites, [14 February 1860]), but this letter did not arrive at Down until late in March (see letter to G. H. K. Thwaites, 21 March [1860]).

[8] Probably Joseph Reay Greene, an Irish naturalist, and Daniel Oliver, assistant in the herbarium of the Royal Botanic Gardens, Kew.

[9] Letter from Francis Boott, 29 February 1860.

[10] John Lindley was professor of botany at University College, London, and editor of the *Gardeners' Chronicle.*

[11] T. H. Huxley 1860a. In a letter to Asa Gray dated 16 March 1860, Hooker reported that 'Huxley made a failure of the RI. Institution Lecture which was a great pity, as he intended to have backed the book but unfortunately managed to damage it.' (Gray Herbarium, Harvard University). A report of the lecture appeared in the *Athenæum*, 3 March 1860, p. 308.

[12] The letter from Hewett Cottrell Watson has not been found.

[13] William Henry Harvey's publications were almost exclusively taxonomic or descriptive works. CD had read Harvey's more general treatise (Harvey 1854) but did not find it of great interest (see *Correspondence* vol. 6, letter to S. P. Woodward, 18 July 1856).

[14] Charles Lyell, Andrew Crombie Ramsay, Joseph Beete Jukes, and Henry Darwin Rogers.

[15] Thomas Henry Huxley, John Lubbock, Leonard Jenyns, and Searles Valentine Wood.

[16] William Benjamin Carpenter and Henry Holland.

[17] Hooker, Watson, Asa Gray, Francis Boott, and Thwaites.

To a Bookseller 4 March [1860]

Down Bromley Kent
March 4th

Dear Sir

Please procure & send to 57 Queen Anne St. "Jukes' Students Manual of Geology", published two or three years ago.[1]

Allso please procure *& send here by Post*, Macmillan's Magazine Nor *2, for December 1859*[2] published at Cambridge & 23. Henrietta Str.[3]

Yours faithfully | C. Darwin

Endorsement: '1860 | March 5.'
Pierpont Morgan Library

[1] Jukes 1857. CD recorded reading this work in 1860 (*Correspondence* vol. 4, Appendix IV, 128: 26). His annotated copy is in the Darwin Library–CUL.

[2] The December 1859 issue of *Macmillan's Magazine*, pp. 142–8, contained Thomas Henry Huxley's review 'Time and life. Mr. Darwin's "Origin of species"'. There is an annotated copy in the Darwin Pamphlet Collection–CUL.

[3] The locations of the offices of Williams and Norgate, publisher of *Macmillan's Magazine.*

To William Erasmus Darwin [4 March 1860][1]

Sunday night

My dear William

You seem to have set about finding out what was best about your degree in a downright sensible manner;[2] & as you will have two years mathematicks, & as you fancy reading for Law, I quite agree that you had better so decide. I do not, however, think that studying mathematicks & History or Law can have at all the same training effect on the mind. The concentrated attention required for a problem is just a faculty which could hardly be got without some training in very peculiar studies.—

But I do not mean at all by this, that you have not judged wisely: no doubt a lawyer besides his law-knowledge, some oratory, good judgment, also is greatly profited by diversified knowledge (a nice little catalogue of qualities requisite for a famous Lawyer; & you know I mean you to be Lord Chancellor & buy Holwood Park!!)[3] & you will, I suppose have rather more time to pick up knowledge of all kinds.— If you do not dislike it, had you not better consult Abdy about reading aloud or oratory?[4] & likewise about entering at Lincoln's Inn or the Temple?—[5] So much for the Law: I think you had certainly better write to Hemmings; but it will be a difficult letter to write.—[6]

How I sh^d^ like to see you in your Uniform: I do not hear much about the Rifles here.[7] M^rs^. Lubbock says that except Montagu all are very idle in drilling.[8] Mamma's visit to London answered pretty well,—at least the latter part.[9] As we came home we called at a Picture's dealers & bought a view on sea-coast of S. Wales in Oil for £12. ^s^10.[10] I think it looks well in the Dining Room: I shall be curious to hear what you will think about it— I lean for more oil-paintings; as they will be sure to last.— The House to day is rather bad: Etty very poorly & Lenny with very bad cough. I have had bad cold, & all the House has been poorly more or less; & Parslow has been in bed for a week.—[11]

I go to London tomorrow for one night to see my Doctor;[12] but I am not very hopeful of his doing me much good. On Friday the Lyells & Tollets come here.[13] A deluge of letters still keep pouring in about my Book, till I am got weary of praise. It is making a great row in America, where it has been reprinted & sells for 5^s^. A German Edition is preparing by very good Naturalist:[14] & I hear of some foreign Reviews coming out.— It has been very well reviewed in N. America.— There was rather fierce attack on it in Annals of Nat. History, by my friend Wollaston,[15] & I hear it has been cut up in a paper read before Royal Soc. of Edinburgh.—[16] By Jove the Book has made row enough, & I sh^d^ now like to get on quietly with my work.

It will be jolly having you here at Easter; & I long for some Billiards I have not had a game for a week or more.

My dear old Guillielmus | Your affect Father | C. Darwin

DAR 210.6

[1] Dated by the reference to the expected visit of the Lyells on 'Friday'. See n. 13, below.

[2] William was in his second year at Christ's College, Cambridge University, reading for the mathematical tripos.

[3] Holwood Park, near Keston, almost two miles from Down, was the seat of Robert Monsey Rolfe, Baron Cranworth, who was lord chancellor of England from 1852 to 1858.

[4] John Thomas Abdy was regius professor of civil law at Cambridge University.

[5] Lincoln's Inn, the Middle Temple, Inner Temple, and Gray's Inn prepared law students for admission to the English bar. William was admitted to Lincoln's Inn in 1861.

[6] CD refers to George Wirgman Hemming, the mathematician and law reporter. He was a fellow of St John's College, Cambridge.

[7] The volunteer movement of 1859 was sparked by a general fear of a French invasion following upon the outbreak of France's war with Austria over hegemony in Lombardy. Many gentlemen joined local, semi-official military units that held monthly meetings to encourage rifle shooting. John William Lubbock had founded the Down Rifles unit in 1859. Joseph Parslow, the Darwin's butler, was a member (see *Correspondence* vol. 7, letter to John Lubbock, 17 December [1859]). William served in the University Volunteers and was also an officer in the Down Corps (F. Darwin 1914, p. 20). On 29 January 1860, CD bought a commission for William in the volunteers (CD's Account book, Down House MS).

[8] CD refers to John Lubbock's wife, Ellen Frances Lubbock, and to Lubbock's brother Montagu, who was 18 years old.

[9] Emma Darwin's diary records that she went to London on 24 February 1860.

[10] The sum mentioned in the letter was recorded in CD's Account book (Down House MS) on 17 February 1860. The picture was a view of the Mumbles, near Swansea, and is on display in the Darwin Museum, Down House.

[11] CD refers to Henrietta Emma Darwin and Leonard Darwin, and to Joseph Parslow.

[12] In 1860, CD began consulting the London physician Frederick William Headland.

[13] Emma Darwin's diary records that Charles and Mary Elizabeth Lyell came to stay on 9 March 1860 and that the 'Tollets' came on 10 March. Emma probably refers to her childhood friends, Georgina and Ellen Harriet Tollet.

[14] See letter to H. G. Bronn, [*c.* 25 February 1860].

[15] [Wollaston] 1860.

[16] Andrew Murray read a paper giving his criticisms of *Origin* at a meeting of the Royal Society of Edinburgh on 20 February 1860 (Murray 1860a). See also letters to Andrew Murray, 28 April [1860] and 28 [April 1860]. There are two copies of Murray's review in the Darwin Pamphlet Collection–CUL, one of which is heavily annotated.

To T. H. Huxley 4 March [1860][1]

Down Bromley Kent
March 4th

My dear Huxley

You poor Londoners I daresay never see the Gardeners' Chronicle so I write to say that in to days number, there appears the following.

"We borrow from the Times the following brilliant notice of the work &c &c"

Then comes your whole Article—[2] I never knew G. Ch. do such a thing before see you best of "general agents" what capital service you have done in the good cause of removing prejudice &c.

Yours Most sincerely | C. Darwin

By the way, a fortnight ago Harvey attacked my book in rather weak article, & Hooker made really admirable rejoinder in last week's number.—[3]

Imperial College of Science, Technology and Medicine Archives (Huxley 5: 109)

[1] Dated by the reference to the *Gardeners' Chronicle and Agricultural Gazette* (see n. 2, below).
[2] Huxley's anonymous review of *Origin* in *The Times*, 26 December 1859, p. 8, was reprinted in the *Gardeners' Chronicle and Agricultural Gazette*, 3 March 1860, pp. 192–3.
[3] *Gardeners' Chronicle and Agricultural Gazette*, 18 February 1860, p. 145, and 25 February 1860, pp. 170–1. See letter to J. D. Hooker, 26 [February 1860].

To James Lamont 5 March [1860][1]

Down. | Bromley. Kent.
March 5th.

Dear Sir.

I am much obliged for your long and interesting letter—[2] You have indeed good right to speak confidently about the habits of wild birds & animals; for I should think no one beside yourself has ever sported in Spitzbergen & Southern Africa![3] It is very curious & interesting that you should have arrived at the conclusion that so called "Natural Selection" had been efficient in giving their peculiar colours to our grouse— I shall probably use your authority on the similar habits of our grouse & the Norwegian Species.[4]

I am particularly obliged for your very curious fact of the effect produced by the introduction of the lowland grouse on the wildness of the grouse in your neighbourhood— It is a very striking instance of what crossing will do in affecting the character of a breed. Have you ever seen it stated in any sporting work that Game has become wilder in this country? I wish I could get any sort of proof of the fact, for your explanation seems to me equally ingenious & probable. I have myself witnessed in S. America a nearly parallel [case][5] with that which you mention in regard to the reindeer in Spitzbergen, with the *C. campestris* of La Plata;[6] it feared neither man nor the sound of shot of a Rifle, but was terrified at the sight of a man on horseback; every one in that country always riding.—

As you are so great a sportsman perhaps you will kindly look to one very trifling point for me, as my neighbours here think it too absurd to notice— Namely whether the feet of birds are dirty, whether a few grains of dirt do not adhere occasionally to their feet. I especially want to know how this is in the case of birds like herons & waders which stalk in the mud—You will guess that this relates to dispersal of seeds—which is one of my greatest difficulties—[7]

My health is very indifferent & I am seldom able to attend the scientific meetings, but I sincerely hope that I may sometime have the pleasure of meeting you—[8]

Pray accept my cordial thanks for your very kind letter, & believe me. Dear Sir. | Your obliged servant | Charles Darwin.

Copy
DAR 146

[1] Dated by the relationship to the letter from James Lamont, [23 February 1860].
[2] See letter from James Lamont, [23 February 1860].
[3] Lamont travelled to Spitzbergen on a hunting expedition in 1858 and again in 1859. He had gone big-game hunting in South Africa in 1851. See Lamont 1950, pp. 9, 11.
[4] Lamont is not cited on this point in subsequent editions of *Origin* or in *Variation.*
[5] The copyist omitted a word here: '[case]' has been added in the margin, presumably by Francis Darwin.
[6] *Colaptes campestris* is a species of woodpecker found in southern Brazil and the pampas of Argentina that feeds solely on the ground. See *Origin*, p. 184, and *Natural selection*, p. 344.
[7] For CD's earlier attempts to gain information on the means of dispersal of seeds and ova, see *Correspondence* vols. 5, 6, and 7.
[8] Lamont was a fellow of the Geological Society of London. In 1859 he read a paper on the geology of Spitzbergen at a meeting of the society.

To Albrecht Carl Ludwig Gotthilf Günther 6 March [1860][1]

Down Bromley Kent
March 6th

Dear Sir

I trouble you with one line to say that I have looked to my original catalogue, & find that I collected 5 specimens of Snakes at the Galapagos from Charles & James Islands.—[2] These snakes I briefly describe as far as colour & stripes are concerned in my Catalogue; & as they differ somewhat in colour, & as when I published the 1st. Edition no Erpetologist had looked at the snakes, I was led into the blunder of supposing that there were several species.[3] This blunder I corrected as soon as M. Bibron had provisionally looked over my collection.— Perhaps Mr Bell may have some specimens of mine from the Galapagos.[4]

Dear Sir | Yours very faithfully | C. Darwin

Shrewsbury School Library

[1] Dated by the reference to CD's snake specimens from the Galápagos, which Günther discussed in a paper read to the Zoological Society of London on 24 January 1860 (see n. 2, below).
[2] CD's catalogue of reptile specimens collected during the *Beagle* voyage is in the British Museum (Natural History). His 'Diary of observations on zoology' (DAR 30 and 31) does not contain descriptions of a Galápagos snake. The letter is a response to a paper Günther had read to the Zoological Society of London in January (Günther 1860). CD may have attended the meeting: he was in London on 24 January (Emma Darwin's diary) and was keen to discuss the distribution of Galápagos birds with Philip Lutley Sclater, the secretary of the Zoological Society (see letter from P. L. Sclater, [3? February 1860], and letter to P. L. Sclater, 4 February [1860]). In the paper, Günther described a new species of snake (*Herpetodryas biserialis*) from the Galápagos Islands and stated that it was the only snake found there. Between the date of the meeting and publication of the paper, Günther's attention was drawn to CD's record of a different species of snake on the Galápagos (Günther 1860). CD's letter supplied additional information that Günther appended in a note to his paper (Günther 1860, p. 97 n.):

> Darwin says in his Journ. of Research., p. 381, speaking on the Zoology of the Galapagos Islands:—"There is one snake which is numerous; it is identical, as I am informed by M. Bibron, with the *Psammophis temminckii* from Chile." Although subsequently, in the 'Erpétologie

Générale,' nothing is mentioned by Duméril and Bibron about the occurrence of *P. temminckii*, or of any other snake, in these islands, that determination of Bibron may possibly be correct. If such be the case, there are two species of Snakes in that group of islands.

3 When CD published *Journal of researches* in 1839, he did not have access to the detailed descriptions of his reptile collection. Thomas Bell, who had agreed to describe the collection for *Reptiles*, passed the snakes on to Gabriel Bibron (*Reptiles*, p. vi). Bibron, however, was unable to complete the final part of his *Erpétologie générale* (Duméril and Bibron 1834–54), and CD's collection remained undescribed. A pencil note in the back of CD's copy of *Journal of researches* 2d ed. (Cambridge University Library) reads: 'Dr Günter has shown. Zoolog Soc. Jan 24 60 that the snake from Galapagos is distinct from that of St Domingo: American form'. On page 381, in the margin next to the description of the snake, CD noted in ink: 'No | Dr Günter says endemic species.' CD managed to insert a note about the snake in the revised 'postscript' added to the new issue of *Journal of researches* (1860), p. vii:

> The snake mentioned at page 381, as being, on the authority of M. Bibron, the same with a Chilian species, is stated by Dr. Günter (Zoolog. Soc., Jan. 24th, 1859) to be a peculiar species, not known to inhabit any other country.

CD erred in the spelling of Günther's name and in giving 1859 as the year of publication of Günther 1860.

4 The specimens in Bell's possession (see n. 3, above) were given to the British Museum in 1845 (Porter 1985, p. 104). According to Donoso-Barros 1975, p. iii, Günther may have unknowingly described CD's specimen which, by then, had been for some years in the British Museum collection.

To Williams and Norgate? 6 March [1860][1]

Down Bromley Kent
March 6th

Dear Sir

Please order & send to 57 Queen Anne St.—"La *première* partie du troisième volume de Histoire Nat. générale des regnes organiques" by Is. Geoffroy St. Hilaire.[2]

I see that it is just published.

Yours faithfully | C. Darwin

Brown University Library

1 Dated by the reference to the third volume of I. Geoffroy Saint-Hilaire 1854–62.

2 The third volume of I. Geoffroy Saint-Hilaire 1854–62 was issued in two parts, the first of which was published in 1860. The volume requested by CD includes discussions of the origin of domestic animals, teratology, and hybridisation. In the back of his copy, CD noted: 'I have selected only for my 1st volume *on Dom Animals [*interl blue crayon*]— Must be all gone over again & Indexed Especially for Hybridisation very good.—' CD's annotated copy of the entire work is in the Darwin Library–CUL.

To Samuel Pickworth Woodward 6 March [1860][1]

Down. | Bromley. Kent.
March 6th.

My dear Sir.

I will not fail to speak what I think about your high merits if consulted by any of the Senate.[2]

I should be very glad of any criticisms on the origin—I imagine you have not finished it. I suspect that we are at cross purposes about "typical" & "specialisation"—I look at the latter (ie. specialisation of organs) with all physiologists as a great advantage to each being & therefore as *not* likely to lead to their extinction.— And with respect to Typical—I observe that Naturalists use it in two very different senses; hence I have almost entirely or entirely avoided its use—

You will see something about the preservation of old former forms in islands & in isolated situations—a subject on which I could have much enlarged.— I doubt whether you understand my notions on extinction; but when we meet you can explain what you mean by "Specialisation" & "typical".

The fair way to view the argument of my book, I think, is to look at Natural Selection as a mere hypothesis (though rendered in some degree probable by the analogy of method of production of domestic races; & by what we know of the struggle for existence) & then to judge whether the mere hypothesis explains a large body of facts in Geographical Distribution, Geological Succession, & more especially in Classification, Homology, Embryology, Rudimentary Organs The hypothesis to me does seem to explain several independent large classes of facts; & this being so, I view the hypothesis as a theory having a high degree of probability of truth. All turns on whether the above classes of facts seem to you satisfactorily explained or not.—The difficulties are great; but they concern the "imperfection of the Geological record," "means of distribution" & "possibility of transitions of organs" And on these classes of facts we are confessedly ignorant, & we do not know how ignorant.—I simply believe that we are **far** more ignorant than any one supposed—

Forgive me for troubling you with this harangue, for I should very much like to *stagger* you,—to *pervert* you or any good man ought & must require months of self reflection.[3]

Yours very sincerely | C Darwin.

It may be a vain & silly thing to say, but I believe my Book must be read twice carefully to be fully understood.—You will perhaps think it by no means worth the labour.

Copy
DAR 148

[1] Dated by the reference to *Origin*.

[2] Woodward may have been a candidate for one of the two examinerships in geology and palaeontology vacant in the University of London. The election by the Senate was to be held on 25 April 1860 (*Athenæum*, 10 March 1860, p. 321). Woodward's obituary notices allude to the fact that he was, at some unspecified time, appointed examiner in geology and palaeontology for the University of London.

[3] CD and others wrongly suspected that Woodward was the author of an anonymous critical review of *Origin* published in the *Athenæum*, 19 November 1859, pp. 659–60. See *Correspondence* vol. 7, letter to J. D. Hooker, [20 November 1859], and letter from J. D. Hooker, [21 November 1859]. Woodward was, nevertheless, among those who found it impossible to accept CD's views (see H. B. Woodward 1882).

To Asa Gray 8 March [1860][1]

Down Bromley Kent
March 8th.—

My dear Gray

I now send some more M.S. & I shd. be extremely glad if they are not too late for the Edition.[2] If the long addition for p. 126 is too late, that at p. 336 cannot be inserted, as they hang together—unless indeed the M.S for p. 126 were given as a supplement.[3] These additions seem to me, judging from number of letters which I have received on the subject of some little importance.— I will for future keep to my resolution of not wasting more time on the origin. So that you need not fear my giving more trouble. Most sincerely do I thank you from my heart for all your generous kindness, & interest about my book. If there be any profit I am not so foolish as to despise it, but you are perfectly right that I care *more* for a good & corrected American Edition; & it is the simple truth that I shd. be *infinitely* pleased to share profits with you; & this would be fair in many ways. Your truly admirable Review would make a most useful prefix,[4] so I hope the Publishers will agree. Believe me that I fully appreciate your motives in taking such great trouble about my Book.—

I have not yet been able to read coolly a second time all your Review; I have had to lend it to so many persons.— I have had a long letter from Dr. Boott, full of the most noble love of truth & candour.[5] He goes far with me, but cannot swallow all. No one could until he had enlarged his gullet by years ⟨of⟩ practice, as in my own case.— You ask about Zoonomia Da⟨rwin⟩ I am his Grandson.—[6] How extraordinary is ⟨the⟩ state of mind of Agassiz: it is, as you say, a case of "science run mad." Owen after much shuffling & secresy; with bitter sneers to some & modified very slight praise to others, has just spoken out that he rejects my views on the ground of the imperfection of the geological record. Now this is just the subject on which he knows *nothing*; for in his life he has never examined a single stratum. I can count in England 4 geological converts & 8 or 9 other Naturalists. I hear that Thwaites of Ceylon is one.—[7] Did you see Harvey in G. Chronicle versus the Origin: the article, I thought rather weak; & Hooker's answer admirably good.[8] I send Macmillan, not that it is worth sending; look at the verses at the end of following article![9]

Farewell my kind friend. Believe me—Yours most sincerely | C. Darwin

If possible I shd. like out of curiosity to receive through Murray of Albemarle St a copy of the American Edition—[10]

P.S. Since the above was written I have received Appletons letter to you with his P.S.—[11] I much fear that the additions herewith sent will be too late.

You might tell the Appletons that I mean to leave all future Editions of the "Origin" (as far as I can tell) in its present state with the additions now sent.—

Hearty thanks about the money.[12] I should have thought the £50 very handsome. & a bird in Hand worth two in the Bush.— —

With renewed thanks | Ever yours | C. D.

Gray Herbarium of Harvard University (31)

[1] Dated by the reference to a revised American edition of *Origin.*

[2] Gray was superintending a revised and augmented edition of *Origin* in the United States (see letters from Asa Gray, [10 January 1860], [17 January 1860], and 23 January 1860). CD had already sent Gray new material and alterations to be incorporated into the edition (see letters to Asa Gray, 28 January [1860], 1 February [1860], and [8 or 9 February 1860]). See also letter to Asa Gray, 2 March [1860].

[3] The additional material for the chapter on natural selection was added to the new edition in a separately paginated section inserted between pages 116 and 117 (*Origin* US ed., pp. 116*–21*). See Appendix IV. The alterations to the text on p. 336 of *Origin*, 'On the state of development of ancient forms', were the same as those made to *Origin* 2d ed. (*Origin* US ed., pp. 293–4). This material was substantially revised in the third edition (see Peckham ed. 1959, pp. 547–8).

[4] CD was strongly in favour of prefixing Gray's review of *Origin* ([Gray] 1860a) to the new American edition (see letter to Asa Gray, [8 or 9 February 1860]). CD had received sheets of the review in February (see letters to Asa Gray, 18 February [1860] and 24 February [1860]).

[5] Letter from Francis Boott, 29 February 1860. Boott was a close friend of Gray's. He had written to Gray about his favourable reaction to CD's theory on 5 January 1860 (Gray Herbarium, Harvard University).

[6] CD appended a note to his discussion of Jean Baptiste Pierre Antoine de Monet de Lamarck's views on species change in the historical sketch for the authorised American edition of *Origin* that reads: 'It is curious how completely my grandfather, Dr. Erasmus Darwin, anticipated these erroneous views in his Zoonomia (vol. i. pp. 500–510), published in 1794.' (*Origin* US ed., p. vi n.). The reference is to E. Darwin 1794–6.

[7] See letter from G. H. K. Thwaites, [14 February 1860].

[8] See letters to Charles Lyell, 18 [and 19 February 1860], and to J. D. Hooker, 26 [February 1860]. William Henry Harvey had also written a long letter to Gray on 9 February 1860 itemising his objections to CD's theory (Gray Herbarium, Harvard University).

[9] The December 1859 issue of *Macmillan's Magazine* contained Thomas Henry Huxley's review 'Time and life: Mr Darwin's *Origin of species*'. The verses in 'Colloquy of the Round Table', attributed to Thomas Hughes and David Masson (*Wellesley index* 1: 557), read (*Macmillan's Magazine* 1 (1859–60): 157):

> Rumours there were and ominous whisperings | Of the great bomb-shell book from Murray's press, | Threatening flagration in the common mind, | And scientific warring not less loud | Than that which followed the "Vestiges"— I mean | Charles Darwin's "Origin of Species | By Natural Selection;" in which words | The author indicates this principle | As principle-in-chief of all his work, | That Nature is a progress of conditions | Still tightening, tightening, tightening through the ages, | Growing more strict and difficult, and thus | Strangling the worser breeds, and letting through | The finer, subtler, stronger, who escape | Minus their clumsier parts, and modified | On and still on for ever.

[10] CD's copy of the revised and augmented American edition of *Origin* is in the Cambridge University Library.

[11] See letter from Asa Gray, 20 February 1860, and also Dupree 1959, pp. 271–2.

[12] See letter from Asa Gray, 20 February 1860.

From Frederick Wollaston Hutton 8 March 1860

get you some more if you wish for them; If they are of no use to you please throw them away as I can get more if I want them.

I may mention also, on the chance of its being useful, that, on Nov 9th. 1852, I caught on board the Alfred in Lat 2°"20′ South and Long 88°"40′ East, a species of Land rail.[1]

Hoping that you will excuse me for taking the liberty of writing to you | I remain | Yours truly | F. W. Hutton | 23rd. R. W. Fusiliers

March 8th. | 1860

Diadema Lasinassa
(female)[2]

Butterfly caught
in
Lattitude 4"16 N.
&
Longitude 88.35 E

Camberwell beauty
Caught Nov 14th. 1852.

probably
Libellula viridula

Caught in
Lat 9°"40′ N
Long 24°"57′ W

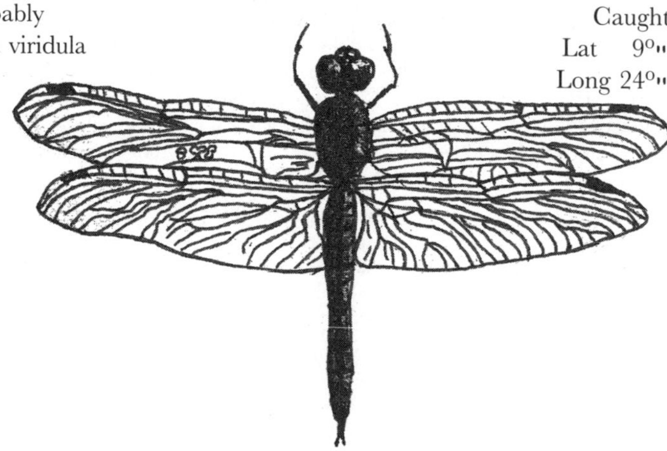

Loo-Choo | Ceylon | Port Natal

Incomplete
DAR 205.2 (Letters)

CD ANNOTATIONS
1.1 get you . . . them. 1.2] *crossed ink*
2.1 I may] '(18)'[3] *brown crayon*
Diagram 1: 'F. W. Hutton | 23^d R. W. Fusiliers.' *ink*; '375 miles from N. Point of Sumatra & rather more distant from Ceylon' *pencil*
Diagram 2: 'F. W. Hutton | 23. R. W. Fusiliers' *ink*; 'Nearly 600 miles from W. c. of Africa near Sierra Leone' *pencil*; 'There were about 100 of these Dragon-flies' *ink*

[1] Hutton was a lieutenant in the Royal Welsh Fusiliers who had served in the Crimean War and during the Indian Mutiny. In 1852, he was attached to the Indian mercantile marine. Having since returned to England, Hutton was studying geology and mineralogy at the Sandhurst and Woolwich military academies. He may have recently read *Origin* (see Hutton 1899, pp. 32–3, and Stenhouse 1990, pp. 415–18).
[2] For CD's notes on the diagrams, see CD's annotations.
[3] The number of CD's portfolio of notes on the means of distribution of animals and plants.

To J. D. Hooker 12 March [1860]

Down Bromley Kent
March 12th

My dear Hooker

I ought to have written sooner to say, with hearty thanks, that the two Plants arrived quite safely on Friday morning.[1]

The Lyells went this morning.[2] I had much talk most interesting to me. & it did not kill me to the extent which I expected. We talked over your Essays & agreed about the Book which you ought to make. What fine materials in all combined, including as Lyell remarked, the Galapagos papers![3] But I see in the Gardeners' Chron: that you have started on a gigantic task with Bentham.[4] By the way I now quite understand Bentham's silence on the modification of species.—[5]

When will you come here? Pray do not forget what pleasure it would give us: & I would get Huxley to come, so that if I were knocked up, you would have a companion.— Do not forget us & do not put off for an indefinite period coming here. There are lots of things I want to talk over.—

Ever yours affec^ly | C. Darwin

Endorsement: '1860'
DAR 115.1: 46

[1] Hooker sent CD living specimens of *Leschenaultia formosa* from the Royal Botanic Gardens, Kew. CD had requested specimens of the genus *Goodenia*, a closely related group of plants with a two-lipped corolla, in order to investigate the plants' mechanism of pollination (see letter to J. D. Hooker, [26 February or 4 March 1860]). CD's experiments on the plants that Hooker supplied are recorded in his Experimental book, pp. 50–1 (DAR 157a). The first entry is headed: 'Ap. 18/60/ Leschenaultia

formosa from Kew (Goodnenaciaciæ)'. The correct spelling of the family name is Goodeniaceae.

[2] Emma Darwin's diary records that Mary Elizabeth Lyell and Charles Lyell visited Down from 9 to 12 March 1860. Lyell made notes on several of his discussions with CD in his scientific journal (Wilson ed. 1970, pp. 357–64). Some of Lyell's entries are dated 13 March 1860. Among the subjects discussed, as indicated by Lyell, were progression and specialisation in animals, the geology of the Weald, various criticisms of *Origin*, Asa Gray's review of *Origin* ([Gray] 1860a), and Hooker's essay on the plants of the Galápagos Islands (see n. 3, below). CD evidently showed Lyell several letters he had received about *Origin*, for they were copied into Lyell's journal during the visit.

[3] Lyell had recorded in his journal: 'Hooker's Galapagos paper good' (Wilson ed. 1970, p. 362). The reference is probably to Hooker 1847a and 1847b. He may also have read an earlier paper by Hooker on a new genus of Portulacaceae from the Galápagos Islands (Hooker 1846). CD had suggested that Hooker compile his published materials into a new volume (see letter to J. D. Hooker, 8 February [1860]).

[4] An announcement in the *Gardeners' Chronicle and Agricultural Gazette*, 3 March 1860, p. 189, stated that Hooker and George Bentham intended to produce a new *Genera plantarum* that would list all the known genera of plants with their synonyms. According to the editor of the *Gardeners' Chronicle*, John Lindley, 'No work in natural history is more needed'. The first part of the first volume was issued in 1862 (Bentham and Hooker 1862–83).

[5] Hooker's letter, in which he discussed Bentham's reaction to *Origin*, has not been found, but see the letter to J. D. Hooker, 3 March [1860].

To Charles Lyell 12 [March 1860]

Down
12th

My dear Lyell

Thinking over what we talked about the high state of Intellectual development of the old Græcians with the little or no subsequent improvement being an apparent difficulty, it has just occurred to me that in fact the case harmonises perfectly with our views.—[1]

The case would be a decided difficulty on the Lamarckian or Vestigian doctrine of necessary progression, but on the view which I hold of progression depending on the conditions, it is no objection at all & harmonises with the other facts of progression in corporeal structure of other animals.— For in a state of anarchy or despotism or bad government or after irruption of Barbarians force strength or ferocity & not intellect would be apt to gain the day.—

We have so enjoyed your & Lady Lyell's visit—[2] Good Night | C. Darwin

By an odd chance (for I had not alluded even to subject) the Ladies attacked me this evening & threw the high state of old Græcians into my teeth, as an unanswerable difficulty; but by good chance I had my answer all pat & silenced them.[3] Hence I have thought it worth scribbling to you.—

Endorsement: 'March | 1860'
American Philosophical Society (203)

[1] Lyell recorded part of his discussion with CD in his scientific journal (Wilson ed. 1970, pp. 364–5).

[2] Charles and Mary Elizabeth Lyell had visited Down from 9 to 12 March 1860 (Emma Darwin's diary).

[3] Probably a reference to Georgina and Ellen Harriet Tollet, childhood friends of Emma Darwin. Emma recorded in her diary on 10 March 1860 that the 'Tollets' came and stayed until 14 March. CD's sister Emily Catherine Darwin was also visiting at the time.

To Joseph Prestwich 12 March [1860][1]

Down. Bromley Kent.
March 12th

My dear Sir.

The recent important changes in taxation[2] will probably have absorbed much of your attention but at some future time when you have a little leisure and when you have read my "Origin of Species" I should esteem it a *singular* favour if you would send me any general criticisms—[3] I do not mean of unreasonable length, but such as you could include in a letter. I have always admired your various memoirs so much that I should be eminently glad to receive your opinion, which might be of real service to me.[4]

Pray do not suppose that I expect to *con*vert or *per*vert you; if I could stagger you in ever so slight a degree I should be satisfied; nor fear to annoy me by severe criticisms for I have had some hearty kicks from some of my best friends. If it would not be disagreeable to you to send me your opinion I certainly should be truly obliged—

I find that I have not put clearly how, as it seems to me, the general argument ought fairly to be viewed; namely, natural selection as a mere hypothesis (rendered in some slight degree probable by the analogy of domestication and the struggle for existence.) which hypothesis has to be judged of by whether it explains a number of facts in Geographical Distribution, Geological Succession Classification, Homology, Embryology &c. If it does explain such facts then the hypothesis seems to me to rise in rank to a theory.

Pray forgive me troubling you with this note, and believe me | My dear Sir | Yours sincerely | C. Darwin.

Copy
DAR 147

[1] Dated by the reference to *Origin*.

[2] Prestwich was the proprietor of a family wine business in London. New tax laws affecting the wine trade were currently being instituted (see *Hansard* vol. 156).

[3] CD sent Prestwich a presentation copy of *Origin* (see *Correspondence* vol. 7, letter to Thomas Henry Huxley, 27 November [1859]).

[4] Prestwich had written important papers on the geology of the coalfields of Shropshire and on the water-bearing strata of the south of England. In 1859 and 1860, he turned his attention to a study of the flint implements found in France and England and their significance for the question of the antiquity of man.

To J. D. Hooker 18 [March 1860]

Down Bromley Kent
18th

My dear Hooker

I am truly glad that you will come here on Thursday 5th. I shall most thoroughily enjoy it, if I can but be even moderately well.— Cannot you persuade Mrs Hooker to come; Emma & myself shd be very glad to see her.— After April 1st I will write & tell you the afternoon Trains & in all probability we shall be able to send & meet you at Station. I will ask Huxley to come.—

I am very glad to hear that you are cogitating about your Book.[1]

Adios, | C. Darwin

Endorsement: 'March 60'
DAR 115.1: 47

[1] CD had encouraged Hooker to compile his materials into a general work on botany (see letters to J. D. Hooker, 8 February [1860] and 12 March [1860]). Hooker wrote on this subject to John Stevens Henslow in March 1860: 'Murray and others are very anxious, I understand, that I should bring out a Darwinian book on Botany—a sort of elementary book on Classification, Distribution, and origin of species. I am dubious and considering. I think I could make it a good instructional one with woodcuts illustrating all sorts of transitional forms, independent of all theory.' (L. Huxley ed. 1918, 1: 535).

To T. H. Huxley 20 March [1860][1]

Down Bromley Kent
March 20th

My dear Huxley

Hooker comes here on Thursday April 5th & stays till Monday[2] Will you come with him; it would give me great pleasure if you will.

Mrs. Darwin desires me to say how happy we should be if you could persuade Mrs. Huxley to come also. I believe our Trains change on April 1st. & I will let Hooker know on 1st or 2d & he will inform you, if I hear that you can come. I hope to be able to send my carriage to Bromley to meet you.— Supposing that you are much engaged & cannot spare time would you come down on Saturday & spend the Sunday with us.—

Do come on Thursday if you conveniently can

Ever most truly yours | C. Darwin

Perhaps you would bring my M.S. Skulls Drawings & Book with you.—[3]

I am so poor a devil that you must make up your mind to be dull here.—

Imperial College of Science, Technology and Medicine Archives (Huxley 5: 160)

[1] The year is given by the invitation to stay at Down. Emma Darwin's diary records that 'Mr & Mrs Huxley' came on 7 April 1860.

[2] See preceding letter.

[3] CD had lent Huxley notes and other material to use in the preparation of his lecture at the Royal Institution on 10 February 1860 (T. H. Huxley 1860a). See *Correspondence* vol. 7, letters to T. H. Huxley, 25 November [1859], 13 December [1859], and 16 December [1859].

To H. G. Bronn 21 March [1860][1]

Down. | Bromley Kent
March 21st.

Dear & Honoured Sir.

I thank you for your kind letter. I daresay your kind present may have arrived & may be at my brother's house—I will send & enquire if I do not go to London in a few days— I feel that I am trespassing on your kindness, but shall be truly obliged for your "Morphological Studies"[2] I will read the passage which you point out; but I am a very poor German Scholar, & to my grief—I do not improve, having a very bad head for languages. I possess your Geschichte der Natur & *very useful* it will be for my larger work.—[3] I have often been astonished at the amount of knowledge which it displays I am very much pleased & surprised at the extraordinary quickness with which the translation has been made![4] It is very good news that p. 126 has not been printed off. I send the desired explanations which I hope will be clear—[5]

Your remarks shew me how clearly you understand my book, I shall ever lie under *deep* obligation to you, & remain Dear Sir. Yours sincerely—obliged— | C. Darwin.

Copy
DAR 143

[1] Dated by the reference to the German translation of *Origin*.

[2] Bronn 1858a. A copy is in the Darwin Library–CUL.

[3] Bronn 1841–9. CD's annotated copy is in the Darwin Library–CUL. CD cited the work repeatedly in *Variation*.

[4] The first part of Bronn trans. 1860 was published on 4 April 1860 (*Börsenblatt für den Deutschen Buchhandel* 24 (1860): 683).

[5] CD refers to a long amendment intended for *Origin*, p. 126, responding to early criticisms of natural selection, that he wished to insert in the German translation being prepared by Bronn. CD had just sent the same amendment to Asa Gray to be inserted into a revised American edition (see letter to Asa Gray, 8 March [1860]). The section appeared in Bronn trans. 1860, pp. 131–7.

To G. H. K. Thwaites 21 March [1860][1]

Down Bromley Kent
March 21st

My dear Mr. Thwaites

I thank you very sincerely for your letter & am **much** pleased that you go a little way with me.[2] You will think it presumptuous, but I am well convinced from my own mental experience, that if you keep the subject at all before your mind, you will ultimately go further. The present vol. is mere abstract & there are great omissions— one main one, which I have rectified in the foreign Editions is an explanation (which has satisfied Lyell, who made same objection with you) why many forms do not progress or advance.[3] (& I quite agree about some retrograding)— I have also M.S. discussion on beauty—but do you really suppose that for instance Diatomaceæ were created beautiful that man after millions of generations sh^d. admire them through the microscope?—[4] I sh^d. attribute most of such structures to quite unknown laws of growth; & mere repetition of parts is to our eyes one main element of beauty. When any structure is of use (& I can show what curiously minute particulars are often of *highest* use) I can see with my prejudiced eyes no limit to the perfection of the coadaptations which could be effected by natural selection.—

I rather doubt whether you see how far, as it seems to me, the argument from Homology & Embryology may be carried. I do not look at this as mere analogy. I would as soon believe that fossil shells were mere mockeries of real shells, as that the same bones in foot of dog & wing of bat—or the similar embryo of mammal & bird—had not a direct signification & that signification can be unity of descent or nothing. But I venture to repeat how much pleased I am that you go some little way with me. I find a number of naturalists do the same, & as their halting places are various & I must think arbitrary, I believe they will all go further. As for changing at once ones opinion; I would not value the opinion of a man who could do so: it must be a slow process.—

Thank you for telling me about the Lantana;[5] & I sh^d. at any time be most grateful for any information which you think would be of use to me.—I hope that you will publish a list of all naturalised plants in Ceylon, as far as known, carefully distinguishing those confined to cultivated soils *alone*. I feel sure that this most important subject has been *greatly* undervalued.

With hearty thanks for your letter | Believe me | Yours very sincerely | C. Darwin

Smithsonian Institution Libraries (Special collections)

CD note:[6]
March 21/60 Thwaites objects about the beauty of Diatomaceæ— was it created for man after millions of generations to admire through microscope. NB. mere symmetrical repetition is to us ['one' *del*] one main element of beauty & that no difficulty—mere vegetative repetition

[1] The year is given by the relationship to the letter from G. H. K. Thwaites, [14 February 1860].

[2] Letter from G. H. K. Thwaites, [14 February 1860].

[3] See letter to Charles Lyell, 18 [and 19 February 1860]. CD and Charles Lyell had discussed this topic during Lyell's recent visit to Down. Lyell recorded in his scientific journal (Wilson ed. 1970, p. 359):

> Why so many simple beings still? Lamarck introd^d. spontaneous generations, but Darwin's natural selection implies no innate tendency to perfection & development is necessary. Rhizopods, or infusoria, foraminifera, have remained for several geolog^l. periods unchanged.

[4] Thwaites was an expert on the Diatomaceae. In *Origin*, p. 199, CD briefly discussed the question of beauty as it related to selection, stating that some naturalists 'believe that very many structures have

been created for beauty in the eyes of man, or for mere variety. This doctrine, if true, would be absolutely fatal to my theory.' Beautiful, though apparently useless, structures were discussed in detail in *Descent*.

5 Thwaites had told CD that *Lantana aculeata*, a species of Verbenaceae introduced to Ceylon, was spreading rapidly there (see letter from G. H. K. Thwaites, [14 February 1860]. See also *ML* 1: 145 n. 2.

6 The note is in DAR 48: 42. It is marked 'Ch 8' in brown crayon, a reference to chapter 8 of CD's 'big book' on species (*Natural selection*), 'Difficulties on the theory'.

To William Darwin Fox 22 [March 1860]

Down Bromley Kent
22d

My dear Fox

The hybrid exhibited some years ago was examined by Owen & showed in its filed hoofs clear proof of artifice. If I go to London I will look at that now shown; but nothing less than dissection & proof that the internal organs & bones were in some degree intermediate would convince me of the possibility of so astounding a hybrid.—[1]

You ask about myself; but I cannot give a good account of myself: I am trying under a Mr Headland a course of nitro-muriatic acid, eating no sweet things & drinking some wine:[2] but it has done nothing for me as yet & I shall go to my grave, I suppose, grumbling & groaning with daily, almost hourly, discomfort. I have begun slowly at my larger book; & this reminds me that I have found a memorandum to the effect that I *thought* I remembered your saying that the common gander does not *always* turn white. If you have any positive knowledge will you send me a line; & if I do not hear, I will understand that you have no positive case.—[3] Linnæus says the goslings of both dusky & white geese are similar yellow;[4] Do you know how this is?—

The "origin of species" has made quite a commotion amongst naturalists & I am well contented with its reception. A German edition will appear immediately & *two* (!) American Editions. So that what is right & what wrong in my Book will soon be sifted & known.—

Like a very bad man that you are, you do not say one word about yourself

Farewell my old friend | Yours affectly | C. Darwin

Erasmus, I grieve to say, has not yet quite lost his ague.—[5]

Postmark: MR 22 60
Christ's College Library, Cambridge (Fox 127)

1 Fox had recommended to CD that he examine an alleged hybrid deer being exhibited in London. See letter to W. D. Fox, 18 May [1860].

2 According to Frederick William Headland's theory of the action of medicines, 'nitro-muriatic acid' (nitro-hydrochloric acid) was valuable in the treatment of blood disorders such as gout. He recommended its usage specifically in cases of oxaluria, a condition that was attributed to 'some fault in

the complicated processes of assimilation and nutrition' (Headland 1852, p. 184). Headland may have diagnosed CD as suffering from oxaluria, in which there is an abnormal formation of oxalic acid in the blood and an excess of urea in the urine. The treatment described by CD in the letter was intended to interrupt and neutralise the formation of oxalic acid and urea during the digestive process. See Headland 1852, pp. 186–90.

[3] CD stated in *Variation* 1: 288 that ganders of the species *Anser ferus* do not always become white, but he did not cite Fox as his source.

[4] Linnaeus 1811, 1: 5. CD recorded this fact in his abstract of the work (DAR 71: 68). He had read the book in 1846 (see *Correspondence* vol. 4, Appendix IV, 119: 17a).

[5] Erasmus Alvey Darwin, CD's brother.

To Charles Lyell [24 March – 3 April 1860][1]

[Down]

My dear L.

What wonderful zeal & kindness you show! I hope enclosed will do.—[2] I have thought myself compelled not to speak of *personal* qualifications, of which I know *nothing*.— I could not allude to precedent under L.d Auckland,[3] for I know nothing about it,—not even whether there was naturalist.

I will write to Blyth this afternoon.—C. D.

I now feel certain that Sedgwick is author of article in Spectator.[4] No one else would use such abusive terms. And what misrepresentation of my notions! Any ignoramus would suppose that I had *first* broached the doctrine that the breaks between successive formations marked long intervals of time. It is very unfair.—But poor dear old Sedgwick seems rabid on question.— **Demoralised** understanding!![5] If ever I talk with him, I will tell him that I never could believe that an inquisitor could be a good man, but now I know that a man may roast another & yet have as kind & noble a heart as Sedgwick's.—

American Philosophical Society (204)

[1] The date range is set by the publication of Adam Sedgwick's review of *Origin* (see n. 4, below) and by CD's letter to Asa Gray, 3 April [1860], which comments on it.

[2] Lyell and CD were apparently writing a testimonial for Edward Blyth, curator of the museum of the Royal Asiatic Society of Bengal, who was seeking a position as naturalist to the Chinese military expedition of 1860. See *Correspondence* vol. 7, letters to W. H. Sykes, 20 December [1859], and to Charles Lyell, 29 [December 1859].

[3] George Eden, Earl of Auckland, was governor-general of India from 1836 to 1842. He had also served as president of the Royal Asiatic Society of Bengal and was vice-president of the Zoological Society of London at the time of his death in 1849. CD perhaps alludes to Eden's Afghan campaign of 1839.

[4] Adam Sedgwick's anonymous review of *Origin* ([Sedgwick] 1860) was published in the *Spectator* on 24 March 1860 and reprinted on 7 April in order to correct two mistakes.

[5] CD refers to a passage in Sedgwick's review ([Sedgwick] 1860, p. 286):

> But I cannot conclude without expressing my detestation of the theory, because of its unflinching materialism; . . . because it utterly repudiates final causes, and thereby indicates a demoralised understanding on the part of its advocates.

From Williams & Norgate 29 March 1860

14, Henrietta-Street, Covent-Garden. | *London, W.C.*
March 29 *18*60

Dear Sir

The translation of origin of species from Stuttgart have not yet arrived.—[1] The Academy has not published a new edition of their dictionary,[2] we believe Poitevin's new french dictionary 2 vols 4º (60/ $\frac{1}{2}$ morocco)[3] is superior to the Academy's English french the best dict. is Flemmings (2 vols 4º 50/)[4] we shall take care to address your parcels and letters according to directions receivd.[5]

Yours very oblig | fpa Williams & Norgate | Ernest Greenoak[6]

Charles Darwin Esquire

DAR 91: 82

[1] The first part of the German translation of *Origin* (Bronn trans. 1860) was published by E. Schweizerbart'sche Verlag, Stuttgart, on 4 April 1860 (*Börsenblatt für den Deutschen Buchhandel* 27 (1860): 683). CD received copies of the first part on 10 April (see letter to H. G. Bronn, 10 April [1860]).

[2] The sixteenth edition of the *Dictionnaire de l'Académie Français* was published in 1835. The next edition was not published until 1878.

[3] Poitevin 1856–60.

[4] Fleming and Tibbins 1841–4.

[5] CD had a printed slip made for him giving the correct postal addresses for parcels and letters to Down. For a transcript, see *Correspondence* vol. 7, letter to Richard Owen, 10 December [1859], n. 5.

[6] Greenoak was presumably a clerk in the London office of Williams and Norgate.

To J. L. A. de Quatrefages de Bréau 30 March [1860][1]

Down. Bromley Kent.
March 30th

My dear Sir

I have taken a long time to thank you for your extremely kind present on the "Maladies" of the Silkworm.[2] But I was not able to read it until lately—and I did not wish to thank you until I had. I have found now quite a large number of facts *very* interesting to me with respect to my notions on Species.—[3] I have been astonished at the variety of diseases to which the unfortunate caterpillars are subject—

The Gentleman who wished to translate my "Origin of Species" has failed in getting a publisher:—Bailliere, Masson & Hachette all rejected it with contempt[4] It was foolish & presumptious in me, hoping to appear in a French dress; but the idea would not have entered my head had it not been suggested to me. It is a great loss— I must console myself with the German Edition which Prof. Bronn is bringing out.

I sincerely hope that your health is quite reestablished, and with my sincere thanks for all your kindness. I remain with much respect, | Yours sincerely, | C. Darwin.

Copy
DAR 147

[1] Dated by the reference to the German translation of *Origin.*
[2] Quatrefages de Bréau 1859. There is an annotated copy in the Darwin Library–CUL. CD also owned a copy of the second edition, Quatrefages de Bréau 1860.
[3] CD cited Quatrefages de Bréau 1859 and 1860 on several occasions in *Variation.*
[4] CD refers to Pierre Theodore Alfred Talandier (see letter to Edward Cresy, 15 January [1860]). CD had previously written to Quatrefages de Bréau concerning a possible French translation of *Origin.* See *Correspondence* vol. 7, letter to J. L. A. de Quatrefages de Bréau, 5 December [1859]; and this volume, letter to J. L. A. de Quatrefages de Bréau, 21 January [1860]). J. B. Baillière et fils, Masson, and L. Hachette et cie, were prominent Parisian publishing houses.

To Miss Holland[1] [April 1860][2]

Down Bromley Kent

My dear Miss Holland—

About a year ago I heard indirectly from you about a Mountain-Ash, of which the berries were greedily sought by Birds.—[3] I sh^{d} be *particularly* obliged for more precise observations made this Autumn. Namely dates of how much sooner this tree is completely or almost completely stripped of berries before the other trees in same garden.—

What birds are the eaters? Is there any difference in appearance of tree or berries? Do the berries taste differently to us? Are the berries ripe earlier than in other trees.—And any other particulars

Perhaps you would be so kind as to keep this note & read over my queries at time.—[4]

Pray believe me | My dear Miss Holland | Yours very truly | Charles Darwin

American Philosophical Society (Getz 2395)

[1] It has not been possible to identify Miss Holland (see n. 3, below). Mary, Lucy, Elizabeth, and Hannah Holland are listed as living in Knutsford, Cheshire (*Post Office directory of Cheshire* 1857).
[2] The date is suggested by a note dated April 1859 in which CD recorded the information referred to in the letter (see n. 3, below).
[3] A note by CD in DAR 46.1: 47 reads:

> Ap. 28/59/ Amy tells me that at Knutsford she saw in Miss Hollands Garden the thrushes eating the fruit of one particular Mountain-ash, before that of any other tree; & was assured that this happened every year: this tree being cleared of its fruit, which no doubt was better; this tree is largely sowed by Birds: & so w^{d} be favoured by being devoured.

CD refers to his niece Amy Wedgwood, the daughter of Francis (Frank) Wedgwood, and Frances Mosley Wedgwood.
[4] CD used this, or very similar, information in *Variation* 2: 230, although Miss Holland was not cited.

To F. J. Pictet de la Rive 1 April [1860][1]

Down Bromley Kent
Ap. 1st

Dear Sir

I received this morning your Review & have just read it.[2] I thank you most cordially for it. There have been many reviews in England, opposed to me, but yours is the single one which seems to me *perfectly* fair & just & candid. I literally agree to every word you say. I admit there are no *direct* proofs of the greater modifications which I believe in.— I most fully admit that I by no means explain away all the vast difficulties. The only difference between us is that I attach much more weight to the explanation of facts, & somewhat less weight to the difficulties than you do.— I am conscious that I always jump at any theory which groups & explains facts; & attach too little weight to unexplained difficulties. Your mind is more cautious & I fear that the world would say more philosophical. The first part of your Review gives a really quite admirable condensation of my views.

Your fifth objection (p. 21) shows me that you think my idea of the spreading of the dominant species & their subsequent multiplication not satisfactory.—[3]

Allow me again to express to you my cordial thanks. I never thought that I shd read an opposed Review *perfectly* fair & just! I shall send it to Lyell, Hooker & Huxley to read.

With sincere respect | I remain yours very faithfully | C. Darwin

Would you like to possess a copy of my Journal of Researches during the voyage of H.M.S. Beagle I shd be truly *proud* to send it you?— How should I send it.?

Bibliothèque Publique et Universitaire de Genève (MS. fr. 1651, f. 10–11)

[1] The year is given by the reference to Pictet de la Rive 1860.

[2] There is an annotated copy of Pictet de la Rive 1860 in the Darwin Pamphlet Collection–CUL.

[3] 'Les formes zoologiques se sont modifiées de la même manière sur toute la surface de la terre . . . Comment le choix naturel aurait-il amené ce parallélisme.' (Pictet de la Rive 1860, pp. 252–3). [The zoological forms are modified in the same manner all over the face of the earth . . . How could natural selection bring about these parallelisms?]

To G. R. Waterhouse 1 April [1860][1]

Down Bromley Kent
Ap. 1st

My dear Waterhouse

I have looked through my collection of combs & have not a piece with drone cells! I had plenty, & one *choice* piece, but some months ago I threw away a lot, & I suppose I did not notice that this was a selected piece. Nothing easier, of course, than to replace the piece. I am very sorry that I cannot answer your query. From memory I am convinced that the shorter diameter of oval Queen cell is larger than Drone cell.—

I cannot see how the relation which you wish to establish will help you, as you have to account for so fine a gradation in the two sizes.— I did indeed most carefully think over subject & came most deliberately to conclusion that distance at which each Bee stands from the others must be the governing element.[2] How they instinctively judge this I cannot conjecture. I got comb & Bees from W. Indies solely for this object.—[3] I think you overrate the difficulty of this judgment of distance.[4] The jumping spiders can judge distance accurately at which they spring on prey.— One has only to perfect this instinct. I am convinced you are not right in speaking of cell-construction as due to excavation *alone*; the Bees can certainly build rough wall in proper position for any particular side of cell. They do not require to make whole cell or sphere, but begin to work at *one* face of the pyramidal basis alone. All that is required is for each Bee to stand in proper relative position & distance with respect to the other Bees.—

I venture to caution you (not from my own knowledge) about the mathematical part of the question; for on the theory of spheres (& *spheres* must be used to make the pyramidal basis) not one angle or side of the hexagon touches the sides of spheres; the points of intersection lie quite within— I have had models made. —[5] The intersection of cylinders to make simple hexagon does not suffice as guide with respect to the intersection of spheres.—

The tenacity with which your Osmia stuck to same position[6] does not apply to Hive Bee for a score of different individuals work one after another at the very commencement of any one cell.—

I much wish I could have answered your query.—

My dear Waterhouse | Yours very sincerely | C. Darwin

The Natural History Museum (General Library, MSS DAR)

[1] The year is suggested by the reference to a 'comb & Bees from W. Indies' (see n. 3) and by the relation to the letter from Frederick Smith, 3 April 1860, and the letter to Frederick Smith, 4 April [1860].

[2] The difference in sizes of the drone, worker, and queen cells in a comb complicated the problem of how bees could achieve the hexagonal structure of the cells and yet keep the thickness of walls uniform. CD's explanation was that the bees instinctively worked at the distances from each other that were required to produce the eventual size and thickness. See *Origin*, pp. 232–3.

[3] See *Correspondence* vol. 7, letter from Richard Hill, 26 November 1859.

[4] A note in DAR 48 (ser. 2): 44 discusses Waterhouse's objection:

> March 31. 1860. Waterhouse seems to dispute that insects can judge of distance— look at springing spider they must judge— we only want to perfect that.— So Melipona judges.— How they judge—whether guided by size of body or by touching each other (not by eyesight), but there is only $\frac{}{1000}$ difference in diameter of common & Drone cells & this is graduated away by 5 or 6 rows of cells, so that the [*over* 'its'] judgment cannot be made by particular parts of body

CD presumably meant to write '$\frac{1}{1000}$' rather than '$\frac{}{1000}$'.

[5] See *Correspondence* vol. 7, letter from W. H. Miller, [14 May 1858], and letters from Erasmus Darwin, [8 June 1858] and [after 8 June 1858].

[6] See *Correspondence* vol. 7, letter from G. R. Waterhouse, 10 February 1858.

To Williams and Norgate 1 April [1860][1]

Down Bromley Kent
Ap 1.

Dear Sir

I am much obliged for information about French Dic^{y}.,[2] when next in London, I will call & speak on subject.—

Please send me **by Post** the number of Westminster Review which will be published tomorrow or immediately afterwards[3]

Yours faithfully | C. Darwin

Hans Rhyn collection, Berne

[1] Dated by the relationship to the letter from Williams and Norgate, 29 March 1860.

[2] See letter from Williams and Norgate, 29 March 1860.

[3] The April 1860 issue of the *Westminster Review* carried a lengthy, unsigned review of *Origin* by Thomas Henry Huxley ([T. H. Huxley] 1860b).

To J. S. Henslow 2 April [1860]

Down Bromley Kent
Ap. 2^{d}.

My dear Henslow

I write one line to remind you to be so kind as to give me a sketch & account of the wasp's comb in transitional state from horizontal to vertical, & the country whence procured, &c., &c.—[1]

As I am in the way of reminding, I will remind about Anacharis—viz how far it has travelled up the Cherry Hinton-Brook—how long the journey has consumed, the means &c.—[2] I have since you were here found statement by Babington that this plant is now not so common as it was at first introduction.[3] I sh^{d}. very much like to have this confirmed, & what creatures destroy it.— *Lastly what plants in any particular spot or pond it has nearly or quite exterminated by taking their places.?* Here is a goodly list of queries!

Many thanks for your Sermon & for copies of the Examination Papers received some time since.[4] Sedgwick's was not very fair towards the Students;[5] but Murray, the Publisher, thought it splendid for selling copies to the unfortunate Students.—

My dear Henslow | Yours very truly | C. Darwin

Endorsement: '*1860*'
DAR 93: 65–6

[1] Henslow had visited Down from 14 to 16 February 1860 (Emma Darwin's diary), during which time he and CD may have discussed wasps' nests. Henslow had previously sent CD part of a wasps' nest (see *Correspondence* vol. 7, letter to J. S. Henslow, 10 January [1859]).

[2] *Anacharis* or *Elodea*, an introduced water-weed, had flourished in English streams to such an extent that rivers were being choked by it. Cherry Hinton is a village near Cambridge. Soon after the introduction of *Anacharis alsinastrum* to the Cambridge Botanic Gardens, the plant 'made its escape through a waste

pipe, found its way into the Cam, and in 1852 impeded navigation and threatened to injure the drainage of the fen country' (A. M. Babington ed. 1897, pp. lxxiii–lxxiv).

[3] The source of Charles Cardale Babington's remarks has not been traced. Babington was the first to describe *A. alsinastrum* (C. C. Babington 1848).

[4] Henslow served on the examination boards for Cambridge and London Universities (see Russell-Gebbett 1977, pp. 69–73). In February 1860, the Cambridge University Senate enacted new regulations whereby those students who passed the Natural Sciences Tripos with honours qualified for admission to the degree of Bachelor of Arts. A Board of Natural Sciences Study was directed to establish the examination requirements and appoint examiners. See Winstanley 1947, pp. 187–8.

[5] Adam Sedgwick. One of the aims of the new regulations was to raise the level of achievement of graduates by forcing candidates to concentrate more on individual subjects for the examination (Winstanley 1947, p. 188).

To Asa Gray 3 April [1860][1]

Down Bromley Kent
Ap. 3d.—

My dear Gray

Although I have nothing particular to say I must thank you for your pleasant letter of March 19th.—[2] But let me say—I know what a busy man you are, & pray do not waste more time over me. My book, your Review, & letters &c &c must have consumed an awful amount. In one sense the time spent on Review has not been wasted; for I feel sure (& I have again (3d time) read it all consecutively) that it will produce great effect, in leading people to think, & that is all I wish. Hooker indeed tells me he knows cases where your article has had this effect & has greatly mollified opposition to my Book.— I agree largely to what you say about "vera causa" "theory"; "hypothesis"; indeed on reading your whole review in one read I saw that some of my remarks were rather superfluous.[3]

It is curious that I remember well time when the thought of the eye made me cold all over, but I have got over this stage of the complaint, & now small trifling particulars of structure often make me very uncomfortable. The sight of a feather in a peacock's tail, whenever I gaze at it, makes me sick! Under this point of view your story of the Black Pigs in the Everglades *delights* me,[4] & supports other cases, which though founded on very good evidence I could hardly digest.—

Pray keep Prof. Wyman up to the mark about writing to me; I shd, also, look at it a great honour & favour, *If you possibly can find out name positively of the red nuts.*—[5]

I shall be very curious to see Agassiz's remarks:[6] I met a few days ago Prof Cooke of your Cambridge[7] & he brought me direct from Agassiz all sorts of very civil speeches. What can this mean? I hope to God A. is a sincere man; I had always fancied that he was so.—

You may like to hear about Reviews on my Book. Sedgwick (as I & Lyell feel *certain* from internal evidence) has reviewed me savagely & unfairly in the Spectator.[8] The notice includes much abuse & is hardly fair in several respects. He would actually lead anyone, who was ignorant of geology, to suppose that I had invented the great gaps between successive geological formations; instead of its being an almost universally admitted dogma. But my dear old friend Sedgwick with his noble heart is old & is rabid with indignation.— It is hard to please everyone; you may remember that in my last letter I asked you to leave out about the Weald denudation:[9] I told Jukes this (who is head man of Irish geological survey) & he blamed me much for he believed every word of it, & thought it not at all exaggerated![10] In fact geologists have no means of gauging the infinitude of past time. There has been one prodigy of a Review, namely an *opposed* one by Pictet the palæontologist in the Bib. Univers. of Geneva,[11] which is *perfectly* fair & just & I agree to every word he says; our only difference being that he attaches less weight to argument in favour & more to argument opposed, than I do. Of all the opposed reviews I think this the only quite fair one, & I never expected to see one. Please observe that I do not class your Review by any means as opposed, though you think so yourself! it has done me *much* too good service ever to appear in that rank in my eyes. But I fear I shall weary you with so much about my Book. I sh^{d} rather think there was a good chance of my becoming the most egotistical man in all Europe.! What a proud preeminence!— Well you have helped to make me so, & therefore you must forgive me if you can.—

My dear Gray | Ever yours most gratefully | C. Darwin

I posted letter to Vilmorin.[12]

Gray Herbarium of Harvard University (47)

[1] The year is given by the references to [Gray] 1860a.

[2] The letter has not been found.

[3] CD refers to [Gray] 1860a. See letter to Asa Gray, 24 February [1860].

[4] Apparently Gray included in his now-missing letter of 19 March 1860 an example, given him by Jeffries Wyman, of the correlation between an animal's colour and its immunity to poison (see letter from Jeffries Wyman, [*c.* 15] September 1860). The black pigs were the only varieties that could eat the paint-root, 'which caused the hoofs of all but the black varieties to fall off'. CD added the account to *Origin* 3d ed., p. 12.

[5] See letter from Jeffries Wyman, [*c.* 15] September 1860.

[6] CD may be referring either to Louis Agassiz's review of *Origin* (Agassiz 1860), which appeared in June 1860, or to Agassiz's contribution to special meetings of the American Academy of Arts and Sciences in March and April, which focused on discussions of CD's views on evolution by natural selection. A report of these meetings appeared in the *Proceedings of the American Academy of Arts and Sciences* 4 (1860): 410–16, 424–31. CD's annotated copy of this volume is in the Cambridge University Library. See also Dupree 1959, pp. 285–8.

[7] Josiah Parsons Cooke was professor of chemistry at Harvard University.

[8] [Sedgwick] 1860.

[9] See the enclosure with the letter to Asa Gray, 1 February [1860] and n. 16. In *Origin*, pp. 285–7, CD had estimated that erosion of the Weald formations would require 306,667,400 years. The estimate was halved in the second edition, and in the third edition the discussion was omitted entirely. CD had been convinced by a criticism in the *Saturday Review*, 24 December 1859, that his reckoning was erroneous on several counts.

[10] Joseph Beete Jukes. The figures on which CD based his calculations had come from Jukes's friend and colleague on the Geological Survey, Andrew Crombie Ramsay (see *Correspondence* vol. 7, letter from A. C. Ramsay, 29 December 1858).

[11] Pictet de la Rive 1860. See letter to F. J. Pictet de la Rive, 1 April [1860].

[12] Pierre Louis François Léveque de Vilmorin published works on economic botany and plant breeding. He died in Paris on 21 March 1860.

From Frederick Smith 3 April 1860

British Museum
3rd. April 1860

My dear Sir

I have carefully examined your Hive-bees from Jamaica[1] and can detect not the slighest difference between them & A mollifica dont accept this as my final determination because I intend to have another investigation— You mentioned when here last that y⟨ou⟩ had a piece of honey-comb—⟨or⟩ had seen one—in which a beautiful gradation of enlargement in the hexagonal cells was shown—i.e. between worker cells & drone cells— I said that I had seen the same in wasp combs— at that time I was not aware that I had preserved a piece—which I will show you in which by a repeated widening of the cells laterally

in about four rows they became large enough for the Queens or females Qy—are the *small Queens*—alias *large workers*—reared in these transition cells?—

I am bothering you with my queries &c——but exc⟨use⟩ me I pray.— There was a very animated discussion here ⟨yes⟩terday speakers— Dr. Gray. Mr W⟨ollaston⟩ & Mr. P⟨*as*⟩coe[2]—or rather outpouring of opinion respecting the apterous or subapterous ⟨Bruchoi⟩ds— this brought to my mind that the same thing occurs in different genera of Insects—I dont think I have pointed it out & dont know whether you have noted it in any way— I will only mention one genus in which it occurs— **Pompilus** I believe every gradation of abbrevation may be found between the ample winged genus *Macroceris* & *Brachypterus brevissimis*—instances also occur in Coleoptera in which the same species is sometimes winged sometimes apterous or subapterous—Example—*Calathus melanocephalus*—

excuse my troubling you if all this is a twice told tale | and believe me | Yours very truly | Fredk Smith

Chas. Darwin Esqre

DAR 177 (fragile)

CD ANNOTATIONS
0.1 My . . . cells—1.6] *crossed pencil*
1.9 Qy . . . cells?—1.10] *scored brown crayon*; 'This refers to Wasp' *ink*
2.7 **Pompilus**] 'Yes' *ink*; 'Pompilidæ??' *pencil over ink*
Top of first page: '10'[3] *brown crayon*

[1] The bees were sent to CD by Richard Hill. See *Correspondence* vol. 7, letter from Richard Hill, 26 November 1859.

[2] Francis Polkinghorne Pascoe was a member of the Entomological Society of London. Smith refers to a meeting of the society on 2 April 1860 at which John William Douglas and John Obadiah Westwood exhibited specimens of weevils of the genus *Bruchus* that infested rice and palm nuts (*Transactions of the Entomological Society* n.s. 5: (1858–61), Proceedings, pp. 113–14). Presumably the discussion concerned the apterous or near-apterous state of weevils and other insects. The genus *Pompilus* referred to in the letter comprises the digging-wasps, which provision their nests with spiders. Smith also refers to John Edward Gray and Thomas Vernon Wollaston.

[3] The number of CD's portfolio of notes on instinct.

To John Murray 4 April [1860][1]

Down Bromley Kent
Ap. 4th

My dear Sir

I enclose with many thanks the receipt for your Bills.—[2]

The Review in Westminster, I have not yet read: it is by Huxley.[3] Prof. Pictet has published an excellent Review, though opposed to me, in the Bibliotheque Univer. of Geneva.[4]

My dear Sir | Yours very sincerely | C. Darwin

John Murray (publishers)

[1] Dated by the bills mentioned the letter (see n. 2, below).

[2] An entry dated 28 May 1860 in CD's Account book (Down House MS) records 'Murray's Bill' for £180. In December 1859 Murray had sent CD a cheque for this amount, due for payment on 27 May 1860, as his share of the profits on the first edition of *Origin* (*Correspondence* vol. 7, letter to John Murray, 10 December 1859).

[3] [T. H. Huxley] 1860b.

[4] Pictet de la Rive 1860.

To Frederick Smith 4 April [1860][1]

Down Bromley Kent
Ap. 4th

My dear Sir

I am much obliged for your note.—Whenever you have finally determined about the Bee, not that it is likely that you will find cause to change your opinion, I will carefully measure the diameter of the cells; & if, as I fully believe, they are sensibly larger, the fact will be curious. The more curious, as Reaumur thought the size so fixed, that he proposed this as a universal standard of measurements for mankind![2]

Thanks for information about the genera allied to Pompilus, which I did not know.—I had heard of Calathus & wish I knew where I could obtain more special information,—viz whether all the gradations in size of wing occur in the same place, or whether the variations are local, &c &c.—

With sincere thanks for your never-failing kindness, believe me | My dear Sir | Yours sincerely | C. Darwin

P.S Waterhouse wants much to see the gradation between common & Drone cells.—[3]

H. R. Glennie

[1] Dated by the relationship to the letter from Frederick Smith, 3 April 1860.
[2] Réaumur 1734–42, 5: 398–9.
[3] See letter to G. R. Waterhouse, 1 April [1860].

To W. B. Carpenter 6 April [1860][1]

Down Bromley Kent
Ap. 6th

My dear Carpenter

I have this minute finished your Review in the Med. Chirurg. Review.[2] You must let me express my admiration at this most able Essay, & I hope to God it will be largely read for it must produce a great effect.— I ought not, however, to express such warm admiration, for you give my Book, I fear, far too much praise. But you have gratified me extremely; & though I hope I do not care very much for the approbation of the non-scientific readers, I cannot say that this is at all so with respect to such few men as yourself.— I have not a criticism to make for I object to not a word; & I admire all, so that I cannot pick out one part as better than the rest. It is all so well balanced.—But it is impossible not to be struck with your extent of knowledge in geology, Botany, & zoology. The extracts which you give from Hooker seem to me *excellently* chosen & most forcible.[3] I am so much pleased in what you say, also, about Lyell.—[4]

In fact I am in a fit of enthusiaism & had better write no more. With cordial thanks | Yours very sincerely | C. Darwin

Down House (MS 6: 5)

[1] Dated by the reference to Carpenter's review of *Origin* (see n. 2, below).
[2] [Carpenter] 1860b. There is a heavily annotated copy of the review in the Darwin Pamphlet Collection –CUL. At the head, CD wrote in pencil: '(no new facts to me)'.
[3] Carpenter also reviewed Hooker 1853 and 1859 in the same article. See [Carpenter] 1860b, pp. 383–94.
[4] [Carpenter] 1860b, p. 399, mentioned 'the mastery of Sir Charles Lyell's admirable reasonings upon the subject, in that grand work on the 'Principles of Geology''.

From J. S. Henslow 7 April 1860

HRS.[1]
7 April 1860

My dear Darwin,

I send you all I can well manage—a mere sketch of what you might expect to find *if* you think the specimen sufficiently interesting to have it forwarded—[2]

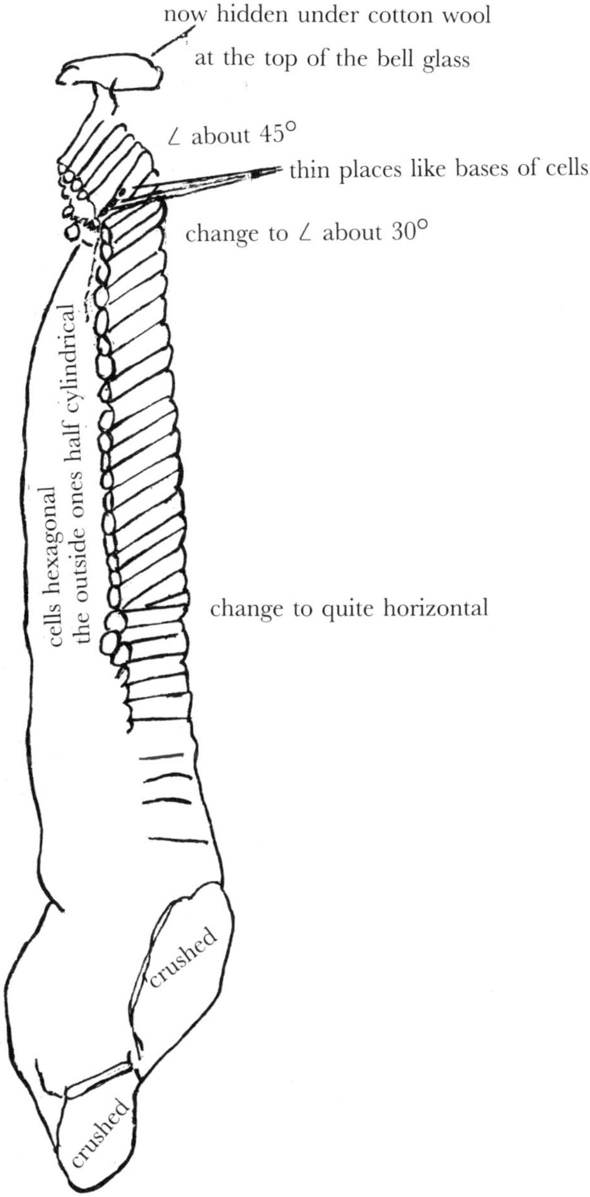

It is suspended from inside of a bell glass by liquid glue & cotton wool—I can easily detach it by Alcohol— The edges are somewhat crushed here & there, & altogether it is beyond my opportunities to examine it so thoroughly as to give you any very precise details— I find it came from *Cuba* not Mexico, as I told you. The material resembles that of the nests of V. vulgaris & V. holsatica—*firm* fibre —not the *herbaceous* materials which become so rotten in the nests of V. crabro & V. rufa—& look as if these were composed of rotten wood— Perhaps they may be *in part*—but certainly not entirely, for I have found the material, when freshly laid on, quite green in some cases—

My Curate is away for a fortnight, & I have my hands full—but I did not like to defer this scanty account any longer— If you prefer propounding *Questions*, to seeing the specimen I will endeavour to answer them—but not feelin⟨g sure⟩ of what you *exactly* want I am at a loss—

Ever Y^rs affec, | J S Henslow

DAR 166.1:180

CD ANNOTATIONS
Diagram: 'Speaks of cells vertical & horizontal' *pencil*

CD note:[3]
Ap. 23^d. 1860[4] Mr Smith showed me a supposed paper wasp nest from Siam also vertical—with cells at top inclined downwards at about 45° & then horizontal; but oddly at lower end of several narrow strips of comb they bend inwards & become horizontal—so that cells at lower end are vertical.— Some of cells at outside are quite cylindrical.— Mr Smith showed me wasps nest with common worker & Queen cells with 4 rows of intermedial gradation—with side-walls elongated, which looks as if distance of building was increased.—**Great** difference in size of **Queen** & Worker cells.—

Waterhouse remarked Queen cell normal form—and not lower half $\frac{1}{2}$ sphere, for this would suffice for all to work

[1] The abbreviation stands for Hitcham Rectory, Suffolk, where Henslow lived.
[2] In the letter to J. S. Henslow, 2 April [1860], CD had requested information about the form of cells in a wasps' nest in Henslow's possession. The diagram has been reduced: the original measures 12 inches and was drawn in ink over pencil.
[3] The note was written on the verso of Henslow's letter.
[4] CD was in London from 21 to 24 April 1860 (Emma Darwin's diary). He apparently visited Frederick Smith and George Robert Waterhouse at the British Museum, where both were employed.

To Maxwell Tylden Masters 7 April [1860][1]

Down Bromley Kent
Ap. 7th

Dear Sir

I hope that you will excuse the liberty which I take in writing to you & begging a favour.— I have been very much interested by the abstract (too brief) of your

Lecture at Royal Institution.[2] Many of the facts alluded to are full of interest for me. But on one point I sh^{d} be infinitely obliged if you could procure me any information; namely with respect to Sweet Peas.[3] I am a great believer in the natural crossing of individuals of the *same* species. But I have been assured by Mr Cattell of Westerham, that the several vars. of Sweet Pea can be raised close together for a number of years without intercrossing.[4] But on other hand he stated that they go over the beds, & pull up any false plant which they very naturally attribute to wrong seeds getting mixed in the lot.— After many failures I succeeded in artificially crossing two vars. & the offspring out of the same pod, instead of being intermediate, were very nearly like the two *pure* parents; yet in one, there was a trace of the cross & these crossed peas in the next generation showed still more plainly their mongrel origin.—

Now what I want to know whether there is much variation in Sweet Peas which might be owing to natural crosses. What I sh^{d} expect would be that they would keep true for many years, but that occasionally, perhaps at long intervals, there would be a considerable amount of crossing of the varieties grown close together. Can you give or obtain from your Father any information on this head & allow me to quote your authority?[5] It would really be a very great favour & kindness.—

I hope that you will forgive me asking you thus to oblige me, & believe me dear Sir | Yours very faithfully | C. Darwin

P.S. | As it seems you have read my "Origin", I need not say that, if you have leisure, how infinitely I sh^{d} be obliged for any facts bearing on any of the points discussed.—[6]

New York Public Library (Berg collection; Astor, Lenox and Tilden Foundations)

[1] Dated by the reference to Masters's lecture at the Royal Institution (see n. 2, below).

[2] Masters was a lecturer in botany at St George's Hospital, London. He read a paper on the relation between normal and abnormal formations in plants on 16 March 1860 at the Royal Institution (Masters 1860). There is an annotated copy of the paper in the Darwin Pamphlet Collection–CUL. In the lecture, Masters specifically addressed CD's views in relation to the question of whether variability could be considered a natural phenomenon in plants.

[3] In the conclusion of his lecture, Masters stated, on the authority of his father William Masters, that white flowering sweet peas seldom, if ever, vary. He went on to say that 'in proportion as the flower becomes darker in colour, so is the liability to vary greater: and these changes are not confined to the colour merely, but affect the pods and other organs.' (Masters 1860, p. 227). The passage is marked in CD's copy of the paper.

[4] John Cattell was a nurseryman in Westerham, Kent, from whom CD frequently ordered seeds. CD cited Cattell's information about sweet peas (*Lathyrus odoratus*) in the manuscript of his 'big species book' (*Natural selection*, p. 70).

[5] Masters's father, William Masters, was a nurseryman in Canterbury. CD had made a similar inquiry to William Masters in 1850 (*Correspondence* vol. 4, letter from William Masters, 3 April 1850) and cited the information in *Natural selection*, p. 70. Additional information provided by William Masters is given in *Variation* 1: 329 and 2: 20.

[6] See letter from William Masters, [after 7 April 1860], and letter to M. T. Masters, 13 April [1860].

To Albert Way 7 April [1860]

Down Bromley Kent
April 7th

My dear Way

Will you forgive me for troubling you with a question, which I presume it is extremely unlikely that you can answer; but I imagine you would be more likely to answer it than anyone.—[1] I am anxious to ascertain what little I can of the history of some of our most strongly marked breeds of domestic animals; & it has occurred to me as possible that archæologists may know when our gigantic dray or waggon Horses were first recorded or noticed. There might be old drawings of heavy carts with horses not so heavy as our present ones or some such evidence, that *heavy* cart-horses did not exist at a given date.— I have read that the old knights in armour rode very powerful horses, but I presume they were not waggon-horses.

Do you know anyone, whom you meet at the archæological Societies, who could throw any light on this point, if it has never occurred to you?[2] Would it be of any use to put a question in Notes & Queries?—though that is a periodical I have no means of seeing.—[3] If you cannot aid me, do not trouble yourself by answering this; as I write it on the bare chance of your being able to give me some information.

It is a very long time since we met.— Eheu Eheu, the old Crux Major days are long past.[4] I sincerely hope that you are well in health.

Believe me Dear Way | Yours very sincerely | C. Darwin

American Philosophical Society (205)

[1] Way was an antiquary and the founder of the Archaeological Institute. He had been a close friend of CD's when they were both undergraduates at Cambridge University. See *Correspondence* vol. 1.

[2] Way was apparently unable to answer CD's query. The chapter on the horse in *Variation* has no information on the first appearance of the dray horse in England.

[3] The first issue of *Notes and Queries*, 3 November 1849, included the statement that 'some cheap and frequent means for the interchange of thought is certainly wanted by those who are engaged in literature, art, and science, and we only hope to persuade the best men in all, that we offer them the best medium of communication with each other.'

[4] Way and CD's cousin William Darwin Fox had introduced him to insect collecting at Cambridge (see *Correspondence* vol. 1). CD often mentioned the beetle *Panageus crux-major* when he recalled his undergraduate entomological expeditions. See also *Autobiography*, pp. 62–4.

From William Masters [after 7 April 1860][1]

Facts and inferences obtained from the observation of the varieties of the Sweet Pea under culture for more than a quarter of a century—

—The following varieties of the Sweet Pea have been largely cultivated by me— viz —White *Painted Lady*—Scarlet—*Purple*—Black and Striped or Powdered (Blotched)

During more than 20 years the *White* never changed at all—

The *Painted Lady* never became *white* or indeed presented any great amount of change—but when a change *was* manifested it was as an advance in color—the *pink or scarlet variety* as it is sometimes called, being produced: And here a marked change occurred—the habit of the plant was altered and the form of the "alæ" or lateral petals was slightly changed—The advance having begun; transition now became more frequent, the general tendency still being shown in the increasing depth of color in the petals so that the *purple* variety was produced. This occurred so frequently that annual selection became necessary to keep the two sorts (pink and purple) distinct—[2]

The *purple and black* varieties seem to bring up the series—they are lighter or darker varieties of the same form and the black so liable to revert to purple that I ceased to grow them as distinct kinds—

In no one instance do I remember to have seen a *white* or *Painted Lady* become purple but the pink or scarlet variety presented this change by no means infrequently.

The other variety—the *striped or powdered*, I obtained from France— Here the flowers instead of having their colors distinct had purple blotches on a grey ground.— This was the most difficult variety of any to keep "true" its tendency being to return to the black variety, but I have no doubt but that by selection, in a few generations, I should have had both pure white and pure black from the same stock—[3]

From these observations I infer that all Sweet Peas were originally white, and as variation is induced so the vigor of the Variety becomes manifest in the intensity of its color, hence the natural deepening of color till we arrive at the last—the striped or powdered variety.— This latter I believe to be a cross between the white and the deep purple and hence its deviation from the other kinds enumerated, all of which I believe to be natural variations produced without hybridization—

That "Natural Selection" is the cause of such general uniformity I cannot doubt for each variety is fertilized by its own pollen although surrounded on all sides by the flowers of the other varieties.

In the whole course of these observations not one variation was produced worthy of selection as constituting a new *sub* tribe

William Masters

DAR 77: 39–40

CD ANNOTATIONS

1.2 *Painted Lady*] *underl brown crayon*

1.2 *Purple*] *underl brown crayon*

3.3 *pink or scarlet variety*] *underl brown crayon*

3.6 shown . . . produced. 3.7] *scored brown crayon*

6.3 This was . . . "true"] *scored brown crayon*

[1] Dated by the relationship to the letters to M. T. Masters, 7 April [1860] and 13 April [1860]. The draft of CD's reply to this letter is headed 'April 13 1860 to M[r] Masters' (DAR 77: 27). See letter to M. T. Masters, 13 April [1860] and CD note.

[2] The letter is a response to a query posed by CD in his letter to M. T. Masters, 7 April [1860].

[3] See letter to M. T. Masters, 13 April [1860] and CD note.

To T. H. Huxley 9 April [1860][1]

Down Bromley Kent
Ap. 9th

My dear Huxley

It is hardly worth troubling you, but I shd like just to tell you that I have looked at Owen's "Invertebrata" 1st Edit, published before my volume on cirripedes,[2] & he admits that the so-called Branchiæ in Balanidæ are respiratory;[3] & I feel almost sure that they are so denominated in case of Coronula or Tubicinella in Cat. of Coll. of Surgeons, on I think, Hunter's belief.—[4]

I never saw such an amount of misrepresentation.[5] At p 530 he says we are called on to accept the hypothesis on the plea of ignorance, whereas I think I could not have made it clearer that I admit the imperfection of geological record as a great difficulty.—

The quotation at p. 512 of Review about "young & rising naturalists with plastic minds", attributed to "nature of limbs" is a false quotation, as I do not use words "plastic minds" At p. 501 the quotation is garbled, for I only ask whether naturalists believe about elemental atoms flashing &c, & he changes it into that I state that they do believe.

At p. 500 It is very false to say that I imply by "blindness of preconceived opinion" the simple belief of creation.— And so on in other cases.—But I beg pardon for troubling you.—I am heartily sorry that in your unselfish endeavours to spread what you believe to be truth, you shd have incurred so brutal an attack.[6] And now I will not think any more of this false & malignant attack

Ever yours | C. Darwin

I most thoroughily enjoyed your visit[7] & have had a start which will last me for some good time.—

Of course this requires no answer.—

Imperial College of Science, Technology and Medicine Archives (Huxley 5: 111)

[1] Dated by the reference to Huxley's recent visit to Down (see n. 7, below).

[2] *Living Cirripedia* (1854).

[3] In R. Owen 1843, p. 158, Richard Owen stated: 'The ordinal distinction between the pedunculated and sessile Cirripedes is not less strongly manifested by their outward forms than by the branchial organs, which, in the Balanoids, consist of two or more broad, transversly plicated, vascular membranes, attached to the inner surface of the mantle.' CD's copy of the second edition of this work is in the Darwin Library–CUL.

[4] Owen had compiled a catalogue of specimens held in the Royal College of Surgeons's museum, which was formed around the collection of the anatomist John Hunter. The two cirripede genera mentioned by CD are described in [R. Owen] 1833–40, 2: 73–4.

[5] CD refers to Owen's review of *Origin* ([R. Owen] 1860a), in which Owen criticised CD's understanding of the anatomy of the branchia of Cirripedia and attacked him on other points. According to Owen, CD had not demonstrated the relation of the folded membranes (branchiae) in Balanidae with the heart or vascular system, 'which could alone prove the respiratory function of such membranes' ([R. Owen] 1860a, p. 489). There is an annotated copy of the review in the Darwin Pamphlet Collection –CUL. In the margin beside these passages, CD wrote in pencil: 'in Owen lecture 1st Edition, p. 158 these organs are admitted to be Branchiæ & I am almost certain are so admitted by Hunter.' CD had

stated that this organ in sessile cirripedes was homologous with the ovigerous frena in pedunculated forms and pointed to this case as a possible illustration of the transition in function gradually assumed by an organ (*Origin*, pp. 191–2).

[6] Owen's review included a scathing attack on Huxley's Royal Institution lecture of 10 February 1860 in which Huxley discussed CD's theory (T. H. Huxley 1860a).

[7] Huxley and his wife Henrietta Anne Huxley visited Down on 7 April 1860 (Emma Darwin's diary).

To William Marshall 9 April [1860][1]

Down, Bromley, Kent.
April 9th.

Dear Sir

My old friend Prof. Henslow has told me that I might use his name in introducing myself to you, and in venturing to ask from you a little information with respect to the Anacharis if you have the leisure and the inclination to give it.[2] Of course I do not by any means expect full answers or information, but any at all on any of the following points, would greatly oblige me.

(1) Is it true that since the Anacharis spread so largely over the waters of your country, that it has become reduced in frequency.[3] If this be so can you give me any dates of maximum increase and subsequent decrease.

(2). Do water-molluscs or other animals feed on it? or can you in any way account for its decrease, if it be true that it has decreased.

(3) Do you know in the places where it now or formerly most abounded, whether it took the place of, and reduced the numbers of, or *locally* extirpated any other kinds of plants; and what plants?

(4) In those cases in which the Anacharis has travelled up non-navigable brooks, have you any idea how it has been transported up the stream?

If you give me any information, will you permit me to publish it at some future time, with due caution, on your authority.[4] Hoping that you will excuse the liberty which I have taken, I remain, Dear Sir | Yours faithfully | C. Darwin

Copy
DAR 146

[1] Dated by CD's interest in the plant *Anacharis* (see n. 2, below).

[2] See letter to J. S. Henslow, 2 April [1860]. Marshall had written two notices giving information about the introduction and geographical spread of the water-weed *Anacharis* (Marshall 1852 and 1857–8).

[3] CD gave Charles Cardale Babington as the source of this information in the letter to J. S. Henslow, 2 April [1860].

[4] No reply from Marshall has been found.

To John Murray 9 April [1860]

Down Bromley Kent
Ap. 9th

My dear Sir

Will you be so good as to send a copy of the Origin addressed as follows, marked from Author

Prof. Kölliker[1] &c &c
care of Prof. Huxley
Museum of Geology
Jermyn Str—

Have you seen the Edinburgh Review?[2] I am thrashed in every possible way to the full content of my bitterest opposers. The article is very venemous, & is manifestly by Owen. I wish for auld lang syne's sake he had been a little less bitter.—

My dear Sir | Yours very sincerely | C. Darwin

Sedgwick has been very fierce in the Spectator; but his blows are fair & open.[3] There is some misrepresentation, but no doubt unintentional. Upon the whole I have as yet escaped pretty well.—

Please also to send a copy of Origin to M^essrs^ Williams & Norgate, with enclosed memorandum.—[4] Marking copy "from Author"

Endorsement: 'April 9 1860'
John Murray (publishers)

[1] CD had asked Thomas Henry Huxley to write to Rudolf Albert von Kölliker inquiring about a possible German translation of *Origin*. See *Correspondence* vol. 7, letter to T. H. Huxley, [9 December 1859], and this vol., letter to T. H. Huxley, 4 February [1860].

[2] [R. Owen] 1860a.

[3] [Sedgwick] 1860.

[4] The memorandum has not been found. CD used the booksellers and publishers Williams and Norgate to send presentation copies of *Origin* to foreign naturalists. See *Correspondence* vol. 7, letter to J. D. Hooker, [27 October or 3 November 1859].

To H. G. Bronn 10 April [1860][1]

Down. | Bromley. Kent.
Ap. 10th

My dear Sir.

I received this morning 4 Copies of the translation[2] and I must trouble you with one line to say how much pleased I am with their appearance.

I have read some pages and my sense seems very clearly given; for poor German Scholar as I am, I could read it with some facility—

M Schweizerbart has sent me a copy of your Geschichte;[3] but in thanking him I have not thought it necessary to tell him that I already possessed a copy—; but the copy shall not be wasted for I will give it to the Library of the Royal or Linnean or some other Society which I may find does not possess a copy—The work on the development of organisms[4] which you so kindly said you would give me was not in the parcel which contained only the translation & the Geschichte—[5]

Once again permit me to return you my heartfelt thanks for the undertaking of the translation I feel it almost the greatest honour ever paid me; & what is of

Richard Owen. Photograph by Maull & Polyblank, 1859.
(From Edward Walford, ed. *Photographic portraits of living celebrities*. London, 1859.
By permission of the Syndics of Cambridge University Library.)

William Benjamin Carpenter.
Photograph by Ernest Edwards, 1864.
(From Lovell Reeve, ed. *Portraits of men of eminence in literature, science, and art*. London, 1863–1864. By permission of the Syndics of Cambridge University Library.)

Thomas Vernon Wollaston
(Courtesy of the Library of the New York Botanical Gardens, Bronx, New York.)

more consequence your well known name will almost ensure the views contained in my work being fairly considered, & that is all that I can wish for—

With sincere respect & gratitude, | I remain, | My dear Sir, | Yours truly obliged, | Ch Darwin.

Prof. Pictet has published a most candid & fair Review in the Bib. Univers of Geneva—[6] I agree to every word which he says, Our only difference is that he attaches less weight to my argument in favour of modification; & I attach less weight to the *great* difficulties.

The German Circular or Advertisement does me much honour.

Copy
DAR 143

[1] Dated by the reference to the German translation of *Origin.*

[2] Bronn sent CD the first part (pp. 1–160) of the German translation of *Origin* (Bronn trans. 1860), published on 4 April (*Börsenblatt für den Deutschen Buchhandel* 27 (1860): 683). CD's copy is in the Darwin Library–CUL, but it is imperfect. The last 56 pages of the book, comprising most of the final chapter of *Origin* and all of Bronn's epilogue, are missing. The pages of the second and third parts and some of those in the first are uncut. According to the list of presentation copies of *Origin* (Appendix III), CD sent the remaining three copies of the German edition to Thomas Henry Huxley, Jan van der Hoeven, and Jakob Moleschott.

[3] Bronn 1841–9. CD had read (and annotated) volume 2 in 1846 (*Correspondence* vol. 4, Appendix IV, 119: 16a). His copy of the work is in the Darwin Library– CUL.

[4] Bronn 1858b.

[5] The parcel had come from E. Schweizerbart of Stuttgart, the publisher of Bronn 1841–9, Bronn 1858b, and the German translation of *Origin.*

[6] Pictet de la Rive 1860.

To Charles Lyell 10 April [1860][1]

Down Bromley Kent.
Ap. 10th

My dear Lyell

Thank you much for your note of the 4th.[2] I am very glad to hear that you are at Torquay. I shd. have amused myself earlier by writing to you; but I have had Hooker & Huxley staying here, & they have fully occupied my time;[3] as a little of anything is a full dose for me.— I have got "The Future", but cannot clearly make head or tail of it.—[4] There has been a plethora of Reviews, & I am really quite sick of myself.— There is very long Review by Carpenter in Med.-Chirurg. Review:[5] *very good* & well-balanced but not brilliant. He discusses Hookers book at as great length as mine, & makes *excellent* extracts; but I could not get Hooker to feel the least interest in being praised.

Carpenter speaks of you in thoroughily proper terms.[6] There is a *brilliant* review by Huxley, with capital hits;[7] but I do not know that he much advances subject: I *think* I have convinced him that he has hardly allowed weight enough to the cases of varieties of plants being in some degree sterile.—[8]

To diverge from Reviews Asa Gray sends me from Wyman (who will write) a good case of all the pigs being black in the Everglades of Virginia;[9] on asking about cause, it seems (I have got capital analogous cases) that when the *black* pigs eat a certain nut, their bones become red & they suffer to certain extent, but that the *white* pigs lose their hoofs & perish; "& we aid by *selection* for we kill most of the young white pigs" This was said by man who could hardly read.—

By the way it is a great blow to me that you cannot admit to potency of natural selection; the more I think of it, the less I doubt its power for great & small changes.—

I have just read the Edinburgh, which without doubt is by Owen.[10] It is extremely malignant, clever & I fear will be very damaging. He is atrociously severe on Huxley's lecture, & very bitter against Hooker. So we three *enjoyed* it together: not that I really enjoyed it, for it made me uncomfortable for one night; but I have got quite over it today. It requires much study to appreciate all the bitter spite of many of the remarks against me; indeed I did not discover all myself.— It scandalously misrepresents many parts. He misquotes some passages altering words within inverted commas. Makes me say that the dorsal vertebræ of pigeons vary & refers to page where the word dorsal does not appear. Sneers at my saying a certain organ is the branchiæ of Balanidæ; whilst in his own "Invertebrata" published before I published on cirripedes, he calls them organs without doubt branchiæ.—[11]

It is painful to be hated in the intense degree with which Owen hates me.—

Now for a curious thing about my Book, & then I have done. In last Saturday Gardeners' Chronicle, a Mr Patrick Matthews publishes long extract from his work on "Naval Timber & Arboriculture" published in 1831, in which he briefly but completely anticipates the theory of Nat. Selection.—[12] I have ordered the Book, as some few passages are rather obscure but it, is certainly, I think, a complete but not developed anticipation![13] Erasmus always said that surely this would be shown to be the case someday. Anyhow one may be excused in not having discovered the fact in a work on "Naval Timber".

I heartily hope that your Torquay work may be successful.[14] Give my kindest remembrances to Falconer, & I hope he is pretty well.— Hooker & Huxley (with Mrs. Huxley) were extremely pleasant, but poor dear Hooker is tired to death of my Book; & it is a marvel & prodigy if you are not worse tired, if that be possible.

Farewell my dear Lyell | Yours affect | C. Darwin

How hard it is to please everyone. I told Jukes that I sh^d.^ leave out in any next Edition about the Weald, & he demurred greatly & said "he almost fancied he had written himself", as he bravely told Murchison.—[15]

American Philosophical Society (206)

[1] Dated by the reference to William Benjamin Carpenter's review of *Origin* ([Carpenter] 1860b).

[2] Lyell's letter has not been found.

[3] Joseph Dalton Hooker, Thomas Henry Huxley, and Henrietta Anne Huxley visited Down over the weekend of 7 and 8 April 1860 (Emma Darwin's diary).

[4] The *Future* was a London journal of 'philosophical research and criticism'. The first number, issued in April 1860, included an anonymous review of *Origin.* There is a copy in the Darwin Pamphlet Collection–CUL.

[5] [Carpenter] 1860b. See letter to W. B. Carpenter, 6 April [1860].

[6] See letter to W. B. Carpenter, 6 April [1860].

[7] [T. H. Huxley] 1860b.

[8] Thomas Henry Huxley maintained that CD's theory could not be fully acknowledged until it was shown how sterility could arise between interbreeding hybrids (T. H. Huxley 1860a, p. 198).

[9] Asa Gray's letter is missing, but see the letter to Asa Gray, 3 April [1860]. Jeffries Wyman wrote to CD later in the year (letter from Jeffries Wyman, [*c.* 15] September 1860).

[10] [R. Owen] 1860a.

[11] See letter to T. H. Huxley, 9 April [1860].

[12] Patrick Matthew's article was published in the *Gardeners' Chronicle and Agricultural Gazette*, 7 April 1860, pp. 312–13. It included long extracts from Matthew 1831. See Appendix V.

[13] CD's annotated copy of Matthew 1831, inscribed 'Ap. 13th 1860', is in the Darwin Library–CUL.

[14] Brixham Cave, near Torquay, was being excavated by Hugh Falconer and William Pengelly. Lyell had joined the group to inspect the fossil remains and flint implements they had uncovered. Lyell described the results of the excavations in C. Lyell 1863, pp. 96–103. For an account of Lyell's work on the antiquity of man, see Bynum 1984.

[15] The letter has not been found, but see the letter to Asa Gray, 3 April [1860]. CD refers to Joseph Beete Jukes and Roderick Impey Murchison.

To T. H. Huxley 11 April [1860][1]

Down Bromley Kent
Ap. 11th

My dear Huxley

Hooker sent me this morning your Lecture, at R. In. & I have just read it. I must have the pleasure of telling you that I think the whole conclusion one of the most eloquent productions which I ever read in my life.—[2]

In my last note I became so full of Owen, that I quite forgot what I meant to have written about! viz that I have told Murray to send to you for Kölliker a copy of the Origin, if you will give it to him with my very sincere respects.—[3]

Since then I have forwarded to you by Post the first part of the German Translation.[4]

Adios | C. Darwin

Imperial College of Science, Technology and Medicine Archives (Huxley 5: 113)

[1] Dated by the reference to Huxley's lecture at the Royal Institution on 10 February 1860.

[2] T. H. Huxley 1860a. There is an annotated copy in the Darwin Pamphlet Collection–CUL.

[3] See letters to T. H. Huxley, 9 April [1860], and to John Murray, 9 April [1860].

[4] CD refers to Heinrich Georg Bronn's German translation of *Origin* (Bronn trans. 1860). CD had received copies of the first part on 10 April (see letter to H. G. Bronn, 10 April [1860]).

To H. G. Bronn 13 April [1860][1]

Down Bromley Kent
Ap. 13

Dear & Honoured Sir

I write in haste to catch the Post just to say that I received this morning your "Untersuchungen uber die Entwickelung"[2] & **two** copies of your "Morphologische Studien"[3] (one of which I will give to some good man)[4] & likewise your Portrait.

For all these *most* kind gifts I most heartily thank you.—

With sincere respect | Yours very truly obliged | Ch. Darwin

American Philosophical Society (207)

[1] Dated by the relationship to the letter to H. G. Bronn, 10 April [1860].
[2] Bronn 1858b. See letter to H. G. Bronn, 10 April [1860]. There is a copy in the Darwin Library–CUL.
[3] Bronn 1858a. See letter to H. G. Bronn, 21 March [1860]. There is only one copy of the work in the Darwin Library–CUL.
[4] See letter to T. H. Huxley, 14 April [1860].

To *Gardeners' Chronicle* [13 April 1860][1]

I have been much interested by Mr. Patrick Matthew's communication in the Number of your Paper, dated April 7th.[2] I freely acknowledge that Mr. Matthew has anticipated by many years the explanation which I have offered of the origin of species, under the name of natural selection. I think that no one will feel surprised that neither I, nor apparently any other naturalist, had heard of Mr. Matthew's views, considering how briefly they are given, and that they appeared in the appendix to a work on Naval Timber and Arboriculture. I can do no more than offer my apologies to Mr. Matthew for my entire ignorance of his publication. If another edition of my work is called for, I will insert a notice to the foregoing effect.[3]

Charles Darwin, Down, Bromley, Kent.

Gardeners' Chronicle and Agricultural Gazette, 21 April 1860, pp. 362–3

[1] The date is given by the relationship to the following letter. The letter was not published in the *Gardeners' Chronicle and Agricultural Gazette* until 21 April 1860.
[2] *Gardeners' Chronicle and Agricultural Gazette*, 7 April 1860, pp. 312–13. See following letter and letter to Charles Lyell, 10 April [1860]. For a transcription of Patrick Matthew's letter, see Appendix V.
[3] CD discussed Matthew's work in the augmented historical sketch included in the third edition of *Origin* (pp. xiv–xv).

To J. D. Hooker 13 [April 1860]

Down Bromley Kent
13th

My dear Hooker

Questions of priority so often lead to odious quarrels, that I sh.d esteem it a great favour if you would read enclosed.[1] If you think it proper that I sh.d send it (& of this there can hardly be question) & if you think it full & ample enough, please alter date to day on which you post it & let that be soon.— The case in G. Chronicle seems a *little* stronger than in Mr Matthews book, for the passages are therein scattered in 3 places.[2] But it would be mere hair-splitting to notice that.—If you object to my letter please return it; but I do not expect that you will, but I thought that you would not object to run your eye over it.— My dear Hooker it is a great thing for me to have so good, true, & old a friend as you. I owe much to science for my friends.—

Many thanks for Huxley's lecture: the latter part seemed to be grandly eloquent.[3] Thanks for your pleasant little note.— I am very glad that you are so much pleased with your Boy Willy,— that is worth everything.—[4]

I have had most obliging & kind communication from Masters. His Father sends me some facts which partly remove a weight about crossing which has pressed me down for years.—[5]

Ever my dear friend | Yours affect | C. Darwin

I have had nice long letter from Lyell.—[6]

PS. I have gone over Owen's Review again & compared passages, & I am astonished at the baseness of the misreprentations. But I am glad I resolved not to answer. Perhaps it is selfish—But to answer & think more on subject is too unpleasant. I am so sorry that Huxley by my means has been thus atrociously attacked.—[7] I do not suppose you much care about the gratuitous attack on you.—

Lyell in his letter remarked that you seemed to him as if you were overworked. Do pray be cautious, & remember how many & many a man has done this, who thought it absurd till too late. I have often thought the same,—you know that you were bad enough before your Indian Journey.—[8]

Endorsement: 'April/60'
DAR 115.1: 48

[1] See preceding letter.

[2] The sections that made up Patrick Matthew's letter to the *Gardeners' Chronicle* were drawn from Matthew 1831, pp. 364–5, 381–8, and 106–8. See Appendix V.

[3] CD refers to T. H. Huxley 1860a. See letter to T. H. Huxley, 11 April [1860].

[4] William Henslow Hooker was Hooker's oldest son, born in 1853.

[5] See letter from M. T. Masters, [after 7 April 1860], and the following letter.

[6] Charles Lyell's letter has not been found, but see CD's reply (letter to Charles Lyell, 10 April [1860]).

[7] Richard Owen attacked Thomas Henry Huxley's lecture on CD's theory given at the Royal Institution (T. H. Huxley 1860a) in his review of *Origin* ([R. Owen] 1860a).

[8] Hooker had travelled in India and the Himalayas from 1847 to 1851. He had been unwell before setting off. See *Correspondence* vol. 4.

To M. T. Masters 13 April [1860][1]

Down Bromley Kent
Ap. 13th.

My dear Sir

I thank you very sincerely for your two kind notes. The next time you write to your Father I beg you to give him from me my best thanks, but I am sorry that he should have had the trouble of writing when ill. I have been much interested by the facts given by him.[2] If you think he would in the least care to hear the result of an artificial cross of two Sweet Peas; you can send the enclosed;[3] if it will only trouble him, tear it up. There seems to me so much parallelism in the kind of variation from my experiment which was certainly a cross and what Mr Masters has observed, that I cannot help suspecting that his Peas were crossed by Bees, which I have seen well dusted with the pollen of the Sweet Pea; but then I wish this, and how hard it is to prevent one's wish biassing one's judgment!

I was struck with your remark about the Compositæ &c.—[4] I do not see that it bears much against me, and whether it does or not is of course of not the slightest importance. Although I fully agree that no definition can be drawn between monstrosities and slight variations (such as my theory requires) yet I suspect there is some distinction. Some facts lead me to think that monstrosities supervene generally at an early age; and after attending to the subject I have great doubts whether species in a state of nature ever become modified by such sudden jumps, as would result from the natural selection of monstrosities. You cannot do me a greater service than by pointing out errors. I sincerely hope that your work on monstrosities will soon appear, for I am sure it will be highly instructive.—[5]

Now for your notes, for which let me again thank you.—[6]

(1) Your conclusion about parts developed &c. &c: not being extra variable agrees with Hooker's: You will see that I have stated that the rule apparently does not hold with plants, though it *ought*, if true, to hold good with them.[7]

(2) I cannot now remember in what work I saw the statement about Peloria affecting the axis;[8] but I know it was one which I thought might be trusted. I consulted also Dr Falconer[9] and I think that he agreed to the truth of it; but I cannot now tell where to look for my notes.— I had been much struck with finding a **Laburnum** tree with the terminal flowers alone in each raceme peloric though not perfectly regular.[10] The *Pelargonium* case in the "Origin" seems to point in the same direction.[11]

(3) Thanks for the correction about furze; I found the seedlings just sprouting, and was so much surprised at their appearance that I sent them to Hooker; but I never plainly asked myself whether they were cotyledons or first leaves.[12]

(4) That is a curious fact about the seeds of the furze,—the more curious as I found with some Leguminosæ that immersion in plain cold water for a very few days killed some kinds.—[13]

If at any time anything should occur to you illustrating or opposing my notions, and you have leisure to inform me, I should be truly grateful, for I can plainly see that you have wealth of knowledge.

My dear Sir | Yours truly obliged | C. Darwin

With respect to advancement or retrogression in organisation in monstrosities of the Compositæ &c. do you not find it very difficult to define which is which.

Anyhow most Botanists seem to differ as widely as possible on this head.

Copy
DAR 146

CD note:[14]
April 13 1860 to Mr Masters
[*del* 'The'] I know the 6 kinds of Sweet Peas named by Mr Masters. I castrated 2 flowers of the *Purple* & fertilised with pollen of *Painted Lady*: one pod produced seedlings [*interl*] *all* closely [*interl*] like the Painted Lady; the other had produced some [*above del* 'one' *pencil*] plants ['s' *added pencil*] not distinguished from the mother or Purple, & [*above del* 'all'] the other peas *in the* [*interl*] *same pod* produced plants like Painted Lady. The early flowers were indistinguishable from *those of [*interl*] the Painted Lady; but later in season the wing-petals became slightly streaked & blotched with *dark or pale [*above del* 'pale purplish'] Pink.—

I raised *many* [*del* 'grandchild'] children from these latter mongrels (ie grandchildren of the Purple ♀ Painted Lady ♂) & one single plant most closely resembled the Purple with a mere [*interl*] *trace* of paler streaks on the [*interl*] patch.[15] All the other *many* plants were *more or less intermediate between the [*interl above del* 'like'] Painted Lady, *& Scarlet in tint, but were all more or less [*interl above del* 'but varying in intensity of colour & more or less blotched'] blotched & streaked [*del* ', but none [*del* 'were'] had any purple on them.—'] with darker or lighter pink, but not with true purple.— see p. 69 for gr. grandchildren

[1] Dated by the relationship to the letter to M. T. Masters, 7 April [1860].
[2] See letter from William Masters, [after 7 April 1860].
[3] The enclosure has not been found, but it was probably a fair copy of a note in DAR 77: 27 (see CD note transcribed following the letter).
[4] Masters had stated that there seemed to be limits to variation, citing as evidence the fact that malformations of complex, specialised plants like the Compositae did not alter or improve the type. In simple plants, however, malformations appeared to make the organism more advanced (Masters 1860, p. 226).
[5] Masters 1869. There is a copy in the Darwin Library–CUL.
[6] The notes from M. T. Masters have not been found.
[7] *Origin*, p. 150.
[8] *Origin*, p. 145. CD cited Moquin-Tandon 1841 on this point in *Variation*. There is a copy of the work in the Darwin Library–CUL.
[9] Hugh Falconer.
[10] See *Correspondence* vol. 4, letters to J. D. Hooker, [2 June 1847], [10 June 1847], and [12 June 1847].
[11] *Origin*, p. 145. See also *Correspondence* vol. 7, letter to John Lubbock, [September–October 1858].
[12] In *Origin*, p. 439, CD called the early leaves of the furze 'embryonic leaves'. The sentence was changed in the third edition to read: 'first leaves'.
[13] For CD's work on the immersion of seeds and plants in salt-water, see *Correspondence* vols. 5 and 6. He was experimenting on the ability of plants to be transported over long stretches of ocean. The topic was discussed in *Origin*, pp. 356–65.
[14] The note is in DAR 77: 27. CD also wrote on it in pencil: 'So there may be crossing'.
[15] CD scored this sentence twice in the margin in red crayon and underlined '*trace*' twice in red crayon.

To T. H. Huxley 14 April [1860][1]

Down Bromley Kent
Ap. 14th

My dear Huxley

Many thanks for your kind & pleasant letter. I have been much interested by

Deep Sea Soundings & will return it by this Post or soon as soon as I have copied a few sentences.—[2] I see Dayman thinks the deposit thin & that it was penetrated by the Plunger.[3] It ought to have been stated by him whether any of the fragments of rock seemed as if recently broken off.— I think you said that some one was investigating the Soundings. I *earnestly* hope that you will ask the someone to carefully observe, whether any considerable number of the calcareous organisms are more or less friable, or corroded, or scaling So that one might form some crude notion whether the deposition is so rapid that the foraminifera are preserved from decay & thus are forming strata at this profound depth. This is a subject which seems to me to have been much neglected in examining soundings.—

Bronn has sent me 2 copies of his "Morphologische Studien über die Gestaltungs Gesetze".[4] It looks elementary. If you will write you shall have the copy; if not, I will give it to Linnean Library

Yours Most sincerely | C. Darwin

I quite agree with letter from Lyell that your extinguished Theologians laying about the cradle of each new science &c &c is *splendid*[5]

Imperial College of Science, Technology and Medicine Archives (Huxley 5: 115)

[1] Dated by the reference to [T. H. Huxley] 1860b.

[2] Huxley had contributed an account of the deposits of the sea-bed to a report on an Admiralty survey of deep-sea soundings in the North Atlantic (T. H. Huxley 1858).

[3] Joseph Dayman was lieutenant-commander of the *Cyclops*, commissioned for the survey. He was the author of the Admiralty report.

[4] Bronn 1858a. See letter to H. G. Bronn, 13 April [1860].

[5] The letter from Charles Lyell has not been found. The reference is to a passage in Huxley's anonymous review of *Origin* in the *Westminster Review*: 'Extinguished theologians lie about the cradle of every science as the strangled snakes besides that of Hercules.' ([T. H. Huxley] 1860b, p. 556).

To Charles Lyell 15 April [1860][1]

Down Bromley Kent
Ap. 15th

My dear Lyell

I was very glad to get your nice long letter from Torquay. A press of letters prevented me writing to Wells.—[2] I was particularly glad to hear what you thought about not noticing Owen's Review.[3] Hooker & Huxley thought it a sort of duty to point out alterations of quoted citations; & there is truth in this remark, but I so hated the thought, that I resolved not to do so. I shall come up to London on Saturday 24th for Sir B. Brodie's party,[4] as I have an accumulation of things to do in London, & will (if I do not hear to contrary) call about $\frac{1}{4}$ before 10 on Sunday morning, & sit with you at breakfast, but will not sit long & so take up much of your time.

I must say one more word about our quasi-theological controversy about natural Selection, & let me have your opinion when we meet in London.— Do you consider that the successive variations in size of the crop of the Pouter Pigeon, which man has accumulated to please his caprice, have been due to "the creative & sustaining powers of Brahma".[5] In the sense that an omnipotent & omniscient Deity must order & know everything, this must be admitted; yet in honest truth I can hardly admit it. It seems preposterous that a maker of Universes sh^{d} care about the crop of a Pigeon solely to please men's silly fancies. But if you agree with me in thinking such an interposition of the Deity uncalled for, I can see no reason whatever for believing in such interpositions in the case of natural beings, in which strange & admirable peculiarities have been naturally selected for the creature's own benefit. Imagine a Pouter in a state of nature wading into the water & then being buoyed up by its inflated crop, sailing about in search of food. What admiration this would have excited,—adaptation to laws of hydrostatic pressure &c &c &c.—

For the life of me I cannot see any difficulty in Natural selection producing the most exquisite structure, *if such structure can be arrived at by gradation*; & I know from experience how hard it is to name any structure towards which at least some gradations are not known.—

Ever yours | C. Darwin

The conclusion at which I have come, as I have told Asa Gray,[6] is that such question, as is touched on in this note, is beyond the human intellect, like "predestination & free will" or "the origin of evil".

American Philosophical Society (208)

[1] Dated by CD's reference to a forthcoming trip to London (see n. 4, below).

[2] Probably the town of Wells in Somerset. Lyell may have indicated that he was staying there en route to Torquay.

[3] CD was upset by Richard Owen's misrepresentation in [R. Owen] 1860a of several of his views. See letters to T. H. Huxley, 9 April [1860], and to Charles Lyell, 10 April [1860].

[4] CD was mistaken about the date: Saturday was 21 April 1860. Emma Darwin's diary records that CD went to London on this date. Benjamin Collins Brodie, president of the Royal Society, held a soirée in the society's rooms at Burlington House on 21 April (*Athenæum*, 28 April 1860, p. 584).

[5] This is possibly a repetition of an expression used in conversation with Lyell during his recent visit to Down, 9 to 12 March 1860. See also letter to Charles Lyell, 12 March [1860]. In Lyell's scientific journal, there is a note written late in March headed 'Natural Selection deified', in which Lyell refers to 'the Creator, Brahma' (Wilson ed. 1970, p. 369).

[6] See letter to Asa Gray, 24 February [1860].

To W. B. Tegetmeier 17 April [1860][1]

Down Bromley Kent
Ap. 17th

My dear Sir

As you have so often aided me, will you do so again? The enclosed queries require only "yes" or "no" to be written on the paper & returned to me.—(& the answers will be nearly all "no"[2] I daresay you will meet Fanciers at some club, &

could obtain me answers. Only please do not trust anyone who speaks at random, & only those who have kept the breeds.— Consult your own leisure.—

There is one other point on which I sh^d be particularly grateful for information; it relates to perpetual interbreeding. When the Queen-Bee takes her flight in order to be fertilised, are there always Drone-bees ready in the *same* Hive? Mr Knight thought that the Queen was generally impregnated by the Drones of another Hive; but he gives no satisfactory evidence.—[3]

When a bee-keeper has long kept the same stock of Bees have you ever heard it advised to get a new Hive from some other district, in order to cross the stock?

Some time ago you said you would let me examine your Fowl Skulls. In the course of two or three months I sh^d be very glad to see them.[4] I shall publish nothing on subject for 18 months, or more likely 30 months, as I work so slowly & so sh^d not interfere with anything further than that already published you might wish to publish, & I heartily wish you would publish on the subject.—[5]

Do not hurry yourself to answer this, for I know that you are a busy man.—

My dear Sir | Yours sincerely | C. Darwin

New York Botanical Garden Library (Charles Finney Cox collection)

[1] The year is suggested by CD's reference to preparing his volumes on variation (see n. 4, below).

[2] The queries have not been found.

[3] Knight 1828, p. 319. See letters to W. B. Tegetmeier, 20 July [1860] and 30 July [1860], and to *Cottage Gardener*, [after 8 May 1860].

[4] CD was writing the chapters on pigeons for *Variation*. He finished the manuscript on 10 June. The next chapter discussed variation in fowls. See 'Journal' (Appendix II).

[5] *Variation* was not published until 1868. It appears that Tegetmeier did not publish any notice of the osteology of fowls before this time. In *Variation* 1: 260, CD stated that of the fifty-three fowl skulls he examined, 'nearly half of these skulls I owe to the kindness of Mr. Tegetmeier'.

To J. D. Hooker 18 [April 1860]

Down Bromley Kent
18th

My dear Hooker

Hearty thanks about G. Chronicle.[1] I return Owen's letter:[2] what a base dog he is. Some of my relations say it cannot *possibly* be Owen's article, because the Reviewer speaks so very highly of Prof. Owen.[3] Poor dear simple folk! My clever neighbour Mr Norman,[4] says the article is so badly written with no definite object that no one will read it.—

What a narrow minded man Balfour must be to bring so poor an argument as St Peter, with so many good arguments which he might use.—[5] Asa Gray has sent me an article from U. States; clever & dead against me.[6] But one argument funny. The Reviewer says that if the doctrine were true, geological strata would be full of monsters "tentative attempts", which had failed.— A very clear view this writer had of the Struggle for existence!

I am delighted to hear that you have begun experimentising you will be glad of it hereafter & I daresay will go on largely.—

I am glad you like Adam Bede so much: I was charmed with it.—[7]

We think you must by mistake have taken with your own No^{r} of National R. my previous Number.[8] I wish you would look.—

I shall be curious to see Huxley's answer.—I could have pointed out falsifications of my writing, but am glad I did not.—[9]

I sh^{d} be infinitely obliged if you would look at a flower of the Goodenia, & look at my *splendid* diagram.[10] I am confounded by it; for it seems to me that the stigmatic surface is *outside* indusium above the brush of Hairs on *one* side. Anyhow there were pollen-grains in one flower there immersed in humid secretion; but it is beyond me to know whether they then penetrate the surface. If this is stigmatic surface, what on earth is use of indusium? Do for Heaven-sake look at it.—

C. Darwin

[Enclosure]

Flower *not* nearly open. (Leschenaultia formosa.—)

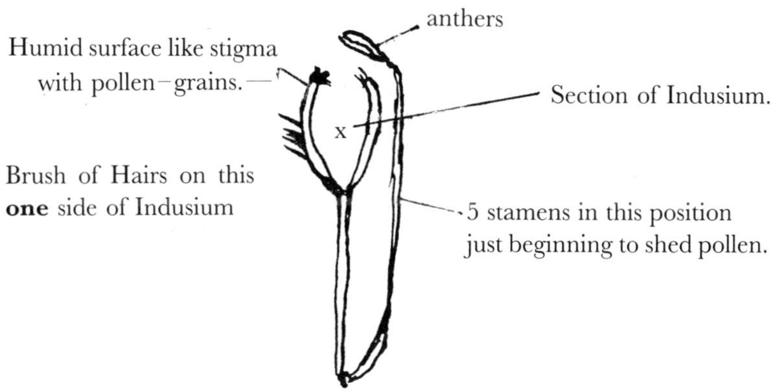

(In a flower more open, the indusium was closed.—)

I must look when flowers more open & see whether pollen gets enclosed within Indusium, as seemed to be case.

My *suspicion* is that the closing of the indusium is merely an action to brush pollen out of anthers.—

Endorsement: 'April /60'
DAR 115.1: 49

[1] CD had sent Hooker the draft of a letter to the *Gardeners' Chronicle* to see whether he thought he should send it to the journal for publication. See letters to *Gardeners' Chronicle*, [13 April 1860], and to J. D. Hooker, 13 [April 1860].

[2] The letter from Richard Owen to Hooker is also mentioned in CD's letter to Asa Gray, 25 April [1860].

[3] Owen reviewed *Origin*, Hooker 1859, and three of his own publications in the *Edinburgh Review* ([R. Owen] 1860a).

[4] George Warde Norman resided at the Rookery in Bromley, Kent.

[5] John Hutton Balfour criticised CD's views in several later publications. Hooker knew that Balfour was strongly against Darwin's views (see L. Huxley ed. 1918, 1: 515).

[6] An unsigned review of *Origin* appeared in the *North American Review* 90 (April 1860): 474–506. The author was Francis Bowen, Alvord professor of natural religion at Harvard University. CD's copy of the review is in the Darwin Pamphlet Collection–CUL. See also letter to Asa Gray, 25 April [1860].

[7] Eliot 1859. For CD's response on reading the work, see *Correspondence* vol. 7, letter to J. D. Hooker, [26 May 1859] and n. 3.

[8] The January number of the *National Review* contained an unsigned review of *Origin* by William Benjamin Carpenter ([Carpenter] 1860a). Hooker had brought the April issue of this periodical when he visited Down (see following letter).

[9] CD refers to Owen's attack on Thomas Henry Huxley in [R. Owen] 1860a, pp. 522–4.

[10] Hooker had sent specimens of *Leschenaultia formosa* for CD's investigation of its pollination mechanism.

From J. D. Hooker [20 April 1860][1]

Kew
Friday.

Dr Darwin

I have no National Review but my own N° XX which I had at Down.[2] I will look at *Goodenia* flowers the first spare moment.[3] I think you will find a similar viscid? surface above the hairs in *Lobelia* & *Campanula*, where also the pollen sticks abundantly, but never penetrates.

Bell told me yesterday that Owen avows the Review,[4] I can hardly believe it.

Your observation about curvature of style reads good, I will bear it in mind— There must of course be a morphological rationale of the coincidence— Curved styles generally imply oblique flowers.— Oblique flowers have generally unilateral nectaries— Unilateral nectaries are usually posticous (or anticous by torsion of pedicel)— From æquilateral flowers with curved styles nothing can be deduced E.G. Rhododendron, for there is no side more advantageous for passage of insect than another as far as form of Corolla is concerned. The curvature is I think in all flowers?? always *towards axis*

Melianthus

Rhodod

5 symmetr.
nects.

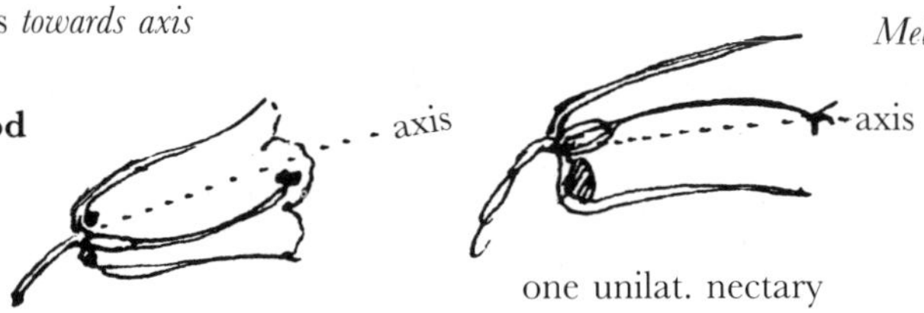

one unilat. nectary

I must clarify my ideas on the subject, which are muddled— It is in leaving the flower that the insect does the trick.— Perhaps I do not quite understand you.

I go to Cambridge tomorrow for 2 days.—I have had an awfully narrow escape lately. HRH. wanted me to lecture the young R^l. family on Botany—at B^m. Palace.[5] I have effected a compromise by shoving the lectures off on Henslow,[6] & saving my credit by offering to teach the children a little practical Botany afterwards. I hope Henslow will not repudiate Owen is lecturing them on Zoology[7] but I cannot endure the thoughts of getting up Lectures.— I shudder at even teaching them with specimens, but think I could scrape through that—To refuse is of course impossible in an Assist: Director of Kew! & I am only too thankful to have jockeyed out of the *Lectures.* Pray say nothing of all this beyond your own house

Ever Yrs affy | Jos D Hooker

I wonder who wrote article on Testimony of Geology in National I like it very much.[8]

DAR 100: 139–40

CD ANNOTATIONS
1.1 I have . . . mind—3.1] *crossed ink*
5.1 I go . . . could 5.7] *crossed ink*
Top of first page: 'Ch. 3'[9] *brown crayon, underl brown crayon*; 'April 20 1860' *ink*

[1] The date is given by CD's annotation.

[2] See preceding letter.

[3] There is a diagram drawn by Hooker of the stigma and style of *Leschenaultia formosa*, with a few accompanying annotations, in CD's Experimental book (DAR 157a). The diagram, dated 19 April 1860, was presumably drawn after Hooker received CD's queries contained in the preceding letter and was possibly sent to CD enclosed with Hooker's letter.

[4] Thomas Bell was president of the Linnean Society. The reference is to Richard Owen's review of *Origin* ([R. Owen] 1860a).

[5] Prince Albert arranged for the royal children to be taught scientific subjects by London experts during the late 1850s. The lecture was to take place at Buckingham Palace, Queen Victoria's London residence.

[6] John Stevens Henslow delivered the botany lecture at Buckingham Palace in June (letter from J. S. Henslow to J. D. Hooker, 5 May 1860, Royal Botanic Gardens, Kew Archive).

[7] Richard Owen gave lectures on natural history to the royal children in April 1860 (R. S. Owen ed. 1894, 2: 98–100).

[8] The article, entitled 'The testimony of geology to the age of the human race', appeared in the *National Review* 10: 279–312. This was the April number that Hooker had taken to Down with him (see preceding letter). The *Wellesley index* does not list the author of the article.

[9] CD's annotation refers to chapter 3 of his 'big book' on species, 'On the possibility of all organic beings crossing' (*Natural selection*, pp. 33–91).

To T. H. Huxley [after 20 April 1860][1]

[Down]

P.S. I lost about the time when you were at Down,[2] the National Review with Carpenter's Review.—[3] Did you take it away by mistake? Hooker says he did not.—[4]

Imperial College of Science, Technology and Medicine Archives (Huxley 5: 255)

[1] Dated by the relationship to the preceding letter.

[2] Huxley and his wife Henrietta Anne Huxley had visited Down on the weekend of 7 April 1860 (Emma Darwin's diary). Joseph Dalton Hooker was also a guest at the time.

[3] [Carpenter] 1860a appeared in the January number of the *National Review*.
[4] See preceding letter and letter to J. D. Hooker, 18 [April 1860].

To Asa Gray 25 April [1860][1]

Down Bromley Kent
Ap. 25th

My dear Gray

I have no doubt I have to thank you for copy of a Review on the "Origin" in N. American Review.[2] It seems to me clever, & I do not doubt will damage my Book. I had meant to have made some remarks on it; but Lyell wished much to keep it, & my head is quite confused between the many Reviews which I have lately read. I am sure the Reviewer is wrong about Bees cells, ie about the distance; any lesser distance would do, or even greater distance, but then some of the planes would lie outside the generative spheres, but this would not add much difficulty to work—[3]

The Reviewer takes a strange view of Instinct: he seems to regard intelligence as developed instinct; which I believe is wholly false.—[4] I suspect he has never much attended to Instinct & minds of animals, except perhaps by reading.

My chief object is to ask you if you could procure for me a copy of the New York Times for Wednesday March 28th: it contains *a very striking* review of my Book, which I shd. much like to keep.[5] How curious that about the two most striking Reviews (ie yours & this) shd. have appeared in America! This review is not really useful, but somehow is impressive.— There was good Review in the Revue des Deux Mondes April 1st. by M. Laugel, said to be a very clever man.—[6]

Have you seen how I have been thrashed by Owen in last Edinburgh: he misquotes & misrepresents me badly, & how he lauds himself.— But the manner in which he sneers at Hooker is scandalous,—to speak of his Essay & never allude to his work on Geograph. Distribution is scandalous.[7] When Hooker's Essay appeared Owen wrote a note, which I have seen, full of strongest praise! What a strange man he is. All say his malignity is merely envy because my Book has made a little noise. How strange it is that he can be envious about a naturalist, like myself, immeasurably his inferior! But it has annoyed me a good deal to be treated thus by a friend of 25 years duration. He might have been just as severe without being so spiteful.— Owen consoles himself by saying that the whole subject will be forgotten in ten years.—

Have you noticed how completely I have been anticipated by a Mr. P. Matthew, in Gardeners Chronicle.[8]

Hooker about a fortnight ago staid here a few days & was very pleasant; but I think he overworks himself. He is beginning to experimentise on various subjects at Kew, & if he can but spare time I think he will be glad of it. He is going to try to degenerate our culinary vegetables. What a gigantic undertaking, I imagine, his & Benthams Genera Plantarum will be![9] I hope he will not get too much immersed in it, so as not to spare some time for Geograph. Distribution & other such questions.

I have begun to work steadily, but very slowly as usual, at details on Variation under domestication.

My dear Gray | Yours always truly & gratefully | C. Darwin

Gray Herbarium of Harvard University (13)

[1] Dated by the reference to the review of *Origin* in the *North American Review* ([Bowen] 1860a).

[2] Francis Bowen, professor of natural religion at Harvard University and a friend of Louis Agassiz's, was the author of the review. CD's copy is in the Darwin Pamphlet Collection–CUL. See also letter to J. D. Hooker, 18 [April 1860].

[3] [Bowen] 1860a, pp. 494–9. In his copy of the review, CD wrote in pencil next to the discussion of bees' methods of constructing cells: 'All this mistaken' ([Bowen] 1860a, p. 496).

[4] Speaking of instinct, the reviewer stated that 'in the infinite number of variations which he requires us to believe this faculty to have undergone, why has it never overstepped the clearly marked line which still divides it from reason?' ([Bowen] 1860, p. 497). CD scored this passage in his copy of the review and added an exclamation mark in the margin.

[5] The *New York Times*, 28 March 1860, p. 3, carried a long and generally favourable review of *Origin*. The review was unsigned, but the text suggests that it may have been written by James Hall, who was palaeontologist to the New York State Geological Survey. Referring to the most recent work on the earliest fossils in the geological record as support for CD's theory, the reviewer stated that 'the present writer had an opportunity last Summer, in company with Sir Wm. Logan, of examining a fossil from the Laurentian rocks of Canada—far older than the Silurian strata—which if it holds its authenticity, must be the intimation of a new and elder world of organic forms, into which research and induction may yet be prolonged.' William Edmond Logan was head of the Geological Survey of Canada. Although criticising some of CD's views, the reviewer nonetheless stated:

> It is clear that here is one of the most important contributions ever made to philosophic science; and it is at least behooving on scientists, in the light of the accumulation of evidence which the author has summoned in support of his theory, to reconsider the grounds on which their present doctrine of the origin of species is based.

[6] Laugel 1860. There is an annotated copy in the Darwin Pamphlet Collection–CUL.

[7] [R. Owen] 1860a. The article reviewed Hooker 1859 in addition to *Origin*.

[8] *Gardeners' Chronicle and Agricultural Gazette*, 7 April 1860, pp. 312–13. See Appendix V. For CD's reply to Patrick Matthew, see letter to *Gardeners' Chronicle*, [13 April 1860].

[9] Bentham and Hooker 1862–83. See letter to J. D. Hooker, 12 March [1860].

To ? 25 [April 1860?][1]

Down.
25

My dear Sir

One line to say that Mr Laugel has sent me a copy of the Revue des Deux Mondes.[2]

Yours very sincerely | C. Darwin

Quaritch (dealers)

[1] The date is suggested by the reference to Laugel 1860 and by the relationship to the preceding letter. The letter is bound into a copy of the first edition of *Origin*.

[2] There is a copy of Antoine Auguste Laugel's review of *Origin* (Laugel 1860) in the Darwin Pamphlet Collection–CUL. It is lightly annotated.

To J. D. Hooker 26 April [1860]

Down Bromley Kent
Ap. 26

My dear Hooker

It is a horrid shame to add in slightest degree to your labour; but I certainly am very curious about Leschenaultia formosa.— I have had hunt for pollen tubes!! but could not find any, but my greenhouse is too cold for flowers to open well.— I cannot help continuing to suspect that surface indicated is stigma.—when the indusium closes, the humid surface comes to summit of pistil.—[1] Would not Mr Bentham look for me, if you have not time??—[2]

I hope you will tell me, if you hear again distinctly from Bell that Owen avows the Review.—[3] I have come to conclusion (& Lyell says he came to same) that Owen treats you more ungenerously, though perhaps not more spitefully, than he does me. But I now clearly see that the Review does not really signify.

With respect to curved style, I will try & find my few notes; & then you will see what few facts I have got.— The cases which have struck me most are those in which flower is regular or nearly regular, but in which honey is secreted by only two of the petals, & then *if style bends*, it bends so that stigma must be brushed by insect in alighting & departing.— In the Fraxinella Dictamnus??? the pistil is at first straight, but after some days when pollen is shedding, & nectar secreting by two upper petals, then & not till then it bends upwards thus.

There is a Corydalis I think tuberosa, which offers good case; as there is an aborted & really secreting nectary, & pistil bends to real nectary.—

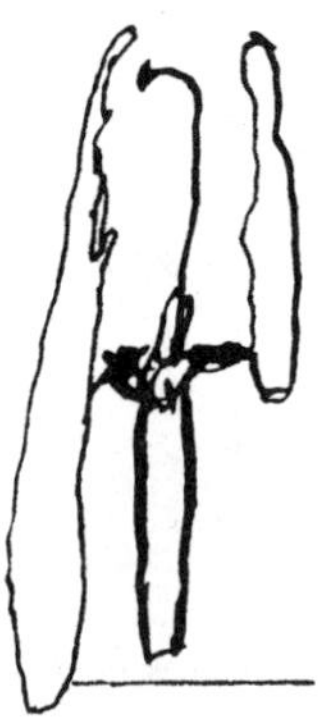

When there is a nectary on both sides, as in Dielytra spectabilis pistil is straight. These are sort of facts which have struck me.— If I remember in Rhododendron honey is secreted by only two petals?? viz the two towards which pistil bends. Insects always know at once where nectar lies, & have not to search for it.—

I must say it seems to me a piece of impertinence in Royalty asking such men as you & Owen to Lecture to their children.—[4] I will not mention it to a soul.—

Your affect | C. Darwin

Endorsement: '/60'
DAR 115.1: 50

[1] See letter to J. D. Hooker, 18 [April 1860], and letter from J. D. Hooker, [20 April 1860]. The female organs of *Leschenaultia formosa* are shaped in such a way that it is difficult to tell which part is which: the style ends in a cup-like expansion, the indusium, enclosing the stigma.

[2] George Bentham lived in London and pursued his taxonomic investigations at the Royal Botanic Gardens, Kew.

[3] Thomas Bell and Richard Owen. See letter from J. D. Hooker, [20 April 1860].

[4] See letter from J. D. Hooker, [20 April 1860].

To J. D. Hooker 27 April [1860][1]

[Down]
Ap. 27th

My dear Hooker

Will you run your eye over other side & then tear up or keep as you like.— What I want are facts to upset this apparent rule or (which of course wd please me most) any strong cases to support it. Any **one** species in a group, with nectar-secreting surface on *one* side of flower with pistil bent towards that side, wd be, I shd think, good case.—Do you think Mr Oliver would keep & decipher my miserable notes, & observe a few flowers for me this summer. —[2] Why I care about it, is that it shows that visits of insects are so important, that these visits have led to changed structure.

C. D.

Leguminosæ; in vast number of species pistil bent rectangularly upwards & Bees get nectar at base of standard; but in Kidney-Bean & garden Lathyrus, they always alight on left-hand petal & the stigma by curving points to this same side.

Mem. case of Corydalis & Dielytra alluded to, & Fraxinella—

Alstrœmeria (common orange-flowered kind) 2 upper petals secrete honey, & the pistil is bent up to this side

Gladiolus (orange) much honey on lower side of flower, where the filaments are attached; pistil *when mature* bends down to this lower side

Delphinus grandiflorus, pistil bent rectangularly so that stigma lies in gangway to nectary. (contrast with columbine)

Tropæolum, stigma *when mature slightly* bent into gangway

Several Scrophulariaceæ, as Antirrhinum & Pedicularis; some Labiatæ, as the fine Blue Salvia have pistils more or less bent into gangway to nectary

Lobelia fulgens stigma slightly bent down towards gangway

Polygala (common English species) has pistil directed at right angles to where Bees suck the flower

Rhododendron, Viola tricolor, Horse-chesnut all have pistils more or less bent in same manner, ie in path which Bees must brush over in sucking the nectar[3]

C. Darwin

Down House (MS 10: 67)

[1] The year is given by the relationship to the preceding letter.

[2] Daniel Oliver was an assistant in the herbarium of the Royal Botanic Gardens, Kew.

[3] None of these plants is included in the discussion 'On the relation between the structure and conspicuousness of flowers, the visits of insects, and the advantages of cross-fertilization' in *Cross and self fertilisation*, pp. 383–4.

To Charles Lyell 27 and 28 April [1860][1]

Down Bromley Kent
Ap. 27th.—

My dear Lyell

Thanks for Newberry:[2] I will read it in about a week's time & will keep it, if useful to me; otherwise will return it to you.— I quite agree that the non-comittal men do not always most help on science.

I sent you a day or two ago, the clever review by Laugel[3] & President of Tyneside Naturalists' Address.—[4]

With respect to Dogs (& several other domestic animals) I have always thought the case must remain doubtful, but I think the argument strongly preponderate in favour of the multiple origin.[5] I shd. not like to commit myself to names until I can weigh all the evidence in mass. I do not at all believe that Owen did not know *perfectly* well some of the wild Canidæ to which I alluded.[6] The case has been too often discussed from time of Pallas to present day for him not to know.[7] Looking to Dogs of world, I believe in their blood, more or less mingled, there is the blood of the European Wolf—*two* very distinct N. American wolves; probably the Guyana dog or wolf; probably (for I think argument of Pallas & lately of Isidore Geoffroy St. Hilaire have much weight) of several so-called species or wild races of Jackall.—[8]

I will compile facts in my present volume;[9] but it seems to me more prudent not to enter on detail on this doubtful subject in the Origin.—

Dawson's remark on variability of Canidæ may be true, but I suspect he would find it very hard to prove.—[10]

My dear Lyell | Yours very sincerely | C. Darwin

Please return the 3 Reviews whenever you have quite done with them.

P.S.—Apr. 28th | I have received your various budgets. I am *extremely* obliged for drawings & have sent them to Paris, with strictest injunctions about care & their return.—[11]

I presume that Carpenter would call the Vertebrata a class, & the Birds an order; but this certainly is not usual.[12] But to estimate the values of groups has always been found hopelessly difficult.—

The case of Spitz Dog is (from me probably) originally from Bechstein,[13] but no particulars are given.— I think the Sheep & Goat in Chile is nearest case of reversion to one pure parent by repeated crosses.—But no one can doubt that this could be easily effected, seeing that the number of generations in which it can be done with various plants was well ascertained by Gartner & I think by Kolreuter.—[14] By the way it has been effected with the very distinct Phasianus colchicus & versicolor,—not that I have compared the reverted breed with the original.—[15]

I see that I have misunderstood you, & you mean reversion when the hybrids are bred inter se. The case you allude to seems very wonderful & improbable, & I must endeavour to find out where described. It almost passes my belief!

I have been much interested by your closing remarks. I cannot explain why, but to me it would be an infinite satisfaction to believe that mankind will progress to such a pitch, that we sh^d. be looked back at as mere Barbarians.—[16] I have received proof-sheets (with a wonderfully nice letter) of very hostile review by Andrew Murray, read before Royal Soc. of Edinburgh.—[17] But I am tired with answering it. Indeed I have done nothing this whole day but write letters, so no more.—

My dear Lyell | Ever Yours | C. D.

American Philosophical Society (209)

[1] Dated by the reference to John Strong Newberry's paper on North American fossil plants (see n. 2, below).

[2] An annotated copy of Newberry 1860 is in the Darwin Pamphlet Collection–CUL. It is inscribed: 'Sir Chas Lyell with the respects of J. S. Newberry'. Newberry was an American geologist who had studied the Cretaceous flora and fauna of the south-western part of the United States.

[3] Laugel 1860.

[4] Henry Baker Tristram was president of the Tyneside Naturalists' Field Club in 1860. In his presidential address of 29 March 1860, Tristram criticised *Origin* (Tristram 1860, pp. 218–28).

[5] CD and Lyell had carried on a lengthy discussion about the origin of the different breeds of domestic dogs in October and November 1858. See *Correspondence* vol. 7. Lyell inclined towards upholding descent from a single ancestor, whereas CD favoured multiple origin. CD discussed the issue in *Variation* 1: 15–33. CD consulted Lyell about the topic again in September 1860 (see letters to Charles Lyell, 23 [September 1860] and 26 [September 1860]).

[6] In his review of *Origin*, Richard Owen had referred to CD's belief that domestic dogs had descended from several wild species ([R. Owen] 1860a, p. 528):

> It may be so; but what are the species here referred to? Are they known, or named, or can they be defined? If so, why are they not indicated, so that the naturalist might have some means of judging of the degree of probability, or value of the surmise, and of its bearing on the hypothesis?

[7] Pyotr Simon Pallas had written extensively on the origin and variation of domestic animals. CD had studied his writings closely (see *Correspondence* vol. 4, Appendix IV, 119: 7a, 10a, 13a).

[8] Pallas 1780 and I. Geoffroy Saint-Hilaire 1854–62, 3: 107, are cited on this point in *Variation* 1: 16 n. 2. CD had recently obtained the first part of the third volume of I. Geoffroy Saint-Hilaire 1854–62 (see letter to Williams and Norgate, 6 March [1860]).

[9] CD began work on *Variation* in March 1860 (see 'Journal'; Appendix II).

[10] Lyell had been reading John William Dawson's *Archaia* (Dawson 1860a), a treatise on the relationship between cosmogony and natural history (see Wilson ed. 1970, p. 371). In the chapter on the unity and antiquity of man, Dawson discussed the topic of variability within species; stating his belief that wolves and foxes were very prone to vary, he went on to say (Dawson 1860a, pp. 254–5):

> Constitutional liability to vary is sometimes connected with or dependent on extreme simplicity of structure, in other cases on a high degree of intelligence and consequent adaptation to various modes of subsistence. Those minute, simply organised, and very variable creatures, the Foraminifera, exemplify the first of these apparent causes; the crafty wolves furnish examples of the second.

Lyell subsequently wrote to Dawson about this work and defended CD's view of the unlimited variability of species (K. M. Lyell ed. 1881, 2: 332–3).

[11] Budget: 'The contents of a bag or wallet; a bundle, a collection or stock.' (*OED*). The drawings that Lyell sent CD have not been identified.

[12] This point relates to a discussion included in William Benjamin Carpenter's recent review of *Origin* ([Carpenter] 1860b). Carpenter queried the extent to which natural selection could operate. He allowed that all birds could have arisen from one progenitor since they showed little divergence from a common type, but not that birds could have originated from the same line as reptiles or mammals. He advocated separate creation for 'each of the great types' and for humans (see [Carpenter] 1860b, p. 404). Lyell made notes on these points in his scientific journal (Wilson ed. 1970, p. 378). Carpenter's review is also mentioned in an entry dated 25 April 1860 that is headed 'With Darwin' (Wilson ed. 1970, pp. 372–3). CD visited Lyell in London on 22 April 1860 (see letter to Charles Lyell, 15 April [1860]).

[13] Bechstein 1793–1805 is cited in *Variation* 1: 31 as the authority for statements on the Spitz (Pomeranian) dog. CD's annotated copy of the work is in the Darwin Library–CUL.

[14] Karl Friedrich von Gärtner and Joseph Gottlieb Kölreuter. CD's annotated copies of Gärtner 1844 and 1849 are in the Darwin Library–CUL. His abstracts of Kölreuter's papers are in DAR 116.

[15] See *Natural selection*, p. 432 and n. 2. *Phasianus colchicus* is the common pheasant; *P. versicolor* is known as Diard's pheasant.

[16] Lyell discussed this point in his scientific journal (see Wilson ed. 1970, pp. 374–5, 379).

[17] Murray 1860a. See letter to Andrew Murray, 28 [April 1860].

From J. D. Hooker [28 April 1860][1]

Kew
Saturday

Dear D$^{r.}$

I have had a pretty good examination of Leschenaultia, but have not found pollen-tubes in any part of stigma or style. I have however I think satisfied myself that the external viscid surfaces (there are 2, one on each lip) have nothing whatever to do with the stigmatic surfaces,

1 They have not the structure.
2. They are *outside* the bundles of nutrient Vessells of style. ∴
3 They do not & cannot communicate with conducting tissue of style.
4 There is a perfectly good & normal stigmatic surface directly communicating

with the conducting tissue, where it ought to be, at the base of the cup *inside*, on each lip, but most developed on the smaller lip. These surfaces are bona fide structurally, ie. histologically "stigmatic"—& I have no doubt, had we fine weather & impregnated flowers, we should find pollen tubes in them.

Brown remarks that some species are impregnated *after* Expansion, by pollen of other flowers,—I know not on what facts he goes—others he says by pollen retained in the indusium on its closing.[2]

I see no flaw in your style & nectary conclusions— In most (all?) irregular flowers the greatest development of perianth & stamens is towards axis of plant (or turned away by resupination of flower) & is balanced by greater development of nectary on opposite side of flower. This appears to me to throw the curvature of the style necessarily in the position you indicate.—it is as it were pushed towards the side where the development of perianth &c is greatest & this is in

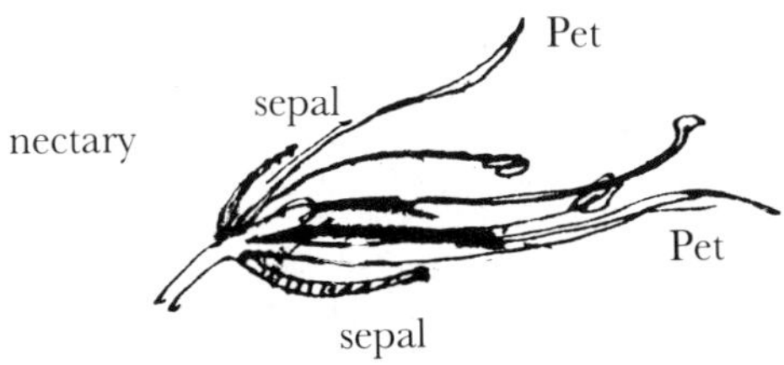

part counteracted by the endeeavor of style as central organ to keep straight—which results in bend of stigma to nectary side. I think this is so in all plants whether they have nectaries or no— It is a case of force & direction of growth.

Ever yrs affy | Jos D Hooker

P.S. No record of Digitalis with 5th. rud. stamen.—ask Masters.[3]

[Enclosure][4]

There is in Herb. Hook. a specimen of a *Primula* collected at Over, Cheshire May 1824 by W. Wilson Esq.[5] of which that excellent Brit. Botanist says,

"Primula elatior, with simple & umbellate flower-stalks,— I think this specimen (having the leaves of the proper shape assumed by Sir J. E. Smith) as a character for *P. elatior* would have convinced him of that being only a variety of *P. vulgaris*"

In Hook Herb. there seem to be all sorts of intermediates in foliage & inflorescence.

There is a good remark by H.C.W. in Hook & Arn. Brit. Flora, that all the forms of Cowslip & Cowslip-oxlip have short tomentum—[6] all of Primrose & Primrose-oxlip have long villous hairs (on scape &c). I find this holds *pretty well* in all the Herb. Hook specimens: but is at best rather hazy character.

DAR 100: 150–1, DAR 166.2: 262

CD ANNOTATIONS
0.1 Dear . . . closing. 2.3] *crossed pencil*
Top of first page 'Ch. 3'[7] *brown crayon, underl brown crayon*
Enclosure: 'April 28/1860' *pencil*; 'Ch. 4.—'[8] *pencil*

CD note:[9]
Mucuna, some particulars.— *Did it ever set seeds before. [*added*][10]
Primrose & Cowslip on same root[11]
Thesium[12]
Bentham on S. or N. limit [*above del* 'range'] of plants, how abrupt?[13]
Bentham, whether in genus Digitalis, rudiment of 5th stamen, ever occurs.[14]

1 Dated by the relationship to the letters to Joseph Dalton Hooker, 26 April [1860] and 27 April [1860]. In 1860, 28 April was a Saturday.
2 Brown 1814, pp. 559–61.
3 Maxwell Tylden Masters was an expert on plant teratology. CD had expressed interest in Masters's recent lecture on the subject (Masters 1860), given at the Royal Institution on 16 March (see letters to M. T. Masters, 7 April [1860] and 13 April [1860]). See CD note and n. 14, below.
4 There were apparently two enclosures sent with the letter, each containing answers to some of the questions posed by CD in the letters to J. D. Hooker, 26 April [1860] and 27 April [1860]. The first enclosure is in DAR 166.2: 262. It was dated by CD 'April 28/1860' and further marked in pencil: 'Ch. 4.—'. The second enclosure (see CD note) was a list of queries in CD's hand, perhaps given to Hooker during his recent visit to Down and now returned by Hooker with his answers added. See nn. 9–14, below.
5 William Wilson had described several mosses for Hooker 1844–7; he had also contributed to the second edition of James Edward Smith and James Sowerby's *English botany* (1832–46).
6 Hooker refers to Hewett Cottrell Watson, as cited in William Jackson Hooker and George Arnott Walker Arnott's *British flora* (1850).
7 CD's annotation refers to chapter 3 of *Natural selection*, 'On the possibility of all organic beings occasionally crossing'.
8 CD's annotation refers to chapter 4 of *Natural selection*, on variation under nature.
9 The note is in DAR 100: 157. On it Hooker wrote his response to CD's queries. The first page was crossed by CD in pencil. A paragraph in Hooker's hand relating to the northern and southern limits of plants (see n. 13, below) was also crossed in pencil. CD marked the bottom of the note 'Ch. 3' in brown crayon. See n. 7, above.
10 Hooker's answer to the first question reads: 'see back *Mucuna* never set any seed till impregnated, by us, in our great Palm Stove when fruit ripened & seeds were raised.' *Mucuna* is a tropical and subtropical genus of leguminous vines.
11 Hooker's answer reads: '—See note in letter'. Hooker refers to the first enclosure. CD had come across a report of this abnormal flower stem in Masters 1860, p. 224.
12 Hooker's answer reads: '— translation of quotation enclosed'. In Masters 1860, p. 225, Masters stated that Siegfried Reissek (in Reissek 1843) had been led 'in some measure, to anticipate the views of Mr. Darwin'. The translation to which Hooker refers was one prepared by Frances Harriet Hooker of a passage from Reissek 1843, pp. 653–4. It is bound in CD's copy of Masters 1860 (Darwin Pamphlet Collection–CUL). See also letter to J. D. Hooker, 30 April [1860] and n. 3. *Thesium* is a tropical parasitic plant.
13 Hooker's answer reads: 'see back of this Bentham's observations on limits of plants did not refer to N. & S. limits, but to abrupt termination of certain Spanish plants in S. France towards Cevennes— The more I think, the more I feel satisfied that N. & S. limits of plants are not at all abrupt like of animals.' The reference is to Bentham 1826, p. 6.
14 Hooker's answer reads: 'see letter'.

To Andrew Murray 28 April [1860][1]

Down Bromley Kent
April 28th

My dear Sir

I thank you from my heart for your most kind letter. I never knew or heard of a hostile Reviewer doing so kind & generous an action.—[2] I have read your Review, & I am sure I have no grounds whatever to complain, but on the contrary, to thank you for the general manner of speaking of me. I will scribble down a few remarks & corrections; but there is only one of much importance.—[3]

Of course on the weight of general argument it would be superfluous in me to make any remarks.— If you think my explanations of rudimentary organs or laws of embryology—succession of same forms on same areas &c &c are not true or of no value, there is nothing more to be said.—[4] So my few remarks will apply to details.— Believe me that I feel most truly grateful to you for your kindness

My dear Sir | Yours very sincerely | C. Darwin

I have put little pencil crosses to your Review, to call your attention to passages if you have patience to read my too lengthy notes.

I shd be very much obliged for a copy when published.—[5]

[Enclosure]

Review | *p. 276.*—[6] I presume that you think my discussion (p 173–179) on forms now being distinct is of no value (nor do I value it much; I look at it as my only way of getting out of difficulty) but your Reader would imagine that I had slurred over the difficulty.—

p. 277 Is it right to give your idea of opposites marrying, without any allusion to plants & to the many marine hermaphrodite animals? I **heartily** wish I could believe in such fine sexual selection that a big & dark Carabus preferred for his wife a small & light-coloured female; or conversely—[7]

p 278. Have you any grounds for believing that a plant's own pollen is prepotent over that of another **variety** (please observe not another species).[8] Again I rather wish I could believe this: but it is utterly opposed to all that I have observed & experimentised on.— Is it fair to state this without at least some facts?

p 278. Will you please read my p. 14, 15, & consider whether I speak "defiantly" against reversion.—[9] I state I rather believe in it in such case as cabbage; but I earnestly desire some good facts.— I have some, but they are not very strong —(rabbits turned out reverting in colour &c) I think 'defiantly' is rather strong, when I give the most striking case from experiments, as far as I know, on record of Reversion in colour at least, in Pigeons.— I have since been experimentising for *this sole purpose*, & have got grand case in Fowls.—

p 279. For above reasons I think "total or almost total negation of reversion" is rather strong.—[10] I simply do not believe so strongly in it as you do.— What I believe is, & as it seems to me all experience shows, that with a very little selection, we can keep our cart-horses, greyhounds & fantail pigeons &c true. To improve them or keep them up to *highest* standard of merit no doubt very careful selection is requisite; but I know of no facts whatever to lead me to believe that a lot of fantails, without any selection exposed to the same conditions under which they have been reared, would not remain fantails for the next 10, thousand years. It may not be so, but I want some shadow of evidence to make me give up this belief—

p. 284.— With respect to your discussion on absence of links in formations, I suppose it is as fair as it could be made so briefly. It would be hardly possible for you to allude to all the points, such as the impossibility of knowing a link if really found; & the impossibility unless every intermediate gradation were found of our distinguishing a linking variety & species.— I have been much pleased & surprised at finding so many *practical* geologists, attaching no or very little weight to absence of links in our geological formations, for instance, Lyell, Ramsay, Jukes Rogers.—[11] That admirable palæontologist Mr Salter of the Geolog. Survey showed me 2 or 3 days ago, the Spirifers of Devonian, Lower & Upper Carboniferous formations arranged (not at my request, for I thought he was dead against me) after my diagram in the "Origin", & it astonished me what a beautiful branching gradation he made by intercalating the varieties & species according to geological age.—[12] Pray remember how many of Al. d'Orbigny's fine species of shells in successive stages are now sunk into the rank of successive varieties in the estimation of good palæontologists,—for instance of Woodward.—[13]

p. 285. Here is the only grave misapprehension of my views in your Review.— I by no means believe that the mouth of the Bear (how often that abominable animal has been made to worry me!!) might be increased by use; I referred here exclusively to the "natural selection" of bigger & bigger mouths because advantageous, in this foolish & imaginary illustration.— So if you read my pages on the Bat, you will see that I never dreamed of Galeopithecus gaining its membrane, or the seal its flipper, by *use*, but by natural selection.[14]

p 285. I should not rank a **treeless** plain under *physical* conditions; the presence of trees, ie of organic beings, I have always ranked under "organic conditions"—[15] you will find the paramount importance of "organic conditions" repeated ad nauseam. For organic conditions, as I believe, mainly determine what variations are profitable & what are accordingly naturally selected. This sentence gives a very wrong idea of my views.—

p. 286 You will see at p. 346 347 that I do not speak of S. America & Africa having same conditions, **as close as the same species generally require**, but only of "certain large tracts" within certain latitudes.—[16]

p. 286. I wish much you would insert that I never meant to confine my remarks to the *insects* of the caves. Taking a general view of all inhabitants I doubt whether I am wrong.— I grieve much if I have misunderstood, (God knows it was unintentionally) Schiodte, but I am glad to see that you admit passage is vague.—[17] I cannot collect my mind to consider subject now.— You are, of course, very much more likely to be right than I.[18] I have, since I published, thought it more probable that as

Adelops is blind & is yet found sometimes not in caves, that Anophthalmus might have formerly been a genus fitted for dusky places, & that consequently it was more easily fitted for the caves than any other carabidous insect. I understood, in sense in which you do, Schiödte remark about the gradation in the degree of blindness, & meant no more.—[19] Would it be asking you too great a favour to refer me to papers on the several other cases specified by you? It is a load removed from my mind to hear that the species of Anophthalmus are different in the different caves.—[20]

As I am asking questions, can you tell me how the blind Pau**ss**us (or Pau**s**us I forget which) in same country gets from one ants' nest to another?—[21]

p. 288. That is a good hit about eyeless state & physical conditions—yet of course I admit in this case the effect of physical conditions. I cannot point out distinction between the effect of physical conditions, as commonly used, & "Disuse", yet it strikes me that there is some difference.[22]

p. 289.— Azara most fully confirms my account of the habits of the La Plata Colaptes (p. 184).—[23] I was well aware of difference between Colaptes & the typical woodpeckers—but I thought in saying that it agrees in "*every* **essential** part of its organisation" I had sufficiently guarded myself. Its undoubted position amongst the woodpeckers seems to me show that it agrees in its essential organisation.[24]

Dana Biomedical Library, Dartmouth College, and R. D. Pyrah

[1] Dated by the reference to Murray's review of *Origin* (Murray 1860a).

[2] Murray sent CD the proof-sheets of his review of *Origin* before it was published (Murray 1860a). See letter to Charles Lyell, 27 and 28 April [1860]. There are two copies of the review in the Darwin Pamphlet Collection–CUL; one is an offprint from the *Proceedings of the Royal Society of Edinburgh*, and the other is a separately paginated reprint. Both were annotated by CD.

[3] CD refers to the whale–bear case described in *Origin*, p. 184. See enclosure.

[4] Murray stated that the strongest points in favour of CD's theory were the taxonomic relationships between groups of organisms and the existence of homologous organs in different animals, some rudimentary and some highly developed (Murray 1860a, p. 289). Nevertheless, he concluded that 'Mr Darwin's theory is unsound' (*ibid.*, p. 291).

[5] See n. 2, above.

[6] The page numbers given by CD do not exactly correspond with those of the published review. CD refers to Murray 1860a, pp. 276–7, in which Murray claimed that if variation were unlimited, 'there would be nothing but an indiscriminate mass of creatures running all into each other'. In the published version, possibly as a result of showing CD the proof-sheets, Murray added a note at this point (Murray 1860a, p. 277 n.):

> One of Mr Darwin's explanations of the absence of intermediate forms may be taken as his answer to this objection—viz., that these forms are, in point of fact, numerically weaker than the forms on each side which they link together, and thus are liable to be exterminated sooner than them. But, admitting the fact to be that they are less numerous, why should they be so under Mr Darwin's theory? With unlimited powers of modification, why should the intermediate forms always be *originally* fewer.

[7] Murray 1860a, p. 277, refers to an 'instinctive inclination which induces individuals of the same species by preference to intercross with those possessing the qualities which they themselves want, so as to preserve the purity or equilibrium of the breed.'

[8] Murray 1860a, p. 278. In his separate offprint of the paper (Darwin Pamphlet Collection–CUL), CD wrote next to this passage in pencil: 'oh Cabbages'.

[9] Murray may have altered his proofs in acknowledgment of CD's point. The printed version reads: 'This phase of reversion to type is slightly alluded to, and slightingly admitted as an element by Darwin.' (Murray 1860a, p. 277). Later in the article he also stated: 'Mr Darwin of course disputes it [Reversion], or at least does not admit it, and desiderates the evidence on which the statement has been so often made, that our domestic varieties, when run wild, gradually but certainly revert in character to their aboriginal stocks.' (Murray 1860a, p. 278).

[10] Murray 1860a, p. 279, reads: 'my first objection to the principles on which Mr Darwin's theory rests is, that it is founded on exaggerated and undue estimate of the one—the power of modification; and if not a negation, at least an inadequate concession of the other, viz. the principle of reversion to type.'

[11] CD refers to Charles Lyell, Andrew Crombie Ramsay, Joseph Beete Jukes, and Henry Darwin Rogers. See Murray 1860a, pp. 283–4.

[12] CD had learned of the close relationships between species of fossil *Spirifer* from Jukes earlier in the year (see letter from J. B. Jukes, 27 February 1860). John William Salter was a colleague of Jukes's at the Museum of Practical Geology. CD had visited London from 21 to 23 April 1860 ('Journal'; Appendix II).

[13] Alcide Charles Victor Dessalines d'Orbigny and Samuel Pickworth Woodward.

[14] Murray added a note to his citation of CD's whale–bear case (Murray 1860a, p. 275 n.):

> In quoting this, I do not at all mean to give it as a fair illustration of Mr Darwin's views. I only refer to it as indicating the extent to which he is prepared to go. The example here given I look upon (as I have reason to know Mr Darwin does himself) merely as an extreme and somewhat extravagant illustration, imagined expressly to show in a forcible way how "natural selection" would operate in making a mouth bigger and bigger, because more advantageous.

For CD's previous discussion of this case, see *Correspondence* vol. 7, letter to Richard Owen, 10 December [1859]. *Galeopithecus* is the flying lemur.

[15] On p. 285 of his review (Murray 1860a), Murray presented a series of arguments against CD's view that physical conditions could not effect change in a species. In *Origin*, p. 184, CD used woodpeckers living on the plains of La Plata, 'where not a tree grows', as an illustration of how species change might occur through natural selection. Although Murray removed the reference to 'treeless plains' from p. 285 (see letter from Andrew Murray, 3 May 1860), he nonetheless stated that the birds CD referred to were an example of alteration in structure 'consequent upon different physical conditions of life' (*ibid.*, p. 259). In his separate offprint of Murray's paper (Darwin Pamphlet Collection–CUL), CD underlined the expression 'physical conditions' in pencil and added two exclamation marks in the margin.

[16] Murray 1860a, pp. 285–6.

[17] Murray devoted a long section of his article to a discussion of the origin of cave fauna, on which he was an expert (Murray 1860a, pp. 286–8). He considered that CD's theory could not explain the existence of closely related species of blind insects in caves geographically far removed from one another. CD maintained that the species under consideration were not closely related (*Origin*, p. 138). In the published version of the article, Murray added a note addressing CD's point (Murray 1860a, p. 286):

> Although Mr Darwin here uses the observations of Schiodte upon blind insects as an illustration, his remarks (as he himself has had the kindness to inform me) are not meant to be confined to them, but also to be applied to the whole of the animals found in caves. But as his theory, if true, should meet every case, a clear flaw in even one would be fatal to the whole, and I would have tested it with these insects, whether they had been referred to by Mr Darwin himself or not.

The reference is to Schiödte [1849]. CD had studied this and other works on cave fauna in 1856 (see *Correspondence* vol. 6, letter from J. O. Westwood, 23 November 1856). See also letter from John Lubbock, [after 28 April 1860].

[18] Murray had published a paper on the comparative anatomy of insect eyes, in which he likened the absence of eyes in larval forms to the absence of eyes in cave insects. Both, he maintained, had no need for them in their search for food (Murray 1857).

[19] Murray 1860a, p. 288.

[20] Murray 1857, pp. 132–4,

[21] Murray discussed the recent identification of blind insects living in ants' nests in Murray 1857, p. 130. *Paussus*, however, was not mentioned. The Paussidae are a family of ground beetles that inhabit ants' nests in tropical countries (Leftwich 1976).

[22] Murray offered a hypothetical case of an ancestral member of the Trechidae wandering into a cave and, by a process of natural selection, eventually turning into the eyeless *Anophthalmus*. Murray went on to state: 'But if he thus saves his theory, what becomes of his disbelief in the effects of physical conditions? If it has no effect, why have they all turned into *Anophthalmi*?' (Murray 1860a, p. 288).

[23] Murray 1860a, p. 289. CD refers to Azara 1809. There is an annotated copy of the work in the Darwin Library–CUL.

[24] See n. 15, above.

To Andrew Murray 28 [April 1860][1]

Down Bromley Kent
28th

My dear Sir

I omitted one little criticism on your explanation of the use of Hybridisation on the supposition (singular enough as it seems to me) of all the males of a rare species dying.[2] Do you not think that you ought to say that this applies only to the unisexual animals & to but few plants—unless indeed you choose to suppose the male organs to fail in acting, during same season in all the individuals of a species. Such astonishing precautions to prevent extinction, seeing that the extinction of every form of life in the course of time is a law of nature, seems to me rather improbable.—

I rather suspect that Owen & many of my other reviewers would have hit me hard had I advanced a view of this nature without facts in support.—[3] Forgive me for troubling you; for I well know that you wish to make your Review as true, as well as as just, as possible.—

My dear Sir | Yours very sincerely | C. Darwin

P.S. Forgive me for adding one word. I rather dispute that Oken, Lamarck & Co. throw some light on Classification, Embryology & Rudimentary organs &c.—, as the theory of Nat. Selection seems to me to do.—[4] In the case of Embryology there must be introduced the principle of variations not supervening at *very* early age & being inherited at corresponding age.— So with Rudimentary organs combined with disuse & Nat. Selection.—

In case of classification, descent alone, as I believe I have shown, will not do; you must combine principle of divergence of character & descent from dominant & increasing forms.— But as you do not believe in my notions it is folly in me pointing out how they differ from the notions of my predecessors—

R. D. Pyrah

[1] Dated by the relationship to the preceding letter.

[2] Murray asserted that if all the males of a species died, extinction would ensue unless the females bred with the males of another species (Murray 1860a, p. 276).

[3] Richard Owen was the author of an unsigned review of *Origin* ([R. Owen] 1860a).

[4] Lorenz Oken and Jean Baptiste Pierre Antoine de Monet de Lamarck. Murray stated, with respect to explaining the principles of taxonomy and embryology, that CD's theory was no more successful than those of earlier naturalists who had espoused the 'doctrine of progressive development' (Murray 1860a, p. 290). CD was anxious to distinguish his evolutionary theory from the views of those who advocated linear, progressive development in nature.

From John Lubbock [after 28 April 1860?][1]

My dear Mr Darwin

I send you Murray's Paper which you are quite welcome to keep.[2]

Haeckel on the eyes of Starfishes[3] & Claparede on those of Insects[4] are both in Siebold & Kollikers Arch.[5] for 1859. Zweites Heft.

I can lend it to you whenever you like to have it.

Yours most faithfully | John Lubbock

DAR 48: 68

CD ANNOTATIONS

End of letter: 'Star-fish eye with crystalline lenses'; 'Claparede disputes Mullers view on origin of eyes—'[6] *pencil*

[1] The date is suggested by CD's interest in the structure and origin of eyes in connection with *Origin*. In the letter to Andrew Murray, 28 April [1860], CD responds to criticisms of his views on blind cave insects made in Murray's review of *Origin* (Murray 1860a). In the same section of his review, Murray cited his paper on the disguises of nature (Murray 1860b); this may be the paper that Lubbock enclosed with the letter (see n. 2, below).

[2] The paper referred to may be Murray 1857, in which Andrew Murray described the anatomy of insect eyes and specifically discussed the subject of blind insects (see letter to Andrew Murray, 28 April [1860]). It could also be Murray 1860b, an annotated copy of which is in the Darwin Pamphlet Collection–CUL. In the latter paper, read at the 1859 meeting of the British Association for the Advancement of Science, Murray discussed the generic identity of the blind insects found in caves in Europe and in the United States.

[3] Haeckel 1860.

[4] Claparède 1860.

[5] The *Archiv für wissenschaftliche Zoologie* was co-edited by Karl Theodor Ernst von Siebold and Rudolf Albert von Kölliker. The second number of the tenth volume containing the papers by Ernst Heinrich Philipp August Haeckel and Édouard Claparède was issued on 12 December 1859, but the volume itself is dated 1860.

[6] Claparède's study of arthropod eyes had led him to disagree with the commonly accepted theory of vision, particularly as it related to the compound eyes of insects and Crustacea, first suggested by Johannes Peter Müller in J. P. Müller 1826 and refined in J. P. Müller 1834–7. CD owned a copy of the English translation of this work (J. P. Müller 1838–42), which is in the Darwin Library–CUL. All three volumes contain annotations; the second chapter in volume 2, which discusses the anatomy and physiology of the eye, is heavily annotated. CD discussed the structure of the eye in a passage added to the American edition of *Origin*, Supplement, pp. 426–7. See also Appendix IV.

To J. D. Hooker 30 April [1860][1]

Down Bromley Kent
Ap. 30th

My dear Hooker

I am not going to ask any more questions; but only to thank you much for your most full answers to all my many questions.—[2] Pray thank Mrs. Hooker for Extract.—[3] Especial thanks for Leschenaultia, which, if you had not settled the case, would have always haunted me with doubts. Ill luck to it, the case is a great difficulty to me; but when we meet (& I shall enter it in my note-book to ask you) I must get you tell me where R. Brown talks about insects & Goodenia.[4] If the species referred to has "indusium", it is extra odd; & the case becomes exactly parallel to Bee orchis (in respect to insect impregnation) for though this Bee orchis has sticky gland to pollen apparently to aid transport by insects, it never is transported as far as I can make out.

I do not care much about the curvature of style towards nectary if it be a necessary result of growth; but if it be a special adaptation it would be very valuable to me.— I shall ask you when we meet, whether the curvature of pistil, when the pollen is shedding does not look more like special adaptation, rather than a necessary result of kind development of flower.—

The *rectangular* bend in some cases, as in some Delphiniums, looks to me too much to depend on the irregularity of the flower & the consequent effect of force of growth.—

Haughton has been down on us with awful force.[5] His frame of mind is exactly that of Jemmy Button,[6] when he saw his countrymen, & when he remarked "d—d fools, no sabe nothing".—

With hearty thanks | Your affect | C. Darwin

I shall be up in London about end of May, when I daresay we may meet.—

Do not kill yourself with work.—

DAR 115.1: 51

[1] Dated by the reference to *Leschenaultia* and its pollination by insects. See letter from J. D. Hooker, [28 April 1860].

[2] See letter from J. D. Hooker, [28 April 1860].

[3] CD had requested a translation of an extract from Reissek 1843, which Frances Harriet Hooker prepared. See letter from J. D. Hooker, [28 April 1860], CD note and note 12.

[4] Robert Brown. See letter from J. D. Hooker, [28 April 1860] and n. 2.

[5] Samuel Haughton, professor of geology at Dublin University, had publicly criticised CD's theory (see *Correspondence* vol. 7, letter to J. D. Hooker, 3 May [1859]). His anonymous review of *Origin* was published early in June 1860 ([Haughton] 1860b). See letters to J. D. Hooker, 5 June [1860], and to Asa Gray, 8 June [1860].

[6] Jemmy Button was a Fuegian who was brought to England in 1830 by Robert FitzRoy and returned to Tierra del Fuego on the *Beagle* in 1833.

From B. P. Brent [May–June 1860?][1]

Dear Sir,

I have to apolo⟨gise for⟩ not answering your enquiry ⟨about⟩ Canaries, throwing back, I ⟨am⟩ sorry I cannot furnish yo⟨u⟩ any authentic case, thoug⟨h I⟩ do not doubt but such of⟨ten⟩ do occur.[2]

You express a wish to see some of the latter volu⟨m⟩es of The Cottage Gardener.[3] I have ⟨ ⟩s by me ⟨ ⟩ ⟨w⟩hich ⟨ ⟩ lend you ⟨ ⟩ but ⟨ ⟩ not bound. Should you ⟨wish it⟩ I would ta⟨ke o⟩ff the ⟨adver⟩tisment sheets, and forward ⟨ ⟩ as you may direct.

⟨I ha⟩ve just read your work on ⟨the⟩ Origin of Species through a second time, in your chapter on instinct you refer to the Cuck⟨oo⟩ laying in other birds nest[4] I thought ⟨ ⟩ Pige⟨on⟩ ⟨ ⟩ interested ⟨ ⟩ notice that when a ⟨ ⟩ lays only a single egg, ⟨and⟩ the first has been broken, or she ⟨has⟩ been much disturbed, she ⟨w⟩ill frequently lay in another's nest, a case in Point. A Scotch house Tumbler hen was about to lay having built and the cock was driving her to nest; April 24th. could feel the egg hard and ready to be deposited, toward evening ⟨ ⟩ ⟨t⟩he nest, ⟨ ⟩ laid ⟨ ⟩ and it ⟨ ⟩ with ⟨ ⟩ of young ⟨ ⟩ different part of the ⟨ ⟩ house, I immediately ⟨conc⟩luded she would not lay a se⟨cond⟩, which proved correct. She laid again on the 3rd & 5th. of May, sat and hatched a pair, I sat the odd egg under another pair but it did not hatch, I had kept it too long or perhaps it was never fertile—

I am | Dear Sir | Yours respectfully | B P Brent

To C Darwin Esqe

DAR 160.3: 297

[1] The date is suggested by the publication of Brent's articles on canaries (see nn. 2 and 3, below).

[2] Brent was an expert on canaries and on poultry. CD had consulted him about canaries on several occasions (see *Correspondence* vol. 6, letter from B. P. Brent, 23 October 1857; and vol. 7, Supplement, letter from B. P. Brent, [after August 1856]). Between April and September 1860, Brent published a series of articles in the *Cottage Gardener* describing the different canary breeds. CD may have inquired about reversion in canaries in response to these articles.

[3] In the 10 April issue of the *Cottage Gardener* 24 (1860): 25–6, Brent discussed the citril finch, supposed by some authorities to be the ancestral form of the European canary. On 17 April (*ibid.*, p. 43), he discussed the wild canary of Madeira, referring to a specimen lent to him by CD.

[4] *Origin*, pp. 216–19. CD used the example of the cuckoo laying its eggs in the nest of another species as an illustration of an instinct being modified by natural selection. He also cited instances of other birds that occasionally laid their eggs in the nests of other species.

To Roderick Impey Murchison 1 May [1860][1]

Down Bromley Kent
May 1

Dear Sir Roderick Murchison

I am much obliged for your kind note & for forwarding to me Count Keyserling's

note[2] I am pleased by it, for I look at any good geologist & palæontologist going one inch with me as more important than a naturalist going two or three inches
Pray believe me | Yours sincerely | Charles Darwin

Copy
DAR 146

[1] The year is given by the reference to Alexandr Andreevich Keyserling (see n. 2, below).
[2] CD had sent the Russian palaeontologist Keyserling a presentation copy of *Origin* (see Appendix III and *Correspondence* vol. 7, letter to T. H. Huxley, 15 October [1859]). Keyserling and Murchison had travelled through Russia together in 1844. For Keyserling's views concerning CD's theory, see the letter to Charles Lyell, 4 May [1860].

From Charles Lyell 2 May 1860[1]

May 2d. 1860.

It is small comfort or consolation to one who feels that Lamarck or Darwin have lessened the dignity of their ancestry making them out to be without souls, to be told 'Never mind you will be succeeded in unbroken lineal descent by angels who like the 'Superior Beings' spoken of by Pope

Will show a Newton as we show an ape[2]

The case of the theologian may be desperate but is there not some view by which the philosopher aspiring to a higher destiny may come to look upon the progressive system as more consonant than any other with his hopes—

It may perhaps show a tendency in all spiritual natures to advance & even at a geometrical ratio & thus if individuality be not a dream, if it be something real which cannot be lost & annihilated by death & on that the aspirations of all men must depend then a system of decay & degradation or even a stationary aspect of the Universe wd. be less promising & pleasing than one of endless variety & improvement. Taking the inner light & intuition as guiding Man to a belief in the immortality of the soul the picture of progressive advancement in a long series of generations & species with (occasionally only) retrograde movements would not be unconsolatory— It involves no doubt the link of common descent with inferior & soulless beings but this exists around us & is no new enigma— To those alone who have been in the habit of closing their eyes to the phenomena of every day life & of human nature can this riddle be for the first time revealed by Geology. Imbecility, idiotcy, madness—all the grades between the imbecile & the man of same but limited capacity—inferior races—the fact of hundreds of millions in each century dying before they have passed in infancy the line which clearly can be claimed as parting responsible from irresponsible beings, whether they have passed from this life as responsible beings or not, undeveloped—the case from the mere embryo to the new born—

Let every geologist freely grant that these are astounding facts, who has not anxiously reflected on them if he has thought at all—who ought to be surprised if in lifting up the veil & seeing more of Nature's past history the same enigma reappears— it is what you sh[d] have expected more of the same kind of perplexing doubts

In doubt to deem himself a God or beast.[3]

It is worthy of remark that, when the poet writing Essay on Man a century ago, enumerated the various sciences which that wondrous creature Man had entered upon, go measure earth weigh air & state the tides[4] he omits the insight gained into the past history of the earth & its inhabitants, but he represent Man nevertheless as being as much as now

"In doubt to deem himself a God or Beast"

It did not require that his remote genealogy sh[d] be pried into in order to raise the question—

No mention is made of attempts to unfold Natures laws in the remote past—but the realities of the present period of the world were enough without any speculation on past genealogies to make him sketch man as a being 'In doubt &c'

Draft
Kinnordy MS, Charles Lyell's journal V, pp. 176–9

[1] The letter is transcribed from a draft in Lyell's scientific journal. The entry is headed with the date and 'Perfectibility | Letter of C. L. to Darwin'. It is printed in Wilson ed. 1970, pp. 382–3.
[2] Pope 1733–4, Epistle II: 31–4:

Superior Beings, when of late they saw
A mortal Man unfold all Nature's Law,
Admir'd such Wisdom in an earthy Shape,
And show'd a Newton as we show an *Ape*

See letter to Charles Lyell, 27 and 28 April [1860].
[3] Pope 1733–4, Epistle II: 7–8:

in doubt to act, or rest,
To deem himself a Part of God, or Beast;

[4] Pope 1733–4, Epistle II: 20.

From Henry Doubleday 3 May 1860

Epping
May 3rd 1860

My dear Sir,

I have read with real pleasure your very interesting work "On the Origen of Species" and I may, some day, send you a few remarks upon portions of it—

My object in now writing is to say that I have tried a great many experiments upon the *Primrose*, *Oxlip* and *Cowslip*[1] all of which tended to confirm my opinion that these three plants are distinct species—or what are called species in all our Botanical works— I never could raise a *primrose* from the seed of the *Cowslip* nor a *Cowslip* from the seed of the *Primrose*—The true *Primula elatior* or *Oxlip* with pale, drooping, scentless flowers appears to be a very local plant in Britain—but it grows in great profusion in swampy meadows at Bardfield in Essex—[2] In wet winters these meadows are always flooded— The *Primrose* does not exist in the Parish of Bardfield although the woods and lanes there seem very favourable situation for it— The *Cowslip* is plentiful in the dry fields—

Twenty years since I brought a number of roots of *Oxlips* from Bardfield and planted them in my garden in a border under a north wall— they have grown luxuriantly there and seed profusely every year— Many thousands of seedlings have flowered and all have been similar to the original plants— There has not been a primrose or cowslip among them— I will send one or two seedlings of last year with this— I tried the experiment of crossing one of these plants with a primrose and the seedlings are now in flower and are just like primroses except in being on a raised flower stem— I will enclose a bunch of flowers—

On the edges of our woods and forests where the primrose and Cowslip grow together we often find plants—apparently hybrids between the two— these are commonly called oxlips— I believe they are never found except where the *cowslip* and *primrose* grow together— I have brought home many of these plants and have grown them in my garden for years *but in no instance have I known one to produce seed*—[3] I will send you a small piece of one to show you what I mean— you will know it by the deep colour—

I must ask you to excuse this hurried note and with best wishes in every way believe me | My dear Sir | Yours most sincerely | Henry Doubleday

C Darwin Esq

DAR 162.2: 237

CD ANNOTATION
3.6 I tried the experiment] 'I tried the experiment' *added ink*

[1] Doubleday was a naturalist and entomologist. CD cited Doubleday's paper on oxlips (Doubleday 1842) several times in his study of the relationships between cowslips, primroses, and oxlips in his 'big book' on species (*Natural selection*, pp. 128–33; the author of the paper was wrongly identified as Edward Doubleday by the editor of *Natural selection*).

[2] In the first edition of *Origin*, p. 49, CD mistakenly called the cowslip *Primula elatior*; in subsequent editions he changed his reference in accordance with information given to him by Hewett Cottrell Watson (see *Correspondence* vol. 7, letter from H. C. Watson, 30 November [1859] and n. 3). CD cited Doubleday in *Forms of flowers*, p. 32, as the first to call attention to the existence of *P. elatior* in England. See also letter from Henry Doubleday, 16 May 1860 and n. 2.

[3] In *Origin*, p. 49, CD referred to the primrose and cowslip as examples of strongly marked varieties that are usually classed as separate species because they are crossed only with difficulty and then produce sterile hybrids.

From Andrew Murray 3 May 1860

Eding. 1 Scotland S^t^
3 May 1860.—

My dear Sir

I have been from home & only returned yesterday—so that I have been unable sooner to thank you for your kind letters.— I was sure you would read my remarks in the spirit with which I offered them to you.—[1]

You are quite correct in supposing that want of space has prevented me alluding to many parts of your argument wh. bear upon my objections.— I did allude to them—but my paper has been so sadly cut down to bring it within the space allowed me that I am often compelled to do injustice both to you & to myself— that you bear so patiently the omission of what you must think a triumphant refutation of what I say, only makes me regret the more the dismemberment wh. my paper has undergone.— Want of space also prevents me enlarging upon & explaining some of my own crude notions—For instance as to "marrying opposites"[2] I imagine the seminal fluid of a hermaphrodite may be more readily received or taken up by the typical joint than by an abnormal one, in the same way that I hold the pollen of the typical plant to be prepotent over other varieties.— You dispute this & think it not fair to make the statet. so broadly unless I have facts to support it & that I should give them— I think I have facts, but I have not space to give them— For instance, Mr Isaac Anderson[3] a great hybridiser here—(that is *great here* as well as a hybridizer here) while working at making white calceolarias found the greatest difficulty in what he calls "washing out" the yellow— He had (by working in the dark & under blue glass) succeeded in getting a calceolaria with a white ground —at least that was the means by wh. he imagined he had succeeded—and he bred white calceolarias from that plant:—but in all his crossings the slightest tinge of yellow in any of the parents always carried the day—the native or typical yellow was prepotent.—

I have endeavoured to avail myself of your remarks so as to make the paper quite consistent with your views.— I have struck out wholly the passage about use & disuse & have appended a note to the "abominable animal" saying that I only quote it (as I have reason to believe you intended it) as an extreme & somewhat extravagant illustration intended to show forcibly the operation of Natural Selection—[4] I have cancelled the ostriches & the treeless plains—& considering that these passages are the only ones which approach to what is called smart writing the sacrifice is good evidence of my sincerity— "Defiantly" no longer rears his head—and I have stated as regards the cave animals that altho your illustration is taken from Schiodtes paper on insects, your remarks as you yourself have had the kindness to inform me were not intended to be Confined to insects.—[5]

I have been very much interested by your account of the result brought out by Mr Salters grouping the spirifers.[6]—and so far as this goes to prove descent I cannot help going very much along with you—but descent by what you would call a leap is all that I see in it— However without actually testing this & other evidences by ocular inspection it is very much like dogmatizing without knowledge to give an opinion upon it—

You wish a reference to the papers where the caves I mention are treated of.— There is not much of a paper about any of them— My information is taken from some descriptions of new insects found in these caves.—

Betharram Low Pyrenees—	Annales de la Soc: Ent: de France 1857. (Bulletin p. xciii)[7]	
Trou des Fades— pres d'Hyères.	D^o 1859.	p. 67[8]
Gargas— High Pyrenees	D^o. "	p. 71[9]
Grotto des Ours. Ligurie Orientale	D^o.	p. 26[10]
Monte Viso	D^o.	p. 27[11]

The Carniolan Cave Anophthalmi are described by Sturm[12]—but Schiodte is the only author so far as I know who has given anything like an acco^t. of the Caves (except Muller perhaps)

Laibach (Müller) Stettiner Ent: Zeitung (1856— 17th. year) p. 310[13]
D^o. Carniola—(Müller) D^o. 1857— (18th. y^r.). p. 65[14]
This is a good paper— Ueber die Lebensweise der Augenlosen Käfer in den Krainer Höhlen—

You ask me how the blind Paussi in foreign Countries get from one ants nest to another— Your question is pertinent enough as regards some insects such as Claviger, Batrisus, &^c., which are blind & live in ants nests—But the Paussi are not blind (at least I know of no blind ones).— On the contrary if we are to believe Afzelius[15] (one species is) provident with a phosphorescent globe to the antennae which it bears before it like a coach lamp giving light to good large prominent Eyes while traversing the dark galleries of the ants nests— I would *like* to believe this.— But it is scarcely less curious to find Paussi (different species mind—up to nearly 100 in number) found in ants nests all over the world, each country having its peculiar species— Physical condition I say— Natural Selection you say— It is a case very like that of the Cave insects—

I shall send you a copy of the paper when it is thrown off.

Believe me | My dear Sir | Yours very sincerely | And^w. Murray

DAR 47: 153–153a

CD ANNOTATIONS
1.1 I have . . . upon it—4.6] *crossed pencil*
Top of first page: '16'[16]
End of letter: 'May 5—1860' *pencil*; 'Ch 7'[17] *brown crayon*

1 See letters to Andrew Murray, 28 April [1860] and 28 [April 1860].
2 Murray 1860a, p. 277. See the enclosure to the letter to Andrew Murray, 28 April [1860].
3 Isaac Anderson-Henry was an Edinburgh solicitor and horticulturist (R. Desmond 1977).
4 Murray 1860a, p. 275 n. See the enclosure to the letter to Andrew Murray, 28 April [1860] and n. 14.
5 Schiödte [1849] was referred to in Murray 1860a, p. 287. See the enclosure to the letter to Andrew Murray, 28 April [1860] and n. 17.
6 John William Salter. See letter from Andrew Murray, 28 April [1860].
7 Murray refers to the report of a meeting of the Société Entomologique de France, at which Charles Delarouzée exhibited cave insects from Bétharram in the Hautes-Pyrénées (*Annales de la Société Entomologique de France* (1857), Bulletin, pp. xciii–xcv).
8 Delarouzée 1859, p. 67.

[9] Linder 1859, p. 71.
[10] Fairmaire 1859, p. 26.
[11] Fairmaire 1859, p. 27.
[12] Sturm 1844 and 1847.
[13] H. Müller 1856.
[14] H. Müller 1857.
[15] Afzelius 1798.
[16] The number of CD's portfolio of notes on hybridism.
[17] A reference to chapter 7 of *Natural selection*, in which blind animals and the effects of use and disuse are discussed.

To Charles Lyell 4 May [1860][1]

Down Bromley Kent
May 4th

My dear Lyell

I have had promised me an arrowhead found by Col. Erskine on his property in peat in Aberdeenshire;[2] which I will send you; but the only point of interest about the case, is that vast numbers were found at one place where there were stones fitted for the manufacture.— I daresay you know about it, but I may mention that John Lubbock tells me that the flint tools in France are found in such vast numbers, **in Peat** that M. Boucher de Perthes told him that he might take as many as he liked.—[3]

These facts, to my mind, remove one of the greatest difficulties of the case of the gravel-beds—celts,—namely their surprising numbers.— I do hope that you will go to France again, & give us lots of Sections.—[4] I found that until J. Lubbock drew me a rough section I did not *in the least* understand their position; & hardly anything seems known about the extension of the beds of gravel, clay &c or their manner of formation. The case seems to me to deserve not day's but month's of work—

I will keep Newberry's paper, as it is very interesting:[5] By the way I was much pleased to see how strongly he put the supposed fact that the continent of America has stood there since Palæozoic times.— I suspect my crude notion of the cause of deposits anterior to the Palæozoic times not being known, will hereafter be found to have truth in it—

I have had brief note from Keyserling,[6] but not worth sending you: he believes in change of species,—grants that natural selection explains well adaptation of forms, but thinks species change too regularly, as if by some chemical law, for natural selection to be sole cause of change.— I can hardly understand his brief note, but this is, I think, the upshot.—

Pray do not forget to look at the Spirifers arranged by Salter.—[7]

I will send A. Murray's paper whenever published: it includes speculations (which perhaps he will modify) so rash & without a single fact in support, that had I advanced them, he or other Reviewers would have hit me very hard.—[8]

I am sorry to say that I have no "consolotory view" on the dignity of man; I am content that man will probably advance & care not much whether we are looked at as mere savages in a remotely distant future.— Many thanks for your last note—[9]

Yours affect[ly] | C. Darwin

I have received in a Manchester Newspaper a rather a good squib, showing that I have proved "might is right", & therefore that Napoleon is right & every cheating Tradesman is also right—[10]

Endorsement: '1860'
American Philosophical Society (210)

[1] The date is incorrectly given as 'January 4th? 1860' in *LL* 2: 261.

[2] Henry Knight Erskine was lieutenant-colonel commandant of the Royal Aberdeenshire Highlanders (*Army list* 1860).

[3] Early in April 1860, John Lubbock travelled with Joseph Prestwich and others to Abbeville, France. They visited Jacques Boucher de Crèvecoeur de Perthes and examined the flint implements he had collected as well as the gravel-beds of the Somme Valley (Hutchinson ed. 1914, 1: 51).

[4] Lyell visited several of the sites that contained worked flints in 1859 and announced his belief that these were implements of early man at the 1859 meeting of the British Association for the Advancement of Science (see *Correspondence* vol. 7, letter to Charles Lyell, 20 September [1859]). He revisited France and also toured sites in Germany in September 1860 (K. M. Lyell ed. 1881, 2: 336).

[5] Newberry 1860. See letter to Charles Lyell, 27 and 28 April [1860].

[6] The note from Alexandr Andreevich Keyserling has not been found, but see the letter to R. I. Murchison, 1 May [1860].

[7] John William Salter was palaeontologist to the Geological Survey of Great Britain. At the Museum of Practical Geology, London, he had arranged specimens of the brachiopod genus *Spirifer* from three Palaeozoic stages, the Devonian, Lower and Upper Carboniferous, using CD's illustration of the principle of divergence (*Origin*, facing p. 117) as his model. He glued the spirifers onto a board and arranged them in single, branching lines, with horizontal lines marking the formations. See letter to Andrew Murray, 28 April [1860], and *Correspondence* vol. 9, letter to T. W. St C. Davidson, 26 April 1861. CD had seen Salter's arrangement in April 1860 (see letter to Andrew Murray, 28 April [1860]). See also letter from J. B. Jukes, 27 February 1860 and n. 5.

[8] See preceding letter and the letters to Andrew Murray, 28 April [1860] and 28 [April 1860].

[9] Letter from Charles Lyell, 2 May 1860.

[10] The *Manchester Guardian*, 20 April 1860, p. 4, carried a commentary on *Origin* entitled 'National and individual rapacity vindicated by the law of nature'. The article mentioned Louis Napoleon's recent seizure of Savoy. Only the author's initials T. H. W. were given.

From J. S. Henslow 5 May 1860

7 Downing Terrace | Cambridge
5 May 1860

My dear Darwin,

I read your wishes to my Class—& yesterday after lecture a few of us walked to Cherry Hinton—[1] I found the Elodea had made great progress since I saw it a year or two ago— It is now quite up to the source of th⟨e⟩ stream & fills the ditches, in Cherry Hinton itself— But I find this stream is not connected with the water course that runs past the Botanic Garden (as I had supposed) but runs down to the paper mills beyond Barnwell— It seems therefore to have travelled up from the river itself which is full of it— So far as mere recollection guides me it seems to have greatly diminished the quantity of Ranunculus aquatilis var fluitans which used to abound in the stream at the part we visited— It decidedly preponderates over every other aquatic, but I found it associated with Potamogeton densus (in small quantity) & Ranunculus aquatilis, & intermixed with plants of the Lemna trisulca—

Sedgwick is to illuminate us on Monday at the Philosophical Society in regard to your supposed errors![2] How can Owen be so savage with your views when his own are to a certain extent of the same character—[3] If I understand him, he thinks the "Becoming" of species (I suppose he means the *producing* of species), a somewhat rapid & not a slow process—but he seems to think them *progressive* organised out of previously organized beings— {analogous (?) to minerals (simple & compound) out of some ± 60 Elements} I don't think it is at all *becoming* in one Naturalist to be bitter against another—any more than for one sect to burn the members of another—

Kind regards to Mrs D.—&c— Y^rs affect^ly | J S Henslow

DAR 186.1

CD ANNOTATIONS

1.2 the Elodea] '(or Anacharis)' *added pencil*
2.1 Sedgwick . . . Henslow 3.1] *crossed pencil*
Top of first page: 'Ch 5 (& 20)'[4] *brown crayon*

[1] See letter to J. S. Henslow, 2 April [1860].

[2] Adam Sedgwick read a paper criticising *Origin* at a meeting of the Cambridge Philosophical Society on 7 May 1860.

[3] In November 1859, Owen had written that he was 'disposed to believe' in some form of transmutation (see *Correspondence* vol. 7, letter from Richard Owen, 12 November 1859). However, his review of *Origin* ([R. Owen] 1860a) was highly critical (see letters to T. H. Huxley, 9 April [1860], and to Charles Lyell, 10 April [1860]).

[4] 'Ch 5' was CD's chapter on the struggle for existence from his 'big book' on species (*Natural selection*, pp. 172–212). '20' was the number of his portfolio of notes on the geographical distribution of plants. CD was interested in the spread of plants like *Elodea* (or *Anacharis*) as examples of introduced species competing successfully with comparable native forms.

To Andrew Murray 5 May [1860][1]

Down Bromley Kent
May 5th

My dear Sir

I must thank you again for your kindness,—or rather I should say generosity. I find that I did not understand you about "prepotency";[2] I thought that you referred to fertility,— with respect to resemblance I fully believe there is often the greatest difference both in crossed species (as has been proved) & in crossed varieties in impressing likeness on offspring. I do not yet understand about marrying hermaphrodite opposites.

It is very kind in you to put in a note about the "abominable animal"—[3] the case has been struck out in my 2d Edition,—the main fact left in as illustration of *diversified habits*.—

With respect to the cave animals, reflect on the cave-Rat, the fish Amblyopsis & Astacus in America—the Proteus in caves of Europe,[4] & you will admit that on creation doctrine, there has been surprising diversity for such similar habitations.— Thank you very much for famous list of references.—

I blundered about the Ant-insect— I remembered that some were blind & thought it was Paussus. What a noble thing would natural science be, if all its followers were actuated by your kind & liberal spirit. If ever I feel inclined to write spitefully about an antagonist, I will try & remember & learn a lesson from you, my antagonist.

My dear Sir | Yours very sincerely | Charles Darwin

Since last writing I have heard from capital Geologist Count Keyserling & he makes no difficulty about imperfection of Geological Record.—[5]

R. D. Pyrah

[1] Dated by the relationship to the letter from Andrew Murray, 3 May 1860.

[2] See letter to Andrew Murray, 28 April [1860], and letter from Andrew Murray, 3 May 1860.

[3] CD refers to the bear–whale example given in *Origin*, p. 184. See the enclosure with the letter to Andrew Murray, 28 April [1860].

[4] *Proteus* is the blind cave salamander.

[5] Murray referred to the imperfection of the fossil record as a great difficulty for CD's theory in his review of *Origin* (Murray 1860a, pp. 281–4). CD refers to Alexandr Andreevich Keyserling. See letters to R. I. Murchison, 1 May [1860], and to Charles Lyell, 4 May [1860].

To J. D. Hooker 7 May [1860][1]

Down Bromley Kent
May 7th.—

My dear Hooker

Do you chance to know any observing man living in the country (whatever country that may be) inhabited by the Leschenaultia formosa?[2] I am very unhappy about this flower; & I can "wriggle out" of my difficulty only by supposing that Bees go from flower to flower & open the indusium. I know that Bees will do more difficult work for pollen. Our English Bees will not look at the flower. I would write & supplicate anyone (& use your name as Introduction), if any naturalist lives in country of Leschenaultia, to watch the Bees at work.—[3]

I have this morning been looking at my experimental Cowslips & I find some plants have all flowers with long stamens & short pistils which I will call "male plants"—others with short stamens & long pistils, which I will call "female plants".[4]

This I have somewhere seen noticed, I think by Henslow;[5] but I find (after looking at only two sets of plants) that the stigma of male & female is of slightly different shape & certainly different degree of roughness, & what has astonished me the pollen of the so-called female plant, though very abundant, is more transparent & each granule is exactly only $\frac{2}{3}$ of size of pollen of the so-called male plant.— Has this been observed?? I cannot help suspecting that the cowslip is in fact dioicous— but it may turn out all blunder, but anyhow I will mark with sticks the so-called male & female plants & watch their seeding. It would be fine case of gradation between an hermaphrodite & unisexual condition.— Likewise a sort of case of Balancement of long & short pistils & stamens.— Likewise perhaps throws light on Oxlips.—[6]

Yours affect. | C. Darwin

You will be sorry to hear that poor Etty has a remittent Fever, it is now become slight, but will run out for 14 or 21 days, the Doctor tells us.— It reduces her & exhausts her much.—[7]

I have now examined Primroses & find exactly same difference in size of pollen, correlated with same difference in length of style & roughness of stigma!—

Cowslips & Primroses

Male plants.	long stamens,	big pollen-grains,	short style,	smooth stigma,
Female plants	short stamens,	small " ",	long ",	rough " ,

I measured pollen-grains in water & dry by Micrometer.—

Endorsement: '/60'
DAR 115.1: 52

[1] The endorsement is confirmed by the relationship to the letter from J. D. Hooker, [28 April 1860].

[2] *Leschenaultia formosa* is a native of Australia. Hooker had supplied specimens of this species for CD's investigation of the plant's pollination mechanism (see letters to J. D. Hooker, [26 February or 4 March 1860] and 12 March [1860]). CD's experiments are recorded in his Experimental book, pp. 50–2 (DAR 157a).

[3] See letters to J. D. Hooker, 15 [May 1860], and to James Drummond, 16 May 1860.

[4] CD's observations on cowslips, dated 8 May 1860, are recorded in his Experimental book, p. 53. His work and the identification of two kinds of cowslip flower were eventually described in a paper 'On the two forms, or dimorphic condition, in the species of *Primula*, and on their remarkable sexual relations', published in the *Journal of the Proceedings of the Linnean Society (Botany)* 6 (1862): 77–96 (*Collected papers* 2: 45–63).

[5] Henslow 1830. CD cited John Stevens Henslow's work on primroses and cowslips in his 'big book' on species (*Natural selection*, pp. 129–30).

[6] Oxlips were commonly thought to be the hybrid offspring of primroses and cowslips. CD, however, believed the oxlip to be a form intermediate between the two that retained characters of the common ancestor of cowslips and primroses (see *Natural selection*, pp. 132–3). See also letter from Henry Doubleday, 3 May 1860.

[7] Henrietta Emma Darwin was seriously ill throughout much of 1860. Emma Darwin's diary records the onset of her ill health on 28 April.

To T. H. Huxley 7 May [1860][1]

Down Bromley Kent
May 7th

My dear Huxley

As you seemed somewhat interested about changes in proportions of Pigeons, I will tell you the general results after reworking the whole subject in every possible manner & by the aid of about (more or less) ten thousands rule of three sums.—[2]

Sternum, scapulæ & furcula invariably reduced in size. Feet & tarsi generally reduced, in size, but are more plainly governed by correlation with length of Beak.—

Wings, measured from end to end of radius, generally reduced in length—viz in 17 Birds (of different races); but in 8 other birds rather increased in length, but then 5 of these 8 are long-beaked birds. But the wings if measured from extreme tip to tip are generally much increased in length—viz in 21 birds longer & in only 5 shorter.—

So that the proportion of birds with longer or shorter wings proportionally with the wild rock-pigeon are reversed, according as the measurement is made from tip to tip of wing or from end to end of radius.—

The increased length of wing, when measured from tip to tip, seems wholly due to greater length of the primary wing feathers, & may not be more related to use & disuse than the quicker or slower growth of nails on our fingers.—

Have you seen last Saturday Review?[3] I am very glad of the defence of you & of myself.— I wish the Reviewer had noticed Hooker.—[4] The Reviewer, whoever he is, is a jolly good fellow, as this Review & the last on me, showed. He writes capitally & understands well his subject.— I wish he had slapped Owen a little bit harder[5]

Ever yours | C. Darwin

Do not answer this.—

Imperial College of Science, Technology and Medicine Archives (Huxley 5: 117)

[1] Dated by the reference to the *Saturday Review* (see n. 3, below).

[2] CD was writing up the results of his study of pigeons for his larger work on species ('Journal'; Appendix II). He refers to the measurements he made comparing the bone dimensions of wild birds and domestic breeds. In most cases he used the common arithmetical 'rule of three' to find the quotient, against which he assessed the actual measurements of domestic breeds. The measurements are given in *Variation* 1: 162–79.

[3] An anonymous review of Richard Owen's *Palæontology* (R. Owen 1860b) appeared in the *Saturday Review*, 5 May 1860, pp. 573–4. The reviewer suggested that Owen was the author of the unsigned, critical review of *Origin* in the *Edinburgh Review* ([R. Owen] 1860a), as in fact he was. See also letter to Asa Gray, 22 May [1860].

[4] CD refers to Hooker 1859.

[5] The review defended CD's right to propose hypotheses and criticised Owen's anonymous attack on Huxley's Royal Institution lecture ([R. Owen] 1860a, pp. 521–3).

From Charles Lyell 7 May 1860

53, Harley St. London. W.
May 7. 1860

My dear Darwin

I saw Salters spirifers, a very good proof of the fructifying nature of your indefinite modifiability principle.[1] I also asked him if he did not think that Barrandes "primordial" (the Cambrian of Manual 5th Edn.) was getting more & more separated from Lower Silurian—[2] He replied certainly— I am beginning to think the gap was enormous between them. And then there are the Huronian & Laurentian each with rippled sands 1000ds of feet thick unconformable to the Potsdam or "primordial".[3] The latter I hear has just been found by Cassiano del Prado in some part of Spain.[4]

Suppose the gap as I begin to suspect to be as great as between Cretaceous & Eocene what a grand discovery the Cambrian type is—

On Wednesday May 16th we expect a large meeting of the Geol. Socy in the hall in Burlington House when Lartet's paper[5] on contemporaneity of

AL incomplete
DAR 205.9 (Letters)

CD ANNOTATIONS
3.1 On . . . contemporaneity of 3.2] *crossed brown crayon*
Top of first page: '22'[6] *brown crayon*

[1] See letter to Charles Lyell, 4 May [1860].

[2] Joachim Barrande believed that the lower Silurian formations around Prague constituted a new system, which he named the 'Primordial'. Comparable rocks in Britain were the subject of heated debate between Adam Sedgwick and Roderick Impey Murchison. Sedgwick insisted, like Barrande, that the formations were sufficiently distinct to merit a separate name (the 'Cambrian'); Murchison, on the other hand, maintained that they were merely part of the lower Silurian. In his *Manual of geology* (C. Lyell 1855, pp. 452–5), Lyell had given his support to Sedgwick's view. For a study of the debate, see Secord 1986.

[3] Huronian and Laurentian formations, which were being actively studied by members of the Geological Survey of Canada at the time, were later identified as pre-Cambrian. See Zaslow 1975.

[4] Casiano de Prado began a geological survey of Spain in 1848.

[5] Édouard Amant Isidore Hippolyte Lartet's excavations of flint-bearing sites helped to establish the contemporaneity of man and extinct animals (*DSB*). His paper 'On the coexistence of man with certain extinct quadrupeds' (Lartet 1860) was read at a meeting of the Geological Society on 16 May 1860. Leonard Horner, president of the society, contributed remarks that were printed with the paper.

[6] The number of CD's portfolio of notes on palaeontology and extinction.

To J. S. Henslow 8 May [1860]

Down Bromley Kent
May 8th

My dear Henslow

Very many thanks about the Elodea,—which case interests me much. I wrote to Mr Marshall at Ely & in due time he says he will send me whatever information he can procure.—[1]

Owen is indeed very spiteful. He misrepresents & alters what I say very unfairly. But I think his conduct towards Hooker most ungenerous, viz to allude to his Essay, & not to notice the magnificent results on geographical distribution.[2]

The Londoners says he is mad with envy because my book has been talked about: what a strange man to be envious of a naturalist like myself, immeasurably his inferior! From one conversation with him I really suspect he goes at the bottom of his hidden soul as far as I do!—

I wonder whether Sedgwick noticed in the Edinburgh Review, about the "Sacerdotal revilers"[3]—so the revilers are tearing each other to pieces.— I suppose Sedgwick will be very fierce against me at the Phil. Soc.—[4] Judging from his notice in the Spectator he will misrepresent me, but it will *certainly* be unintentionally done.—[5] In a letter to me & in the above notice he talks much about my departing from the spirit of inductive philosophy.—[6] I wish, if you ever talk on subject to him, you would ask him whether it was not allowable (& a great step) to invent the undulatory theory of Light—ie hypothetical undulations in a hypothetical substance the ether. And if this be so, why may I not invent hypothesis of natural selection (which from analogy of domestic productions, & from what we know of the struggle of existence & of the variability of organic beings, is in some very slight degree in itself probable) & try whether this hypothesis of natural selection does not explain (as I think it does) a large number of facts in geographical distribution—geological succession—classification—morphology, embryology &c. &c.— I shd really much like to know why such an hypothesis as the undulation of the ether may be invented, & why I may not invent (not that I did *invent* it, for I was led to it by studying domestic varieties) any hypothesis, such as natural selection.

Pray forgive me & my pen for running away with me & scribbling on at such length.—

My dear old Master | Yours affectly | C. Darwin

I can perfectly understand Sedgwick or any one saying that nat. selection does not explain large classes of facts; but that is very different from saying that I depart from right principles of scientific investigation.—

Endorsement: '1860'
DAR 93: 67–9

[1] Letter to William Marshall, 9 April [1860].

[2] Richard Owen had reviewed Hooker 1859 as well as *Origin* in [R. Owen] 1860a.

[3] A reference to the statement in [R. Owen] 1860a, p. 511:

> We have no sympathy whatever with Biblical objectors to creation by law, or with the sacerdotal revilers of those who would explain such law.

[4] Adam Sedgwick was preparing a public address on *Origin* to be delivered before the Cambridge Philosophical Society. See letter from J. S. Henslow, 5 May 1860, and letter from J. S. Henslow to J. D. Hooker, 10 May 1860.

[5] [Sedgwick] 1860.
[6] See *Correspondence* vol. 7, letter from Adam Sedgwick, 24 November 1859.

To Charles Lyell 8 [May 1860]

Down Bromley Kent
8th

My dear Lyell

Your letter contained much news for me.— It is indeed most important that the Cambrian or Barrandes primordial should be getting so much separated (in spite of Murchison) from Lower Silurian: I did not at all know this.[1]

I fear it is not very likely that I shall be up on 16th (I have written to J. Lubbock)[2] though I shd. much like to hear the paper & discussions, for, I am sure that you will both be sorry to hear that Etty is ill with remittent Fever.[3] There is at present no cause for anxiety, though she is very ill, & the Fever is sure to run on for a fortnight or three weeks, and anxiety always knocks me up.—

I have sent for Canadian Naturalist;[4] if I cannot procure a copy I will borrow yours.—I had letter from Henslow this morning who says that Sedgwick was on last Monday night to open a battery on me at Cambridge Phil. Socy.—[5] Anyhow I am much honoured by being attacked there & at Royal Soc. of Edinburgh.—[6] With respect to Aster, I remember long ago reading curious paper by Asa Gray & another on Aster, in which they give cases of two forms so very distinct that they must consider them specifically distinct, & yet perfectly united by intermediate varieties or links & they admit that in almost every other case that this suffices to upset two species as distinct.—[7] But I do not think it worth while to contradict single cases. Nor is it worth while arguing against those who do not attend to what I state. A moment reflexion will show you that there must be (on our doctrine) large genera not varying—see p. 56 on this subject in 2d Edit of Origin— Though I do not there discuss case in detail—

It may be sheer bigotry for my own notions, but I prefer to the Atlantis my notion of plants & animals having migrated from old to new world or conversely when climate was much hotter by approximately the line of Behring's Straits.[8] It is most important as you say to see living forms of plants going back so far in time. I wonder whether we shall ever discover the flora of dry land of the Coal period & find it not so anomalous as the swamp or coal-making flora!

I am working away over the blessed Pigeon manuscript;[9] but from one cause or another I get on very slowly,—chiefly from my abominable stomach.—

This morning I got letter from Acad. Nat. Sc. of Philadelphia, announcing that I am elected a Correspondent;[10]— I do not suppose that this is much of honour; but it shows that some naturalists there do not think me such a scientific profligate as many think me here.—

Ever | My dear Lyell | Yours gratefully | C. Darwin

What a grand fact about extinct Stags horn worked by man!—[11]

Endorsement: 'May 1860'
American Philosophical Society (211)

[1] See letter from Charles Lyell, 7 May 1860. CD refers to Roderick Impey Murchison.

[2] In the missing portion of his previous letter, Lyell had evidently mentioned several of the topics to be discussed at the next meeting of the Geological Society of London on 16 May 1860 (see letter from Charles Lyell, 7 May 1860). Édouard Amant Isidore Hippolyte Lartet and Leonard Horner spoke about the evidence for humans having existed contemporaneously with animals that had since become extinct. John Lubbock was also deeply interested in the topic.

[3] An entry in Emma Darwin's diary on 7 May 1860 records that Henrietta Emma Darwin was suffering from 'cold & fever'. She had fallen ill on 28 April 1860 (Emma Darwin's diary).

[4] A review entitled 'Darwin on the origin of species by means of natural selection', signed 'J. W. D.', appeared in the *Canadian Naturalist and Geologist* 5 (1860): 100–20. John William Dawson was the editor of the journal. CD's copy of the article (Darwin Pamphlet Collection–CUL) identifies Dawson as the author; CD also wrote on it: 'good geologist & Palaeontologist.'

[5] See preceding letter and letter from J. S. Henslow to J. D. Hooker, 10 May 1860.

[6] A reference to Andrew Murray's critique of *Origin* in a paper read before the Royal Society of Edinburgh on 20 February 1860 (Murray 1860a).

[7] CD may be referring to Torrey and Gray 1838–43, 2: 104, in which the following statement is included in a passage discussing difficulties in the classification of species of *Aster*:

> It must, however, be admitted that, in this as in all large and natural genera, several species which we cannot but consider as distinct (such for instance as A. cordifolius and A. sagittifolius) do frequently present very puzzling intermediate forms;

[8] CD refers to Edward Forbes's proposed former Miocene continent, across which organisms were supposed to have migrated from Europe to the Azores. CD rejected hypothetical continents and other land-bridges for which there was no obvious geological evidence in favour of occasional and accidental transport of seeds and ova across the intervening ocean. He and Lyell had discussed the subject at length in 1856. See *Correspondence* vol. 6, especially letters to Charles Lyell, 16 [June 1856] and 25 June [1856], and letters from Charles Lyell, 17 June 1856 and [1 July 1856].

[9] CD was drafting chapters on pigeons that were eventually published in *Variation* (see 'Journal'; Appendix II).

[10] CD had been elected as a corresponding member of the Academy of Natural Sciences in Philadelphia (see letter to Thomas Stewardson, 8 May 1860). The academy, founded in 1812, was one of the most prestigious scientific societies in the United States.

[11] Lartet 1860, p. 472.

From William Masters 8 May 1860

Canterbury
8 May 1860

Dear Sir,

In reply to your enquiry about Hollyhocks, the sorts were all growing in one bed, and in rows many plants of each kind in a row—[1]

As it regards Cabbages & Savoy &c I have found the purple Cale or Kale to be the most infectious of all Varieties—[2] a few years agone, a Cottager at $\frac{1}{2}$ a mile distance was growing some for Seed and to my knowledge no Kale was nearer—and yet my pure stocks became seriously affected with purple bastards (sic Shakespear) the year following.

In this case insects were the probable Carriers.

Believe me Dear Sir | Your obliged Servant | William Masters

Ch Darwin Esqr

DAR 76 (ser. 2): 166–7

CD ANNOTATIONS

1.1 In reply . . . a row—1.2] *crossed pencil*; 'Distance of pollen crossing' *added red crayon*
2.1 Cabbages] *underl red crayon*
Top of first page: 'Ch 3.' *brown crayon* ; 'Keep for last chap'; 'Also Gartner'[3] *red crayon*

[1] CD had directed queries about variation and crossing in sweetpeas to William Masters earlier in the year through his son Maxwell Tylden Masters. See letters to M. T. Masters, 7 April [1860] and 13 April [1860], and letter from William Masters, [after 7 April 1860]. CD seems to have sent further inquiries between 13 April and 8 May, but these have not been found. William Masters was a seedsman who lived in Canterbury, Kent.

[2] Cabbages are described in *Variation* as plants that cross readily.

[3] CD's annotation refers to chapter 3 of *Natural selection*, 'On the possibility of all organic beings occasionally crossing'. The plant breeding experiments of Karl Friedrich von Gärtner are cited frequently in this chapter.

To Thomas Stewardson 8 May 1860

Down Bromley Kent
May 8. 1860

Sir

I beg leave to acknowledge the receipt of your letter of April 16th, in which you announce to me that I have been elected a Correspondent of the Academy of Natural Sciences of Philadelphia.[1] I assure that I fully appreciate & feel grateful for the great honour thus conferred on me, & I remain

Sir | Your obliged & obedt servant | Charles Darwin

To | Dr. Th. Stewardson | Corr. Secretary

Academy of Natural Sciences of Philadelphia

[1] The letter has not been found. Stewardson was the corresponding secretary of the Academy of Natural Sciences of Philadelphia.

To W. B. Tegetmeier 8 May [1860][1]

Down Bromley Kent
May 8th

My dear Sir

You have answered all my questions as fully, as clearly & as kindly as you have ever given me information on any point on which I was curious. I thank you sincerely.— I am now drawing up an account of the differences of the breeds of pigeons & this morning used on your authority the very curious case about the down of just hatched Pigeons & the colour of the mature bird.—[2]

By the way it just occurs to me, whether any of the Breeders have noticed any difference in length of time of eggs being hatched in such different breeds as Tumbler, Turbit, Pouter, Carriers & Runts.— I do not suppose there would be any difference, but it would be a good fact to give if you could ascertain it without much trouble & permit me to quote it on your authority. I never thought about it, till today or I would have observed when I had pigeons alive. It would be almost equally curious to know *positively* that such differently sized birds as Runts & Tumblers hatched their eggs in same time.—

With cordial thanks | Yours sincerely | C. Darwin

P.S. I hope that you will give my sincere thanks to Mr H. Weir, Mr Esquilant & Mr Jones for their kindness.—[3]

Beinecke Library, Yale University

[1] Dated by CD's reference to writing his manuscript on pigeons (see n. 2, below).

[2] CD's journal indicates that he wrote up his material on pigeons between March and June 1860 ('Journal'; Appendix II). The manuscript was eventually published as part of *Variation*. The case referred to was described in *Variation* 1: 170:

> Mr. Tegetmeier has informed me of a curious and inexplicable case of correlation, namely, that young pigeons of all breeds, which when mature become white, yellow, silver (*i.e.* extremely pale blue), or dun-coloured, are born almost naked; whereas other coloured pigeons are born well clothed with down.

The case was mentioned again in *Variation* 2: 332. See also *Correspondence* vol. 6, letters to W. B. Tegetmeier, 19 November [1856], 12 [May 1857], and [18 June 1857].

[3] CD thanked Harrison William Weir for his assistance in *Variation* 1: 132 n. 2. F. C. Esquilant, secretary of the Philoperisteron Society, a pigeon-fanciers' club, is also cited in support of this point in *Variation* 1: 170. P. H. Jones was a leading member of both the Philoperisteron and the National Columbarian Society and became the first secretary of the amalgamated National Peristeronic Society (Fulton [1874–6]).

To *Cottage Gardener* [after 8 May 1860][1]

[Down]

"A Devonshire Bee-keeper" states (page 94) that he caught a common drone entering one of his hives of the pure Ligurian stock.[2] Will he have the kindness to state at what distance in a straight line there are hives of the common bee? I believe it is not known how far the drones commonly wander from their own hive. Andrew Knight believed, as stated in the "Philosophical Transactions," that the queen was seldom fertilised by her own blood-relations, the drones of her own hive.[3] Does "A Devonshire Bee-keeper," who seems to be so conversant with the habits of bees, believe in this doctrine of Andrew Knight?—[4]

C. D.

Cottage Gardener 24 (1860): 143

[1] The letter is a reply to a note published on 8 May 1860 in the *Cottage Gardener* 24 (1860): 94. CD's letter appeared in the issue of 29 May. There is an undated draft of this letter in the Fitzwilliam Museum library, Cambridge.

[2] The article, signed 'A Devonshire bee-keeper', was published in the *Cottage Gardener* on 8 May 1860 (see n. 1, above). The author stated that two out of his four Ligurian queens appeared to be hybrids and that he had seen an ordinary drone entering the hive of his Ligurian bees.

[3] Knight 1828, p. 307. See also letter to W. B. Tegetmeier, 17 April [1860].

[4] In a reply, dated 24 May, that immediately followed CD's query, 'the Devonshire bee-keeper' reported that he now believed the 'unwelcome stranger' to have been a 'small hybrid drone' rather than 'a straggler from some other colony.' On other evidence, he inferred that drones extended their flight to a greater distance than was generally imagined. Nevertheless, he believed that females were 'very generally fertilised by the offspring of the same mother' (*Cottage Gardener* 24 (1860): 143).

From J. S. Henslow to J. D. Hooker[1] 10 May 1860

7 Downing Terrace | Cambr.
10 May 1860.

My dear Joseph,

I don't know whether you care to hear Phillips, who delivers the Rede Lecture in the Senate House next Tuesday at 2. P.M.[2] It is understood that he means to attack the Darwinian hypothesis of Natural Selection.

Sedgwicks address last Monday[3] was temperate enough for his usual mode of attack, but strong enough to cast a *slur* upon all who substitute hypotheses for strict induction, & as he expressed himself in regard to some of C.Ds suggestions as *revolting* to his own sense of wrong and right & as Dr Clark[4] who followed him, spoke so unnecessarily severely against Darwin's views; I got up, as Sedgwick had alluded to me, and stuck up for Darwin as well as I could, refusing to allow that he was guided by any but truthful motives, and declaring that he himself believed he was exalting & not debasing our views of a Creator, in attributing to him a power of imposing laws on the Organic World by which to do his work, as effectually as his laws imposed upon the inorganic had done it in the Mineral Kingdom—

I believe I succeeded in diminishing, if not entirely removing, the chances of Darwin's being prejudged by many who take their cue in such cases according to views of those they suppose may know something of the matter— Yesterday at my lectures I alluded to the subject,[5] & showed how frequently naturalists were at fault in regarding as *species*, forms which had (in some cases) been shown to be varieties, and how legitimately Darwin had deduced his *inferences* from positive experiment— Indeed I had, on Monday, replied to a sneer (I don't mean from Sedgwick) at his pidgeon results, by declaring that the case *necessitated* an appeal to such *domestic* experiments, & that this was the legitimate & best way of proceeding for the detection of those laws which we all endeavouring to discover—

I do not disguise my own opinion that Darwin has pressed his hypothesis too far—but at the same time I assert my belief that his Book is (as Owen described it to me)[6] the 'Book of the Day'— I suspect the passages I marked in the Edinburgh Review[7] for the illumination of Sedgwick have produced an impression upon him to a certain extent— When I had had my say, Sedgwick got up to explain, in a very few words, his good opinion of Darwin, but that he wished it to be understood that his chief attacks were directed against Powel's late Essay,[8] from which he quoted passages as "from an Oxford Divine" that would astound Cambridge men, as no doubt they do. He showed how greedily, (if I may so speak) Powell has accepted all Darwin had suggested, & applied these suggestions (as if the whole were already proved) to his own views—

I think I have given you a fair, tho' very hasty, view of what happened, & as I have just had a letter from Darwin, & really have not a minute to spare for a reply this morning perhaps you will send this to him, as he may like to know, to some extent, what happened—

As he also wishes to know of all criticisms, pro & con, he will find an *adverse* view in the last N° (just received) of the Dublin Magazine of Natural history—[9] Of course he knows of the reply to Owen in a late Saturday magazine,[10] & also the articles in Spectator—[11]

Let me know, as soon as you conveniently can, whether my *ideas* of the Palace lectures meet yours.[12] Every lecturer must, of course, be guided to a considerable extent by his own— But there may be something or other which he has neglected, or is not aware of, that would induce him to modify his plans— I have only a fortnight more here, & have to *select* such materials as I think may be useful from the Museum here, & pack them up—so I have not much time to spare— A few words will be enough to set me *thinking* & if these lectures are to come off I should wish to make them as instructive or *suggestive* (as well as agreeable) as 4 lectures may admit, & my opportunities allow— I must now have 100 auditors here, the room is so full;—& above 50 seem likely to try for a pass this year— We have had 3 meetings within the week to arrange our new Schemes for Triposes—& so far all is working well—[13] Grace Hawthorn is here for ± 24 hours— Leonard has been for 2 or 3 nights—[14]

Love to F^s^ &c.[15] | Ever affectl^y^ | J. S Henslow

R. A. Hooker

[1] Henslow suggested that Hooker send the letter on to CD. CD discusses it in his letters to J. D. Hooker, 13 [May 1860], and to J. S. Henslow, 14 May [1860].

[2] The provisions of the Cambridge University lectureship endowed in 1524 by Robert Rede, chief justice of the common pleas, had been revised in 1858. The new statutes stipulated that a man of eminence in science or literature be appointed annually to deliver one lecture in the Easter term. The Rede lecture of 1859, the first under the new system, was given by Richard Owen. John Phillips was president of the Geological Society and professor of geology at Oxford University. The title of Phillips's lecture was 'Life on the earth, its origin and succession' (Phillips 1860). Phillips addressed, among other topics, the question of natural selection and the mutability of species (Phillips 1860, pp. 200–17).

[3] Adam Sedgwick read a paper on CD's theory at a meeting of the Cambridge Philosophical Society on 7 May 1860. A synopsis of his remarks was reported in the *Cambridge Chronicle*, 19 May 1860, pp. 4–5. The *Proceedings of the Cambridge Philosophical Society* 1 (1843–63): 223 notes only that Sedgwick's lecture was entitled 'On the succession of organic forms during long geological periods; and on certain theories which profess to account for the origin of new species'.

[4] William Clark was professor of anatomy at Cambridge University. See also letter to Charles Lyell, 18 May [1860] and n. 7.

[5] Henslow was professor of botany at Cambridge University from 1825 to 1861. He resided in Cambridge while giving his annual course of lectures.
[6] Richard Owen.
[7] Henslow refers to Owen's review of *Origin* ([R. Owen] 1860a).
[8] Baden Powell, until his death in June 1860, was Savilian professor of geometry at Oxford University. The 'late Essay' was his contribution to the 1860 volume of *Essays and reviews*, in which he referred to *Origin* as 'a work which must soon bring about an entire revolution of opinion in favour of the grand principle of the self-evolving powers of nature' (Powell 1860, p. 139). See also CD's two letters to Baden Powell, 18 January [1860], and letter to Charles Lyell, 15 [February 1860], in which CD quotes Powell as having told him he had 'never read anything so conclusive' as the passage in *Origin* on the eye.
[9] The first number of the *Natural History Review* for 1860 included an anonymous article entitled 'Biogenesis' that criticised *Origin*. The author was the Dublin geologist Samuel Haughton ([Haughton] 1860b). See letter to a Bookseller, 13 [May 1860].
[10] The *Saturday Review*, 5 May 1860, pp. 573–4, carried an anonymous review of Richard Owen's *Palæontology* (R. Owen 1860b). The reviewer took the opportunity to discuss *Origin*, which he noted was scarcely mentioned in Owen's book. In particular, the reviewer commented favourably on the principle of natural selection:

> Natural selection will, we are persuaded, be henceforward recognised as a *vera causa* which has operated both in modifying animal and vegetable forms and in extinguishing those that have ceased to be fully adapted to the surrounding conditions of existence; but we are equally persuaded that, taken by itself, it is inadequate to explain the entire past history of vital phenomena as developed in our planet.

[11] The *Spectator*, 24 March 1860, pp. 285–6, published remarks on *Origin* communicated to the magazine by Richard Whately, archbishop of Dublin. A 'revised and corrected' reprint of the article also appeared in the issue of 7 April 1860, pp. 234–5. The author primarily attacked CD's geological arguments but also mentioned a number of other points. A letter dated 16 April 1860 from Belfast and signed 'J. J. M.', appeared in the issue of 21 April 1860, p. 380, under the title 'Darwin's theory defended'. The author responded to several of the critical points raised in the earlier article and concluded: 'I do not see how any unprejudiced person can read Darwin's masterly chapters on the geographical distribution of organisms, without seeing that his argument is to a great extent *cumulative*. But the main strength of the transmutation theory is in the facts of morphology, especially in the existence of rudimentary organs.'
[12] Henslow was to deliver botanical lectures to the children of Queen Victoria and Prince Albert at Buckingham Palace. See letter from J. D. Hooker, [20 April 1860].
[13] The tripos is the system of honours examination for Cambridge University undergraduates. In 1859, the university administration set up a committee to consider changes in the natural sciences tripos. As a result of the committee's report, a board of studies was established in March 1860 for the natural sciences. Henslow, the professor of botany, was a member of the new board (see Winstanley 1947, pp. 187–8).
[14] Leonard Ramsay Henslow, also a clergyman, was Henslow's oldest son. It has not been possible to identify Grace Hawthorn.
[15] Henslow's daughter Frances Harriet was Hooker's wife.

From H. C. Watson 10 May 1860

Thames Ditton | Kingston—S.W.
10 May 1860

My dear Sir

I return by same post the four notices of 'Origin' which you kindly sent.—Gradually, the various arguments, objections, cavils, etc. on the grand subject are growing into a sort of conglomerate in my recollections; who said this, & who said that, getting pretty much confused together. I hope *you* can keep them mentally distinct.

Pictet goes a good way,[1]—farther perhaps than he himself clearly knows,— on page 15, where you pencil the no. 12,[2] your difficulty is & will be to lead zoologists & botanists, those who look to existing "species", so far as Pictet "believes". Until a faith in certain impassable barrier between existent species becomes thoroughly shaken, naturalists will resist your views, & hail difficulties as if conclusive arguments on the contra side.

Differently as these unseen barriers are traced or placed, they are believed in about as strongly by almost all. Let the subjoined numerals represent botanical forms, & the lines be the barriers as placed by different botanists;

Jordan[3]   1 | 2 | 3 | 4 | 5 | 6 | 7 | 8 | 9 | 0
Babington[4] 1 2 | 3 4 | 5 6 | 7 8 | 9 0
Bentham[5]   1 2 | 3 4 5 6 | 7 8 9 0
Hooker[6]   1 2 3 4 5 6 | 7 8 9 0
  at any rate,
  previous to 1859.

This represents an approximate uniformity which is not always found; the differences lying only in the more numerous barriers of J e B, as compared with B e H.— But these transpositions sometimes occur;

Hooker 1 2 3 4 — One species
Smith[7] 1 2 | 3 4 — Two species
Bab^{n}. 1 4 | 2 3 — Two (other) species

Shall I bore you with a simple illustration of convergence?[8] You have at any rate the option not to read it.

Take two plants, nore or less unlike in all other respects than these three characters:—

1.	Glabrous —	Leaves linear —	Flowers white.
2.	Pubescent —	Leaves oval —	Flowers blue
1	varies with leaves lanceolate	—	Flowers pink.
2	varies with Leaves lanceolate	—	Flowers pink.

Then let one (1) become pubescent, or the other (2) glabrous; & the convergence is completed so far as the three characters go.—

Accumulate similar convergences, in your way of accumulating divergences, and though you may never actually change two *distant* species into *each other*, you still may counterpoise endless divergence. Now, as facts in nature, the changes from pubescent to glabrous,—from linear or oval to lanceolate,—from white or blue to pink,—are ordinary, & excite no surprize singly. And yet those three characters would be held (as a rule) amply sufficient to justify a separation of 1 e 2 as perfectly distinct species, tho' all other technical characters should agree.

I presume you would say that some ancestors of 1 e 2 had lanceolate leaves & pink flowers, which had diverged into linear & white, oval & blue. But you can at best only *suppose* this lanceolate & pink fore-parent. And if you try to account for all converging variations in the like manner, your original adam or dozen of adams must surely have combined every character in one? Unless you believe that diverging characters are novelties, while converging are restorations only?

Sincerely yours | Hewett C! Watson

C. Darwin | Esqe

DAR 47: 160–1

CD ANNOTATIONS

2.1 Pictet . . . knows,—] *double scored pencil*

5.1 Shall . . . divergence. 8.3] 'The variations wd have to be *[illeg]* else one form wd blend the other if living together; if separate not likely same organic & inorganic conditions'; 'Have 2 varieties ever produced an identical variety.—' *added ink*

Top of first page: 'Ch VI'[9] *brown crayon*

End of letter: 'I think it possible or even probably that offspring of 2 species might & vary & converge so as to be classed as vars. as one.'; 'In cultivated vars I daresay many would be placed in wrong groups from converging variation.—' *ink*

[1] Pictet de la Rive 1860. CD's copy of the review is in the Darwin Pamphlet Collection–CUL.

[2] CD's copy of Pictet de la Rive 1860 has been marked with the numbers 1 to 15, sometimes in an unidentified hand. Two passages are marked '12'. The one on p. 15, indicated in the letter, reads:

> Je crois que la longue série des temps géologiques peut lui avoir donné un peu plus de portée, et lui a permis, par l'accumulation de résultats analogues, de faire naître d'une même espèce quelques espèces très-voisines. [I believe that the long sequence of geological time could have given him a longer period of gestation, making it possible, by the accumulation of analogous results, for one species to give birth to other, closely related species.]

[3] Alexis Jordan was renowned for his tendency to classify plants as distinct species that other botanists might consider varieties of a single species. Watson had made a similar comparison between taxonomic 'splitters' and 'lumpers', also citing Jordan, in letters to CD in 1855. See *Correspondence* vol. 5, letters from H. C. Watson, 13 August 1855, 2 October 1855, and 8 November 1855.

[4] Charles Cardale Babington was also known as a 'splitter'. See *Correspondence* vol. 7, letters to C. C. Babington, 22 February [1858], and to J. D. Hooker, 28 February [1858].

[5] George Bentham.

[6] Joseph Dalton Hooker.

[7] James Edward Smith.

[8] Watson had sent CD a long memorandum on the problem of convergence in species earlier in the year. See letter from H. C. Watson, [3? January 1860].

[9] CD's annotation refers to chapter 6 of *Origin*, entitled 'Difficulties on theory'. CD inserted a discussion of Watson's point in both the revised American edition and in the German translation of *Origin*. See letters to Asa Gray, 8 March [1860], and to H. G. Bronn, 21 March [1860].

To J. D. Hooker 11 May [1860][1]

Down Bromley Kent
May 11th

My dear Hooker

I hope I shall not bore you with my infinitesimally small facts.— I have just opened indusium of 8 flowers (& some before) of the Leschenaultia & in 7 abundance of pollen within, but in no one case was there any pollen where you indicate in your nice drawing the stigmatic surfaces to lie.[2] Therefore I believe insect agency to be necessary for **self** impregnation. My Greenhouse is too cold for the Plants, but I will try. But will you get one of your men with fine camel-hair pencil-brush to poke into indusium, so as to force pollen to the bottom & then *mark the flowers* operated & watch whether they set seed or whether they even commence to set seed, or show other signs of fertilisation.—

Do not hate me my dear fellow, | your affect | C. Darwin

I have examined other primroses & cowslips & find what I said as yet universal.[3] I shall be very curious to note which plants produce seed.—

I suppose Mrs Hooker's time is drawing near. I most heartily wish you through it well.—[4]

Etty has perspired & we are in good spirits, the more so as Doctor has looked little grave these two days—says it is a form of Typhus fever—an ugly word.—[5]

Has Lechenaultia ever produced seed at Kew?

Endorsement: '/60'
DAR 115.1: 53

[1] The endorsement is confirmed by the reference to Frances Harriet Hooker's pregnancy.

[2] See letter from J. D. Hooker, [28 April 1860].

[3] See letter to J. D. Hooker, 7 May [1860].

[4] The Hookers' fifth child, Brian Harvey Hodgson Hooker, was born on 27 May 1860.

[5] Typhus fever, a highly contagious disease, is characterised by great prostration and a specific form of skin eruption. The course of the disease is marked by a 'crisis' in which patients perspire freely as the feverish high temperature falls. It is possible, however, that Henrietta Emma Darwin was suffering from typhoid fever, at the time thought to be a variety or derivative of typhus. In *Emma Darwin* 2: 176, Henrietta Litchfield referred to this illness as follows: 'In 1860 my poor mother's thoughts and time were engrossed with the care of me in a long illness (probably typhoid fever) lasting with relapses from May, 1860, till Midsummer, 1861.' The difference between typhoid fever and typhus had been definitively shown by the work of William Jenner at the London Fever Hospital, 1849–51. CD's description of Etty having passed a crisis in the disease marked by slight perspiration fits more with the course of typhus than with typhoid, although she apparently also suffered from abdominal inflammation characteristic of typhoid.

From John Cattell 12 May 1860

Westerham
May 12th/60

Dear Sir,

Pray receive my best thanks for your very kind and sympathysing letter—[1]

I much regret I cannot do more for you in the articles you write about.

Auriculas of the named kinds I am not a grower of, I had in the autumn a nice batch of seedlings but they are nearly all killed My Linum flavum are destroyed, and out of something like a hundred thousand Lettuce plants I have none to send you, the few I have done up are seedlings they will produce seed should we have a fine autumn.

Apples all gone.

I think there is a place on our Chart (Common)[2] where I could get a good bunch of Common Polyanthus flowers, should they be of any service to you I shall have much pleasure in getting them, or anything else at any time that I can send you for your observation I shall be happy in doing so, Please to allow me to remark that we are in the habit of saving several kinds of Lettuce Seeds near at hand but have never observed that they cross naturally[3]

I am | Dear Sir | Yours obed[ly] | John Cattell

C. Darwin Esq

DAR 77: 171–2a

CD ANNOTATIONS
1.1 Pray . . . doing 5.4] *crossed pencil*
5.4 Please . . . naturally 5.6] *scored brown crayon*

CD note:[4]
Ex | Exper. Book p. 8— 2 Lettuces grown together did not cross—but then did not flower together* (Cattell told me they w[d] not).

[1] CD's letter has not been found. Cattell was a nurseryman and seedsman in Westerham, Kent.
[2] Westerham Chart is a large area of common land adjacent to Westerham.
[3] CD was investigating the natural crossing of different varieties of garden plants and vegetables grown in close proximity. See also letters to M. T. Masters, 7 April [1860] and 13 April [1860], and letter from William Masters, 8 May 1860.
[4] CD's note is in DAR 77: 172b. It refers to an experiment he carried out at Down and recorded in his Experimental book, p. 8 (DAR 157a).

To a Bookseller 13 [May 1860]

Down—
13th

Dear Sir

Be so good as to send to 6 Queen Anne St.[1] the numbers last published of the "North British Review" & of "Dublin Magazine of Natural History."—[2]

Please enquire if a French Translation has ever appeared of "Unger's, Versuch einer Geschichte der Pflanzen-Welt 1852"[3] And if such Translation has appeared, please order it for me.—

Yours | C. Darwin

kindly inform me if there be no Translation—

Endorsement: '1860'
American Philosophical Society (Getz 2797)

[1] The street numbers in Queen Anne Street were changed in the spring of 1860. The number of Erasmus Alvey Darwin's house had formerly been 57.

[2] There were unsigned reviews of *Origin* in the *North British Review* 32 (1860): 455–86 ([Duns] 1860) and in the *Natural History Review* 7 (1860): 23–32 ([Haughton] 1860b). CD wrote the names of the authors on his copies of the reviews, which are in the Darwin Pamphlet Collection–CUL.

[3] Unger 1852. There is a copy of the work in the Darwin Library–CUL. Apparently no French translation of the work was ever issued. The recipient of the letter wrote 'no' next to CD's request. CD had recorded the work in the 'books to be read' section of his reading notebook in 1852 (see *Correspondence* vol. 4, Appendix IV, 128: 182).

To Francis Galton 13 May [1860][1]

Down Bromley Kent
May 13th

Dear Galton

I have in simple truth been quite unreasonable in begging favours of you; but yet I am going to beg one more.— I remember that you said that you knew Mr M. Parkyns.[2] If you know him well enough to supplicate an answer to the enclosed query, will you do so for me?[3] If you do not, tear up the question; I would not trouble you if I were not really anxious for information on the point—

Humbly begging forgiveness, I am | Dear Galton | Yours very sincerely | Charles Darwin

University College London Library (Galton 39)

[1] The year is suggested by CD's work on his pigeon manuscript during 1860 (see n. 3, below).

[2] Mansfield Parkyns had travelled in Abyssinia from 1843 to 1846; he had been an attaché in the British embassy in Constantinople between 1850 and 1852. CD had read Parkyns's description of his travels (Parkyns 1853) in 1855 (see *Correspondence* vol. 4, Appendix IV, 128: 14). Parkyns and Galton were both active members of the Royal Geographical Society of London.

[3] The enclosure has not been found, but a passage in *Variation* 1: 183 indicates that CD received information from Parkyns about doves in Abyssinia. Parkyns informed CD that *Columba guinea* was not domesticated in Abyssinia, contrary to some reports. CD worked on his chapters on pigeons (*Variation* 1: 131–224) from March to June 1860 ('Journal'; Appendix II).

To J. D. Hooker 13 [May 1860][1]

Down
13th

My dear Hooker

I return Henslow, which I was very glad to see. How good of him to defend me.—[2] I will write & thank him.—[3] As you said you were curious to hear Thomson's opinion, I send his kind letter.[4] He is evidently a strong opposer to us; but his arguments rather remind me of an old Lady who defended Christianity to my Father, saying "Doctor Doctor, I know that sugar is sweet, & I know that my Redeemer liveth."—[5]

Your affect | *C. D.*

Endorsement: 'May *1860*'
DAR 115.1: 54

[1] The endorsement is confirmed by the reference to the letter from J. S. Henslow to J. D. Hooker, 10 May 1860.
[2] Letter from J. S. Henslow to J. D. Hooker, 10 May 1860.
[3] See following letter.
[4] Thomas Thomson's letter has not been found. Thomson had collaborated with Hooker on the *Flora Indica* (Hooker and Thomson 1855). He was superintendent of the botanic garden in Calcutta.
[5] This anecdote is given in *Autobiography*, p. 96.

To J. S. Henslow 14 May [1860]

Down Bromley Kent
May 14th

My dear Henslow

I have been greatly interested by your letter to Hooker;[1] & I must thank you from my heart for so generously defending me as far as you could against my powerful attackers.— Nothing which persons say hurts me for long, for I have entire conviction that I have not been influenced by bad feelings in the conclusions at which I have arrived. Nor have I published my conclusions without long deliberation & they were arrived at after far more study than the publick will ever know of or believe in.— I am certain to have erred in many points, but I do not believe so much as Sedgwick & Co. think. Is there any Abstract or Proceedings of the Cambridge Phil. Soc. published? If so & you could get me a copy I shd like to have one.—[2]

Believe me my dear Henslow I feel grateful to you on this occasion & for the multitude of kindnesses you have done me from my earliest days at Cambridge.—

Yours affectionately | C. Darwin

P.S. I think I remember your observing that the pistil in different flowers of cowslips & Primroses varies much in length.[3] From observations which I have been making I have strong suspicion that they (& Auriculas)[4] are dioicous; but I shall know this autumn, for I have marked what I consider the male & female plants.[5] Why I mention this to you, is that I have a vague remembrance of your stating that some other plants varied *greatly* in length of pistil; if so I shd **much** like to know what, that I might carefully observe them. Do not think of writing unless you can tell me of any such plants.—

Endorsement: '1860'
DAR 93: 70–1

[1] Letter from J. S. Henslow to J. D. Hooker, 10 May 1860. Henslow had asked Hooker to send it on to CD.
[2] Adam Sedgwick's paper criticising *Origin* was not published in the *Proceedings of the Cambridge Philosophical Society*, but a report of its contents was given in the *Cambridge Chronicle*, 19 May 1860, pp. 4–5, and in the *Literary Gazette*, 12 May 1860, p. 582. See letter to J. S. Henslow, 17 May [1860].

[3] Henslow 1830.

[4] CD's Experimental book, p. 54 (DAR 157a) contains an entry dated 12 May indicating that CD examined 'some common Auriculas supplied by M[r] Cattell'. See also letter from John Cattell, 12 May 1860.

[5] See the following letter and the letter to J. D. Hooker, 7 May [1860]. In a paper on the dimorphic condition of *Primula* published in 1861, CD suggested that these hermaphroditic plants were possibly in the process of becoming dioecious. Contrary to his expectations, the short-styled flowers produced more seed than the long-styled forms, and hence were more likely to become the future females. See 'On the two forms, or dimorphic condition, in the species of *Primula*, and on their remarkable sexual relations', *Journal of the Proceedings of the Linnean Society (Botany)* 6 (1862): 77–96; *Collected papers* 2: 45–63.

To J. D. Hooker 14 May [1860]

Down Bromley Kent
May 14th

My dear Hooker

I daresay you will be too busy to give instructions about the fertilisation of Leschenaultia;[1] but in case you do, I write to say that I find the way to do it, is to move a camel-pencil, holding handle nearly *parallel* to the pistil & axis of flower, just in same direction & manner as the belly of a Bee would brush over the indusium, whilst crawling in to suck the copious nectar; & then the points of the brush hit against the rough protuberant lip, open the indusium, & some of the hairs enter & stir up the pollen splendidly & generally bring out some grains. So that the operation is done in a second, & I feel nearly sure that every bee which sucks this flower will open the indusium, stir up the pollen, remove some, & often leave pollen from other flowers within the indusium.

In fact I am at ease about Leschenaultia; but sh[d] like *excessively* to prove that stirring up the pollen within the indusium is necessary or favourable to fertilisation. For then I sh[d] look at this curious contrivance, as specially related to visits of insects; as I begin to think is almost universally the case. The evidence about dioicous condition of Cowslips & **Auriculas** seems to me to get clearer & clearer.—

And now I will (honour bright) not bother you with another note for a long time —that is if I can possibly resist

Yours affect[y] | C. Darwin

Poor dear Etty is a little better; not seriously ill, but our Doctor, who comes daily, says the Fever may run on for a considerable time. She eats hardly anything, but does not weaken so rapidly as I sh[d] have feared.[2]

P.S. | Very hearty thanks for your pleasant note (& M[rs] Hooker's) just received. I daresay after all my vows I shall not resist answering it.—

I find before flowers are utterly withered of Cowslips the seeds of "females" are *twice* as big as seeds or ovules in the "Male" flowers— Hence I expect the latter will abort entirely. Out of 522 flower stalks gathered by the children I find 281 males & 241 Females.—[3]

Endorsement: '1860'
DAR 115.1: 55

[1] See letter to J. D. Hooker, 11 May [1860].

[2] Henrietta Emma Darwin had been diagnosed as suffering from a form of typhus fever. See letter to J. D. Hooker, 11 May [1860].

[3] These observations are recorded in CD's Experimental book, pp. 55 and 56 (DAR 157a). CD's paper 'On the two forms, or dimorphic condition, in the species of *Primula*, and on their remarkable sexual relations' discusses these points (*Collected papers* 2: 45–63). See also preceding letter and n. 5.

To J. D. Hooker 15 [May 1860][1]

Down Bromley Kent
15th

My dear Hooker

I have not the least doubt that Lyell was de facto the originator of the importance of geological changes of level with respect to distribution.[2] No doubt he had precursors such as those who speculated on England & France having been joined from presence of Wolf & other Noxious animals in England.— I have passages on subject in my Journal, viz on introduction of old world mammals by Behren's Sts & former more intimate connection of N. & S. America but nothing of consequence, that I can remember & I do remember in one case beginning by saying that the facts were good illustrations of Lyell's doctrines.—[3]

No doubt Forbes de facto originated the doctrine of distribution during glacial period.[4] But Link in his "Urwelt" 1821 speculates on changed climate from greater former height of mountains as explaining identity of alpine & northern plants.—[5] With respect to passages in Asa Gray I wrote to him to say that I had no claim for position in which he places me.—[6] I have no doubt the reason was that he used (slightly modifying it) the view which I sent him of migration from old to new (or conversely) worlds during former warmer periods by present land;[7] & that instead of expressly acknowledging this he gave me undue general credit. I do not say this invidiously, for I have not least doubt, he did it without thought & perhaps because he was bothered, in as much as he had slightly altered my view he could not well acknowledge it.— Of course you will not say anything of this to him.— I may just add that I gave my views at his request, & seeing how he introduced them I naturally thought that he had arrived at analogous view before hearing from me. But he subsequently said in answer to my letter in which I said I had no claim to position assigned me.—he had not; & stated that he had mentioned my name in the way he does as an acknowledgement.— But I have told a very long story about a very small affair.

I most entirely agree to what you say about Müller,[8] I was disgusted. One ought not to judge without hearing all that can be said about Candidates.— But I was disgusted at the list; & fancied I saw the effects of a great Surgeon-President.—[9] Spence Bate & Sir E. Tennent ought to have been elected **long** before R. Palmer & Baring M.P..[10] It seems to me monstrous.—

I will write to Drummond about Leschenaultia;[11] but I am pretty easy now, as I am sure intercrossing is at least *possible*. Would it bother you to make mere outline of flower for me for woodcut of this size & like this? & then perhaps I would give it & your section of indusium.—

How paltry it is in such men as Balfour, Arnott & Co. not reading your Essay.[12] It is incredibly paltry.— They may all attack me to their hearts' content. I am got case-hardened. As for the old fogies in Cambridge it really signifies nothing.[13] I look at their attacks, as a proof that our work is worth the doing. It makes me resolve to buckle on my armour.— I see plainly that it will be a long uphill fight.— But think of Lyell's progress with geology.— One thing I see most plainly that without Lyell, yours, Huxley & Carpenter's aid my book would have been a mere flash in the pan.— But if we all stick to it, we shall surely gain the day. And I now see that the battle is worth fighting. I deeply hope that you think so.— Does Bentham progress at all?[14]

I do not know what to say about Oxford.[15] I sh^{d} like it *much* with you; but it must depend on health.—

Etty goes on the same.[16] The Doctor to day said "well I feel easy about her now"—which shows me that he was more uneasy than he ever expressed.—

Yours most affect | C. Darwin

Endorsement: 'May 1860'
DAR 115.1: 56

[1] The endorsement is confirmed by CD's reference to James Drummond. CD wrote to Drummond on 16 May 1860.

[2] CD refers to Charles Lyell's *Principles of geology*, which discusses the geographical distribution of animals and plants (C. Lyell 1830–3).

[3] *Journal of researches*, pp. 151–4.

[4] E. Forbes 1846.

[5] Heinrich Friedrich Link had discussed the relationship between alpine and Arctic plants in Link 1821. A copy of the first volume of this work is in the Darwin Library–CUL.

[6] See *Correspondence* vol. 7, letter to Asa Gray, 24 December [1859]. CD refers to a passage in Gray 1859 in which Asa Gray implied that CD had been the first naturalist to provide an explanation of the geographical distribution of Arctic and alpine plants (Gray 1859, p. 446): 'considerations which Mr. Darwin first brought to bear upon such questions, and which have been systematically developed and applied by the late Edward Forbes, by Dr. Hooker, and by Alphonse de Candolle.'

[7] See *Correspondence* vol. 7, letter to Asa Gray, 11 August [1858].

[8] Hooker apparently proposed Ferdinand Jakob Heinrich von Mueller for election to the Royal Society as a foreign member, but the recommendation was not approved by the council. Mueller was elected to the society in 1861. The new foreign members were announced at a meeting of the Royal Society on 24 May 1860 (*Athenæum*, 2 June 1860, p. 759).

[9] The surgeon Benjamin Collins Brodie was president of the Royal Society from 1858 to 1861.

[10] CD refers to the annual election of fifteen new fellows of the Royal Society. The list of nominees was decided at a council meeting in May and announced on 7 June 1860. Charles Spence Bate and James Emerson Tennent were not elected members until 1861 and 1862, respectively. CD considered Bate a 'rising' authority on Crustacea (see *Correspondence* vol. 5, letter to J. D. Dana, 6 December [1853]). He had read Tennent's book on Ceylon (Tennent 1859) in 1859 (see *Correspondence* vol. 4, Appendix IV, 128: 155). Roundell Palmer was a lawyer and politician who had botanical interests, and Thomas Baring was a financier and MP for Huntingdon. The results of the election were announced in the *Athenæum*, 9 June 1860, p. 792.

[11] See letter to James Drummond, 16 May 1860.

[12] The 'essay' was Hooker 1859. CD refers to the anti-transmutationist views of the botanists John Hutton Balfour and George Arnott Walker Arnott. See letter to J. D. Hooker, 18 [April 1860]. In a letter written at about this time to Thomas Anderson in Edinburgh, Hooker remarked: 'I hope you have read Owen's review in the *Edinburgh*. I should think it must add gall to the Balfourians' bitterness of spirit, for not content with snubbing me and spitefully entreating Darwin and Huxley, the cool fish hedges for a transmutation view of his own!' (L. Huxley ed. 1918, 1: 515). The reference is to [R. Owen] 1860a.

[13] CD refers to a meeting of the Cambridge Philosophical Society on 7 May 1860 at which Adam Sedgwick and William Clark criticised CD's views. See letter from J. S. Henslow to J. D. Hooker, 10 May 1860.

[14] George Bentham. See letters to J. D. Hooker, 3 March [1860] and 12 March [1860].

[15] The 1860 meeting of the British Association for the Advancement of Science was to be held in June in Oxford. CD and Hooker had both attended the 1847 meeting in Oxford (see *Correspondence* vol. 4).

[16] Henrietta Emma Darwin was recovering from typhus fever. See letter to J. D. Hooker, 11 May [1860].

From Henry Doubleday 16 May 1860

Epping
May 16th 1860

My dear Sir,

I hope you will excuse my apparent neglect in not replying to your letter, but I have been so much engaged the last few days that I really have had no spare time—

I will not however any longer delay answering your queries— In the few remarks I made about *Primroses*, *Cowslips* and *Oxlips* I merely stated what had occurred to me—[1] I am perfectly aware that it is often necessary to repeat the same experiments many times before you can arrive at the real facts of the case.

Since I received your letter yesterday[2] I have examined the plants of the *Bardfield Oxlip* which remain in flower and find they are both "male" and "female" plants—⟨that is plants w⟩ith long pistils and short stamens and others with long stamens and short pistils—I have only two or three plants of our *hybrid oxlip* left and these have long stamens and short pistils but I think I have seen them with long pistils—[3] I have taken no notice of them the last four or five years and the two or three dry hot summers we have had killed most of them— Some years ago I saw a statement in the Gardener's Chronicle that upon the Continent the *primrose* and *Cowslip* never grow together and that the plant commonly called *Oxlip* here does not exist—altho' the true *P. elatior* is common in certain localities— I have never met with these apparently hybrid plants except where the *Primrose* and *Cowslip* were growing together—and they vary exceedingly—some of the flowers being deep in colour and not much larger than cowslips while others are pale and nearly as large as primroses—and many of them produce flowers on single stalks like primroses as well those in umbels— I have never seen this in the Bardfield Oxlip—

If I can find any plants of our Oxlip in flower in the field⟨s⟩ or edges of the forest I will ⟨carefully⟩

AL incomplete
DAR 162.2: 238

CD ANNOTATIONS
3.7 Some years . . . Bardfield Oxlip—3.15] *scored pencil*
3.15 Oxlip] 'H. Doubleday' *added pencil*
Top of first page: 'Ch. 4'[4] *brown crayon*

[1] See letter from Henry Doubleday, 3 May 1860.
[2] CD's letter to Doubleday has not been found.
[3] In *Forms of flowers*, CD referred to the common oxlip as a hybrid between the cowslip and primrose and to the Bardfield oxlip as a distinct species (*Primula elatior*) 'found in England only in two or three of the eastern counties' (*Forms of flowers*, pp. 32, 55, 72). He cited Doubleday as the first to discover this species in England (*ibid.*, p. 32).
[4] A reference to notes pertaining to chapter 4 of *Natural selection*, on 'Variation under nature'.

To James Drummond 16 May 1860

Down, Bromley, Kent
May 16th. 1860.

Dear Sir

I hope that you will excuse the liberty which I take in writing to you and asking you a favour.—[1] Dr. Hooker has told me that I may use his name as an introduction.— I am very curious about the fertilisation of Leschenaultia formosa. I must just allude to its structure, though no doubt you are well acquainted with it. The indusium or cup at the summit of the pistil is open in the bud, but before the flower opens the indusium closes and in closing includes nearly all the pollen.— In my greenhouse the indusium never opens again. The pollen within the indusium is enclosed near the lips, and not a grain lies, as I find, on the stigmatic surface at the bottom of the indusium. How then is the plant fertilised?[2]

My belief is that insects in creeping in to suck the copious nectar brush open the indusium, and the hairs of their abdomens stir up the pollen and push it down on to the stigmatic surface. I find that this easily effected by a camel-hair brush. Now what I want to beg is for you to have the great kindness to watch for a short time the Leschenaultia and see whether Bees visit it; and if they do, to endeavour to observe whether in crawling in or whilst sucking, they do not open the lips of the indusium. As this plant may be visited by nocturnal insects, it would be a very interesting experiment to cover over with bag on frame made of *very open* gauze one or two plants (plucking off any open flowers) and then see whether they seed at all or less freely than plants left to the visits of insects.—[3]

This may appear a trifling enquiry to you; but the subject has been largely discussed by R. Brown, Aug: St. Hilaire[4] and other celebrated Botanists; and Dr Hooker and I are experimenting on the subject.— If any other Goodenaceæ, furnished with an indusium, grow near you, I should be infinitely obliged for any notices on their structure in relation to the following point, "how in one flower is it possible that pollen from another flower or plant of the same species could get into the indusium," I have strong reason to believe that this is a universal possibility. In the Leschenaultia, if insects open the indusium in the manner in which I suspect, their abdomens dusted with pollen from one flower might easily carry grains into the indusium of another flower.

I hope that you will excuse the liberty I take in begging this favour and | I remain, Dear Sir | Your obliged Servant | Charles Darwin

LS(A)
Library Board of Western Australia, State Archives, Accession 2275A

[1] Drummond was superintendent of the government botanic gardens of Western Australia (R. Desmond 1977).

[2] For Drummond's reply, see the letter from James Drummond, 17 September 1860.

[3] CD had conducted similar experiments to ascertain the role of insects in facilitating the pollination of clover and kidney-beans, the results of which are recorded in his Experimental book (DAR 157a). See also *Correspondence* vol. 6, letter to *Gardeners' Chronicle*, 18 October [1857].

[4] Robert Brown and Augustin François César Prouvençal de Saint-Hilaire. CD refers to Brown 1814, pp. 559–60, and to Saint-Hilaire 1841, pp. 572–3, in which the fertilisation of *Goodenia* (a plant closely related to *Leschenaultia*) is discussed. The passage in Saint-Hilaire 1841 is marked in CD's copy of the work (Darwin Library–CUL).

To J. S. Henslow 17 May [1860]

Down Bromley Kent
May 17th

My dear Henslow

Many thanks for your note & information about my opposers.—I have sent for Literary Gazette.— [1] For the chance of your caring I send the characters by which I can divide *all* Primroses & cowslips into what I suspect will turn out Male & Female Plants. These two forms exist in about equal numbers.— I have marked a set of both forms to see about seeding. The difference in state of pollen is very clear & invariable.[2]

I suspect it will turn out fine case for me: the first gradation in the formation of a dioicous plant.— The Holly forms a second step, for here the male plant has anthers but no pollen.— The male cowslips have *abundant* pollen, but all grains small-sized.—

It will be curious if I can show that so common a plant is dioicous or nearly so. Auriculas are in same state as far as I have seen.—

This is reason why I wanted to know whether you had observed long & short pistils in any other flowers.

Yours affect^ly | C. Darwin

[Enclosure]

Cowslips & Primroses

Male Plants. Tube of corolla long, throat short— Stamens long— pollen in water about $\frac{9}{6000}$ of inch in diameter. Pistil short, stigma far beneath anthers,— surface of stigma smoother

Female plants:— Tube of corolla short, throat long.— Stamens short, pollen in water about $\frac{6}{6000}$ of in inch in diameter.—[3] Pistil long, stigma far above anthers, surface of stigma rougher.—

Endorsement: '1860'
DAR 93: 72–3, 116

[1] The *Literary Gazette*, 12 May 1860, p. 582, included a report on Adam Sedgwick's and William Clark's remarks about *Origin* at the Cambridge Philosophical Society meeting on 7 May 1860.

[2] See letter to J. S. Henslow, 14 May [1860].

[3] In CD's paper on the dimorphic condition of *Primula*, read on 21 November 1861, the measurements were changed to $\frac{10-11}{7000}$ of an inch for the pollen from short-styled plants and $\frac{7}{7000}$ of an inch for that from the long-styled plants. By the time CD wrote this paper, he had found that the short-styled plants produced more seeds than the long-styled, and so he designated the short-styled ones 'females'. See *Collected papers* 2: 49–51.

To W. D. Fox 18 May [1860]

Down Bromley Kent
May 18th

My dear Fox

It was very kind of you to take so much trouble about the curious dog; but unless one is on the spot it is hopelessly difficult to understand the amount & origin of difference. The dog, however, must be a most curious creature.— I am glad to hear of a naturalist taking up so difficult & neglected a subject as spiders.—

If I have time & if in London moderately soon, I will look at the deer-Hybrid— But the fact is, if it appeared ever so intermediate I sh^d. not believe in its hybrid nature, *unless after careful dissection* of its internal organs; & one might easily be deceived by hoofs being filed &c &c.[1]

You will be sorry to hear that our eldest girl, Etty, has now been 3 weeks ill with odd fever, partly remittent partly typhoid;[2] but I have great hopes we see signs of abatement. It has harassed us much, though not exactly frightened us. At one time, however, the Doctors seemed rather anxious.— But I think it must end soon.— What a household you must have had with seven with Hooping Cough!

My health has been better of late, which I am inclined to attribute to mineral acids, no sugar, & drinking wine.—[3]

I do not know whether you ever see various Reviews, but the attacks have been falling thick & heavy on my now case-hardened hide.— Sedgwick & Clarke opened regular battery on me lately at Cambridge Phil. Soc.y & dear old Henslow defended me in grand style, saying that my investigations were perfectly legitimate.[4] I have begun my bigger Book, but make very very slow progress.—

My dear old friend, | Yours affect | C. Darwin

Postmark: MY 19 60
Christ's College Library, Cambridge (Fox 128)

[1] See letter to W. D. Fox, 22 [March 1860].
[2] Henrietta Emma Darwin had been diagnosed as having 'a form of Typhus fever' (letter to J. D. Hooker, 11 May [1860]).
[3] For a description of CD's new dietary regimen, see the letter to W. D. Fox, 22 [March 1860].
[4] See letter from J. S. Henslow to J. D. Hooker, 10 May 1860.

To Asa Gray 18 May [1860][1]

Down Bromley Kent
May 18th

My dear Gray

I thank you for your pleasant letter received this morning.[2] I return Messrs. Appleton's letter: you refer to *two* letters from them. I have habit of filing everything, & have looked through my whole file & can find no other.[3] It is provoking that he does not publish 2d. Edit. & how kind it is in you to go on troubling yourself on the subject.—[4] It seems very true that as you have put your hand in it, it is not so easy to draw it out. What a battle you do seem to have been fighting on the origin of species.— I can most plainly see that whatever amount of truth my book may contain, the saving of it (as I told Hooker the other day) will surely be wholly due to a very small body of men. Had it not been for Lyell, Hooker, yourself & two or three others, I am well convinced my book & the whole subject would have been mere flash in the pan. For the attacks are now here incessant & very bitter. For instance Sedgwick & Prof. Clarke attacked me savagely at Cambridge Phil. Soc. but dear old Henslow (though he goes but little way with me) stood up manfully for the subject as legitimately within bounds of science, & produced excellent effect.— Since then Prof. Phillips the Geologist has lectured against me at Cambridge—[5] A. Murray the Entomologist has read paper at Royal Soc. of Edinburgh against me.[6] Dr Dawson in Canada Nat. Mag. do.[7] Some one in Dublin Nat. Mag.[8] Mr Haughton at Dublin Geolog. Soc.[9] And thus I could go on for many more!!

But the effect on me is that I will buckle on my armour & fight my best. You seem to have done so allready in grand style. And I believe Hooker will, as certainly will Lyell & Huxley. But it will be a long fight. By myself I shd. be powerless. I feel my weak health acutely, as I cannot work hard.— So bitter is the feeling of some that Hooker tells me that neither Harvey, Balfour, or Arnott have ever read his Australian Essay![10] Is this not incredibly paltry? It makes me savage to think of the slighting way in which Owen alludes to Hooker's Essay in Edinburgh Review.[11] Lyell is working very hard at geological history of Man; & it is really marvellous how rapidly curious facts are turning up. I expect his Essay, in which he will discuss origin of species, will make a great commotion with the reading public.

Thanks about red "*roots*".—[12] I most sincerely hope that you may publish your stunner of an answer to Bowen Agassiz & Co.—[13]

With hearty thanks | Yours most sincerely | C. Darwin

Gray Herbarium of Harvard University (14)

[1] The year is given by the references to reviews of *Origin.*

[2] Gray's letter has not been found, but see the letter from Asa Gray, 20 February 1860.

[3] Gray had forwarded to CD a letter he had received from D. Appleton & Co. of New York dated 17 February 1860 agreeing to grant CD copyright over the American edition of *Origin* that they published (see enclosure to letter from Asa Gray, 20 February 1860). The second letter to which CD refers may be that of 7 February 1860 from D. Appleton to Gray (Gray Herbarium Archives). The text of this letter reads as follows:

> We return you as requested the contents &c of Darwin's work and must apologise for the soiled state but the printers who supplied the article for the new ed[n] did not understand you wished it again.
>
> We suppose before long we shall require a new ed[n] & we thought we would write to learn if any additions could be made for a second edition

Gray had apparently thought he had sent this letter on to CD as well. See also letter from Asa Gray, 23 January 1860, n. 2, and letter to Asa Gray, 22 May [1860].

[4] The first three printings of *Origin* published in the United States by D. Appleton & Co. of New York (issued in January, February, and March 1860) all had identical texts taken from the first English edition. The fourth printing, however, was 'revised and augmented', including a 'Historical preface', a number of corrections to the text, and a supplement of seven pages of additional changes that CD sent to Asa Gray (see letters to Asa Gray, 28 January [1860], 1 February [1860], [8 or 9 February 1860], and 8 March [1860]). The revised American edition thus incorporated most of the changes made to the second English edition of *Origin* as well as several subsequently added to the third English edition (1861). See Freeman 1977, p. 83.

[5] John Phillips. See letter from J. S. Henslow to J. D. Hooker, 10 May 1860.

[6] Murray 1860a.

[7] Dawson 1860b.

[8] [Haughton] 1860b.

[9] Haughton 1860a.

[10] See letter to J. D. Hooker, 15 [May 1860].

[11] [R. Owen] 1860a.

[12] See letter to Asa Gray, 3 April [1860] and nn. 4 and 5.

[13] CD refers to Francis Bowen, Louis Agassiz, and John Amory Lowell. Gray defended CD's theory at a meeting of the American Academy of Arts and Sciences in Boston on 10 April 1860. Bowen, Agassiz, and Lowell had criticised CD's views at a previous meeting on 27 March. The remarks were later published in the *Proceedings of the American Academy of Arts and Sciences* 4 (1860): 410–16, 424–31. CD's annotated copy of an offprint of these pages is in the Darwin Pamphlet Collection–CUL. The discussion also formed the basis for Gray's article on *Origin* in the *Atlantic Monthly* ([Gray] 1860b).

To Charles Lyell 18 May [1860]

Down Bromley Kent
May 18th

My dear Lyell

I send letter from Asa Gray to show how hotly battle rages there.— Also one from Wallace, very just in his remarks, though too laudatory & too modest, & how admirably free from envy or jealousy.— He must be a good fellow. Perhaps I will enclose letter from Thomson of Calcutta, not that it is much, but Hooker thinks so highly of him.—[1]

Many thanks for your interesting letter of 15th.[2] What interesting facts are turning up about man!— With respect to Rabbit & Hares; Lewis takes the account from Isid. G. St. Hilaire:[3] I have not yet looked at original account; such good men are concerned that I think there can be no mistake. Yet I shall not be satisfied till I hear how he got his *first* hybrids.— There is an old variety called Hare-rabbit— A *good* naturalist once assured me he had found hybrid pheasant-fowls perfectly fertile int se; after a time I found out that he had bought the *first* cross, & this supposed cross was the variety called "pheasant-fowl" & which does resemble the pheasant considerably.— Some time when you have done with it I shd much like to see Schaafhausen pamphlet on natural selection.—[4]

I have ordered Canadian Mag;[5] but if I cannot get it, will borrow yours.—

Henslow informs me that Sedgwick & then Prof. Clarke made a regular & savage onslaught on my Book lately at Cambridge Phil. Soc. but Henslow seems to have defended me well, & maintained that the subject was a quite legitimate one for investigation.—[6] Since then Phillips has given lecture at Cambridge on same subject, but treated it very fairly.[7] How splendidly Asa Gray is fighting the battle.[8] The effect on me of these multiplied attacks is simply to show me that the subject is worth fighting for, & assuredly I will do my best. But ill-health & multifarious interrutions make my progress excessively slow.— I can very plainly see, as I lately told Hooker, that my Book would have been & be a mere flash in the pan, were it not for you, Hooker & a few others.— I hope all the attacks make you keep up your courage; & courage you assuredly will require.

Etty to day has been a shade better.[9] Tomorrow or Sunday makes 3 weeks.— It has been dreadfully harassing to us. She has not suffered much, except the pitiable weariness of so long a fever.—

My dear Lyell | Yours affect | C. Darwin

I have read some of Godron.—[10] He strikes me as rather common-place, which surprises me as he wrote capital paper on means of Distribution of plants.—[11] He puts the old case well that mere physical conditions do very little in modifying organic forms.—

I have just received 8 Pages of verses, anonymous, apparently well done, quizzing & lauding my Book & me, but I have hardly read them. They have been copied so I cannot recognise hand-writing.—[12]

I forget whether I told you that there was very good & long quiz in Manchester newspaper some time ago, showing that I had proved that "might was right" was the universal law of nature.—[13]

Endorsement: '1860'
American Philosophical Society (212)

[1] The letters from Asa Gray, Alfred Russel Wallace, and Thomas Thomson have not been found.

[2] Lyell's letter has not been found.

[3] CD refers to a description of the hybrid progeny of rabbits and hares in a paper by George Henry Lewes (Lewes 1860, p. 604). Lewes stated that over 1000 hybrids had been reared and sold at market in Angoulême in France and cited I. Geoffroy Saint-Hilaire 1854–62, vol. 3, as his authority. Lewes's article was reprinted in book form in 1862 under the title *Studies in animal life*. For Lyell's response to this information, see the letter from Charles Lyell, 19 June 1860; see also Wilson ed. 1970, p. 450. When CD discussed the case in *Variation*, he referred to Pierre Paul Broca's 'interesting memoir on this subject' (*Variation* 1: 105 and n. 7). The reference is to Broca 1858–9.

[4] Schaaffhausen 1853. There is an annotated copy in the Darwin Pamphlet Collection–CUL. Bound with the pamphlet is a manuscript translation of the section discussing the development of organic life (Schaaffhausen 1853, pp. 423–4), possibly prepared for CD by the Darwins' governess, Camilla Ludwig.

[5] Dawson 1860b.

[6] CD refers to John Stevens Henslow, Adam Sedgwick, and William Clark. See letter from J. S. Henslow to J. D. Hooker, 10 May 1860.

[7] Phillips 1860.

[8] See preceding letter and n. 13.

[9] Emma Darwin recorded in her diary on 28 April 1860 that Henrietta Emma Darwin was 'poorly'. Her fever had been diagnosed as a form of typhus (letter to J. D. Hooker, 11 May [1860]).

[10] Godron 1859. This book contains reprints of two of Dominique Alexandre Godron's papers on species that CD had particularly wanted to read in 1855 (see *Correspondence* vol. 5). He had apparently been unsuccessful in obtaining copies of the papers. There is an annotated copy of the book in the Darwin Library–CUL.

[11] Godron 1853.

[12] The verses have not been found. CD suspected that the author was the botanist Francis Boott (see letter to J. D. Hooker, 20 May [1860]).

[13] See letter to Charles Lyell, 4 May [1860].

To John Murray 18 May [1860][1]

Down Bromley Kent
May 18th

My dear Sir

I write one line to thank you very sincerely for your too magnificent present of six copies of my Journal.— It looks excellently, & I am greatly pleased at my creditable appearance in a green coat.—[2] With very many thanks

My dear Sir | Yours sincerely | C. Darwin

Dunedin Public Library, Reed collection

[1] The year is given by the publication of a reprint edition of *Journal of researches*.

[2] The 1860 edition of *Journal of researches*, the tenth-thousand printed, was published in the same green cloth binding as *Origin* (Freeman 1977, p. 36). Previous editions published by Murray had been bound in red cloth.

To A. R. Wallace 18 May 1860

Down Bromley Kent
May 18th 1860

My dear Mr Wallace

I received this morning your letter from Amboyna dated Feb. 16th, containing some remarks & your too high approbation of my book.[1] Your letter has pleased me very much, & I most completely agree with you on the parts which are strongest & which are weakest. The imperfection of Geolog. Record is, as you say, the weakest of all; but yet I am pleased to find that there are almost more Geological converts than of pursuers of other branches of natural science. I may mention Lyell, Ramsay, Jukes, Rogers, Keyserling, all good men & true—[2] Pictet of Geneva is not a convert, but is evidently staggered (as I think is Bronn of Heidelberg) & he has written a perfectly fair review in the Bib. Universelle of Geneva.—[3] Old Bronn has translated my book, well-done also, into German & his well-known name will give it circulation.—[4]

I think geologists are more converted than simple naturalists because more accustomed to reasoning. Before telling you about progress of opinion on subject, you must let me say how I admire the generous manner in which you speak of my Book: most persons would in your position have felt some envy or jealousy. How nobly free you seem to be of this common failing of mankind.— But you speak far too modestly of yourself;—you would, if you had had my leisure done the work just as well, perhaps better, than I have done it.—

Talking of envy, you never read anything more envious & spiteful (with numerous misrepresentations) than Owen is in the Edinburgh Review. I must give one instance he throws doubts & sneers at my saying that the ovigerous frena of cirripedes have been converted into Branchiæ, because I have not proved them to be Branchiæ; whereas *he himself* admits, before I wrote, on cirripedes, without the least hesitation that these organs are Branchiæ.—[5]

The attacks have been heavy & incessant of late. Sedgwick & Prof. Clarke attacked me savagely at Cambridge Phil. Soc. but Henslow defended me well, though not a convert.—[6] Phillips has since attacked me in Lecture at Cambridge.[7] Sir W. Jardine in Eding. New Phil. Journal.—[8] Wollaston in Annal of Nat. History.—[9] A. Murray before Royal Soc. of Edinburgh—[10] Haughton at Geolog. Soc. of Dublin—[11] Dawson in Canadian Nat. Magazine,[12] And *many others*.[13] But I am got case-hardened, & all these attacks will make me only more determinately fight. Agassiz sends me personal civil messages but incessantly attacks me; but Asa Gray fights like a hero in defence.—.— Lyell keeps as firm as a tower, & this autumn will publish on Geological History of Man, & will there declare his conversion, which now is universally known.—[14] I hope that you have received Hooker's splendid Essay.—[15] So far is bigotry carried, that I can name 3 Botanists who will not even read Hooker's Essay!![16]

Here is a curious thing, a Mr. Pat. Matthew, a Scotchman, published in 1830 a work on Naval Timber & Arboriculture, & in appendix to this, he gives *most clearly* but very briefly in half-dozen paragraphs our view of natural selection. It is most complete case of anticipitation. He published extracts in G. Chronicle:[17] I got Book, & have since published letter, acknowledging that I am fairly forestalled.—[18] Yesterday I heard from Lyell that a German Dr Schaffhausen has sent him a pamphet published some years ago, in which same View is nearly anticipated but I have not yet seen this pamphet.—[19] My Brother, who is very sagacious man, always said you will find that some one will have been before you.—

I am at work at my larger work which I shall publish in separate volumes.— But from ill-health & swarms of letters, I get on very very slowly.— I hope that I shall not have wearied you with these details.—

With sincere thanks for your letter, & with most deeply-felt wishes for your success in science & in every way believe me,[20] | Your sincere well-wisher | C. Darwin

British Library (Add. 46434: 21–23v.)

[1] Wallace's letter has not been found. The only extant record of Wallace's initial response to *Origin* is given in two letters he wrote to friends at the end of 1860. In a letter to George Silk, dated 1 September 1860, Wallace wrote about *Origin* (Wallace 1905, 1: 372):

> I have read it through five or six times, each time with increasing admiration. It will live as long as the 'Principia' of Newton. It shows that nature is, as I before remarked to you, a study that yields to none in grandeur and immensity. The cycles of astronomy or even the periods of geology will alone enable us to appreciate the vast depths of time we have to contemplate in the endeavour to understand the slow growth of life upon the earth. The most intricate effects of the law of gravitation, the mutual disturbances of all the bodies of the solar system, are simplicity itself compared with the intricate relations and complicated struggle which have determined what forms of life shall exist and in what proportions. Mr. Darwin has given the world a *new science*, and his name should, in my opinion, stand above that of every philosopher of ancient or modern times. The force of admiration can no further go!!!

In a letter to Henry Walter Bates, dated 24 December 1860, Wallace again gave his reaction to the book (Wallace 1905, 1: 374):

> I know not how, or to whom, to express fully my admiration of Darwin's book. To *him* it would seem flattery, to others self-praise; but I do honestly believe that with however much patience I had worked and experimented on the subject, I could *never have approached* the completeness of his book, its vast accumulation of evidence, its overwhelming argument, and its admirable tone and spirit. I really feel thankful that it has *not* been left to me to give the theory to the world. Mr. Darwin has created a new science and a new philosophy; and I believe that never has such a complete illustration of a new branch of human knowledge been due to the labours and researches of a single man. Never have such vast masses of widely scattered and hitherto quite unconnected facts been combined into a system and brought to bear upon the establishment of such a grand and new and simple philosophy.

[2] Charles Lyell, Andrew Crombie Ramsay, Joseph Beete Jukes, Henry Darwin Rogers, and Alexandr Andreevich Keyserling.

[3] Pictet de la Rive 1860.

[4] Bronn trans. 1860.

[5] See letter to T. H. Huxley, 9 April [1860].

[6] See letter from J. S. Henslow to J. D. Hooker, 10 May 1860.

[7] Phillips 1860.

[8] [Jardine] 1860.

[9] [Wollaston] 1860.

[10] Murray 1860a.

[11] Haughton 1860a.

[12] Dawson 1860b.

[13] For a list of reviews of *Origin* that were published in 1859 and 1860, see Appendix VII.
[14] Lyell intended to treat the history of humankind in a new edition of his *Elements of geology*, but in fact he published his work on this subject as a separate volume in 1863 (C. Lyell 1863).
[15] Hooker 1859.
[16] See letters to J. D. Hooker, 15 [May 1860], and to Asa Gray, 18 May [1860]. The three botanists were William Henry Harvey, John Hutton Balfour, and George Arnott Walker Arnott.
[17] For a transcription of Patrick Mathew's letter, see Appendix V.
[18] See letters to *Gardeners' Chronicle*, [13 April 1860], and to J. D. Hooker, 13 [April 1860].
[19] Schaaffhausen 1853. See letter to Charles Lyell, 18 May [1860].
[20] The letter has deteriorated, and the final paragraph is now missing. The text has been taken from a transcript made by the editors at an earlier date.

To J. D. Hooker 20 May [1860][1]

Down Bromley Kent
May 20th

My dear Hooker

The following, I believe, are all the chief references on experiments on Cowslips &c. for Harvey.—[2]

W. Herbert.	Transact. Hort. Soc.	Vol. IV.	p. 19[3]
Henslow	Loudon's Mag. of Nat. Hist.	Vol. 3. 1830.	p. 409
He succeeded in getting primrose from cowslip[4]			
H. C. Watson.	Phytologist	Vol. 2.	p. 217 & p. 852.
		Vol 3.	p. 43.—[5]
Sidebotham	Phytologist	vol. 3.	p. 703.[6]

Notwithstanding, the amount of evidence from witnesses who experimentised with hostile bias, I cannot avoid sometimes suspecting error.— No experiment is good unless plants protected by gauze or some other similar protection.— I think Mr Sidebotham alone took such pains; yet for some unknown cause H. C. Watson sneers at his experiments.— Mr Doubleday has often tried, & has found no variation.[7] I reared lots of Primroses from manured primroses with no variation.— I have now got cowslips under gauze, which I fertilize with care. Nevertheless looking through all the above accounts (& I have some loose corroborative facts) the evidence does seem overwhelming that they are varieties. The case does not half please me, as I remember saying to you, it does not work well in with nat. selection—it is rather too sudden a jump.—especially the adaption to different climatal ranges.— I am inclined by conjecture to look at cowslip as the modified descendant of Primrose; & that the seeds from the cowslip do not simply vary, but revert back in different degrees to primrose. Most of the experiments have been made on cowslips & oxlips.— on this notion, the case would be more intelligible.

Yours affect | C. Darwin

I have received long copy of M.S. verses from some anonymous author, highly laudatory, yet partly quizzing the Origin.— They are from some Botanist. I suspect Dr Boott.—[8] Does he write verses?

Etty keeps just the same, poor little thing, now 3 week, wearing Fever. What a world it is—[9]

New York Botanical Garden Library, Cox collection

[1] Dated by the reference to Henrietta Emma Darwin's continuing fever.

[2] William Henry Harvey had asked Hooker to obtain information from CD about hybridisation among *Primula* species (see L. Huxley ed. 1918, 1: 516–20). This phenomenon had been cited as support for CD's view of intermediate gradations between varieties becoming sufficiently distinct that the forms could be judged as different species (see *Spectator*, 21 April 1860, p. 380).

[3] Herbert 1822. CD cited this paper frequently in discussions of hybridisation in the manuscript of his 'big book' on species (*Natural selection*).

[4] Henslow 1830.

[5] Watson 1847.

[6] Sidebotham 1849. CD discussed both this paper and Watson 1847 in *Natural selection*, pp. 130–3.

[7] See letters from Henry Doubleday, 3 May 1860 and 16 May 1860.

[8] The verses have not been found. Francis Boott's theological beliefs had been shaken following his reading of *Origin* (see letter from Francis Boott, 29 February 1860). For CD's opinion of Boott's response to *Origin*, see the letter to J. D. Hooker, 3 March [1860].

[9] Henrietta Emma Darwin had been very ill since 28 April 1860 (Emma Darwin's diary).

To Asa Gray 22 May [1860][1]

Down Bromley Kent
May 22^d

My dear Gray.

Again I have to thank you for one of your very pleasant letters (of May 7th), enclosing a very pleasant remittance of 22£.—[2] I am in simple truth astonished at all the kind trouble you have taken for me. I return Appleton's account.—[3] For the chance of your wishing for a formal acknowledgment, I send one.— If you have any further communication to the Appletons pray express my acknowledgment for his generosity; for it is generosity in my opinion. I am not at all surprised at sale diminishing: my extreme surprise is at greatness of sale. No doubt the public has been *shamefully* imposed on! for they bought the book, thinking that it would be nice easy reading. I expect the sale to stop soon in England: yet Lyell wrote to me the other day that calling at Murrays he heard that 50 copies had gone in previous 48 hours.— I am extremely glad that you will notice in Silliman additions in the Origin.[4] Judging from letters (& I have just seen one from Thwaites to Hooker) & from remarks, the most serious omission in my book was not explaining how it is, as I believe, that all forms do not necessarily advance,—how there can now be *simple* organisms still existing.—[5] The article in Med. & Chirurg. Review is by Carpenter.—[6] I would send Pictet's, if you cannot see it, but I sh^d require it back.—[7] Sedgwick has been firing broadsides into me, but exclusively on geological grounds.— Prof. Clarke of Cambridge says publickly that the chief characteristic of such books as mine is their "consummate impudence".—[8]

I hear there is **very** severe review on me in North British by a Revd Mr Dunns,[9] a free-Kirk minister & dabbler in Nat. Hist^y I sh^d be very glad to see any good American Reviews,—as they are all more or less useful.— You say that you shall touch on other Reviews.—[10] Huxley told me some time ago that after a time he would write review on all Reviews, whether he will I know not.— If you allude to Edinburgh, pray notice *some* of the points which I will point out on separate slip.[11] In "Saturday Review" (one of our cleverest periodicals) of May 5th. p. 573 there is a nice article on Owen's Review, defending Huxley, but not Hooker; & the latter I think Owen treats most ungenerously.—[12] But surely you will get sick unto death of me & my Reviewers.—

With respect to the theological view of the question; this is always painful to me.—I am bewildered.— I had no intention to write atheistically. But I own that I cannot see, as plainly as others do, & as I shd wish to do, evidence of design & beneficence on all sides of us. There seems to me too much misery in the world. I cannot persuade myself that a beneficent & omnipotent God would have designedly created the Ichneumonidæ with the express intention of their feeding within the living bodies of caterpillars, or that a cat should play with mice. Not believing this, I see no necessity in the belief that the eye was expressly designed. On the other hand I cannot anyhow be contented to view this wonderful universe & especially the nature of man, & to conclude that everything is the result of brute force. I am inclined to look at everything as resulting from designed laws, with the details, whether good or bad, left to the working out of what we may call chance. Not that this notion *at all* satisfies me. I feel most deeply that the whole subject is too profound for the human intellect. A dog might as well speculate on the mind of Newton.— Let each man hope & believe what he can.—

Certainly I agree with you that my views are not at all necessarily atheistical. The lightning kills a man, whether a good one or bad one, owing to the excessively complex action of natural laws,—a child (who may turn out an idiot) is born by action of even more complex laws,—and I can see no reason, why a man, or other animal, may not have been aboriginally produced by other laws; & that all these laws may have been expressly designed by an omniscient Creator, who foresaw every future event & consequence. But the more I think the more bewildered I become; as indeed I have probably shown by this letter.

Most deeply do I feel your generous kindness & interest.—

Yours sincerely & cordially | Charles Darwin

[Enclosure 1][13]

Asa Gray for Mr. Darwin

Statement of the Sale of Darwin's "Origin of Species" *to* May *1st, 18*60

On hand last account,		—	*On hand this date,*	250	
Printed since	Jany/60	1500	*In hands of Booksellers,*	300	550
	Feby—/60	500	*Given away,*		200
	Mch/60	500	*Sold to date,*		1750
		2500			2500

1750 *Sold*, at 5% on $1.25 *Copyright amounting to* $109.37

[Enclosure 2][14]

Edinburgh Review

p. 489. I think I told you of the illiberality of Owen about the Branchiæ; he having in independent works fully admitted that the Branchiæ were Branchiæ[15]

496 what a quibble to pretend he did not understand what I meant by *inhabitants* of S. America: & anyone would suppose that I had not throughout my volume touched on Geograph. Distrib. He ignores also everything which I have said on Classification, Geological Succession. *Homologies*, Embryology & Rudimentary Organs.

500 He falsely applies what I said (too rudely) about "blindness of preconceived opinions" to those who believe in creation; whereas I exclusively apply the remark to those who give up multitudes of species as true species, but believe in remainder.—

p. 501. He slightly alters what I say.— I **ask** whether creationist really believe that elemental atoms have flashed into life. He says I describe them as so believing; & this surely is a difference.—

501. He speaks of my "clamouring against" all who believe in creation: & this seems to me an unjust accusation.

512. It is a good joke that the Reviewer is so full of Owens work that he refers to "On the Nature of the Limbs", when he means to refer to my book & give right page; but alters the quotation.

522— He makes me say that the Dorsal vertebræ vary— this is simply false: I nowhere say a word about dorsal vertebræ.

525. What an illiberal sentence that is about "any pretension to candour" & about my rushing through barriers which stopped Cuvier: such an argument would stop any progress in science.

530 How disingenuous to quote from my remarks to you about my *brief* letter, as if it applied to whole subject.

530. How disingenuous to say that we are called on to accept the theory from the imperfection of geological Record, when I over & over again how grave a difficulty the imperfection offers.—

519. How often that abominable arctic Bear has been made to worry me; but Owen is false, when he says that the case has no application, as put in my 2d. Edit, for it certainly well illustrates "diversified & changed habits" of life.—

Gray Herbarium of Harvard University (26 and 37a)

1 Dated by an entry in CD's journal recording the payment mentioned in the letter (see n. 2, below).

2 Gray's letter has not been found. CD recorded receiving a payment of £21 17*s.* 6*d.* in May 1860 as his share of the profits from the sale of the first American edition of *Origin* ('Journal'; Appendix II).

3 See the first enclosure. The first American edition of *Origin* had been published in January 1860 by the New York firm D. Appleton & Co. (Freeman 1977, pp. 83, 85). Gray had acted on CD's behalf in negotiating the terms regarding the author's share of the profit from sales of the American edition of *Origin*. See letter from Asa Gray, 20 February 1860 and the enclosure.

4 The fourth printing of *Origin* in the United States incorporated, through Gray's exertions, substantial additions and corrections to certain sections of the text. See letters to Asa Gray, 28 January [1860],

1 February [1860], [8 or 9 February 1860], and 8 March [1860]. Gray was an associate editor of the *American Journal of Science and Arts*, also known as 'Silliman's journal'. However, the article that Gray was preparing (see n. 10, below) appeared in the *Atlantic Monthly* (Gray 1860b). Gray's notice about the changes in the revised and augmented American edition of *Origin* is on p. 116.

[5] See following letter. CD replied to this point in his letter to G. H. K. Thwaites, 21 March [1860]. He rectified the omission of this discussion in the third edition of *Origin* (*Origin* 3d ed., pp. 134–7).

[6] [Carpenter] 1860b.

[7] Pictet de la Rive 1860.

[8] CD refers to an account of a meeting of the Cambridge Philosophical Society on 7 May 1860 reported in the *Literary Gazette*, 12 May 1860, p. 582. See also letter from J. S. Henslow to J. D. Hooker, 10 May 1860.

[9] The anonymous review of *Origin* in the *North British Review* was written by the editor, John Duns, a Scottish divine. See *LL* 2: 311 and *Wellesley index* 5: 236. CD had recently ordered a copy of the journal (letter to a Bookseller, 13 [May 1860]).

[10] Gray published an article in the *Atlantic Monthly* 6 (1860): 406–25 entitled 'Darwin and his reviewers' (Gray 1860b). The article is reprinted in Gray 1876, pp. 72–145.

[11] See the second enclosure. The reference is to [R. Owen] 1860a.

[12] *Saturday Review*, 5 May 1860, pp. 573–4. See also letter to T. H. Huxley, 7 May [1860].

[13] The first enclosure is a printed sales slip from the firm of D. Appleton & Co. sent to CD by Gray. CD wrote on the bottom of the slip '= £22.00—', which was his share of the profit on sales after converting the sum given in dollars. See nn. 2 and 3, above.

[14] The second enclosure is in the Gray Herbarium of Harvard University (37a).

[15] See letter to T. H. Huxley, 9 April [1860].

To J. D. Hooker 22 [May 1860][1]

Down Bromley Kent
22^d^

My dear Hooker

I was very glad to see Thwaites' letter:[2] he makes same objection almost totidem verbis as Lyell did at first. It was great omission in my part not discussing this subject, (as I will if there be future Edition) for I think I can remove this difficulty.—[3]

That is curious fact about Beetle: something analogous is known in certain crustaceans; but hardly well enough made out in the case of Beetle to quote.— Have you Pyrolas at Kew? if so for **Heaven sake** observe curvature of pistil towards gangway to nectary. If Bees visit this plant, it is best of all to observe them.

You have turned the tables in case of Lavatera with a vengeance against me. Godron in his work "sur l'Espece" gives apparently the best & most carefully worked out cases, which I have met with, of seeds alive & long buried in earth.[4]

I fear by appearance, Leschenaultia will not seed, possibly owing to conditions of life; I can see no sign of incipient fertilisation.—

I have had lately two most kind letters from A. Gray, with 22£ from American Publisher![5] I have had, also, capital letter from Wallace, after reading the Origin.—[6]

Lyell tells me that Binney has published in Proc: of Manchester Soc^y^ paper trying to show that coal plants must have grown in very marine marshes.—[7] Do you remember how savage you were long years ago at my broaching such

I am sincerely sorry to hear about Carpenter's disappointment, but I wonder he cares about it—[9]

Poor Etty keeps nearly same, but we think certainly improves a little.— Your kind expressions about her pleased us much.—

My dear old friend | Your affect | C. Darwin

P.S. | As Horse-chesnuts have male flowers & hermaphrodite flowers I have wished to examine their pollen, & this has made me observe a thing which has surprised me.—**All** the flowers now open on my *several* trees are *male* with rudimentary pistil with *pollen shedding*; so that I began to think my memory had deceived me & that the pistil was never well developed; but on opening *buds near end* of each little lateral twig of the flower-truss, I find plenty of hermaphrodite flowers with pistils well developed. So that on all my trees there has been a gigantic crop of *quite useless* male flowers, with millions of pollen-grains wasted, for there is not a female flower nearly open.— Now will you just look if you think fact odd enough at Kew at Horse chesnuts; only you must remember that your trees will be 7 or 10 days earlier than mine; so that I daresay your hermaphrodite flowers & males from **upper flower** of each side twig will be open; but I daresay you will be able to see whether the **lower** flowers on each little side twig are exclusively male.— If this is case elsewhere & in native country, it w^{d} be great & curious blunder in dame nature.[10] I shall look to pollen hereafter.— The pollen of useless males seems good & swells in water.—

P.S. 2^{d} I think I shall make out good case of want of *perfection* in Horse-chesnut— I find 140 flowers on a *small* truss— I think each truss produces only one or sometimes two seeds—& I have got vague notion that in each ovary there are at first many ovules.— I might put it, "why should we wonder at thousands of wasted *male* flowers, when we see such waste of flowers of all kind?"— I have, also, somewhere got some notes on the prodigious waste of insect life—by the sticky scales on the buds buds, —by the thousands, & all uselessly killed.—

Endorsement: 'May/60'
DAR 115.1: 57

[1] The endorsement is confirmed by the reference to CD having received £22 from the publisher of the American edition of *Origin* ('Journal'; Appendix II).

[2] The letter from George Henry Kendrick Thwaites to Hooker giving his opinion of *Origin* has not been found. However, see the letter to G. H. K. Thwaites, 21 March [1860].

[3] See following letter and letter to Asa Gray, 22 May [1860]. CD added to the third edition of *Origin* a brief explanation of why organisms, particularly simple ones, do not necessarily have to advance (*Origin* 3d ed., pp. 133–43).

[4] Godron 1859, 1: 147–54. For CD's interest in the vitality of long-buried seeds, see *Correspondence* vols. 2, 5, and 6.

[5] See letter to Asa Gray, 22 May [1860].

[6] Alfred Russel Wallace's letter has not been found, but see CD's reply (letter to A. R. Wallace, 18 May 1860).

[7] Binney 1847. See following letter and n. 10.

[8] See *Correspondence* vol. 4, letters to J. D. Hooker, [1 May 1847], [6 May 1847], and [12 May 1847]. CD referred to this incident in his *Autobiography*, p. 105.

[9] The reason for William Benjamin Carpenter's 'disappointment' has not been identified.

[10] CD and Hooker had discussed the imperfection of nature on a number of occasions. See, for example, *Correspondence* vol. 6, letters to J. D. Hooker, 13 July [1856] and 11 September [1857]. See also *Origin*, pp. 202–3.

To Charles Lyell 22 May [1860]

Down Bromley Kent
May 22[d]

My dear Lyell

Perhaps you w[d] like see another of A. Gray's letter received this morning. The Appletons are gentlemen, though payment not large.— The Edition was 2500 of the Origin, & I am far from surprised that sale slackens.—[1]

I send by this Post Isidore G. H. on Hare-rabbit p. 222:[2] you can send it afterwards to my Brothers. I have not yet seen N. British; though I have copy at my Brothers, & am very glad to know who is Author.—[3] I hear that it is very savage.—

The Medical Review referred to by Asa Gray is only that (I now find from Williams & Norgate) which Carpenter wrote in Medical & Chirurg. Review;[4] it is at my Brothers & you can take it, if you have not already seen it.—

I am sorry that I troubled you with Sedgwick in Cambridge paper.—[5]

Hooker has sent me letter of Thwaites of Ceylon, who makes exactly same objection which you did at first about the necessity of all forms advancing & therefore the difficulty of simple forms still existing.[6] There was no worse omission than this in my Book & I had discussion all ready. I am extremely glad to hear that you intend adding new argument about imperfection of Geological Record:[7] I always feel this acutely & am surprised that such men as Ramsay & Jukes do not feel it more.—[8] I quite agree no sufficient evidence about Mummy wheat.[9]

When you can spare it, I sh[d] like (but out of *mere* curiosity) to see Binney on coal-*marine* marshes:[10] I once made Hooker very savage by saying that I believed Coal-plants grew in sea like mangroves.—[11]

Etty keeps nearly the same, but rather improves: it is now more than 3 weeks.— We have just moved her to fresh room for change.—

I have much to write, so goodBye | Yours affect. | C. Darwin

What a fact about the Coal Land Shells!!![12]

Endorsement: '1860'
American Philosophical Society (213)

[1] Asa Gray's letter has not been found, but see letter to Asa Gray, 22 May [1860]. CD recorded the American sales of *Origin* in a journal entry dated 22 May 1860 ('Journal'; Appendix II).

[2] I. Geoffroy Saint-Hilaire 1854–62, 3: 222. CD's annotated copy is in the Darwin Library–CUL. See letter to Charles Lyell, 18 May [1860].

[3] [Duns] 1860. There is a copy of the review in the Darwin Pamphlet Collection–CUL.
[4] [Carpenter] 1860b was published in the *British and Foreign Medical-Chirurgical Review*.
[5] Adam Sedgwick. See letter to Charles Lyell, 18 May [1860].
[6] See preceding letter and letter to Asa Gray, 22 May [1860]. Lyell had raised this point in October 1859. See *Correspondence* vol. 7, letter to Charles Lyell, 11 October [1859].
[7] Lyell was intending to publish an extensively revised edition of his *Elements of geology*. Instead, he included this material in his book on the antiquity of man (C. Lyell 1863, pp. 448–50).
[8] Andrew Crombie Ramsay and Joseph Beete Jukes. See letter from J. B. Jukes, 27 February 1860.
[9] CD refers to widespread accounts that wheat grains found in Egyptian tombs had germinated when sown.
[10] Binney 1847. Lyell had recently read the volume of the *Memoirs of the Literary and Philosophical Society of Manchester* that included both this paper and Binney 1848 (see Wilson ed. 1970, p. 404). CD evidently thought that Lyell was referring to a more recent paper: he was already familiar with Binney 1847 (see *Correspondence* vol. 4, letter to J. D. Hooker, [1 May 1847]). See also preceding letter.
[11] See preceding letter.
[12] In a paper read at a meeting of the Geological Society on 14 December 1859, the Canadian geologist John William Dawson described the discovery of a fossilised terrestrial mollusc found in the coal formations of Nova Scotia (Dawson 1859). Lyell had formerly visited this site with Dawson and described fossils found there (Lyell and Dawson 1853 and Dawson 1859, pp. 268–9). The occurrence of fossilised land shells in coal was of particular interest in connection with the question of the origin of coal. See *Correspondence* vol. 4, letter to J. D. Hooker, [1 May 1847].

To John Lubbock 25 May [1860]

Down.
May 25th

Dear Lubbock.

I have no comment to make on your Report.— I have forwarded it to Innes[1] with your query about E. Town,[2] which I cannot answer.— I have suggested to Innes that the Report ought to be sent to Mr Stephens[3] for his approval, as he is ex officio on Committee.[4] Remembering that you asked Mr Stephen to take the account (as well as asking me) on account of your manifold employment & that Mr S. refused I have asked Innes whether he had not better suggest to Mr Stephens to undertake this slight labour & thus relieve you.— Whether he will of course I know not.— I am pretty sure (from manner alone) that Mr. S. was a little sore at not being appointed to the School committee, so that some time ago I mentioned this to Mr Innes, & he said that he had no doubt that he was ex officio on committee & I informed Mr Stephens of this; & I suspect that this soreness led him to refuse to take the accounts.—

Many thanks for your enquiries about Etty; her fever runs on & is mild, but it is terribly wearing to her as it has now lasted 4 weeks tomorrow.—

Dear Lubbock | Yours very truly | C. Darwin

Endorsement: '1860'
Down House

[1] John Innes was the perpetual curate of Down. Innes no longer resided in Down owing to his inability to secure suitable accommodation for his family.

[2] There is a Daniel Town, carpenter, listed as residing in Down in the *Post Office directory of the six home counties* 1859. E. Town may have been a relative.

[3] Probably Thomas Selwood Stephens, who is listed in CD's Address Book (Down House MS) as residing in Down. His name appears (misspelled) in the 1862 *Post Office directory* as curate of Down. Only in 1865, however, was he first listed as such in the *Clergy list.*

[4] CD refers to the committee overseeing the national school in Down, of which John Lubbock and his father, John William Lubbock, were benefactors. The school was funded by charitable donations from the local gentry and administered through a committee appointed by the parish vestry.

To J. D. Hooker 29 [May 1860][1]

Down Bromley Kent
29th

My dear Hooker

Your letter contained two capital pieces of news, that Mrs. Hooker's troubles are safely over which is a real blessing,[2] & that the Admiralty should have been liberal.[3] I truly rejoice at this, though of course the remuneration is as nothing to the labour expended.— Many thanks for Harvey's letter,[4] which I will keep a little longer & then return.— I will write to him & **try** to make clear, from analogy of domestic productions, the part which I believe Selection has played.[5] I have been reworking my pigeons & other domestic animals & I am sure that anyone is right in saying that Selection is **the** efficient cause, though as you truly say variation is the base of all.—

Why I do not believe so much as you do in Physical agencies, is that I see in almost every organism (though far more clearly in animals than in plants) *adaptation*, & this except in rare instance must, I shd. think, be due to selection.—

Do not forget the Pyrolas when in flower. My blessed little Scævola has come into flower & I will try artificial fertilisation on it.—

Yours ever affect | C. Darwin

I have looked over Harvey's letter & have assumed (I hope rightly) that he could not object to knowing that you had forwarded it to me.—

Endorsement: 'May/60'
DAR 115.1: 58

[1] The endorsement is confirmed by the reference to the Hookers' new baby (see n. 2, below).

[2] Brian Harvey Hodgson Hooker was born on 27 May 1860.

[3] The reference has not been traced. It may possibly relate to the publishing of Hooker 1859, the costs of which Hooker had primarily borne. See letter to J. D. Hooker, 8 February [1860].

[4] William Henry Harvey corresponded with Hooker in May and June 1860 about the theological implications of CD's theory. A letter from Hooker to Harvey dated 26 May 1860 indicates that Harvey had written a long letter to Hooker discussing natural selection, which is probably the one referred to here. See L. Huxley ed. 1918, 1: 516–20.

[5] CD's letter to Harvey has not been found. See, however, letter to W. H. Harvey, [20–4 September 1860].

To John Lubbock 29 [May 1860][1]

Down.—
29th

Dear Lubbock

I enclose Innes answer. It is evident that he will not speak to Mr Stephens.[2] If on reflexion you still much wish to give up accounts I would ask Mr Stephens for you; for I could make good reason, saying that you had asked me, & that I refused as I keep the Friendly Club (which is no joke) Coal Club & Saving Bank accounts.—[3] Or you might say this.—This fact, by the way, is an answer to Mr Innes' remark on all being thrown on the Clergyman's back.—

You will see query about E. Town answered & I quite agree.—[4]

I hope that you agree that the Report shd be sent to Mr Stephens for approval & signature

Etty keeps very ill, & though we were anxious yesterday, I think she is better to day.—

Dear Lubbock | Yours most truly | C. Darwin

P.S. 10s is due from you this day for annual Subscription to Friendly Club.— I am glad to say Mr Stephens, without asking has become Subscriber to Friendly Club.—

Endorsement: '1860'
Down House

[1] Dated by the relationship to the letter to John Lubbock, 25 May [1860].

[2] Probably Thomas Selwood Stephens, a resident of Down who was apparently acting as curate. See letter to John Lubbock, 25 May [1860].

[3] CD had been treasurer of the Down Coal and Clothing Club since 1848 (see *Correspondence* vol. 4, letter to John Innes, [8 May 1848]). In 1850, the administration was expanded to include a Friendly Society, a mutual insurance and benefit club, of which CD was treasurer for thirty years (*LL* 1: 142).

[4] Possibly a relative of Daniel Town, carpenter, who is listed as residing in Down in the *Post Office directory of the six home counties* 1859.

To J. D. Hooker 30 May [1860][1]

Down Bromley Kent
May 30th

My dear Hooker

I return Harvey's letter: I have been very glad to see the reason why he has not read your Essay;[2] I feared it was bigotry. And I am glad to see that he goes a little way (*very much* further than I supposed) with us on Nat. Selection.— I was not sorry for a natural opportunity of writing just to show that I was not piqued at his turning me & my book into ridicule,—not that I think it was proceeding which I deserved or worthy of him.—[3]

It delights me that you are interested in watching progress of opinion on change of species; I feared that you were weary of subject; & therefore did not send A.

Gray's letters. The battle rages furiously in U. States. Gray says he was preparing a speech which would take $1\frac{1}{2}$ hour to deliver, & which he "fondly hoped would be a stunner". He is fighting splendidly & there seem to have been many discussions with Agassiz & others at the meetings. Agassiz pities me much at being so deluded.—As for progress of opinion, I clearly see that it will be excessively slow, almost as slow as change of species. In fact it will, I believe, be insensible. I am getting wearied at the storm of hostile Reviews; & hardly any useful.—

Did you see in Literary Gazette that Prof. Clarke of Cambridge says the chief characteristic of such Books as mine is "their consummate impudence"—mild & gentleman-like language!—[4]

I am in great doubt about Oxford;[5]—assuming that Etty, is then quite well (she is better today, but the weary fever drags on), my doubt being chiefly from my own health, whether it would answer.— I have written to ask whether I c.[d] have rooms in a college.—[6] If, however, everything is transcendantly well, my wife says she will go,—but all is dark yet.— I hope M.[rs] Hooker goes on well—

Yours affect | C. D.—

Blyth writes from Calcutta that his ideas on species are quite revolutionised.—[7]

Endorsement: '/60'
DAR 115.1: 59

[1] The endorsement is confirmed by CD's reference to the meeting of the British Association for the Advancement of Science to be held in Oxford in June 1860.

[2] William Henry Harvey had been so occupied with delivering his course of lectures at Trinity College, Dublin, that he had not found time to read Hooker's introductory essay (Hooker 1859) and had only recently read *Origin*. See L. Huxley ed. 1918, 1: 517, 519.

[3] Harvey had presented a 'serio-comic squib on Darwin' at a meeting of the Dublin University Zoological and Botanical Association on 17 February 1860 (L. Huxley ed. 1918, 1: 515–16). He later sent it to CD 'with the writer's repentance' (see letter from W. H. Harvey, 8 October 1860). The paper is in the Darwin Pamphlet Collection—CUL.

[4] See letters to J. S. Henslow, 17 May [1860], and to Asa Gray, 22 May [1860].

[5] See n. 1, above.

[6] See letter to George Rolleston, 1 June [1860].

[7] Edward Blyth, curator of the museum of the Asiatic Society of Bengal, had corresponded with CD since 1855. He became an adherent of natural selection in 1860 and defended the theory at meetings of the society in November (*DSB*). Unfortunately, none of the letters from Blyth to CD during the years 1859 to 1861 have survived. Blyth's views on *Origin* were published in the *Journal of the Asiatic Society of Bengal* 29 (1860): 436–7 and in the *Calcutta Review* 35 (1860): 64–88. The latter paper is in the Darwin Pamphlet Collection–CUL.

To Charles Lyell 1 [June 1860]

Down Bromley Kent
Friday night—1.st

My dear Lyell

Poor Etty's fever drags on its weary course & Sunday will be five weeks.— I believe there is no cause yet for serious anxiety, but we have sent for Sir H. Holland[1] to aid our local Doctor.—

I sent yesterday A. Murray's review:[2] the speculative parts seem to me weak. I am not convinced that I have misrepresented Schiödte.—[3] Have you seen Hopkins in the new Fraser?[4] the public will I sh[d] think, find it heavy. He will be dead against me as you prophesied; but he is generously civil to me personally. On his standard of proof, *natural* science would never progress; for without the making of theories, I am convinced there would be no observation.[5] I have ordered the Future,[6] & have begun reading the N. British,[7] which so far strikes me as clever.

Phillips' Lecture at Cambridge is to be published.[8]

All these reiterated attacks will tell heavily; there will be no more converts & probably some will go back. I hope you do not grow disheartened. I am determined to fight to the last.— I hear, however, that the *great* Buckle highly approves of my Book.—[9]

I have had note from poor Blyth of Calcutta, who is much disappointed at hearing that L. Canning will not grant any money;[10] so I much fear that all your great pains will be thrown away.[11] Blyth says (& he in many respect a very good judge) that his ideas on Species are quite revolutionised.

I send Asa Gray's letter,[12] though it is really not worth sending; & do not return it till you have occasion to write; though if I were not the most honourable of men, I should jump at the excuse of getting a note from you & require his letter back soon.—

I have never attended to gestation of dog question,[13] thinking it's importance overrated, as the period is known to be variable in many of our domestic animals.— But I have begun to make some enquiries, & have got *accurate* return from excellent stud of Grey-hounds—period varies from 60—or 61 (according as 1st day reckoned)—to 65 or 66 days.

I do not think lowness of Rodents has any bearing on the question of fertility of Hybrids, as plants, & even Fuci, follow same general laws in these respects with animals.—

My dear Lyell | Ever yours | C. Darwin

American Philosophical Society (214)

[1] Henry Holland, a distant relative of both CD and Emma Darwin, was a physician in ordinary to Queen Victoria. CD had consulted him on previous occasions when health problems in the family were particularly severe (see *Correspondence* vol. 4). The Darwins' local doctor had diagnosed her illness as a form of typhus fever (letter to J. D. Hooker, 11 May [1860]).

[2] Murray 1860a.

[3] CD had queried this point after reading proof-sheets of Andrew Murray's review (see letter to Andrew Murray, 28 April [1860]). CD had apparently just received a printed copy of the review. Murray criticised CD's discussion of the improbability of the separate creation of blind insects in caves located on separate continents (Murray 1860a, p. 286):

> It is quite correct for him [CD] to say that we should expect close similarity in the caves in question, but it is incorrect to say that "this is not the case"; for the similarity in some is marvellously close; and it is also incorrect to say that Schiödte and others have remarked that "this is not the case."

The reference is to Schiödte 1849.

[4] Hopkins 1860, which appeared in *Fraser's Magazine*. William Hopkins, a mathematician and geologist, was renowned for his work in physical geology. He had corresponded with CD in the late 1840s (see *Correspondence* vol. 3).

[5] Hopkins argued that CD's theory should meet the same standards of evidence as physical theories.

[6] A review of *Origin* appeared in the first number of the new and short-lived publication entitled *The Future: a Monthly Journal of Research and Criticism in the Physical and Historical Sciences*, April 1860, pp. 10–16. A copy of the review is in the Darwin Pamphlet Collection–CUL. A second article discussing *Origin* appeared in the May issue; a copy of this piece is also in the Pamphlet Collection. At the head, CD wrote: 'By a Mr Berke' and made a number of comments about the author's arguments. Luke Berke was the editor of *The Future*.

[7] [Duns] 1860. See letter to Asa Gray, 22 May [1860].

[8] Phillips 1860.

[9] CD had met Henry Thomas Buckle, the historian, at a party at the home of Hensleigh Wedgwood in 1858 (see *Correspondence* vol. 7, letter to J. D. Hooker, 23 February [1858]). The source of information concerning Buckle's remark has not been traced, but Buckle wrote favourably about *Origin* in his correspondence. In 1859 he told a friend that the book was 'full of thought, and of original matter' (Huth 1880, 2: 28).

[10] Edward Blyth's name had been put forward by CD for the post of naturalist to the forthcoming expedition to China (see *Correspondence* vol. 7, letters to W. H. Sykes, 20 December [1859], and to Charles Lyell, 29 [December 1859]). Charles John Canning had been made viceroy of India in 1858 after the East India Company transferred power to the British Crown. He objected to the appointment of a naturalist to the China expedition on the ground that scientific observations in a hostile country would entail personal danger. Despite Blyth's assurances that he was 'quite willing to encounter the danger', no naturalist was appointed (Grote 1875).

[11] Lyell also provided a testimonial for Blyth (see *Correspondence* vol. 7, letter to Charles Lyell, 29 [December 1859]).

[12] The letter has not been found, but for CD's reply, see the letter to Asa Gray, 22 May [1860].

[13] Lyell questioned whether domestic breeds of dog were descended from one or several wild species and suggested that the question might be settled by comparing the gestation period in the domestic dog with that of the supposed aboriginal species. See *Correspondence* vol. 7, letters from Charles Lyell, 28 October 1859 and 21 November 1859. Lyell consulted Thomas Henry Huxley on the matter on 6 June 1860 (see Wilson ed. 1970, pp. 439–40).

To George Rolleston 1 June [1860][1]

Down Bromley Kent
June 1st

Mr Darwin presents his compliments to Dr Rolleston & is extremely much obliged for his very kind invitation,[2] which he would have had very great pleasure in accepting, had not the state of his health, rendered it almost necessary for Mr Darwin to have private lodgings so as to avoid fatigue.—

Wellcome Institute for the History of Medicine Library, London (WMS. 6119/1)

[1] Dated by the relationship to the letter to J. D. Hooker, 30 May [1860], in which CD states that he has written to inquire about rooms at Oxford (see n. 2, below).

[2] Rolleston was professor of anatomy and physiology at Oxford University and president of the subsection on physiology at the forthcoming meeting of the British Association for the Advancement of Science in Oxford.

To *Gardeners' Chronicle* [4–5 June 1860][1]

I should be extremely much obliged to any person living where the Bee or Fly Orchis is tolerably common,[2] if he will have the kindness to make a few simple observations on their manner of fertilisation. To render the subject clear to those who know nothing of botany, I must briefly describe what takes place in our common British Orchids. The pollen-grains form two pear-shaped masses; each borne on a foot-stalk, with a sticky gland at the end. The pollen masses are hidden in little pouches open in front. When an insect visits a flower, it almost necessarily, owing to the position of the parts, uncovers and touches the sticky glands. These firmly adhere to the head or body of the insect, and thus the pollen-masses are drawn out of their pouches, are dragged over the humid stigmatic surface, and the plant is fertilised. So beautifully are the relative degrees of adhesiveness of the gland, and of the grains of pollen to each other and to the stigmatic surface mutually adapted, that an insect with an adherent pollen-mass will drag it over the stigmas of several flowers, and leave granules of pollen on each. The contrivance by which the sticky glands are prevented from drying, and so kept always viscid and ready for action, is even still more curious; they lie suspended (at least in the two species which I have examined) in a little hemispherical cup, full of liquid, and formed of such delicate membrance, that the side projecting over the gangway into the nectary is ruptured transversely and depressed by the slightest touch; and then the glands, sticky and fresh out of their bath, immediately and almost inevitably come into contact with and adhere to the body which has just ruptured the cup. It is certain that with most of our common Orchids insects are absolutely necessary for their fertilisation; for without their agency, the pollen-masses are never removed and wither within their pouches. I have proved this in the case of *Orchis morio* and *mascula* by covering up plants under a bell-glass, leaving other adjoining plants uncovered;[3] in the latter I found every morning, as the flowers became fully expanded, some of the pollen-masses removed, whereas in the plants under the glass all the pollen-masses remained enclosed in their pouches.

Robert Brown, however, has remarked that the fact of all the capsules in a dense spike of certain Orchids producing seed seems hardly reconcileable with their fertilisation having been accidentally effected by insects.[4] But I could give many facts showing how effectually insects do their work; two cases will here suffice; in a plant of *Orchis maculata* with 44 flowers open, the 12 upper ones, which were not quite mature, had not one pollen-mass removed, whereas every one of the 32 lower flowers had one or both pollen-masses removed; in a plant of *Gymnadenia conopsea* with 54 open flowers, 52 had their pollen-masses removed. I have repeatedly observed in various Orchids grains of pollen, and in one case *three* whole pollen-masses on the stigmatic surface of a flower, which still retained its own two pollen-masses; and as often, or even oftener, I have found flowers with the pollen-masses removed, but with no pollen on their stigmas. These facts clearly show that each flower is often, or even generally, fertilised by the pollen brought by insects from another flower or plant. I may add that after observing our Orchids during many years, I have never seen a bee or any other diurnal insect (excepting once a butterfly) visit them; therefore I have no doubt that moths are the priests who perform the marriage ceremony. The structure, indeed, of some Orchids leads to this same conclusion; for no insect without a very long and extremely fine proboscis could possibly reach the nectar at the bottom of the extremely long and narrow nectary of the Butterfly-Orchis; and entomologists have occasionally captured moths with pollen-masses adhering to them. If any entomologist reads this, and can remember positively having caught a moth thus furnished, I hope he will give its name, and describe exactly to which part of the moth's body the sticky gland adhered.[5]

We may now turn to the genus Ophrys; in the Fly Orchis (*Ophrys muscifera*), the pollen-masses, furnished with sticky glands, do not naturally fall out of their pouches, nor can they be shaken out; so that insect-agency is necessary, as with the species of the other genera, for their fertilisation. But insects here do their work far less effectually than with common Orchids; during several years, previously to 1858, I kept a record of the state of the pollen-masses in well-opened flowers of those plants which I examined, and out of 102 flowers I found either one or both pollen-masses removed in only 13 flowers. But in 1858 I found 17 plants growing near each other and bearing 57 flowers and of these 30 flowers had one or both pollen-masses removed; and as all the remaining 27 flowers were the upper and younger flowers, they probably would subsequently have had most of their pollen-masses removed, and thus have been fertilised. I should much like to hear how the case stands with the Fly Orchis in other districts; for it seems a strange fact that a plant should grow pretty well, as it does in this part of Kent, and yet during several years seldom be fertilised.

We now come to the Bee Orchis (*Ophrys apifera*), which presents a very different case; the pollen masses are furnished with sticky glands, but differently from in all the foregoing Orchids, they naturally fall out of their pouches; and from being of the proper length, though still retained at the gland-end, they fall on the stigmatic surface, and the plant is thus self-fertilised. During several years I have examined many flowers, and never in a single instance found even one of the pollen-masses carried away by insects, or ever saw the flower's own pollen-masses fail to fall on the stigma. Robert Brown consequently believed that the visits of insects would be injurious to the fertilisation of this Orchis; and rather fancifully imagined that the flower resembled a bee in order to deter their visits. We must admit that the natural falling out of the pollen-masses of this Orchis is a special contrivance for its self-fertilisation; and as far as my experience goes, a perfectly successful contrivance, for I have always found this plant self-fertilised; nevertheless a long course of observation has made me greatly doubt whether the flowers of any kind of plant are for a perpetuity of generations fertilised by their own pollen. And what are we to say with respect to the sticky glands of the Bee Orchis, the use and efficiency of which glands in all other British Orchids are so manifest? Are we to conclude that this one species is provided with these organs for no use? I cannot think so; but would rather infer that, during some years or in some other districts, insects do visit the Bee Orchis and occasionally transport pollen from one flower to another, and thus give it the advantage of an occasional cross. We have seen that the Fly Orchis is not in this part of the country by any means sufficiently often visited by insects, though the visits of insects are indispensable to its fertilisation. So with the Bee Orchis, though its self-fertilisation is specially provided for, it may not exist here under the most favourable conditions of life; and in other districts or during particular seasons it may be visited by insects, and in this case, as its pollen masses are furnished with sticky glands, it would almost certainly receive the benefit of an occasional cross impregnation. It is this curious apparent contradiction in the structure of the Bee Orchis—one part, namely the sticky glands, being adapted for fertilisation by insect agency—another part, namely the natural falling out of the pollen-masses, being adapted for self-fertilisation without insect agency—which makes me anxious to hear what happens to the pollen-masses of the Bee Orchis in other districts or parts of England. I should be extremely much obliged to any one who will take the trouble to observe this point and to communicate the result to the *Gardeners' Chronicle* or to me.

Charles Darwin, Down, Bromley, Kent.

Gardeners' Chronicle and Agricultural Gazette, 9 June 1860, p. 528

[1] Dated by the relationship to the following letter, in which CD states that he has 'written & sent notice for Gardeners' Ch.' The letter was published in the *Gardeners' Chronicle and Agricultural Gazette*, 9 June 1860, p. 528, and also in the *Entomologist's Weekly Intelligencer*, 23 June 1860, pp. 93–4, and 30 June 1860, pp. 102–3 (see letter to H. T. Stainton, 11 June [1860]). See also *Collected papers* 2: 32–5.

[2] The bee orchis (*Ophrys apifera*) and the fly orchis (*O. muscifera*) are only found on chalky soil like that in the vicinity of Down.

[3] This experiment is recorded in CD's Experimental book, p. 59 (DAR 157a). *Orchis morio* and *O. mascula* were then common in English woodlands. CD had discussed the problem of cross-fertilisation in these orchids in his 'big book' on species (*Natural selection*, pp. 65–6).

[4] Brown 1832. In this paper, Robert Brown suggested that the floral structure of the bee orchis was an adaptation for self-fertilisation. CD agreed with Brown that insects do not visit this particular species, but he believed that cross-fertilisation must sometimes occur. He was never able, however, to discern a mechanism for such crossing (see *Orchids*, pp. 63–72).

[5] See letter to H. T. Stainton, 11 June [1860].

To J. D. Hooker 5 June [1860]

Down Bromley Kent
June 5th

My dear Hooker

It is a pleasure to me to write to you, as I have no one to talk to about such matters as we write on.— But I seriously beg you not to write to me, unless so inclined; for busy as you are & seeing many people, the case is very different between us.—

Sir H. Holland came down on Sunday, & consoled us about no immediate cause of anxiety, but feared her recovery would be very slow.[1] And on Monday she was so weak & exhausted that we were very unhappy & anxious; but our Bromley Doctor thinks her rather better today.[2] It is a miserable long illness.—

I enclose letters (for I had unfortunately written to Harvey before your last letter came, but I begged him *not* to answer me) from Harvey & from (received today) A. Gray, for the chance of your liking to see them.[3] In the note which you forwarded to me from A. Gray, he begs you & me to give him any hint about Owen's article in Edinburgh, or other Review.—[4] You will not be the man to do that. I had already made comments on Owens review, & remarked how basely ungenerous, I thought his treatment of you.—

Have you seen Haughton's coarsely-abusive article of me in Dublin Mag. of Nat. History.[5] It outdoes even N. British & Edinburgh in misapprehension & misrepresentation.[6] I never knew anything so unfair as in discussing cells of Bees, his ignoring the case of Melipona which builds combs almost exactly intermediate between Hive & Humble-bee.—[7] What has Haughton done that he feels so immeasurably superior to all us wretched naturalists & to all political economists, including the great philosopher Malthus?[8]

This Review, however, & Harveys letter have convinced me that I must be a very bad explainer. Neither really understand what I mean by natural selection.— I am inclined to give up attempt as hopeless.— Those who do not understand, it seems, cannot be made to understand.—

By the way, I think, we entirely agree, except perhaps that I use too forcible language about selection. I entirely agree, indeed would go almost further than you, when you say that climate (ie variability from all unknown causes) is "an active handmaid influencing its mistress most materially."— Indeed I have never hinted that Natural selection is "the efficient cause to the exclusion of the other" ie variability from climate &c. The very term *selection* implies something, ie variations or differences, to be selected.—

You speak of adaptation being rarely *visible* though present in plants: I have just recently been looking at common Orchis, & I declare I think its adaptations in every part of flower quite as beautiful & plain, or even more beautiful, than in Woodpecker. I have written & sent notice for Gardeners' Ch. on curious difficulty in Bee Orchis,[9] & sh^d^ much like to hear what you think of the case. In this article I have incidentally touched on adaptation to visits of insects; but the contrivance to keep the sticky glands fresh & sticky beats almost everything in nature. I never remember having seen it described; but it must have been, & as I ought not in my book to give the observation as my own, I sh^d^ be very glad to know where this beautiful contrivance is described.—[10]

A Revd Mr Dunns wrote N. British R^w^ as I hear from Lyell, who heard from Chambers.—[11]

Hopkins of Cambridge has article in Fraser just published;[12] he speaks of me in kindest manner personally, but evidently thinks me an illogical & rash blunderer.

According to his standard of proof all we naturalists had better shut up shop, & never attempt to reason. He says there is no difference between my view & Lamarcks.—

I am much puzzled about the upshot of the cowslip case; some *few* of the capsules on the so-called male plants, judging from external feeling, are swelling as well as on female plants.[13] It would be best possible case of transition to have the male plants producing a few, *but markedly fewer* seeds than females; but the case w^d. require very good evidence. It would be catching a plant in actual transition from hermaphrodite to unisexual condition.— If both so-called males & females plants produce an equal or nearly equal number of seeds, how very strange the specified differences will be, especially that of size of pollen-grains:—[14]

How does your book progress (I mean your general sort of book on Plants);[15] I hope to God you will be more successful than I have been in making people understand your meaning. I should begin to think myself wholly in wrong, & that I was an utter fool, but then I cannot yet persuade myself that Lyell, & you & Huxley, Carpenter, Asa Gray & Watson &c are all fools together. Well time will show, & nothing but time. Farewell.

I hope M^rs. Hooker goes on well[16] | Yours affect | C. Darwin

Endorsement: '/60'
DAR 115.1: 60

1 The London physician Henry Holland had been called in to examine Henrietta Emma Darwin, who had been ill with 'remittent Fever' since late in April (see letter to J. D. Hooker, 7 May [1860]).

2 Probably Edward Augustus Williams, a surgeon in Bromley. He is listed in CD's Address book (Down House MS). See also letter to J. D. Hooker, 11 May [1860].

3 Neither the letter from William Henry Harvey nor that from Asa Gray has been found. The one from Harvey was a reply to information forwarded to Harvey via Hooker (see letter to J. D. Hooker, 20 May [1860]).

4 Gray was writing an article about CD's book and its reviewers (see letter to Asa Gray, 22 May [1860]). CD refers to [R. Owen] 1860a.

5 [Haughton] 1860b.

6 [Duns] 1860 and [R. Owen] 1860a.

7 A long section in Samuel Haughton's anonymous review, headed 'The causes of variation assigned are not adequate to produce the effects ascribed to them', criticised CD's explanation of the formation of bees' cells ([Haughton] 1860b, pp. 30–1). Haughton referred to the absence of any apparently intermediate stage between the cells of the humble-bee and the hive-bee, ignoring the case of *Melipona* discussed by CD in *Origin*, pp. 225–8. CD marked this section extensively in his copy of [Haughton] 1860b (Darwin Pamphlet Collection–CUL).

8 Haughton, referring to natural selection, stated: 'this notable argument is borrowed from Malthus' doctrine of Population, and will, no doubt, find acceptance with those Political Economists and Pseudo-Philosophers who reduce all the laws of action and human thought habitually to the lowest and most sordid motives.' ([Haughton] 1860b, p. 27). Haughton refers to Thomas Robert Malthus.

9 See preceding letter.

10 CD briefly described his observations of the 'sticky glands' of *Ophrys apifera* in his 'big book' on species (*Natural selection*, pp. 67–8). Although this species seemed to offer one of the best examples of continual self-fertilisation, the presence of viscid discs in the flower of the bee orchis indicated to CD that insects might occasionally effect cross-pollination. This point, however, was not developed in *Orchids*.

[11] [Duns] 1860. CD wrote 'Rev^{d} M Dunns' on his copy of the review (Darwin Pamphlet Collection –CUL). CD also refers to Charles Lyell and Robert Chambers.

[12] Hopkins 1860. See also letter to Charles Lyell, 1 June [1860].

[13] See letters to J. D. Hooker, 7 May [1860], 11 May [1860], and 14 May [1860]. The results of CD's experiments are recorded in his Experimental book, pp. 55–7 (DAR 157a).

[14] Further observations led CD to reverse his designation of 'male' and 'female' plants when he found that the short-styled plants, which he originally labelled 'male', produced more seeds than the long-styled. See his paper 'On the two forms, or dimorphic condition, in the species of *Primula*, and on their remarkable sexual relations', *Journal of the Proceedings of the Linnean Society (Botany)* 6 (1862): 77–96, reprinted in *Collected papers* 2: 45–63.

[15] See letters to J. D. Hooker, 12 March [1860] and 12 July [1860].

[16] Frances Harriet Hooker had given birth to the Hookers' fifth child, Brian Harvey Hodgson Hooker, on 27 May 1860.

To William Hallowes Miller 5 June [1860][1]

Down Bromley Kent
June 5

My dear Miller

I thank you much for your letter.[2] Had I seen the interest of my remark I would have made many more measurements though I did make several. I stated the facts *merely* to give the general reader idea of thickness of walls.[3]

Especially if I had seen that the fact had any general bearing I should have stated that as far as I could measure, the walls are by no means perfectly of the same thickness. Also I should have stated that the chief difference is when the thickness of walls of the upper part of the hexagon and of the pyramidal basal plates are contrasted Will you oblige me by looking with a strong lens at the bit of comb, brushing off with a knife the upper thickened edges, and then compare, by eye alone, the thickness of walls there with the thickness of the basal plates, as seen in any cross section. I should very much like to hear whether, even in this way, the difference is not perceptible. It is generally thus perceptible by comparing the thickness of the walls of the hexagon (if not taken very close to the angle) near to the basal plates, where the comparison by eye is of course easier. Your letter *actually* turned me sick with *panic*, from not seeing any great importance of fact, till I looked at my notes, I did not remember that I made several measurements.[4] I have now repeated some[5] measurements *roughly* with same general results but difference I think is hardly double.[6]

I should not have mentioned the thickness of the basal plates at all, had I not thought it would give an unfair notion of the thickness of the walls, to state the lesser measurements alone—

With cordial and sincere thanks | Yours very truly | C. Darwin

Copy
DAR 146

[1] Dated by the reference to CD's discussion of bees' cells in *Origin*.

[2] Miller was professor of mineralogy at Cambridge University. In 1858 he had helped CD with his study of the geometry of bees' cells (see *Correspondence* vol. 7, letters to J. D. Hooker, 10 April [1858], and to W. H. Miller, [15 April 1858], and letter from W. H. Miller, [14 May 1858]). Miller presumably wrote to CD after reading the discussion of the subject in *Origin*, pp. 224–35, in which CD cited Miller as his authority.

[3] CD described his experiments to determine the mode of formation of the cell walls in *Origin*, pp. 228–31. He found that bees at first build thick walls and then gnaw them down to 'an excessively thin finished wall' (*ibid.*, p. 231).

[4] CD's notes on bees' cells are in DAR 48 (ser. 2): 5–77.

[5] Francis Darwin was not certain whether this word was 'some' or 'same' and wrote a note to this effect on the copy of the letter.

[6] CD refers to the sentence in *Origin*, p. 231, where he estimates that the plates of the pyramidal basis are about twice as thick as the hexagonal walls.

To W. H. Miller [after 5 June 1860][1]

[Down]

My dear Miller,

I had no thought that you would *measure* thickness of walls of cells; but if you will, and allow me to give your measurements it will be immense advantage.[2] As it is no trouble, I send more specimens. If you measure please observe that I measured thickness of walls of hexagonal prisms not very near base: but from your very interesting remarks the lower part of walls ought to be measured—

Thank you for suggestion about how bees judge of angles and distances.[3] I will keep it in mind. It is a complete perplexity to me, and yet certainly insects can rudely somehow judge of distances. There are special difficulties on account of gradation in size between worker-cells and the larger drone-cells. I am trying to test the case practically by getting combs of different species, and of our own bee from different climates. I have lately had some from W. Indies of our common bee,[4] but cells *seem* certainly to be larger; but they have not yet been carefully measured. I will keep your suggestion in mind whenever I return to experiments on living Bees; but that will not be soon.

As you have been considering my little discussion in relation to Ld Brougham,[5] and as I have been more vituperated for this part than for almost any other, I should like just to tell you how I think the case stands. The discussion *viewed by itself* is worth little more than the paper on which it is printed, exepting in so far as it contains 3 or 4 certainly new facts. But to those who are inclined to believe the general truth of the conclusion that species and their instincts are slowly modified by what I call natural selection, I think my discussion nearly removes a very great difficulty. I believe in its truth chiefly from the existence of the *Melipona*, which makes a comb so intermediate in structure between that of Humble and Hive bee; and especially from the new and curious fact of the bees making smooth cups or saucers, when they excavated in a thick wax, which saucers stood so close that hexagons were built on their intersecting edges. And lastly because when they excavated on their slip of wax, the excavation on both sides of similar smooth basins was stopped, and flat planes left between the nearly opposed basins. If my view were wholly false these cases would I think never have occurred. Sedgwick and C° may abuse me to their hearts content,[6] but I shall as yet continue to think that mine is a rational explanation (as far as it goes) of their method of work.

With *cordial* thanks for all your great kindness. | Yours very sincerely | C. Darwin

Copy
DAR 146

[1] Dated by the relationship to the preceding letter.

[2] CD cited Miller's measurements of the thickness of the cell wall of hive-bees in *Origin* 3d ed., p. 252. See letters to W. H. Miller, 1 December [1860], 27 December [1860], and 31 December [1860].

[3] See *Origin*, p. 227.

[4] See *Correspondence* vol. 7, letters from Richard Hill, 10 January 1859 and 26 November 1859.

[5] CD prefaced his discussion of bees' cells in the first edition of *Origin* with an allusion to Henry Peter Brougham's work on the same subject. The passage reads (*Origin*, p. 224):

> We hear from mathematicians that bees have practically solved a recondite problem, and have made their cells of the proper shape to hold the greatest possible amount of honey, with the least possible consumption of precious wax in their construction.

CD refers to Brougham 1839 and 1858. See also *Correspondence* vol. 7, letter to W. B. Tegetmeier, 8 [June 1858], n. 3.

[6] Adam Sedgwick criticised CD's explanation of the construction of the cells of honey-bees, stating that his views made 'demands on our credulity, that are utterly beyond endurance, and do not give us one true natural step towards an explanation of the phenomena—viz., the perfection of the structures, and their adaptation to their office.' ([Sedgwick] 1860, p. 335).

To Charles Lyell 6 June [1860]

Down Bromley Kent
June 6

My dear Lyell

We think Etty is rather better today; her pulse has fallen & she seems stronger. We had Sir H. Holland down on Sunday, & though he thought there was no immediate cause for anxiety, he discouraged us in hopes of speedy recovery.[1]

Did you read Haughton in Dublin Mag. of Nat. Hist.[2] He is more coarsely contemptuous than even Mr Dunns in N. British[3] & overdoes everyone else in misrepresentation. I never knew anything so unfair as his ignoring in his remarks on Bee's cells the almost exactly intermediate comb of Melipona; & so in many other cases. It consoles me that he sneers at Malthus, for that clearly shows, mathematician though he may be, he cannot understand common reasoning.[4] By the way what a discouraging example Malthus is to show during what long years the plainest case may be misrepresented & misunderstood.

I have read the "Future":[5] how curious it is that several of my reviewers sh^d^ advance such wild arguments as that vars. of dogs & cats do not mingle; & sh^d^ bring up the old exploded doctrine of definite analogies or Quinarianism.—[6] I am beginning to despair of ever making the majority understand my notions. Even Hopkins does not thoroughly. (by the way I have been so much pleased by the way he personally alludes to me)[7] I must be a very bad explainer. I hope to Heaven that you will succeed better. Several Reviews, & several letters have shown me too clearly how little I am understood. I suppose "natural selection" was bad term; but to change it now, I think, would make confusion worse confounded. Nor can I think of better; "Natural preservation" would not imply a preservation of particular varieties & would seem a truism; & would not bring man's & nature's selection under one point of view. I can only hope by reiterated explanations finally to make matter clearer.

If my M.S. spreads out, I think I shall publish one volume exclusively on "Variation of animals & plants under domestication". I want to show that I have not been quite so rash as many suppose.—

Though weary of Reviews, I sh.d like to see Lowell sometime,[8] & likewise (*if you can spare*) Binney on Coal & Schaafhausen or some such name on natural selection.—[9] I suppose Lowell's difficulty about instinct is same as Bowens;[10] but it seems to me wholly to rest on assumption that instincts cannot graduate as finely as structure. I have stated in my volume, that it is hardly possible to know which i.e. whether instinct or structure changes first by insensible steps.— Probably sometimes instinct, sometimes structure.— When a British insect feeds on exotic plant, instinct has changed. by very small step & then structure might change so as to fully profit by new food. Or structure might change first, as the direction of tusks in one var. of Indian Elephant, which leads it to attack the tiger in different manner from other kinds of Elephants.—

Thanks for your letter of 2d, chiefly about Murray[11] (N.B Harvey of Dublin gives me in letter the argument of tall men marrying short women, as one of great weight!)[12]

I attributed in Origin blindness of cave animals not, (as you say) to "selection of million of chance varieties" but *exclusively* to disuse. But about insects I was hasty, I knew (but overlooked bearing of fact) that one genus Adelops is found under moss **out** of caves & yet blind. It is *possible* that the genus Anophthalmus (a Carabidous Beetle) was similarly an extra-cavernal blind insect. It seems not unlikely that a blind insect would be less inconvenienced in dark cave than other insects & would become tenant. There were several passages in Murray's article which I could not clearly understand.

I do not quite understand what you mean by saying "that the more they prove that you underrate physical conditions, the better for you, as geology comes into your aid."—

Godron puts well the little effect of *climate*,[13] which always becomes stronger & stronger conviction on my mind. I do not say confidently food.— I see in Murray & many others, one incessant fallacy when alluding to slight differences of physical conditions as being very important; namely oblivion of fact that all species, except very local ones, range over a considerable area, & though exposed to what they would call considerable diversities, yet keep constant. I have just alluded to this in Origin in comparing productions of Old & New Worlds.[14]

Farewell. Shall you be at Oxford?[15] If Etty gets quite well; perhaps I shall go there.— Yours affect | C. Darwin

Endorsement: '1860'
American Philosophical Society (215)

[1] See letter to J. D. Hooker, 5 June [1860] and n. 1.
[2] [Haughton] 1860b.
[3] [Duns] 1860.
[4] See letter to J. D. Hooker, 5 June [1860] and n. 8.
[5] *Origin* was criticised in articles in both the April and May 1860 issues of *The Future: a Monthly Journal of Research and Criticism in the Physical and Historical Sciences*. Copies of these reviews are in the Darwin Pamphlet Collection–CUL.
[6] The author of the May 1860 review in *The Future*, p. 29, stated 'that order and system are the rule, and disorder and indeterminateness the comparatively rare exceptions, even within the limits of species', and he advanced the general absence of mongrelism among dogs and cats as a case in point. On his copy of the article (Darwin Pamphlet Collection–CUL), CD attributed the article to 'a Mr Berke', referring to the journal's editor Luke Berke. CD also wrote: 'His main argument against me is Swainson's analogies.' CD refers to William Swainson's idealistic system of classification. For CD's opinion of Swainson's system, see *Correspondence* vol. 3, letter to J. D. Hooker, 31 March [1844].
[7] William Hopkins had expressed his 'full participation in the high respect in which the author is universally held, both as a man and a naturalist' (Hopkins 1860, 61: 752).
[8] [Lowell] 1860.
[9] Binney 1847 and Schaaffhausen 1853. See letter to Charles Lyell, 22 May [1860].
[10] Francis Bowen had reviewed *Origin* in the *North American Review* 49 (1860): 474–506. There is an annotated copy of this review ([Bowen] 1860a) in the Darwin Pamphlet Collection–CUL. In another article on CD's theory written in 1860 but published the following year, Bowen acknowledged that he had written the review (Bowen 1860b, p. 115 n.). There is also an annotated copy of Bowen 1860b in the Darwin Pamphlet Collection–CUL. Both Lowell and Bowen attacked CD for his 'denials' of design. See also letter to Asa Gray, 18 May [1860].
[11] Lyell's letter discussing Andrew Murray's criticisms of *Origin* has not been located. CD had sent Lyell a copy of Murray 1860a (see letter to Charles Lyell, 1 [June 1860]).
[12] William Henry Harvey's letter has not been found. Murray had also raised this point in his criticism of natural selection (Murray 1860a, p. 277). See letter to Andrew Murray, 28 April [1860].
[13] Godron 1859, 2: 6–11. See letter to Charles Lyell, 18 May [1860].
[14] *Origin*, pp. 346–7.
[15] The 1860 meeting of the British Association for the Advancement of Science was to be held in Oxford. See letter to Charles Lyell, 25 [June 1860].

To George Rolleston 6 June [1860][1]

Down Bromley Kent
June 6th

Dear Sir

I thank you sincerely for your very kind note & offer of assistance.[2] My plans may be unfortunately altered, for my daughter has a low fever, & unless every shade of anxiety is removed by her almost entire recovery, I should not like to leave home.—

If it would not be selfish, I think the plan which would cause least trouble would be for you, if not asking too great a favour, to reserve me rooms in a near college; & then if I am prevented coming I will promise to write 2 or 3 days before the meeting, so that the rooms would be at your disposal for some one else.—[3]

Believe me with many thanks for all your kindness & most obliging invitation | Dear Sir | Yours truly obliged | ⟨C. Darwin⟩[4]

AL
Wellcome Institute for the History of Medicine Library, London (WMS. 6119/2)

[1] Dated by the relationship to the letter to George Rolleston, 1 June [1860].
[2] Rolleston's letter has not been found. CD had written to inquire about lodgings in Oxford during the forthcoming British Association meeting (letter to George Rolleston, 1 June [1860]).
[3] CD was unable to attend the meeting. See letter to Charles Lyell, 25 [June 1860].
[4] The signature has been excised from the letter.

To J. D. Hooker 7 June [1860]

Down Bromley Kent
June 7th

My dear Hooker

Best & most beloved of men, I supplicate & entreat you to observe one point for me. Remember that the Goodeniaceæ have weighed like an incubus for years on my soul.—[1] It relates to Scævola microcarpa. I find that in bud the indusium collects all the pollen splendidly, but differently from Leschenaultia cannot be afterwards easily opened.[2] Further I find that at early stage, when flower first opens a boat-shaped stigma lies at bottom of indusium; & further that this stigma after flower sometime expanded grows very rapidly, *when plant is kept hot*, & pushes out of indusium a mass of pollen & at same time two horns project at corners of the indusium; now the appearance of these horns makes me suppose that these are the stigmatic surfaces. Will you look to this? for if they be, you will see by relative position of parts, with indusium & stigmas bent at right angles to style, that an insect entering a flower could not fail to have whole back (at the period when as I have seen, a whole mass of pollen is pushed out) covered with pollen which would almost certainly get rubbed on the two horns.—

Indeed I doubt whether, without this aid whether pollen would get on to the horns.— What interests me in this case is the analogy in result with Lobelia but by very different means. In Lobelia the stigma before it is mature pushes by its circular brush of hairs the pollen out of conjoined anthers; here the indusium collects pollen & then the growth of stigma pushes it out.— In course of about 1½ hour, I found an indusium with hairs on outer edge perfectly clogged with pollen, & horns protruded, which before the 1½ hour had not one grain of pollen outside the indusium. & no trace of protruding horns. So you will see how I wish to know whether the horns are the true stigmatic surfaces.— I could try the case experimentally by putting pollen on the horns, but my greenhouse is so cold & my plant so small & in such little pot, that I suppose it would not seed.—

Forgive me.— Your affect | C. Darwin

Etty is rather decidedly better today & we are all in jolly spirits.[3]

The little length of stigmatic horns at the moment when pollen is forced out of indusium, compared to what they ultimately attain, makes me fancy that they are not then mature or ready & if so, as in Lobelia, each flower must be fertilised by pollen from another & earlier flower. How curious that indusium sh^{d} first so cleverly collect pollen & then afterwards push it out! Yet how closely analogous the Campanulas brushing pollen out of the anthers & retaining it on hairs till stigma is ready.— I am going to try whether Campanula sets seeds without insect-agency.—

Endorsement: '/60'
DAR 115.1: 61

[1] The Goodeniaceae apparently contradicted CD's belief that all organic beings must occasionally cross-fertilise. He had referred to the problem in his 'big book' on species (*Natural selection*, pp. 63–4), explaining how the pollen and stigma in *Goodenia* were completely enclosed by an indusium in the flower. He had subsequently studied the structure of the related genus *Leschenaultia* (see letters to J. D. Hooker, [26 February or 4 March 1860], 12 March [1860], and 7 May [1860]).

[2] CD's observations and experiments on *Scaevola* and *Leschenaultia* are recorded in his Experimental book (DAR 157a). The observations described in the letter are dated 1 June 1860 (Experimental book, p. 64).

[3] Henrietta Emma Darwin had been ill since the end of April 1860.

To John Murray 7 June [1860][1]

Down Bromley Kent
June 7th

My dear Sir

Will you be so kind at your convenience to tell me how the "Origin" goes off.[2] & whether it is likely there will be wanted soon, ever, or never another Edition. I want to know partly out of curiosity, & partly, as I am come to a break in my present work, because it will decide what my next job shall be.—[3]

My dear Sir | Yours sincerely | C. Darwin

John Murray (publishers)

[1] The year is given by CD's reference to coming 'to a break' in his work (see n. 3, below).

[2] The second edition of *Origin*, issued on 7 January 1860, had a print-run of 3000. A note on the letter, probably written by one of Murray's clerks, reads '1000 on hand'. There was no further issue or reprint of *Origin* until April 1861, when the third edition was published (Freeman 1977, p. 78).

[3] On 10 June 1860, CD recorded in his 'Journal' (Appendix II): 'Finished **2^{d}** Ch. on Pigeons. arranged papers for next Ch. & Experiments.' He was preparing the first part of his longer work on natural selection, ultimately published as *Variation*.

To Asa Gray 8 June [1860]

Down Bromley Kent
June 8th

My dear Gray

I have to thank you for two notes, one through Hooker, & one with some letters to be posted, which was done.—[1] I anticipated your request by making a few remarks on Owen's Review.[2] Hooker is so weary of Reviews; that I do not think you will get any hint from him.—[3] I have lately had many "more kicks than half-pence". A review in last Dublin Nat. Hist. Review. is the most unfair thing which has appeared,—one mass of misrepresentation. It is evidently by Haughton, the geologist, chemist & mathematician. It shows immeasurable conceit & contempt of all, who are not mathematicians. He discusses Bees-cells, & puts a series, which I have never alluded to, & wholly ignores the intermediate comb of Melipona which alone led me to my notions. The article is a curiosity of unfairness & arrogance. But as he sneers at Malthus, I am content, for it is clear he cannot reason.[4]

He is a friend of Harvey, with whom I have had some correspondence.[5] Your article has clearly as he admits influenced him: he admits to certain extent natural selection, yet I am sure does not understand me. It is strange that very few do, & I am become quite convinced that I must be an extremely bad explainer. To recur for moment to Owen. He grossly misrepresents & is very unfair to Huxley.[6] You say that you think the article must be by pupil of Owen; but no one fact tells so strongly against Owen, considering his former position at Coll. of Surgeons, as that he has never reared one pupil or follower.—

In the number just out of Fraser's magazine, there is article or Review on Lamarck & me by W. Hopkins, the mathematician; who like Haughton despises the reasoning powers of all naturalists. Personally he is extremely kind towards me; but he evidently in following number means to blow me into atoms.[7] He does not in least appreciate the difference in my views & Lamarck's, as explaining adaptation, the principle of divergence, the increase of dominant groups, & the almost necessary extinction of the less dominant & smaller groups &c. &c.—

Talking of adaptation, I have lately been looking at our common orchids, & I daresay the facts are as old & well-known as the hills, but I have been so struck with admiration at the contrivances, that I have sent notice to Gardeners Chronicle.[8] The Ophrys Apifera, offers, as you will see, a curious contradiction in structure.— I get on very very slowly with my larger work; & of late from a very unhappy cause, my poor eldest daughter about 16 years old, has now been 40 days ill with low fever: we have had a good deal of anxiety, but I hope & think she has at last turned the corner, though the fever has not gone.—[9]

Farewell. When so inclined, pray write;—it is a great pleasure to me to hear from you.— Yours most truly & gratefully | C. Darwin

P.S. I have been making some little trifling observations which have interested & perplexed me much. I find with Primroses & Cowslips, that about an equal number of plants are thus characterised

So called (by me) *male* plants Pistil much shorter than stamens; stigma rather smooth,—*pollen grains large*, throat of corolla short.—

So-called female plants Pistil much longer than stamens, stigma rougher—*pollen-grains smaller*,—throat of corolla long.

I have marked a lot of plants, & expected to find so-called male plants barren; but judging from the feel of the capsules, this is not the case, & I am very much surprised at difference in the size of pollen, which I measured many times & always found 2 to 3/6000 of an inch less in diameter in the "female" than in the "male" plants.— If it should prove that the so-called male plants produce less seed than the so-called females, what a beautiful case of gradation from hermaphrodite to unisexual condition it will be! If they produce about equal number of seed, how perplexing it will be.—[10]

Endorsement: '1860'
Gray Herbarium of Harvard University (40)

[1] Neither of Gray's letters has been located.

[2] See letter to J. D. Hooker, 5 June [1860]. CD refers to [R. Owen] 1860a.

[3] In a letter to John Stevens Henslow written in June 1860, Hooker commented that he was 'getting very tired of Darwinian Reviews; there is wonderfully little to the purpose in any but Gray's and Owen's, Huxley's and Carpenter's. All the rest seem ignorant prejudice.' (L. Huxley ed. 1918, 1: 514).

[4] [Haughton] 1860b. See letters to J. D. Hooker, 5 June [1860], and to Charles Lyell, 6 June [1860].

[5] William Henry Harvey was a close friend of Gray's and had corresponded with him about *Origin* (Dupree 1959, p. 278).

[6] [R. Owen] 1860a, which reviewed *Origin* and Hooker 1859, attacked Thomas Henry Huxley's February 1860 lecture on species at the Royal Institution (T. H. Huxley 1860a).

[7] Hopkins 1860. The review was published in *Fraser's Magazine* in two parts: the first part appeared in the June issue, no. 61 (1860): 739–52; the second in the July number, no. 62 (1860): 74–90.

[8] See letter to *Gardeners' Chronicle*, [4–5 June 1860].

[9] For CD's relief at Henrietta Emma Darwin's improving health, see letter to J. D. Hooker, 7 June [1860].

[10] See letters to J. S. Henslow, 14 May [1860] and 17 May [1860], and to J. D. Hooker, 20 May [1860]. CD's observations are recorded in his Experimental book, pp. 53–7 (DAR 157a).

From J. D. Hooker 8 June 1860

Kew
June 8 | /60

My dear Darwin

We are much relieved by your good news of dear Etty What a long trial the poor thing has had. I suppose you look to change of air as the next measure when practicable. I shall be anxious to know where you will go.

Your observations on *Scævola* are capital.[1] I am sure you are right about the horns being the stigmata, I know them well to be so, but I will look to *S. microcarpa* again with especial interest— The Indusium collecting the pollen is a splendid fact, as it is undoubtedly the homologue of the hairs of Campanula, & indeed I consider

Goodeniaceæ only a tribe of Campanulaceæ including Lobeliaceæ. Our *Leschenaultia* does nothing how should it in this miserable weather—so different from K^{g} G^{s} Sound.[2] The genus is almost confined to S.W. Australia if not altogether so.

I expect to find that the horns are protruded *from* the boat shaped organ; or rather from the bottom of it: for if my memory serves me right the boat shaped organ is a *second indusium* whose base inside is stigmatic surface & afterwards protrudes as horns.— Just as base inside of cup of Leschenaultia is stigmatic & the glandular fold *outside* indusium of Leschenaultia is I take it a rudimentary breaking up of indusium into two concentric ones.

I am, as you rightly suppose, extremely interested in the progress of opinion on Species, it amuses & instructs me. I do not despair of Harvey— I may tell you that I have purposely kept his printed affair from you till he has settled a little—[3] he will I expect take up the subject at the *Gooseberry season*—[4] You know he never had any Education in Science at all; & calls himself an Amateur only,—lays no claim to be philosophical or scientific— he was brought up a Quaker & his MD is *purely honorary* he took up Botany as another takes up coins, or seals—without a notion of ever being scientific— his science has *grown on him* he has not raised it. If you consider this you will agree with me that his letter is an interesting & not despicable one.[5] He is all wrong in his reasoning on biform fruit of Algæ.— how can it tend to preserve species? I am inclined to agree with his meaning at lines 2–6 of p. 3, but there again he is not clear N.S.[6] is not *dependent* on variability in any logical sense— As to the question of *personality* it is simply ridiculous— I agree with him & you too that the *real* primary agency is as great a mystery as ever, call it *mystery* or *theistic element* or *God* or *nature* or what he will, *unseen Power* if he likes. How odd that people should not

Incomplete
DAR 157a

CD ANNOTATIONS
0.1 My . . . will go. 1.3] *crossed pencil*
4.1 I am, . . . should not 4.16] *crossed pencil*
Top of letter: 'June 11/1860' *pencil*

[1] See letter to J. D. Hooker, 7 June [1860].

[2] *Leschenaultia* is a genus of Australian shrubs commonly found in south-west Australia near King George's Sound.

[3] Hooker refers to William Henry Harvey's privately printed pamphlet, *On the origin of the human animal* (1860). CD's annotated copy is in the Darwin Pamphlet Collection–CUL. In the preface, dated 21 February 1860, Harvey refers to it as 'a serio-comic squib'. Harvey sent it to CD in October (see letter from W. H. Harvey, 8 October 1860).

[4] Gooseberry season: 'the time of year when newspapers have plenty of space to record trifles' (*OED*). Harvey had apparently used this expression in a letter to CD that is now missing. See L. Huxley ed. 1918, 1: 518.

[5] The letter from Harvey has not been found. For the correspondence between Hooker and Harvey about CD's theory in May 1860, see L. Huxley ed. 1918, 1: 516–20.

[6] Natural selection.

To Emma Gärtner 9 June [1860][1]

Down Bromley Kent
June 9th

Madam

I thank you sincerely for your most kind note & for your presents,[2] which I shall value much & shall no doubt receive in due time. I have long venerated your Father & estimated his two great works most highly.[3] I can perfectly well believe that the "search after truth was the leading principle of his life & labours"; I was thoroughily impressed with this, after reading them; & strict conscientiousness seemed to me their noble & prevailing character.— I shall be very glad to read something of his life; for I remember some few years ago enquiring on the subject, as I had got to feel so much interest in his great labours; but did not then hear of his life.—

Your Father's name is pretty well known in this country, but not so universally as it ought to be, & not nearly well enough known in France, judging from such French Books as I read. If I can in ever so little degree make it more generally known I shall be pleased; & in my larger work I shall have to quote him far oftener. But your Father's works have enduring merit, & will certainly be of high service to science many years after I am dead & gone.—

With sincere thanks for your letter, which has given me much satisfaction, & with my gratitude for your most kind presents, believe me Madam, with much respect, | Yours sincerely & obliged | Charles Darwin

Medical Center Library, Duke University (Trent collection)

[1] Dated by the relationship to the letter to Emma Gärtner, 14 July [1860].

[2] Emma Gärtner was the daughter of the late German botanist Karl Friedrich von Gärtner, whose works CD frequently cited in his species manuscript (*Natural selection*) and in *Origin* and *Variation*. The 'presents' included engravings of Gärtner and a copy of Gustav Jäger's obituary notice (Jäger 1851). Jäger's biographical notice, inscribed by Emma Gärtner, is in the Darwin Pamphlet Collection–CUL. See letter to Emma Gärtner, 14 July [1860].

[3] CD studied Gärtner's works on hybridisation very closely during the 1850s. His annotated copies of Gärtner 1844 and 1849 are in the Darwin Library–CUL; further notes on these works are in DAR 116. See *Correspondence* vol. 5.

To Henry Tibbats Stainton 11 June [1860]

Down Bromley Kent
June 11th

Dear Sir

I take the liberty to send the enclosed.—[1] I fear it is quite out of the question your being able to insert the whole in the Intelligencer;[2] but perhaps you would oblige me by putting in the paragraph in which I ask for information on what kinds of moths the pollen-masses of Orchids have been found adhering.—[3] My best & almost sole chance will be through the columns of the Intelligencer.

Dear Sir | Yours faithfully | C. Darwin

P.S. I have just received your extremely kind & to me **most** interesting note.—[4] As you have Gardeners Ch. I will not send the Duplicate copy, which I got to send you, but think I will send it to M[r] Parfitt & enquire whether he is certain that he recognises pollen-masses of Bee Orchis.[5] The *green* pollen masses I sh[d] think were those of true Orchis.—[6] Again I thank you sincerely. It is extremely curious the same moth having been found with pollen-masses in two parts of England.

I c[d] recognise pollen-masses of Butterfly Orchis, & I daresay with a little care of most of our orchids.—

Endorsement: '1860'
The Natural History Museum (General Library, MSS DAR)

[1] See the postscript to the letter.

[2] CD's letter to the *Gardeners' Chronicle*, [4–5 June 1860], was reprinted in the *Entomologist's Weekly Intelligencer*, 23 June 1860, pp. 93–4; and 30 June 1860, pp. 102–3. Stainton was the editor of the publication.

[3] See letter to *Gardeners' Chronicle*, [4–5 June 1860].

[4] Stainton's letter has not been found.

[5] Edward Parfitt was a noted entomologist and botanist. He was currently the curator of the Archaeological and Natural History Society of Somerset.

[6] In a short notice in the *Entomologist's Weekly Intelligencer*, 3 October 1857, pp. 3–4, Parfitt described the adherence of pollinia from the bee orchis, *Ophrys apifera*, to the proboscis of a moth. Parfitt apparently did reply to CD's query: a note in *Orchids*, p. 36, states that he sent CD specimens of the moth *Anthrocera trifolii* having pollen masses attached to them. CD also stated that Parfitt was mistaken in believing that they had come from the bee orchis, as CD suggests in the letter. See also following letter.

To J. D. Hooker 12 [June 1860]

Down Bromley Kent
12th

My dear Hooker

Many thanks for your long & pleasant letter & for the enclosures herewith returned.— I am sorry Thomson does not budge an inch about permanence of species;[1] but how capitally he appreciates the Geographical part of your Essay,[2]— he ranks it as it deserves.— I am pleased to observe that Thwaites advances a little.—[3]

I enclose two slips, with points on which I want observations; if you can persuade M[r] Oliver kindly to observe;[4] mind that I know that you are much too busy.—

Have you ever thought of getting that done, which you said you thought you could, viz a rough approximation of proportions of plants in some one greenhouse or Hot-house—or even out-of-doors, which set seeds.— This would be invaluable for me—In Gardeners Ch. I have said from sticky gland that I thought that pollen-masses of Bee-orchis would be sometimes removed by insects—[5] I have now received account of numerous moths caught with pollen-masses on proboscis—(in one case **3 pair** on one!) & the writer asserts that they were pollen-masses of *Bee* orchis; but I have written to know how he knew this.—[6]

Thank you for telling me about homologies of stigma &c of Leschenaultia & Scævola; I was curious on this head.—[7]

I have long meant to experimentise (as far as insects are concerned) on the imperfect flowers of Viola & certain Leguminosæ; but I do not know whether I shall have time.[8]

I am sorry to hear about bronchitis & your little boy;[9] we know what that is with children.—

We were grievously disappointed yesterday by decided return of Fever-fit with Etty.— It is extremely doubtful whether I shall get to Oxford.

Yours affect | C. Darwin

P.S. The following metaphor gives good view of my notion of relative importance of Variability & Selection.— Squared stones, bricks or timber are indispensable for construction of a building; & their Nature will to certain extent influence character of building, but selection I look at, as the architect; & in admiring a well-contrived or splendid building one speaks of the architect alone & not of the brick-maker.—

Endorsement: '/June 60'
DAR 115.1: 62

[1] Thomas Thomson had written to CD expressing his opposition to the theory of species change through natural selection. The letter has not been found, but see letter to J. D. Hooker, 13 [May 1860].

[2] Hooker 1859.

[3] George Henry Kendrick Thwaites had been corresponding with Hooker about *Origin*. See letter to J. D. Hooker, 22 [May 1860].

[4] Daniel Oliver was Hooker's assistant in the herbarium of the Royal Botanic Gardens, Kew.

[5] Letter to *Gardeners' Chronicle*, [4–5 June 1860].

[6] CD refers to information obtained from Edward Parfitt. Parfitt probably wrote to CD after reading his letter to the *Gardeners' Chronicle*, [4–5 June 1860]. There would not have been enough time for Parfitt to have responded to CD's inquiry mentioned in the preceding letter.

[7] See letter to J. D. Hooker, 7 June [1860], and letter from J. D. Hooker, 8 June 1860. CD subsequently published Hooker's observations in a communication entitled 'Fertilisation of Leschenaultia' printed in the *Gardeners' Chronicle and Agricultural Gazette*, 9 September 1871, p. 1166 (see also *Collected papers* 2: 162–5).

[8] There is only one experiment on *Viola* recorded in CD's Experimental book, p. 59 (DAR 157a), and this is dated 24 May, 24 June, and 8 July 1860.

[9] Probably Charles Paget Hooker, aged 5.

To John Higgins[1] 13 June [1860]

Down Bromley Kent
June 13th

My dear Sir

You need not fear my changing my mind, I have £18,000 as good as ready money (ie £16,800 on Grt. N., & N.W. Railway Bonds; & the remainder on Deposit account at 7 days notice) & I shd not object to sell Railway shares for another £2000. or even £3000. Beyond this £21,000, I shd. not much like to go.—[2] I have no objection to the situation of the Farm; as I want it as a safe investment for my sons, it will depend whether land good—Buildings in fair repair & price what you consider reasonable. On all these points I must rely on your kindness & judgment. If what you hear is promising I trust to your being so good as to inspect the land; for this is for me an important purchase.—

I may mention that my Bonds have about $1\frac{1}{2}$ year to run, & pay me good interest $4\frac{1}{2}$ & $4\frac{3}{4}$, so that I am in no hurry & my Deposit account will slowly increase.— So please not to purchase unless you think the investment a fairly good one. I shall certainly invest in land to above amount.[3]

I do not know how you will manage if you buy at auction about paying Deposit money, until titles are examined by my solicitors.[4] I may mention that I have about £1400 or £1500 current account & on emergency I could draw nearly all out.—

With my sincere thanks for your uniform kindness, I remain My dear Sir | Yours very faithfully | C. Darwin

P.S. | As soon as you have decided that the Land seems promising, kindly inform me as my eldest son,[5] would like to see it & will start immediately; & will after looking at it, probably come on to Alford & look at Beesby & Claythorpe.—[6]

Lincolnshire Archives Office

[1] Higgins was the Darwin family's agent who managed their investments in Lincolnshire farms. CD owned farms near Alford.

[2] CD had written to Higgins asking him to look for another farm for him to purchase. See *Correspondence* vol. 7, letter to John Higgins, 26 December 1859.

[3] In 1854, CD had asked Higgins to look for land to purchase and then changed his mind (see *Correspondence* vol. 5, letter to John Higgins, 9 April [1854]).

[4] CD's solicitor was Thomas Salt of the Shrewsbury law firm of Dukes and Salt.

[5] The land was intended to form an investment to be divided among CD's sons (see *Correspondence* vol. 7, letter to John Higgins, 26 December 1859). William Erasmus Darwin was CD's oldest son.

[6] Beesby farm had been bought for CD by his father, Robert Waring Darwin, in 1845 (see *Correspondence* vol. 3). Claythorpe farm was owned by CD's sister Susan Elizabeth Darwin.

To Charles Lyell 14 [June 1860]

Down Bromley Kent
14th

My dear Lyell

We have been grievously disappointed in Etty having retrograded a little.[1] I cannot work, so will pass a little time in scribbling to you.— I copy extract from letter of Blyth, who expresses much gratitude to you for interest about the Chinese Expedition.—[2] I enclose letter of Hopkins, thinking that you might like to see it.— I fancy he is horrified about man; I have told him that I thought man must be included under same category with animals.[3]

I return by this Post 4 Pamphlets (Binney sent today by mistake in duplicate).[4] I do not see much in Binney. Lowels review is pleasantly written; but it is clear he is not naturalist.[5] He quite overlooks the importance of accumulation of mere individual differences; & which I think I can show is great agency of change under domestication. It is no wonder Agassiz denies varieties in animals, when he calls even the *same* forms in two distant countries, two species.—[6]

I have not finished Schaafhausen,[7] as I read German so badly: I have ordered copy for self & sh^d. like to keep yours till my own arrives; but will return it to you instantly if wanted.— He admits statements rather rashly, as I daresay I do.— I see only one sentence as yet at all approaching nat. selection.[8]

There is notice of me in penultimate no^r of "All the Year Round", but not worth consulting; chiefly a well-done hash of my own words.—[9] Your last note was very interesting & consolatory to me.[10]

I have expressly stated that I believe physical conditions have more direct effect on Plants than on animals.— But the more I study the more I am led to think that natural selection regulates in a state of nature most trifling differences.— As squared stones, or bricks, or timber are the indispensable materials for a building & influence its character; so is variability not only indispensable, but influential; Yet, in same manner, as the architect is the *all*-important person in a Building, so is Selection with organic bodies.—

I do not in least understand what A. Murray meant by Agassiz & ab ovo; but I did not much puzzle my brains on subject.—[11]

My dear Lyell | Yours affect. | C. Darwin

Blyth says in Sabines Translation of Wrangell's Voyage Introduction p. 117 there is account of Flint tool f^d. in ice.—[12] Bears perhaps or Mastodon in ice &c. Worth looking to, I sh^d think Blyth thinks Esquimaux had no iron tools, when first discovered, refers to some Essay by Sir J. Richardson[13]

Endorsement: 'June | 1860'
American Philosophical Society (216)

[1] Henrietta Emma Darwin.

[2] Edward Blyth's letter has not been found, but see the letter to Charles Lyell, 1 [June 1860].

[3] William Hopkins's letter has not been found. In the second part of his review of *Origin* (Hopkins 1860, 62: 88), Hopkins stated that he considered 'the most difficult and most important point' of any transformist theory was 'the transition in passing up to man from the animals next beneath him, not to man considered merely as a physical organism, but to man as an intellectual and moral being.'

[4] Binney 1847. See letter to Charles Lyell, 22 May [1860].

[5] [Lowell] 1860. The wealthy Bostonian John Amory Lowell was a patron of the sciences, particularly of botany. He was an active member of the American Academy of Arts and Sciences (Dupree 1959).

[6] Louis Agassiz often described as two species forms that would today be identified as geographical varieties or subspecies. See Winsor 1979, p. 105.

[7] Schaaffhausen 1853.

[8] CD's copy of Schaaffhausen 1853 is in the Darwin Pamphlet Collection–CUL. Bound with it is a manuscript translation of pp. 423–4, possibly in the hand of Camilla Ludwig, the Darwins' governess. CD marked on the translation the following passage: 'Many species may have kept true during a long period, only a few individuals, with a gradually changing formation, may have separated from them, and thus new varieties or species may have been formed beside the original species. Species are independent, without connecting links, (gradual transitions) because these have not been preserved'. CD quoted this statement in the revised historical sketch he added to the third edition of *Origin* (1861).

[9] *All the Year Round* was a popular weekly magazine edited by Charles Dickens. The unsigned notice appeared in the issue of 2 June 1860 (*All the Year Round* 3 (1860): 175–8, 293–9).

[10] Lyell's letter has not been found.

[11] The passage reads (Murray 1860a, p. 290):

> The distinctive character of Mr Darwin's theory is not development *ex ovo*; that is the theory of Oken, of Agassiz, of the author of the "Vestiges of Creation"; . . . of Bonnet and of Priestley, who . . . held "that all the germs of future plants, organical bodies of all kinds, and the reproducible parts of them, were really contained in the first germ."

Andrew Murray refers to Lorenz Oken, Louis Agassiz, Robert Chambers, Charles Bonnet, and Joseph Priestley.

[12] Sabine ed. 1840, p. cxvii. The book had been translated from Wrangel 1839 by Elizabeth Juliana Sabine. CD first read this work in 1840 (see *Correspondence* vol. 4, Appendix IV, 119: 10a).

[13] Probably J. Richardson 1852, p. 322, in which John Richardson mentioned the use by the Eskimo of wood or bones for tools and other implements.

From Charles Lyell 15 June 1860

June 15—1860

Your comparison of Selection to the Architect—variations to the stones, is what I deduced from some passages but cannot accept.[1] The architect who plans beforehand & executes his thoughts & invents the Corinthian & other styles of architecture & then by means of machinery, living & inanimate, cranes & horses & even sometimes intelligent men (forman allowed some discretion & power of choosing) such an architect must not be confounded in his functions with the humble office of the most sagacious of breeders. As I have said all along it is the deification of Natural Selection.

Copy
Kinnordy MS, Charles Lyell's journal VI, pp. 108–9

[1] See preceding letter and letter to J. D. Hooker, 12 [June 1860]. The transcript has been taken from Lyell's scientific journal. It is headed by the note 'Extract from letter of C. L. to Darwin'. It is also printed in Wilson ed. 1970, p. 445.

From Frederick Bond [16? June 1860][1]

to them, viz. Polia Occulta and Spilodes Palealis, the first a night & the other an evening flyer, I will send them if you wish it—[2] I have frequently seen Plusia Gamma, H. Dentina, Mamestra Anceps and two or three of the Pyralidæ, on Clover, Trefoil & Vetches. I have seen field beans' almost smothered with moths (when in bloom) at night, & Believe me to be | Yours truly | F^k Bond

C. Darwin. Esqr

Incomplete
DAR 76 (ser. 2): 168

CD ANNOTATIONS
1.1 to them . . . wish it—1.2] *crossed pencil*
1.2 I have] *square bracket before, pencil*
1.3 H. Dentina,] 'Hadina' *added ink*
1.3 Pyralidæ,] 'Pyralidæ' *added ink*
1.4 Trefoil] 'Trefoil' *added ink*
1.4 beans'] 'Beans' *added pencil*
Top of letter: '3'[3] *brown crayon, circled brown crayon*
End of letter: 'June 18 1860' *ink*

[1] Although CD marked the letter 'June 18 1860' (see CD annotations), he had apparently received the letter before this date. In his letter to J. D. Hooker, 17 June [1860], CD states that he had received 'good facts about moths & orchids.—'
[2] Bond is cited in *Orchids*, p. 35, as having sent CD a large number of moths with orchid pollinia attached to their proboscises. A list of the species of moths received is given in *Orchids*, p. 36, and includes two of the species mentioned in the extant portion of the letter.
[3] The number of CD's portfolio of notes on variation and varieties.

To W. B. Carpenter 17 June [1860][1]

Down Bromley Kent
June 17th

My dear Carpenter

I should have written to you long before this to have reminded you of your kind promise of running down here for a Sunday; but I have for last *seven* weeks been hoping against hope that my daughter would rapidly recover.— She has now been in her bed with low fever for exactly this period; & is only just now beginning very slowly to recover; whenever she is strong enough to get her out of bed, we shall have to move somewhere for change of air; so that I am sorry to say we must defer your little visit here till the latter part of summer or autumn. I know that you go to your sea-side home in Scotland every year;[2] & I suppose we shall have to go, also, to sea-side for my daughters sake.—

I have been of late sufficiently well pitched into to please anybody, about my Book.— But I care very little, which I entirely & absolutely owe to the generous & kind support of a very few men.— When I reflect, as I often do, that such men as Lyell, yourself, Hooker, & Huxley go a certain way with me, nothing will persuade me that I am so wholly & egregiously in error as many of my reviewers think.—

Pray do not trouble yourself to answer this.— My dear Carpenter | Yours very sincerely | C. Darwin

Down House (MS 6: 6)

[1] The year is given by the reference to Henrietta Emma Darwin's continuing ill health (see 'Journal'; Appendix II).
[2] Carpenter and his family spent their summers on Holy Island, in the Firth of Clyde, off the western coast of Scotland. See *Correspondence* vol. 7, letter to W. B. Carpenter, 26 January [1858].

To J. D. Hooker 17 June [1860][1]

Down Bromley Kent
June 17th

My dear Hooker

I have reread your paper on Listera[2] & have been looking as specimens, & have hardly ever been more amused. In Neottia nidus-avis, the rostellum is thicker, & the summit is so wonderfully sensitive, that a touch from a child's hair, not held close to point was sufficient to cause the ejection of the fluid!! & what delighted me was that this fluid sets so quick, becoming slightly opake, that in 2 seconds or even less when the touching object was withdrawn, a mass of pollen was withdrawn with it.— I send the hair in quil, with which I touched top of rostellum of Neottia, & almost instantly withdrew it & you will see (if you think it worth cutting open quil) pollen glued to end. So that insect by this wondrous contrivance would be sure to carry pollen from flower to flower.— The same thing happens with Listera, but in this case *whole* pollen-mass in a flower only lately expanded, is with drawn.—

I cannot believe from what I have seen that pollen mass can be ejected by the ejection of fluid matter; for during first second it is not very viscid. A more important difference is that in Neottia pollen is shed in bud & is cast irregularly on to the rostellum & does not adhere to it till the fluid is ejected; & the adhesion is indefinite ie to any point of the pollen-masses, & not to the apices of the pollen-mass as in Listera.— I suspect, from *one* case that I **saw** that pollen-masses are ejected from anthers of Listera with force, *before* & independently of ejection of fluid.— I can see that this viscid fluid is *analogous* to sticky gland of orchis, but it seems to me, **in my ignorance**, that in Listera the opposite end of pollen-masses are attached, to what is the case in Orchis.—[3]

In Cephalanthera grandiflora the pollen-masses are shed in bud & stand close in two friable column close to upper edge of stigma; & this upper edge, (which I presume answers to rostellum) whilst in bud-state, ejects fluid (without touch) on each side & glues the pollen-masses there; as the rim with the glue turns reddish, it is probably homologous to the viscid secretion of Listera & Neottia.—[4] I am sure I do not know whether you care enough about Listera to hear this about Neottia.

Yours affect | C. Darwin

It requires no sort of answer.

I can see *externally* to loculi in rostellum of Neottia.—

I have got good facts about moths & orchids.—*good* evidence thank Heaven of pollen-masses of **Bee Orchis** adhering to proboscis of moth; & one moth sent me with 13 pollen-masses of some orchis adhering to its proboscis, rendering it quite arborescent.—[5] I could recognise pollen-mass of Butterfly-orchis on 2 other moths.—

Down House (MS 10: 68)

[1] Dated by the relationship to the letters to the *Gardeners' Chronicle*, [4–5 June 1860], and to J. D. Hooker, 12 [June 1860].

[2] Hooker 1854b.

[3] CD described the pollen-masses of *Listera* and *Neottia* in *Orchids*, pp. 152–5. He wrote: 'I found in 1860 so many flowers exploded, with a bead of the purplish hardened cement attached to the crest of the rostellum and to the unremoved pollinia, that I suspect that it does explode, after a time, spontaneously, without the excitement of a touch.' (*Orchids*, p. 153).

[4] CD eventually considered the mechanisms for cross-fertilisation in *Neottia* to be intermediate between those of *Epipactis* and *Cephalanthera* (*Orchids*, p. 155).

[5] See letter to J. D. Hooker, 12 [June 1860], and letter from Frederick Bond, [16? June 1860]. The specimen was probably one of several moths that Bond sent to CD having large numbers of pollinia attached to their mouth-parts (see *Orchids*, p. 35).

To Charles Lyell 17 June [1860]

Down Bromley Kent
June 17th

My dear Lyell

One word more upon the "Deification" of Natural Selection.[1] Attributing so much weight to it, does not exclude still more general laws i.e. the ordering of the whole universe. I have said that nat. selection is to the structure of organised beings, what the human architect is to a building. The very existence of the human architect shows the existence of more general laws; but no one in giving credit for a building to the human architect, thinks it necessary to refer to the laws by which man has appeared. No astronomer in showing how movements of Planets are due to gravity, thinks it necessary to say that the law of gravity was designed that the planets shd pursue the courses which they pursue.— I cannot believe that there is a bit more interference by the Creator in the construction of each species, than in the course of the planets.—

It is only owing to Paley & Co,[2] as I believe, that this more special interference is thought necessary with living bodies.— But we shall never agree, so do not trouble yourself to answer.—

I shd think your remarks were very just about mathematicians not being better enabled to judge of probabilities, than other men of common sense.—

I have just got more returns about gestation of Hounds,—period differs at *least* from 61 to 74 days; just as I expected.—[3]

Etty has been decidedly better these two or three last days.[4] This day makes 7 weeks in bed!

Yours affect | C. Darwin

I was thinking of sending G. Chronicle to you on account of a paper by me on fertilisation of Orchids by insects, as it involves a curious point; & as you cared about my paper on Kidney Beans; but as you are so busy, I will not.—[5]

Endorsement: '1860'
American Philosophical Society (217)

[1] See letter to Charles Lyell, 14 [June 1860], and letter from Charles Lyell, 15 June 1860.

2 William Paley and other natural theologians argued that the organisation and adaptations of living organisms indicated the existence of an intelligent creator.

3 The source of CD's information has not been identified. See letters to Charles Lyell, 1 [June 1860] and 25 [June 1860].

4 Henrietta Emma Darwin.

5 See letter to *Gardeners' Chronicle*, [4–5 June 1860]. For CD's earlier paper on the fertilisation of kidney-beans, see *Correspondence* vol. 6, letter to *Gardeners' Chronicle*, 18 October [1857] and *Collected papers* 1: 275–7.

To Thomas Campbell Eyton 18 June [1860][1]

Down Bromley Kent
June 18th

Dear Eyton

Can you give me any authentic cases of duration of gestation in any breeds of Dogs,—Hounds, Otter-Hounds &c &c. &c.—[2] If you can I sh^{d} esteem it a great favour.— Pray do not trouble yourself to write unless you can & are willing to aid me.—[3]

Yours very sincerely | C. Darwin

American Philosophical Society (218)

1 The stationery is watermarked 1860.

2 CD's interest in this question had been stimulated by queries raised by Charles Lyell. See preceding letter and letter to Charles Lyell, 1 [June 1860].

3 Eyton's reply has not been found, but in *Variation* 1: 30, CD stated: 'Mr. Eyton of Eyton, who has had much experience with dogs, also informs me that the time is apt to be longer with large than with small dogs.'

To W. D. Fox 18 June [1860]

Down Bromley Kent
June 18th

My dear Fox

I have been looking over old letters of yours, & I want to ask a question: but I do so on condition, that you just answer it, at your leisure as briefly as possible.— Dixon says that his Musk Ducks often took to roost & perch on high places or ridge of Barn, or perch along side Hens &c.—[1] I see you used to keep Musk ducks—did you ever observe this or any tendency to lay eggs on high places? Case bears on retention of aboriginal habits— I see in your letter capital case of a cross between white & some other var. of Musk-duck *reverting* to black & white or aboriginal colour of Musk Duck.—[2]

Can you give me any authoritive case with exact dates of period of gestation of any breeds of Dog?[3] How I wish I knew for instance about Blood Hounds— I suppose no chance of Galton's knowing; indeed I do not know whether Mr H. Galton is alive.—[4]

Poor Etty has now been in bed for exactly 7 weeks with low fever; but at last, thank God, she is beginning to gain little strength.—[5]

Yours affect. | C. Darwin

Postmark: JU 18 60
Christ's College Library, Cambridge (Fox 129)

[1] Dixon 1848, p. 66. The passage is marked in CD's copy (Darwin Library–CUL).

[2] Fox's letter discussing this point has not been found. CD cited Fox's responses in *Variation* 2: 40: 'the Rev. W. D. Fox tells me that, by putting a white drake to a slate-coloured duck, black birds, pied with white, like the wild musk-duck, were always produced.'

[3] In *Variation* 1: 30, CD stated: 'Rev. W. D. Fox has given me three carefully recorded cases of [the period of gestation of] retrievers'.

[4] CD refers to John Howard Galton, a relative of both Fox and CD, with whom Fox had corresponded about bloodhounds in 1838. See *Correspondence* vol. 2, letter from W. D. Fox, [*c.* November 1838].

[5] Henrietta Emma Darwin.

From Charles Lyell 19 June 1860

19. June 1860[1]

Did I refer you to a passage which I told Huxley I objected to in his review, in the *West^r^* p. 546.[2] "Matter & Force are the two names of the one artist who fashions the living as well as the lifeless." This I presume is a specimen of the reaction against what you call Paley & Co[3] who in search of some higher power outside of the machinery called Nature by us & capable it is thought, of adding new causes to it from time to time, such as vital powers introduced into our inorganic planet & then instinct & then reason have instead of deifying Matter & Force or Natural Selection likened the Unknown Cause to the Mind & Soul of Man, this being the loftiest conception, they could form, they have striven to intensify man's intellectual & moral attributes & seem to me far more philosophical in so doing & to keep nearest to analogy. For the free will of Man which however inconsistent with belief in constant laws you must admit or give up your source of all knowledge & ignore the constitution of your own mind, must I think have some counterpart in the Deity or First Cause according to the highest conception I can make of him or it. Volition or Free Will in Man is a new cause which in the time of the Deinosaurians had no action, did not interfere with the course of vital action in the globe. The idea of such an anomaly being ever allowed to play such pranks as the breeder has played & to sport with God's creatures & the laws of reproduction so as to perpetuate pouter pigeons, & other monstrosities, would have been scouted by a philosopher of the Wealden Period, if you can suppose some human reasoner disembodied, or not yet in corporeal existence, to have studied the organic phenomena as displayed so long before the human era. If you say that in such matters we are beyond our depth this objection naturally arises when too much is attributed to any secondary cause whether it be such a "unity in duality" as Huxley's "one artist," or your "Selection". But perhaps I misunderstood Huxley.

What you say of the gestation of the hound is very remarkable.[4] How comes it that in the human race there is such regularity?

Why have we not had years ago many specimens of the fertile hybrids called leporines between the rabbit & hare in the Zoo! gardens? Do speak to the Council.[5] Lewes talks of a thousand having been sold from the Angoulême stock.[6] It is the greatest fact yet asserted. Can it be true? It would help one to believe Pallas' theory of the multiple origin of dogs from jackall, wolf &c. which I suppose I must reluctantly embrace.[7] Perhaps variation in time of gestation is a consequence of such multiple origin whereas if Man belongs to a species which came from one original stock, (or one area) & is not a mixture of several species this might explain why the Negro & European have the same period of gestation. (Have they?)

Has Hooker written to you about the absence of peculiar forms in extra-arctic Greenland & his explanation?[8] It confirms my notion that the glacial period tho' it may have required half a million of years was a brief episode in the last geological epoch, not above 1 or 2 per cent of difference in the shells—so that the species making power had not time to produce new plants.

Copy
Kinnordy MS, Charles Lyell's journal VI, pp. 117–23

[1] The text has been taken from a copy in Lyell's scientific journal with the heading 'Letter to Darwin'. It is also printed in Wilson ed. 1970, pp. 449–50.

[2] [T. H. Huxley] 1860b.

[3] See letter to Charles Lyell, 17 June [1860].

[4] See the letter to Charles Lyell, 17 June [1860], in which CD states that the gestation period of dogs varies widely.

[5] CD was a fellow of the Zoological Society of London, which owned and managed the zoological gardens in Regent's Park.

[6] See letter to Charles Lyell, 18 May [1860]. Lyell refers to Lewes 1860, p. 604.

[7] Pallas 1780. For Lyell's concern about this question, see *Correspondence* vol. 7, especially the letters written in October and November 1859.

[8] Joseph Dalton Hooker read a paper on the flora of Greenland and Arctic America at a meeting of the Linnean Society on 21 June 1860 (Hooker 1861). The paper did not, however, include his explanation for the absence of peculiar forms.

To *Entomologist's Weekly Intelligencer*[1] [20 June 1860][2]

I once saw several individuals of a small moth *apparently* eating the pollen of the *Mercurialis*; is this physically possible? I have during several years watched the smaller clovers, such as *Trifolium procumbens*, and the *Vicia hirsuta* which has such extremely minute flowers, and I never saw a bee visit them. I am, however, aware from experience that it is very difficult to assert that bees do not visit any particular kind of plant. As Mr. F. Bond informs me that he has often seen moths visiting papilionaceous flowers,[3] even such small ones as those of the trefoil, it has occurred to me that small moths may suck the flowers of *T. procumbens* and of *V. hirsuta*. From analogy we must believe that the smaller clovers secrete nectar; and it does not seem probable that the nectar would be wasted. I should esteem it a great favour if any Lepidopterists would communicate their experience on this point.—[4]

Charles Darwin,

Down, Bromley, Kent.

Entomologist's Weekly Intelligencer, 30 June 1860, p. 103

[1] The letter, which was enclosed with the letter to H. T. Stainton, 20 June [1860], was intended for publication in the *Entomologist's Weekly Intelligencer*, of which Stainton was editor. The published version is preceded by the heading: 'Do the Tineina or other small moths suck Flowers, and if so what Flowers?—'

[2] Dated by the relationship to the letter to H. T. Stainton, 20 June [1860].

[3] This information was probably in the section of the letter from Frederick Bond, [16 June 1860], that is now missing.

[4] When this letter was published, it was followed by an editorial comment (*Entomologist's Weekly Intelligencer*, 30 June 1860, pp. 103–4):

> In reply to Mr. Darwin's enquiry we may observe that very many of the Tineina are provided with tongues, and that these appendages are naturally used in extracting the sweets of flowers. It is no uncommon sight to see an Umbellifer swarming with the pretty little *Glyphipteryx Fischeriella*, each with its proboscis extended sucking at the flowers. The *Depressariæ*, as is notorious to every collector of Noctuæ, come very freely to sugar, and no doubt naturally visit flowers.
>
> But the fertilization of flowers may be accomplished by insects in another way. Many species oviposit on the blooming flowers; they do not deposit all their eggs on a single plant, but sparingly a few here and a few there; a female protruding her ovipositor down the corolla of a flower, and then flying off to repeat the operation elsewhere may herself be "the priest who performs the marriage ceremony.

CD's copy of the issue of the *Entomologist's Weekly Intelligencer* in which his letter appeared is in DAR 76: 74.

To Charles Lyell 20 [June 1860]

Down Bromley Kent
20th

My dear Lyell

I send Blyth—[1] it is a dreadful hand-writing—the passage is on p. 4.— In former note he told me he feared there was hardly chance of getting money for Chinese Expedition & spoke of your kindness.—

Many thanks for your long & interesting letter. I wonder, admire & thank you for your patience in writing so much.—[2] I rather demur to Deinosaurus not having "free will", as surely as we have.— I demur, also, to your putting Huxleys "force & matter" in same category with natural selection. The latter may of course be quite false view, but surely it is not getting beyond our depths to first causes.—

It is truly very remarkable that the gestation of hound shd. vary so much, whilst that of man does not.[3] It may be from multiple origins.— The eggs from Musk & common Duck take intermediate period in hatching. But I shd. rather look at it, as one of ten-thousand cases, which we cannot explain—namely when one part or function varies in one species & not in another.—

Hooker has told me nothing about his explanation of few Arctic forms; I knew the fact before.— I had speculated on what I presume from what you say, his explanation; but there must have been at all times an Arctic region.— I f^{d} speculation got too complex, as it seemed to me, to be worth following out.— I have been doing some most interesting work with Orchids.— Talk of adaptation in Woodpeckers—some of the Orchids beat it.— I showed the case to Elizabeth Wedgwood[4] & her remark was "now you have upset your own book, for you won't persuade me that this could be effected by nat. selection".—

Yours affect | C. Darwin

Endorsement: 'June 1860'
American Philosophical Society (219)

[1] The letter from Edward Blyth has not been found. CD had enclosed a copy of a passage from it in the letter to Charles Lyell, 14 [June 1860]. Lyell had apparently asked to see the whole letter.
[2] Letter from Charles Lyell, 19 June 1860.
[3] See letter to Charles Lyell, 17 June [1860], and letter from Charles Lyell, 19 June 1860.
[4] Sarah Elizabeth (Elizabeth) Wedgwood was Emma Darwin's older sister. She lived in Hartfield, Sussex.

To H. T. Stainton 20 June [1860]

Down Bromley Kent
June 20th

Dear Sir

I am very much obliged for your second note. I have had a very satisfactory answer from Mr Parfitt, who is evidently a careful & conscientious observer.—[1]

Would you *kindly* insert enclosed question in the Intelligencer, & answer it yourself, if it lies in your power.—[2]

Dear Sir | Yours truly obliged | C. Darwin

Endorsement: '1860'
The Natural History Museum (General Library, MSS DAR)

[1] The letter from Edward Parfitt has not been found. See letter to H. T. Stainton, 11 June [1860].
[2] See letter to *Entomologist's Weekly Intelligencer*, [20 June 1860]. The letter was published in the issue of 30 June 1860, p. 103. See also *Collected papers* 2: 35–6.

To John Higgins 21 June [1860]

Down Bromley Kent
June 21.

My dear Sir

I have just received your letter of the 19th (the Post taking two days here) & thank you for it.[1] I place the affair entirely in your hands, fully believing that you will act in the best way possible for me.—

As the time is so near it will be impossible for me to hear from my Son, after seeing the farm, in time to communicate with you.[2] My son will defer his visit for a week or two, but whether or no Anwick be purchased he will probably pay Alford a visit in the course of a week or two.[3] And I am much obliged for your very kind offer of hospitality & of showing him the Farms.—

But I know how much occupied your time is & he must not trespass on your kindness.—

My dear Sir | Yours sincerely | C. Darwin

Please to rememmber my statements about money in Hand & on Deposit at 7 days, & act as well as you can about payment.—

Post takes two days here.—

P.S. | If possible I sh^d be glad of **one** line by Post on Monday, on account of money, as it possible, though not probable, that I might wish to go to British Assoc^n: at Oxford on Wednesday afternoon.[4]

Post from here **to** Alford ought to take only *one* day.

Endorsement: '1860'
Lincolnshire Archives Office

[1] Higgins's letter has not been found. It was a reply to the letter to John Higgins, 13 June [1860].
[2] William Erasmus Darwin intended to visit the farm before CD decided whether to purchase it (see letter to John Higgins, 13 June [1860]).
[3] Anwick village, near Sleaford, Lincolnshire, is about 35 miles from Alford, where Higgins lived.
[4] See letter to Charles Lyell, 25 [June 1860].

To Alexander Goodman More 24 June 1860

Down, Bromley, Kent
June 24, 1860

Dear Sir

I hope that you will forgive the liberty which I take in writing to you and requesting a favour.[1] Mr H. C. Watson[2] has given me your address and has told me that he thought that you would be willing to oblige me. Will you please to read the enclosed;[3] and then you will understand what I wish observed with respect to the Bee-Orchis. What I especially wish, from information which I have received since publishing the enclosed, is that the state of the pollen-masses should be noted in flowers just beginning to wither, in a district *where the Bee-Orchis is extremely common.* I have been assured that in parts of Isle of Wight, viz, Freshwater Gate, that *numbers* occur almost crowded together: whether anything of this kind occurs in your vicinity, I know not; but if in your power, I should be infinitely obliged for any information. As I am writing I will venture to mention another wish which I have namely to examine *fresh* flowers and buds of the Aceras, Spiranthes, marsh Epipactis and any other rare Orchis. The point which I wish to examine is really very curious but it would take too long space to explain. Could you oblige me by taking the great trouble to send me in an old tin cannister any of these orchids, permitting me of course to repay postage.[4] It would be a great kindness, but perhaps I am unreasonable to make such a request. If you will inform me whether you have leisure so far to oblige me, I would tell you my movements, for on account of my own health and that of my daughter, I shall be on the move for next 2 or 3 weeks.

I am sure I have much cause to apologise for the liberty which I have taken, and I beg leave to remain | Dear Sir | Yours faithfully | Charles Darwin

I have examined

Ophrys apifera and muscifera
Musk Orchis
Listera ovata
Neottia Nidus-avis
Butterfly orchis
Cephalanthera grandiflora
Orchis mascula, maculata, morio, conopsea, pyramidalis.

None others grow here.

Copy
DAR 146

[1] More was a naturalist who lived on the Isle of Wight. In 1860, he contributed the chapters on zoology and botany for Edmund Venables's guide to the Isle of Wight (Venables 1860). In 1858 he had compiled a manuscript catalogue of the island's plants (see Moffat ed. 1898, p. 101).

[2] More was influenced by Hewett Cottrell Watson's work, in particular by his *Cybele Britannica* (Watson 1847–59). More had spent some days with Watson in June 1859. See Moffat ed. 1898, pp. 26, 103.

[3] CD probably enclosed a copy of his letter to the *Gardeners' Chronicle* published in the issue for 9 June 1860 (see letter to *Gardeners' Chronicle*, [4–5 June 1860]).

[4] See letter to A. G. More, 3 July [1860].

To Charles Lyell 25 [June 1860]

Down. Bromley Kent
25th

My dear Lyell

I have thought that you might like to add the enclosed very pretty arrow-heads to your Collection. They were sent me by Mrs Moir,—the mother-in-law of Col. Erskine.—[1] The place is given in note within; & they may be considered authentic & will make good contrast with the Cets.—[2]

You need not believe one word of what I said about gestation of dogs;[3] since writing to you I have had more correspondence with the master of Hounds, & I see his record is worth nothing— it may of course be correct, but cannot be trusted.— I find also different statements about wolf—in fact I am all abroad.—

I have given up Oxford;[4] for my stomach has utterly broken down & I am forced to go on Thursday for a little water-cure, to "Dr Lanes Sudbrook Park,

Richmond Surrey",[5] where I shall stay a week, & sh^d. stay rather longer had it not been for Etty. Etty improves slightly, but so slightly that it takes weeks to perceive any difference; she cannot sit up in bed for more than few minutes.—[6]

Farewell— I hope that you will have pleasant time at Oxford; I much wished to have been there, but I could not stand it, or indeed anything.—

Ever yours | C. Darwin

P.S. Many thanks for letter just received.[7] I return Dawson, which I was glad to see.—[8] I must borrow his Review sometime, for my Bookseller has never had copy.—[9] I do not think much of Dawson's letter. It would be insanity to compare evidence of organic change with geological change, at present, as far as strength of evidence goes.

But what inches of elevation on coast of Sweden are to great mountains so are the numerous varieties & endless doubt what to call species & what varieties, to undoubted species.— I entirely deny that there is no evidence of change. But time alone will bring naturalists round, when they find that they can explain many facts on such views as mine, & cannot on view of creation.—

Farewell.—

Endorsement: 'June | 1860'
American Philosophical Society (220)

[1] Henry Knight Erskine. See letter to Charles Lyell, 4 May [1860].

[2] CD refers to the shaped stones known as celts that had been found in alluvial deposits and identified as arrowheads or hatchets by Jacques Boucher de Crèvecoeur de Perthes. Recently celts had been found in association with the remains of extinct animals. Lyell, Hugh Falconer, Joseph Prestwich, and others were engaged in examining the objects and their locations in an effort to verify their age and origin. See Bynum 1984.

[3] See letters to Charles Lyell, 1 [June 1860] and 17 June [1860].

[4] The British Association for the Advancement of Science was to meet in Oxford from 27 June to 4 July 1860. Lyell was vice-president of the geology section of the meeting.

[5] Edward Wickstead Lane purchased a hydropathic establishment at Sudbrook Park, Surrey, in 1860. He had formerly owned a similar establishment at Moor Park, Surrey, which CD had visited for treatment on many occasions beginning in April 1857. CD greatly admired Lane and his family (see *Correspondence* vols. 6 and 7).

[6] Henrietta Emma Darwin had been ill since the end of April.

[7] Lyell's letter has not been found.

[8] CD refers to a letter from John William Dawson to Lyell. Lyell discussed the substance of the letter in an entry in his scientific journal dated 25 June 1860 (see Wilson ed. 1970, pp. 457–8). It reads in part:

> J. W. Dawson says that Lyell could point to changes in progress which support his theory of modern causes, such as elevation and subsidence, volcanic ejections, river-sediments, now actually in progress—whereas Darwin has to invoke new & unheard of powers & to produce new units of existence.

[9] Dawson 1860b. The review appeared in the *Canadian Naturalist.* There is a lightly annotated copy in the Darwin Pamphlet Collection–CUL.

To John Obadiah Westwood 25 June [1860][1]

Down Bromley Kent
June 25th

My dear Sir

I see in last Gardeners Ch. by signature "W", that some one has sent you Hive-Bees & Humble-bees with pollen-masses of Orchids, sticking to them;[2] as I have so long watched orchids & never seen a Bee visit one,[3] I shd much like to see these specimens, if you would lend them to me & not think that I am giving you too much trouble.—[4] I know the pollen-masses of most orchids, & think that I could recognise the genus, & perhaps species of orchis.— I shd. like to know locality, where taken.—

Hoping that you will excuse, me, | I remain | My dear Sir | Yours very faithfully | Ch. Darwin

I hope that you will have a pleasant meeting—[5] my health is too bad to come.—

I shall soon go from Home for a week,[6] so that your specimens, if kindly sent, will not be returned immediately.—

Postmark: JU 26 60
Oxford University Museum (Hope Entomological collections)

[1] Dated by the reference to a notice in the *Gardeners' Chronicle* (see n. 2, below).

[2] In the *Gardeners' Chronicle and Agricultural Gazette*, 23 June 1860, p. 580, in the section entitled 'Notices to correspondents', there is an editorial note that reads: '*W W.* The appendages on the nose of the honey and humble bees sent are not fungoid. They are the pollen masses on some Orchis. See Mr. Darwin's late paper on this subject in *Gardeners' Chronicle. W.*' The correspondent 'W. W.' has not been identified. The author of the note was presumably Westwood, who frequently contributed entomological articles to the *Gardeners' Chronicle.*

[3] Although CD only rarely observed insects visiting the flowers of orchids, he was convinced that they effected the transfer of pollinia between flowers. In *Orchids*, p. 35, he cited information from Westwood confirming that bees sometimes visit orchids.

[4] Westwood sent CD the specimens (see letter to J. O. Westwood, 9 July [1860], and *Orchids*, p. 35).

[5] CD refers to the forthcoming meeting of the British Association for the Advancement of Science in Oxford. Westwood read two papers on entomological topics at the meeting.

[6] CD's journal records that he visited Edward Wickstead Lane's hydropathic establishment at Sudbrook Park in Surrey from 28 June to 7 July 1860 ('Journal'; Appendix II).

From Frederick Bond 26 June 1860

Cavendish Rd
June 26th. 1860

Dear Sir

The box was registered & arrived here on Saturday night but my being away it was not acknowledged.[1] I regret that it was so. I hope you kept it as long as you wished. They are very careful at the post office generally I do not recollect losing a box & have had very few injured & those have been badly packed or bad boxes.

Should I have an opportunity this season I will attend to your wishes as to the small clover's &c[2] I know there are several small moths very fond of some of them

& are hardly to be found anywhere but on the flower's & | Believe me to be | Yours very truly | F^{k} Bond

C. Darwin. Esqr

DAR 76 (ser. 2): 169

CD ANNOTATIONS
1.1 The box . . . wishes 2.1] *crossed pencil*
2.1 as to] 'A' *over* 'a', *pencil*
2.2 clover's] 'Clovers' *added pencil*
2.2 moths] *crossbar added to* 't', *pencil*
2.3 hardly to] 'ly' *added pencil*; 'o' *added pencil*
2.3 but] *crossbar added to* 't', *pencil*
2.3 on the] *crossbar added to* 't', *pencil*

[1] The box that CD returned to Bond contained specimens of Lepidoptera that had pollinia attached to their mouth-parts. See letter from Frederick Bond, [16? June 1860].
[2] See letter to *Entomologist's Weekly Intelligencer*, [20 June 1860].

To J. D. Hooker 26 [June 1860]

Down Bromley Kent
26th

My dear Hooker

Many thanks for your most interesting letter. I have often speculated loosely about Greenland, so shall be most curious to see your paper;[1] as yet I have not digested your note.—

Etty makes a little progress certainly, but terribly slowly.—

My stomach has utterly failed; & I cannot think of Oxford; on Thursday I go for week of water-cure to "Dr Lanes, Sudbrook Park, Richmond Surrey": please let me have line to say whether you will be at Kew on Thursday week July 5, or 6th or 7th; for on one of those days I shall return home & would come to Kew & spend *one* hour with you, with list of things to talk over & then by Railway home.— I should so enjoy this.—

I am *very* proud at being called a "spoiled Botanist".

I have had such luck to day saw two small Hymenopterous insects (Sphegidæ?) *licking* labellum & when they crawled in they touched the rostellum & came out with the pollen masses fixed on forehead.—[2] On one of them when dead I saw mark where anther had been fixed.— One very minute insect of same Family, which had no business to attempt to walk was caught & fixed to rostellum & was dying there!

I think I have made out whole process as far as insects are concerned in Listera.—[3]

I was so proud, I dissected the stigmatic horns in Gymnanedia[4] & found it penetrated by hundreds of pollen-tubes, which I traced up to pollen.—

Farewell I will scribble no more.— I am very curious about homologies of anthers of Listera, & must get you to explain further.—

Farewell | C. Darwin

Endorsement: 'June /60'
DAR 115.1: 63

[1] Hooker's letter has not been found. CD refers to Hooker 1862a. Hooker read this paper on the distribution of Arctic plants at a meeting of the Linnean Society of London on 21 June 1860.

[2] CD's observations are given in *Orchids*, p. 146.

[3] See *Orchids*, pp. 139–52.

[4] CD's observations on the pollen tubes of *Gymnadenia* are described in *Orchids*, p. 81.

To A. G. More [30 June 1860][1]

Sudbrook Park | Richmond
Saturday,

Dear Sir

I write a line to thank you much for the kind manner with which you have received my rather unreasonable request.[2] If you find any pollen-masses removed, will you watch a group of the Bee-orchis for $\frac{1}{4}$ or $\frac{1}{2}$ an hour, and see what sort of insect visits them. I have received account of pollen-masses of this plant having been seen on proboscis of a day-moth; but I cannot help feeling sceptical about the identification.[3] If you send any other Orchids perhaps you would kindly enclose a Bee-orchis, (especially if you find one or more with pollen-masses removed) for this summer I have as yet searched in vain for specimen near my home; and I want to have pollen-masses for standard of comparison with those observed on the probosces of moths. The Spiranthes would be *especially* valuable to me and the Epipactus.

I shall return home on next Thursday (5th) (to Down, Bromley, Kent) and on Monday 9th or 10th I shall go to Rev. C. Langton's, Hartfield, Tonbridge Wells.[4]

I will venture to tell you my movements later as I fear I shall not be stationary at home for some little time.

With very sincere thanks I remain | Dear Sir | Yours truly obliged | Ch. Darwin

Copy
DAR 146

[1] This date was the only Saturday during CD's stay at Sudbrook Park (see 'Journal'; Appendix II). The copy from which the letter is transcribed is incorrectly dated '1862'.

[2] See letter to A. G. More, 24 June 1860.

[3] See letters to H. T. Stainton, 11 June [1860] and 20 June [1860].

[4] CD returned to Down on 7 July 1860. On 10 July, he left to visit Charlotte and Charles Langton who lived in Hartfield, Sussex ('Journal'; Appendix II). Charlotte Langton was Emma Darwin's sister.

From J. D. Hooker 2 July 1860

Botanic Gardens Oxford
July 2/60

Dear Darwin

I have just come in from my last moonlight saunter at Oxford & been sililoquizing over the Ratcliffe & our old rooms at the corner & cannot go to bed without inditing a few lines to you my dear old Darwin.[1] I came here on Thursday afternoon & immediately fell into a lengthened revirie: without you & my wife I was as dull as ditch water & crept about the once familiar streets feeling like a fish out of water—[2] I swore I would not go near a Section & did not for two days—but amused myself with the Colleges buildings & alternate sleeps in the sleepy gardens & rejoiced in my indolence. Huxley & Owen had had a furious battle over Darwins absent body at Section D.,[3] before my arrival,—of which more anon. H. was triumphant—You & your book forthwith became the topics of the day, & I d—d the days & double d—d the topics too, & like a craven felt bored out of my life by being woke out of my reveries to become referee on Natural Selection &c &c &c— On Saturday I walked with my old friend of the Erebus Capt Dayman[4] to the Sections & swore as usual I would not go in; but getting equally bored of doing nothing I did. A paper of a yankee donkey called Draper on "civilization according to the Darwinian hypothesis" or some such title was being read,[5] & it did not mend my temper; for of all the flatulent stuff and all the self sufficient stuffers—these were the greatest, it was all a pie of Herb[t] Spenser & Buckle without the seasoning of either—[6] however hearing that Soapy Sam[7] was to answer I waited to hear the end. The meeting was so large that they had adjourned to the Library[8] which was crammed with between 700 & 1000 people, for all the world was there to hear Sam Oxon— Well Sam Oxon got up & spouted for half an hour with inimitable spirit uglyness & emptyness & unfairness, I saw he was coached up by Owen & knew nothing & he said not a syllable but what was in the Reviews— he ridiculed you badly & Huxley savagely— Huxley answered admirably & turned the tables,[9] but he could not throw his voice over so large an assembly, nor command the audience; & he did not allude to *Sam's* weak points nor put the matter in a form or way that carried the audience. The battle waxed hot. Lady Brewster fainted,[10] the excitement increased as others spoke—my blood boiled, I felt myself a dastard; now I saw my advantage—I swore to myself I would smite that Amalekite Sam[11] hip & thigh if my heart jumped out of my mouth & I handed my name up to the President (Henslow) as ready to throw down the gauntlet—[12] I must tell you that Henslow as president would have none speak but those who had *arguments* to use, & 4 persons had been burked[13] by the audience & President for mere declamation: it moreover became necessary for each speaker to mount the platform & so there I was cocked up with Sam at my right elbow, & there & then I smashed him amid rounds of aplause— I hit him in the wind at the first shot in 10 words taken from his own ugly mouth—& then proceeded to demonstrate in as few more 1 that he

Oxford University Museum. Engraving from the *Oxford Almanack* (1860).
(Courtesy of the Oxford Museum of the History of Science.)

Samuel Wilberforce, Bishop of Oxford.
Photograph by Charles Lutwidge Dodgson
(Lewis Carroll), 1860.
(Courtesy of the M. L. Parrish Collection,
Princeton University Library.)

Thomas Henry Huxley
After a photograph taken at
the British Association meeting in Oxford, 1860.
(From William Tuckwell, *Reminiscences of Oxford*. London, 1900.
By permission of the Syndics of Cambridge University Library.)

could never have read your book & 2 that he was absolutely ignorant of the rudiments of Bot. Science— I said a few more on the subject of my own experience, & conversion & wound up with a very few observations on the relative position of the old & new hypotheses, & with some words of caution to the audience— Sam was shut up—had not one word to say in reply & the meeting *was dissolved forthwith* leaving you master of the field after 4 hours battle. Huxley who had borne all the previous brunt of the battle & who never before (thank God) praised me to my face, told me it was splendid, & that he did not know before what stuff I was made of— I have been congratulated & thanked by the blackest coats & whitest stocks in Oxford (for they hate their Bishop quite *[section illeg]* love) & plenty of ladies too have flattered me—but eheu & alas never is[14]

Incomplete
DAR 100: 141–2

CD ANNOTATIONS
Top of letter: '(Keep)' *pencil*

[1] Hooker alludes to the 1847 meeting of the British Association for the Advancement of Science in Oxford, which both he and CD attended (see *Correspondence* vol. 4).

[2] Hooker had become engaged to Frances Harriet Henslow during the 1847 meeting (see L. Huxley ed. 1918, 1: 219, and *Correspondence* vol. 4).

[3] Hooker refers to an exchange between Thomas Henry Huxley and Richard Owen that occurred during the Thursday session of the botanical and zoological section (Section D). For a report of the discussion, see Appendix VI. See also L. Huxley ed. 1900, 1: 180–1, and F. Darwin ed. 1892, pp. 236–42.

[4] Joseph Dayman had served as mate on HMS *Erebus*, on which Hooker served as assistant-surgeon and botanist during the Antarctic voyage commanded by James Clark Ross (1839–43).

[5] John William Draper was professor of chemistry and dean of the medical school of the University of the City of New York. His belief that chemical forces were the source of elementary organisation led him to accept the inheritance of acquired characteristics and transmutation. His paper applied these ideas, first enunciated in his leading textbook of physiology (Draper 1856), to the development of society (Draper 1860). See Fleming 1950. For a summary of Draper's paper, see Appendix VI.

[6] Hooker refers to the developmentalist views of Herbert Spencer and Henry Thomas Buckle.

[7] Samuel Wilberforce, bishop of Oxford, acquired the nickname 'Soapy Sam' as a result of his prolonged vacillation over the appointment of Renn Dickson Hampden to the see of Hereford in 1847 (*DNB*). As a vice-president of the British Association, Wilberforce was entitled to attend and contribute to sessions of the Oxford meeting.

[8] Hooker refers to the library of the new Oxford University Museum, in which Section D held its meeting. The museum was a product of the university's decision to create a central teaching facility for natural science and to unite its various natural history collections (Simcock 1984).

[9] The precise wording of Huxley's response to Wilberforce has never been established. In one version of the story, the bishop had asked Huxley whether he preferred to be descended from an ape on his grandfather's or his grandmother's side. Huxley gave his own account of the proceedings in a letter to Frederic Daniel Dyster about two months later: 'If then, said I, the question is put to me would I rather have a miserable ape for a grandfather or a man highly endowed by nature and possessed of great means and influence and yet who employs those faculties for the mere purpose of introducing

ridicule into a grave scientific discussion—I unhesitantly affirm my preference for the ape' (Jensen 1988, p. 168). See also Appendix VI. The events of the session have been documented in *LL* 2: 320–3; F. Darwin ed. 1892, pp. 236–42; L. Huxley ed. 1900, 1: 179–89; L. Huxley ed. 1918, 1: 521–7; and Jensen 1988.

10 Jane Kirk Purnell became David Brewster's second wife in 1857.

11 The Amalekites were the arch-enemies of Israel (Gen. 24:12–16; 1 Sam. 15).

12 John Stevens Henslow was president of Section D for zoology and botany, including physiology.

13 To burke: 'To smother, "hush up", suppress quietly' (*OED*). After William Burke, the criminal executed in Edinburgh in 1829 for smothering persons in order to sell their bodies for medical dissection (Onions ed. 1966).

14 The passage '(for they . . . never is' has been heavily inked over.

To J. D. Hooker [2 July 1860][1]

Sudbrook Park
Monday Night—

My dear Hooker

I have just received your letter. I have been very poorly with almost continuous bad headach for 48 hours, & I was low enough & thinking what a useless burthen I was to myself & all others, when your letter came & it has so cheered me. Your kindness & affection brought tears into my eyes. Talk of fame, honour, pleasure, wealth, all are dirt compared with affection; & this is a doctrine with which, I know from your letter, that you will agree from the bottom of your heart. It is indeed most true there is nothing in this world at all like a good wife.—[2] How I sh^{d} have liked to have wandered about Oxford with you, if I had been well enough; & how still more I sh^{d} have liked to have heard you triumphing over the Bishop.— I am astounded at your success & audacity. It is something unintelligible to me how anyone can argue in public like orators do. I had no idea you had this power.

I have read lately so many hostile reviews, that I was beginning to think that perhaps I was wholly in wrong & that Owen was right when he said whole subject would be forgotten in ten years;[3] but now that I hear that you & Huxley will fight publicly (which I am sure I never could do) I fully believe that our cause will in the long run prevail. I am glad I was not in Oxford, for I sh^{d} have been overwhelmed, with my stomach in its present state.—

Poor Etty has had one little return of Fever since I left home: but it has not pulled her much back, & I think we shall move her soon.—[4]

I leave this place on Thursday;[5] & would come for **one** hour between 10 & 11 if that would suit you, but I almost fear from your letter that Thursday will not suit— If not I would drive over on Wednesday afternoon;[6] this would not be quite so good, as I sh^{d} lose $\frac{1}{2}$ a day of idleness & baths.— Let me have one line to say whether you will be at home on Thursday morning. I do not think it would be worth your coming here, for unless I improve much, I am up to very little talking. After seeing you on Thursday I sh^{d} go on by Rail to London & Home.—

God Bless you my dear & most kind friend.— | Your affect, | C. Darwin

P.S | If you cannot have me on Thursday morning, I think my strongest time for Wednesday will be soon after breakfast, so that I would be with you for about an hour between 10 & eleven; ie if I can receive an answer in time.— I sh^d be very sorry to miss seeing you, but you must not think of altering any arrangement on Thursday for my pleasure.— You cannot tell how your letter has warmed my soul.—

Endorsement: 'July 2/60'
DAR 115.1: 64

[1] The letter is clearly a response to the preceding letter, also dated 2 July 1860. Hooker's letter must have been dispatched early in the morning on Monday, 2 July.
[2] See preceding letter.
[3] The occasion on which Richard Owen made this remark has not been identified. CD first mentioned Owen's comment in March 1860 (see letter to J. D. Hooker, 3 March [1860]).
[4] The Darwins were planning to leave for Hartfield, Surrey, where two of Emma Darwin's sisters lived, shortly after CD returned from taking the water-cure at Sudbrook Park. See 'Journal' (Appendix II).
[5] CD returned to Down on Saturday, 7 July 1860 ('Journal'; Appendix II). See letter to J. D. Hooker, [3 July 1860].
[6] Sudbrook Park was in the village of Richmond, Surrey, only a short distance from Kew, where Hooker lived.

To Asa Gray 3 July [1860]

Down Bromley Kent [Sudbrook Park]
July 3^d

My dear Gray

I received the mathematical papers,[1] for which many thanks, but have not yet had time to try to understand them: Your letter of June 17th has been forwarded to me here (for I am at Water-Cure).[2] I have had an unhappy household of late: my eldest girl has had low fever, 9 weeks in bed, but at last we have moved her: & my health quite broke down. We shall soon have to take her to sea-side & I want a considerable change; for from anxiety & consequent ill-health have done hardly anything for last six weeks.— I am very sorry to hear how extremely hard you are pressed with work. It is a pity that you should spend more time over Reviews of my Book:[3] you have done an immensity & been of incalculable service. I feel very grateful; though I know well that this is not a personal affair: you wish that the subject sh^d be fairly treated & discussed. Nevertheless I cannot help feeling deeply obliged to you.— I can now very plainly see from many late Reviews, that I sh^d have been fairly *annihilated*, had it not been for 4 or 5 men, including yourself. The *early* favourable reviews were highly important in preventing the subject from being "burked."—

Yesterday I had letter from Hooker at B. Assoc^n at Oxford;[4] & he tells me that there was one day a savage fight on my Book between Owen & Huxley; & subsequently a discussion of utmost warmth of 4 hours duration(!) on a paper by Draper of U. States on some subject,[5] in which somehow (I know not how) my book became subject: Bishop of Oxford, one of most eloquent men in England, ridiculed me at great length & with much spirit; & Hooker answered him, I imagine, with wonderful spirit & success.—

Owen will not prove right, when he said that the whole *subject* would be forgotten in 10 years.[6] My book has stirred up the mud with a vengeance; & it will be a blessing to me if all my friends do not get to hate me. But I look at it as certain, if I had not stirred up the mud some one else would very soon; so that the sooner the battle is fought the sooner it will be settled,—not that the subject will be settled in our lives' times. It will be an immense gain, if the question becomes a fairly open one; so that each man may try his new facts on it pro & contra.—

Very many thanks about N.Y. Times;[7] I daresay you will be disappointed with the Article; & I cannot for the life of me tell what it is that struck Lyell & me on it— I hope I may find it at home when I return there in 3 or 4 days.— I will order the two numbers of the Atlantic when I know which months contain your Articles;[8] as I shall be very anxious to see them.

I have just reread your letter: in truth I am myself quite conscious that my mind is in simple muddle about "designed laws" & "undesigned consequences".—[9] Does not Kant say that there are several subjects on which directly opposite conclusions can be proved true?![10]

I forgot to say that in Fraser's Magazine for July, there is strong article by our great man W. Hopkins. It is written with very fair spirit & with not more of the arrogance of a mathematician, than might have been expected. I have remonstrated with him for so coolly saying that I base my views on what I rank as great difficulties.—[11]

Anyone by taking these difficulties alone can make a most strong case against me. I could myself write a more damning Review than has as yet appeared! On question of Hybridity he passes over the insensibly fine gradation from utter sterility to complete fertility—,—the fertility (W. Herbert) of some hybrids,—& the sterility of the vars. of Verbascum (Gärtner) & of the vars. of Tobacco (Kölreuter), which latter facts you, by the way, never notice.—[12] Your letters are great pleasure to me, but **pray** do not write whilst so overworked.—

I have this minute had letter from Lyell; who is just starting for continent; on his return he is going to investigate reported cases of Hippopotamus *subsequent* to Glacial epoch: he finds Falconer & Prestwich now believe in this remarkable fact; if so there has probably been in Europe a period subsequent to Glacial warmer.[13] Do you remember my saying that I hoped I sh^{d} be proved wrong to punish me for disbelieving in you; & it seems that my punishment is at hand!—[14] Not being able lately to work I have amused myself about Orchids. I have been struck with amazement at beauty of contrivances with respect to fertilisation by insects. The insects led me to find that two horns (I subsequently dissected out pollen-tube) in Gymnadenia conopsea are the stigmas & you never saw anything so pretty as the contrivances are.

Do you know Hookers paper in Phil. Transact on Listera;[15] he misapprehended meaning of his facts. I find the rostellum so delicate that the explosion takes place by touch of human-hair, & the fluid sets hard in under two seconds. It was really beautiful to see a little Sphex licking the labellum, & as soon as its head touched the rostellum the explosion took-place, & the insect crawled out with the 2 pollen-masses cemented to its forehead, ready to impregnate next flower into which it crawled.

One word more on "designed laws" & "undesigned results". I see a bird which I want for food, take my gun & kill it, I do this *designedly*.— An innocent & good man stands under tree & is killed by flash of lightning. Do you believe (& I really sh^d^ like to hear) that God *designedly* killed this man? Many or most persons do believe this; I can't & don't.— If you believe so, do you believe that when a swallow snaps up a gnat that God designed that that particular swallow sh^d^ snap up that particular gnat at that particular instant? I believe that the man & the gnat are in same predicament.— If the death of neither man or gnat are designed, I see no good reason to believe that their *first* birth or production sh^d^ be necessarily designed. Yet, as I said before, I cannot persuade myself that electricity acts, that the tree grows, that man aspires to loftiest conceptions all from blind, brute force.

Your *muddled* & affectionate friend | Ch. Darwin

Endorsement: '[1860]'
Gray Herbarium of Harvard University (41)

1 Gray had sent CD the June 1860 issue of a New York periodical, the *Mathematical Monthly*. It contained an article by the mathematician Chauncey Wright entitled 'The economy and symmetry of the honey-bees' cells' (Wright 1860). This issue of the *Mathematical Monthly* is in DAR 48: 48–61; Wright's paper was annotated by CD.

2 Gray's letter has not been found. CD took the water-cure at Sudbrook Park, Richmond, Surrey, from 28 June to 7 July 1860 ('Journal'; Appendix II).

3 Gray was preparing an essay surveying the reviews of *Origin* (see letter to Asa Gray, 8 June [1860]). It was published as the last part of his anonymous three-part article in the *Atlantic Monthly* ([Gray] 1860b).

4 See letter from J. D. Hooker, 2 July 1860.

5 Draper 1860.

6 See preceding letter and n. 3.

7 CD had asked Gray to procure a copy of the *New York Times*, 28 March [1860], which contained a review of *Origin*. See letter to Asa Gray, 25 April [1860] and n. 5.

8 [Gray] 1860b comprised three articles that were published in the July, August, and October issues of the *Atlantic Monthly*. Annotated copies of the articles are in the Darwin Pamphlet Collection–CUL.

9 In the third of his articles, Gray addressed this point ([Gray] 1860b, p. 414), stating that CD used:

> expressions which seem to imply that the natural forms which surround us, because they have a history or natural sequence, could have been only generally, but not particularly designed,—a view at once superficial and contradictory; whereas his true line should be, that his hypothesis concerns the order and not the cause, the *how* and not the *why* of the phenomena, and so leaves the question of design just where it was before.

The passage is marked in CD's copy of the work.

[10] In the *Critique of pure reason* (Kant 1781), Immanuel Kant called certain transcendental ideas 'antinomies of pure reason', or pairs of contradictory metaphysical ideas, each of which could be conclusively proved. CD's knowledge of Kant's writings was probably drawn from Whewell 1840.

[11] Hopkins 1860. CD's letter to William Hopkins has not been found, but see letter to Charles Lyell, 5 [July 1860].

[12] CD discussed the work of William Herbert, Karl Friedrich von Gärtner, and Joseph Gottlieb Kölreuter in *Origin*, pp. 246–50. Their respective investigations into hybridism were also extensively discussed in his 'big book' on species (*Natural selection*).

[13] Charles Lyell's letter has not been found. He left England on 6 July 1860 for a tour in France, Belgium, and Germany. His purpose was to investigate the flint implements and fossils first discovered by Jacques Boucher de Crèvecoeur de Perthes. Before his departure, Lyell had discussed the possibility of a second post-glacial warm period with Hugh Falconer and Joseph Prestwich (see Wilson ed. 1970, pp. 452–5, 460). The subject had arisen after fossilised bones of the extant African elephant were found in association with those of extinct species in caves near Palermo. See letter from Hugh Falconer, 9 July 1860, and K. M. Lyell ed. 1881, 2: 334–5.

[14] See *Correspondence* vol. 7, letters to Asa Gray, 11 November [1859] and 24 December [1859]; see also letter to Asa Gray, 7 January [1860].

[15] Hooker 1854b.

To J. D. Hooker [3 July 1860]

Sudbrook Park
Tuesday morning

My dear Hooker

I have heard this morning from home that Emma hopes to move Etty today;[1] if so it would be folly in me not to try whether the Baths would not do me some good: so that I think I shall stay here till Saturday; but a bad account would take me home on Thursday; so I must leave my visit to you rather uncertain.—[2] If I do **not** hear I shall understand that you *probably* will be at home on Saturday morning between 10 & 11— But I shall judge by your note to certain extent.— Do not trouble yourself to write, unless you are certain that you will not be at home on Saturday.[3]

Yours affect | C. Darwin

I have reread your letter again this morning with infinite satisfaction.—[4]

I see Trains do not work in very well for me, for I cannot leave (to catch Down Train) before 1º 14′ from Kew so I must leave your house at 1º 5′. Will you give me a sandwich at 12º 45′? I will not come till past eleven; & if tired I can take a stroll in garden & save myself & you time.—

Endorsement: 'July 3ᵈ 1860'
DAR 115.1: 66

[1] Emma Darwin wrote in her diary on 3 July 1860 'came to Hartfield.' The house of Emma's sister Sarah Elizabeth Wedgwood was known in the family as 'the kindly hospital for all who are sick or sorry' (*Emma Darwin* 2: 176).

[2] See letter to J. D. Hooker, [2 July 1860].

[3] CD stayed at Sudbrook Park until Saturday, 7 July ('Journal'; Appendix II). He visited Hooker on his return to Down (see letter to J. D. Hooker, 12 July [1860]).
[4] Letter from J. D. Hooker, 2 July 1860.

To T. H. Huxley 3 July [1860][1]

Sudbrook Park | Richmond | (I return to Down on Saturday)
July 3^d

My dear Huxley

I have been water-curing here for a few days at first with very little success, but now with better hope; for I was quite worn out by anxiety about my girl & my stomach had got into dreadful state. My girl is a little better & I hope this day was moved first stage from home.[2]

I had letter from Oxford written by Hooker late on Sunday night, giving me some account of the awful battles which have raged about "species" at Oxford. He tells me you fought nobly with Owen, (but I have heard no particulars) & that you answered the B. of O. capitally.—[3] I often think that my friends (& you far beyond others) have good cause to hate me, for having stirred up so much mud, & led them into so much odious trouble.— If I had been a friend of myself, I should have hated me. (how to make that sentence good English I know not.) But remember if I had not stirred up the mud some one else certainly soon would.— I honour your pluck; I would as soon have died as tried to answer the Bishop in such an assembly. Was Owen very blackguard? H. says that the Bishop turned me into ridicule & was very savage against you.— I hardly like to ask you to write, for I know how you are overworked; but I sh^d rather like to hear a bit about the battle. I did not imagine that it would have turned up at Oxford; but I am now glad that I had no choice about going for I was utterly unfit.— The world surely will soon get weary of subject & let us have some peace. Though, on other hand, I do believe this row is best thing for subject.— As I am never weary of saying I sh^d have been **utterly** smashed had it not been for you & three others.

I can now see (which I did not at first) how **very** important the few early favourable reviews were.— Have you seen Hopkins in last Fraser:[4] it is well done & good spirit shown, (except about the soul)[5] but nothing new; the weak points are well put.— I have attacked him in letter for saying that I base my view on the difficulties of the subject; & for ignoring the fertility of some hybrids. & infertility of the Verbascum & Tobacco vars.—[6]

I fancy from what Hooker says he must have answered the Bishop well.— God knows, I honour & thank you both.

Ever yours | C. D.

Imperial College of Science, Technology and Medicine Archives (Huxley 5: 121)

[1] CD was at Sudbrook Park hydropathic establishment from 28 June to 7 July 1860 ('Journal'; Appendix II).

[2] Emma Darwin recorded in her diary that she and the children left for Hartfield on 3 July 1860.

[3] See letter from J. D. Hooker, 2 July 1860. CD refers to Richard Owen and Samuel Wilberforce, the bishop of Oxford, both of whom attacked *Origin* at separate sessions at the Oxford meeting of the British Association for the Advancement of Science. See Appendix VI.

[4] In the second part of his review of *Origin* published in *Fraser's Magazine* (Hopkins 1860), William Hopkins discussed the religious implications of CD's theory.

[5] 'We believe that man has an immortal soul, and that the beasts of the field have not. If any one deny this, we can have no common ground of argument with him. Now we would ask, at what point of his progressive improvement did man acquire this spiritual part of his being, endowed with the awful attribute of immortality? Was it an "accidental variety" seized upon by the power of "natural selection" and made permanent?' (Hopkins 1860, 62: 88).

[6] CD's letter to Hopkins has not been found. See also letter to Asa Gray, 3 July [1860].

To A. G. More 3 July 1860

Sudbrook Park, Richmond
July 3, 1860

My dear Sir

I thank you heartily for all your great kindness. I received this morning the specimens *quite fresh* and was very glad to see them.[1] The Bee-orchis ought to be looked to when flowers are beginning to wither. I shall be most grateful for the E. palustris and it will be all the better for me in 10 days time.[2] Please see that there are *some* buds; as these are the best in some respects for points of structure which I am examining. It is a shame that you so kindly will mulct yourself of sundry red stamps.

On the 10th or 11th my address will be at

Rev. C. Langton's
Hartfield.
Tonbridge Wells[3]

and I will take with me my microscope and dissecting tools, which unfortunately I have not now with me.

In haste and with cordial thanks, believe me | My dear Sir | Yours truly obliged | C. Darwin

Copy
DAR 146

[1] See letters to A. G. More, 24 June 1860 and [30 June 1860].

[2] CD thanked More for sending him specimens of *Epipactis palustris* in *Orchids*, p. 95 n.

[3] Charles Langton had married Emma Darwin's sister Charlotte in 1832. Their home at Hartfield Grove was only a quarter of a mile from The Ridge, where Sarah Elizabeth Wedgwood lived.

To J. D. Hooker [4 July 1860][1]

Sudbrook—
Wednesday

My dear Hooker

I have just received your note saying that you could see me on Thursday, but were going to London at 11 or 12. But I have just this minute heard joyful news that Etty has been moved to Hartfield—& was safe & comfortable in bed, though much exhausted.[2] Hence I shall stay here till Saturday Morning. If I do **not** hear I shall understand that you will be at home on that morng between 11 & 1. If you cannot see me then, I would come earlier on Friday

God Bless you | My dear Friend | C. Darwin

DAR 115.1: 65

[1] This was the Wednesday following CD's letter to J. D. Hooker, [3 July 1860].

[2] Emma Darwin took Henrietta Emma Darwin and the other children to Hartfield on 3 July 1860 (Emma Darwin's diary).

To John Lubbock [4 July 1860][1]

Sudbrook Park | Richmond.
Wednesday

Dear Lubbock

I have just seen in the Times that Mrs Lubbock has been confined, & I most heartily congratulate you on the anxiety being over.—[2]

I was very sorry that I was quite unable to go to Oxford;[3] I have been bad enough here, but I began yesterday to revive.—I had the joyful news this morning that Etty has been safely moved to Hartfield in Sussex. I return home on Saturday evening & go to Hartfield on Tuesday or Wednesday We shall stay a week or so there & then for Etty's sake go to sea-side.—

If you have nothing better to do will you pay me half-an-hour visit on Sunday or Monday at any hour. It is so long since I have seen you that I sh^d *very* much like just to have a little chat with you before going away, if you can so manage it. I would have come to High Elms to call on you; but everything knocks me up so easily & so completely— I most truly hope that Mrs Lubbock is going on well. I shall hear at Down.— Do not trouble yourself to write, but come if you can.

Dear Lubbock | Believe me Yours very sincerely | C. Darwin

Endorsement: '1860'
Down House

[1] Dated by the reference to the Lubbocks' new baby (see n. 2, below).

[2] Ellen Frances Lubbock gave birth to Constance Mary Lubbock on 2 July 1860. The birth was reported in the *The Times*, 4 July 1860.

[3] Lubbock had attended the Oxford meeting of the British Association for the Advancement of Science. He spoke in support of CD's theory at the Saturday, 30 June, session at which Thomas Henry Huxley confronted the bishop of Oxford, Samuel Wilberforce (see Appendix VI). In a letter to Francis Darwin dated 2 January 1896 (DAR 106/7 (ser. 4): 30), Lubbock recalled that he was 'proud of having born my part in the great Oxford debate . . . I took the point in reply to Wilberforce & [Robert] FitzRoy that the embryology of the individual in many cases represents the past history of the Species.' See also Hutchinson ed. 1914, 1: 50, and F. Darwin ed. 1892, p. 239.

To T. H. Huxley [5 July 1860][1]

Sudbrook Park | Richmond
Thursday

My dear Huxley

I must send you a line to say what a good fellow you are to send me so long an account of the Oxford doings.[2] I have read it twice & sent it to my wife & when I get home shall read it again: it has so much interested me.— But how durst you attack a live Bishop in that fashion? I am quite ashamed of you! Have you no reverence for fine lawn sleeves? By Jove, you seem to have done it well. If anyone were to ridicule any belief of the Bishop's, would he not blandly shrug his shoulders & be inexpressibly shocked?

I am very very sorry to hear that you are not well; but am not surprised after all your self-imposed labour: I hope you will soon have an outing; & that will do you real good.—

I am glad to hear about J. Lubbock, whom I hope to see soon, & shall tell him what you have said.[3] Have you read Hopkins in last Fraser[4]—well put, in good spirit, except soul-discussion, bad, as I have told him.— —nothing actually new —takes the weak points alone, & leaves out all other considerations—

I heard from Asa Gray yesterday;[5] he goes on fighting like a Trojan

God Bless you— get well, be idle & always, reverence a Bishop.—

Ever yours most truly | C. Darwin

Imperial College of Science, Technology and Medicine Archives (Huxley 5: 123)

[1] CD took the water-cure at Sudbrook Park from 28 June to 7 July 1860 ('Journal'; Appendix II).
[2] Huxley's letter has not been found.
[3] See preceding letter and n. 3.
[4] Hopkins 1860. CD's letter to William Hopkins has not been found.
[5] Asa Gray's letter has not been found.

To Charles Lyell 5 [July 1860]

Sudbrook Park | Richmond
Thursday 5th.

My dear Lyell

I thank you for your farewell note,—the last, for the present, of the very many most interesting letters, which you have so kindly written to me.— I had no idea that you were going for so long;[1] I am glad to hear that you are going to Amiens: do for goodness sake invent some theory to explain how the flint gravel was spread out high & low. I am, also, glad to hear that you are going to investigate post-glacial period— it is a most interesting question with respect to distribution & many other things.[2]

I was rejoiced to hear yesterday that Etty was safely moved to Hartfield on Tuesday,—of course much exhausted. I leave this place on Saturday; somewhat improved in stomach for I was very bad, & am rejoiced that I did not attempt Oxford. Huxley, Hooker & J. Lubbock (as I am pleased to hear) seem to have stuck up for modification of Species like Trojans.[3] Asa Gray, as I hear to day, also goes on fighting well.— I am in good heart on the question & think we shall in long run conquer.—

Have you seen Hopkins in last Fraser?[4] it is very well done & in good spirit; but I regret the soul-discussion, as I have told him. He very coolly puts it as if I had *based* my belief on what I rank as greatest difficulties. Anyone who will put forward the difficulties *alone* may make a very damaging review.—

I am particularly glad to hear about the Hare-rabbits:[5] your facts have convinced me that they are hybrids, & I never was before *quite* convinced.— I forget whether it is stated that the half & half breed inter se.— The case is the more odd as I have at least two authorities who tried with perseverence & could never get the Hare & Rabbit to breed.

Ever my dear Lyell | Yours gratefully & affectionately | C. Darwin

Endorsement: 'July 1860'
American Philosophical Society (221)

[1] Lyell's European tour lasted from 6 July until 23 September 1860 (Wilson ed. 1970, p. 490 n. 88).

[2] Lyell's letter has not been found. For a discussion of some of its contents, see the letter to Asa Gray, 3 July [1860].

[3] See letter from J. D. Hooker, 2 July 1860, and letters to J. D. Hooker, [2 July 1860], to T. H. Huxley, 3 July [1860], and to John Lubbock, [4 July 1860].

[4] The second part of Hopkins 1860 appeared in the July issue of *Fraser's Magazine*.

[5] In his scientific journal (Wilson ed. 1970, p. 465), Lyell recorded that he had seen the so-called 'leporines' during a visit to the zoological gardens in London on 1 July 1860. He further noted that the gestation periods of the two parent species differed, a point that related to the question of the fertility of hybrids.

From Hugh Falconer 9 July [1860][1]

The Athenæum
9th July

My Dear Darwin

Hooker—Huxley & others of your correspondents will have informed you about the doings at Oxford—and how the Saponaceous Bishop got basted and larded by Huxley. Owen also came in for such a set down by Huxley, as I have never witnessed within my experience of Scientific discussion.[2] Your interests I assure you were most tenderly watched over by your devoted Elèves.

What I want to tell you now is quite a different affair—but one which I am sure will interest you very much. Baron Anca a Sicilian Friend, who followed up my inquiries in the Sicilian caves, has brought over from Sicily two molars of the *Existing African Elephant* and upwards of 20 jaws of the Existing Spotted Hyæna (Hyæna Crocuta), of the Cape—from the Caves![3]

Admiral Smyth laid down "Adventure Bank" a shoal with a narrow channel, between Trapani the Western End of Sicily & Capo Bono—the promontory of Tunis.[4]

We can now show that the division of Sicily from the African Continent is quite as late—if not later than the separation of England from France.[5]

This is a great point gained—and I think the dispersion of the *Hyæna Crocuta* from the Cape of Good Hope to Sicily, will fructify into some congruity with your doctrines

My Dear Darwin | Yours Ever Sinly | H Falconer

DAR 164.1: 5

CD ANNOTATIONS
1.3 Owen . . . discussion. 1.4] *scored blue crayon*
Top of first page: 'Keep Geograph Distribution' *pencil*; '19 & 20'[6] *brown crayon*

[1] Dated by the reference to fossil remains found by Francesco Anca (see n. 3, below).

[2] See letter from J. D. Hooker, 2 July 1860, and letters to J. D. Hooker, [2 July 1860], to T. H. Huxley, 3 July [1860], and to John Lubbock, [4 July 1860].

[3] Anca announced his discovery at the Oxford meeting of the British Association for the Advancement of Science (Anca 1860). The fossils, found near Palermo, were given to Hugh Falconer for identification (see K. M. Lyell ed. 1881, 2: 334). The fact that the remains of living African species had been found in association with the remains of extinct species suggested that Sicily had been connected with Africa during a recent geological period.

[4] William Henry Smyth, who retired from the navy in 1846, had engaged in naval operations with the Sicilian flotilla during the Napoleonic wars. During that time and for several years afterwards, he surveyed the Sicilian coast. His findings were published in Smyth 1824. The information given in the letter was published in Anca 1860.

[5] Falconer discussed the implications of the fossils from Palermo in Falconer 1860.

[6] Respectively, the numbers of CD's portfolios of notes on the geographical distribution of animals and of plants.

To J. O. Westwood 9 July [1860]

Down Bromley Kent
July 9th.—

My dear Sir

I am very much obliged to your kindness in sending the Bees;[1] & I am glad to have seen the pollen-masses attached. I am sorry to say the Hive-Bee came with body loose & what is much worse the pollen-masses were off; I cannot conceive how this could occur for the pin was not loose.— I can see place where pollen-masses were attached; but I have searched in vain for the pollen-masses themselves. I suppose they must have rolled out of Box, whilst I was looking at the Bee, but I do not think that I inclined the Box sufficiently for this— I am *extremely* sorry for this accident.

I do not know of Morren's paper about Orchids & insects.[2] I was much interested by one on the part which insects play in the fertilisation of the (I think) Stylideæ,[3] & sh^{d} be glad to have any other reference if you will take the trouble to send it.—

With my best thanks & sincere regret for the accident to your Bee which is unaccountable to me, I remain, My dear Sir | Yours sincerely | C. Darwin

I have spent so much money lately that I am not willing to subscribe for the purchase of Wollaston's collections for your Museum.—[4]

Endorsement: '1860'
Oxford University Museum (Hope Entomological collections)

[1] See letter to J. O. Westwood, 25 June [1860], in which CD requested from Westwood specimens of bees with pollinia attached.

[2] Morren 1836. The paper is not cited in *Orchids*. For CD's further correspondence with Westwood about this paper, see *Correspondence* vol. 9.

[3] Morren 1838.

[4] Westwood was the curator of the Hope Entomological collection, housed in the new Oxford University Museum. Frederick William Hope presented his collection of insects and other natural history specimens, as well as portraits and prints, to the university in 1849. He also endowed a chair of zoology, to which Westwood was appointed in December 1860. See A. Z. Smith 1986. Westwood was soliciting subscriptions toward the purchase of Thomas Vernon Wollaston's collection of land and freshwater shells from Madeira, which the museum acquired in 1861 (A. Z. Smith 1986, p. 161).

To S. P. Woodward 9 [July 1860][1]

Down Bromley Kent
9th

My dear Sir

I am very sorry to say that I cannot answer one of your questions.—[2] Von Buch or Johnston's Phys. Atlas would probably tell number of Volcanos.—[3] I am nearly certain there are the best remarks of anybody's on antiquity of sub-aerial volcano in Lyell Principles or Elements on *Auvergne*;[4] I remember being astonished at their antiquity.— Degradation would prevent any sub-aerial volcano *in a recognizable condition* being of excessive antiquity. St. Helena of which upper part at least is sub-aerial, struck me as of extreme antiquity. The Canary isl^{d} from *littoral* miocene shells must as a volcanic group be very old.—[5]

I entirely & absolutely disagree with Von Buch's elevation-crater-theory—indeed I think it proved false.[6] As there are so-considered volcanic ash beds in oldest formations, viz Silurian, no doubt *sub-aerial* Volcanos has always existed,—or at least *probably*, for on reflexion I do not suppose volcanic mud & ashes when stratified under water could be distinguished.—

I am very glad to hear about your Brother.[7]

I am sorry that I cannot answer your queries.—

In Haste | Yours very sincerely | C. Darwin

The Cordillera were volcanic with islands since early part of Cretaceous period or older; but I do not say that the volcanoes were sub-aerial—some stream however flowed in *not* deep water—[8]

Iceland from Miocene plants must have been volcano since that period— I suspect that this oldest ascertained volcano.—

The Natural History Museum (General Library, MSS DAR)

[1] The date is given by the reference to Woodward's brother (see n. 7, below) and by a note written on the manuscript in an unidentified hand that reads: 'To S. P. Woodward | 1860 | British Museum'.

[2] Woodward was preparing an article on volcanoes for the eighth edition of the *Encyclopædia Britannica* (S. P. Woodward 1860). In the article, he cited information from CD's *Volcanic islands*. He also discussed CD's views on the elevation and subsidence of segments of the earth's crust (S. P. Woodward 1860, pp. 606–7).

[3] The reference is to Buch 1836 (the French translation of Buch 1825) and to Johnston ed. 1856. Christian Leopold von Buch described in this work his theory of the formation of volcanic craters and gave a list of the principal volcanoes of the globe. Alexander Keith Johnston's edition of the *Physikalischer Atlas* of Heinrich Karl Wilhelm Berghaus included a double-page plate of volcanoes. Woodward referred to both these works in S. P. Woodward 1860, p. 603. He also cited CD's map of active volcanic vents published in *Coral reefs*.

[4] CD probably refers to the ninth edition of Charles Lyell's *Principles of geology* (C. Lyell 1853), a copy of which is in the Darwin Library–CUL. Lyell's *Manual of geology* discussed the volcanoes of the Auvergne in some detail (C. Lyell 1855, pp. 550–9). A copy of this work is also in the Darwin Library–CUL.

[5] Woodward discussed the antiquity of volcanic vents in S. P. Woodward 1860, p. 606.

[6] CD had expressed his disbelief in Buch's theory as early as 1844, when Buch and Lyell had first debated the formation of volcanic craters. See *Correspondence* vols. 2 and 4. In his article, Woodward stated that Buch's theory had 'lost its popularity' owing to Lyell's and CD's criticism, but he found it difficult to believe that huge craters could have formed the base of volcanic cones, as Lyell's theory required (S. P. Woodward 1860, p. 605).

[7] Bernard Bolingbroke Woodward, S. P. Woodward's older brother, was appointed librarian in ordinary to Queen Victoria at a ceremony at Windsor Castle on 2 July 1860 (*DNB*).

[8] CD's point was discussed in S. P. Woodward 1860, p. 606.

From Asa Gray [10 July 1860][1]

Your observation on Primroses relates to a subject with the general facts which I am very familiar, though I never thought of the *accurate* investigation by measurements &c you undertake. I have long ago convinced myself that the two forms are *both fertile*, but always looked at it as the first step toward bisexuality. When your notion about the probable necessity of cross-fertilization came to my knowledge, these *diœcio-dimorphous* flowers (See Torr. & Gray, Fl. N. Amer. 2. p. 38)[2] occurred to me as a pretty case of arrangement for it. The case is *extremely common, in very various families*,—tho' I did not know it in *Primulaceæ*. The Rubiaceæ, Borragineæ,

Labiatæ, &c &c. are full of it. I *think* cases can be found of its gliding into structural unisexuality.

Adieu, dear Darwin, | Truly Yours | A. Gray

Incomplete
DAR 110 (ser. 2): 77

CD ANNOTATIONS
End of letter: 'See Manual of Bot | p. 171'[3] *blue crayon*

[1] Dated by CD's statement in the letter to Asa Gray, 11 August [1860]: 'I received your note of July 10th'. See also n. 3, below.

[2] Torrey and Gray 1838–43.

[3] CD did not own a copy of Torrey and Gray 1838–43. The expression 'diœciously dimorphous' is discussed in Gray's *Manual of botany* (Gray 1856, p. 171 n.). In CD's annotated copy of this work (Darwin Library–CUL), this expression is underlined in pencil. CD thanked Gray for these 'valuable hints' in the letter to Asa Gray, 11 August [1860].

To Hugh Falconer 12 July [1860][1]

The Ridge, Hartfield, | Tonbridge Wells
July 12th

My dear Falconer

I thank you for your *most* interesting note.[2] The facts are really grand. We have had a melancholy household for some time; as my eldest girl has been very ill for a long time and we have brought her here for a change.[3] I have been bad enough of late, and have not been in London for an age. Whenever I do come I must pay you a visit and have one of those chats with you, which I so much enjoy. But our plans are all uncertain being dependent on my daughter's health.

I must say I do heartily enjoy Owen having had a good setting down—his arrogance and malignity are too bad.

My good old friend | Yours most truly | C. Darwin

Copy
DAR 144

[1] Dated by the relationship to the letter from Hugh Falconer, 9 July [1860].

[2] Letter from Hugh Falconer, 9 July [1860].

[3] The Darwins had taken Henrietta Emma Darwin to convalesce at the home of Sarah Elizabeth Wedgwood in Hartfield, Sussex ('Journal'; Appendix II).

To J. D. Hooker 12 July [1860]

Miss Wedgwood's | Hartfield | Tonbridge Wells
July 12th

My dear Hooker

Many thanks for newspaper. I have had a little, but very interesting, note from Falconer;[1] he tells me how capitally things went on at Oxford.

Poor Etty, I find, very little better; but what shade of difference there is, is certainly on the right side. It is most disheartening to us.—

I did so thoroughily & greatly enjoy my little visit to you.—[2] I want two questions answered; I know how busy you are. Do you think Mr Oliver would think me unreasonable, if I asked him to answer them?[3] If he would, pray tell him, I sh^d^ be very much obliged.

Firstly name of enclosed Saxifrage it is often kept in pot in Houses[4] on account of slightly curved pistil towards nectar-secreting surface.—& impregnation by Diptera

Secondly the name of that wonderful hot-house flower, with rectangularly-bent pistil, which has been falsely said to turn round. This flower[5] has stuck in my throat ever since: I was a fool not to have accepted your offer of taking it home to examine.—[6] I want to know whether a sort of fold at base of each petal secretes nectar: a closely similar fold at base of 2 upper petals in an orange-coloured Alstrœmeria does secrete nectar. I care for this case because it is strongest apparent exception to my rule of bent pistil & gangway into nectary.

Is not this position of parts? Instead of the pistil revolving, would not large insect successively visiting the supposed nectaries, brush over anthers & stigma; so that the pistil is bent into the **circular** gangway to the nectaries?

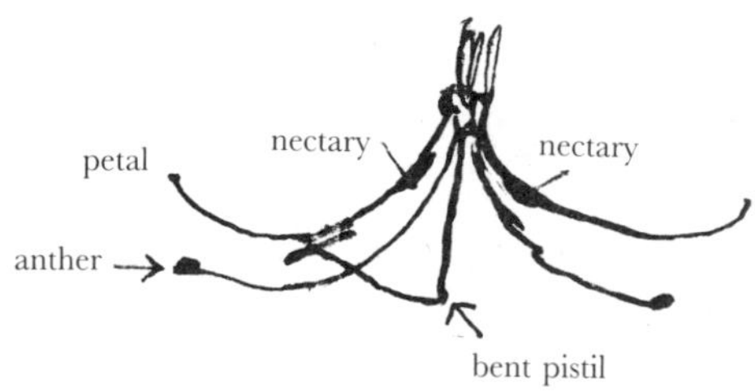

(N.B I find common Martagon Lily with curved pistil so like in all respects that I can make out above points.)[7]

I have been examining Orchis pyramidalis, & it almost equals, perhaps even beats, your Listera case:[8] the sticky glands are congenitally united into saddle-shaped organ, which has great power of movement & seizes hold of bristle (or proboscis) in admirable manner, & then another movement takes place in pollen-masses, by which they are beautifully adapted to leave pollen on the two *lateral* stigmatic surface.— I never saw anything so beautiful.—

Ask your scientific gardener to have another try at fertilising Leschenaultia & **mark the flowers**;[9] for the ovaria of two flowers of which I stirred up the pollen in indusium before going to Sudbrook have swollen to *twice diameter* of any other ovaria: hence I feel pretty sure that I am on the right road.—

I have been thinking a bit about your Book & the more I think of it the more awfully difficult it seems, & therefore the more worthy of your attempting.[10] One of the first points seems naturally to occur viz difference between plant & animal! And then, as I suppose you will allude to unicellular plants, what makes an individual!! And thirdly the difference between propagation by gemmation & sexual generation! Nice simple little subjects to discuss!—

Yours ever affectionately | C. Darwin

Do not you answer; if you do not like to ask M^r. Oliver, let this note sleep, for everything will keep.

Endorsement: '/60'
DAR 115.1: 67

[1] See letter from Hugh Falconer, 9 July [1860].

[2] CD visited Hooker on his way to Down from Edward Wickstead Lane's hydropathic establishment at Sudbrook Park, Richmond, Surrey. See letters to J. D. Hooker, [2 July 1860], [3 July 1860], and [4 July 1860].

[3] Daniel Oliver was an assistant botanist at the Royal Botanic Gardens, Kew.

[4] CD added the phrase 'it is often kept in pot in Houses' in pencil above the line (see Manuscript alterations and comments).

[5] The text from 'This flower' to 'the **circular** gangway to the nectaries?' was deleted by CD. It has been included in the transcription because the following sentence '(N.B I find common Martagon Lily . . . above points.)' relates to it.

[6] CD refers to his visit to Hooker at Kew (see n. 2, above).

[7] CD presumably deleted the passage preceding this sentence when he realised that his questions had been answered through his examination of the martagon lily (*Lilium martagon*).

[8] CD refers to Hooker's paper on *Listera* (Hooker 1854b). For a summary of CD's findings, see the letter to Asa Gray, 3 July [1860].

[9] See letter to J. D. Hooker, 26 April [1860].

[10] CD was encouraging Hooker to write a general book on botany. See letters to J. D. Hooker, 8 February [1860], 14 February [1860], and 12 March [1860].

To H. G. Bronn 14 July [1860][1]

Down. | Bromley Kent. [Hartfield][2]
July 14th.

Dear & Honoured Sir.

On my return home after an absence of some time I found the Translation of the third part of the Origin,[3] and I have been delighted to see a final chapter of Criticisms by yourself.[4] I have read the first few paragraphs & final paragraph and am perfectly contented, indeed more than contented, with the generous & candid spirit with which you have considered my views— You speak with too much praise of my work.—[5] I shall of course carefully read the whole chapter; but though I can read descriptive books like Gäertners pretty easily,[6] when any reasoning comes in I find German excessively difficult to understand.[7] At some *future* time I should very much like to hear how my book has been received in Germany & I most sincerely hope M. Schweizerbart will not lose money by the publication—[8] Most of the Reviews have been bitterly opposed to me in England; yet I have made some converts & *several* naturalists who would not believe in a word of it, are now coming slightly round & admit that Natural Selection may have done something. This gives me hopes that more will ultimately come round to a certain extent to my views.

I shall ever consider myself deeply indebted to you for the immense service & honour which you have conferred on me in making the excellent translation of my book.

Pray believe me with most sincere respect | Dear Sir. | Yours gratefully.[9] | Charles Darwin.

Copy
DAR 143

[1] The year is given by the reference to Bronn's translation of *Origin* into German (Bronn trans. 1860).

[2] CD visited Sarah Elizabeth Wedgwood in Hartfield, Sussex, from 10 July 1860 to 2 August ('Journal'; Appendix II).

[3] The third and final part of the German translation of *Origin* was published on 11 June 1860 (*Börsenblatt für den Deutschen Buchhandel* 27 (1860): 1166). The first and second parts had appeared on 4 April and 2 May.

[4] CD had suggested to Bronn at the outset that he might like to add his own comments about CD's theory to the translation of *Origin* (letter to H. G. Bronn, 4 February [1860]). Bronn did so in a concluding fifteenth chapter entitled 'Schlusswort des Übersetzers'. CD's copy of the translation in the Darwin Library–CUL is incomplete: it lacks pages 465–520, including most of the fourteenth chapter as well as Bronn's concluding remarks. Associated with the book are two pages of notes by CD headed 'Bronn's criticism for New Edit of Origin'. For a transcript of these notes, see letter to H. G. Bronn, 5 October [1860].

[5] Bronn prefaced his remarks with praise for CD, calling him 'a genuine naturalist who regards in an ingenious and penetrating manner old facts that he has collected and considered for twenty years'. He went on to say (Bronn trans. 1860, p. 495):

> Tief in seinem Gegenstand versenkt, von der Wahrheit der gewonnenen Resultate unerschütterlich überzeugt, trägt er sie mit so bewältigender Klarheit vor, beleuchtet er sie mit so viel Geist, vertheidigt er sie mit so scharfer Logik, zieht er so wichtige Schlüsse daraus, dass wir, was auch unsre bisherige Überzeugung gewesen seyn mag, uns eben so wenig ihrem Eindrucke entziehen, als unsre Anerkennung der Aufrichtigkeit versagen können, womit er selbst alle Einreden, die man ihm entgegen-halten kann, herbeisucht und nach ihrem Gewichte anerkennt. [Deeply immersed in his objective, unshakably convinced of the truth of his results, he writes with such overpowering clarity, elucidates his material with so much spirit, defends it with such sharp logic, draws from it such important conclusions, that we ourselves, whatever our previous views may have been, are just as little able to avoid the impression that they make on us as we are able to deny our appreciation of the sincerity with which he himself seeks out and acknowledges, according to their importance, all objections that can oppose him.]

[6] The reference is to Gärtner 1844 and 1849, both of which CD had carefully studied in the preparation of his work on species (see *Correspondence* vols. 5 and 6).

[7] CD first began to learn German in 1839 (see *Correspondence* vol. 2, letter to Caroline Wedgwood, [27 October 1839]). According to Francis Darwin, CD read German slowly and with great effort. 'He himself learnt German simply by hammering away with a dictionary; he would say that his only way was to read a sentence a great many times over, and at last the meaning occurred to him . . . In spite of his want of grammar, he managed to get on wonderfully with German, and the sentences that he failed to make out were generally really difficult ones.' (*LL* 1: 126).

[8] The German translation of *Origin* was published by E. Schweizerbart'sche Verlagshandlung, Stuttgart.

[9] The copyist wrote 'faithfully', which has been corrected to read 'gratefully' in an unidentified hand.

To Emma Gärtner 14 July [1860]

Down Bromley Kent [Hartfield]
July 14th

Dear Madam

I have received & am grateful for your extremely kind letter. After an absence from home of a fortnight I found on my return the memoir of your honoured Father[1] (which I have not had time to read yet, as I was obliged to leave home again immediately on account of the illness of one of my children) & the two Engravings.— I value much that of your Father & the expression of countenance is most intellectual & delightful.—

You generously offer me a collection of your Father's specimens of Hybrid plants. The offer is a most valuable one; but I do not collect plants, & I am not likely to have time to experimentise much myself, on Hybrid plants.— If on reflexion you should still think that a collection of these plants would be well bestowed on England, I would suggest your giving them to the Herbarium at the Royal Gardens at Kew near London.[2] The Collection is under the care of Sir William Hooker & of Dr. Hooker & is by far the best collection in Britain; indeed I believe one of the very best in the world. I asked Dr. Hooker & he said if presented, the collection would be much valued & taken care of for perpetuity.

Should you think of this, I would advise you to consult some German Botanists, & see whether my account of Kew is not correct.

Pray believe me dear Madam, with much respect & gratitude | Yours very truly obliged | Charles Darwin

Postmark: JY 15 60
Kenneth Spencer Research Library, Department of Special Collections (MSS P87: 1), University of Kansas

[1] See letter to Emma Gärtner, 9 June [1860]. The presentation copy of the memoir of Karl Friedrich von Gärtner (Jäger 1851), inscribed by Emma Gärtner, is in the Darwin Pamphlet Collection–CUL.
[2] Gärtner's herbarium was not presented to the Royal Botanic Gardens, Kew.

To J. S. Henslow 16 July [1860]

at Miss Wedgwoods | Hartfield | Tonbridge Wells
July 16th

My dear Henslow

I thought it wd be better to write direct to Daubeny & tell him there was nothing *direct* on subject in my Book.—[1] I did not see much in his paper.—[2] The frequency of "Sports" ie modified buds in plants goes direct against his view.[3] There is no greater mystery in the whole world, as it seems to me, than the existence of sexes,— more especially since the discovery of Parthenogenesis.[4] The origination of sexes seems beyond all speculation.— Nevertheless I quite agree with your remarks in answer to Daubeny.—[5]

I am glad to hear M[rs] Barnard's affair is safely over.—[6]

How successful your Fete seems to have been!—

In Haste | My dear old Master | Yours affect. | C. Darwin

I did not hear of poor FitzRoy with the Bible at the Geographical Section—[7] I think his mind is often on Verge of insanity.—

Endorsement: '1860'
DAR 93: 74–5

[1] CD's letter to Charles Giles Bridle Daubeny has not been found. Daubeny discussed the sexuality of plants in relation to CD's views in a paper delivered at the Oxford meeting of the British Association for the Advancement of Science (Daubeny 1860). See Appendix VI.

[2] Daubeny presented a copy of his paper (Daubeny 1860) to CD; it is in the Darwin Pamphlet Collection –CUL. CD noted that it contained 'excellent facts new to me' on p. 30, where Daubeny gave examples of monstrous plants that, in their abnormal state, resemble plants of quite different genera. These passages were marked by CD.

[3] Daubeny suggested that sexual reproduction in plants exists to prevent uniformity; vegetative reproduction, he stated, could only produce a 'mere counterpart' of the parent (Daubeny 1860, p. 110).

[4] CD refers to Karl Theodor Ernst von Siebold's research on parthenogenesis (Siebold 1856), which demonstrated that the ova of bees and certain butterflies are capable of developing without being fertilised. For CD's views on Siebold's work, see *Correspondence* vol. 6.

[5] Henslow's remarks may have been made at the British Association meeting. Henslow was president of Section D, in which Daubeny read his paper.

[6] Henslow's daughter Anne Barnard gave birth to a daughter on 16 July 1860 (*Gentleman's Magazine* n.s. 9 (1860): 312).

[7] Robert FitzRoy attended the British Association meeting in Oxford, at which he delivered a paper on the prediction of storms to improve the safety of shipping (FitzRoy 1860). No record of his having spoken in the geographical section of the meeting has been found, but he had made comments during the discussion of CD's theory in Section D (see Appendix VI and also F. Darwin ed. 1892, p. 239).

To J. D. Hooker [17 July 1860][1]

at Miss Wedgwoods | Hartfield | Tonbridge Wells.
Tuesday Night

My dear Hooker

Asa Gray has sent me Proc. American Acad. p. 409 to 416 and 425 to 434. So that 417 to 424 are missing.[2] If you have these pages & can lay your hand on them, pray put them in blank cover & send them to me,—as I much wish to see all.— He seems to me to discuss subject quite **excellently**.—[3] I have been turning in my mind whether I would not send **my** copy to the Athenæum with request that they would reprint this discussion— But Heaven knows whether they would.— I shall think a bit more & reread it, before deciding[4]

Yours affect. | C. Darwin

Etty has improved a little more these two last days, & has twice sat in arm-chair for above $\frac{1}{2}$ Hour.— I can see Doctors are afraid of mesenteric mischief.—[5]

Endorsement: 'July 22/60'
DAR 115.1: 69

[1] Although the letter is endorsed 'July 22/60', CD wrote it on the Tuesday before the letter to Asa Gray, 22 July [1860].

[2] CD refers to the *Proceedings of the American Academy of Arts and Sciences* 4 (1860): 410–15, 424–31, in which a debate on *Origin* between Asa Gray, John Amory Lowell, Louis Agassiz, and Francis Bowen was reported. The pages that CD considered missing were in fact not concerned with *Origin*. Both parts of the article are in the Darwin Pamphlet Collection–CUL.

[3] In the report of the academy's meeting held on 10 April 1860, Gray defended CD's views from criticisms raised by Lowell, Bowen, and Agassiz at the previous session. See *Proceedings of the American Academy of Arts and Sciences* 4 (1860): 411–15. Gray's comments are marked in CD's copy.

[4] See letter to Asa Gray, 22 July [1860].

[5] For a fuller account of Henrietta Emma Darwin's condition, see the letter to J. D. Hooker, 29 July [1860].

To John Innes 18 July [1860]

at Miss Wedgwoods | Hartfield | Tonbridge Wells
July 18th

Dear Innes.

Thanks for Coal-bill which I will pay as soon as I get home in about 10 days time.—[1] I am sincerely sorry to hear so poor an account of Mrs. Innes.[2]

We moved Etty here with much difficulty a fortnight ago, & she has improved very little, but thank God she has improved a little. She now can generally sit up for above $\frac{1}{2}$ hour twice a day. I have been a good deal knocked up of late & have had to recur to Water-cure; but all our anxiety with Etty ill for 12 weeks has been enough to knock us up. But my wife has stood the incessant nursing wonderfully. Women are so good & unselfish, that helping others seems to do them good.—

Our plans are utterly Uncertain; we hope to take Etty to sea, whenever she is strong enough; but where & when are both quite unknown to us.—

What a wandering life you have been leading!

We shall be very glad to see you at Down.— I enquired (not alluding to you) about Mr Ainslie's house,[3] but can hear nothing whatever. Parslow asked Mr Baxter,[4] but he knew nothing.

Farewell with our kind remembrances to Mrs. Innes—believe me | Dear Innes | Yours very sincerely | C. Darwin

I hope from your note that Johnny is tolerably well.—[5]

Postmark: JY 18 60
Cleveland Health Sciences Library (Stecher collection)

[1] The bill probably related to the affairs of the Down Coal and Clothing Club, of which CD was treasurer.

[2] Eliza Mary Innes.

[3] Robert Ainslie, of Tromer Lodge, Down, had left the village in 1858 (see *Correspondence* vol. 7, letter to W. E. Darwin, 14 [May 1858]).

[4] Joseph Parslow was the Darwins' butler. CD possibly refers to William Baxter, the dispensing chemist in Bromley, Kent (*Post Office directory of the six home counties* 1859). Innes, the perpetual curate of Down,

had been trying for a number of years to purchase or rent a house in Down (see J. R. Moore 1985, p. 469).

[5] John William Brodie Innes.

To J. D. Hooker 19 [July 1860][1]

at Miss Wedgwoods | Hartfield | Tonbridge Wells
Thursday 19th

My dear Hooker

I am very sorry to hear about your Baby & that Mrs Hooker is prevented going as yet to sea-side; for it would have done her good. Milk disagreed with one of our Babs; but it ultimately got strong.—

Many thanks for Review: it **is** by Asa Gray:[2] I have sent for copy to America.: it was intended to be strictly anonymous; but publisher let name escape to A. Gray's vexation. Hence he wrote in the no-care-devil style; but it is very clever & pleasant. What a first-rate man he is.—

It is a real good joke my discussing homologies of Orchids with you, after examining only 3 or 4 genera: & this very fact makes me feel positive I am right!! I do not quite understand some of your terms; but sometime I must get to explain the homologies; for I am intensely interested on subject, just as at a game of chess.—[3] I must look to your description of the viscid fluid.— *Practically* it may be said then pollen-masses are *never* retained on the rostellum except by accident.

I have been comparing parts of fructification again of the Habenaria chlorantha & bifolia: I do not know whether you compared minutely; but there is *vast* difference in stalk of pollen-mass—in the important junction (cause of movement) of footstalk & sticky surface—in size of pollen-grains, & shape of stigmatic surface. By Jove if all this is variation it is wonderful case: the difference is fully as great as between Bee & Fly Ophrys.—[4]

I have seen Temple's sermon & like first part very much;[5] in latter part there are, I think, some terrific assumptions & illogical remarks.—

Ever yours affect.— | C. Darwin

I could write on lots more subjects & shd. enjoy it, but refrain out of simple virtue to save your time.—

I am very glad that you have ordered Lechenaultia to be fertilised.—[6]

Endorsement: 'July/60'
DAR 115.1: 68

[1] The endorsement is confirmed by CD's visit to Sarah Elizabeth Wedgwood in Hartfield and by the reference to [Gray] 1860b.

[2] CD refers to the first part of Asa Gray's unsigned review of *Origin*, published in three parts in the July, August, and October 1860 issues of the *Atlantic Monthly* ([Gray] 1860b). Annotated copies of all three parts are in the Darwin Pamphlet Collection–CUL. CD noted in pencil on the first part: 'Nothing in the 1st Article'.

[3] CD discussed the homologies of orchids in *Orchids*, pp. 286–307. Hooker's advice about how to trace the homologies of the spiral vessels was mentioned by CD several times in *Orchids*, pp. 290–301.

[4] CD repeated these remarks in *Orchids*, p. 89, explaining that the structural differences between the two species of *Habenaria* were greater than those usually found among members of the same genus.
[5] Possibly Temple 1860, a sermon on the relation between science and religion. Frederick Temple was headmaster of Rugby School.
[6] See letter to J. D. Hooker, 12 July [1860].

To J. D. Hooker [20? July 1860][1]

Tunbridge Wells

P.S. I have just read Quarterly R.[2] It is uncommonly clever; picks out with skill all the most conjectural parts, & brings forwards well all difficulties.— It quizzes me quite splendidly by quoting the Anti-Jacobin versus my grandfather.—[3] You are not alluded to; nor, strange to say, Huxley, & I can plainly see here & there Owen's hand.—[4] The concluding pages will make Lyell shake in his shoes. By Jove if he sticks to us he will be a real Hero.—[5]

Good night—your well-quizzed, but not sorrowful & affectionate friend. C. D.

I can see there has been some queer tampering with the Review—for a page has been cut out & reprinted.—[6]

Endorsement: 'July 23 [or 26]/60'
Incomplete
DAR 115.1: 33a

[1] Although the note was endorsed 'July 23/60' and then corrected in the same hand to read 'July 26/60', it seems likely that both dates are incorrect. The note was probably written before the letters to T. H. Huxley, 20 July [1860], and to John Lubbock, 20 July [1860]. It could possibly have been enclosed with the preceding letter, but then CD would have had little time to obtain and read the review to which he refers. The *Quarterly Review* in question (see n. 2, below) was published on Wednesday, 18 July 1860 (*Publisher's Circular*, 17 July 1860, p. 339).
[2] The July issue of the *Quarterly Review* carried an anonymous review of *Origin*. The author was Samuel Wilberforce, bishop of Oxford (Wilberforce 1874; *Wellesley index* 1: 743). CD's annotated offprint of the review is in the Darwin Pamphlet Collection–CUL. Errors noted on an erratum sheet in the *Quarterly Review* have been corrected in the offprint. CD wrote in ink at the head: 'Bishop of Oxford aided by Owen & Murchison(?)'. He refers to Richard Owen and Roderick Impey Murchison.
[3] CD refers to the following passage in the review ([Wilberforce] 1860, pp. 254–5):

> We do not think that, with all his matchless ingenuity, Mr. Darwin has found any instance which so well illustrates his own theory of the improved descendant under the elevating influences of natural selection exterminating the progenitor whose specialities he has exaggerated as he himself affords us in this work. For if we go back two generations we find the ingenious grandsire of the author of the 'Origin of Species' speculating on the same subject, and almost in the same manner with his more daring descendant . . . Many of our readers will remember the humour with which Frere and Canning, in the 'Poetry of the Anti-Jacobin,' exposed these philosophical arguments of the last generation.

The reference is to 'Loves of the triangles' a parody of Erasmus Darwin's 'Loves of the plants', published in *Poetry of the Anti-Jacobin* (1799), edited by John Hookham Frere, George Canning, and others.
[4] The reviewer quoted extensively from the works of Richard Owen, whose philosophy he found 'just as much truer to nature as it is to revelation than all these speculations of the transmutationist.' ([Wilberforce] 1860, p. 260).

[5] Charles Lyell's attack on transmutation in his *Principles of geology* (C. Lyell 1830–3) was mentioned in the concluding pages of the review: 'We trust that he [CD] is mistaken in believing that he may count Sir C. Lyell as one of his converts . . . no man has been more distinct and more logical in the denial of the transmutation of species than Sir C. Lyell, and that not in the infancy of his scientific life, but in its full vigour and maturity.' ([Wilberforce] 1860, p. 263). Several of Lyell's remarks against transmutation were quoted.

[6] It has not been possible to determine which page CD thought had been reprinted.

To T. H. Huxley 20 July [1860]

at Miss Wedgwoods | Hartfield | Tonbridge Wells
July 20th

My dear Huxley

Many thanks for your pleasant letter. I agree to every word you say about Fraser & the Quarterly.[1] I have had some really admirable letters from Hopkins. I do not suppose he has ever troubled his head about Geograph. Distrib. Classification, morphologies &c &c. & it is only those who have that will feel any relief in having some sort of rational explanation of such facts. Is it not grand the way in which the Bishop asserts that all such facts are explained by ideas in God's mind?![2] The Quarterly is uncommonly clever; & I chuckled much at the way my grandfather & self are quizzed. I could here & there see Owen's hand.— By the way how comes it that you were not attacked? Does Owen begin to find it more prudent to leave you alone?[3] I would give five shillings to know what tremendous blunder the Bishop made; for I see that a page has been cancelled & new page gummed in.—

I am indeed most thoroughly contented with progress of opinion. From all that I hear from several quarters, it seems that Oxford did the subject great good.— It is of enormous importance the showing the world that a few first-rate men are not afraid of expressing their opinion. I see daily more & more plainly that my unaided Book would have done *absolutely* nothing. Asa Gray is fighting *admirably* in U. States. He is thorough master of subject,—which cannot be said by any means of such men, as even Hopkins.

I have been thinking over what you allude to about a Natural Hist. Review.[4] I suppose you mean really a *Review* & not Journal for original communications in Nat. History. Of the latter there is now superabundance.— With respect to a good Review, there can be no doubt of its value & utility: nevertheless, if not too late, I hope you will consider deliberately before you decide.— Remember what a deal of work you have on your shoulders; & though you can do much yet there is limit to even the hardest worker's power of working.— I should deeply regret to see you sacrificing much time which could be given to *original* research.

I fear to one who can review as well as you do, there would be the same temptation to waste time, as there notoriously is for those who can speak well.— A Review is only temporary; your work shd be perennial. I know well that you may say that unless good men will review there will be no good Reviews. And this is true. Would you not do more good by an *occasional* Review in some well established Review, than by giving up *much* time to the editing, or largely aiding if not editing, a Review which from being confined to one subject would not have a very large circulation?— But I must return to the chief idea which strikes me, viz that it would lessen the amount of original & perennial work which you could do.— Reflect how few men there are in England who can do original work in the several lines, in which you are excellently fitted.[5] Lyell, I remember, on analogous grounds many years ago resolved he would write no more Reviews.—

I am an old, slow coach & your scheme makes me tremble. God knows in one sense I am about the last man in England, who ought to throw cold water on any review, in which you would be concerned,—as I have so immensely profited by your labours in this line.—[6]

With respect to Reviewing myself; I never tried:[7] any work of that kind stops me doing anything else, as I cannot possibly work at odds & ends of time; I have, moreover, an insane hatred of stopping my regular current of work. I have now materials for a little paper or two; but I know I shall never work them up. So I will not promise to help; though not to help, if I could, would make me feel very ungrateful to you.— You have no idea during how short a time daily I am able to work. If I had any regular duties like you, & Hooker, I sh[d] do *absolutely* nothing in Science.

I am heartily glad to hear that you are better; but how such labour as volunteer-soldiering (all honour to you) does not kill you, I cannot understand.[8]

My poor girl improves a little;[9] but terribly little. I fear some organic mischief. We shall go home in about 10 days.—

Farewell my good friend | Yours most sincerely | Ch. Darwin

For God's sake remember that your field of labour is original research in the highest & most difficult branches of Natural History. Not that I wish to underrate the importance of clever & solid Reviews.—

Bronn at end of Translation, gives a Ch. of Criticism: but the German is so difficult I have not read it yet.—[10]

Endorsement: '[1860]'
Imperial College of Science, Technology and Medicine Archives (Huxley 5: 125)

[1] Hopkins 1860 and [Wilberforce] 1860.

[2] In his review of *Origin* in the *Quarterly Review*, Samuel Wilberforce stated ([Wilberforce] 1860, p. 259):

> How can we account for all this? By the simplest and yet the most comprehensive answer. By declaring the stupendous fact that all creation is the transcript in matter of ideas eternally existing in the mind of the Most High . . .

CD's copy of [Wilberforce] 1860 is in the Darwin Pamphlet Collection–CUL. Beside this passage he wrote: 'Mere words'.

[3] In his review of *Origin* ([R. Owen] 1860a), Richard Owen had attacked Huxley as being a supporter of CD's theory. See also preceding letter.

[4] The *Natural History Review* was founded in 1854 to publish the transactions of a number of Irish natural history societies. By 1860, the editors found it hard to continue publishing, having only about a hundred subscribers in Ireland and fewer in England, and they were compelled to make

the *Review* more 'cosmopolitan' in a new series (*Natural History Review* 7 (1860): 65–6). The editors of the first series were William Henry Harvey, Samuel Haughton, Arthur Riky Hogan, and E. Perceval Wright, all members of Dublin University. Wright had recently discussed with Huxley the possibility of transferring the general editorship to him, and Huxley was receptive to the proposal. Huxley wrote to Joseph Dalton Hooker on 17 July 1860 that the 'tone of the *Review* will be mildly episcopophagous, and you and Darwin and Lyell will have a fine opportunity if you wish of slaying your adversaries.—' (L. Huxley ed. 1900, 1: 210).

5 Huxley mentioned this in a subsequent letter to Hooker: 'Darwin wrote me a very kind expostulation about it, telling me I ought not to waste myself on other than original work. In reply, however, I assured him that I *must* waste myself willy-nilly, and that the *Review* was only a save all.' (L. Huxley ed. 1900, 1: 210).

6 Huxley decided to accept the general editorship of the new series of the *Natural History Review*, which had a board of eleven editors in all. The first volume of the new series was issued in January 1861.

7 This is not strictly true. CD had reviewed the first volume of Waterhouse 1846–8 in the *Annals and Magazine of Natural History* 19 (1847): 53–6 (*Collected papers* 1: 214–17).

8 Huxley had apparently joined one of the volunteer rifle units being organised as part of a popular movement to defend Britain against possible invasion by France.

9 Henrietta Emma Darwin.

10 Bronn trans. 1860, pp. 495–520. See letters to H. G. Bronn, 14 July [1860] and 5 October [1860].

To John Lubbock 20 July [1860]

at Miss Wedgwoods | Hartfield | Tonbridge Wells
July 20th

My dear Lubbock

I am puzzled what to think about the Review.—[1] Good periodicals undoubtedly do very good service in cause of Science.— Unless good men will devote some time to them, they will never be good; yet I cannot avoid regretting that men like Huxley & yourself shd give up much of your precious time to such work. If you had not the Bank work, I shd be strong for your assisting.—[2] I suppose you could not do such work in Banking hours? If not I must say I doubt about wisdom.— Do not quote me to Huxley, as it would be so ungracious.— I have heard from Huxley himself today & I shall tell him that I doubt much the wisdom of his undertaking the work.[3]

On other hand it is wonderful the amount of work which you hard-working men continue to do.— In fact, the more I think the more puzzled I feel.— On one point I am clear, that if it would stop your doing original work you ought not, even pro bono publico, undertake this new work.— If it would not stop you, I feel sure your name & help would be highly serviceable in a good cause.— This note is a collection of platitudes, but I can say nothing else.— I feel that the subject is of importance; for I am convinced that you are one of the very few philosophical naturalists which this country contains; & therefore you are bound to do the best service which you can in the good cause of Natural Science.—

Poor dear Etty does mend a little, but it is fearfully little, only just perceptibly— The Doctors fear some Mesenteric mischief; but they hope not permanent. Her pulse still keeps high.— It is wretched work—

GoodBye, my dear friend | Yours affectly | C. Darwin

Have you seen Quarterly R.? Very clever & amusing, & quizzes me so capitally, that I chuckled again, & read it with thorough enjoyment.—[4]

Endorsement: '1860'
Down House

[1] Thomas Henry Huxley had written to Lubbock to solicit his assistance were he to take over the editorship of the *Natural History Review*. See preceding letter and L. Huxley ed. 1900, 1: 209. Lubbock presumably asked CD's opinion about this matter in a letter that is now missing.

[2] Lubbock was a partner in his father's bank, Robarts, Lubbock & Co.

[3] See preceding letter.

[4] [Wilberforce] 1860. See the two preceding letters.

To W. B. Tegetmeier 20 July [1860]

Down Bromley Kent [Hartfield]
July 20th

My dear Sir

I am sure that you have no cause to apologise to me; but rather I to you for so often plaguing you with questions.— I did *not* ask about young pigeon-down; but I am *especially* glad of this additional information, which I will give on your authority.[1] What I asked about was whether you had any facts leading you to believe that it was any advantage with Hive-Bees to cross the breed, when the same stock has been long kept by same person in same place.[2] But do not think of troubling yourself to answer.[3] I am sorry that I did not bring in your name more clearly about the Bees-cells; but my volume was merely an abstract, & I quote no one in detail.[4] If I live I hope to do more justice to all whom I quote.

Very sincere thanks for your kind offer of Poultry skulls, which I will most gladly avail myself of;[5] but I am making **very** slow progress in my larger work, chiefly owing to a terribly long illness of my eldest daughter & my own poor health; so that I shall not attack poultry for some time.—[6] I *have* seen notices of your most curious observations on the eggs of the Worker-Bees.—[7] I am sorry to say that I am entirely ignorant with respect to the native country of the Italian Alp-Bee.—[8]

Pray believe me, My dear Sir, with thanks for all your kindness. | Yours very sincerely | C. Darwin

P.S. I have just remembered, what my other question was, namely whether it would be possible to procure for me trustworthy information, whether the duration of incubation is exactly the same in breeds of Pigeons so different as short-faced Tumblers, Runts, Carriers, or Pouters. Any fact on this head would be very valuable.—[9]

Endorsement: '1860'
New York Botanical Garden Library (Charles Finney Cox collection)

[1] Tegetmeier had supplied CD with information about pigeon-down a few months earlier. See letter to W. B. Tegetmeier, 8 May [1860]. CD cited Tegetmeier's observations in *Variation* 1: 170.

[2] See letters to W. B. Tegetmeier, 17 April [1860], and to *Cottage Gardener*, [after 8 May 1860].

[3] Tegetmeier investigated the topic suggested by CD. On 8 August 1860, he presented a short paper entitled 'Natural cross breeding in bees' at a meeting of the Entomological Society (Tegetmeier 1860). See also letter to W. B. Tegetmeier, 30 July [1860].

[4] Tegetmeier assisted CD in his study of the construction of bees' cells (see *Correspondence* vol. 7). Tegetmeier was cited twice in the discussions of this subject in *Origin*, pp. 228 and 233.

[5] See letter to W. B. Tegetmeier, 17 April [1860].

[6] CD recorded in his 'Journal' in 1861: 'May 16. Finished Fowls (8 weeks)' (see de Beer ed. 1959).

[7] At a meeting of the Entomological Society on 4 June 1860, Tegetmeier exhibited specimens of fertile workers of the honey-bee *Apis mellifica* that had come from a hive lacking a queen. He believed this disproved Jean Pierre Huber's statement that such workers were only produced if they had fed on royal jelly reserved for the queen (*Transactions of the Entomological Society of London* n.s. 5 (1858–61), Proceedings, p. 118). A notice of this paper was also published in the *Gardeners' Chronicle and Agricultural Gazette*, 14 July 1860, p. 650.

[8] According to a notice in the *Gardeners' Chronicle and Agricultural Gazette*, 5 May 1860, p. 411, the 'newspapers have lately been much occupied by discussions respecting a new sort of bee which apiarians have discovered to be domesticated in the Swiss Alps, which entomologists call *Apis helvetica* or *Apis ligustica*, and which also bears the name of the Ligurian Bee.' The bee was also sometimes called the Italian Alp-bee. Tegetmeier's inquiry was related to his attempt to obtain a stock of these bees. In a letter to the *Gardeners' Chronicle*, 11 August 1860, pp. 734–5, Tegetmeier stated that he had 'after many discouragements, succeeded in obtaining a thriving and populous stock of Ligurian bees' and would be pleased to show them to any bee-keeper. He exhibited specimens of these bees at the Entomological Society on 8 August 1860 (see n. 3, above).

[9] See letter to W. B. Tegetmeier, 8 May [1860].

To Asa Gray 22 July [1860][1]

Down Bromley Kent [Hartfield]
July 22[d]

My dear Gray

Owing to absence from home at Water-cure & then having to move my sick girl to whence I am now writing,[2] I have only lately read the discussion in Proc.[s] American Acad.[3] And I now I cannot resist expressing my sincere admiration of your most clear powers of reasoning. As Hooker lately said in a note to me, you are more than *anyone* else the thorough master of the subject.[4] I declare that you know my Book as well as I do myself; & bring to the question new lines of illustration & argument, in a manner which excites my astonishment & almost my *envy*! I admire these discussions, I think, almost more than your article in Silliman's Journal.[5] Every single word seems weighed carefully, & tells like a 32-pound shot. It makes me much wish (but I know that you have not time) that you could write more in detail; & give, *for instance*, the facts on the variability of the American wild fruits.—

The Athenæum has largest circulation & I have sent my copy to Editor with request that he would republish the first discussion: I much fear he will not, as he reviewed subject in so hostile a spirit & is not a liberal man.—[6] I fear that I shall lose my copy.— I have received p. 410 to 416—and pp. 425 to 434—, so that pages 417 to 424 are missing[7] I wish you would tell me whether I could buy these sheets of Proceedings.—

New York Times had not arrived when I left home.— Hooker has lent me (*& was much puzzled about authorship*!) the July Monthly Atlantic:[8] it is uncommonly pleasantly written, & will tell well on public.— I shall be curious (& will order) the August number, as soon as I know that it contains your Review of Reviews.—[9] My conclusion is that you have made a mistake in being a Botanist, you ought to have been a Lawyer, & you would have rolled in wealth by perverting the truth, instead of studying the living truths of this world.—

Of Reviews I have to report one in Quarterly by the Bishop of Oxford (Wilberforce) aided by Owen:[10] it is uncommonly clever, not worth anything scientifically, but quizzes me in splendid style till I chuckled with laughter at myself— From all that I gather, the B. Assoc. at Oxford advanced subject greatly, owing to effort of Hooker, Huxley & Lubbock, not withstanding incessant attacks. Henslow & Daubeny are shaken.—[11] I hear from Hooker that he hears from Hochstetter that my views are making very considerable progress in Germany;[12] & that good workers are discussing the question. Bronn at end of his Translation has a chapter of Criticism; but it is such difficult German, that I have not yet read it.—[13] Hopkins' Review in Fraser is thought the best which has appeared against us;[14] I believe that Hopkins is so much opposed because his course of study has never led him to reflect much on such subjects as Geograph. Distribution, Classification, Homologies &c &c; so that he does not feel it a relief to have some kind of explanation.—

I see most clearly that my Book would have been a dead failure, had it not been for all the generous labour bestowed on it (not for my sake, but for subject sake) by yourself, Hooker, Huxley & Carpenter; & to these names I hope soon Lyell's may be added.— But I know that the Quarterly will cause Lyell to shake in his shoes.[15] Considering his age, his former views & position in Society, I think his conduct has been heroic on this subject.—

I am tired, so will write no more. Farewell my kind & good friend. | Yours most truly | C. Darwin

Gray Herbarium of Harvard University (30)

[1] CD visited Hartfield, Sussex, where Sarah Elizabeth Wedgwood lived, from 10 July 1860 to 2 August ('Journal'; Appendix II).

[2] CD had taken the water-cure at Edward Wickstead Lane's establishment at Sudbrook Park, Richmond, Surrey, from 28 June 1860 to 7 July ('Journal'; Appendix II). He refers to the long-continued illness of Henrietta Emma Darwin.

[3] *Proceedings of the American Academy of Arts and Sciences* 4 (1860): 410–15, 424–31. See letter to J. D. Hooker, [17 July 1860].

[4] Joseph Dalton Hooker's remark was probably made in a letter, now missing, that was written in the interval between receiving the letters to J. D. Hooker, [17 July 1860] and 19 [July 1860].

[5] [Gray] 1860a.

[6] William Hepworth Dixon was editor of the *Athenæum* from 1853 to 1869. The review of *Origin* in the *Athenæum*, 15 November 1859, pp. 659–60, was written by John R. Leifchild. It upset CD, who believed he had been 'cut up in Athenæum, and under a theological point of view, unfairly' (see *Correspondence* vol. 7, letter to Caroline Sarah Wedgwood, [after 21 November 1859]). An extract

from the material CD sent to Dixon (*Proceedings of the American Academy of Arts and Sciences* 4 (1860): 411–16) was published in the *Athenæum*, 4 August 1860, p. 161. See also letter to J. D. Hooker, 7 August [1860].

7 The 'missing' pages did not include any discussions relating to Darwin. See also letter to J. D. Hooker, [17 July 1860].

8 [Gray] 1860b. See letter to J. D. Hooker, 19 [July 1860].

9 CD's annotated copies of all three articles from the *Atlantic Monthly* ([Gray] 1860b) are in the Darwin Pamphlet Collection–CUL. Gray's article 'Darwin and his reviewers' appeared in the October issue.

10 [Wilberforce] 1860. CD believed that Richard Owen had helped Samuel Wilberforce write the review (see letters to J. D. Hooker, [20? July 1860], and to T. H. Huxley, 20 July [1860]).

11 See letter from J. D. Hooker, 2 July 1860, and Appendix VI.

12 Ferdinand Hochstetter, who was appointed professor of mineralogy and geology at the Imperial Polytechnic Institute in Vienna in 1860, attended the 1860 meeting of the British Association for the Advancement of Science in Oxford.

13 Bronn trans. 1860, pp. 495–520.

14 Hopkins 1860.

15 See letter to J. D. Hooker, [20? July 1860] and n. 5.

From Charles Hardy[1] 23 July 1860

Hayling Vicarage | Havant
23 July 1860.

Sir,

I venture to mention to you in connexion with a remark made by you at p. 73 of your "Origin of Species" that "humble bees alone visit the red clover (Trifolium pratense) as other bees cannot reach the nectar"—that I was at Bury near Arundel last year in the months of July August & September. There was a field of red clover at the back of the house which was always frequented by myriads of bees.[2] Speaking moderately I think a single dash of the net in any part of the field would have taken fifty of them. I called the attention of my children to the fact as contradicting what I told them was a popular error—which circumstance they perfectly recollect. I have repeatedly observed the same thing at other times & in other places.

I ought to say that the clover in question was a second growth (& I believe that it is from the second growth exclusively that the seed is saved) & that the flowers are somewhat smaller than those produced by a first growth. I have never particularly observed whether the latter are equally frequented by bees. If my life is spared for another year I will endeavour to ascertain that point.

I remain Sir | faithfully yours | Charles Hardy.

C. Darwin Esq[re]

DAR 76 (ser. 2): 170

CD ANNOTATIONS
Top of letter: 'For *Origin* | *good*' *ink*; 'Ch. 3'[3] *brown crayon*

[1] Hardy was vicar of North and South Hayling, Hampshire.

[2] Hardy apparently means that he saw hive-bees sucking at the clover. In *Origin*, pp. 94–5, CD stated that hive-bees could not reach into the corolla of the red clover, whereas humble-bees could.

[3] See following letter. CD's annotation 'Ch. 3' refers to the third chapter, 'On the possibility of all organic beings crossing', of his 'big book' on species (see *Natural selection*, pp. 56, 68).

To Charles Hardy 27 July [1860][1]

(Down Bromley Kent) | (now staying at Miss Wedgwoods | Hartfield | Tonbridge Wells.)
July 27th

Dear Sir

Absence from home has prevented my answering your note & thanking you very sincerely for it.[2] The correction of a blunder is a greater kindness than a new fact.— I did not speak rashly, for I watched a large clover-field for *many* days, & never saw any Hive Bees at work.— An experienced Apiarian also assured me of the truth of the statement; & the statement is made in Maunder's(?) American Bee Keeper.—[3] It has been a great evil to me publishing in abstract: for it was not possible for me to guard my statements sufficiently.—

I have twice seen Hive-Bee sucking single *stunted* plant of T. pratense. In the large clover field above alluded to which I had watched, was joined by a field of Sanfoin, & which was visited by thousands of Bees, when this was cut one morning, the Bees came to the Clover & tried to suck it & succeeded with some of the withered flowers & those at the bottom of the heads. So that if the flowers are smaller of the second-growth I can well believe that they could be sucked. This would be a very interesting little fact for me, & I will take the liberty of quoting your note[4] & shd be extremely much obliged if you would make & communicate any further observations on subject to me—

Pray accept my sincere thanks for your kindness. & believe me | Dear Sir | Yours truly obliged | Charles Darwin

If Mrs. Hardy has not forgotten me, pray give her my best compliments.[5]

Smithsonian Institution Libraries (Special collections)

[1] Dated by the relationship to the preceding letter.

[2] See preceding letter.

[3] CD probably refers to T. B. Miner's *American bee keeper's manual*, in which it is stated that red clover is useless to hive-bees (Miner 1849, p. 243).

[4] Hardy's point was incorporated into the third edition of *Origin*, p. 100: 'I have been informed, that when the red clover has been mown, the flowers of the second crop are somewhat smaller, and that these are abundantly visited by hive-bees.' See also Peckham ed. 1959, p. 184.

[5] It is not known when CD made the acquaintance of Charlotte Hardy. Charles Hardy had been a student at Christ's College, Cambridge, between 1822 and 1826 (*Alum. Cantab.*).

To J. D. Hooker 29 July [1860]

at Miss Wedgwoods | Hartfield | Tonbridge Wells
July 29th

My dear Hooker

We have had Sir H. Holland here to see Etty again,[1] & to our great grief he tells us the fever has left mischief in the abdomen with some fluid accumulated, but thank God not much. He is tolerably sanguine for her ultimate recovery; but it will be very long. We have had a miserable time of it, but are now more composed & hopeful. I know well how sincerely you will sympathise with us. But do not write: there is nothing to be said. We can only hope.

I trust your Baby is better & that Mrs. Hooker has got to Worthing.[2]

I return home with our children on Thursday, & Emma follows with Etty in two or three days.

Latterly I have done nothing here; but at first I amused myself with a few observations on the insect-catching power of Drosera; & I must consult you some time whether my "twaddle" is worth communicating to Linnean Soc.—[3]

I have made some good observations, extending your Listera work in case of Malaxis,—curious modification of same process.[4]

In 3d nor of "London Review" very good geological article on my Book; & I shd. *very much* like to know who author is.—[5]

I have had note from A. Gray: he is fighting away like a Trojan.— The Athenæum will insert A. Gray's discussion.—[6]

My dear Hooker | Yours affect | C. Darwin

Endorsement: '/1860'
DAR 115.1: 70

1 Henry Holland was called to Hartfield on 26 July 1860 (Emma Darwin's diary). Emma Darwin described his visit in a letter to Mary Elizabeth Lyell, written on 28 August 1860: 'We had the bad luck at Hartfield to fall into the hands of a desponding medical man, and it really was a great injury to us. We had a visit from Sir Henry Holland, who cheered us again, and I fully believe his view is the true one. He has been so constantly kind, and taken so much trouble, that we feel very grateful.' (*Emma Darwin* 2: 177). The 'desponding medical man' was probably William Wallis of Hartfield (see n. 4, below). When Henrietta first fell ill in April, she was diagnosed as suffering from 'a form of Typhus fever' (letter to J. D. Hooker, 11 May [1860]). Subsequently, CD described the fever as being 'partly remittent partly typhoid' (letter to W. D. Fox, 18 May [1860]). One of the symptoms of patients with typhoid fever is the enlargement of the mesenteric glands of the abdomen (*EB*).

2 See letter to J. D. Hooker, 19 [July 1860].

3 The common sundew (*Drosera rotundifolia*) is an insectivorous plant that catches insects by the movement of sticky hairs on its leaves. CD's notes on his observations are in DAR 60.1: 15–23. He published an account of them in *Insectivorous plants*, pp. 1–3:

> During the summer of 1860, I was surprised by finding how large a number of insects were caught by the leaves of the common sun-dew (*Drosera rotundifolia*) on a heath in Sussex. I had heard that insects were thus caught, but knew nothing further on the subject. I gathered by chance a dozen plants, bearing fifty-six fully expanded leaves, and on thirty-one of these dead insects or remnants of them adhered . . . Many plants cause the death of insects, for instance the sticky buds of the horse-chestnut (*Æsculus hippocastanum*), without thereby receiving, as far as we

can perceive, any advantage; but it was soon evident that Drosera was excellently adapted for the special purpose of catching insects, so that the subject seemed well worthy of investigation.

Emma Darwin described CD's work in a letter to Mary Elizabeth Lyell (see n. 1, above). She wrote: 'At present he is treating Drosera just like a living creature, and I suppose he hopes to end in proving it to be an animal.' (*Emma Darwin* 2: 177).

4 *Malaxis* is a genus of orchids that resembles *Listera* in not having the rostellum permanently attached. Hooker had published a paper on *Listera* several years earlier (Hooker 1854b). CD described the anatomy of the flower of *M. paludosa* in *Orchids*, pp. 130–9, having dissected specimens provided by 'Mr. Wallis, of Hartfield, in Sussex' (*Orchids*, p. 130 n.). William Wallis was a surgeon in Hartfield (*Post Office directory of the six home counties* 1859). CD found that the flowers of *Malaxis* secreted a drop of viscid fluid on the rostellum, which served to catch the pollinia and attach them to an insect's proboscis (*Orchids*, pp. 136–7). Hooker had described a similar phenomenon in *Listera* (Hooker 1854b, p. 259, and *Orchids*, p. 142).

5 The author of the review of *Origin*, which appeared in the *London Review and Weekly Journal of Politics, Literature, Art and Society* 1 (1860): 11–12, 32–3, 58–9, has not been identified.

6 See letters to J. D. Hooker, [17 July 1860] and 7 August [1860], and to Asa Gray, 22 July [1860].

To James Dwight Dana 30 July [1860]

Down Bromley Kent [Hartfield]
July 30th

My dear Sir

I received several weeks ago your note telling me that you could not visit England, which I sincerely regretted, as I should most heartily have liked to have made your personal acquaintance.—[1] You gave me an improved, but not very good, account of your health. I shd at some time be grateful for a line to tell me how you are.—

We have had a miserable summer owing to a terribly long & severe illness of my eldest girl, who improves slightly but is still in a precarious condition.—

I have been able to do nothing in science of late. My kind friend Asa Gray often writes to me & tells me of the warm discussions on origin of species in the U. States.— Whenever you are strong enough to read it, I know you will be dead against me, but I know equally well that your opposition will be liberal & philosophical. And this is a good deal more than I can say of all my opponents in this country. I have not yet seen Agassiz's attack;[2] but I hope to find it at home, when I return in few days, for I have been for several weeks away from home on my daughter's account. Prof. Silliman sent me an extremely kind message by Asa Gray that your Journal wd be open to a reply by me;[3] I cannot decide till I see it, but on principle I have resolved to avoid answering anything, as it consumes much time, often temper, & I have said my say in the Origin.— No one person understands my views & has defended them so well as A. Gray;—though he does not by any means go all the way with me.— There was much discussion on subject at B. Assoc. at Oxford; & I had many defenders & my side seems (for I was not there) almost to have got the best of the battle.— Your correspondent & my neighbour J. Lubbock goes on working at such spare time as he has.—

This is an egotistical note; but I have not seen a naturalist for months. Most sincerely & deeply do I hope that this note may find you almost recovered.

Pray believe me | Yours very truly | C. Darwin

Endorsement: 'July | 60'
Dana family papers, Manuscripts and Archives, Yale University Library

[1] Dana was convalescing in Italy following a breakdown in his health (*DAB*). CD had sent Dana a presentation copy of *Origin* (see Appendix III and *Correspondence* vol. 7, letter to J. D. Dana, 11 November [1859]). For CD's concern about Dana's health and his wish that Dana should not read *Origin* until he was well enough, see *ibid.*, letter to J. D. Dana, 30 December [1859].

[2] Louis Agassiz criticised CD's views in the third volume of his *Contributions to the natural history of the United States of America* (Agassiz 1857–62). The discussion, taken from advance sheets of the volume, was printed in the *American Journal of Science and Arts* 2d ser. 30 (1860): 142–55; it was also printed in *Annals and Magazine of Natural History* 3d ser. 6 (1860): 219–32. There is an annotated copy of the *American Journal of Science and Arts* article in the Darwin Pamphlet Collection–CUL. For CD's opinion of Agassiz's 'attack', see the letter to T. H. Huxley, 8 August [1860].

[3] Benjamin Silliman Jr, professor of chemistry at Yale University, was Dana's brother-in-law. He and Dana co-edited the *American Journal of Science and Arts*.

To W. E. Darwin [30 July 1860][1]

at Miss Wedgwoods | Hartfield | Tonbridge Wells
Monday

My dear William.

It was extremely considerate & kind of you to think of offering to come here, my dear old fellow; but all the males of family, old & young, return home on Thursday, & that will relieve poor Gingo's ennui.[2] Miss Ludwig,[3] Lizzie[4] & Anne[5] on Friday— Mamma & Etty & Jane[6] on Monday or Tuesday. Etty has had two good days & is improving. Mr Wallace frightened us extremely,[7] but Sir H. H has comforted us: poor Etty will long be an invalid, but we are now too happy even at that poor prospect.—[8]

Your letter has amused us all extremely & was read with roars of laughter. Etty has not yet heard it; but you cannot think what a pleasure your letters are to her; they amuse & cheer her so nicely. I shall copy your account of dialogue before the Bishop & send it to Hooker & Huxley.—[9] I daresay I will send some queries to your friend the Cook. You may tell the Gardener that I have seen an ant's nest in a tree, but it is rare.—[10]

You seem to have a quiet sort of pleasant time at Cambridge & I am glad that you have got into full swing. I sh^d^ have thought that Justinian would have been a choker.—[11] You would find a visit to Skipworth very pleasant I sh^d^ think.[12] I have had very civil letter from old Higgins really regretting he did not see you.—[13] I shall have to write before long to Union Bank, & then I will repay you.— Your trip was very moderate in cost I think.—

All our plans for the future are necessarily utterly vague on account of Etty.

Whenever she can move we shall go to sea-side.— Mamma sends some "corrections of the press."

The Review by Bishop of Oxford + Owen in last Quarterly is worth looking at.— I am splendidly quizzed by a quotation from Anti-Jacobin.—[14] The Naturalists are fighting about the Origin in N. America even more than here, as I see by the printed Reports—

My dear old fellow | Your affect. Father | C. Darwin

DAR 210.6

[1] The date is inferred from the reference to returning home 'on Thursday'. CD's 'Journal' records that he returned to Down from Hartfield on Thursday, 2 August 1860 (Appendix II). There is an accompanying cover in DAR 210.6 with the postmark: 'JY 30 60'.

[2] A nickname for George Howard Darwin, aged 15.

[3] Camilla Ludwig, the Darwin children's governess.

[4] Elizabeth Darwin.

[5] Anne was one of the Darwins' nursery maids.

[6] Jane was also a maid at Down.

[7] Probably William Wallis, the surgeon and parish registrar in Hartfield, Sussex (*Post Office directory for the six home counties* 1859). Wallis supplied CD with orchid specimens during his visit. See letters to J. D. Hooker, 29 July [1860], n. 4, and 7 August [1860].

[8] The Darwins had consulted Henry Holland about Henrietta Emma Darwin's health (see letter to J. D. Hooker, 29 July [1860]).

[9] See following letter.

[10] CD made several important observations on ants' nests and behaviour during his stay at Hartfield. Notes on his observations are in DAR 205.11: 99–102.

[11] Possibly a reference to William's studies in the coming year, his final year at Cambridge University. For William's intention to read law, see the letter to W. E. Darwin, [4 March 1860].

[12] CD may be referring to William's plans for the approaching summer vacation. Septimus Patrick Skipworth, a friend of William's from Christ's College, lived in Caistor, Lincolnshire.

[13] John Higgins was CD's land agent for the farms he owned in Lincolnshire. William had intended to visit the farms (see letters to John Higgins, 13 June [1860] and 21 June [1860]).

[14] [Wilberforce] 1860, p. 255. See letter to J. D. Hooker, [20? July 1860] and n. 3.

To T. H. Huxley [30? July 1860][1]

[Hartfield]

Wolstenholme[2] (tutor of Christ's Coll) told *me* (i.e. my son William) last night;[3] that two Cambridge men, one a blind man named Fawcett,[4] were at the Brit. Assoc.; and after the meeting they happened to be near the B. of Oxford; and the one asked Fawcett whether he thought the Bishop had ever read the Origin; and the blind man shouted out in a loud voice "Oh no, I would swear he has never read a word of it." The Bishop bounced round with an awful scowl and was just going to pitch into him, when he saw that he was blind, and said nothing.—

Copy
DAR 145

[1] Dated by the relationship to the preceding letter.

[2] Joseph Wolstenholme, a fellow of Christ's College, Cambridge, was William Erasmus Darwin's mathematics tutor.

[3] See preceding letter.

[4] Henry Fawcett, a fellow of Trinity Hall, Cambridge, lost his sight in a shooting accident in 1858 (*DNB*). The reference is to Samuel Wilberforce, the bishop of Oxford. See also Appendix VI.

To Charles Lyell 30 July [1860]

Hartfield.
July 30th

My dear Lyell

We have been here for 3 weeks on Etty's account, but return home in a few days.— We had Sir H. Holland here a few days ago, as we got much alarmed about poor dear Etty. He gives us hope for future, but says the fever has left mischief in internal organs, & that her recovery will be excessively slow. We have had an unhappy time of it. All this trouble has prevented my writing, which I shd. have done sooner, though I have nothing to say, as I have seen no one (except indeed Hooker for an hour or two at Kew) for an age. I had lots of pleasant letters about Brit. Assoc. & our side seems to have got on very well. There has been as much discussion on other side of Atlantic as on this. No one I think understands whole case better than Asa Gray, & he has been fighting nobly. He is capital reasoner; I have sent one of his printed discussions to our Athenæum, & Editor says he will print it.[1] The Quarterly has been out sometime; it contains no malice, which is wonderful considering Owen's aid (by the way it seems generally admitted that Huxley smashed Owen at Oxford); it quizzes me really in capital style; it makes me say many things which I do not say.— At the end it quotes all your conclusions against Lamarck & makes solemn appeal to you to keep firm in the true faith.—[2] I fancy it will make you quake a little. Owen has ingeniously primed the Bishop with Murchison against you as head of uniformitarians.[3] Owen is really wonderfully clever in his malevolence.—

The only other review worth mentioning, which I can think of is in 3d Nor of "London Review"; by some geologist & favourable for a wonder; it is very ably done, & I shd. like much to know who is author.—[4] I shall be very curious to hear on your return, whether Bronns German Translation of the Origin has drawn any attention to subject.—[5] Huxley is eager about a Natural History Review, which he & others are going to Edit, & he has got so many first-rate assistants, that I really believe he will make it a first-rate production.—[6] I have been doing nothing, except a little Botanical work as amusement.— I shall hereafter be very curious to hear how your tour has answered.— I expect your Book on Geological History of man will with a vengeance be a bomb-shell:[7] *I hope it will not be very long delayed.*

Our kindest remembrances to Lady Lyell.—[8] This is not worth sending, but I have nothing better to say. Yours affectionately | C. Darwin

Endorsement: '1860'
American Philosophical Society (222)

[1] See letters to Asa Gray, 22 July [1860], and to J. D. Hooker, 7 August [1860].
[2] [Wilberforce] 1860, pp. 263–4. See letter to J. D. Hooker, [20? July 1860] and n. 5.
[3] In his review of *Origin* ([Wilberforce] 1860), Samuel Wilberforce cited Roderick Impey Murchison's views and implied that he was a prominent advocate for gradual geological change. CD believed that Richard Owen had supplied Wilberforce with much of the scientific material discussed in the review.
[4] *London Review and Weekly Journal of Politics, Literature, Art and Society* 1 (1860): 11–12, 32–3, 58–9. See also letter to J. D. Hooker, 29 July [1860].
[5] Lyell was travelling on the Continent (see letter to Charles Lyell, 5 [July 1860]). CD refers to Bronn trans. 1860.
[6] See letters to T. H. Huxley, 20 July [1860], and to John Lubbock, 20 July [1860]. Thomas Henry Huxley had recently agreed to become general editor of the new series of the *Natural History Review*; the board of eleven editors also included John Lubbock, George Busk, William Benjamin Carpenter, Philip Lutley Sclater, and Daniel Oliver.
[7] C. Lyell 1863.
[8] Mary Elizabeth Lyell.

To W. B. Tegetmeier 30 July [1860]

Down | Bromley, Kent [Hartfield][1]
July 30th

My dear Sir

One line to thank you for your information on Pigeon-Hatching & on Drones.—[2] Your case is much to the point. I believe, as many believe, that an occasional cross is almost indispensable;[3] Mr Westwood in G. Chronicle threw in my teeth Hive-Bees, & assumed that with them there was no intercrossing, but perpetual intermarrying.[4]

With many thanks | My dear Sir | Yours sincerely | C. Darwin

My daughter still keeps very ill, but improves a little.—

Endorsement: '1860'
New York Botanical Garden Library (Charles Finney Cox collection)

[1] CD was in Hartfield, Sussex, until 2 August 1860 ('Journal'; Appendix II).
[2] See letter to W. B. Tegetmeier, 20 July [1860].
[3] In his paper on crossing among bees, Tegetmeier described the way in which drones (fertile male bees) could move from hive to hive without being attacked. He believed that the bees of any one hive probably included a number of 'foreign' drones, thereby facilitating some cross-breeding between different stocks. This finding contradicted the popular belief that queen bees only mated with the drones of their own hive. See Tegetmeier 1860. CD cited Tegetmeier's information in *Variation* 2: 126.
[4] CD refers to two letters written by John Obadiah Westwood to the *Gardeners' Chronicle* in 1860. The first, published in the *Gardeners' Chronicle and Agricultural Gazette*, 14 January 1860, p. 26, addressed the question of the 'doubtful permanence of cross-breeds'. Westwood there alluded to the fact that CD's theory appeared to be refuted by the hive-bees, in which, he stated, 'no crossing is necessary, even if but a single hive were to be kept from year to year.' CD responded to this point in his letter to the *Gardeners' Chronicle*, [before 21 January 1860]. Westwood wrote again (*Gardeners' Chronicle and Agricultural Gazette*, 11 February 1860, p. 122), reiterating the point that: 'Extensive bee-keepers do not find it necessary to import hives from a distance to keep up their establishments, and thus the species would

keep true, immaterial whether the queens paired with their own subjects or with those of adjacent hives.' CD was attempting to gather further information on the crossing of hive-bees from Tegetmeier and others in order to dispute Westwood's claim (see letters to W. B. Tegetmeier, 17 April [1860], and to *Cottage Gardener*, [before 8 May 1860]).

To Frederick Watkins[1] 30 July [1860][2]

Down, Bromley, Kent [Hartfield]
July 30

My Dear Watkins.

Your note gave me real pleasure[3] Leading the retired life which I do with bad health I oftener think of old times than most men probably do; and your face now rises before me, with the pleasant old expression as vividly as if I saw you.

My book has been well abused, praised and splendidly quizzed by the Bishop of Oxford,[4] but from what I see of its influence on really good workers in science, I feel confident that *in the main* I am on the right road— With respect to your question I think the arguments are valid, showing that all animals have descended from four or five primordial forms; and that analogy & **weak** reasons go to show that all have descended from some single prototype.

Farewell my old friend. I look back to old Cambridge days with unallowed pleasure.

Believe me | Yours most sincerely. | Charles Darwin.

Copy
DAR 148

[1] Watkins had attended Shrewsbury School and Cambridge University with CD (see *Correspondence* vol. 1, letter from Frederick Watkins, 18 September 1831, and letter to Frederick Watkins, 18 August 1832). After graduating Watkins took holy orders and served as Her Majesty's Inspector of Schools in Rotherham, Yorkshire until 1874, when he became archdeacon of York.

[2] Dated by the reference to Samuel Wilberforce's review of *Origin* ([Wilberforce] 1860).

[3] Watkins's letter has not been found.

[4] [Wilberforce] 1860.

To A. G. More 3 August 1860

Down, Bromley, Kent
Aug. 3, 1860

My dear Sir

I thank you most sincerely for sending me the Epipactis:[1] you can hardly imagine what an interesting morning's work you have given me; as the rostellum exhibited a quite new modification of structure. It has been extremely kind of you to take so *very* much trouble for me. Have you looked at the pollen-masses of the Bee-Ophrys?[2] I do not know whether the Epipactis grows near to your house; if it does and any object takes you to the place (pray do not for a moment think me so very unreasonable as to ask you to go on purpose) would you be so kind to watch the flowers for quarter of an hour, and mark whether any insects (and what?) visit these flowers:

Trifling as this may appear I have reason to believe important in fertilisation.

I should suppose they would crawl in by depressing the terminal portion of the labellum; and that when within the flower this terminal portion would resume its former position; and lastly that the insect in crawling *out*, would *not* depress the labellum but would crawl out at back of flower. An insect crawling out of a *recently opened flower* would I believe, have parts of the pollen-masses adhering to the back or shoulder. I have seen this in Listera. How I should like to watch the Epipactis.

With cordial thanks. | My dear Sir, Yours sincerely | C. Darwin

If you can at any time send me Spiranthes or Acera or O. ustulata you would complete your work of kindness.

P.S. *If* you should visit the Epipactis again would you gather a few of the lower flowers, which have been opened for some time and have begun to *wither a little*, and observe whether pollen is *well cleared* out of anther-case. I have been struck with surprise that in nearly all the *lower* flowers sent by you, though much of the pollen has been removed yet a good deal of pollen is left wasted within the anthers. I observed something of this kind in Cephalanthera grandiflora.[3] But I fear that you will think me an intolerable bore.

Copy
DAR 146

[1] CD had asked More to send him specimens of this and other orchids (see letter to A. G. More, 24 June 1860). He thanked More in *Orchids*, p. 95 n., for 'repeatedly sending me fresh specimens of this beautiful Orchis.'

[2] See letters to A. G. More, 24 June 1860 and [30 June 1860].

[3] CD compared the fertilisation mechanisms of *Epipactis* and *Cephalanthera* in *Orchids*, pp. 93–112.

To John Murray 3 August [1860][1]

Down Bromley Kent
Aug. 3[d]

My dear Sir

I suppose that I have to thank you for a copy of the Quarterly, which I found here on my return home.—[2] The article on the Origin seems to me very clever & I am quizzed splendidly; I really believe that I enjoyed it as much as if I had not been the unfortunate butt. There is hardly any malice in it, which is wonderful considering the source whence many of the suggestions came.[3] The Bishop makes me say several things which I do not say, but these very clever men think they can write a review with a very slight knowledge of the Book reviewed or subject in question.—

With my thanks for your kind present | Pray believe me | Yours very sincerely | C. Darwin

I see there is a cancelled page, which I presume contained some great blunder; what sweet revenge it would have been, had but the page been left in!—

Did you ever read such magnificent nonsense as the "strong shudder which ran through all this the world"![4]

John Murray (publishers)

[1] The year is given by the reference to the review of *Origin* in the *Quarterly Review* (see n. 2, below). CD returned to Down from Hartfield, Sussex, on 2 August 1860 (Emma Darwin's diary).

[2] The July issue of the *Quarterly Review* carried Samuel Wilberforce's anonymous review of *Origin* ([Wilberforce] 1860). John Murray, who published *Origin*, was also the publisher of the *Quarterly*. CD's annotated copy of Wilberforce's review is in the Darwin Pamphlet Collection–CUL.

[3] CD believed that Richard Owen had supplied Wilberforce with the scientific details mentioned in the review.

[4] [Wilberforce] 1860, p. 260. The passage reads:

> But we can give him a simpler solution still for the presence of these strange forms of imperfection and suffering amongst the works of God.
>
> We can tell him of the strong shudder which ran through all this world when its head and ruler fell.

To A. G. More 5 August 1860

Down, Bromley, Kent.
Aug. 5, 1860.

My dear Sir

I am infinitely obliged for your most clearly stated observations on the Bee-Orchis.[1] It is now perfectly clear that something removes the pollen-masses far more with you than in this neighbourhood. But I am utterly puzzled about the foot-stalk being so often cut through.[2] I should suspect snails. I yesterday found 39 flowers and of them only 1 pollen-mass in 3 flowers had been removed, and as these were extremely much withered flowers I am not quite sure of the truth of this. The wind again is new element of doubt. Your observations will aid me *extremely* in coming to some conclusion. I hope in a day or two to receive some day moths, on the probosces of which I am assured the pollen-masses of the Bee-Orchis still adhere.

Did you ever read Dr Hooker's paper in Philosph. Transactions, 1854, p. 259, on the curious structure of the Rostellum in Listera.[3] Dr H. is considerably mistaken about the use of the parts, but his facts are accurate; and it is following out the remarkable modification in the structure and function of the rostellum, that I am so anxious to examine Spiranthes and Epipactis and indeed all Orchids. The paper is really worth reading, though the facts in several respects are far more curious than Hooker suspected. I wrote yesterday to thank you for the *Epipactis*.[4] For the chance of your liking to look at what I have found; take a recently opened flower, drag gently up the stigmatic surface, almost any object (the *side* of a hooked needle) and you will find the cap of the hemispherical rostellum comes off with a touch, and being viscid on under surface, clings to needle, and as pollen-masses are *already* attached to the back of rostellum, the needle drags out much pollen. But to do this, the curiously projecting, and fleshy summits of anther-cases must at some time be pushed back slightly. Now when an insect's head gets into the flower, when the flap of the labellum has closed by its elasticity, the insect would naturally creep out by the back-side of the flower. And mark when the insect flies to another flower with the pollen-masses adhering to it, if the flap of labellum did not easily open and allow free ingress to the insect, it would surely rub off the pollen on the upper petals and so not leave it on stigma. It is to know whether I have rightly interpreted the structure of this whole flower that I am so curious to see how insects act. Small insects I daresay would crawl in and out and do nothing. I hope that I shall not have wearied you with these details.

With my cordial thanks Believe me | My dear Sir | Yours truly obliged | Charles Darwin

If you would like to see a pretty and curious little sight look to *Orchis pyramidalis* and you will see that the sticky glands are congenitally united into a saddle-shaped organ. Remove this under microscope by *pincers applied to footstalk* of pollen-mass and look quickly at the spontaneous movement of the saddle-shaped organs, and see how beautifully adapted to seize proboscis of moth.

Copy
DAR 146

[1] In *Orchids*, pp. 67–8, CD described an experiment performed by More on the bee-orchis (*Ophrys apifera*). More observed that 'in plants growing singly both pollinia were invariably present. But on taking home several plants, from a large number growing in two places, and *selecting plants* which seemed to have had some pollinia removed, he examined 136 flowers: of these ten had lost both pollinia, and fourteen had lost one; so here there seems at first evidence of the pollinia having been removed by their adhesion to insects'.

[2] 'Mr. More found no less than eleven pollinia . . . with their caudicles cut or gnawed through, but with their viscid discs still in their pouches, and this proves that some other animals, not insects, probably slugs, had been at work.' (*Orchids*, p. 68).

[3] Hooker 1854b.

[4] See letter to A. G. More, 3 August 1860.

From T. H. Huxley 6 August 1860

My dear Darwin

I have to announce a new and great ally for you worth all the Owens & Bishops that ever were pupped

Von Bär[1] writes to me thus[2] "Et outre cela je trouve que vous ecrivez encore des redactions Vous avez ecrit sur l'ouvrage de M. Darwin une critique dont je n'ai trouvè que des débris dans un Journal allemand[3] J'ai oublié le nom terrible du journal Anglais dans lequel se trouve votre recension.[4] En tout cas aussi je ne peux pas trouver le journal ici. Comme je m'interesse beaucoup pour les idées de M. Darwin sur lesquelles j'ai parlé publiquement et sur lesquelles je ferai peutêtre imprimer quelque chose—[5] vous m'obligeriez infiniment si vous pourriez me faire parvenir ce que vous avez ecrit sur ces idées

J'ai enoncé les mêmes idées sur la transformation des types ou origine d'espèces que M. Darwin. Mais c'est seulement sur la geographie zoologique que je m'appuie Vous trouverez dans le dernier chapitre du traité "Ueber Papuan und Alfuras" que j'en parle très decidemment sans savoir que M. Darwin s'occupait de cet objet"[6]

The treatise to which Von Bär refers he gave me when over here but I have not been able to lay hands on it since this letter reached me two days ago—

When I find it I will let you know what there is in it[7]

Ever yours faithfully | T H Huxley

Augt. 6th. 1860.

DAR 98 (ser. 2): 31–2

CD ANNOTATIONS
End of letter: 'Rudolph Wagner | Agassiz.'[8] *pencil*

[1] Karl Ernst von Baer, the renowned embryologist, was professor of anatomy at the St Petersburg Academy of Sciences. He had visited Huxley during a trip to England late in the summer of 1859 prior to the publication of *Origin*. Huxley, who had learned of 'the value of development as the criterion of morphological views' from Baer's writings, had translated into English several of Baer's most important embryological texts (T. H. Huxley trans. 1853). See L. Huxley ed. 1900, 1: 163.

[2] 'And besides, I find that you are still writing reviews. You have written a critique of Mr. Darwin's work of which I have found only fragments in a German journal. I have forgotten the terrible name of the English journal in which your review appeared. In any case I cannot find the journal here. As I am much interested in Mr. Darwin's ideas, on which I have spoken publicly and on which I shall perhaps publish something, you would oblige me infinitely if you would send me what you have written on these ideas. I have expressed the same ideas on the transformation of types or origin of species as Mr. Darwin. But it is only on zoological geography that I rely. You will find in the last chapter of the treatise "Ueber Papuas und Alfuren" that I speak of this very positively without knowing that Mr. Darwin was concerning himself with this subject.' (The translation has been modified from that given in Oppenheimer 1959, pp. 295–6.)

[3] Baer may be referring to a notice of *Origin* that appeared in the *Archiv für Naturgeschichte* in 1860. It appeared in a report on recent works in general zoology and the natural history of man prepared by Rudolph Wagner. In addition to CD's book, Wagner also discussed Louis Agassiz's *Essay on classification* (Agassiz 1859) and Baer's essay 'Über Papuas und Alfuren' (Baer 1859), noting that each work dealt with the question of the origin of species (Wagner 1860a, pp. 2–7).

[4] Baer perhaps refers to Huxley's review of *Origin* published in the *Westminster Review* ([T. H. Huxley] 1860b).

[5] Baer repeated his intention to publish something on CD's theory in a subsequent letter to Huxley (Huxley papers MS 10.188 (Imperial College of Science, Technology and Medicine Archives)). See also letter to T. H. Huxley, 2 December [1860]. He did not publish a full account of his views on *Origin*, however, until after the publication of *Descent* in 1871. Baer, who conceived of nature in teleological terms, became an opponent of CD's theory. See Oppenheimer 1959 and Ospovat 1981.

[6] Baer 1859. In this ethnographic work, Baer addressed the species question, stating that the geographical distribution of certain species seemed to indicate a relationship of ancestry. He wrote that 'it seems

apparent from the distribution of animals that also many such species that are now separated and reproduce were not originally separate, that therefore from varieties they have become, systematically speaking, specifically different species.' (translated from Baer 1859, p. 343).

[7] Huxley was apparently unable to find Baer's treatise. In a letter written to an unidentified correspondent on 2 August [1866] (*Calendar* no. 5170), CD stated that he had only seen an extract of this work.

[8] CD's annotations relate to his reply to this letter. See letter to T. H. Huxley, 8 August [1860]. See also n. 3, above.

To J. D. Hooker 7 August [1860][1]

Down Bromley Kent
Aug. 7th

My dear Hooker

Your last letter pleased & cheered me much, as indeed yours always do.— Etty came home last night, not much tired: her abdomen my wife thinks certainly less hard & swollen, which of course is our dreadful fear; & certainly she gains a very little strength.—

I return Watsons nice letter not at all cantankerous:[2] I have not yet read Cybele-supplement.[3] The "conspious zoologist", who says nat. selection does not apply to plants must be J. E. Gray.—[4] By the way Owen has sent me a copy of one of his Reports, which he never did before![5] so he does not want to split with me, only to have malignant pleasure of sneering at me anonymously.

The Editor of Athenæum has inserted Asa Grays capital answers to Agassiz & Co.—[6] Rudolph Wagner, has sent me German pamphet in which he says truth probably lies half-way between Agassiz & self.—[7] It is curious how many will go half-way or quarter way,—all such I look at as booked to go further; for I observe none even hint why they do not go further.

I will gratefully send my notes on Drosera when copied by my copyer: the subject amused me much, when I had nothing to do.—[8]

I am convinced that your Listera discovery is the key-stone to understand the structure of many orchids.—[9] I enclose abstract of facts of two cases. I never so well realised M. Edwards saying "niggard in innovation & prodigal variety"—[10] observe what little change would convert Epipactis to Orchis: the temporary or non-congenital point of attachment in working down to base of pollinia; & to viscous matter (at least greater part) does not come out of the rostellum.[11] It is very idle of me thus to amuse myself; but how I sh^d like to examine a few forms in other divisions. I suspect that the slow movement in one direction which I find in all pollinia of Orchis, would explain the springing out of pollinia in Catasetum. Could you *lend* me a plant in bud & bloom of Catasetum? Have you any orchids, (not very rare) with irritable labellum? Malaxis is rare at Hartfield, but when there again I could get you some plants; & if you wished *much* for them, I could perhaps ask a surgeon there (who found my specimens) to search for you;[12] but if you do not care much perhaps it w^d be better to wait & not ask this favour— Would you like to see my curious movement in pollinia of O. pyramidalis? if so I could send some in tin—by Post: it would not take five minutes to see the movements.—

If you keep to your good intention of answering the Bishop,[13] I have put a few notes about blunders in detail—though whether they will be of any use I know not.—

I agree to all what you say about Daubeny's paper.[14]

I am sorry about your ear; but can hardly sympathise with you **not** having tooth out.—

Farewell | C. Darwin

I hope your poor Baby will improve.—

Can you tell me name of plant, which grew a foot or two in height in my Father's garden, with little white flowers on branching stem, & *the flowers of which used to catch a multitude of flies by their proboscis.* I must try & get this plant to observe it, with respect to intercrossing.—[15]

Please answer this sometime

On reflexion I fear the Catasetum do not grow in pots & would not bear even for short time temperature of sitting room & would be too precious to *lend* me.— Could I buy any *Green-house* orchids in flower do you suppose?

I am sillily & very idly interested in them.—

I have just had note from Huxley that Von Baer goes a great way with me.[16] It seems that he has written on subject, from Geograph. Distribution grounds, before having read my Book. Von Baer counterweighs Owen + Agassiz. The latter has attacked me fiercely but not well.—[17]

DAR 115.1: 72

[1] The year is given by the reference to the *Athenæum* (see n. 6, below).

[2] The letter from Hewett Cottrell Watson has not been found.

[3] Watson 1860. The book was a supplement to Watson's *Cybele Britannica* (Watson 1847–59). CD's copy of the entire work is in the Darwin Library–CUL.

[4] John Edward Gray, keeper of zoology in the British Museum, did not admit the possibility of transmutation, linking CD's theory with the discredited views of Jean Baptiste Pierre Antoine de Monet de Lamarck (Gunther 1975, pp. 453–5). CD reported Gray's initial, sarcastic response to his views in a letter to Hooker in November 1859 (see *Correspondence* vol. 7, letter to J. D. Hooker, 14 December [1859]).

[5] Probably R. Owen 1859b, a report delivered at a meeting of the British Association for the Advancement of Science entitled 'On the orders of fossil and recent Reptilia, and their distribution in time'. There is an annotated copy in the Darwin Pamphlet Collection–CUL, bound in sequence with other pamphlets received during 1860.

[6] CD had forwarded an extract from the *Proceedings of the American Academy of Arts and Sciences* 4 (1860): 411–15, sent to him by Asa Gray, to William Hepworth Dixon, the editor of the *Athenæum.* The extract was printed in the *Athenæum*, 4 August 1860, p. 161. See also letters to J. D. Hooker, [17 July 1860], to Asa Gray, 22 July [1860], and to Charles Lyell, 30 July [1860]. In the extract, Asa Gray defended CD's views against the critical remarks made at a meeting of the American Academy of Arts and Sciences by John Amory Lowell, Francis Bowen, and Louis Agassiz. Gray stated that 'he had no doubt that variation and natural selection would have to be admitted as operative in nature,

but were probably inadequate to the work they had been put to.' (*Proceedings of the American Academy of Arts and Sciences* 4 (1860): 411). CD's copy of the offprint is in the Darwin Pamphlet Collection–CUL and is marked: 'This has been printed in Athenæum'.

[7] Wagner 1860b. There is a copy of the pamphlet in the Darwin Pamphlet Collection–CUL.

[8] See letter to J. D. Hooker, 29 July [1860].

[9] CD hoped to show that there was a gradation in the structure of the reproductive organs of orchids from the simplest to the most complex. Hooker, in his paper on the rostellum of *Listera*, had pointed out the close structural relationship and apparent gradation of form between these structures in different orchid species (Hooker 1854b, p. 263).

[10] CD cited this remark by Henri Milne-Edwards in his 'big book' on species (*Natural selection*, p. 354) and in *Origin*, p. 194. His source was Milne-Edwards 1851, p. 9. An annotated copy of this work is in the Darwin Library–CUL.

[11] CD stated that different species of orchid differ as to whether or not the pollinia are congenitally attached to the rostellum (see *Orchids*, pp. 314–15). CD described the gradations and alterations in the forms of attachment of the pollinia in *Orchids*, pp. 314–33, concluding his discussion with the observation that: 'An ancient form, combining most of the characters, but in a less developed state, of Goodyera, Epipactis, and Spiranthes, would, by further slight modifications, give rise to the whole tribe of the Ophreæ.' (*Orchids*, p. 333). The variability of *Epipactis* and its possible gradation towards the Ophreae is further discussed in *Orchids*, pp. 97–8.

[12] William Wallis, a surgeon in Hartfield, Sussex, had provided CD with orchid specimens. See letters to J. D. Hooker, 29 July [1860], and to W. E. Darwin, [30 July 1860].

[13] [Wilberforce] 1860. Hooker apparently abandoned his plan to answer Samuel Wilberforce in print.

[14] Daubeny 1860. See letter to J. S. Henslow, 16 July [1860], and Appendix VI.

[15] This and the following paragraphs were written on a separate slip of paper and enclosed with the letter. CD's allusion to Karl Ernst von Baer confirms that the slip was a continuation of his letter (see preceding letter).

[16] See preceding letter.

[17] See following letter.

To T. H. Huxley 8 August [1860]

Down Bromley Kent
Aug 8th

My dear Huxley

Your note contained magnificient news & thank you heartily for sending it me.—[1] Von Baer weighs down with a vengeance all the virulence of Owen & weak arguments of Agassiz. If you write to Von Baer for Heaven sake tell him that we should think one nod of approbation on our side of the greatest value; & if he does write anything beg him to send us a copy;[2] for I would try & get it translated & published in the Athenæum & in Silliman to touch up Agassiz.— By Jove how it would rile Owen! I am getting very spiteful towards that grand Seigneur. The other day he sent me a copy of one of his Reports!!![3]

Have you seen Agassiz's weak metaphysical & theological attack on the Origin in last Silliman;[4] I would send it you, but apprehend it would be less trouble for you to look at it in London than return it to me.— R. Wagner has sent me a German Pamphlet, giving an Abstract of Agassiz Essay on Classification "Mit Rucksicht auf Darwin's Ansichten &c &c".—[5] He wont go vy "dangerous lengths" but thinks the truth lies half way between Agassiz & the Origin.—[6] As he goes thus far he will nolens & volens have to go further. He says he is going to review me in yearly Report.[7]

My good & kind agent for the propagation of the Gospel ie the Devil's gospel.— Ever yours | C. Darwin

Endorsement: '[1860]'
Imperial College of Science, Technology and Medicine Archives (Huxley 5: 133)

[1] Letter from T. H. Huxley, 6 August 1860.
[2] See letter to T. H. Huxley, 2 December [1860].
[3] R. Owen 1859b. See preceding letter.
[4] Louis Agassiz reviewed *Origin* in the *American Journal of Science and Arts*, popularly known as 'Silliman's journal' after the founding editor Benjamin Silliman. There is an annotated copy of the review (Agassiz 1860) in the Darwin Pamphlet Collection–CUL, presented to CD by Benjamin Silliman Jr. Agassiz's critique was drawn from a discussion of the species question in his forthcoming third volume of *Contributions to the natural history of the United States* (Agassiz 1857–62).
[5] Wagner 1860b. A separately paginated offprint of the article, which was first published in the *Göttingische gelehrte Anzeigen*, is in the Darwin Pamphlet Collection–CUL. In the article, Rudolph Wagner discussed points relating to the species question brought forward by Agassiz in the first volume of Agassiz 1857–62, reprinted separately as Agassiz 1859. Wagner received a letter from CD in response to his pamphlet, a passage from which Wagner later published (Wagner 1862, p. 167): 'Although You are far from agreeing with me I thank You by heart for the liberal and most kind way in which You allude to it. All that I can hope and expect is, that my views should be fairly considered'.
[6] Wagner 1860b, p. 795. The passage is marked in CD's copy.
[7] Wagner contributed a report to the 1861 volume of the *Archiv für Naturgeschichte* on recently published works in general zoology and the natural history of man; he devoted a considerable portion of this report to a discussion of *Origin*. See Wagner 1861.

To A. G. More 9 August 1860

Down, Bromley, Kent.
Aug. 9, 1860.

My dear Sir

In case you visit Epipactis, will you make one other trifling observation for me. Though the distal end of labellum was nearly closed in flower sent me, perhaps it may be fully open in sunshine; and if so would you try whether irritation with stalk of grass on the jointed part or on the odd crumpled orange central part causes any movement, slow or quick. I know that it is ten thousand to one against any such movement, but as the labellum in some orchids is irritable, it would be worth trial. It is possible that the labellum may be irritable as well as elastic.[1] The only chance of seeing insects at work would be the first bright day after this miserable weather or a bright gleam of few hours in middle of one of our gloomy days.[2]

I see that you have lately published on the Flora of the I. of Wight.[3]

My dear Sir | Yours truly obliged | Charles Darwin

Copy
DAR 146

[1] The elasticity of the labellum of *Epipactis*, which compels insects to brush against the rostellum when retreating, is described in *Orchids*, pp. 98–100. More is cited as having performed experiments on the movements of the labellum, although this particular inquiry about irritability is not mentioned (*Orchids*, pp. 99, 101–2 n.). CD believed that the labellum of *Epipactis* might be irritable by analogy with other species of orchid: 'In certain exotic Orchids the labellum suddenly moves and catches insects as if in a box' (*Orchids*, p. 342).

[2] The summer of 1860 was notorious for its extended period of bad weather. See *Orchids*, p. 39, in which CD writes of the 'injurious effects of the extraordinary cold and wet season of 1860'.

[3] More contributed the chapter on 'Flowering plants and ferns' to Edmund Venables's guide to the Isle of Wight (Venables 1860, pp. 466–502).

To Asa Gray 11 August [1860][1]

Down Bromley Kent
Aug. 11th.—

My dear Gray

On my return home from Sussex about a week ago I found several articles sent by you.— The 1st article from the Atlantic which I am very glad to possess.[2] By the way the Editor of Athenæum has inserted your answer to Agassiz, Bowen & Co. & where I therein read them;[3] I admired them even more than at first. They really seem to me admirable in their condensation, force, clearness & novelty.

I am surprised that Agassiz did not succeed in writing something better. How absurd that logical quibble;—"if species do not exist how can they vary?"[4] As if anyone doubted their temporary existence. How coolly he assumes that there is some clearly defined distinction between individual differences & varieties. It is no wonder that a man who calls identical forms when found in two countries distinct species, cannot find variation in nature.[5] Again how unreasonable to suppose that domestic varieties selected by man for his own fancy (p. 147) shd. resemble natural varieties or species. The whole article seems to me poor: it seems to me hardly worth a detailed answer (even if I could do it, & I much doubt whether I possess your skill in picking out salient points & driving a nail into them) & indeed you have already answered several points. Agassiz's name, no doubt, is a heavy weight against us; but yesterday I heard that a man, whom I believe to be greater than Agassiz, viz Von Baer goes a long way (how far I know not) with me, & has spoken out publickly & will probably publish.[6] R. Wagner has published, also, in Germany an abstract of Agassiz's Essay on Classification, & says he believe the truth lies between us two;[7] & this will make A. very savage, I shd. think.—

If you see Prof. Parsons, will you thank him for the extremely liberal & fair spirit in which his Essay is written.—[8] Please tell him that I reflected much on chance of favourable monstrosities (ie great & sudden variations) arising.[9] I have, of course, no objection to them; indeed it wd. be great aid; but I did not allude to subject, for after much labour I could find nothing which satisfied me of the probability of such occurrences. There seems to me in almost every case too much, too complex, & too beautiful adaptation in every structure to believe in its sudden production. I have alluded under head of beautifully hooked seeds to such possibility.[10] Monsters are apt to be sterile, or *not* to transmit monstrous peculiarities. Look at fineness of gradation in the shells of successive *sub-stages* of same great formation. I could give many other considerations which made me doubt such view.— It holds to certain extent with domestic productions no doubt, where man preserves some abrupt change in structure. It amused me to see Sir R. Murchison quoted as a judge of affinities of animals;[11] & it gave me a "cold shudder" to hear of anyone speculating about a true Crustacean giving birth to a true Fish.![12]

Since I wrote last I received your note of July 10th, in which you confirm the capital pig story:[13] (I fear Prof. Wyman will never give me his facts about striped Horses)[14] Also you give me valuable hints about diœcio-dimorphous flowers.—[15] I was forced to gather my cowslip seed too soon, & I am all at sea about the differences in fertility; but I shall try & work out case better next summer.—

My poor daughter is decidedly better, though still very ill & weak. But I hope the organic mischief suspected by the Doctors, consequent on the fever, is slowly righting. There was fluid in the abdomen, but this seems to have absorbed; but there is still some hardness. We have had a miserable time of it.—

Farewell my kind friend | Yours most truly | C. Darwin

Gray Herbarium of Harvard University (35)

[1] Dated by the reference to reviews of *Origin*.

[2] [Gray] 1860b, pp. 109–16. There is an annotated copy in the Darwin Pamphlet Collection–CUL.

[3] *Athenæum*, 4 August 1860, p. 161. See letters to Asa Gray, 22 July [1860], and to J. D. Hooker, 7 August [1860].

[4] Agassiz 1860, p. 143. The sentence reads: 'If species do not exist at all, as the supporters of the transmutation theory maintain, how can they vary? and if individuals alone exist, how can the differences which may be observed among them prove the variability of species?' In his copy of the review (Darwin Pamphlet Collection–CUL), CD marked the passage and added in pencil: 'exist only temporarily'.

[5] See Winsor 1979.

[6] See letter from T. H. Huxley, 6 August 1860.

[7] Wagner 1860b. See letter to T. H. Huxley, 8 August [1860] and n. 5.

[8] Theophilus Parsons, professor of law at Harvard University, reviewed *Origin* in the *American Journal of Science and Arts* (Parsons 1860). Gray apparently sent the review to CD: there is an annotated copy in the Darwin Pamphlet Collection–CUL.

[9] In his review, Parsons asked whether it was necessary for CD to limit himself to small, gradual changes in organisms (Parsons 1860, p. 2). CD marked the passage in his copy and added: 'I cd find no facts— Monsters apt to be sterile Such fine gradation in many groups—look at shells in closely related formations—look at recent species.—'

[10] *Origin*, p. 392.

[11] Roderick Impey Murchison's expertise was primarily in structural geology. The passage in which Parsons quoted Murchison is marked in CD's copy with two exclamation marks (Parsons 1860, p. 5).

[12] Parsons cited a remark from Murchison in which he suggested that fossil flying fishes (*Pterichthys*) and 'the curious genus Cephalaspis *form the connecting links* between crustaceans and fishes.' (Parsons 1860, p. 5).

[13] Gray had sent CD information from Jeffries Wyman on the black pigs of Florida (see letter to Asa Gray, 3 April [1860]). The 'story' was given in *Origin* 3d ed., p. 12.

[14] See letter from Jeffries Wyman, [*c.* 15] September 1860.
[15] See letter from Asa Gray, [10 July 1860].

To Charles Lyell 11 August [1860]

Down Bromley Kent
August 11th

My dear Lyell.

I was very glad to get your letter.[1] We have returned home about a week.[2] Etty stood the journey well & has decidedly improved a little. One of her most dangerous symptoms has disappeared; but the pulse keeps sadly too high. The Doctors, however, all keep sanguine. They say her face does not look like serious organic mischief.— Hope on is our motto; but of course all this anxiety has much interrupted my work.—

I have laughed at Woodward thinking that you were a man who could be influenced in your judgment by the voice of the public;[3] & yet after mentally sneering at him, I was obliged to confess to myself that I had had fears, what the effect might be of so many heavy guns fired by great men.—[4] As I have (sent by Murray) a spare Quarterly R. I send it by this post; as it may amuse you.—[5] The anti-jacobin part amused me.— It is full of errors; & Hooker is thinking of answering it.[6] There has been a cancelled page; I sh^d like to know what gigantic blunder it contained. Hooker says that Owen has played on the Bishop & made him strike whatever note he liked; he has wished to make the article as disagreeable to you as possible. I will send the Athenæum in a day or two.[7]

There has been a second discussion before the American Academy, in which A. Gray argued capitally, but my copy was imperfect so I could not send it to the Athenæum.[8]

By the way the other day Owen sent me a copy of one of his Reports,[9] so he does not wish to come to quarrel with me.

As you wish to hear what Reviews have appeared, I may mention that Agassiz has fired off shot in last Silliman—not good at all—denies variations & rests on perfection of geological evidence.[10] Asa Gray tells me that a very clever friend has been almost converted to our side by this Review of Agassiz's.[11] Rudolph Wagner has published in Germany abstract of Agassiz Essay on classification in relations to "Darwins ansichten", & concludes that the truth lies between us;[12] & this will make Agassiz savage. Talking of Germany Huxley tells me the grand news that the truly great Von Baer is much interested with the Origin & goes a long way with us; & that he has spoken publicly & in letter to Huxley says he will perhaps publish on subject.—[13]

Prof. Parsons has published in same Silliman a speculative paper correcting my notions—worth nothing.—[14] In Highland Agricult Journal, there is Review by some entomologist—not worth much.—[15] This is all that I can remember; I forget whether I said that there was a third article *very good* & geological & favourable in

London Review;[16] I cannot think who author can be.— As Huxley says the platoon-firing must soon cease. Hooker & Huxley & Asa Gray, I see, are determined to stick to the battle, & not give in.— I am fully convinced that whenever you publish, it will produce great effect on all trimmers, & on many others.— By way I forgot to mention Daubeny's pamphlet—**very** liberal & candid; but scientifically weak.—[17]

I believe Hooker is going nowhere this summer: he is excessively busy. M[rs] H. is at Worthing, & their Baby has been very delicate & in precarious health. He has written me many most nice letters.— I shall be very curious to hear on your return some news of your Geological doings. Talking of geology, you used to be interested about the "pipes" in the chalk.—[18] About 3 years ago a perfectly circular hole suddenly appeared in a flat grass field to everyone's astonishment & was filled up with many waggon loads of earth; & now 2 or 3 days ago, again it has circularly subsided about two feet more.— How clearly this shows what is still slowly going on.—

This morning I recommenced work & am at dogs— when I have written my short discussion on them; I will have it copied & if you like you can then see how the argument stands about their multiple origin. As you seemed to think this important; it might be worth your reading; though I do not feel sure that you will come to same probable conclusion that I have done.—[19]

By the way, the Bishop makes a very telling case against me by accumulating several instances, where I speak very doubtfully; but this is very unfair, as in such cases, as this of the Dog the evidence is & must be very doubtful.—

Good night with kindest remembrances to Lady Lyell & all your party— | Believe me my dear Lyell | Ever yours most truly | C. Darwin

Endorsement: '1860'
American Philosophical Society (223)

[1] Lyell's letter has not been found.

[2] CD returned to Down from Hartfield, Sussex, on 2 August 1860 ('Journal'; Appendix II).

[3] Samuel Pickworth Woodward's remark may have been made in a letter to Lyell, who was travelling on the Continent. Woodward, who reviewed scientific works for the *Critic*, was known to be a firm believer in the idea of design in nature. Lyell later stated that Woodward was 'the best arguer I have met with against natural selection and variation. He puts conchological difficulties against it very forcibly.' (K. M. Lyell ed. 1881, 2: 364).

[4] CD refers to comments made in the closing pages of Samuel Wilberforce's review in the *Quarterly Review* ([Wilberforce] 1860) regarding Lyell's published opposition to transmutation. See letters to J. D. Hooker, [20? July 1860], and to Charles Lyell, 30 July [1860].

[5] CD read the issue of the *Quarterly Review* that included [Wilberforce] 1860 while he was visiting Hartfield (see letters to J. D. Hooker, [20? July 1860], to T. H. Huxley, 20 July [1860], and to Asa Gray, 22 July [1860]). John Murray, the publisher of the *Quarterly Review* and of *Origin*, had also sent CD a copy (see letter to John Murray, 3 August [1860]).

[6] See letter to J. D. Hooker, 7 August [1860].

[7] CD refers to the 4 August issue of the *Athenæum*, which carried an extract from Asa Gray's defence of CD's views published in the *Proceedings of the American Academy of Arts and Sciences*. The piece had appeared in the *Athenæum* at CD's request. See preceding letter and letter to Asa Gray, 22 July [1860].

[8] See letters to J. D. Hooker, [17 July 1860], and to Asa Gray, 22 July [1860].
[9] R. Owen 1859b. See letters to J. D. Hooker, 7 August [1860], and to T. H. Huxley, 8 August [1860].
[10] Agassiz 1860.
[11] Probably Jeffries Wyman. See preceding letter.
[12] Wagner 1860b. See also letter to T. H. Huxley, 8 August [1860] and n. 5.
[13] See letter from T. H. Huxley, 6 August 1860.
[14] Parsons 1860. See preceding letter.
[15] The reference is to an anonymous review of *Origin* in the *Journal of Agriculture and the Transactions of the Highland and Agricultural Society of Scotland* n.s. 69 (1860): 333–53. There is an annotated copy of the article in the Darwin Pamphlet Collection–CUL.
[16] The unsigned review was published in the *London Review and Weekly Journal of Politics, Literature, Art and Society* 1 (1860): 11–12, 32–3, 58–9. See also letter to J. D. Hooker, 29 July [1860].
[17] Daubeny 1860.
[18] Lyell had been interested in the origin of sandpipes in Chalk deposits since 1839, believing they were formed by the corroding action of water containing carbonic acid (C. Lyell 1839). He described his views in succeeding editions of his *Principles of geology* and *Manual of geology*.
[19] CD refers to his chapter on domestic dogs for *Variation*. Lyell had found it difficult to accept CD's view that the various breeds of dogs had descended from several distinct species rather than from a single progenitor. See the correspondence between the two in October and November 1859 (*Correspondence* vol. 7).

To Robert Monsey Rolfe?[1] 15 August [1860]

Down Bromley Kent
Aug.[t] 15th.—

My dear Lord

We are very much obliged for your Lordship's kind invitation to dinner on Saturday, but I am sorry to say that my health still prevents me going out to a distance in the evening;—and a very great loss it is to me in this instance.—

With our best thanks, I beg leave to remain | Your Lordships | Truly obliged | Charles Darwin

Endorsement: '1860'
Paul C. Richards (dealer)

[1] The recipient is inferred from the reference to 'your Lordship'. Rolfe was created Baron Cranworth in 1850. He lived at Holwood House in Keston, Kent, a few miles from Down. See also *Correspondence* vol. 7, letter to R. M. Rolfe, 10 November [1858].

From Charles Théophile Gaudin[1] [15 August 1860][2]

On a depuis peu d'années une nouvelle race d'abeilles qui ont une trompe plus longue et peuvent sucer le treffle rouge. Une ruche de miel pese jusqu'à 90 lb tandis qu'une bonne ruche de l'abeille ordinaire ne pèse que 60 lb. Cette abeille ne pique pas; elle aime l'homme On l'appelle ligurienne. Elle est cultivée dans les Grisons[3] et un Suisse en a emporté cet été une cargaison aux Etats Unis—

Je serais heureux de procurer à M. Darwin tous les renseignemens qui pourraient lui être agréables sur ce sujet[4]
Charles Th Gaudin

Eglantine, Lausanne

Incomplete
DAR 47: 164

[1] For a translation of this letter, see Appendix I. This incomplete note, written in pencil, was given to CD by the original recipient of the letter, probably Hugh Falconer (see nn. 2 and 3, below).

[2] The date is written at the top of the letter in Hugh Falconer's hand. The letter may have been given to CD when he visited Falconer in London (see letter to Charles Lyell, 28 August [1860]). According to Emma Darwin's diary, CD went to London on 21 August 1860.

[3] Grisons is a canton in south-east Switzerland. At this point, the letter was torn into two parts. The parts were subsequently glued together and attached to another piece of paper, on which notes about the sequence of fossil elephants were written in Falconer's hand. Gaudin, a Swiss palaeobotanist, studied the fossils of the Arno River valley and incorporated some of Falconer's findings in Gaudin 1859. Gaudin refers to comments made about bees and clover in *Origin*, pp. 94–5. The information given in the note was used, without attribution, in the fourth edition of *Origin* (see Peckham ed. 1959, p. 184).

[4] The information in the note relates to CD's researches on the role of insects in the cross-fertilisation of leguminous plants such as clover; it also includes information about a new kind of bee, called the Ligurian bee, that had recently been introduced into Britain. On the first point, see the letter to the *Entomologist's Weekly Intelligencer*, [20 June 1860]; on the second, see the letters to the *Cottage Gardener*, [after 8 May 1860], and to W. B. Tegetmeier, 20 July [1860] and n. 8.

From William Henry Harvey 24 August 1860

Trin. Coll. Dublin—
Aug. 24. 1860

My dear Sir

I have taken advantage of the summer vacation to finish the reading of your book, & considering what has already passed between us on the subject, I feel bound to tell you that my opinions have been, at least modified.[1] It is true that I cannot as yet (probably never shall) receive the theory of Natural Selection as a satisfying explanation of the Origin of species—but I am willing to admit that it explains several facts which are not otherwise easily to be accounted for. Until however something more is known of the inciting causes of the *Variation* & *Correlation of Organs*, which in Nature ever go hand in hand, I can only regard Natural Selection as one Agent out of several;—a handmaid or wetnurse—so to say—but neither the *housekeeper*, nor the mistress of the house.

So long as you deal in generalities I can usually follow your arguments assentingly. Many of the facts brought forward illustrate your general principles & fit well into the place assigned them. But unhappily, my assent is most ready in cases where your data are *fewest*! and I find that particular instances, intended to illustrate how Natural Selection has acted or might be supposed to act in certain cases, almost always incline me to withdraw my assent to the previous argument.

Thus, when—following your argument—I am half inclined to admit the successive development of species from species, I come to your *Illustrations*, instead of my faith in my teacher being confirmed, I am absolutely repelled, & forced to suspect some undetected flaw in his line of reasoning. For instance, the speculation on the bear & the whale, of which I dare say you have heard enough, simply made me laugh.[2] But that on the Ostrich & a bustardlike-bird (p. 134), though less improbable, is almost as *fanciful*; for, if your previous statement of the reason why ground feeding birds have imperfect wings be a *vera causa*, it is just as likely that the Ostrich may be a decreasingly corpulent Dinornis[3] *acquiring* wings in its efforts to escape from a recently selected Lion. The modern Ostrich assuredly does use its wings constantly, both in running, & in flapping its sides in hot weather.— Again, the imagined adaptation of the tube of the red-clover to the proboscis of an improved domestic bee (p. 95) seems to me referable to the times of the "may-bees" only;—and lastly (for it is useless to say more, since the result on my mind has been *uniform*), what you say (p. 392) of insular trees belonging to orders which elsewhere include only herbaceous species, seems to me to be unsupported by sufficient evidence. You cite no particular trees, & I may therefore be wrong in *guessing* that the Orders you allude to are *Scrophularineæ* & *Compositæ* ; & the insular trees the Antarctic *Veronicas* & the arborescent Compositæ of S. Helena, Tasmania &c. But—in S. Africa, *Halleria* (a Scrophularinea) is often as large & woody as an apple tree; & there are several S. African arborescent Compositæ (Senecio & Oldenburghia).[4] Besides, in Tasmania at least, the arborescent Composites are not found "competing with herbaceous plants *alone*, & growing taller & taller by overtopping them" & so being converted "first into bushes & ultimately into trees"; for the most *arborescent of them all (Eurybia argophylla*, the Musk-tree) grows, not among bushes & herbaceous plants but in *Eucalyptus* forests, where it forms an undergrowth to trees so hopelessly taller than itself that competition is impossible.[5] And so of the S. African *Halleria*, which is a tree, among trees. What the conditions of the arborescent *Gerania* of the Sandwich Islands may be I am unable to say, or whether you allude to them. I cannot remember any other instances; nor can I accept your *explanation* in any of the cases I have cited.

The Chapters of your Essay that weigh most with me are the latter ones.— The fact of an orderly succession of forms throughout geological eras suggests the notion of an orderly transmutation of species successively from epoch to epoch.—The facts of geographical distribution, making allowance for residua, seem to me *favourable* on the whole to a theory of successive mutation; particularly as regards the biology of outlying, small islands.— The facts of Embryology & of morphology are favourable to belief in an intimate relationship of organism to organism throughout nature; & this relationship is explicable on the hypothesis of a community of descent: & though these facts are not contrary to a notion of separate creation they are unexplained by it.— But if Creation be the finished work of a single Divine Author a *common-plan* may be anticipated throughout; there will be a "wheel within a wheel", the same idea endlessly repeated & endlessly diversified; the same results arrived at by innumerable different contrivances. There will thus arise, between the different organisms, graduated similarities & diversities, & these will suggest relationship; but the *real* blood-relationship may be no more than that between different patent contrivances for accomplishing the same end. Watt's steam-engine is not necessarily the *parent* of all steamengines.—[6] The major facts of classification therefore, the subordination of group to group, appear to me to be perfectly consonant with a theory of separate creation.

But there are other facts of classification which certainly favour the view of *mutation*, & these are what you have most enlarged on in the early chapters. The tendency of forms to vary; the impossibility of distinguishing between species & variety; in many cases the difficulty of drawing a clear line between genus & genus, order & order, class & class, & even between the vegetable & animal kingdoms; —these are difficulties to the believer in separate acts of creation & are perhaps the strongest evidence in favour of a doctrine of *mutation*. But it does not therefore necessarily follow that the mutation has been effected through "natural selection" *alone*.

For some years back I have contented myself in believing in the *absolute* creation of "natural species" as distinguished from "book-species"; allowing "book-species" to have originated by variation within limits. But I must freely confess that my "natural species" are purely *ideal*, & never can be absolutely made plain to the senses. It is for such species as these that Mr Hopkins contends in Frazers Magazine.[7]

At p. 484. you state your belief "that animals have descended from at most only 4 or 5 progenitors, & plants from an equal or lesser number";—and you admit that analogy suggests that "all organic beings may have descended from some one primordial form, into which life was first breathed by the Creator". Now, if you admit 4 or 5, or even but one, primal organisms, you admit so many "natural species", in the same ideal sense that I have been accustomed to do. The difference between us at starting is merely as to the *number* of originally created forms; a difference in *degree* & not one in *kind*. We both believe that Life was first given by the Creator. Hitherto I have believed in thousands of "natural species"; nor do I yet see any absurdity in so believing. I may yet be driven back, by evidence, step by step, as you have been, into the narrows, & back even to one primordial form; but so long as we hold that form to have been "*created*", we cannot be said to be infidels. To my mind there is no more difficulty in believing in 50, or 500 or 5000 absolute acts of creation, than in believing in one solitary creative act;—nor do I see greater difficulty in believing in many successive acts than in few or many simultaneous acts; but you seem to feel a difficulty in both these cases. What has *once* occurred, may occur more than once.

Again, at p. 483, you ask, "Were all the infinitely numerous kinds of animals & plants created as eggs or seed, or as full grown"? To this it is sufficient to reply; was your primordial organism, or were your 4 or 5 progenitors created as egg, seed or full grown? Neither theory attempts to solve this riddle, nor yet the riddle of the Omphalos.

In a former letter I troubled you with a few remarks on the *geological* difficulties of the Theory,[8] & I have still something to say under this head.

Granting the highly developed condition of Silurian fossils we are driven to seek for the "primordial form" in some much earlier bed, "long before the first bed of the Silurian was deposited". At p. 285, you calculate the denudation of the Weald at 306,662,400 years, or at the lowest 150 to 100 millions of years. We shall not trouble ourselves to compute thereby the enormous length of the *post-silurian* epoch (from the present year, back to the date of the first silurian bed);— but let us endeavour to conceive, guided by palæontological evidence, the probable length of the *pre-silurian*-organic period, from the lowest bed, back to the bed of the Primordial Form. Little as we know of Silurian Animals we know at least that they were greatly diversified, many of them highly organised, and probably very far removed from a "primordial form", & if we could fully unveil the buried world of that era, it would probably compel us to push back our "primordial" some millions of centuries further than the known evidence now obliges us to do.

But how is palæontological evidence to be applied so as to compute a probable lapse of time? Granting that species & genera divaricate in the manner explained by your diagram of divarication,[9] we may observe the number of fossiliferous beds through which *well marked* & *nearly allied* genera extend; remaining mutually distinct below, and undivaricated above. Taking the group of Mollusca, I find, among the fragmentary evidence collected by the palæontologist, that between 20 & 30 molluscan genera which have numerous living species, are represented by fossil species in Silurian strata, & considering our very small acquaintance with Silurian fossils I think this a *large* number. I cannot doubt that many other genera which at present seem to fail a little above the Silurian will yet be traced down into it. Every fresh discovery will add to the lapse of time, required to gather in the divarications to a primordial.

But let us take *Patella* & *Chiton*, two well-marked & nearly allied Silurian genera, having numerous modern as well as ancient species. The fossil Chitons have the same number of valves, similarly arranged & sculptured as their modern congeners, & except the subgenera *Chitonellus* & *Cryptochiton*, neither very different from the type, I know of no generic divarication. Now if *Chiton* & *Patella*, littoral shells, living side by side on the same rocks, have continued with the same habits, the same organization, generically undivaricated since the first bed of the silurian was deposited; may not these same two genera go down through lower strata with the same habits, the same organization & generically disunited to some indefinitely distant primal sea, before a tendency to coalesce exhibit itself; & may they not then go down through an equally indefinite but probably much longer period before an animal half-patella & half-chiton, the common progenitor of both will be found? But suppose us to have reached this *Chitona-patella*: can we believe it to be the common progenitor of all Mollusca? Certainly not, for *Chiton* & *Patella* are far removed from the lowest of the Molluscan series.— But to hunt back for a common progenitor of Molluscs through supposed forks of an imaginary family-tree would be a very hopeless task, and when at length, after fresh myriads of centuries we had traced this parent of Mollusca, we should have to go still further down, & probably to an equally remote era before the common progenitor of Mollusca & Articulata were found;— & this low organism might still be myriads of centuries removed from your ultimate *Protozoon.*

By your theory, the changes from generation to generation are infinitessimal; & to the continued accumulation of such changes through vast periods is due the present diversified condition of the organic world. But till I began this conjectural calculation I had no conception of the smallness of an infinitessimal generic difference. Judging now by the *differences* which may have modified the organization of Chiton & other molluscan genera during the whole post-silurian epoch, I can only liken each small *cyclical* divergence to the distance between two of the striæ on some microscopic test-object! Well may those naturalists who believe in the *cyclical fixity* of a *variable type* be forgiven when they see so large, so ancient & so widely diffused a genus as Chiton is so persistently slow in its progressive development. I am tempted to think that "*festina lente*" ought still to be our motto, in receiving a theory of *unlimited* divarication.

Moreover, we must bear in mind, that as, in imagination, we descend through pre-silurian beds toward our first organism, many circumstances conspire to render every successive step slower & slower. For you tell us that higher & more improved forms *alter quicker*, but low, unimproved forms *endure longest & alter most slowly*. We know how long *Chiton* & other molluscan genera have endured & flourished, & how slowly they have altered, *if altered at all*, in organization. In every step below them therefore we must allow still longer periods of stability, still slower cycles of change.— Besides, at every step downwards, we encounter fewer & fewer forms; hence, there will be less & less "struggle for life" in a continually thinner & thinner population. "Natural Selection" consequently will become feebler & feebler, as the "Struggle for life" diminishes; & this latter, it seems to me, must cease altogether long before the era of your Protozoon be arrived at.

But granting that we are at length arrived at the bottom of the well & have secured our "primordial form", just as it was "created". What was it like?— It is plain that it must have belonged to the *Protozoa*; for it is the original *protozoon* itself. Besides, as it united in itself the *undeveloped* Animal & Vegetable *ideas*, it could not be higher in organization than the simplest animal & vegetable soldered together. A pair of nucleated cells, combined into an atom of sarcode, would answer to this description. But the *size* of each cell might be as large as you please. The largest cells known in the Vegetable Kingdom occur among the Algæ; those of *Valonia Forbesii* are sometimes as big as a pigeon's egg. But remembering the enormous bulk of the animals of the early world, & that the *earliest* types of each group are the *most gigantic*, I think we may (if necessary) make the cells of our primal organism as big as plum-puddings—or as balloons. There is nothing necessarily to limit the size but the tenacity of cellulose & sarcode, & that might vary almost indefinitely.

But probably the enormous size suggested may be objected to, & there is nothing compelling us to any size. The organization *must* have been simple, but the size is an open question. Whatever was the size, large or small, the production of the first organism was a stupendous miracle. The "flashing up of elemental atoms into living tissue" (p. 483) which then occurred is doubtless as wonderful as the creation of a new world.[10] For, however brought about, this is what took place. Certain atoms of carbon, oxygen, hydrogen & azote, changing into cellulose & protoplasm, became a living body, endowed with growth; capable, by feeding, of changing mineral atoms into more cellulose & protoplasm; having *an instinct* to turn to the light & perhaps the germs of indefinitely progressive instincts; and above all, possessing a procreative power, enabling it to transmit from generation to generation of similar organisms, throughout all time, similar powers & capacities. This first organism in fact possessed distinct *personality*; hence (my reason assures me) the Power that called it into being & endowed it with secondary powers must be a **Personality**, & not merely "a law or laws acting around us." And as *every* organism in Nature has its *personality*, so, whether every one were separately created, or the higher derived through the lower, we require a *personal*, creative or moulding Power alike in all cases. Secondary (or physical) laws suffice for all the phenomena of the inorganic world, but life is made up of contingencies which physical laws, unmodified by personal Agency, will not always meet.

And here I must observe that if we allow a Self-originated, Supernatural Power to have *once* acted as above described, we cannot presume to limit the further working of that Power. If the derived generative force of animal or plant may act an indefinite number of times, must not the primary Generating Force be held capable of acting an indefinite number of times? Hence I see no theoretical reason for limiting the number of originally *created* forms, either as to their variety among themselves or as to the number of individuals of each kind that may have been simultaneously or successively called into being. Neither is it of any theoretical consequence whether they were created as eggs or as adults; for the Power that could call *a living tissue* from the dust, must be held capable of, at will, calling up an egg or an adult.

To return to our primordial form; let us see what becomes of him on the theory of natural selection. By theory we have made him the simplest in structure; the feeblest in divaricating tendency (because "unimproved") & there is a probability in favour of his being of small size. By theory also he is alone in the world, because every species starts from a solitary individual or pair. Now granting a high rate of increase to such a being we may safely allow myriads of years or perhaps of centuries to roll by before the world could have been so fully stocked with undivaricated protozoa that there should have arisen any "struggle for life". Millions of ordinary protozoa may congregate without crowding in a cubic foot of water; how many million millions would it require to overstock a primæval, universally extended Ocean?—"Variability" if solely depending on "conditions of life" could effect but little, for protozoic forms flourish equally well on arctic-snows, in temperate & tropical rivers

& seas & in thermal springs. They are the *simplest* of organisms, & the simpler an organism is, the less is it affected by "conditions of life". Any attempt therefore to set "natural selection" in action under such circumstances appears to me to be impossible.

I have endeavoured, by theory, to overcome this difficulty; but granting that "Growth", & "Reproduction" are ordinarily in each animal or plant fixed-average quantities; and considering that "Variability" arising from conditions of life, from use & disuse, is feeble or wholly inoperative in a being placed and constituted as our protozoon; and that a "Struggle for life" is all but impossible to conceive; I have exhausted your Productive "Laws" (p. 490) & have no resource left but to call up the *Unknown Laws of Variation*, those namely which cause an Organism "*to sport*" or diversify itself *unexpectedly.*— Natural Selection, no doubt, is ready to take advantage of such *contingent variations*, but cannot be said either to *explain*, or to *originate* them. If therefore these are *necessary*, at starting, to set natural selection in motion, we are surely calling up a wholly different Agency to any set forth in your theory.

For once that we call in *Unknown* Laws of Variation, "Growth" will no longer be a fixed-average quantity, but may be supposed to vary indefinitely either in the same being or in its immediate progeny, as we see illustrated by sponges & the lower algæ. The Protozoon may be supposed as plastic as an *Amoeba.* Hence divarication from the original Protozoon may immediately commence, & even be rapid, but would be disorderly or monstrous but for another *unexplained* Law, "the correlation of Organs".

By the help of Divarication & Correlation acting together we arrive at the symmetrical diversity of organisms, so obvious in all around us; and with "Reproduction" ever multiplying individuals, & "Inheritance" seizing on & perpetuating new variations, we may conceive the Earth to be rapidly peopled with diversified & diversifying forms. And these may have been *originated*, & have gone on divaricating for long ages before a "Struggle for Life" shall have brought Natural Selection into being. Whilst we have Instruments at command so powerful as "Unknown Laws" (of Variation & Correlation) we no longer feel hopelessly crushed by the weight of 50 or 500 post-silurian epochs, for we see that, time being allowed, we shall at length emerge from the abyss. But without calling in these Unknown & Unexplained laws, organic nature, if it originated in a single "primordial form" appears to me to be doomed to perennial sameness. The primordial form, like the *Protococcus nivalis*, would have reproduced its own likeness from the dawn of life to the present day. Mere lapse of time can effect no change whatever.

But what are these "Unknown Laws of Variation"; what this "Correlation of Organs"? My reason tells me that they are probably fresh revelations of the same Supernatural Power which *originated* our "primordial form", and endowed it with its reproductive, & all its other *powers*. Organic nature, from beginning to end, is a continued miracle, & this is true on both theories; whether each species has been separately created, or whether the past & present conditions of the organic world have arisen by successive divarications of a primæval germ. Every living atom bears its witness to an Everliving, Superintending, Upholding, & Contriving Intelligence. In every variation from type, in every correlation of parts, there are evidences of Creative Power, such as no secondary agency, like that of Natural selection accounts for; and I think also there are evidences of *one* Great Design, beginning at the simplest elementary form, & culminating in the "fearfully & wonderfully made" framework of man.

Once or twice in your essay you object to our attributing to the Creator a way of working similar to that of intelligent man. At p. 188, speaking of the structure of the eye, you say,—"Have we any right to assume that the Creator works by intelligent powers like those of man"?— We have at least some reason for "assuming" that He possesses such powers; for how has man acquired them but through His gift? Nor do I think it presumption to believe that the Infinite Intelligence is ever acting, everywhere throughout creation. And if not, whence arise those wondrous correlations, beginning with the balanced movements of the spheres, & penetrating every particle of matter? Are they merely coincidences? Or, do they come through natural-selection?—

Do not suppose me to mean that I consider you to deny a Superintending Providence. Your admission of a *created* germ of life, & the quotation from Bishop Butler,[11] explanatory of the word *natural* ("what is natural requires & presupposes an intelligent agent to render it so") show that you admit a Supernatural Agency. But it seems to me that in developing the theory of natural selection & referring *every* operation in nature to it, the underworking Power is too generally lost sight of as an element in the problem, & that the continual mention of "usefulness" is the only evidence of theism that the work supplies;—because, an appreciation & perpetuation (or *selection*) of the "useful" presupposes a *vigilant* & *intelligent* agent. I do not suppose you to say, for instance, (p. 235) that "*nature*" is no more conscious than the reforming bees of the saving of wax effected by "sweeping equal spheres at a given distance from each other in a double layer" &c. There is abundant evidence of *intelligence* somewhere; either among the bees (which we do not think) or *Above* & around the bees. But it is surely no *explanation* to say, as you often do: "If so & so were *useful* to any animal in its struggle for life, *it would be easy for natural selection*, by slow degrees, more & more perfectly, to fit the creature to its peculiar habits, & endow it with such & such instincts;"—as in the case of the Ant & the Bee, whose instincts, supposed to be acquired *through inheritance*, are *transmitted* through successive communities of *neuters*! You seem to me to be merely asserting this trueism; Nature or the Author of Nature, can easily effect any preconceived design, & form & endow any creature, as He pleases. Nothing is really *explained* by saying "it would be easy" "if it were useful". And when we strike off such "explanations" as these, with which the book abounds, the facts of organisation *explained* by natural selection will be very seriously diminished.—

If there be any part of the theory of natural selection more firmly established than another, it is what we may call *the law of the strongest*. Dominant races throughout nature will gradually drive out feeble races; "the weakest will go to the wall".[12] In applying this universally recognised principle you strongly insist that *improved* races will continually supplant *unimproved*, & hence you deduce the gradual progress of organic nature from the simple primordial forms up to man. We know what an "improved" race among domestic animals means; we know how it is manufactured, how disseminated & how preserved from degenerating; but we also know that it is feebler in constitution & less fitted to conquer in the struggle for life than its *unimproved* progenitor. Artificially "improved" races could never become dominant, if neglected by man. It is of importance to your theory therefore to determine whether, in Nature, the most *dominant* races are also the **organically** *improved* races? Unless I mistake your meaning you seem very generally, so to argue, & many of your readers understand you to contend for *one continued progress* onward, from the monad to the man, accomplished by the continual extinction of *less improved* forms, by dominant *improved* varieties.

Such appears to me to be the *general* drift of your argument; but the chapters on "use & disuse" allow for occasional retrogression. They teach us, that if an animal, through profitable use of its limbs has risen, in successive generations, from a very low to a very high organization; its degenerating descendants, subjected to changed conditions of life, which would cause them gradually, in successive generations, to lose the formerly acquired *habit*, & the *organs* fitted to those habits, may descend in the scale of being to the state of the unimproved original from which the race ascended. The blind animals, vertebrate & invertebrate of the Kentucky caverns illustrate such retrogressions. You suppose these animals to have "slowly migrated, by successive generations, from the outer world into the deeper & deeper recesses of the caves" (p. 138); and we see that they have already gone so far back as to have *lost* the eyes which their progenitors had *acquired* (through natural selection, as *all* eyes have arisen (p. 136)), and to have had their bodies considerably correlated with their new habitats. Now, they have only to go still further back, and after successive migrations & generations their descendants may lose every *acquired* organ, one by one, till they revert to a protozoic form. But how can we certainly tell that they are migrating inwards, deeper & deeper into the caves? May they not, on the contrary, be migrating *outwards*, *seeking* the light, as almost all known organisms instinctively do;—and if so, they are animals in progress, *acquiring*, not losing organs. Their imperfect eyes are *nascent*, not obliterated eyes. And in the case of the blind-rat, with its well formed though sightless eyeballs we have an animal almost ready for emergence?[13]

But to return to the dominant races: are these the *improved* or the unimproved in *organization*? Viewing organic nature in its widest aspect I think it is unquestionable that the truly dominant races are not those of high, but those of low organization. The simpler any animal's structure is, the less is it dependent on outward conditions; hence, the better is it fitted to conquer in the struggle for life, & to perpetuate its kind to a remote posterity. Cryptogamic plants & invertebrate animals, natures scavengers, are the most dominant and persistent races of the world. The *Protococcus nivalis* is probably one of the oldest inhabitants of the world, younger only than the snow; and the *P. pluvialis*, which scarcely differs, is equally persistent wherever rain water habitually rests, & these two may be called dominant. But as they are harmlessly dominant, we do not heed them. Other animals & plants of low organization by the ravages they cause, teach man himself that his domination over nature is by sufferance of a Higher Power. The *Botrytis* which so nearly destroyed the potatoe; the *Oidium* which has ruined the vinegrowers throughout the world; these, and the locust & the caterpillar, the tapeworm & its kindred, with the innumerable army of intestinal worms & of animal & vegetable parasites, these are the true *dominant* races; the Lords of creation that never struggle for life and that fear no evil. They have nothing to lose by any chance, short of the anihilation of the world, & if the theory of natural selection be true, they have every thing to win.

The law of the strongest therefore most forcibly applies to the wholly unimproved races; in a less degree to those higher in the scale; and gradually diminishes upwards as organization, & sensibility to outward conditions encrease. I object therefore to its being made a vera causa of the *evolution* of organization. It only applies where a peculiar organization, combined with *strength of constitution & adaptability to circumstances*, gives an animal an advantage over its fellows: but in this case I think the dominancy will depend more on the strength of constitution & power of adaptation than on the organization. For we see improved & unimproved races (like the humble & domestic bees) flourishing together without one supplanting the other; & if the humble bee, notwithstanding its wasteful habits, have greater strength of constitution & less susceptibility to varying climate than the domestic bee, it will eventually win.

I have already far transgressed the proper bounds of a letter, & yet have a few more words to say.

When you suppose one species to pass, by insensible degrees into another, so many facts of variation support your view that it does not seem very improbable; but where a generic limit has to be passed, bearing in mind how *persistent* generic differences are, I think we require a *saltus* (it may be a small one) or real break in the chain, namely, a *sudden* divarication.[14] I know you account for genera by the dropping out of supposed intermediate infinitessimals. But we know also that *sudden* divarications do sometimes occur in nature; and it is *possible* that they may be even necessary consequences of repeated & long-continued infinitessimal changes. The tower of Pisa may be infinitessimally receding from its perpendicular for centuries, & may at last, the point of balance being passed, come down bodily with a crash. I have sometimes amused myself, when looking through a kaleidoscope, by turning it so as to make it indicate *genera* & *species*. Starting with any pattern of figure, by very slow turning of the tube you may get successively a great many *modifications* of the figure, without any radical change in the pattern, thus illustrating several species of one genus. But when you have turned the tube so far as to cause such a displacement of the fragments of glass as makes them topple over, a perfectly new pattern will *suddenly* start up & may then by further slow turning be modified, till it in turn shall topple over. Of course this is merely a possible illustration of the gradual succession of species & genera, if they arise by *smaller* & *greater* divarications.

You object, I know, very strongly to *abrupt* changes. But as there are volcanoes & sudden displacements in the mineral world, there may also be abrupt divarications in the organic. Some of the facts of metamorphosis, particularly that of the Cirrhipedes (p. 441) which for a special purpose suddenly acquire & as suddenly lose their compound eyes & natatory legs; and every similar fact that indicates a *special* interference of Creative intelligence implies at least a *break* in the series of changes. Nor can I divest myself of the belief that there is a real *break* between every established species (however originated), & that that break implies the interference of a First Cause. The line of unfertile hybridity offers us some evidence of such breaks; anatomical structure, in cases where the plan of organization is reversed, will probably furnish more exact evidence. The change from hairs to feathers seems a sudden one, inexplicable on the hypothesis that intermediate links have dropped out. The change from coniferous wood-tissue & no ducts to ordinary wood-tissue with ducts, seems also sudden. And so do other changes that imply contrariety or polarity.

I am therefore still strongly of the opinion that, whether all organisms *originated* in but a few primæval types or in a definite, but considerable, number of "natural species"—a question that at present must be left open—that the present aspect of nature is due, not to a succession of minute differences accumulated through inheritance & preserved by "natural selection", but to renewed *interferences* with ordinary generation, which may be attributed either to "Variation & Correlation of Organs", or to a superintending & moulding Intelligence:—& that these interferences may have caused either *sudden* or *gradual* changes in the previously moulded organisms. By admitting the possibility of *sudden* divarication we get rid of those perfectly innumerable forms of life which your hypothesis requires us to believe in, but of whose existence there is so little evidence either in existing nature or among fossils. But the whole subject is at present obscured by difficulties that no proposed hypothesis fully gets rid of. By directing attention to one set of phenomena your theory of natural selection finds favour, but turning to another set I am driven back nearly to my old ground; and then "natural selection" is to me but a new phrase signifying "the order of nature", which may be further translated into "the will of God", & then the theory dissolves into thin air.

As yet therefore I cannot subscribe myself your disciple, but I remain as heretofore | My dear Sir | very faithfully yours | W. H. Harvey

To | Charles Darwin Esq, F.R.S. | &c. &c &c

DAR 98 (ser. 2): 33–40

First CD note:[15]

Sept. 20th 1860 | Prof. Harvey in letter (p. 13) says ['dominant' *del*] the lowest forms, as protozoa, intestinal worms, & parasitic cryptogams are the dominant forms.— These are most [*interl*] widely spread, ['which' *del*] & I imagine have many species in groups, & are infinitely numerous in individuals which truly shows that [*interl*] they are dominant. *in fact they are so, otherwise they would not exist in numbers. [*added*] As far as number of individuals are concerned, I ought to have put in caution "relatively to size of individuals". They probably are really highly dominant, for no doubt they keep down numbers of higher productions; or rather probably *aid* in keeping down certain forms.— To say that cryptogamic forms such as Botrytis are dominant over phanerogams, implies that if they did not exist, the number of phanerogams, would be larger in any country.

I doubt this, for the land is so thickly clothed with plants they could hardly exist. I daresay there would be change in proportions. Under artic climate, on mountains & barren heath where the higher cryptogams abound, they are to certain extent dominant & more heath or poor grasses would grow. So when he speaks of Protococcus being dominant, it can only be dominant under the conditions in which it can live. A **few** fish of one species is yet dominant over millions of infusoria & worms.— I must somehow show that when I talk of plants being dominant I refer only to those other plants, with which it comes in competition. But by dominant I in fact imply that the single species or group is increasing in numbers; can we believe this of low organisms? Again with respect to parasitic worms being dominant.; in a country fully stocked with mammals, would the number increase if no parasitic worms, perhaps they would, for then they would gain an advantage over insects & other vegetable feeders. No doubt parasitic worms are one check; especially when numbers increase. But we know of no case of animal not increasing in numbers, owing to such causes, if food abounds & there be no struggle for life *for food at certain seasons or [*interl*] with beasts of prey

Hence wide distribution, large groups—or /unlimit/ numbers of individuals (disregarding size) do not make a group dominant looking to world or even to a single country, we must include under closely similar conditions, as inhabitants of water—or plants /drawing nourishment/ from soil & breathing air —

The number of intestinal worms beat that of the species on which parasitic—many belong to large group— ['ma' *del*] ranges as widely & if inhabiting other species, may range wider; yet can hardly be said to be dominant over that species unless it exterminated it & then it would cause its own extermination, or unless it greatly reduced its numbers, irrespectively of other grass-feeders or carnivores with which it came into competition.

So grasshoppers ['are' *del*] may be dominant over herbaceous quadrupeds: & insects [*above del* 'they'] no doubt are a highly dominant group.— The expression as used by me is relatively to allied forms; if of two species one c[d] resist intestinal worms better than other species, or live on food on which grass hoppers did not eat *as roots [*interl*], it would be in fact so far dominant.—

I suspect I have used expression, the dominant forms within each class.—

Second CD note:

Every group of organisms which exists in great numbers may be said to be a dominant group.— In Australia Marsupials are dominant mammals—not so ['Ornithorynchus' *del*] in world. My remarks on dominant forms chiefly relate to dominance over other members of the same group, or to those which come into direct competition with ['it' *del*] them.— Under simple conditions of life the lowest forms are often dominant.

Add ['when' *del*] dominant over other members of the same class which live in same region under ['con' *del*] similar conditions of life—or members of other classes with which the dominant forms come into competition or struggle for life

[1] See letters to J. D. Hooker, 20 May [1860], 29 [May 1860], and 30 May [1860].

[2] *Origin*, p. 184. CD deleted the passage from the second edition of *Origin*, although he allowed it to remain in the American edition. Richard Owen, who at first expressed interest in this example (see *Correspondence* vol. 7, letter to Richard Owen, 10 December [1859]), ridiculed it in his review of *Origin* ([R. Owen] 1860a, pp. 517–18).

[3] *Dinornis* is a large extinct bird, remains of which had been found in New Zealand.

[4] Harvey was an expert on South African flora, having resided in Cape Town from 1835 to 1842. During this time he probably also became familiar with the habits of ostriches, as discussed in the

letter. He and Otto Wilhelm Sonder were at work on a flora of South Africa, the first volume of which was published in 1860 (Harvey and Sonder 1860–5).

[5] Harvey had botanised in Australia between 1853 and 1856.

[6] James Watt, the engineer and inventor, had developed and patented a steam-engine in 1769.

[7] Hopkins 1860.

[8] Harvey's letter has not been found.

[9] CD's explanation of the divergence of organisms is in *Origin*, pp. 112, 127–8. The diagram illustrating the principle faces p. 117.

[10] Next to this passage in his copy of the second edition of *Origin*, CD wrote:

> Harvey *& Bronn [*interl*] rightly object, I cannot answer such queries from my primordial form—yet perhaps less difficult for a mere cell—asexual—& generally by division.—

For Heinrich Georg Bronn's discussion of this point, see the letter from H. G. Bronn, 5 October 1860.

[11] CD prefaced the first edition of *Origin* with quotations from Francis Bacon and William Whewell. For the second edition, published in January 1860, he added an additional quotation taken from Joseph Butler's *Analogy of religion* (Butler 1736). The quotation reads (*Origin* 2d ed., p. ii):

> The only distinct meaning of the word 'natural' is *stated, fixed*, or *settled*; since what is natural as much requires and presupposes an intelligent agent to render it so, *i.e.*, to effect it continually or at stated times, as what is supernatural or miraculous does to effect it for once.

[12] Shakespeare, *Romeo and Juliet*, 1.1.18.

[13] For CD's views on the blind animals inhabiting caves in Kentucky and elsewhere, see *Origin*, p. 137. See also letters to Andrew Murray, 28 April [1860] and 28 [April 1860].

[14] Harvey had argued for the operation of saltation in nature, in opposition to CD's theory, in a letter to the *Gardeners' Chronicle and Agricultural Gazette*, 18 February 1860, pp. 145–6. See letters to J. D. Hooker, [20 February 1860], and to Charles Lyell, 18 [February 1860] and 23 February [1860]. In his communication, Harvey cited the case of a monstrous new *Begonia* that had grown in the Royal Botanic Gardens, Kew. He suggested that new species could perhaps similarly arise in an abrupt fashion.

[15] Although the first note is dated 'Sept. 20th. 1860', it clearly relates to this letter. It was marked by CD in brown crayon 'Ch 5 & 6', which refers to the chapters 'On the struggle for existence as bearing on natural selection' and 'On natural selection' in his 'big book' on species (*Natural selection*, pp. 173–274). The second note is on different paper but its content appears to be closely related to the first. The two notes are in DAR 98 (ser. 2): 41–3. See also CD's response to Harvey (letter to W. H. Harvey, [20–4 September 1860]).

To Edward Cresy 25 August [1860][1]

Down Bromley Kent
Aug. 25th

My dear Sir

Thank you for your note.— My object in calling on you was chiefly selfish; for it was to ask a little advice on education of one of my Boys,—on which subject I remembered to have heard you once talk as if you had thought over the subject.—[2] I want to propose to Mrs Cresy & yourself, instead of walking over here, to take us on your way home & come to dinner & sleep here, & in all probability we can send you on to the station at Bromley.[3] But I am bound to tell you that we are a dismal house: my daughter has been very ill for 15 weeks, but is very slowly recovering, yet Mrs. Darwin has to be with her much of her time, & I am on most days a poor wretch unable to sit a whole evening even with my nearest relations; but if you will put up with this & accept good will, it will give us great pleasure if you will come here.—

M[rs] Darwin would have written to M[rs] Cresy, but I have said that I felt sure she would excuse her doing so.—

I hope that you will adopt my plan, instead of walking here & believe me | My dear Sir | Yours very sincerely | Charles Darwin

Private collection

[1] The year is given by an entry in Emma Darwin's diary (see n. 3, below).

[2] CD may have wanted to discuss the future education of Leonard Darwin, aged ten, whom he considered 'rather slow & backward (in part owing to loss of time from ill-health)' (letter to G. V. Reed, 15 September [1860]). He intended to send Leonard to George Varenne Reed for tutoring before entering school. Francis Darwin, who had been tutored by Reed, entered Clapham Grammar School, London, in the summer of 1860. See J. R. Moore 1977.

[3] Emma Darwin's diary records that Edward Cresy and his wife arrived at Down on 18 September 1860. According to CD's Address book (Down House MS), the Cresys lived in Horton Kirby, near Farningham, Kent, about nine miles from Down.

To Charles Lyell 28 August [1860]

Down Bromley Kent
Aug 28th

My dear Lyell.

As Emma is writing to Lady Lyell,[1] I will add a few line; though I have not much to say. I was in London lately for a couple of days,[2] & heard a piece of news, which if you have not heard of it, you will be sincerely sorry to hear. Sir G. Grey on his passage out,[3] having some suspicion, after he was well engaged at whist, suddenly went to his wife's cabin & found her in bed with Capt. (or Admiral?) Keppell![4] The latter has staid at Rio & what a despatch he must have written to explain to the Admiralty why he did not take on the Governor to the Cape. What a wreck of happiness for poor Sir George. No doubt she was formerly guilty in N. Zealand, when that examination about some Officer took place.—

The only scientific man I saw in London was Falconer, who goes to Sicily in October.[5] He was very full about a tiny new species of Elephant from Malta not much larger than a calf!

The only Review worth notice, since I wrote is an *excellent* argumentative one in the *August* Monthly atlantic by Asa Gray.[6] He is assuredly a first-rate arguer & most completely understands the subject. I have been well abused in a Catholic Journal;[7] but the writer assures his readers that the progress of Science does not in the least shake the authority of Moses. I was very glad to hear in London that the Rajah Sir J. Brooke is about again & dining out,—a sort of resurrection.—[8] I have been getting on a bit better with my work lately & to day finished with Dogs;—& it is rather a curious subject; & I am confirmed in my belief that they have descended from several wild stocks.—

I sent you the Athenæum & Quarterly Review, which I hope you received.—

Farewell. I fear that you will not be settled in London for some considerable time.— Ever yours | C. Darwin

Endorsement: '1860'
American Philosophical Society (224)

[1] Emma Darwin's letter to Mary Elizabeth Lyell, dated 28 August [1860], is printed in *Emma Darwin* 2: 177.

[2] Emma Darwin's diary records that CD went to London on 21 August 1860.

[3] George Grey, who had formerly served as the governor of South Australia, of New Zealand, and of the Cape Colony, was reinstated as governor of the Cape in 1859 after having been recalled from the post a few months earlier. He was on his way back to Cape Town from England (*Aust. dict. biog.*).

[4] Eliza Lucy Grey, née Spencer, had married Grey in 1839. CD refers to Henry Keppel, the third son of the Earl of Albermarle and admiral of the fleet. Keppel's invalid wife, Katherine Louisa Keppel, had died in 1859. Although a rumour circulated in London that the pair had been caught in the act of adultery, Grey had in fact only discovered notes indicating a romance. See Dalton 1974.

[5] Hugh Falconer intended to examine sites in Sicily where fossils had been found by Francesco Anca.

[6] [Gray] 1860b, pp. 229–39.

[7] CD probably refers to the *Rambler*, which carried a review of *Origin* in the March 1860 issue, pp. 361–76. According to the *Wellesley index* 2: 779, the review was by the editor, Richard Simpson.

[8] James Brooke, Rajah of Saráwak, had sent CD specimens of domestic animals from the Malay Archipelago (see *Correspondence* vol. 6, letter to W. B. Tegetmeier, 11 February [1857]). Brooke was in England from 1857 to November 1860.

From Charles Lyell[1] 28 August 1860

Augt 28. 1860

The grand argument from absence of mammalia & batrachians in Oceanic islands is probably felt to be strong by Owen as he has not ventured to impugn it and therefore declined to bring it into notice.[2]

It is a very telling argument & I think just & true, tho' if I had been helping Owen & the Bishop I fancy I could have suggested objections but I have not thought them out yet, & have not therefore troubled you with them, thinking I sh^{d} be able to answer them myself.

I remember however it struck me that the putting it even as possible that the Galapagos or Cape de Verds could have been joined to the nearest Continent was dangerous, as in the case (p. 390) they must have retained some of the smaller mammalia. Also if atolls be remnants of sunk continents would not some of the old continental land quadrupeds have survived or have been slowly modified into amphibious & volant species.[3]

Also, if as I believe the Canaries & Madeiras have been islands ever since the Miocene period (& the Azores), would not the theory of Nat. Selection have time to modify any vertebrate animal into the lowest of the Lyencephala in such a lapse of ages.[4] Or if not, does it not give us a sort of scale for the vast geological time required for such a transmutation. No Galapagos saurian, having risen to a mammal. Perhaps you will say that this would be a greater puzzle to the out-&-out

progressionists who may be converted by you to transmutation. Bronn,[5] e.g., for you do not believe in that constant advance, as an ever working tendency, & allow for stationary & retrograde movements.

You would prefer I conceive to derive Eocene mammalia from the Microlestes of the Trias rather from any reptilian of however high a grade e.g. a deinosaurian. You w^d rather conceive a bird to be turned into a mammal than derive the latter from a reptile. The Microlestes, or its nearest living analogues (marsupials) w^d in its embryology go through the bird-state or likeness before it became even a non-placental mammifer?

But if an island be very ancient the bats would be modified into mammalia of other genera & orders long before any bird could be turned into a mammal. Here comes the question, how old is the oldest isolated Oceanic island? The miocene littoral shells of Grand Canary & of Porto Santo give a miocene date to those lands, already in all probability separated from the Continent as now. But there are no bats there.

New Zealand is so large it probably existed from Eocene times as land?? Why did not bats getting there diverge into a dominant mammalian volant fauna & some lose their wings & become non-volant mammalia? Or why did not seals & marine mammalia turn terrestrial in their habits if they had such a geological period as the post-miocene for nat^l selection to work in. The seals swarmed in Madeira before Man drove them away. This difficulty strikes me as greater in as much as the rate of change in mammalia is more rapid than in inferior grades & there have been in Europe several changes of mammalia since the Upper Miocene period. The adaptation of a seal which is already half terrestrial, requires less time than the conversion of Hearn's bear into a whale which they are never tired of quizzing you for.[6] I shall be glad if you will tell me whether these objections are of any value. Probably you have not thought the evidence of the antiquity of any islands good for much. Yet it seems necessary to grant a good deal of time to cause so many endemic shells in Madeira & Porto Santo & land shells would change much more slowly than mammalia. Still I grant that the conversion of a phoca into a terrestrial quadruped is very different from one helix into another, but ought not the seals to have begun in ancient islands to keep somewhat longer out of water. These are perhaps unfair, imaginary difficulties & ought not to weigh against an hypothesis which explains so much as does transmutation in the case of absence of mammalia & frogs from remote islands while they swarm in islands which we know have been united in newer pliocene times.

If there was time in Galapagos to convert a marine Amblyrhincus into a land species or vice versa. why not a phoca into a land species or other cetaceae into terrestrial. Did the land Amblyrhincus come first into being & of what S. American Saurian was he or his aquatic congener a modification? If all these questions apply equally against independent acts of Creation they will not be very damaging. But when the free agency of the First Cause is allowed to come into play, it may be said that no mammalia were wanted because it was foreseen that Man would come with his attendant train of useful mammals, taking even the camel with success into the Canaries.

Copy
Kinnordy MS, Charles Lyell's journal IV, pp. 164–71

[1] The text of the letter has been taken from a draft, in Mary Elizabeth Lyell's hand, in Lyell's scientific journal. It is also printed in Wilson ed. 1970, pp. 467–9. The copy is headed: 'Letter to Darwin from Rudolstadt'. Lyell was travelling on the Continent: from 2 to 8 August 1860 he was in Rudolstadt, Germany (*ibid.*, p. 490 n. 88).

[2] CD discussed the geographical distribution of Mammalia and batrachians in *Origin*, pp. 393–5. Lyell refers to Richard Owen's assistance in the writing of Samuel Wilberforce's review of *Origin* ([Wilberforce] 1860).

[3] Lyell refers to CD's theory of the origin of coral reefs, according to which coral atolls had been formed as the seabed subsided (see *Coral reefs*). Apparently Lyell had forgotten that CD no longer believed that groups of atolls indicated the location of a former landmass (see *Correspondence* vol. 6, letter from Charles Lyell, 17 June 1856, and letter to Charles Lyell, 25 June [1856]).

[4] According to Richard Owen, Lyencephala, including the monotremes and marsupials, were the most primitive mammals (R. Owen 1859a).

[5] Heinrich Georg Bronn, who prepared the German translation of *Origin*, gave his reasons for believing in the progressive introduction of species in Bronn 1858b. See also Junker 1991.

[6] The bear–whale story in the first edition of *Origin*, p. 184, was derived from a remark in a book by Samuel Hearne (Hearne 1795).

To J. D. Hooker 31 [August 1860][1]

Down Bromley Kent
31st

My dear Hooker

I know you are frightfully busy; but summer is passing, & I am very curious on point in enclosed paper, would you get your scientific gardener or some one to make the small enclosed observation.[2] It is merely the *rate* of closing of leaves of Australian Drosera; if they do close.—[3] I am still working at an imported plantation of Drosera;[4] & really one point turns out very curious. The leaves are first rate chemists & can distinguish even an incredibly small quantity of any nitrogenised substance from non=nitrogenised substances.[5] I won't write more, as I know how busy you are.—I hope you are *all* well.—

C. D.

Etty progress steadily, but very very slowly.—

☞ Do not answer this till you have leisure.—

DAR 115.1: 71

[1] Dated by the relationship to the letter to J. D. Hooker, 2 September [1860].

[2] Daniel Oliver was Hooker's assistant in the herbarium at the Royal Botanic Gardens, Kew. See also letters to J. D. Hooker, 2 September [1860], and to Daniel Oliver, 11 September [1860].

[3] During his research on *Drosera rotundifolia* carried out in Hartfield in July 1860, CD ascertained that the sticky tentacles borne on the upper surface of the leaf bend over to trap insects. He found that the

rate of inflection for the common sundew varied between one and five hours. At this early stage in his study, CD probably did not know whether the Australian sundew (*D. spathulata*), with long narrow leaves, also caught insects. Later Oliver sent CD a living specimen for his observation (letter to Daniel Oliver, [22–3 September 1860]; see also *Insectivorous plants*, pp. 280–1). CD found that it took 18 hours for the tentacles of the Australian sundew to curve over a trapped fly (*ibid.*, p. 280).

[4] CD evidently brought several living specimens of *D. rotundifolia* back with him from Hartfield.

[5] See also letter to J. D. Hooker, 6 September [1860].

To Charles Lyell 1 September [1860]

Down Bromley Kent
Sept 1.

My dear Lyell

I have been much interested by your letter of 28th, received this morning. It has *delighted* me, because it demonstrates that you have thought a good deal lately on Nat. selection. Few things have surprised me more than the entire paucity of objections & difficulties new to me in the published reviews. Your remarks are of a different stamp & new to me. I will run through them & make a few pleadings such as occur to me.

I put in the possibility of Galapagos having been *continuously* joined to America out of mere subservience to the many who believe in Forbes' doctrine,[1] & did not see the danger of admission; about small mammals surviving there in such case. The case of Galapagos, from certain facts on littoral sea-shells, (viz Pacific ocean & S. American littoral species) in fact convinced me more than in other case of other islands that Galapagos had never been continuously united with the mainland: it was mere base subservience & terror of Hooker & Co.—[2]

With respect to atolls I do not think mammals wd. hardly survive *very long* even if main islands (for as I have said in Coral-Book outline of groups of atolls do not look like a former *continent*)[3] had been tenanted by mammals, from the extremely small area, the very peculiar conditions, & the probability that during subsidence all or nearly all atolls have been breached & flooded by sea many times during their existence as atolls.—

I cannot conceive any existing reptile being converted into a Mammal. From homologies I shd. look at it as certain that all Mammals have descended from some single progenitor. What its nature was, it is impossible to speculate. More like, probably, the Ornithorhynchus or Echidna than any known form as these animals combine Reptilian characters (& in lesser degree Bird character) with Mammalian. We must imagine some form as intermediate as is Lepidosiren now in between Reptiles & Fish, between Mammals & Birds on the one hand (for they retain longer the same embryological character) & Reptiles on the other hand. With respect to a mammal not being developed on any island, besides want of time for so prodigious a development, there must have arrived on the island, the necessary & peculiar progenitor having characters like the embryo of a mammal, & not an *already developed* reptile, Bird or Fish.— We might give to a Bird the habits of a mammal, but inheritance would retain almost for eternity some of the bird-like structure, & prevent our new creation ranking as a true mammal.—

I have often speculated on antiquity of islands, but not with your precision or *at all under the point of view of Natural selection not* having done, what might have been anticipated. The argument of *littoral* miocene shells at Canary Is[d] is new to me.— I was deeply impressed from amount of denudation of antiquity of St. Helena, & its age agrees with peculiarity of Flora.—

With respect to Bats at N. Zealand (N.B there are 2 or 3 Europeans Bats in Madeira & I think in Canary Islands) not having given rise to group of non-volant &c Bats, is, now you put the case surprising, more especially as the genus of Bats is in N. Zealand very peculiar, & therefore has probably been long introduced, & they now speak of Cretacean fossils there. But the first necessary step has to be shown, namely of a Bat taking to feed on the ground or anyhow & anywhere except in the air: I am bound to confess I do know one single such fact. viz of an Indian species killing frogs.

Observe that in my wretched polar bear case I do show the first step by which conversion into a whale "would be easy"—"would offer no difficulty."!! So with Seals, I know of no fact showing any the least incipient variation of seals feeding on the shore. Moreover seals wander much: I searched in vain & could not find *one* case of any species of Seal confined to any island; & hence wanderers would be apt to cross with individuals undergoing any change on an island, as in case of land-birds of Madeira & Bermuda. The same remark applies perhaps even to Bats, as they frequently come to Bermuda from the mainland, though about 600 miles distant.

With respect to the Amblyrhynchus of Galapagos; one may infer as probable, from marine habit being so rare with Saurians & from the terrestrial species being confined to a few central islets, that its progenitor first arrived at the Galapagos; from what country it is impossible to say as its affinity (I believe) is not very close to any known species. The offspring of the terrestrial species was probably rendered marine. Now in this case I do not pretend I can show variation in habits; but we have in the terrestrial species a vegetable feeder (in itself a rather unusual circumstance) largely on **lichens**, & it would not be a great change for its offspring to feed first on littoral algæ & then on submarine algæ.

I have said what I can in defence; but yours is a good line of attack. We should, however, always remember that no change will ever be effected till a variation in habits or structure or of both *chance* to occur in right direction so as to give the organism in question an advantage over other already established occupants of land or water; & this may be in any particular case indefinitely long.— I am very glad you will read my Dogs M.S; for it will be important to me to see what you think of balance of evidence.[4] After long pondering on a subject it is often hard to judge.—

With hearty thanks for your most interesting letter. Farewell | My dear old Master | C. Darwin

Emma & self wrote a few days ago to you.—

Endorsement: '1860'
American Philosophical Society (225)

[1] Edward Forbes postulated the existence of former land-bridges to explain geographical distribution patterns between islands and neighbouring areas of Europe (E. Forbes 1846). The possibility of the Galápagos and Cape Verde Islands having been joined to the nearest continents by continuous land is mentioned in *Origin*, pp. 398–9. In the sixth edition, CD inserted the parenthetical phrase 'though I do not believe in this doctrine' (Peckham ed. 1959, p. 631).

[2] CD stated his objections to Forbes's theory in 1846 and again in 1856 when he vigorously debated the question with both Joseph Dalton Hooker and Lyell. CD believed there was insufficient geological evidence to be able to assume the existence of land-bridges and that Forbes and Hooker were misguided in laying so much emphasis on them. Between 1856 and 1858, CD conducted a long series of experiments on the means by which organisms could be transported across oceans to reduce the theoretical necessity for land connections. He discussed his findings at length with Hooker. See *Correspondence* vols. 3–6.

[3] *Coral reefs*, pp. 91–2. See letter from Charles Lyell, 28 August 1860 and n. 3.

[4] See letter to Charles Lyell, 11 August [1860].

To W. B. Tegetmeier 1 September [1860]

Down Bromley Kent
Sept. I.

My dear Sir

I am very much obliged for the Field (returned by this Post), I have copied the passage, which is extremely satisfactory to me.—[1] You have settled the point right well.—[2] You have so repeatedly & so kindly taken much trouble to oblige me, that I am loth to refuse anything which you ask. Not to mention that you *greatly* overvalue anything which I could contribute, I really do not know what I could, I sh[d] so much dislike you to think me churlish, that permit me to be a little prolix.— I have wasted, chiefly owing to my daughter's illness, more than two months this summer, but I am now got to my work again.— I *never* am able to work more than 3 hours per diem. The subject of Bees-cells has almost gone out of my head, & I assure you on my word that it would take me a week or 10 days to get it all up, so as to write anything on the subject.[3] I really cannot spare so much time; so much behind hand I am in all my work. I trust that you will excuse me. If I had anything ready & worth sending I sh[d] feel myself bound to send it you; but I have not & could not prepare anything without going over all my rough notes & specimens.

Pray forgive me & believe me | My dear Sir | Yours sincerely | Charles Darwin

Endorsement: '1860'
New York Botanical Garden Library (Charles Finney Cox collection)

[1] Tegetmeier was the editor of *The Field, the Farm, the Garden, the Country Gentleman's Newspaper*. The passage to which CD refers has not been traced, but it presumably discussed the intercrossing of bees (see n. 2, below).

[2] CD had asked Tegetmeier for facts relating to the crossing of different stocks of hive-bee (letter to W. B. Tegetmeier, 20 July [1860]). Tegetmeier reported some of the results of his experiments on

the movements between different hives of workers and drones of the hive-bee to the Entomological Society of London on 6 August 1860. CD cited Tegetmeier's finding that drones are permitted to enter any hive in *Variation*. CD consequently concluded that 'there is no *à priori* improbability of a queen receiving a foreign drone.' (*Variation* 2: 126).

[3] In 1858 CD had investigated the means by which bees construct their cells in connection with his study of the origin of instinct (see *Correspondence* vol. 7). He published some of the results of this work in *Origin*, pp. 224–35. CD's notes on bees' cells are in DAR 48 (ser. 2): 5–75.

To J. D. Hooker 2 September [1860]

Down Bromley Kent
Sept. 2^d

My dear Hooker

I am astounded at your news received this morning.[1] I am become such an old fogy, that I am amazed at your spirit. For God's sake do not go & get your throat cut.[2] Bless my soul, I think you must be a little insane. I must confess it will be a most interesting tour; & if you get to top of Lebanon, I suppose extremely interesting.— You ought to collect any Beetles under stones there; but the Entomologists are such slow coaches, I daresay no result could be made out of them. The stupid beasts have never worked the Alpines of Britain.—

If you come across any Brine lakes do attend their minute Flora & Fauna; I have often been surprised how little this has been attended to.—

I have had long letter from Lyell, who starts *ingenious* difficulties opposed to nat. selection, because it has not done more than it has.[3] This is very good, as it shows he has thoroughily mastered the subject; & shows he is in earnest. Very striking letter altogether, & it rejoices the cockles of my heart.— I mentioned to him that you had been thinking of answering Q.R.. but he does not think it worth your while.[4]

If you come across many mules or Asses look out for *double* or treble shoulder stripe—or *trace* of forked shoulder stripes;

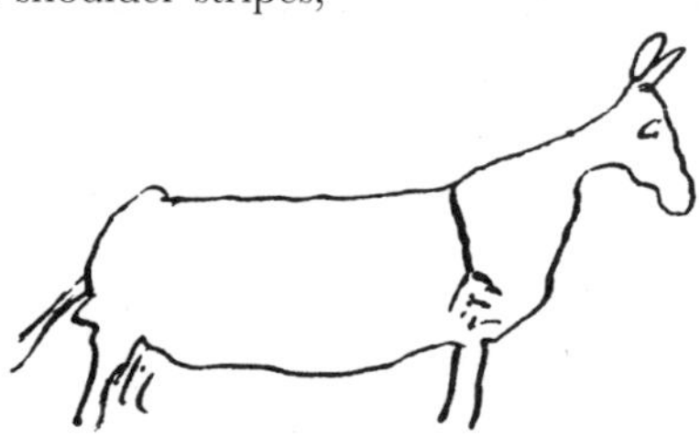

for transverse bars on legs & trace of stripes on side of face: especially forked shoulder stripe, even trace of.— If you see, make slightest sketch,—that will be best & shortest way.—[5]

I am glad to hear pretty good account of Baby. Etty steadily crawls on.

How I shall miss you my best & kindest of friends.— God Bless you | Yours ever affectionly | C. Darwin

I have got nothing to suggest about Arctic Flora.—[6]
I wrote yesterday about Drosera, give it Mr Oliver.[7]
I had not heard of Review in "Once a week"[8]

Endorsement: '60.'
DAR 115.1: 73

[1] The letter has not been found. John Washington, a naval hydrographer, had invited Hooker to take part in a scientific expedition to Syria. Hooker left for Palestine in mid-September in the company of Washington and Daniel Hanbury (L. Huxley ed. 1918, 1: 528). Hooker's objective was to ascend Mount Lebanon and examine the cedars, with particular reference to establishing their relationship to the cedars of the Himalayas.

[2] In his biography of Hooker, Leonard Huxley described the perils of the journey as follows: 'With wars and rumours of wars on every side, the journey promised to be more than a little hazardous; Italy was still engaged in the struggle for liberation from Austria; in Syria Muslim and Christian were at daggers drawn; the French as Protectors of the Faith . . . were chief in organising the Powers' campaign against the Druses.' (L. Huxley ed. 1918, 1: 528–9).

[3] Letter from Charles Lyell, 28 August 1860.

[4] CD refers to the possibility of Hooker answering Samuel Wilberforce's arguments against *Origin* published in the *Quarterly Review* ([Wilberforce] 1860).

[5] There is a note in CD's hand about stripes on horses bound with the letter. The note had been sent to Hooker in 1857 with reference to CD's inquiries about the colour of Norwegian horses (see *Correspondence* vol. 6, enclosure with letter to J. D. Hooker, 11 September [1857]). Hooker apparently located the note and kept it with the present letter. The note was incorrectly endorsed '1860' by Hooker.

[6] Hooker may have sent CD a draft of Hooker 1862a, which he had read at the Linnean Society of London on 21 June (see letter to J. D. Hooker, 26 [June 1860]).

[7] In fact, CD refers to the letter to J. D. Hooker, 31 [August 1860]. See also letter to Daniel Oliver, 11 September [1860].

[8] No review of *Origin* appeared in the journal *Once a Week*.

To A. G. More 5 September 1860

Down, Bromley, Kent.
Sept. 5. 1860.

My dear Sir

The Spiranthes came in splendid condition and I send you hearty thanks, for I have had a most interesting day's work, and fine new type of structure in rostellum.[1] May I understand that you *saw* Dipterous insects insert their *probosces* into the flowers? If this is *not* so, will you please inform me, as otherwise I will say that you saw this.[2] If you had seen them visit the earlier flowers you would have probably seen pollen-masses withdrawn (by curious contrivance) for they were all removed from older flowers. If you happen to walk to place again pray have a look at the Flies. Could you catch one? and send it me to name. I concluded that the dreadful weather prevented you visiting E. palustris: I suppose now it is too late, but I shall be very much obliged for any observations next year. I am extremely curious to know what position the distal half of labellum holds in a fresh flower, as I could hardly judge of this in those sent. Since writing I have examined E. latifolia, which

I have in my garden, and am become more doubtful on the probability of my conjectures about the half of labellum closing after insect has crawled into flower of E. palustris; for the cap of the rostellum is so far easier removed by outward movement than by inward, that this perhaps alone would suffice. But I shall be intensely curious to hear how case really is.[3]

I am very glad you looked at O. pyramidalis;[4] for though I have mentioned fact to some few I am not aware that anyone has looked at the movement. As I did not know whether you would care about the little fact I did not mention a second movement which takes place *after* the saddle of the pollen-mass has seized the needle:

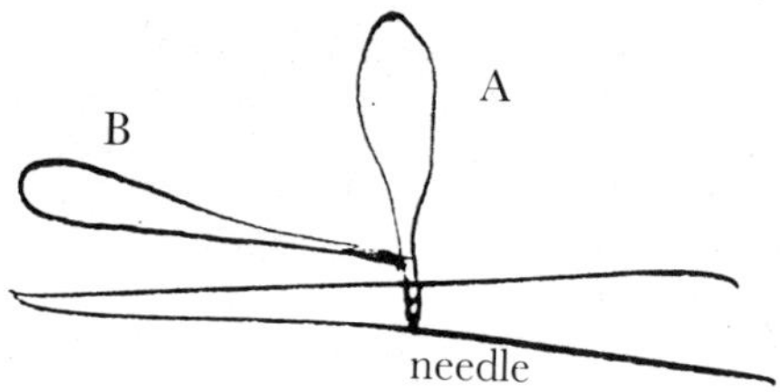

at first it holds the position at A, it then moves slowly (*always*) into position B, and the *two* pollen-masses diverge a little. Now if you poke the needle into the nectary of another flower, the pollen-masses hold such a position with respect to the needle (or proboscis of moth) that their apices infallibly strike against the two *lateral* stigmatic surfaces and leave grains of pollen on them.[5] I am not aware that Botanists have observed that the stigmas of this species are lateral and double quite different from all other species (except O. conopsea). Is not all this a pretty relation to visits of insects?

By the way my correspondent sent me a moth with the supposed pollen-masses of Bee-Orchis affixed to proboscis, and as he remarked it was very odd that the moth had bored through the footstalk of the pollen-masses!! But alas the pollen-masses turned out to belong to O. pyramidalis; and the closely fitting saddle had been mistaken for the bored footstalk![6]

I hope that I have not wearied you with these details.

With very sincere thanks. | Yours very faithfully | C. Darwin

Copy
DAR 146

[1] CD had asked More to send him living specimens of *Spiranthes autumnalis* (see letters to A. G. More, 24 June 1860 and 3 August 1860). CD acknowledged More's assistance in *Orchids*, p. 116 n. The rostellum in this species presented peculiarities that interested CD: he observed a rigid 'boat-formed disc' projecting from the middle of the rostellum that was filled with thick adhesive fluid. The structure and function of this disc are described in *Orchids*, pp. 116–21. See also following letter.

[2] See letters to A. G. More, 3 August 1860 and 9 August 1860, in which CD asks More to observe the activities of insects on *Epipactis*. More apparently gave CD some information relating to Diptera and *Spiranthes* in his reply. The case is not mentioned in *Orchids*.

[3] See letter to A. G. More, 9 August 1860 and n. 1.

[4] CD had suggested that More might like to repeat some of the observations CD had made on *Orchis pyramidalis* (letter to A. G. More, 5 August 1860). CD especially recommended that More note the spontaneous movements of the sticky glands when removed from the flower 'and see how beautifully adapted to seize proboscis of moth'.

[5] CD described these movements in *Orchids*, pp. 25–8, 337–9.

[6] The case is described in *Orchids*, p. 26.

To J. D. Hooker 6 September [1860][1]

Down Bromley Kent
Sept 6th

My dear Hooker

I was surprised & pleased to get your last note; for I did not at all expect another.—[2] We were amused at your original views on insanity.— My Father thought something the same; at least he thought almost everyone was insane on some point. Thank you much for thinking about Asses & Mules: the forked stripe on shoulder is a very curious point (& I have seen traces of it) for the meaning of the fork is plain when the legs are transversely barred, but is an absurdity (except on doctrine of inheritance) with an animal with unstriped legs.—[3] If you go up lofty mountains compare general state of vigour of the *first* individuals of any species of plant, which you meet, with the last, with respect to struggle for life, as alluded to in Origin—;[4] also the degree of abruptness of limitation of the species.—

The rest of this letter I write to amuse myself & not for your sake, as I have a weakness, that I can never enjoy my work, till I have told you.— You are my public.— But before I begin, I must say how entirely agree about the detestable article in Athenæum about Tyndall—sneering at his veracity—& very disagreeable about Huxley.—[5] I look at the Editor as a spiteful old woman, who has taken, what my Brother calls her D.B. degree (ie damned bitch)—[6]

With respect to Drosera, there is no doubt that the plant is acute enough & knows what is good. It releases from its gripe an inorganic body far sooner than an organic body. But fluids show the difference better than the act of mere releasing. I have tried only the following 14 fluids, made as nearly as possible the same **thin** consistency, & in most case each fluid on several leaves. Water, syrup of sugar, starch, gum, olive-oil—sherry wine, carbonate of soda; these contain no nitrogen & never cause any inflection of the leaves or hairs. On other hand, milk cold infusion of raw meat, urine, raw white of egg, thin gelatine, *thin* mucus from lungs, & saliva all cause most powerful contraction & all contain nitrogen.— Saliva contains less than one percent of animal matter. But here comes the really curious case, a solution in proportion of 1 grain of Carbonate of Ammonia to 1 oz of water (ie 1 to 480 parts by weight) & of this solution $\frac{1}{2}$ minim being taken, positively & certainly produces great effect. Hence $\frac{1}{960}$ of one grain of C. of Ammonia causes contraction. This evening I rerepeated experiment on 5 leaves (on some of which

I had put drop of distilled water in morning with *no* effect) & all are now contracted. Is this not curious? Nor am I got to minimum strength of solution yet.—[7]

I daresay you had no time to read my note of action of rostellum in Malaxis & Epipactis.[8] I have now examined Spiranthes autumnalis (for God's sake if you find any odd genus of Orchis *if you can*, put it in weak spirits *with buds* & *old flowers* for me, for I am fairly mad on subject) & in Spiranthes we have new & curious modification, which I must publish to *you* my public.[9] *Within* the rostellum, there is formed a hard canoe-shaped object; placed vertically, ie parallel to axis of flower;

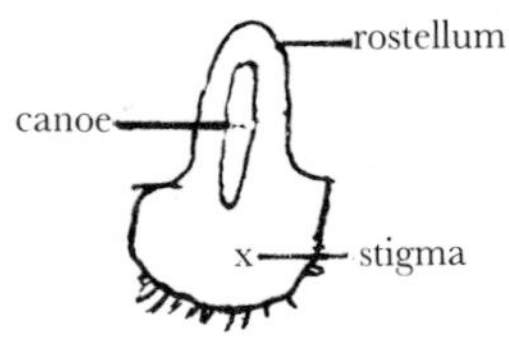

the anthers shrivel & leave pollinia lying at back of rostellum (as in Listera) in bud, & the rostellum splits behind, & the pollinia become permanently attached to the now naked bottom of "canoe". Now the flower opens. If you then touch most delicately the rostellum in front (ie on side over the stigmatic surface,) the external membrane by a vital action instantly splits longitudinally down its whole length, & exposes the open part of canoe, which canoe is filled with extremely viscid matter; & this viscid matter sets instantly hard (as in Listera), so that the touching object—(a bristle or probosis of insect) is glued firmly to the "canoe" & this being already glued to the pollinia, all are withdrawn together. The rostellum then presents a split appearance. The membranous edges of the now split rostellum form the two foliaceous projections to the rostellum described by some authors![10] If you can understand this, you will think it is cerious modification of the Listera type of structure.

Farewell my dear Hooker. Do take care of yourself.— I am very glad your family approve of your Expedition.[11] It is wonderfully spirited.

Farewell may you prosper | Yours affectionately | C. Darwin

I have written all this for my enjoyment & not a bit for yours.—

DAR 115.1: 74

[1] Dated by the reference to Hooker's forthcoming expedition to Syria (see n. 11, below).

[2] Neither of the letters Hooker wrote to CD before his departure for Syria has been found. See also letter to J. D. Hooker, 2 September [1860].

[3] CD asked Hooker to look out for asses with shoulder- or leg-stripes during his trip to Syria (letter to J. D. Hooker, 2 September [1860]). Hooker observed five asses in Syria in which the shoulder-stripe was 'plainly forked over the fore leg.' (*Variation* 1: 63).

[4] *Origin*, p. 69.

[5] The *Athenæum*, 1 September 1860, pp. 280–2, carried an anonymous review of John Tyndall's *Glaciers of the Alps* (Tyndall 1860). The book described some of Tyndall's alpine climbs, on one of which he was accompanied by Thomas Henry Huxley. The reviewer wrote in such a way as to make the climbers' attempts seem foolish.

[6] The editor of the *Athenæum* was William Hepworth Dixon. See also letter to Asa Gray, 22 July [1860]. CD refers to Erasmus Alvey Darwin.

[7] CD performed a series of experiments on *Drosera* after returning to Down from Hartfield on 2 August and before leaving for Eastbourne on 22 September. His notes are in DAR 60.1.

[8] CD probably refers to the 'abstract' mentioned in the letter to J. D. Hooker, 7 August [1860].

[9] The *Spiranthes autumnalis* had been supplied by Alexander Goodman More (see preceding letter).

[10] CD refers to Hooker's paper on the structure of *Listera ovata* (Hooker 1854b). The similarities between the fertilisation of this species and of *S. autumnalis* are discussed in *Orchids*, p. 151.

[11] Hooker left England for Smyrna (Izmir), Turkey, on 15 September 1860. He returned in mid-November (L. Huxley ed. 1918, 1: 528).

To John Innes 6 September [1860]

Down Bromley Kent
Sept. 6th

Dear Innes

Many thanks for your kind enquiries about Etty. I am glad to be able to give a decidedly better account, though her progress is excessively slow. She now sits up several hours every day & has taken two or three very short drives. What is best of all is that the Doctors are now convinced there is no organic mischief.— We have had an unhappy Summer; but I hope the worst is over.— I am glad to have a pretty good account of your son; & I hope Mrs Innes is fairly well.[1] My wife joins me in very kind remembrances to her.—

I hope you have not given up thinking about Down; though what house you could get, I cannot tell. I hear dreadful reports on the state of Mr. Ainslie's house.—[2] We were away from Down for an unusual time this summer, namely, six weeks; & have hardly seen a soul since our return;—except J. Lubbock to talk Natural History with. Every thing goes on much as usual. ⟨I some⟩times see Mr Phillips, ⟨ ⟩ of ⟨ ⟩.—

⟨ ⟩ ⟨re⟩membering my hobby of striped asses.[3] I must say that I am a complete skeptic about the powers of rooks,—curious as your stories are.

What stories one hears about the spirit-rapping now-a-days— the old saying to believe nothing one hears & only half of what one sees is a golden rule.

Farewell | with every good wish. | ⟨ ⟩[4]

Postmark: SP 7 60
AL⟨S⟩
Cleveland Health Sciences Library (Stecher collection)

[1] John William Brodie Innes and Eliza Mary Innes.

[2] Innes, perpetual curate of Down, had been unable to secure a house for his family in Down. Robert Ainslie, the owner of Tromer Lodge, had left the village late in 1858 (see *Correspondence* vol. 7, letter to W. E. Darwin, 14 [May 1858] and n. 5).

[3] William Walker Phillips lived at Down Hall farm (*Post Office directory for the six home counties* 1859). CD discussed the stripes of asses in *Variation* 1: 62–4, but Phillips is not cited.

[4] CD's signature has been excised.

From Charles Lyell[1] 8 September 1860

8th Septr. 1860—

Your argument from absence of mammalia in islands excepting those which we know to have been joined to mainland in Pliocene times is a favourite one of mine when I wish to prove that y^{r} bond of descent is required for the coming in of new species.[2] The older the island, the stronger yr case against special creation without a closely allied antetype as Wallace would say.[3]

The land shells of Madeira and the Canaries are so endemic as to show that none were united with each other or Africa. I believe all were separate from Upper Miocene times; if not, small African species of quadrupeds would have been retained.

But the antiquity of islands void of mammals checks the hypothesis of the easy adaptability of one species of a genus to new conditions. On a new volcanic island before all the best places are seized upon, seals or walrus's, manatees, dolphins & other cetacea, when pressed hard for food w^{d} go up the rivers or if amphibious devour the eggs or young of land-birds. Bats, though not primates, but here Lissencephala ought to get the better in some stations of marine birds if higher classes tend to become dominant.[4]

The long reign of the *Dinornis* family in New Zealand implies that for ages there had been room for mammalia if bats & rodentia & aquatic placentals being at hand had been convertible even into allied structures fitted for land habits. And why bats & rodents should have peopled all Australia without having been developed into something higher in the placental line seeing that the representatives of the latter were imported from Europe, can run wild there, is strange, for why sh^{d} it take more time for a bat to work up into a lemur than for a Myrmecobius to improve into a Thylacinus.[5]

As to the land shells of Madeira & Porto Santo 95 per cent of them have remained absolutely unchanged for a period sufficiently long to allow the form & size of the islands to be altered materially by the waves of the Atlantic, long enough for some species which once very common to become almost extinct—in a few quite lost

The only speculation I have been able to make is this that by far the majority $\frac{9}{10}$^{ths} perhaps are immutable, & when changes come, must die rather than yield. The hereditary power has got too strong a set, after a million of years, in one way It cannot go in any other. But there are some eminently metamorphic species in the Madeiras & these are producing the allied species & sub-species of Lowe in each island.[6] & after a long period there might be found, in spite of extinctions, just as many true species of Helices in the Archipelago as now, supposing Man would let it alone— So I suspect it has always been— Most species immutable & true to the death, as I maintained in the Principles that all were, but a few of the whole plastic & becoming the centers of new generic & ultimately higher groups as you maintain. So it seems to me to hold good in the tertiary series Some shells once abundant in the Mediterranean formation & in the Crag, still survive unaltered though excessively persecuted & rare.—

Y^{r} antagonists are driven to great straits when they try to set up any machinery other than that of normal generation— The greater we can show the role of extinction, the worse is their case because the reactivating power ought to keep pace, if as Bronn pretends, the creation is always growing richer & more varied,[7] which however I doubt since Eocene days.

According to the progressive theory, why sh^{d} there be a living platypus or Ornithorhynchus, for we have nothing synthetical or elementary in the Coal Strata, & if they sh^{d} be found there, how have they escaped being altered, improved and specialized in 30 periods.

Ammonites in uppermost part of the Maestricht, of a Cretaceous type, if not species of Texas are against the doctrine of wide gaps. Between Maestricht & Thanet sands.

What you say of the power of preoccupancy is good[8]—so, as to atolls, but if they subsided very slowly, the absence of volant & amphibious forms of reptile & mammifer in such a region proves rate of transmutation slow, even as compared to revolutions in physical geography.

Copy
Kinnordy MS, Charles Lyell's journal VI, pp. 179–86

[1] The text of the letter has been taken from a copy in Lyell's scientific journal. It is also printed in Wilson ed. 1970, pp. 472–4. The copy is headed: 'C. Lyell to C. Darwin 8th Septr 1860— Coburg'.

[2] See letter to Charles Lyell, 1 September [1860].

[3] Alfred Russel Wallace referred to the original form from which a variety or new species developed as the 'antitype' (see Darwin and Wallace 1858, p. 61, and Wallace 1860, p. 181).

[4] Lissencephala, in Richard Owen's classification of the Mammalia, was the division that included the rodents, insectivores, bats, and edentates. See R. Owen 1859a.

[5] Lyell refers to the numbat or banded ant-eater, an Australian marsupial, and to the so-called Tasmanian wolf, a carnivorous marsupial.

[6] Richard Thomas Lowe, an expert on the natural history of Madeira, published a description of the recent and fossil land and fresh-water shells of the island (Lowe 1854).

[7] Bronn 1858a. See Junker 1991.

[8] CD made this point in the concluding paragraphs of his letter to Charles Lyell, 1 September [1860]. See also *Origin*, p. 403.

To Asa Gray 10 September [1860][1]

Down Bromley Kent
Sept 10th

My dear Gray

On receipt (through Hooker) of your letter of Aug. 26th.,[2] I ordered my Bookseller to send you Hopkins' 2^{d} article in Fraser; it was very stupid in me not to have thought of sending it & the Quarterly.[3] But I presume the Quarterly will have long ago arrived in the States. I have ordered Murray to send the "Origin" to the American Acad. of Sciences, to your care, as I did not know proper address; I shall be grateful for volumes. Many thanks for the sheets replacing those sent to Athenæum.[4] I sent the copy of this Athenæum at Lyell's request to him in Germany (where he is fossil-man hunting) & he writes to me that he thinks the argument by you quite admirable.—[5] I have not yet recived the papers mentioned in your letter sent to Hooker, for H. first lent them to Lady H.—[6] You will have heard of Hooker's astonishing expedition to Syria; if he ascends Lebanon, it may answer scientifically, but I cannot help thinking the expedition rather rash.[7] His absence is so great a loss to me, that I am hardly fair judge of the wisdom of the Journey. I have got, nevertheless, your second article in Atlantic monthly;[8] for a copy of whole number was sent to my Brother-in-law Hensleigh Wedgwood, on account of a Review of his Dictionary.[9] Do you know who the writer is, as my Brother would like to know??

I have ordered another copy of this August number, as I sh[d] like (if I can get it in London) to send copy to Pictet;[10] & I have been thinking & shall consult Huxley, whether I could not get it reprinted in some English Journal.—[11] I have, also, ordered by anticipation two copies of October number; so do not trouble yourself to send it.[12]

By the way I am thinking of taking a very great liberty; but after much consideration I do not think you can object: you said that *it was known that you were the author of the 1st article*;[13] & as the best chance of getting it reprinted in England in a scientific Journal w[d] be to affix your name, I think of doing this & I hope to Heaven that you will not think this an unwarrantable liberty. I think most highly of this Article & I cannot bear to think it sh[d] not be known in England. You will be weary of my praise; but it does strike me as quite admirably argued; & so well & pleasantly written.— Your many metaphors are inimitably good. I said in a former letter that you were a Lawyer; but I made a gross mistake, I am sure that you are a poet. No by Jove I will tell you what you are, a hybrid, a complex cross of Lawyer, Poet, Naturalist, & Theologian!— Was there ever such a monster seen before? I have just looked through the passages which I have marked as appearing to me extra good, but I see that they are too numerous to specify; & this is no exaggeration. My eye just alights on the happy comparison of colours of prism & our artificial groups.—[14] I see one little error of fossil *cattle* in S. America.—[15] It is curious how each one, I suppose weighs arguments in a different balance: embryology is to me by far strongest single class of facts in favour of change of form, & not one, I think, of my reviewers has alluded to this. Variations not coming on at a very early age, & being inherited at not very early corresponding period, explains, as it seems to me, the grandest of all facts in Nat. History, or rather in Zoology. viz the resemblance of Embryos.—[16]

Hensleigh Wedgwood, above alluded to, is a very strong Theist, & I put it to him, whether he thought that each time a fly was snapped up by a swallow, its death was designed; & he admitted he did not believe so, only that God ordered general laws & left the result to what may be so far called chance, that there was no design in the death of each individual Fly.—

Farewell my good Friend | Yours most truly | Charles Darwin

Never write to me when too much pressed by labour,—though I value much your letters.—

Gray Herbarium of Harvard University (34)

[1] Dated by the reference to Gray's articles on *Origin* (see n. 8, below).
[2] Gray's letter has not been found.
[3] CD refers to Hopkins 1860 and [Wilberforce] 1860. Both articles reviewed *Origin*.
[4] The reference is to Gray's defence of *Origin* published in the *Proceedings of the American Academy of Arts and Sciences* 4 (1860): 411–16. See letters to Asa Gray, 22 July [1860], and to J. D. Hooker, 7 August [1860].
[5] See letters to Charles Lyell, 11 August [1860] and 28 August [1860].
[6] Maria Hooker was Joseph Dalton Hooker's mother.
[7] See letters to J. D. Hooker, 2 September [1860] and 6 September [1860]. Hooker and his party had travelled to Syria despite the danger of being caught up in the conflict between Christians and Muslims. They entered Damascus only a day after the Christian quarter of the city had been sacked. See L. Huxley ed. 1918, 1: 529.
[8] [Gray] 1860b, pp. 229–39. There is an annotated copy of all three of Gray's articles for the *Atlantic Monthly* in the Darwin Pamphlet Collection–CUL.
[9] There was a review of the first volume of Hensleigh Wedgwood's *Dictionary of English etymology* (Wedgwood 1859–65) in the *Atlantic Monthly* 6 (1860): 248–54.
[10] In his second article for the *Atlantic Monthly*, Gray discussed the favourable opinion of CD's theory expressed by François Jules Pictet de la Rive in his review of *Origin* (Pictet de la Rive 1860). Gray called the review 'much the fairest and most admirable opposing one that has yet appeared' ([Gray] 1860b, p. 230).
[11] See following letter.
[12] The October issue of the *Atlantic Monthly* carried Gray's concluding article, entitled 'Darwin and his reviewers' ([Gray] 1860b, pp. 406–25).
[13] [Gray] 1860b, pp. 109–16. All three of Gray's articles in the *Atlantic Monthly* were published anonymously.
[14] CD's copy of [Gray] 1860b is heavily marked. The passage referred to is on p. 238:

> Our [classification] systems are nothing, if not definite. They are intended to express differences, and perhaps some of the coarser gradations. But this evinces, not their perfection, but their imperfection. Even the best of them are to the system of Nature what consecutive patches of the seven colors are to the rainbow.

[15] Gray stated that both 'wild horses and cattle' had lived in America during former geological periods ([Gray] 1860b, p. 235). CD circled 'and cattle' and wrote 'no' in the margin of his copy in pencil.
[16] For earlier remarks on this point, see *Correspondence* vol. 7, letter to J. D. Hooker, 14 December [1859], and, in this volume, letter to W. B. Carpenter, 6 January [1860]. See also *Autobiography*, p. 125.

To T. H. Huxley 10 September [1860][1]

Down Bromley Kent
Sept 10th

My dear Huxley

I send by this Post a Review by Asa Gray, so good that I shd like you to see it;[2] I must beg for its return. I want to ask, also, your opinion about getting it reprinted in England. I thought of sending it to Editor of Annals & Mag. of N. Hist (in which two hostile Reviews have appeared)[3] (although I suppose the Annals have very small circulation) & asking them in spirit of fair play to print this, with Asa Gray's name, which I will take responsibility of adding.—[4] Also, as it is long, I would offer to pay expences.—

It is very good in addition as bringing in Pictet so largely.[5] Tell me briefly what you think.—

What an astonishing expedition this of Hooker's to Syria.— God knows whether it is wise.—

How are you & all yours? I hope you are not working too hard. For Heavens Sake, think that you may become such a beast as I am.— How goes on the Nat. Hist. Review?—[6] Talking of Reviews I damned with a good grace the Review in Athenæum on Tyndall with a mean scurvy allusion to you.—[7] It is disgraceful about Tyndall,—in fact doubting his veracity.—

I am very tired, very stomachy & hate nearly the whole world. so good night, & take care of your digestion which means Brain—

My dear fellow | Ever yours | C. Darwin

Etty slowly, very slowly, but steadily improves., & we are thinking of sea-side for her & us all.—

Imperial College of Science, Technology and Medicine Archives (Huxley 5: 135–6)

[1] Dated by the reference to Asa Gray's review of *Origin*.
[2] CD refers to the second part of [Gray] 1860b. See preceding letter.
[3] CD refers to [Wollaston] 1860 and to a reprint of Agassiz 1860, published in the September issue of the *Annals and Magazine of Natural History*. The editors of the journal were Prideaux John Selby, Charles Cardale Babington, William Francis, and John Edward Gray.
[4] See letter to Asa Gray, 26 September [1860].
[5] See preceding letter and n. 10.
[6] See letter to T. H. Huxley, 20 July [1860].
[7] See letter to J. D. Hooker, 6 September [1860] and n. 5.

To George Gordon[1] 11 September [1860]

Down Bromley Kent
Sept 11th

Dear Sir

I hope that you will excuse the great liberty which I take in addressing you & begging a favour.—

Mr H. C. Watson informs me that you are the most likely man to be able to allow & willing to grant this favour.—[2] I have been examining the minute structure of the rostellum of all British Orchids & am in consequence *extremely anxious* to see Goodyera. Is it in your power to send me a few *fresh* specimens? I know that it is a mere chance whether this rare plant grows near you. But if it does, & you would pack up a few specimens, in **bud** & flower, with damp moss round the cut off stems, packed in an old tin cannister & sent by Post (allowing me of course to repay postage) it would be a great favour.

I hope in any case you will forgive the liberty, which I take in troubling you & I beg you not answer this, unless you can kindly send me the specimens.—[3]

With many apologies | I beg leave to remain | Dear Sir | Your obliged servant | Charles Darwin on account of health of my daughter I shall be compelled to go to sea-side in about 9 or 10 days.

Postmark: SP 11 60
Moray District Record Office (DDW 71/860/7)

[1] The recipient is identified by the cover that accompanies the letter. Gordon was the minister of Birnie, near Elgin, Scotland. He was an expert on the flora of Morayshire.

[2] Gordon had contributed catalogues of plants from certain parts of Scotland to Hewett Cottrell Watson's *New botanists' guide* (Watson 1835–7).

[3] See letter to George Gordon, 17 September [1860].

To John Innes 11 September [1860]

Down Bromley Kent
Sept. 11th

Dear Innes.

I am very sorry to say that I fear we shall not see you when you come to Keston,[1] for we have almost made up our minds to be off next week to sea-side for a month to see what the change will do for Etty,—though I have not much hope.

Your account of the Donkey has interested me much & would you be so kind as to have another look at it & observe whether its eyes are pink; but as you say it is rather cream-coloured than white, it probably is not an albino.— Did the owner rear it; if so please ask him whether it was born of same colour & had then no stripes.— Anytime will do for an answer.—[2]

I am sorry to hear about your house difficulties.[3]

Dear Innes | Ever yours very truly | C. Darwin.

Endorsement: '60'
Cleveland Health Sciences Library (Stecher collection)

[1] Keston is a village about three miles from Down.

[2] See letter to John Innes, 6 September [1860]. Innes's information about donkeys is not cited by CD in *Variation.*

[3] See letters to John Innes, 18 July [1860] and 6 September [1860].

To Daniel Oliver 11 September [1860][1]

Down Bromley Kent
Sept 11th

My dear Sir

Dr Hooker told me that you would be so kind as to observe rate of closing of leaf of some Australian Drosera,[2] if it does close its leaf over flies.

There are two other points, on which I shd be infinitely obliged for a little information. Lindley speaks on the closing of the leaf of Drosera lunata of India, in a manner which makes me suppose that it closes in a more marked manner that our Drosera;[3] I see Steudel in his Nomenclator refers to Buchanan & Wight,[4] would it give you much trouble to refer to these Authors & see whether they say anything on the subject.—

Lastly would you look at the Dionæa (if you have living specimen) & observe whether the hairs are viscid, for it almost passes my belief that the leaf can snap so quick as to catch a fly, unless it be in some degree entangled: when a fly is caught it is said to be bathed with a secretion, as in case of our Drosera.—[5]

Pray forgive me troubling you so much & believe me | My dear Sir | Yours very faithfully | C. Darwin

P.S. I believe one of your Gardeners artificially stirred up pollen in the indusium of *Leschenaultia* formosa at my request.[6] Will you ask whether it has made Pods. My plant has made pods on those flowers, which I thus treated. Whether my seeds will ripen I doubt, but I can see seeds through walls of pod.—

P.S. | I have written to Gardeners Chron. to ask for authority of movement of leaves of D. lunata.—[7]

You will think it incredible but it is a fact that genererally $\frac{1}{1920}^{\text{th}}$ part of single grain of Nitrate of Ammonia (nitrogen in both base & acid) causes decided movement of leaf. Even $\frac{1}{2400}^{\text{th}}$ of a single grain sometimes causes movement, & I am myself convinced (though I do not expect any one to believe without ocular evidence) that $\frac{1}{2880}^{\text{th}}$ of a grain occasionally produces an effect on very sensitive & young leaves.— Are not these facts curious, ie if you can swallow them. Believe me that they are not stated without numerous & carefully repeated experiments.—[8]

Down House (MS 10: 9)

[1] Dated by the relationship to the letters to J. D. Hooker, 2 September [1860] and 6 September [1860].
[2] See letter to J. D. Hooker, 2 September [1860].
[3] Lindley 1846, pp. 433–4. See letter to *Gardeners' Chronicle*, 15 [September 1860].
[4] Francis Hamilton Buchanan and Robert Wight were cited in Steudel 1841, p. 531, as the first to identify *Drosera lunata*. CD's copy of Steudel 1841 is in the Darwin Library–CUL. Buchanan and Wight were specialists in the flora of India. The point about which CD was inquiring is not discussed in *Insectivorous plants*.
[5] The Venus's fly-trap (*Dionaea muscipula*) is described in *Insectivorous plants*, pp. 286–320. CD states that the surface of the leaf of this plant is thickly covered with minute glands (*ibid*, p. 287). In addition, the leaves have several rows of sensitive filaments that 'are not viscid, and the capture of insects can be assured only by their sensitiveness to a momentary touch, followed by the rapid closure of the lobes.' (*ibid.*, p. 290).
[6] See letter to J. D. Hooker, 11 May [1860].
[7] See letter to *Gardeners' Chronicle*, 15 [September 1860] and n. 3.
[8] CD's notes on experiments carried out in July, August, and September 1860, in which he tested the effects of various substances on the leaves of *Drosera*, are in DAR 60.1.

To Charles Lyell 12 September [1860]

Down Bromley Kent
Sept 12th

My dear Lyell

I never thought of showing your letter to anyone.—[1] I mentioned in letter to

Hooker that I had been much interested by letter of yours with original objections founded chiefly on natural selection not having done as much as might have been expected.[2] Even if his mind had not been full of Syria he could never have conjectured your precise line of thought.—[3] In your letter just received you have improved your case versus N. Selection;[4] & it would tell with public (do not be tempted by its novelty to make it too strong) yet it seems to me not *really* very killing, though I cannot answer your case, especially why rodents have not become highly developed in Australia. You must assume that they have inhabited Australia for a very long period, & this may or may not be the case. But I feel that our ignorance is so profound, why one form is preserved with nearly the same structure, or advances in organisation, or even retrogrades or becomes extinct, that I cannot put very great weight on the difficulty. Then, as you say often in your letter, we know not how many geological ages it may have taken to make any great advance in organisation; *remember monkeys in Eocene formations.*

I see I misunderstand you[5]

But I admit that you have made out an excellent objection & difficulty: & I can give only unsatisfactory & quite vague answers, such as you have yourself put. I think, however, you hardly put weight enough on the absolute necessity of variations first arising in the right direction, videlicet of seals beginning to feed on shore.—

I entirely agree with what you say about only one species of many becoming modified: I remember this struck me much when tabulating the varieties of plants. & I have a discussion somewhere on point. It is absolutely implied on my ideas of classification & divergence that only one two species of even large genera give birth to new species; & many whole genera become *wholly* extinct, ie none of the species. Please see p. 341 of Origin. But I cannot remember that I have stated in Origin the fact of only very few species in each genus varying. You have put the view much better in your letter. Instead of saying, as I often have that very few species vary at same time, I ought to have said that very few species of a genus *ever* vary so as to become modified; for this is the fundamental explanation of classification & is shown in my engraved diagram.[6]

I can see no way of estimating the number of species "extinguished in a given time".

I do not think passages pp. 168 & 313 are really contradictory; mere variability & variability taken advantage of & selected are widely different considerations.— Look at variability of rudimentary organs,—which from being useless may & do vary greatly, & yet are not selected, & so do not become permanently altered & so make a new specific form.

Am I not right that Ammonites have become wholly extinct in a remarkably sudden manner *relatively* to most other families; I meant only this, but I see I have not been nearly guarded enough.—[7] I cannot find that I have, but I thought I had, alluded to much extinction, (as well as of modification) in the great blank intervals between our formations. (I see I have in regard to Ammonites at bottom of p. 321) But does it not strike you that the extinction of Ammonites has been wonderfully sudden?— I may add that I suspect, & have always suspected, that in southern Chile, near Concepcion, there are beds, which seem to be Tertiary with Ammonites & Baculites.—

I quite agree with you on the strange & inexplicable fact of Ornithorhynchus having been preserved; & Australian Trigonia, or the Silurian Lingula.— I always repeat to myself that we hardly know why any one single species is rare or common in best known countries. I have got a set of notes somewhere on the inhabitants of fresh water; & it is singular how many of these are ancient or intermediate forms, which I think is explained, by the competition having been less severe & the rate of change of organic forms having been slower in small confined areas, such as all the fresh-waters make compared with sea or land.

I see that you do allude in last page as a difficulty, marsupials not having become placentals in Australia; but this, I think, you have no right at all to expect; for we ought to look at marsupials & placentals as having descended from some intermediate & lower form.— The argument of rodents not having become highly developed in Australia (supposing that they have long existed there) is much stronger.— I grieve to see you hint at the creation "of distinct successive types as well as of a certain number of distinct aboriginal types.—"[8] Remember if you admit this, you give up the embryological argument (*the weightiest of all to me*) & the morphological or homological argument.— You cut my throat, & your own throat; & I believe will live to be sorry for it.— So much for Species.—

The striking extract which Emma copied was your own writing!! in a note to me many long years ago, which she copied & sent to Mad. Sismondi; & lately my Aunts in sorting her letters, found Emma's & returned them to her.—[9]

Etty has a wish to try the sea, & we all start there in about a week.— I have been of late shamefully idle; ie observing instead of writing & how much better fun observing is than writing.—

Yours affect | C. Darwin

Endorsement: '1860'
American Philosophical Society (226)

[1] Letter from Charles Lyell, 28 August 1860.

[2] See letter to J. D. Hooker, 2 September [1860].

[3] See letter from Charles Lyell, 25 September 1860.

[4] See letter from Charles Lyell, 8 September 1860.

[5] Before this sentence CD wrote, and then deleted, the following passage: 'One point I vehemently object to, ie your supposing that with any length of time an *existing* marsupial animal would be developed into placental; I quite'. See Manuscript alterations and comments.

[6] CD refers to the diagram in *Origin*, facing p. 117, illustrating divergence.

[7] See *Origin*, pp. 318, 321–2.

[8] This passage does not occur in either of the extant portions of the letters from Charles Lyell, 28 August 1860 and 8 September 1860.

[9] The extract alluded to was probably enclosed in a letter from Emma Darwin to Mary Elizabeth Lyell (see letter to Charles Lyell, 28 August [1860]). The letter from which it was copied was one from Emma to her aunt Jessie de Sismondi, dated 27 August 1845; it is printed in *Emma Darwin* 2: 96. Emma had quoted some remarks Lyell had made about Joseph Blanco White's autobiography (White 1845):

> I would advise every scientific man who is preparing a new edition in any rapidly progressing branch of science, in which he has launched many new speculations and theories, to read over the life of St Blanco the Martyr . . . How ashamed ought every lover of truth to feel if mere self-love or pride makes him adhere obstinately to his views, after seeing the sacrifices which such a man was ready to make for what he believed to be truth.

Lyell's original from which the extract was copied has not been found.

To *Gardeners' Chronicle* 15 September [1860][1]

In Lindley's Vegetable Kingdom (p. 433) it is stated that the leaves of Drosera lunata "close upon flies and other insects that happen to alight upon them."[2] Can you refer me to any published account of the movement of the viscid hairs or leaves of this Indian Drosera?[3]

C. R. Darwin,

Down, Sept. 15.

Gardeners' Chronicle and Agricultural Gazette, 22 September 1860, p. 853

[1] Although the letter is dated 15 September, it was apparently drafted on 11 September. See letter to Daniel Oliver, 11 September [1860].

[2] Lindley 1846, pp. 433–4. See also letter to Daniel Oliver, 11 September [1860].

[3] The letter was followed by a reply, presumably written by John Lindley, editor of the *Gardeners' Chronicle*:

> Dr. Royle is the authority for the statement in question. In his Illustrations of Himalayan Botany there is the following passage:—"D. lunata occurs in the mountains from Silhet to the Sutlej. This I have found in the small valleys enclosed within the different lateral projections of the Mussooree range, where the ground is rather flat, and the soil moist. In such situations it springs up and flowers in considerable quantities, but only during the rainy season, when the thermometer has a range of not more than 10°, between 60° and 70°, and the hygrometer always indicates a degree of moisture approaching that of saturation. This species, which in my MSS. Catalogue I had named D. muscipula, from the glandular ciliæ of its viscous leaves closing upon flies and other insects which happen to light upon them, is remarkable, as in this respect resembling Dionæa muscipula, which is placed in the same natural family."

Lindley refers to Royle 1839, 1: 75.

To Daniel Oliver 15 [September 1860][1]

Down Bromley Kent
15th

My dear Sir

I am very much obliged for your information. I will look to Annales des Sienc.[2] & have ordered the German Book.—[3]

With respect to the Australian Drosera, the *rate or quickness* of closing is the point, on which I am very anxious to have a few observations. I shall be surprised, judging from my observations, if the Australian Drosera can distinguish dry organic & inorganic substances. I found that they closed equally (or nearly so) over any substance, but they *released* vegetable or inorganic substances *much more* quickly than flies or meat.—[4]

The most curious results which I have arrived at, is the recognition by the leaves of *fluids* containing nitrogen & not containing nitrogen. This power of detecting nitrogen in fluids seems to me quite remarkable.—

Lindley does not refer to Buchanan or Wight, but he speaks in Veg. Kingdom p. 433 bottom of the movement of the Indian Drosera lunata; & I see in Steudel's Nomenclator the names of Buchanan & Wight in relation to this species.—[5]

I am surprised, if my instructions were followed by Croker that Leschenaultia has not made pods.—[6]

I am sorry to hear that the plants in the "isosceles beds" have not seeded better. All experiments require a tantalising amount of patience.

With *sincere* thanks for all your kindness | Believe me, My dear Sir | Yours sincerely | C. Darwin

It might be worth if you like the job while to put a *small* drop of milk & of **saliva** on a leaf of the Australian Drosera. Our Drosera likes milk better than any other drink.—[7]

Down House (MS 10: 10)

[1] Dated by the relationship to the letter to Daniel Oliver, 11 September [1860].

[2] The reference is probably to two papers about the leaves and glands of *Drosera rotundifolia* published in the *Annales des Sciences Naturelles (Botanique)* in 1855 (Groenland 1855 and Trécul 1855). These papers are both cited in *Insectivorous plants*, pp. 1 n., 5 n. See also letter to Daniel Oliver, 27 [September 1860].

[3] The reference is probably to the book written by Theodor Rudolf Joseph Nitschke on the physiology of *D. rotundifolia* (Nitschke 1859). CD seems never to have obtained a copy of this work. In *Insectivorous plants*, p. 1 n., he cited papers by Nitschke from the *Botanische Zeitung* for 1860 and 1861. Nitschke wrote two papers on *D. rotundifolia* in each of these volumes (Nitschke 1860a, 1860b, 1861a, and 1861b) that CD believed were 'by far the most important ones' to have been published on *Drosera*.

[4] CD's experiments on Australian *Drosera* are reported in *Insectivorous plants*, pp. 280–1, 281–3.

[5] See letter to Daniel Oliver, 11 September [1860].

[6] Charles William Crocker was foreman of the propagating department at the Royal Botanic Gardens, Kew. See letter to Daniel Oliver, 11 September [1860].

[7] Milk and saliva were among the first nitrogenous fluids that CD used in experiments to test the response of *D. rotundifolia* to various chemicals (*Insectivorous plants*, pp. 79–80). CD stated that he carried out these experiments because he 'was led to believe that the leaves absorbed nutritious matter from the insects which they captured' (*Insectivorous plants*, p. 76). His notes from this period on the effects of various substances on *Drosera* are in DAR 60.1.

To George Varenne Reed 15 September [1860][1]

Down Bromley Kent
Sept 15th

My dear Sir

I am very much obliged for your note & kind enquiries about our poor eldest daughter.— she steadily, but very slowly, improves improves in health; & as she has a wish to try sea-air, we all move to sea-side in a few days' time.—[2] On our return I shall be most glad to take advantage of your kindness & send Leonard twice a week to you.[3] As he is rather slow & backward (in part owing to loss of time from ill-health) it has occurred to us, that *if it did not prove inconvenient to you*, it might be better (until the days became too short,) for him to come to you on any two days in the afternoon a little before 3 oclock after his dinner. For by this means he would not lose his whole mornings, & he has much to make up. Frank, though put in a rather low form, keeps at the top in every subject;[4] & I am rather glad, considering his susceptibility, that he was put rather low, so as to master thoroughily all he has to do.—

I see by my accounts that my two Boys came to you for a fortnight after the last quarter was paid & I suppose it will be fair if I include this time in the next quarter, or ought broken time to rank as a full quarter— Perhaps you will kindly at some future time inform me.— I will let you know when we return in about a month's time; & then you can, if you please, let me hear about the *afternoons* for a month or so.—

With very sincere thanks for all your kindness & with our best compliments to Mrs Reed. Pray believe me | Yours very sincerely | C. Darwin

Buckinghamshire Record Office (D22.39.4)

1 Dated by CD's reference to a trip to the seaside (see n. 2, below).

2 The Darwins took Henrietta Emma Darwin and the other children to Eastbourne, Sussex, in late September 1860 ('Journal'; Appendix II).

3 CD first asked Reed about beginning to tutor Leonard Darwin in July 1859. See *Correspondence* vol. 7, letter to G. V. Reed, 1 July [1859].

4 Francis Darwin, formerly tutored by Reed, entered Clapham Grammar School in the summer of 1860 (J. R. Moore 1977, p. 53).

From Jeffries Wyman [*c.* 15] September 1860[1]

Cambridge
Sept 1860

Dear Sir

During the frequent conversations I have had with our mutual friend Prof Gray, on the Origin of Species, he has sometimes expressed a wish that I would communicate to you such views as I might have on the subject.[2] Entertaining the belief that progressive development is a far more probable theory than progressive creations, I have no desire to trouble you with any notions of my own; for I am sure after all the varied discussions, fair & unfair, which your work has called forth in Journals & Societies, anything I might say would prove superfluous. I think science owes you a huge debt, if for nothing else, for reopening the question in such a masterly manner as to create an universal interest in it both among scientific men & others.

My object in writing is only to say a few words on one or two unrelated points, for which I trust you will excuse me

First of all however permit me to state, that having some 18 months since made an excursion to the Uruguay La Plata & Parana rivers, also across the continent to Valparaiso, how much indebted I am to you for the pleasure, instruction & assistance derived from the "Journal of a Naturalist",[3] which served me as a guide book, wherever my route was the same with yours.— Among the subjects which interested me very much was your description of the Nãta, the curious breed of cattle with imperfectly developed upper jaw.[4] I do not see that you mention it in the Origin of Species, though it seems to me a good instance of a monstrosity becoming transmissible like that of the "otter breed" of sheep Manx cats, Dorking fowls &c. After numerous enquiries I suppose there can be no doubt that it is a *breed.*

What gave me especial interest in it however was, that several years since during an excursion to Labrador I found that a similar monstrosity was occasionally met with in the Cod fish & is sufficiently common to be known among fishermen as the "bull dog cod". I procured two specimens of it & prepared the heads which present deviations analogous to those of the nãta, viz an arrest of the development of the upper jaw. Among the things brought with me from S. America was the skull of a nãta, which shows very well its osteological peculiarities.[5]

I believe Prof Gray communicated to you a statement by me, in regard to the effect of the "paint root" on the Hogs causing the hooves of all but the *black* varieties to drop off.[6] I have every reason to believe the statement true not only from information which I obtained myself when in Florida, but from further enquiries made by Dr A. S. Baldwin of Jacksonville in that state;[7] who informs me that it is not only generally believed but practiced upon. & as one of the "Crackers", (ie a Florida squatter) remarked to me, we "select" the black members of a litter for raising as they alone have a good chance of living.[8]

I was rather sorry to find that the cows living in the swamps *did not* manifest any additional length of legs as Audubon & Bachman assert that the Deer do, when compared with those living on high grounds.—[9]

In the Origin of Sp^s^ you refer to the "rat" from Mammoth cave as illustrating the influence of the absence of light in modifying the eyes. In this case I believe you have been led into an error by Prof Silliman.[10] Prof Baird who is well informed on the subject, & eminently qualified to judge of the matter, tells me that the animal in question is not the Norway rat but belongs to another genus viz the Neotoma, the eyes of which are naturally of larger proportions than those of the Norway rat.[11] It is doubtful therefore whether any change has taken place in them.

In connection with the above I would also refer to what you say in regard to our old friend the tucu-tucu. (Ctenomys.)[12] Would not the blindness be explained in that case in some other way than as the result of accidental injury? I believe it is generally admitted by physiologists that the effects of mechanical injuries are not transmitted by inheritance & in view of the fact that the Jews after so many centuries of mutilation persist in being born with a prepuce, I am inclined to believe it. Is it not the persistence of an embryonic condition? All mammals go through a certain formality as regards their eye lids; viz. after the lids begin to grow they advance towards each other, meet, & adhere quite firmly; after having remained in this condition for a certain time variable in different species they open again, in some before birth as Rum^{nts} & Pachyderms, sometimes about the time of birth, as in Man & Monkeys, & in others not till several days after as in Rodents & Carnivore. The Russian *Rat mole* (Spalax) never opens them but retains through life the adhesive stage of the embryonic condition. I never had the good

AL incomplete
DAR 47: 165–6

CD ANNOTATIONS
0.1 Dear . . . grounds.—5.3] *crossed pencil*
2.1 First] 'Nãta cattle' *added pencil*
3.1 during . . . peculiarities. 3.7] *double scored ink*; *square bracket added after, pencil*
Top of first page: 'Nãta cattle. | Black Hog & Roots | Cave Rat' *ink, del pencil*; 'Keep for cause of Blindness in Tucu-tucu' *pencil*
End of letter: 'Tucutucu' *pencil*

[1] The date is inferred from CD's reply, which is dated 3 October [1860].

[2] Wyman and Asa Gray were both professors at Harvard College in Cambridge, Massachusetts. See letter from Asa Gray, 23 January [1860].

[3] Wyman probably refers to the American edition of CD's *Journal of researches*, published in two volumes in 1846 (Freeman 1977, p. 40).

[4] *Journal of researches* 2d ed., pp. 146–7.

[5] CD later cited this information in *Variation* 1: 89 n. 64. He stated: 'Prof. Wyman, of Cambridge, United States, informs me that the common cod-fish presents a similar monstrosity, called by the fishermen the "bulldog cod." Prof. Wyman also concluded, after making numerous inquiries in La Plata, that the niata cattle transmit their peculiarities or form a race.' Wyman had visited La Plata during an expedition to South America in 1858–9 (Dupree 1951, p. 106 n. 15).

[6] See letter to Asa Gray, 3 April [1860] and n. 4.

[7] A. S. Baldwin was a physician and naturalist who lived in Florida.

[8] CD included Wyman's information on hogs in *Origin* 3d ed., p. 12. In *Variation* 2: 227 and *Origin* 5th ed., the place of habitation was changed from Florida to Virginia.

[9] Audubon and Bachman 1846–54.

[10] CD cited Benjamin Silliman Jr's study of the blind animals of the Mammoth Cave in Kentucky (Silliman 1851) in *Origin*, p. 137, stating: 'In one of the blind animals, namely, the cave-rat, the eyes are of immense size; and Professor Silliman thought that it regained, after living some days in the light, some slight power of vision.'

[11] Wyman refers to the vertebrate anatomist Spencer Fullerton Baird. In fact, Silliman had recognised the difference between the cave-rat and the Norway rat (Silliman 1851, p. 336). CD corrected the third edition of *Origin* to include Wyman's comment that the cave-rats belonged to the genus *Neotoma* (*Origin* 3d ed., p. 154; Peckham ed. 1959, p. 283). See also letter from Benjamin Silliman Jr, 27 October 1860.

[12] CD described *Ctenomys*, a South American burrowing rodent that was occasionally blind, in *Journal of researches* 2d ed., pp. 50–1, and in *Origin*, p. 137.

From Thomas Vernon Wollaston [16 September 1860][1]

The whole of last winter was employed in finishing off Madeiran novelties;[2] so that the 50.000 Canarians are as yet *almost* untouched!—[3] Yet in spite of this, I have once or twice been foolish enough to picture to myself the unknown treasures of the Cape de Verdes (wh. I am aching to know something about), & have even contemplated *the possibility* of making a desperate rush for Fogo,—wh. I feel convinced is but an outlying detached portion of the great Central African Chain, cut off from the mainland by a series of those horrible convulsions to which you have such a violent objection,—tho' heaven only knows why.—[4] However *I believe* that my day-dreams (half-hatched) will (& ought to) end in smoke; & that I shall settle down like a Christian to my winter's work in sober earnest when this Cape de Verdian panic has blown over.—

I have had some curious geographical facts made known within the last few days, touching my Atlantic forms, which have puzzled me beyond measure & await further explanation.— A large & interesting batch of S. African specimens just brought home fr. the *Cape of Good Hope* by my friend Bewicke[5] contains a heap of my Atlantic genera, of the most peculiar types, wh. I had found in Madeira & the Canaries!!—such as **Tarphius** (the most eminently Atlantic form), *Europs* (an anomalous Euphorbia-feeder), & the nearly-blind *Cossyphodes*! (these out of a *single tube*; there is no telling how many more the 22 others will disclose).—— Hence (as far as is known) *Tarphius* has one representative in Sicily, 18 in Madeira, 7 in the Canaries, & 1 at the Cape!— (No doubt there would be 20 more at the C. de Verdes, & 5 or 6 in the Azores).—

I was not aware that Hooker had gone to Syria; he will *certainly* be massacred unless he turns Moslem,—wh perhaps however he intends to do, by way of another step gained? in the high-road of Development. How they have been pitching into you, to be sure!— I only wonder you are alive to tell the tale, & admire your pluck in proportion as I detest your Theory.—[6] I read the Bishop's notice in the Quarterly[7] (it was lent me by his great friend Sir Charles Anderson),[8] & chuckled over it beyond measure,—for I really did not think it ill-natured, though full of fun & caustic allusions which could not do you any harm.

You know I & Sam[9] agree wonderfully well in most things, so that you will not be surprised that I assented to his *reasoning* (in which I hold him very high). A naturalist he is not, & does not profess to be; nevertheless it did not strike me that there was anything even in his facts to which

AL incomplete
DAR 205.3 (Letters)

CD ANNOTATIONS
1.1 The . . . over.—1.11] *crossed brown crayon*
2.1 I have . . . Azores).—2.11] '19'[10] *added brown crayon, circled brown crayon*
2.7 nearly-blind *Cossyphodes*!] *underl brown crayon*

2.8 Hence . . . Azores).—2.11] 'Mem. S African genera of Plants— Fernando Po plants | Highlands of Africa plants'[11] *added pencil*
3.1 I was . . . to which 4.4] *crossed brown crayon*
Top of letter: 'Sept 16th. 1860— Wollaston' *ink*

[1] The date is taken from CD's annotation.

[2] In 1857, Wollaston gave his collection of Madeiran Coleoptera to the British Museum and prepared a catalogue of the collection (Wollaston 1857). From 1858 to 1862, he contributed a series of papers to the *Annals and Magazine of Natural History* on the new species found since the initial catalogue was prepared.

[3] Wollaston described the British Museum's collection of coleopterous insects from the Canary Islands, collected by himself and others, in Wollaston 1864.

[4] CD had criticised Wollaston's assumption in Wollaston 1854 that the islands of the Madeira group were connected to continental Europe in the Miocene period (see *Correspondence* vol. 6, especially letter to Charles Lyell, 16 [June 1856], and letter from T. V. Wollaston, [27 June 1856]).

[5] In Wollaston's papers on Madeiran beetles published in the *Annals and Magazine of Natural History* (see n. 2, above), he stated that many had been collected by his friend C. Bewicke of Palmeira, near Funchal, Madeira.

[6] Wollaston harshly criticised *Origin* in a review published anonymously in the *Annals and Magazine of Natural History* ([Wollaston] 1860). CD's copy of the review is in the Darwin Pamphlet Collection–CUL.

[7] [Wilberforce] 1860.

[8] Charles Henry John Anderson and Samuel Wilberforce had been students together at Oriel College, Oxford.

[9] Samuel Wilberforce.

[10] The number of CD's portfolio of notes on the geographical distribution of animals.

[11] CD's annotation refers to Wollaston's information about the geographical distribution of *Tarphius*. CD apparently intended to consider the problem in relation to the distribution of African plants, particularly the migration of species from north to south during the worldwide cold period.

From James Drummond 17 September 1860

Hawthornden Farm
Sep 17th *1860*

Dear Sir,

In compliance with your request,[1] I have paid particular attention to the *Goodeniaceus* plants which I have met with in flower, since I received your letter, *I met with Lechenaultiae Grandiflora*, in abundance covering the flat top of an ironstone Gravelly hill several acres in extent and so profusely covered with the flowers of this beautiful plant that the whole surface of the ground seems of a blue colour from its blossoms, but I have not been able to learn much relating to this genus which you have not found out by observing L Formosa in your greenhouse, the pollen leaves the anthers long before the flower opens, just when the lips of the indusial cup reaches as high as the filmens which support the anthers, it therefor receives almost their whole contents, when the cup is open to the crown of the Germen it is here no doubt impregnation takes place whenever it does take place but that *comparatively speaking*, is a very rare event in plants of this genus I may safely say that one seed vessel in 500 I would probably be within bounds if I said 5000 ever under any circumstances is fertilized by the pollen, and comes to maturity, the Lechenaultias are all creeping rooted plants at least all the ornamental species and they are altogether independant of seeds for their reproduction[2] I have been more sucessful in other genera of this family the curious Genus Distylis a yellow flowered annual plant growing in this neighbourhood shows clearly the Eoconomy of this family of plants, to understand it the flower but[3] must be taken very young long before the flower opens, I then find that the double cupped Indusium which can only be considered two branches of the style, has just risen to the height of the filaments which support the anthers, the anthers and filaments are closely invested by the unexpanded corolla at this stage therefor the whole of the pollen contained in the anthers is discharged into the two indusial cups and it is evident in this plant that impregnation imedeattely takes place, as the seeds which are many in number and large for the size of the plant very soon begin to enlarge, by the time the flower opens the two indusial cups are greatly altered in appearance they look like two simple ciliated filaments, but by this time the germen is considerably enlarged, I have made observations on another curious *annual* genus of this order, the capsule of which contains two seeds very large for the size of the plant, and resembling grevilleae in the way they lie in the seed vessel, each seed is furnished with a broad wing which goes all round the seeds I have examined 2 species I do not know the name of the genus, I shall try and send you seeds of these two species, and *distylis* the Œconomy with regard to the Anthers and Indusium is exactly the same, as in *distylus*, neither of them are ornamental plants but I think they will clearly explain the Mystery about the impregnation of the Goodeniaceae

I have not yet been able to make any observations on *Brunoniae*, the habbitat I knew of near my residence had been destroyed by a flock of sheep grazin over it, and I have not yet found another

I am Sir Your obedient Humble Servant | James Drummond

DAR 157a

CD ANNOTATION

0.1 Dear . . . colour 1.5] *crossed ink*

[1] See the letter to James Drummond, 16 May 1860, in which CD asks for information about the fertilisation of *Leschenaultia formosa*. Drummond was superintendent of the government botanic gardens in Western Australia, where *Leschenaultia* is common.

[2] According to a later letter, Drummond wrote two letters to CD on or around 17 September (see letter to James Drummond, 22 November [1860]). The other letter has not been found. It appears that Drummond did not provide CD with information about the activities of insects in the fertilisation of species of *Leschenaultia* until the end of the year (see letter to James Drummond, 20 December [1860] and n. 4).

[3] Butt: 'the end or tip of a branch' (*OED*).

[4] See letter to James Drummond, 20 December [1860].

To George Gordon 17 September [1860]

Down Bromley Kent
Sept. 17th

Dear Sir

I thank you sincerely for your great kindness in sending me the famous lot of Goodyera,[1] which arrived in very fair condition, being so excellently packed, & which I have dissected with the greatest interest this morning. The structure of the rostellum is much nearer to that of Epipactis than to that of Spiranthes, though in some degree intermediate between them.[2] I beg leave again to give you my cordial thanks.

Dear Sir | Yours truly obliged | Charles Darwin

I enclose the stamps.

P.S | I have just received your most kind note of the 14th.—[3] I believe I shall leave home on Saturday for the sea-coast; so I fear I shall hardly receive the fresh specimens, if you send them. Those sent were hardly vigorous enough to act perfectly but I feel morally almost certain that the action is the same as in Epipactis. I have planted the roots in peat in shady place & they may flower next year with me. The contrivances for insect fertilisation in Orchids are multiform & truly wonderful & beautiful.

You have added to your kindness in sending me the specimen by taking so much trouble so freely & pleasantly

Endorsement: Sepr *1860*'
Moray District Record Office (DDW 71/860/8)

[1] See letter to George Gordon, 11 September [1860]. CD thanked Gordon in *Orchids*, p. 112, for supplying him with living specimens of 'this rare Highland Orchid'.

[2] CD described *Goodyera repens* as being 'an interesting connecting link between several very distict forms' of orchids (*Orchids*, p. 114). About the rostellum, he recorded: 'In the structure of the rostellum, with the exception of the sloping sides, and in the shape of the labellum, we see the affinity of Goodyera to Epipactis.' He went on to suggest that 'Goodyera probably shows us the state of the organs of fructification in a large group of Orchids, now mostly extinct, but the parents of many living descendants.' (*ibid.*, pp. 115–16).

[3] Gordon's letter has not been found.

To T. H. Huxley 18 September [1860]

Down Bromley Kent
Sept 18th.

My dear Huxley

I was indeed grieved to receive your news this morning.—[1] I cannot resist writing, though there is nothing to be said.

I know well how intolerable is the bitterness of such grief.[2] Yet believe me, that time, & time alone, acts wonderfully. To this day, though so many years have passed away, I cannot think of one child without tears rising in my eyes; but the grief is become tenderer & I can even call up the smile of our lost darling, with something like pleasure. My wife & self deeply sympathise with Mrs Huxley & yourself. Reflect that your poor little fellow cannot have had much suffering.

God Bless you. | Charles Darwin

I have written to John Lubbock

Postmark: SP 19 60
Imperial College of Science, Technology and Medicine Archives (Huxley papers: Family deposit 1987)

[1] Noel Huxley, the Huxleys' oldest child, died from scarlet fever on 15 September 1860. He was nearly four years old. See L. Huxley ed. 1900, 1: 151–2, 216.

[2] CD refers to the death of his oldest daughter, Anne Elizabeth Darwin, aged ten, in 1851. See *Correspondence* vol. 5.

From Charles Lyell[1] 18 September 1860

Sept 18 1860

I found yr letter here & thank you for intending to send me Asa Gray's last article which I will return without delay as also the M.S. on dogs.[2] Agassiz in some one of his Essays leans to Pallas' views,[3] which always struck me as being rash in him who is for the sanctity of species, whereas if any thing could show Nature's entire indifference in regard to the violation of such supposed sanctity it would be that those now existing should be of multiple origin.

You need not be afraid of my starting any theory of successive creation of types. I have never had any leaning towards a compromise and thought Carpenter very inconsistent in going so very far with you, & then concluding that there was probably in the case of Man, an exception, whereas there would be no end of all the morphol! & embryological arguments if such a concession was made.[4]

But what I asked or intended to ask in my last letter was different—perhaps you meant to reply that you had no fixed opinion on the subject—but I wish to be sure.— Can we assume as at all probable that all mammalia came from one original Stock instead of several distinct mammalian-types each developed by small & successive modifications out of lower ornithic reptilian or perhaps monotrematous prototypes. It would greatly simplify matters if single & exclusive areas could be assigned or even speculated upon, on some even slight data, or if single periods could be proposed as those of the first coming in of mammalia. But as I understand your views this is not very probable. If for example we found before the triassic microlestes, some monotremes in Permian or Carbonif^s beds & fixed on them as the probable starting point of mammalia, what influence would the development of a mammal in Asia or Africa have in preventing some other line of monotremes from improving into a totally different genus of mammal in Australia or America & supposing the mammal first formed, say in Asia, to be extinguished, what is there in your system to cause us to despair of the higher or placental grade from ever being evolved because after a geological period, the earlier formed mammalia died out.

I have always expected to find mammals intermediate between Stonesfield & Eocene, but I also looked for them before Stonesfield. When therefore they turned up both in Trias & in Purbeck, it agreed with my anticipations.[5] The Triassic surprised some people more than the Purbeck, for they held the doctrine, propounded by you in p. 316, once begun the group is continued till it dies out.

I wish you could give the slightest reason why it should not begin more than once in more than one place. I incline to think it has not, but why? according to the principle of selection why when once in any one quarter of the globe at any one period the step in advance has been taken are the inferior types elsewhere to be checked, & not to presume to work up into any genus of corresponding grade & class? A frog has a vast deal of one type in him even the double occipital condyle, to say nothing of his limbs & digits & the absence of a tail, tho' he is low in the reptilian scale, (but it is true, he does not as yet go far back in time) but take some triassic reptile or bird I cannot see why they are never to be modified into something higher because a microlestes has entered on the stage.[6] Two independent chains of development would not be discordant with your machinery of selection?

I am not aware of any fossil bats or rodents in Australian caves, so the small antiquity of rodentia may be a reason for their not having given origin as yet to Gyrencephala & does there not seem some connection between the low grade of Lyssencephala as the only occupants of that vast Australian continent?[7]

Can you not refer me to some book or paper where I can find how many plants may have been extirpated in St Helena & may I not presume that as many or more endemic insects have been annihilated there besides some land-shells?

As to Australia, I cannot help thinking that the Marsupials must in the long run (if Man were away) be best fitted for that continent & that after European or Asiatic Gyrencephala had been introduced & then left to themselves, they w.[d] during some 4 years drought, after a century or much more, have all perished & given way before their pouched rivals, who as Owen has somewhere suggested, w[d] have migrated with their young in the pouch, wholly indifferent to the long leaps of the mother.[8] This w[d] not apply to bats, but perhaps you may be able to object that some of the native placental rodents w[d] be no better off than a lion or an Elephant in a 4 years drought?

In y[r] chapter on grafting you might, it struck me, have alluded to those wonderful cases of nat.[l] grafting recorded in the fir-woods of Germany. I think I could find a reference to it, where a totally leafless stump keeps alive by its roots, borrowing from the roots of adjoining fir-trees till the bark closes over.[9] It answers the question why Nature has given such a power & was it for the mere sake of human gardeners.

I am not going to press the objection about the non-variation, adaptation & advance of bats & rodents, so as to destroy the weight of y.[r] argument respecting absence of other mammalia in islands, as I cannot conceive such absence not being due to the want of migratory powers of mammalia, & that alone w^{d} not do, unless the coming in of new mammals was by inheritance from pre-existing types. But it is better to show that one does not overlook objections.

I think I asked you once before whether the admission of the multiple origin of dogs w^{d} not impair one's faith in the single origin of the human race, & as you did not reply, you perhaps meant that you had nothing to hazard in the way of speculation. You remember a passage in Agassiz about negros being black where the Chimpanzee is black, & yellow in Borneo? where the orang is yellow, or something to that effect.[10] I could hunt out the passage if you do not remember it. As he does not connect Man genealogically with any anthropomorph, he says this without any hesitation.

To shorten correspondence I will imagine the best answer I can for you & beg you to try & strengthen it.

It is so many millions to one against the right variation offering itself in the right place & time that many geol^{l} periods must pass away before the first mammal is formed & as an equal lapse of ages w^{d} occur before a 2^{d} offered, the first will in the meantime have spread over the globe & branched out into many genera, w^{h} being of high grade w^{d} keep down any incipient mammals, just as homo sapiens when once evolved w^{d} take good care that no other anthropomorph sh^{d} improve into rationalism & compete with him for the good things of this world. No island has existed for so long a time as that which separates the Lower Miocene from the human epoch, therefore the chances have been far too many against transmutation of any creature of an inferior class into one of a higher class on an island, more especially as bats & rodents w^{d} help to check any signs of an invasion of the prerogatives of the dominant class, they being already there to represent the vertebrate aristocracy.

Incomplete
Copy
Kinnordy MS, Charles Lyell's journal VI, pp. 187–95d

[1] The text of the letter has been taken from a copy in Lyell's scientific journal. It is also printed in Wilson ed. 1970, pp. 474–7. The entry is headed: 'Copy of letter to C. Darwin Bonn Sept 18 1860'.

[2] See the letter to Charles Lyell, 11 August [1860], in which CD offered to send Lyell his manuscript on the origin of domestic dogs. A further letter, in which he mentioned sending the second part of [Gray] 1860b, has not been found. It is not discussed in the letter to Charles Lyell, 12 September [1860].

[3] Louis Agassiz discussed the possible origin of dogs in the first volume of the *Contributions to the natural history of the United States of America*, where he stated (Agassiz 1857–62, 1: 164):

> where is the unprejudiced naturalist who in our days would dare to maintain: 1st, that it is proved that all the domesticated varieties of sheep, of goats, of bulls, of llamas, of horses, of dogs, of fowls, etc., are respectively derived from one common stock; 2d, that the supposition that these varieties have originated from the complete amalgamation of several primitively distinct species is out of the question.

Moreover, Agassiz asserted that such proof, were it ever obtained, was not relevant to the problem of defining species. Pyotr Simon Pallas believed that domestic animals such as dogs were derived from several different species (Pallas 1780, pp. 84, 100). See *Correspondence* vol. 7, letters to Charles Lyell, 25 October [1859] and 31 [October 1859], and letters from Charles Lyell, 28 October 1859 and 25 [November 1859].

4 [Carpenter] 1860b, pp. 403–4. In his review of *Origin*, William Benjamin Carpenter stated that he was 'strongly convinced' of the truth of natural selection, but that it seemed to him probable that 'each of the great types' owed their origin to creative forces. 'So, too, there seems to us so much in the psychical capacity of Man, however degraded, to separate him from the nearest of the Mammalian class, that we can far more easily believe him to have originated by a distinct creation, than suppose him to have had a common ancestry with the Chimpanzee, and to have been separated from it by a series of progressive modifications.' ([Carpenter] 1860b, p. 404). CD's annotated copy of the review is in the Darwin Pamphlet Collection–CUL.

5 Lyell described the mammalian remains from the Upper Trias and Purbeck shale beds in a *Supplement* to his *Manual of elementary geology* (C. Lyell 1857, pp. 17–24). These new fossils were even older than those of the Stonesfield slate formations.

6 *Microlestes* remains in Triassic deposits in Germany were the earliest mammalian fossils to have been found. The discovery of *Microlestes* remains in England had also recently been reported (see C. Moore 1858 and 1860).

7 In Richard Owen's classification of the Mammalia, the Gyrencephala included all the higher mammals except for man; the Lissencephala comprised the lower placental mammals, including rodents, bats, edentates, and insectivores. Lyell apparently refers to Owen's lowest division, the Lyencephala, which included the monotremes and marsupials predominantly found in Australia. See R. Owen 1859a, pp. 24–6, 33.

8 R. Owen 1846a, p. 197.

9 Lyell refers to Göppert 1842. See the letter to Charles Lyell, 23 [September 1860], in which CD asks Lyell for further information on this point, and Lyell's response in the letter from Charles Lyell, 25 September 1860.

10 Lyell probably refers to a remark made by Agassiz in a letter prefixed to Nott and Gliddon eds. 1857, p. xiv: 'I have already alluded, on another occasion, to the identity of color of the Malays and orangs: may we not now remember, also, a similar resemblance between some of the species of Hylobates with the Negrillos and Telingans?' In the same letter, Agassiz referred to his earlier essay on 'The natural provinces of the animal world and their relation to the different types of man' (in Nott and Gliddon 1854), but that work does not contain the allusion mentioned.

From Daniel Oliver 19 September 1860

Kew. W.
19. IX. 60

My dear Sir,

Your observations upon the *sensible* little Drosera, preferring milk & nitrogenous fluids,—also the effect of Nit. Am[a] are very interesting.—[1] Are you satisfied, though, that in respect to fluids the *density* of the drop put upon its leaf has not as much to do with the incurving, &c. as its chemical composition?— Or in the case of the Nit. Am[a] that it is not the same result wh. any soluble irritant might occasion,—say common salt,—destitute of Nitrog[n] I have experimented on one of our *Droseras* with fluids at yr suggestion. I tried (not saliva! but) sol. of gum,—Com. gum diss[d] in water. (perhaps a little sour!) syrup—, & milk. By far the *most marked* result has been afforded by the *gum drops*.[2] Had the syrup been denser it might have equalled it, tho' the result in its case as yet has been immaterial or next to nothing.

But I am by no means sure whether I *can shew* that it is a question of density (or Endosmose &c, rather), for I've thought it possible the gum sol. might, by virtue of its adhesiveness (?) &c. in some way or other deceive me.[3]

On Tuesday P.M. (2–2.15) a drop of milk, of syrup, & of gum placed on 3 several leaves of the Drosera. In each case several of the marginal hairs of the rounded extreme of the leaf were patent or erect.

4.45. Gum drop most *conspicuously*, by far, affected. But 2 glands are left sticking right out. Nearly all the rest are more or less incurved—right into—the drop.— In the milk & syrup not much change noted, perhaps the latter more aff^d than the milk.

4.50 a drop of gum placed on another leaf. All the marginal glands perfect. The leaf supported by a slip of glass.—at 6.8 no incurving notable

Wednesday. 8.36. All the glands of the *last* (4.50) leaf remarkably incurved.— In the old gum-drop no marked *further* change,—but 1 gland now sticks conspic^y out.

Milk, no very marked change,—prob^y a few incline towards the drop.— Syrup no mk^d change.—

I dont know how far you have tried various fluids or whether further experiments are needed to *settle* the Nitrogenous part of the Curving.

The progress of change is not easy to describe without minute observation or perhaps drawings.— A series of wee photographs might do the work nicely.—

Very faithfully yours | Danl Oliver Jr

Chas. Darwin, esq.

DAR 58.1: 12–13

CD ANNOTATIONS

4.1 *4.45.* . . . milk. 4.4] 'Any glue with gum?' *added pencil*

5.1 The leaf . . . notable 5.2] *two marginal crosses added pencil*

[1] See letters to Daniel Oliver, 11 September [1860] and 15 [September 1860].

[2] This was opposed to CD's view that only nitrogenous matter caused the inflection of the leaves.

[3] The term 'endosmose' was coined by René Joachim Henri Dutrochet. He established that 'when two liquids of different densities or of different chemical natures are separated by a membranous partition, two currents are established across this partition, going in opposite directions, and unequal in force'. Dutrochet called the movement of fluid into organic tissue under these conditions 'endosmose'. See Dutrochet 1828, p. 1.

To W. H. Harvey [20–4 September 1860][1]

[Eastbourne]

My dear Sir

I have read your long letter with much interest & I thank you sincerely for your great liberality in sending it me.—[2] But on reflexion I do not wish to attempt answering any part, except to you privately: anything said by myself in defence would have no weight; it is best to be defended by others or not at all.

Parts of your letter seem to me, if I may be permitted to say so, very acute & original; & I feel it a great compliment your giving up so much time to my book. But on the whole I am disappointed; not from your not concurring with me, for I never expected that; & indeed in your remarks on Ch. XII & XIII you go so much further with me, (though a little way) than I ever anticipated & am much pleased at the result. But on the whole I am disappointed, because it seems to me that you do not understand what I mean by Natural Selection, as shown at p. 11 of your letter & by several of your remarks.— As my book has failed to explain my meaning it would be hopeless to attempt it in a letter. You speak in early part of your letter & at p. 9. as if I had said that Natural Selection was the sole agency of modification; whereas I have over & over again, ad nauseam, directly said & by order of precedence implied (what seems to me obvious) that selection can do nothing without previous variability. see p. 80, 108, 127, 468, 469 &c "Nothing can be effected unless favourable variations occur". I consider Natural Selection as of such high importance, because it accumulates successive variations in any profitable direction; & thus adapts each new being to its complex conditions of life.— The term "Selection" I see deceives many persons; though I see no more reason why it should than *elective* affinity, as used by the old chemists. If I had to rewrite my book, I would use "natural preservation" or "naturally preserved". I sh^{d} think you would as soon take an emetic as reread any part of my Book, but if you did & were to erase selection & selected & insert preservation & preserved, possibly the subject would be clearer.

As you are not singular in misunderstanding my Book, I should long before this have concluded that my brains were in a haze, had I not found by published reviews & especially by correspondence that Lyell, Hooker, Asa Gray, H. C. Watson, Huxley & Carpenter & many others *perfectly* comprehend what I mean.[3]

The upshot of your remarks at p. 11 is that my explanations &c & the whole doctrine of natural selection are mere empty words signifying the "order of nature"; as the above named clear-headed men, who do comprehend my views, all go a certain length with me & certainly do not think it all moonshine, I should venture to suggest a little further reflexion on your part. I do not mean by this to imply that the opinion of these men is worth much as showing that I am right, but merely as some evidence that I have clearer ideas, than you think; otherwise these same men must be even more muddled headed than I am; for they have no temptation to deceive themselves. In the forthcoming September no^{r} of the American Journal of Science there is an interesting & short Theological Arcticle (by Asa Gray),[4] which gives incidentally with admirable clearness the theory of Natural Selection, & therefore might be worth your reading: I think that the theological part would interest you.)

You object to all my illustrations: they are all necessarily conjectural, & may be all false; but they were the best I could give. The Bear case has been well laughed at, & disingenuously distorted by some into my saying that a bear could be converted into a whale; as it offended persons I struck it out in 2^{d} Edition; but I still maintain that there is no especial difficulty in a Bear's mouth being enlarged to any degree useful to its changing habits,—no more difficulty than man has found in increasing the crop of the pigeon, by continued selection, until it is literally as big as whole rest of body. If this had not been known, how absurd it would have appeared to say that the crop of a bird might be increased till it became like a balloon.

(With respect to the ostrich, I believe that the wings have been reduced & are not in course of development, because the whole structure of a Bird is essentially formed for flight; & the Ostrich is essentially a bird. You will see at p. 182 of "Origin" a somewhat analogous discussion. At p. 450 of 2^d Edit. I have pointed out the essential distinction between a nascent & rudimentary organ.— If you prefer the more complex view that the progenitor of the Ostrich lost its wings, & that the present ostrich is regaining them, I have nothing to say in opposition.—)

(With respect to Trees on islands: I collected some cases, but took the main facts from Alph. De Candolle, & thought they might be trusted.[5] My explanation may be grossly wrong; but I am not convinced it is so; & I do not see the full force of your argument of certain herbaceous orders having been developed into trees in certain rare cases on continent. The case seems to me to turn altogether on the question whether generally herbaceous orders more frequently afford trees & bushes on islands, than on continents, relatively to these areas.—)

(In p. 4. of your letter you say you give up many Book-species as separate creations; I give up all, & you infer that our difference is only in degree & not in kind. I dissent from this; for I give a distinct reason how far I go in giving up species; I look at all forms, which resemble each other homologically or embryologically as certainly descended from the same parents.)

(You hit me hard & fairly about my question (p. 483 Origin) about creation of eggs or young &c—(but not about mammals with mark of umbilical chord).[6] Yet I still have an *illogical* sort of feeling that there is less difficulty in imagining the creation of an asexual cell, increasing by simple division.)

(Page 5 of your letter,— I agree to every word about antiquity of World; & never saw the case put by any one more strongly or more ably. It makes, however, no more impression on me, as an objection; than does the astronomer when he puts on a few hundred-million miles to the distance of the fixed stars. To compare very small things with great,—Lingula &c remaining nearly unaltered from Silurian epoch to present day is like the dovecot pigeons still being identical with wild rock-pigeons, whereas its "fancy" offspring have been immensely modified & are still being modified by means of artificial Selection.)—)

(You put the difficulty of the first modification of the first protozoon *admirably*: I assure you that immediately after 1st. Edit. was published this occurred to me; & I thought of inserting it in 2^d. Edit. I did not, because we know not in the least what the first germ of life was; nor have we any fact at all to guide us in our speculations on the kind of change which its offspring underwent. I dissent quite from what you say of myriads of years it would take to people world with such imagined protozoon.— In how very short a time Ehrenberg calculated that a single infusorium might make a cube of rock—[7] A single cube on geometrical progression would make the solid globe in (I suppose) under a century. From what *little* I know I, cannot help thinking that you underrate the effect of the physical conditions of life on these low organisms. But I fully admit that I can give no sort of answer to your objection; yet I must add that it would be marvellous if any man ever could, assuming for the moment that my theory is true.— You beg the question, I think, in saying that Protococcus would be doomed to eternal similarity:—nor can you know that the first germ resembled a Protococcus or any other now living form.—)

(Page 12 of your letter. There is nothing in my theory necessitating in each case progression of organisation; though natural selection tends in this line, & has generally thus acted. An animal if it became fitted by selection to live the life, for instance, of a parasite, will generally become degraded. I have much regretted that I did not make this part of the subject clearer; I left out this & many other subjects, which I now see ought to have been introduced. I have inserted a discussion on this subject in the foreign Editions.[8] In no case will any organic being tend to retrograde unless such retrogradation be an advantage to its varying offspring; & it is difficult to see how going back to the structure of the unknown *supposed* original protozoon would ever be an advantage.)

(Page 13 of your letter. I have been more glad to read your discussion on "dominant" forms than any part of your letter:[9] I can now see that I have not been cautious enough in confining my definition & meaning. I cannot say that you have altered my views. If Botrytis had exterminated the wild Potatoe, a low form would have conquered a high; but I cannot remember that I have ever said (I am sure I never thought) that a low form would *never* conquer a high. I have expressly alluded to parasites half exterminating game-animals: & to the struggle for life being sometimes between forms as different as possible, for instance between grasshoppers & herbivorous quadrupeds. Under the many conditions of life which this world affords, any group which is numerous in individuals & species & is widely distributed may properly be called dominant. I never dreamed of considering that any one group, under all conditions & throughout the world, could be *pre*dominant. How could Vertebrata be predominant, under the conditions of life to which parasitic worms live? What good would their perfected senses & their intellect serve under such conditions? When I have spoken of dominant forms, it has been in relation to the multiplication of new specific forms, & the dominance of any one species has been relative generally to other members of the same group, or at least to beings exposed to similar conditions & coming into competition. But I daresay that I have not in the Origin made myself clear, & space has rendered it impossible. But I thank you most sincerely for your valuable remarks, though I do not agree with them.)

(About sudden jumps;[10] I have no objection to them: they would aid me in some cases: all I can say is, that I went into the subject, & found no evidence to make me believe in jumps; & a good deal pointing in the other direction.— You will find it difficult (p. 14 of your letter) to make a marked line of separation between fertile & infertile crosses. I do not see how the *apparently* sudden change (for the suddenness of change in chrysalis is of course largely only apparent) in larvæ during their development throws any light on this subject.—)

I wish I could have made this letter better worth sending to you: I have had it copied to save you at least the intolerable trouble of reading my bad hand-writing. Again I thank you for your liberality & kindness in sending me your criticisms; & I heartily wish we were a little nearer in accord, but we must remain content to be as wide asunder as the poles; but without, thank God, any malice or other ill feeling.—

My dear Sir | Yours very sincerely

Draft
DAR 98 (ser. 2): 45–53

CD note:[11]

Every group of being which is great in number & widely distributed is dominant in Harveys sense & truly so, but *in relation of the multiplication of new species, [*interl*] I have always been referring to dominance over other members of the same group, & which in many, perhaps all [*under del* 'most'] cases; implies the withstanding better the attacks of *various enemies, [*interl*] parasitic animals & plants & *insects too [*interl*] & ultimately tends ['to make, in as far as' *del*] if many new species are produced, to make the groups thus increased by so far more dominant in the region under consideration.

It w[d] be more correct to speak of parasites as enemies, or as unfavourable conditions like cold or heat rather than as competitors or as concerned in struggle for life.—The competition is rather between allied forms which can best withstand such enemies.—

[1] The text has been taken from a draft of the letter in DAR 98 (ser. 2): 45–53. The draft was probably composed over several days. The earliest possible date on which it could have been written was 20 September 1860, when CD composed lengthy notes pertaining to Harvey's letter (see letter from W. H. Harvey, 24 August 1860, CD notes). It was completed by 24 September, for on that day he sent the manuscript to his copyist Ebenezer Norman with instructions to make a clean copy (see letter to Ebenezer Norman, [24 September 1860]).

[2] Letter from W. H. Harvey, 24 August 1860.

[3] At this point in the draft, CD wrote and then deleted: 'These men who are clear-headed men, all go to some length with me; & I would ask you to pause before'.

[4] CD refers to [Gray] 1860c.

[5] In *Origin*, p. 392, CD cited Alphonse de Candolle as the authority for his statement that trees had restricted geographical ranges. CD postulated that trees would not be likely to colonise isolated islands and that those found on islands were usually arborescent forms of herbaceous orders. The reference was taken from A. de Candolle 1855, 1: 527–32. CD's annotated copy of the work is in the Darwin Library–CUL.

[6] In *Origin*, p. 483, CD asked of those who believed in miraculous acts of creation: 'Do they believe that at each supposed act of creation one individual or many were produced? Were all the infinitely numerous kinds of animals and plants created as eggs or seed, or as full grown? and in the case of mammals, were they created bearing the false marks of nourishment from the mother's womb?'

[7] Ehrenberg 1854–6, pp. viii–ix. Christian Gottfried Ehrenberg, the founder of micropalaeontology in Germany, had corresponded with CD from 1844 to 1846 concerning samples of earth CD had taken from various countries visited on the *Beagle* voyage. Ehrenberg described the protozoa he found in these samples in Ehrenberg 1854–6, pp. 289–95.

[8] The discussion was included in the revised American edition of *Origin* (pp. 116*–21*) and in the German translation (Bronn trans. 1860, pp. 133–7). See letters to Asa Gray, 28 January [1860], and to H. G. Bronn, 14 February [1860]. See also Appendix IV.

[9] See CD's note, above.

[10] See letter from W. H. Harvey, 24 August 1860 and n. 13.

[11] The note is in DAR 98 (ser. 2): 44. CD marked it in brown crayon: 'Ch IV Keep—'. He refers to chapter 4, on 'Variation under nature', of his 'big book' on species (*Natural selection*, pp. 95–171). CD also wrote in ink: '& p. 8 of my letter to Harvey'. He marked the first paragraph with square brackets in brown crayon. The page of Harvey's letter to which CD refers discusses the problem of 'dominant' forms.

To Daniel Oliver 21 [September 1860][1]

Down Bromley Kent
21st

My dear Sir

I am **infinitely** obliged for your note.[2] The first thought which occurred to me was endosmose (& I tried olive oil & wine).—[3] I have tried so many experiments that I cannot think I am in error. I rely much in the case of the saliva experiments on the diminishing effect of weaker & weaker doses. Sulphate & Muriate of Ammonia, Nitrate of Potash & N. of Lime produce none or hardly any effect. With respect to the Vegetable fluids I rely on nitrogenous fluid placed on the **same** *leaves* subsequently causing contraction, whereas the vegetable matters produced no effect.—[4]

Nothing ought to be placed under the leaf as a support, for I have ascertained (in our Drosera) that irritation on the smooth under side of leaf causes contraction.

So small a drop ought to be put on as not to entangle the marginal hairs, as the act of drying of a viscid fluid *mechanically* draws the hairs together.—

Are you sure that there is no glue with your gum?[5] I tried *pure* white gum & it produced no action at all. Nor did syrup of white sugar or starch.—For Heaven sake try some pure gum.— You have given me a panic. I tried

milk on	9	leaves	and all strongly contracted. on *different* plants
thin Gelatine on	4		
White of egg on	6		
Saliva on	8		
Urine on	11		
Mucus on	4		
infusion of meat	1		
	43	leaves.—	

I have not tried vegetable fluid so largely only 3 or 4 leaves for each; for absolutely no effect was produced.

Believe me that I am **truly grateful** for your cautions.—

In Haste | Yours very sincerely | C. Darwin

I start for sea-side tomorrow,[6] but letters will be forwarded to me.

P.S | Does not the sourness of your gum prove that it has fermented? And does not fermentation imply that there has been some nitrogenised matter in it? I am nearly sure that yeast is nitrogenised. Most cryptogamic plants certainly contain nitrogen.— For Heaven sake dissolve fresh *pure*, white gum & try again.[7]

The inflection of disc of leaf itself, when it occurs is the best of all evidence.—

I fully believe, I have now ascertained that $\frac{1}{2880}$ of grain of N. of Ammonia occasionally suffices with young & very sensitive leaf—, so that scrupulous care is requisite.—[8]

Down House (MS 10: 11, 14)

[1] Dated by the reference to the Darwins' leaving for the seaside (see n. 6, below).
[2] Letter from Daniel Oliver, 19 September 1860.
[3] See letter from Daniel Oliver, 19 September 1860 and n. 3.
[4] CD's notes on experiments to determine the effects of nitrogenous and non-nitrogenous fluids on the leaves of *Drosera* are in DAR 60.1. The experiments are described in *Insectivorous plants*, pp. 76–84.
[5] When CD discussed this case in *Insectivorous plants*, pp. 77–8, he mentioned the problems of impurities in experimental samples of gum. He probably was referring to Oliver when he stated: 'It is necessary to try pure gum arabic, for a friend tried a solution bought ready prepared, and this caused the tentacles to bend; but he afterwards ascertained that it contained much animal matter, probably glue.'
[6] CD went to Eastbourne on 22 September 1860 ('Journal'; Appendix II).
[7] Oliver wrote on the letter in pencil: 'Some bits of clear gum picked'.
[8] The postscript is on a separate slip of paper. The contents indicate that it belongs with this letter.

To Daniel Oliver [22–3 September 1860][1]

15 Marine Parade | Eastbourne

My dear Sir

This is my address.— I will write soon— I am grateful for your last note.—

I am dying to hear about pure gum[2]

[Yours] | C. Darwin.

P.S.[3] | Will you be so kind as to dry a little of your Gum used on Drosera & burn it; perhaps you would detect smell of animal matter if any gelatine or glue used or size[4] used.—

Any corrosive Sublimate??[5]

Could you send me, when you write, one leaf (& name) of Australian Drosera, that I may see it out of idle curiosity.—[6]

Down House (MS 10: 12, 18)

[1] Dated by CD's reference to having arrived in Eastbourne. CD and the children left Down on 22 September ('Journal'; Appendix II). The letter was written prior to receiving Oliver's reply (see letter from Daniel Oliver, 25 September 1860).
[2] See preceding letter and letter from Daniel Oliver, 19 September 1860.
[3] The postscript is on a separate slip of paper. The contents indicate that it belongs to this letter.
[4] Size: 'an agglutinant consisting of undried glue' (*EB*).

[5] This sentence was added in pencil.

[6] See letter to Daniel Oliver, 11 September [1860]. There is a note headed 'Drosera spathulata Australia from Kew' that is dated 'Sept. 28th—/60/' in DAR 54: 5.

To Charles Lyell 23 [September 1860][1]

15 Marine Parade | Eastbourne
Sunday 23d

My dear Lyell

I got your letter of 18th just before starting here.[2] You speak of saving me trouble in answering: never think of this for I look at every letter of yours as an honour & *pleasure*, which is a pretty deal more than I can say of some of the letters which I receive.— I have now one of 13 *closely written folio* pages to answer on Species![3]

I sent off A. Gray's Review before starting;[4] please tell me in $\frac{1}{2}$ dozen words whether you think it good, for I have been thinking that I would ask Editor of Annals & Mag. of N. H. (who has reprinted Agassiz's article)[5] in fair play to reprint this; & I would attach A. Gray's name to it. From giving so much of Pictet it seems to me very good. I intended to offer to pay for its printing. What think you? The Annals, I fear, has very small circulation.—[6]

Etty has stood her journey excellently:[7] she & Emma drove in one carriage all the way here.—

I suppose you have heard of the Huxleys' terrible loss in sudden death of their little Boy.[8]

I quite misunderstood you on Types. I have a very decided opinion that all Mammals must have descended from a *single* parint. Reflect on the multitude of details, very many of them of extremely little importance to their habits (as number of bones of head &c—covering of hair, identical Embryological development &c &c): now this large amount of similarity I must look at as certainly due to inheritance from a common stock. I am aware that some cases occur in which a similar or nearly similar organ, has been acquired by independent acts of nat. selection, but in most of such cases of these apparently so closely similar organs, some important homological difference may be detected.

Please read p. 193 beginning "The Electric organs" & trust me that the sentence "In all these cases of two very distinct species &c &c" was not put in rashly; for I went carefully into every case.—[9] Apply this argument to the whole frame, internal & external, of mammifers, & you will see why I think so strongly that all have descended from one progenitor.— I have just reread your letter & I am not perfectly sure that I understand your point.

I enclose 2 diaggrams showing the sort of manner I *conjecture* mammals have been developed: I thought a little on this, when writing p. 429 beginning "Mr. Waterhouse".—[10] (please read the paragraph.) I have not knowledge enough to choose between these two diagrams;[11] if the Brain of Marsupials in embryo closely resembles that of placentals, I shd. strongly prefer nor. 2, & this agree with antiquity of microlestes. As a general rule I sh^{d} prefer no^{r} I. diagram. whether or not Marsupials have gone on being developed or rising in rank from a very early period would depend on circumstances too complex for even a conjecture: Lingula has not risen since Silurian epoch, whereas other Molluscs may have risen.— Owen if he chose to attend to such view could work this out.[12]

I at first thought with you that there was no fossil Rodent in Australia; but I then had vague misgivings, so said nothing. with respect to Australia being particularly fitted for Marsupials, I have always thought it a gigantic hallucination of Owens.—[13] Not to mention Rodents; the Dingo has been wild for an *enormous* period before the S. Australian Volcanic outburst.— Then look at number of species of Marsupials in Brazil. But strongest fact is New Guinea, a most humid country; & tenanted by Marsupials as *exclusively* (except one pig, which some doubt being endemic) by marsupials (some belonging to peculiar type, as the tree-kangaroos), as is Australia itself.

I daresay you know evidence of antiquity of Dingo in Australia: you will see references in my Dog-M.S:[14] Ancient as the Dingo is, one cannot help suspecting that it was introduced by man; & if so, it bears on your subject of antiquity of man. If it could be shown that Dingo existed in some other country it would be pretty good evidence that it was not aboriginal in Australia: Lesson states that it is same with the dog of New Ireland:[15] but Lesson is not to be trusted, & confounds the Dingo with the very distinct type of the Canis primæus of India.—

That is a good bit about grafting of roots, & the stumps kept thus alive. I remember reading the paper & sh^{d} much like to have reference to it.[16] But my impression is pretty strong that authors say that the case is confined to Coniferæ.— Perhaps we are thinking of something different: I refer to cut-down stumps (which in the fir-tribe never spring up again) but which are kept alive for years by their roots being inarched or grafted.—

I do not think multiple origin of dogs goes against single origin of man. (Agassiz's remark, in Nott & Gliddon?? on coincidence of colour alone, so fleeting a character, does not go for much in his comparison of man & anthropoid apes.)[17] All the races of man are so infinitely closer together than to any ape, that (as in case of descent of all mammals from one progenitor) I sh^{d} look at all races of man as having certainly descended from single parent.— I should look at it as probable that the races of man were less numerous & less divergent formerly than now; unless indeed some lower & more aberrant race, even than the Hottentot, has become extinct. Supposing, as I for one believe that our dogs have descended from 2 or 3 wolves, jackalls &c; yet these have, *on our view*, descended from single remote unknown progenitor. With domestic dogs the question is simply whether the whole amount of difference has been produced since man domesticated a single species; or whether part of the difference arose in the state of nature.— Agassiz & Co. think the Negro & Caucasian are now distinct species; & it is a mere vain discussion, whether when they were rather less distinct they would, on this standard of specific value, deserve to be called species.

I agree with your answer which you give to yourself on these points; & the simile of man now keeping down any new man which might be developed strikes me as good & new. White man is "improving off the face of the earth" even races nearly his equals.—

With respect to islands, I think I would trust to want[18] time alone, & not to Bats & Rodents (N.B I know of no rodents on Oceanic isld. except my Galapagos mouse which *may* have been introduced by man) keeping down the development of other classes.— Still *much* more weight I shd attribute to there being now neither in islands or elsewhere no known animal of a grade of organisation intermediate between mammals, fish, reptiles &c, whence a new mammal could be developed.— If every Vertebrate were destroyed throughout the world except our *now well established* Reptiles, millions of ages might elapse before Reptiles could become highly developed on a scale equal to mammals; & on principle of inheritance they would make some quite *new Class* & not Mammals,—though *possibly* more intellectual!! I have not an idea, whether you will care for this letter, so speculative.—

Most truly yours | C. Darwin

[Enclosure][19]

A. Unknown form probably intermediate between mammals, Reptiles & Birds as intermediate as Lepidosiren now is between Fish & Batrachians.—— This unknown form probably more closely related to Ornithorhynchus than to any other known form.—

Diagram I.

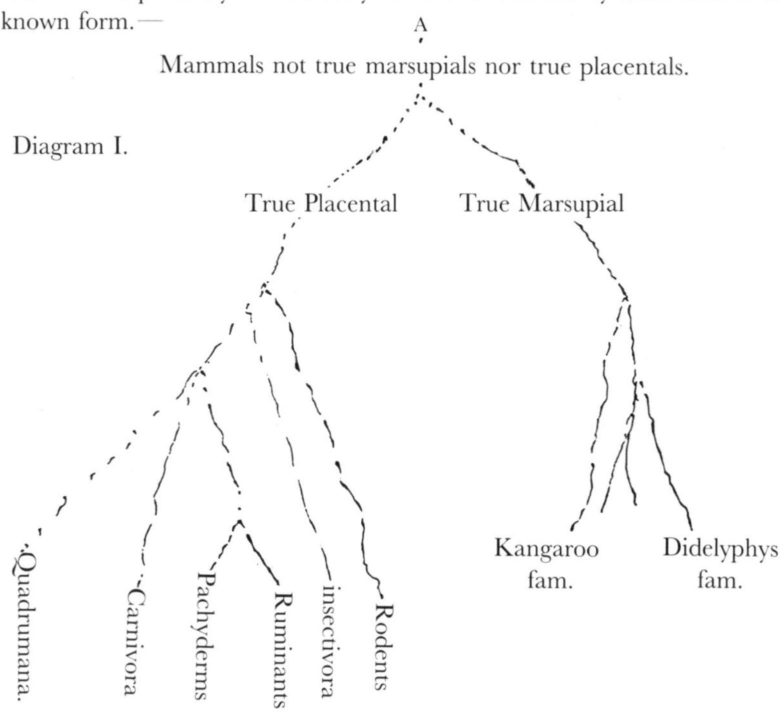

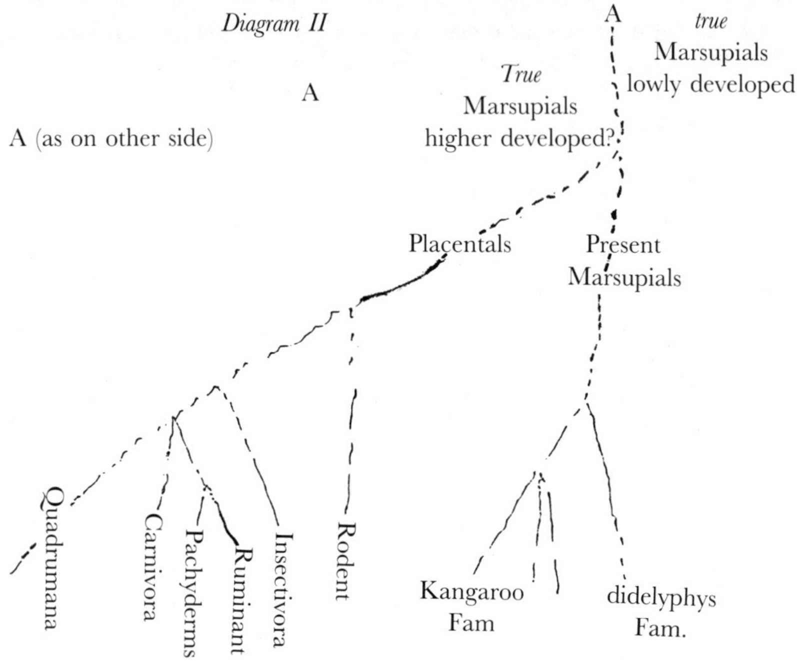

Endorsement: '1860'
American Philosophical Society (227)

[1] Dated by the relationship to the letter from Charles Lyell, 18 September 1860. The first Sunday during CD's stay in Eastbourne was 23 September.

[2] Letter from Charles Lyell, 18 September 1860. CD left for Eastbourne on 22 September ('Journal'; Appendix II).

[3] Letter from W. H. Harvey, 24 August 1860. For CD's reply to Harvey's letter, see letter to W. H. Harvey, [20–4 September 1860].

[4] [Gray] 1860b, pp. 229–39.

[5] Louis Agassiz's criticism of *Origin*, taken from proof-sheets of the third volume of Agassiz 1857–62 and first printed in the *American Journal of Science and Arts*, was reprinted in the August issue of the *Annals and Magazine of Natural History* (Agassiz 1860).

[6] CD had discussed with Asa Gray and with Thomas Henry Huxley the possibility of having Gray's article ([Gray] 1860b, pp. 229–39) reprinted in an English journal (see letters to Asa Gray, 10 September [1860], and to T. H. Huxley, 10 September [1860]). François Jules Pictet de la Rive had spoken favourably about *Origin*, even though he did not fully agree with CD's conclusions (Pictet de la Rive 1860).

[7] Henrietta Emma Darwin had been unwell for many months. The visit to Eastbourne was intended to help her convalesce.

[8] Noel Huxley, who was three years old, died on 15 September 1860. See letter to T. H. Huxley, 18 September [1860].

[9] The sentence reads: 'In all these cases of two very distinct species furnished with apparently the same anomalous organ, it should be observed that, although the general appearance and function of the organ may be the same, yet some fundamental difference can generally be detected.' (*Origin*, p. 193).

[10] The passage gives CD's opinion that when a member of one group of animals exhibits an affinity to others of a quite distinct group, 'this affinity is in most cases general and not special'. He attributed this view to George Robert Waterhouse (*Origin*, pp. 429–30). CD had corresponded with Waterhouse in 1843 about the meaning of affinity and analogy in classification. See *Correspondence* vol. 2.

[11] See the enclosure.

[12] Richard Owen based his classification of the Mammalia on a detailed anatomical comparison of the cerebral characteristics of the various groups of placental and marsupial mammals. See R. Owen 1859a.

[13] See letter from Charles Lyell, 18 September 1860 and n. 7. Owen assumed that the flora and marsupial fauna of Australia represented 'an ancient condition of the earth's surface, which has been superseded in our hemisphere by other strata and a higher type of Mammalian organization.' (R. Owen 1846b, p. 69).

[14] The 'Dog-M.S' CD sent to Lyell for his comments has not been found. It presumably formed the basis for chapter 1 of *Variation.* The references alluded to by CD in the letter were probably Selwyn 1858 and 1860. Both of these papers were cited in relation to the antiquity of the dingo in Australia in *Variation* 1: 26 n. 27.

[15] René Primevère Lesson made this point in his description of the zoological results of the voyage of the *Coquille*, 1822–5 (Duperrey 1826–30, 1, pt 1: 127, 341).

[16] See letter from Charles Lyell, 18 September 1860 and n. 9.

[17] The remark referred to is probably that made in Nott and Gliddon eds. 1857, p. xiv. See letter from Charles Lyell, 18 September 1860 and n. 10.

[18] Lyell inserted 'of' here to make CD's meaning clearer.

[19] The diagrams were drawn on the final leaf of the stationery.

To Ebenezer Norman [24 September 1860][1]

15 Marine Parade | Eastbourne
Monday 23[d]

To M[r] Norman

Dear Sir

Please copy this letter[2] *as soon as you can* on folio page, writing on both sides & send it me by Post, paying postage & charging it to me.—[3]

Yours faithfully | C. Darwin

DAR 98 (ser. 2): 45v

[1] In 1860, 23 September was actually a Sunday. CD made the same error about the date in the following letter, but then corrected it.

[2] CD refers to the draft of a letter to William Henry Harvey (see letter to W. H. Harvey, [20–4 September]). The letter to Norman was written on the verso of the first sheet of the draft.

[3] A payment to Norman for copying, which may have included this transaction, is recorded in CD's Account book (Down House MS) on 26 April 1861.

To Daniel Oliver 24 [September 1860][1]

15 Marine Parade | Eastbourne
Monday 24th.

My dear Sir

I should think you could make an interesting paper about spiny trees.[2] I had a vague notion about plants *varying* by becoming more spinose in dry & hot climates, but I never put two & two together, as you have done, in relation to the *species* of such countries being spinose. This is a point after my own heart.— It is quite new to me about species of far removed orders being spinose in deserts.— The labour would be, I suppose, great; but I can hardly doubt that you might make a very interesting paper on the subject.— The simple fact of bushes & trees being in so marked a manner spinose in deserts alone struck me, from what I have myself seen & read. Livingstone, for instance, was much struck with the contrast in this respect between the plains of the southern parts of Africa & the more humid & intra-tropical parts.[3]

The explanation of the fact, seemed to me to be probably, that where vegetation was scanty, those plants alone could withstand the injury from browsing quadrupeds, which were protected by spines. Even our gorze is an instance of this, as being when bruised & chopped so eminently liked by horses.— I have fancied that desert-plants were often strongly aromatic or strong-tasted for a similar purpose, viz protection.— This struck me much on stony mountains of Chile.—[4]

You allude to another interesting point, about unisexual plants having very fragrant & conspicuous plants.[5] This would be a very curious point, but I sh^{d} suppose very difficult to determine, as fragrancy is so indefinite.— I have imagined or seen stated, that night-blowing flowers are often fragrant to attract insects. & often white. By your excluding orders in which all the species are unisexual you exclude in fact one source of doubt, viz the cases, where wind is sole agent of fertilisation— I have just thought of one strong exception the common-Holly, which is not fragrant or has conspicuous flowers, & is yet essentially unisexual.— Bees can certainly smell any sweet excretion. I fear there would be too many elements of doubts.—

I am not Botanist enough to follow out your ideas about definite & indefinite inflorescence.— I am not sure that I understand "definite" & "indefinite".— I have long thought that dimorphous flowers (& told Dr Hooker so)[6] would be a *very interesting* subject for experiment.— I once got the Russian violet to commence on; but want of time & my poor health prevented me doing anything. This very summer I have had Venus Looking glass[7] under cover partly with this object, but the wet season spoiled this & several other little experiments. I shall never take it up now.— The view which I meant to test was this: that the apetalous forms was *self* fertilised & therefore seeded abundantly; whilst the flowering kind was alone visited by insects, & either required (as in case of Violets & I *believe* all Campanulas) & was much benefitted by the visits of insects for its self-fertilisation; & thus *incidentally* received the benefit of an occasional cross.

This year I tried V. tricolor & it was marvellous the difference in seeding of the flowers, which were visited by Bees or artificially fertilised by me, & those which were untouched.—[8] I think that you would find this an easy & interesting line of experiment.—

Very many thanks for your interesting letter | Yours sincerely | C. Darwin

There is an N. American Campanula with nearly apetalous flower, which seeds largely. How does pollen get on stigma?—

Down House (MS 10: 22)

[1] Dated by the relationship to the letter from Daniel Oliver, 25 September 1860. The only 'Monday 24th' during CD's stay in Eastbourne was in September.

[2] Oliver's letter has not been found. He had been studying families of plants with spines, particularly the Caryophyllaceae. His analysis of the internal structure of plant stems, which in passing mentioned the relation of stems to spines, was published in 1859 (Oliver 1859).

[3] Livingstone 1857, pp. 345–6. CD had been greatly impressed by this work when he read it in 1858 (see *Correspondence* vol. 4, Appendix IV, 128: 23).

[4] *Journal of researches* (1860), p. 327.

[5] Oliver may have mentioned to CD his work on the family Hamamelidaceae, members of which often have unisexual flowers. The family includes many fragrant species. Oliver identified a new genus called *Sycopsis* in a paper read at a meeting of the Linnean Society on 15 March 1860 (Oliver 1860). In this paper, he suggested that the geographical distribution of the family might be elucidated by CD's and Alfred Russel Wallace's views (Oliver 1860, p. 87). There is an annotated copy of the paper in the Darwin Pamphlet Collection–CUL.

[6] See letters to J. D. Hooker, 11 May [1860] and 5 June [1860].

[7] A popular name for *Specularia*, a member of the Campanulaceae. CD's experiments on the fertilisation of plants by insects, performed during the year 1860, are recorded in his Experimental book, pp. 59–61 (DAR 157a). *Specularia*, however, is not mentioned.

[8] CD's experiment on *Viola tricolor* is recorded in his Experimental book, p. 59.

From Charles Lyell[1] 25 September 1860

25. Sept. 1860

I return the M.S. on dogs which I think excellent.[2] The case you make out seems very strong not only of crosses from distinct living species having blended into the dog, but in favour of diff^t savage races having domesticated different canine types, wolves, jackalls &c by domestication in accordance with the hypothesis of Pallas having eliminated the dislike to cross with other species as well as the tendency of such crosses to sterility.[3]

All this helps the doctrine of the several so blendible wild species having themselves come down from a common remote progenitor & I suppose you will say in reply to Quart. Review that the only reason that pigeons have not so mixed is that Man had not the same motive to cross them with other species distinct from the Rock pigeon.[4]

I think the subject too important to bear shortening. As to the antiquity of the Dingo I have turned up to your two references in Quart. G. S. Journal & they by no means bear out your inference.[5] The fossils occurred in a cave in the basalt & were certainly posterior probably very long so, & there is no evidence as caves are gradually choked up that the Dingo was not one of the last creatures whose bones were introduced. I asked Falconer yesterday[6] & he said, 'let him take care he does not get himself into the same scrape which Owen did about Strzelecki's Australian Mastodon;[7] I read the papers about Dingo & thought they had no proof.'

Falconer knows of no fossil, rodent or bat in Australia. He says Strzelecki's Mastodon was from S. America.

It seems strange that the question whether any Newfoundland dogs are semi-web-footed or not sh^{d} be so doubtful, also gestation of jackall. On these and many points y^{r} book will draw forth a number of facts for the 2^{d} Ed^{n}.[8]

Some one sh^{d} go at once to Angoulême & see into that question of the Leporine.[9] If true, it w^{d} take away the antecedent improbability of Pallas' hypothesis of dogs when domesticated losing sterility &c. Bartlett says, at Zool. Gardens, that we must wait till next spring for good experiments.[10] He ought to have as many as possible to start with.

The chapter on dogs makes me wish yr book soon out. That & the pigeons & the Tables of large & small genera w^{d} alone make a useful beginning.[11]

You conclude yr. letter received this morning by doubting whether I care for so speculative a way of dealing with the question. It is just what I wanted & not more conjectural than my letter, much of which I w^{d} not of course touch or venture on in print, nor wish shown to third persons, more especially as one may alter one's speculations the day after. I have referred to the pp. you allude to of Origin with much profit, tho' I remembered them all, but they contain such a condensation of matter that they require to be often re-read.

I have been too deeply interested & taken up with the Dogs to have done with Asa Gray, but see enough to wish it to appear in Annals of N. H. & cannot but think they w^{d} print it for their own sake & not require you to pay.[12] Tell them that the addition of A. Gray's name w^{d} give a wholly new value to it instead of its appearing anonymously I want to buy it, for one.

I think you have understood my point & the idea that if the original type of Mammalia had been lost & the reptilian had been greatly raised in grade, they w^{d} have produced some other great class as high perhaps or higher but not the existing Mammalia, is a grand notion & believing as I do in the infinite Capacity of the creative power, inherent in the organic world, worked out by variation & natural selection, I do not think it an extravagant speculation at all. You might say the same if some monotreme, instead of a reptile, was improved into a mammal, by a series of changes independent of those which gave rise to the first mammal. Possibly birds may be an instance of such a class, though I fear that would not do, because however low the Monotreme, it must perhaps rank typically before birds & birds therefore must have preceded in order of development.

I am glad you reminded me of the N. Guinea Marsupials. I had imagined that there was a larger admixture of placentals. I sh^{d} think still the Australian genera & species were so well fitted for the extraordinary droughts that they w^{d} get the better of the Dingo had he run wild as in Juan Fernandez. The Brazilian case does not tell for much as that country has in the Marsupial line turned out nothing but opossums.

The case of nat^{l} grafting stated by Göppert Überwallung der Tannenstöche, Bonn 1842, where silver fir borrows from roots of Pinus abies is the more analogous to grafting as being one conifer borrowing from another.—[13] Plants of other divisions w[d] do the same probably if near enough to allied trees, which owing to rotation may not be often at hand. If the genus were remote, it w[d] fail.

Do you not think when Man was less advanced 50,000 years or generations back & tribes more isolated there must have been rather more races than now. tho' of course at a remoter period much fewer, & originally only one, & that one lower intellectually than any one now existing. In proportion as more powerful & more cosmopolite races arose they w[d] exterminate inferior ones & also there w[d] be more mixture & a check to divergence into species or the centripetal force of hybridity (excuse the use of the phrase) would come into play.

Falconer has been holding forth today on the diff[t], Mastodons & Elephants not coming in chronologically as they sh[d] do, according to yr views, but when one sees the new Maltese dwarf intermediate between E. antiquus & E. meridionalis & Anca's new Sicilian cave elephant,[14] a modification of the living Indian one leaning towards antiquus, & when one thinks that Falconer can distinguish all American varieties of Mammals from all European fossil species, I confess I attach little value to the objection.[15]

Incomplete
Copy
Kinnordy MS, Charles Lyell's journal VII, pp. 3–12

[1] The text of the letter has been taken from a copy in Lyell's scientific journal. It is also printed in Wilson ed. 1970, pp. 493–5. Lyell returned to London on 23 September following a six-week visit to the Continent (*ibid.*, p. 490 n. 88).

[2] See letter to Charles Lyell, 11 August [1860].

[3] CD described the 'doctrine' of Pyotr Simon Pallas concerning the origin of domestic breeds of dogs in a previous letter to Lyell (*Correspondence* vol. 7, letter to Charles Lyell, 25 October [1859]):

> He *hypothetically* supposes that when long domesticated they lose their tendency to sterility when crossed with the other domesticated *species*; & by their crossing when domesticated he believes that all our domestic races have originated.

[4] The review of *Origin* in the *Quarterly Review* ridiculed CD's example of the power of selection drawn from comparing the breeds of domestic pigeons, stating 'this is all very pleasant writing, especially for pigeon-fanciers; but what step do we really gain in it at all towards establishing the alleged fact that variations are but species in the act of formation, or in establishing Mr. Darwin's position that a well-marked variety may be called an incipient species?' ([Wilberforce] 1860, p. 235). This passage and others relating to pigeons are heavily marked and annotated in CD's copy of the review (Darwin Pamphlet Collection–CUL).

[5] Selwyn 1858 and 1860. See letter to Charles Lyell, 23 [September 1860].

[6] Lyell visited Hugh Falconer soon after returning to London to learn what progress had been made 'on the great question at issue about the relation of certain elephant beds and the glacial epoch.' (K. M. Lyell ed. 1881, 2: 337).

[7] In 1844, Richard Owen had identified a fossilised molar tooth sent to him by Paul Edmund de Strzelecki as that of a mastodon. It was said to have been found in a cave in Australia (R. Owen 1844). Before then, the extinct mammalian fauna of Australia was believed to be exclusively restricted to marsupial forms. Falconer first questioned Owen's identification in 1857; in 1863, he published a full history and refutation of Owen's claim (see Falconer 1857–8, 'Synoptical table of the species of *Mastodon* and elephant', facing p. 319; and Falconer 1863, pp. 96–101).

[8] The edition of *Origin* usually referred to as the second was published in January 1860. Lyell evidently considered this to have been a corrected reprint and rather refers to the major alterations that CD was intending to incorporate into the third edition of *Origin*, published in 1861.

[9] Lyell refers to the supposed hybrid between a hare and a rabbit said to have been bred in Angoulême, France. See letter to Charles Lyell, 5 [July 1860].

[10] Abraham Dee Bartlett was superintendent of the zoological gardens in London. Lyell had seen the so-called leporines at the gardens early in July 1860 (Wilson ed. 1970, p. 465). Bartlett conducted many breeding experiments for CD at the gardens, but an account of this particular cross has not been located in the Darwin Archive.

[11] CD was preparing the first part (*Variation*) of a substantial work, in which he planned to provide all the details and citations he had been unable to include in *Origin*.

[12] See letter to Charles Lyell, 23 [September 1860].

[13] Göppert 1842.

[14] See Anca 1860.

[15] Falconer was engaged in an extensive study of the fossil elephants and mastodons of Europe and America. He published a paper on the British and European species in 1857 and one on the American forms in 1863 (Falconer 1857–8 and 1863). He had recently discovered a 'small elephant, the size of a Shetland pony, in the small island of Malta.' (K. M. Lyell ed. 1881, 2: 339). See also letter to Charles Lyell, 28 [September 1860].

From Daniel Oliver 25 September 1860

Kew. W.
25th. IX/60

Dear Sir,

My results so far with pure gum quite support your opinion.[1] Yesterd^{y} morn^{g} 9.25. a drop was put on a leaf of *D. spathulata* (a leaf of wh. I enclose) at 2.10 P.M. no marked change,—prob^{y} some inflection of inner glands,—but the marginal patent or recurved. 5.55. do. **this day** (Tuesd^{y}) 8.56 AM & 5.48 P.M. No material change as yet. The plant looks unhealthy.

Another drop was placed on a leaf of *D rotundifolia*? 2.13. P.M. (Yester^{y})— No marked result this even^{g}

I must try the old gum again. I have not *burnt* it *yet*. Besides Im not sure if we used it (the new) in equally strong solution. I intend still trying it.

We have another *Drosera* I dont know the species,—(I understand Australian) with narrow linear leaves Its rate of incurving some days ago over a fly was as follows

ab^{t}

—1st day. 3.— P.M.

2— 8.35. AM

— — 2.40 P.M

— — 6.13 P.M

—3$^{\underline{rd}}$ day 8.30. am.
4$^{\underline{th}}$ day. morn^{g}. no mrk^{d}. chge.

I quite look to working up one or two spinous *Floras*[2] I think of, as one, the Flora (Florula!) of *Aden* upon wh. D^{r}. Anderson is just now engaged here.[3] It is to appear in a suppl^{t}. to the Linn: Journal It is a very spiny Flora. Then Delile, for Egypt.[4] *perhaps* Kelaart, for Gibraltar.[5] Munby's list for Algeria—[6] No. West India.—& Some part of Tropical Australia or Mexico would be important. To *contrast* I would take our own Flora,—that of Arctic circle (easy)—upon wh. D^{r}. H. has much material for the Linn Trans. (it will not appear tho' I doubt till tow^{ds}. close of next year)[7]—V. Diemen's Land. (J.D.H.)[8] So. United States (Chapman).[9] Before setting seriously to work upon it I should like to complete a little task I have in hand among the *Aurantiaceae* wh. I am trying (apart from *Citrus*) to rearrange.—[10]

The various sorts of spinosity ought to be separately tabulated,—whether true spines,—epidermal (prickles) & leaf appendages,—hardw^{d}. bracts, &c Often it will be difficult to draw the line between spinose & spineless species,—as among Thistles, Dipsacus, &c. &c.

Apropos of *Drosera* see a footnote in D^{r}. Bromfield's '*Flora Vectensis*' p. 56—"The glands x x x x curving eventually over the flies x x x ensnared by their clammy exudation, but never, so far as I could observe, assisting primarily in their capture by any sudden contraction as in *Dionæa muscipula*"[11]

I put a bit of paper in daubed with *old* gum,— I did not perceive animal odour on burning a piece.

Very faithfully yours | Danl. Oliver. Jr

I doubt if we get good seed this autumn from over *3* or *4* of the isosceles beds![12]

DAR 58.1: 1–3

CD ANNOTATIONS

3.1 I must . . . trying it. 3.2] 'where got?' *added pencil*
4.9 4th . . . chge.] 'morn^{g} no' *underl pencil*; '?' *added pencil*
5.1 I quite . . . I should 5.9] *crossed pencil*
5.9 like . . . *Drosera* 7.1] *crossed ink*
8.1 I put . . . beds! 10.1] *crossed pencil*
Top of first page: '*D Spathulata*', 'D' *red crayon*; 'Spathulata' *pencil, underl red crayon*

[1] CD had suggested that the gum Oliver used in his first experiments on *Drosera* was perhaps contaminated with nitrogenous matter. See letters to Daniel Oliver, 21 [September 1860] and [22–3 September 1860].
[2] See letter to Daniel Oliver, 24 [September 1860].
[3] Thomas Anderson was working at the Royal Botanic Gardens, Kew, on the taxonomy of the flowering plants found in Aden (Anderson 1860). In 1861 he assumed the directorship of the Calcutta Botanic Gardens.
[4] Alire Raffeneau-Delile described the Egyptian plants collected by members of the French Commission d'Egypt (Raffeneau-Delile [1813] and [1824]).
[5] Edward Frederick Kelaart published a flora of Gibraltar in 1846 (Kelaart 1846).
[6] Giles Munby lived in Algeria and published a flora of the native plants of the country in 1859 (Munby 1859).
[7] Oliver refers to Joseph Dalton Hooker, who had read a paper on the 'Outline of the distribution of Arctic plants' at a meeting of the Linnean Society of London on 21 June 1860. The paper was published in 1862 (Hooker 1862a).
[8] Hooker 1855–60.
[9] Alvan Wentworth Chapman was the author of a flora of the southern states of America (Chapman 1860).
[10] Oliver published a study of the Aurantiaceae in 1861 (Oliver 1861).
[11] Bromfield 1856.
[12] See letter to Daniel Oliver, 15 [September 1860].

To Asa Gray 26 September [1860][1]

Down Bromley Kent [Eastbourne]
Sept. 26th.

My dear Gray

I have a multitude of things to thank you for; & hardly know how to begin. I forward & stamped letter to Grisebach in yours of Sept. 2d.[2] Always use me this way, when convenient; as a scrap from you, I consider very good payment. With respect to Von Baer's statement about Guinea-pig; I shd. much like to know where it is;[3] but it can hardly be trusted unless he has brought forward some quite new evidence with respect to the unknown wild type of Guinea-pig. I have seen some of the Hybrids of Rabbit & Harc;[4] & believe case is to be trusted; but I do not know that the exact *half*-bloods are *perfectly* fertile *inter se*: it is a particularly curious case, because many have perseverinly tried & utterly failed even to unite these 2 species.

Your letter of the 2d. was very pleasant, & so was that to Hooker, (just received from Lady Hooker)[5] which made us all (ie my family) laugh heartily. I have received the Dialogue from Silliman's October nor.[6] Also the Proofs of the 3d. Atlantic Article.[7] These both seem to me *very* good; but if forced to choose, I think I shd pick out the 2d article as best.— (I don't know though) I do not pretend to be a good judge, as I have never attended to Logic, Philosophy &c; but it is my opinion that you are best reasoner, of any man, let him who he may, that I ever read. As for your metaphors & similes, they make me envious: I shd like to steal a few: yet you never are tempted to use them prodigally. The two last essays are *far* the best Theistic essays I ever read. But I must return to the metaphors— I like specially that of the woman & the cloth;[8] and that of the rain-drops on the ocean.—[9] I believe my chief difference with you, is that I (unfortunately) think more of the rain-drops on the ocean than on those on the land. All your arguments about Design seem to me excellent; but I must hereafter read all again. I have a feeling that the existence of the multitude of Stars & the motion of the planetary system &c are equally good with living beings to prove a First Cause; & yet if there were no living things, there could hardly be design.— But I well know that I am muddled-headed on this subject.—

I have sent the August no[r] of Atlantic to Annals & Mag. of N. History; & offered to pay for its printing, if its length prevented its insertion: I have affixed your name, as Lyell agreed that it so greatly added to its value:[10] I do most sincerely hope that I have not done wrong. The Saturday Review has lately been discussing Design; so I will send the Dialogue to it, of course without your name; whether they will print, I doubt.—[11] I wish I could think of any Journal to print the October number.—[12]

I have had a letter of 14 folio pages from Harvey against my Book, with some ingenious & new remarks; but it is an extraordinary fact that he does not understand at all what I mean by Nat. Selection.[13] I have begged him to read the Dialogue in next Silliman, as you never touch the subject without making it clearer.[14] I look at it as even more extraordinary that you never say a word or use an epithet which does not express fully my meaning. Now Lyell, Hooker & others, who perfectly understand my book, yet sometimes use expressions to which I demur. If I had to write my book again I would use "Natural Preservation" & drop "Selection", but it is too late now. D[r] Gray of B. Museum says to me "it is, you know, obviously impossible that there can be any *Selection* in case of Plants".[15] I made no answer; for it is hopeless. Well, your extraordinary labour is over; if there is any fair amount of truth in my views, I am well assured that your great labour has not been thrown away. I am perfectly convinced that had it not been for yourself, Hooker, Huxley & Lyell; my Book would Scientifically have been a complete failure. You have been most generous & disinterested in your exertions; & I thank you from my heart.

Farewell, my dear Gray | Yours most truly | C. Darwin

P.S. Please observe that if the Appletons lose by the 2[d] Edition never selling, I sh[d] **prefer** repaying the money they have paid me.—[16]

I am writing this at Eastbourne, where we have come to sea-side for my girls health, & she has certainly profited by it.— Consequently I shall not see Clark (Agassiz's man).[17]

I have been infinitely amused by working at Drosera: the movements are really curious; & the manner in which the leaves detect *certain* nitrogenous compounds is marvellous.— You will laugh; but it is **at present**, my full belief (after endless experiment) that they detect (& move in consequence of) the $\frac{1}{2880}$ part of a single grain of Nitrate of Ammonia; but the Muriate & Sulphate of Ammonia bother their chemical skill, & they cannot make anything of the nitrogen in these salts![18]

I began this work on Drosera in relation to *gradation* as throwing light on Dionæa.—[19]

Does Spiranthes (or any odd genus of Orchids) grow near you; if so, I would ask you next summer to make a few observations, which would not take 10 minutes for me. There is very curious contrivance in our S. autumnalis for insect fertilisation; viz a decked boat-full of viscid matter & on a touch the deck splits & leaves viscid matter exposed. But I would explain if you have a Spiranthes.—

I yet hope, & almost believe, that the time will come when you will go further in believing a very large amnt of modification of species, than you did at first or do now. Can you tell me whether you believe further or more firmly than you did at first? I sh[d] really like to know this. I can perceive in my immense correspondence with Lyell, who objected to much at first, that he has, perhaps unconsciously to himself converted himself very much during last six months.—& I think this is case even with Hooker.— This fact gives me far more confidence than any other fact.—

P.S. 2[d] | How does Dana get on? I wrote to him, a month or two ago.—[20]

Gray Herbarium of Harvard University (28)

[1] The year is given by the reference to [Gray] 1860c.

[2] The letter from Gray mentioned by CD has not been found. August Heinrich Rudolf Grisebach was professor of botany at Göttingen University. CD had used his flora of the Balkans in 1856 (see *Correspondence* vol. 6).

[3] The reference is to Karl Ernst von Baer. See letter to Charles Lyell, 28 [September 1860] and n. 8.

[4] See letter to Charles Lyell, 5 [July 1860], and letter from Charles Lyell, 25 September 1860.

[5] Maria Hooker had apparently sent CD a letter that Gray had written to her son Joseph Dalton Hooker (see letter to Asa Gray, 10 September [1860]). Hooker had just left for Syria (L. Huxley ed. 1918, 1: 528).

[6] CD refers to the article entitled 'Discussion between two readers of Darwin's treatise on the origin of species, upon its natural theology', published in the September issue of the *American Journal of Science and Arts* ([Gray] 1860c). There is a copy in the Darwin Pamphlet Collection–CUL, on which CD wrote: 'Asa Gray(?)'. Gray's authorship is confirmed by his biographer, A. Hunter Dupree (Dupree 1959, pp. 288–9). According to Dupree, the 'Discussion' was a record of an actual debate that took place between Gray and his friend Daniel Treadwell, formerly Rumford professor at Harvard University. Gray described Treadwell as a 'skeptical neighbor of mine, a man who hangs doubtfully on the skirts of atheism, and thinks Darwin's book will give him a firmer hold' (Dupree 1959, p. 289). In the debate, the first reader was Treadwell and the second Gray.

[7] The third and final part of [Gray] 1860b was published in the October issue of the *Atlantic Monthly*. The article was entitled 'Darwin and his reviewers'. There is an annotated copy of Gray's articles for the *Atlantic Monthly* in the Darwin Pamphlet Collection–CUL.

[8] The passage occurs in [Gray] 1860c, p. 239. It describes the way that a 'woman of a past generation' would believe the wool or cotton forming a piece of cloth was carded, spun, and woven by hand rather than by machine.

> If you patiently explained to her the theory of carding machines, spinning jennys, and power-looms, would her reception of your explanation weaken her conviction that the cloth was the result of design? It is certain that she would believe in design as firmly as before, and that this belief would be attended by a higher conception and reverent admiration of a wisdom, skill, and power so greatly beyond anything she had previously conceived possible.

[9] The passage occurs in the third part of [Gray] 1860b, p. 417:

> The whole animate life of a country depends absolutely upon the vegetation; the vegetation

upon the rain. The moisture is furnished by the ocean, is raised by the sun's heat from the ocean's surface, and is wafted inland by the winds. But what multitudes of rain-drops fall back into the ocean, are as much without a final cause as the incipient varieties which come to nothing!

[10] See letter from Charles Lyell, 25 September 1860. The paper was reprinted in the November issue of *Annals and Magazine of Natural History*, pp. 373–86.

[11] See letter to Charles Lyell, 3 October [1860]. The article was not published in the *Saturday Review*.

[12] CD refers to the third part of [Gray] 1860b. It was not reprinted in an English journal.

[13] Letter from W. H. Harvey, 24 August 1860.

[14] See n. 6, above. William Henry Harvey and Gray corresponded about CD's theory. After reading Gray's 'Dialogue' ([Gray] 1860c) and the articles in the *Atlantic Monthly* ([Gray] 1860b), Harvey wrote to Gray on 3 November 1860: '"Almost thou persuadest me to be a Darwinite"—not *quite* so, but thou persuadest me to be a *Grayite*.' (Gray Herbarium, Harvard University).

[15] John Edward Gray's response to *Origin* was negative. See *Correspondence* vol. 7, letter to J. D. Hooker, 14 December [1859] and Gunther 1975, pp. 454–5.

[16] See letter from Asa Gray, 20 February 1860, and letter to Asa Gray, 22 May [1860]. In May 1860, CD recorded in his journal the sum of £21 17*s*. 6*d*., which represented CD's share of the profits of the American edition of *Origin*. See 'Journal' (Appendix II).

[17] Henry James Clark was Louis Agassiz's personal assistant at the Museum of Comparative Zoology, Harvard, and a contributor to Agassiz's *Contributions to the natural history of the United States of America* (Agassiz 1857–62). In June 1860, he was appointed assistant professor of zoology in the Scientific School of Harvard University. Following his appointment, he travelled for his health's sake in England, France, Germany, and Switzerland. See Winsor 1991.

[18] CD continued his experiments on *Drosera* throughout his stay in Eastbourne. His notes on these experiments are in DAR 60.1.

[19] *Dionaea* is a member of the Droseraceae. In *Insectivorous plants*, CD stated of *D. muscipula*: 'This plant, commonly called Venus' fly-trap, from the rapidity and force of its movements, is one of the most wonderful in the world' and 'is one of the most beautifully adapted plants in the vegetable kingdom' (*Insectivorous plants*, pp. 286, 358).

[20] James Dwight Dana returned to the United States in August 1860, having spent over a year in Europe recuperating from a breakdown in his health (*American Journal of Science and Arts* 30 (1860): 308). See letter to J. D. Dana, 30 July [1860].

To Charles Lyell 26 [September 1860][1]

15 Marine Parade | Eastbourne
26th

My dear Lyell

It has just occurred to me that I took no notice of your question on Extinction in St. Helena.[2] I am nearly sure that Hooker has information on extinction of plants:[3] but I cannot remember where I have seen it. One may confidently assume that many insects were exterminated. By the way, I heard lately from Wollaston,[4] who tells me that he has just received eminently Madeira & Canary Isld insect forms from the C. of Good Hope, to which trifling distance, if he is logical, he will have to extend his Atlantis! I have just received your letter; & am very much pleased that you approve. But I am utterly disgusted & ashamed about Dingo; I cannot think how I could have misunderstood the paper so grossly.[5] I hope I have not blundered likewise in its coexistence with extinct species. What horrid blundering.

I am grieved to hear that you think I must work in the notes in text; but you are so much better judge, that I will obey. I am sorry that you had trouble of returning Dog. M.S,—which I suppose I shall receive tomorrow.—

I mean to give *good* woodcuts of all chief races of Pigeons.—

Except the C. oenas (which is partly, indeed almost entirely, a **wood** pigeon) there is no other **rock** pigeon, with which our domestic pigeon could cross,— that is if several *excessively* close geographical races of C. livia which hardly any ornithologists look at as true species, be all grouped under C. livia.—

I am writing higglety-pigglety, as I reread your letter.— I thought that my letter had been much wilder than yours.— I quite feel comfort of writing when one may "alter one's speculations the day after".— It is beyond my knowledge to weigh rank of Birds & monotremes: in the respiratory & circulatory systems & muscular energy I believe birds are ahead of all mammals.—

I knew that you must have known about New Guinea; but in writing to you, I never make myself civil!!.—

After treating of some half-dozen or dozen domestic animals in same manner, as I treat Dogs,—I intended to have a chapter of conclusions.— But Heaven knows when I shall finish, I get on very slowly.— You would be surprised how long it took me to pick out what seemed useful about Dogs out of multitude of details.

I see force of your remark about more isolated races of man in old times, & therefore more in number. It seems to me difficult to weigh probabilities. Perhaps so, if you refer to very slight differences in the races; to make great differences, much time would be required; & then even at earliest period, I sh^{d} have expected one race to have spread, conquered & exterminated others.

With respect to Falconer's series of Elephants, I think case could be answered better than I have in Origin p. 334.—[6] All these new discoveries shows how imperfect the discovered series is, which Falconer thought years ago was nearly perfect.—

I will send today or tomorrow *two* articles by Asa Gray.—[7] The longer one (now **not** finally corrected) will come out in the **October** Atlantic Monthly & they can be got at Trübners.—[8]

Hearty thanks for all your kindness. | Ever yours | C. Darwin

Do not hurry over Asa Gray.— He strikes me as one of the best reasoners & writers, I ever read.— He knows my Book, as well as I do myself.—

Endorsement: 'Septr 1860'
American Philosophical Society (228)

[1] Dated by the relationship to the letter from Charles Lyell, 18 September 1860.

[2] See letter from Charles Lyell, 18 September 1860.

[3] CD discussed the flora of Saint Helena with Joseph Dalton Hooker on several occasions (see particularly *Correspondence* vol. 3). Hooker used some of his work on the flora of the island in assessing the Galápagos plants collected by CD. See Hooker 1847b, pp. 243–4. See also letter from Charles Lyell, [after 3 October 1860].

[4] Letter from T. V. Wollaston, [16 September 1860].

[5] See letter from Charles Lyell, 25 September 1860.
[6] Hugh Falconer's recent work on fossil elephants was mentioned in the letter from Charles Lyell, 25 September 1860. The passage in *Origin* to which CD refers mentions Falconer's finding that fossil mastodons and elephants arranged in two series, 'first according to their mutual affinities and then according to their periods of existence, do not accord in arrangement. The species extreme in character are not the oldest, or the most recent; nor are those which are intermediate in character, intermediate in age.' (*Origin*, p. 334). CD suggested in response that forms successively produced might not always endure as long as more ancient forms, thereby confusing the record of the appearance and disappearance of forms.
[7] [Gray] 1860c and the third part of [Gray] 1860b. See preceding letter.
[8] The London publishing house founded by Johann Nicolaus Trübner was the English agent Asa Gray normally used for the distribution of his publications.

From Charles Lyell[1] 27 September 1860

27th. Septr. 1860

I am haunted with a kind of misgiving, which bye & bye I shall be able to express more clearly—that the multiple origin of the dog will furnish an argument for the multiple origin of a Mammal, or of Man. Do try to consider it that way, for I incline to go far with Hooker (& with you?) in believing that whatever is true in domestication is essentially possible in Nature also, & I am not sure that you confine the multiple origin of the dog to Man's selection.

Incomplete
Copy
Kinnordy MS, Charles Lyell's journal VII, p. 12

[1] The text of the letter is taken from a copy in Lyell's scientific journal. It is also printed in Wilson ed. 1970, p. 496.

To Daniel Oliver 27 [September 1860][1]

15 Marine Parade | Eastbourne
Thursday 27th

My dear Sir

Your note of 25th about the gum has been a great relief to me; for I took a panic, as I know I am very apt to blunder & run away with things. After writing to you I looked through my notes, & found I had been rather more careful, than I thought for I had tried 27 leaves with **non**-nitrogenised substances, *not* counting saline solutions & simple water. The more I reflect on the experiments which I have tried the less I think I am mistaken. Thus I fully expected that 1 gr of gelatine to oz. of distilled water, would have affected the leaves, but it produced no effect, whereas a little stronger solution produced a marked effect, & whereas *1* gr to 1 oz of several salts with nitrogen produced strong effect.— *Thick* syrup on 5 growing leaves produced no effect, whereas a leaf *gathered* & put into the same solution suffered in extraordinary degree from exosmose & the the hairs & disc of leaf collapsed; so that the vital power seems to resist exosmose in the living plant. I think I have made out the simple mechanism of movement.— But why I trouble you with all these details I know not.—

I will give you no more trouble, except that I earnestly hope you will try again the *old* gum; & if it acts, endeavour to find out certainly its composition. Thank you for the pretty leaf of the Australian Drosera; & for paper with gum; but I have a horrid cold & must stop to another day to try it: a large drop or $\frac{1}{2}$ spoon-full dried & then put on heated knife is proper way to try.—

M. Trecul (I am so much obliged to you for telling me of that paper)[2] disbelieves in any movement & accounts for flies being caught by their crawling under the gumy incurved hairs! But he kept the plants in a green-house or hot-house. Could this have paralysed them? I suppose not: anyhow hot-sunshine seems to make them act better. I imagine he looked out for sudden movement. Vapour of Chloroform for 30 seconds paralyses them completely.—

Thank you much for details of the unnamed Australian species. I will with permission quote your observations.[3] Neither D. spathulata or the other species or D. longifolia seem to move quicker than D. rotundifolia.—

I am very glad to hear that you intend to attack spiny plants. Your note shows that it is a complex problem.—[4]

With **many** thanks. My dear Sir | Yours sincerely | C. Darwin

I have just received a cargo by Post of living plants of Drosera from Down & I will try more gum & starch myself.—[5]

In the unnamed Australian Drosera the incurvation of the leaf **itself** was terminal I suppose by your sketch. In D. rotundifolia, sometimes the incurvation is terminal; sometimes lateral; & sometimes termino-lateral, so as to be quite variable: *as far as I saw* in D. longifolia it was always terminal

Down House (MS 10: 23)

[1] Dated by the relationship to the letter from Daniel Oliver, 25 September 1860.

[2] CD cited Trécul 1855 in *Insectivorous plants*, p. 1 n. Oliver supplied the reference in a letter that is now missing (see letter to Daniel Oliver, 15 [September 1860]). Oliver's note giving the reference is in DAR 60.1: 66.

[3] Oliver's observations were given in the letter from Daniel Oliver, 25 September 1860. They were not cited in *Insectivorous plants*.

[4] Letter from Daniel Oliver, 25 September 1860.

[5] Notes on CD's experiment, dated 'Sunday 30th' and headed 'Mr Olivers Gum from Kew, which will have caused contraction.—', are in DAR 60.1: 110.

To J. S. Henslow 28 [September 1860]

15 Marine Parade | Eastbourne.
28th.

My dear Henslow

Just before coming here for our poor girl's health (I am glad to say that she has benefitted decidedly from sea-air) I received your little note, telling me that you were not quite satisfied on antiquity of the Celts.[1]

I fear that this truly dreadful weather will give me no chance of my weed-seeds.—[2]

My object in writing now is to beg for a bit of information; & I cannot think of any one else to apply to, otherwise I would not have troubled you.—

I have been making a great number of observations on the leaves of Drosera, & have come to some curious results about their power of discriminating nitrogenised compounds.—

When the viscid hairs contract or become inflected; they pour out much fluid & the contents of the cells in the footstalks, instead of being a thin pink homogeneous fluid, becomes a broken mass of dark red, thick fluid. When the cells are in this condition, the particles circulate round the cell, *as if* driven by ciliæ. I believe that this is not an uncommon circumstance: is it? But what has surprised me is that the globules & cylinders of the thick dark red fluid or substance keeps on an incessant *slow* contracting & expanding movement: they often coalesce & then separate again; they often send out buds, which rapidly increase at the expence of the larger parent mass; in short endless slow changes in form.[3] These slow incessant movements, which are quite independent of the circulation, resemble the movements of sarcode in the protozoa.[4]

Is any such phenomenon known? it may be quite common, as I am so ignorant of vegetable physiology.— It has surprised me much. Will you be so very kind, whenever you have a little leisure, to let me hear—

I hope you have returned well & strong after your tour.

What a wonderful start this of Hooker's.!—[5]

My dear Henslow | Yours affectionately | C. Darwin

P.S. Though I cannot conceive how it can be effected, I presume that the movement in the red matter must be due to slow absorption of water. Yet how this should make the globules coalesce, divide, coalesce again & bud out into new globules, it is hard to understand.—

Endorsement: 'Sep. 1860—'
DAR 93: 76–8

[1] The letter from Henslow has not been found. Henslow's interest in the question of the antiquity of man had been stimulated by the recent discovery of flint implements known as celts in various deposits in Europe and Britain. He examined the finds at a site in Hoxne, Suffolk, in 1859 and discussed his views in a letter to the *Athenæum*, 19 November 1859, p. 668. He had just returned to England from France, where he visited the celebrated gravel pits near Amiens and Abbeville in which many celts had been found. In a letter to the *Athenæum* dated 14 October 1860, Henslow stated his conviction 'that the facts I have witnessed do not *of necessity* support the hypothesis of a pre-historic antiquity for these works of man.' (*Athenæum*, 20 October 1860, p. 516). See also letters to J. S. Henslow, 26 October [1860] and 10 November [1860].

[2] It is not known for what purpose CD wished to obtain weed seeds. It may have been in connection with his study of the digestive powers of *Drosera* (see *Insectivorous plants*, pp. 127–8). Henslow had previously supplied CD with the seeds of many native plants, some of them collected by the little girls

who attended Henslow's elementary botanical classes in Hitcham (see *Correspondence* vols. 5 and 6). In 1857, CD made observations on weed seedlings naturally appearing in a cleared patch of ground, noting the proportion that perished (see Experimental book, p. 25 (DAR 157a), and *Correspondence* vol. 6, letter to J. D. Hooker, [21 March 1857] and n. 5). He cited his results in *Origin*, pp. 67–8.

[3] This phenomenon, which CD called 'aggregation of protoplasm', is described in *Insectivorous plants*, chapter 3. His notes recording observations made in the autumn of 1860 are in DAR 60.1: 77–81.

[4] The French naturalist Félix Dujardin coined the term 'sarcode' to denote the homogeneous, glutinate substance found in Protozoa. He further noted that this substance was endowed with the ability of self-contraction. See Dujardin 1841, pp. 35–42.

[5] CD refers to Joseph Dalton Hooker's botanical expedition to Syria.

To Charles Lyell 28 [September 1860]

15. Marine Parade | Eastbourne
28 Friday Evening

My dear Lyell

I will now amuse myself by rereading your last letter, & see if I have anything to answer.[1]

The extinction of Ammonites, even after what you tell me about the great breaks in upper challk does yet seem to me a most singular fact; considering how long & how very largely they were developed during the Secondary periods, & considering how high they are in Molluscous scale.—

I do not think I sh^d. save you any trouble (I *have* ordered my own copy) by ordering the Atlantic Monthly for *October*; any Bookseller could order it or get it (for he has copies sent to him) at Trubners: I got lately from him the copy which you have of the August or 2^d. Article.—[2]

I am very glad to hear about the Germans reading my Book.[3] No one will be converted, who has not independently begun to doubt about Species.— Is not Krohn a good fellow?[4] I have long meant to write to him. He has been working at Cirripedes & has detected 2 or 3 gigantic blunders, but in very difficult points. & about which, I thank Heaven, I spoke rather doubtfully, such difficult dissection that even Huxley failed.[5] It is chiefly the interpretation which I put on parts that is so wrong; & not the parts which I describe. But they were gigantic blunders; & why I say all this is, because Krohn, instead of crowing at all pointed out my errors with the utmost gentleness & pleasantness.— I have always meant & write to him & thank him.[6] I suppose "Dr Krohn, Bonn" would reach him.—[7]

With respect to Guinea-pig, I have asked A. Gray where Von Baer makes this statement; unless Von B. has some wonderful new evidence what is the wild parent; the fact goes for little; for I agree with those who entirely deny that the Aperea of La Plata & S. Brazil is the wild parent-stock.[8] It seems that the Guinea-pig was domesticated when America was discovered.—

Did I ever tell you that the great Von Baer goes a long way with us on Species;[9] he has read my Book with much attention.—

I would gladly keep myself the Hybrid Hare-Rabbit; but if I got them from France, what evidence should I have that those which I bought were half & half.?—

Bartlett is setting about it in right way, namely by crossing wild Hare & Rabbit.—[10] If you see him again just suggest that he sh^{d} try several races of the Rabbit with Hare, if one fail; for there is some mystery about the whole case, as several have tried & could never get the two species to pair.—

I cannot see yet how multiple origin of dog can be properly brought as argument for multiple origin of man. Is not your feeling a remnant of that deeply-impressed one on all our minds, that a species is an entity,—something quite distinct from variety? Is it not that the dog case injures the argument from fertility; so that one main argument that the races of man are varieties & not ⟨species, i.e. because they⟩[11] are fertile inter ⟨se is⟩ much weakened?—

I quite agree with what Hooker says that whatever variation is possible under culture is *possible* under nature;—not that the same form would ever be accumulated & arrived at by selection for man's pleasure, & by natural selection for the organism's own good.—

Talking of "Natural Selection", if I had to commence de novo, I would have used ⟨natural preservation⟩;[12] for I find men like Harvey of Dublin cannot understand me; though he has read the Book twice.[13] Dr Gray of B. Museum, remarked to me that "*Selection* was obviously impossible with plants"! "No one could tell him how it could be possible".—[14] And he may now add that the Author did not attempt it to him!—

Yours ever affect | C. Darwin

Endorsement: '28 Sept 1860'
American Philosophical Society (229)

[1] The points CD discusses were apparently mentioned in the letter from Charles Lyell, 27 September 1860, only a part of which has been found.

[2] CD refers to the third and second parts, respectively, of [Gray] 1860b.

[3] Lyell had travelled in Germany during the summer of 1860.

[4] August David Krohn was a Russian-born zoologist who specialised in the anatomy and embryology of marine invertebrates. Lyell may have met him in Bonn during his continental tour. See letter from Charles Lyell, 18 September 1860, and n.7, below.

[5] Krohn published two articles on the anatomy of Cirripedia in the *Archiv für Naturgeschichte* (Krohn 1859 and 1860a), the second of which was translated into English and printed in the *Annals and Magazine of Natural History* 3d ser. 6 (1860): 423–8. In his 1860 paper, Krohn stated that CD was incorrect in *Living Cirripedia* in supposing that the horns of the carapace of the larva metamorphosed into what CD called the 'prehensile antennæ' of the mature individual (Krohn 1860a, p. 7 n.). Krohn had earlier pointed out CD's error in believing that the 'gut-formed glands' were part of a continuous organ that included the cement ducts and the peduncular tubules (Krohn 1859). Thomas Henry Huxley had tentatively confirmed CD's observations on these points in T. H. Huxley 1856–7, 15: 239. See *Correspondence* vols. 4 and 5.

[6] CD had sent a presentation copy of *Living Cirripedia* (1854) to Krohn (see *Correspondence* vol. 5, letter to T. H. Huxley, 2 September [1854], n. 4). Krohn later sent CD a copy of his study of pteropods and heteropods (Krohn 1860b), which is in the Darwin Pamphlet Collection–CUL.

[7] Although Krohn was not officially attached to the University of Bonn, he often worked there when he was not at the seaside studying marine fauna. See Blyakher 1982.

[8] See letter to Asa Gray, 26 September [1860]. The source of Gray's information may have been Baer

1834, p. 15, where Karl Ernst von Baer suggests that the *Cavia aperea* of America is the wild parent-stock of the European domesticated guinea-pig. He went on to say that the domesticated guinea-pig would no longer pair with the wild parent-stock, and so could be considered a new species.

[9] See letter from T. H. Huxley, 6 August 1860, in which Baer's initial response to *Origin* is given. CD had already told Lyell about Baer (letter to Charles Lyell, 11 August [1860]).

[10] Abraham Dee Bartlett. See letter from Charles Lyell, 25 September 1860.

[11] The text was excised by Lyell and the missing words added in another hand at the bottom of the page. See n. 12, below.

[12] The words 'natural preservation' were cut out by Lyell and sent to CD in a later letter because Lyell could not read what CD had written. See letter to Charles Lyell, 3 October [1860].

[13] See letter from W. H. Harvey, 24 August 1860.

[14] CD refers to John Edward Gray. See also letter to Asa Gray, 26 September [1860].

To Daniel Oliver [29 September 1860][1]

15 Marine Parade | Eastbourne
Saturday

My dear Sir

I really thought that I should give no more trouble. But I have strongest wish to observe structure of Dionæa, now that I know Drosera so well.[2] Is it a very precious plant? Could a living plant be packed so as to come here by Railway & could I keep it alive for week or two in sitting room? If so, would you be so kind as to give me address of any nurseryman where I could purchase a plant. Or if there are *several* plants at Kew, would you read this note to Sir William[3] & ask him whether he could **lend** me a plant, which should be returned (carriage free) to Kew; but I should require to gather & dissect some leaves. I want to compare structure of hairs of Dionæa & Drosera, & especially to see whence the fluid is secreted in Dionæa, which is said to bathe the flies caught by it.

The fluid in Drosera is acid & has curious antiseptic powers on meat. Pray forgive me. I would just as soon purchase as borrow: I only mention Kew in case the plant is not to be easily purchased & in case of there being several plants at Kew. How I should like to see a fly caught by it!

Yours very sincerely | C. Darwin

Down House (MS 10: 15)

[1] Dated by the relationship to the letter to Daniel Oliver, 5 October [1860]. The Saturday before 5 October 1860 was 29 September.

[2] CD had inquired about *Dionaea muscipula*, or the Venus's fly-trap, earlier in the month. See letter to Daniel Oliver, 11 September [1860].

[3] William Jackson Hooker was the director of the Royal Botanic Gardens, Kew.

From Charles Lyell[1] 30 September 1860

Sept. 30. 1860.

I expect that when the nondiversification of rodents, bats, manatees seals &c. on remote Miocene islands is fully worked out, it may merely end in satisfying you &

me that the time required for change is longer than was supposed, or without such facts, demonstrable.

It will require renewed enquiry into the antiquity of such islands, also considerations as to the time before first bats & rodents arrived. also the force of preoccupancy—also the arrival of the same species of bats & rodents again & again acting like European colonists into U.S., checking the formation of a new race.

Have you ever speculated in print on turtles changing to land tortoises in remote islands like Galapagos?

From Falunian (Upper Miocene) to Recent is only a part, tho' a large part ($\frac{2}{3}$ ds) of one geol! period measured by change of mollusca.

If as a rule a few species even of large genera vary then if bats & rodents migrate into an island the chances may be a thousand to one that these species are among the unpliant, unvarying ones, & if as you say we know nothing of the law which governs the exceptional cases, we cannot conjecture whether insular conditions would favour such divergence if when the thousandth chance did turn up in some island & an insular species was ready to sport & improve & wd. have done so on land far from sea. Perhaps the chief use to make of my difficulty is this, & it accords with my old notion that species as a rule, or the majority of them, are immutable, & have always been so.

Asa Gray is afraid of argument from imperfection of record.

Falconer observes that the Apteryx has strong clavicles & other bird-like characters without wings.[2] I said, the more useless some of these parts not yet suppressed, the more likely to come down from an ancestor who had need & use of them or to be nascent organs advancing towards perfection, for a posterity which wd. enjoy wings. He only laughed as if the whole was a joke, yet I got him to admit that the hypothesis of limited modifiability was quite as arbitrary as yours.

By the way that reminds me that the keel in the middle of the breast-bone is wholly wanting in some (or all?) of the Dinornis family, tho' I think a little remains in the Apteryx & I am almost sure that H. v. Meyer has lately found this keel in the Pterodactyl.[3]

The long duration & numbers of the Ammonities from trias to chalk & then their extinction, tho' high in the scale, is certainly striking. I presume from the myriads of sepia bones which sometimes strew the Jutland coast that naked cuttle-fish now play their part & are higher in the scale, though not higher than belemnites who accompanied ammonites. The vast number & size of Hippurites & their sudden appearance & disappearance, the only extinct order of Mollusca is also very striking. It would not make much impression here as we have few, but in S. of France, Italy, Sicily & all round Mediterranean.

What you say as to my difficulty is I suspect an explanation up to a certain point, but I hope to put the whole more clearly soon. Dogs of multiple origin & leporines would greatly weaken the objection to regarding the negro as of a different species, in the same sense that the Prairie wolf & common wolf may be. I should like a good naturalist to give me a list of reputed species in Mammalia not more remote than negro & white man. Would they not be many?

Instead of Selection I sh^{d} have said, Variation & Nat. Selection. My only objection is not to the term, but to your assigning to it more work than it can do & the not carefully guarding against confounding it with the creative power to which "variation" & something far higher than mere variation viz. the capacity of ascending in the scale of being, must belong. Most likely you would have chosen some term less worthy of Deification, for 'selection' you had an excellent technical reason.

Copy
Kinnordy MS, Charles Lyell's journal VII, pp. 13–19

[1] The text of the letter has been taken from a copy in Lyell's scientific journal. It is also published in Wilson ed. 1970, pp. 496–8.
[2] Hugh Falconer. The *Apteryx* (Kiwi bird) of New Zealand has rudimentary wings only and is flightless.
[3] Christian Erich Hermann von Meyer was a prominent German palaeontologist. His work on the *Pterodactyl*, a winged fossil reptile, was published in Meyer 1861 and 1862. Lyell had visited Meyer on several occasions and in 1857 learned about his study of *Pterodactyl* (K. M. Lyell ed. 1881, 2: 242–3).

From Thomas Bridges [October 1860 or after][1]

1. The Fuegians do nod their heads vertically, to express assent; and horizontally to express dissent.[2]
2. They blush much, but chiefly in regard to women, but they certainly blush also at their own personal appearance.[3]
3. They certainly, express astonishment by widely openning their eyes, by raising the eyebrows, and openning the mouth.[4]
4. They evince anger with the same signs as we do, with an angry face, which becomes pale. They stamp their feet, and walk about distractedly, and if very angry they cry, and inflict pain on themselves.[5]
5. When out of spirits, they evince it in the very same manner as we do.
6. They express contempt by shooting out the lips, and hissing through them, and by turning up the nose. To spit at one, is the highest mark of contempt[6]
7. They frown greatly when deep in thought, as I have frequently seen.[7]
8. The Fuegians consider a woman a pretty one, who has a round face, who is slender, yet compact, and strong. And it is very certain that the nearer a person approaches to the Caucasian race, the most beautiful, for what European females they have seen, they decidedly looked upon them as beautiful.[8]

⟨*section missing*⟩[9] their noses, and are clever to knock off the shellfish at the first blow. They are excellent swimmers, and ready to bring any bird out of the sea. They are allowed to use the wigwams as their dormitories, and are very good watch dogs. To make their dogs lighter, the Fuegians, whilst they are puppies cut off their tails.[10] The two last questions I cannot answer.[11]

AL incomplete
DAR 85: 39

CD ANNOTATIONS
9.1 their noses . . . answer. 9.5] *crossed ink*
Top of first page: '(America)' *pencil*; '29'[12] *brown crayon*; 'S. America' *brown crayon*
End of letter: 'Informat' *ink, del ink*; 'Answer received through Mr Stirling, from the catechist to the Fuegian Mission, Mr Bridges.' *ink*; 'Information from Mr Bridges, Catechist to Fuegian Mission, through Mr Stirling—'[13] *ink*

[1] The date reflects the minimum time required for Bridges to have received and answered CD's letter of 6 January 1860. Bridges's letter, however, could have been written later in 1860 or in 1861. A missionary working with Fuegians, Bridges was based on Keppel Island in the Falklands. The letter was forwarded to CD by Waite Hockin Stirling, the secretary of the Patagonian Mission Society in London (see CD's annotations).
[2] Cited in *Expression*, p. 276.
[3] This information is cited in *Expression*, p. 318, together with CD's remark that his own experience of Fuegians during the *Beagle* voyage confirmed Bridges's observations.
[4] Cited in *Expression*, p. 279.
[5] Cited in *Expression*, p. 248.
[6] Cited in *Expression*, pp. 260–1.
[7] The letter has been cut in two at this point and the two parts subsequently reunited.
[8] This observation is cited in *Descent* 2: 351 n. 62, followed by CD's comment: 'I cannot but think that this must be a mistake, unless indeed the statement refers to the few Fuegians who have lived for some time with Europeans, and who must consider us as superior beings.'
[9] The missing section of the letter probably addressed the question posed by CD in his letter of 6 January 1860 relating to the breeding and habits of dogs in Tierra del Fuego. See n. 10, below.
[10] This remark about dogs was cited in *Variation* 1: 39. A second reference to Bridges indicates that the missing portion of the letter contained further information about dogs (*ibid.*, 2: 207):

> Hardly any nation is more barbarous than the Fuegians, but I hear from Mr. Bridges, the Catechist to the Mission, that, "when these savages have a large, strong, and active bitch, they take care to put her to a fine dog, and even take care to feed her well, that her young may be strong and well favoured."

[11] The questions concerned the colour of young pigs and calves in the Falkland Islands. See letter to Thomas Bridges, 6 January 1860 and n. 5.
[12] This may be the number of one of CD's portfolios of notes on expression.
[13] The annotations at the end of the letter were duplicated because the letter was at one time cut in two (see n. 7, above). The first remark about Bridges was written on the verso of paragraphs 1–7, which also carries the number '29'; the second was written on the verso of paragraph 8. The letter was subsequently glued together again. In 1867, when CD came to work on his notes for *Expression*, he attempted to acquire further details about the behaviour of Fuegians from Stirling (see n. 1, above), who was by then the senior missionary on Keppel Island. Stirling was in England in 1866 (Macdonald 1929). See *Calendar* nos. 5330 and 5357.

To Charles Lyell 3 October [1860]

15 Marine Parade | Eastbourne
Oct. 3^d

My dear Lyell

Your last letter has interested me much in many ways.—

The Dog M.S is safe at Down.—

I enclose letter of Wyman, which touches on Brain—[1] Wyman is mistaken in supposing that I did not know Cave Rat was an American form: I made special enquiries.[2] He does not know that eye of Tucutucu was carefully dissected.—

With respect to Reviews by A. Gray: I thought of sending the Dialogue to Saturday Review in a week's time or so, as they have lately discussed design.—[3] I have sent the 2^d^ or August Atlantic article to Annals & Mag. of N. History.[4] The copy which you have I want to send to Pictet, as I told A. Gray I would, thinking from what he said he would like this to be done.—[5] I doubt whether it will be possible to get October number reprinted in this country; so that I am in no hurry at all for this.—[6]

I had a letter a few weeks ago from Symonds on imperfection of G. Record, less clear & forcible than I expected. I answered him at length & very civilly, though I could hardly make out what he was driving at.[7] He spoke about you in a way which it did me good to read.—

I am *extremely* glad that you like A. Gray's Reviews. How generous & unselfish he had been in all his labours. Are you not struck by his metaphors & similes?— I have told him he is a poet & not a lawyer.—

I should altogether doubt on turtles being converted into land tortoises on any one island:[8] remember how closely similar tortoises are on all continents, as well as islands; they must have all descended from one ancient progenitor; including the gigantic tortoise of Himalaya.—

I think you must be cautious in not running to convenient doctrine that only one species out of very many ever varies.— Reflect on such cases as the fauna & Flora of Europe. N. America & Japan, which are so similar & yet which have a great majority of their species, either specifically distinct or forming well marked races. We must in such cases incline to belief that a multitude of species were once identically the same in all 3 countries when under warmer climate more in connection; & have varied in all three countries. I am inclined to believe that almost every species (as we see with nearly all our domestic productions) varies sufficiently for natural selection to pick out & accumulate new specific differences, under new organic & inorganic conditions of life; whenever a place is open in the polity of nature. But that looking to *a long lapse of time* & to the whole world or to large parts of the world, I believe only one or a few species of each large genus ultimately becomes victorious & leaves modified descendants. To give imaginary instance, the Jay has become modified in the 3 countries into (I believe) 3 or 4 species; but the Jay genus is not, apparently, so dominant a group as the crows, & in the long run probably all the Jays will be exterminated & be replaced perhaps by some modified crows.—

I merely give this illustration to show what seems to me probable. But oh what work there is before we shall understand the genealogy of organic beings.—

With respect to Apteryx: I know not enough of anatomy, but ask Dr F.[9] whether clavicle &c do not give attachment to some of muscles of respiration. If my views are at all correct; wing of Apteryx cannot be (p. 452 of Origin) a nascent organ, as these wings are useless. I dare not trust to memory, but I know I found whole sternum *always* reduced in size in all the fancy & confined pigeons, relatively to the same bones in the wild rock-pigeon: the keel was generally still further reduced relatively to the reduced length of sternum; but in some breeds it was in a most anomalous manner more prominent.— I have got a lot of facts on reduction of organs of flight in pigeons, which took me weeks to work out, & which Huxley thought curious.—[10]

Yours are interesting remarks on Ammonites—Cuttle-fish & Hippurites.

I can hardly make up my mind to keep the Rabbits; but I will think about it.[11] Isidore G. St. Hilaire evidently knows the case only second-hand.—[12] Do not trust Sclangenweit (the Indian Brothers or some such name) about Yaks, if you come across their statement.[13] Isidore, I have a very good opinion of.— Owen rather sneers at him; & I daresay he is not be trusted on Homologies.—[14]

I have no doubt there are many reputed species (laying on one side question of fertility) not so distinct as Negro & white man.— Ass & Pony? **many** foxes. &c.—

I am utterly ashamed & groan over my hand-writing. It was "Natural Preservation".[15] Natural persecution is what the author ought to suffer. It rejoices me that you do not object to the term. Hooker made same remark that it ought to have been "Variation & nat. selection". Yet with domestic productions, when Selection is spoken of, variation is always implied.— But I entirely agree with your & Hooker's remark.—

Ever yours affect | C. Darwin

Have you begun regularly to write your Book on Antiquity of Man?—

I do **not** agree with your remark that I make N. Selection do too much work.— You will perhaps reply, that every man rides his Hobby-horse to death; & that I am in this galloping state.

Endorsement: '1860'
American Philosophical Society (230)

1 Letter from Jeffries Wyman, [*c.* 15] September 1860. The section of the letter that discussed the brain is now missing. See also following letter.

2 See letter from Jeffries Wyman, [*c.* 15] September 1860 and the following letter. CD had asked James Dwight Dana in 1856 whether the blind rats and other fauna from the Mammoth Cave of Kentucky were American in type. Dana responded by providing information given by Louis Agassiz, who had discussed the cave fauna in Agassiz 1851a. See *Correspondence* vol. 6, letter to J. D. Dana, 14 July [1856], and letter from J. D. Dana, 8 September 1856.

3 [Gray] 1860c. See letter to Asa Gray, 26 September [1860].

4 The second part of [Gray] 1860b.

5 Asa Gray discussed François Jules Pictet de la Rive's critique of *Origin* (Pictet de la Rive 1860) in the second part of [Gray] 1860b.

6 The third and final part of [Gray] 1860b was not reprinted in an English periodical.

7 Neither letter has been found. William Samuel Symonds, rector of Pendock, Worcestershire, was a well-known geologist and a friend of Lyell's. Lyell had received an invitation to visit him in October 1860 (K. M. Lyell ed. 1881 2: 337–8).

[8] See letter from Charles Lyell, 30 September 1860.

[9] Hugh Falconer. See letter from Charles Lyell, 30 September 1860.

[10] CD gave the results of his measurements of pigeon skeletons in *Variation* 1: 162–79. He lent his manuscript on pigeons to Thomas Henry Huxley in January 1860 so that Huxley could use it in preparing his lecture to the Royal Institution on 10 February (T. H. Huxley 1860a). See *Correspondence* vol. 7, letters to T. H. Huxley, 13 December [1859] and 16 December [1859].

[11] CD hinted to Lyell that he might be prepared to conduct cross-breeding experiments between hares and rabbits (letter to Charles Lyell, 28 [September 1860]).

[12] Cited in Lewes 1860. See letter to Charles Lyell, 18 May [1860] and n. 3.

[13] The German naturalists Robert, Adolph, and Hermann Rudolph Alfred Schlagintweit undertook a scientific expedition to India and Tibet between 1854 and 1857. CD corresponded with Robert Schlagintweit about the fertility of the offspring of a cross between a yak and a cow in 1857. See *Correspondence* vol. 6, letter from Robert Schlagintweit, 25 September 1857.

[14] It has not been possible to establish CD's source for his comment about Richard Owen.

[15] Lyell had cut out and sent to CD an illegible phrase from the letter to Charles Lyell, 28 [September 1860].

To Jeffries Wyman 3 October [1860][1]

Down Bromley Kent [Eastbourne]
Oct. 3d

Dear Sir

I thank you sincerely for your letter which has interested & gratified me extremely.[2] If you had had leisure or inclination to have made any general criticisms on the "Origin of Species"; they would have been very far from "superfluous"; for I know hardly anyone whose opinions I should be more inclined to defer to.— I will trouble you with a few remarks on some parts of your letter.

Few facts have interested me more than your case of the black Hogs; it shows such a marvellous relation of mere colour, (generally thought to be so unimportant) & constitution: I have long suspected that such correlation of colour & tendency to fevers, might possibly explain the origin of colour of negros; but I can get no facts to support this crude speculation.[3] I have been the more glad to get your Hog case, as I was hardly able to credit the parallel case of sheep in Sicily.[4]

With respect to the Cave Rat; I knew that it was a strictly American form;[5] but I did not at all know that the genus had large eyes.— I know how busy you are, & that you do not much like writing letters; but if you could get me a little information about this Rat, it would be a *great* kindness.— Firstly, are the eyes of the Cave species not at all larger than in the other species? Secondly, may I trust Prof. Silliman, that the Rat was blind & when kept out of cave, it seemed to acquire some power of vision?[6] And lastly does it inhabit the profoundest depths of caves, or less profound parts?—

As I am asking questions, I will ask one other; viz were you struck with any peculiarity in length of *hind*(?) legs (I am writing away from home, & cannot consult my notes) of the Nãta Cattle? I procured a skull; but it has never been described:[7] would it not be worth your while to insert in some Journal a short description together with the parallel case of the Cod Fish? If you do will you inform me.—[8]

From various sources of information I cannot doubt that the breed is rather ancient, & extremely true: I have copious notes on subject, which I hope some day to use.—

Speaking of inheritance, I was long inclined to entirely disbelieve with you, that Mutilations are ever inherited; (& I got special enquiries made about the Jews); but I have of late been rather staggered; & now Brown-Sequard's case of inherited epilepsy from mutilation seems to almost settle question.[9] I have suspected from some facts that mutilations are inherited only when they produced disease. A clever young Surgeon carefully dissected the eyes of the Tucutucu, which I brought home in Spirits, & assured me that there were traces of inflammation.[10] My Spanish informant had no motive, (for I asked no question) to tell me falsely how often they were blind. Does not the case of Cats with blue eyes being deaf (when one blue eye deaf on one side alone of head) strike you as a very odd case: I have lately observed that all kittens have blue eyes when they first open them; & before they open them, they seem to me after several trials, absolutely deaf;—so that I fancy there is at first some correlation between the blue eyes & deafness, & that the blueness is carried on in the old cats, by a sort of correlation or arrest of development the deafness continues; but my ideas are vague enough.—[11]

With respect to spinal stripe of Horse; I think it is much too common (& characteristic of every species of the genus) to be explained as your American informant supposes: the stripe seems to be commonest with with colts than with old Horses; in same manner as the foals of the E. Hemionus are well striped when first born.—

I have been very glad to hear your remarks on the Brain of Chimpanzee; as will Sir C. Lyell be.—[12] Dr Falconer tells me that Huxley gave Owen the most severe setting down at Oxford on this subject, that ever man received.—[13] I fear the general opinion is true, that Owen truckles to the approbation of those high in church & state. I declare I believe, from some conversation with him that at the bottom of his heart he goes a long way with me on the origin of species, though he has attacked me so severely & not very ingenuously in the Edinburgh Review.—[14] No one other person understands me so thoroughly as Asa Gray. If ever I doubt what I mean myself, I think I shall ask him! His generosity in getting my views a fair hearing, & not caring himself for unpopularity has been most unselfish,—I would say noble.

Forgive me for writing at such length & believe me Dear Sir, with cordial thanks & sincere respect. | Yours very faithfully | C. Darwin

Jeffries Wyman Jr

[1] The year is given by the relationship to the preceding letter and to the letter from Jeffries Wyman, [*c.* 15] September 1860.

[2] Letter from Jeffries Wyman, [*c.* 15] September 1860.

[3] See *Correspondence* vol. 7, letters to Asa Gray, 18 November 1858 and 11 November [1859]. CD had also asked William Freeman Daniell about the possible relation between skin colour and resistance to tropical diseases (see *Correspondence* vol. 6, letter from W. F. Daniell, 8 October – 7 November 1856).

[4] Wyman had given CD information relating to the ability of black pigs to withstand poisons that affected white pigs. CD cited this information in *Origin* 3d ed., p. 12, and in *Variation* 2: 227, where he

also mentioned the case of some races of Sicilian sheep that are similarly affected. See Dupree 1951, p. 106 n. 12.

[5] See preceding letter.

[6] CD cited Benjamin Silliman Jr's study of the blind cave-rat from the Mammoth Cave (Silliman 1851) in *Origin*, p. 137. See also letter from Benjamin Silliman Jr, 27 October 1860.

[7] Richard Owen described the skull of a niata cow in R. Owen 1853, p. 624. CD gave this reference when he came to write *Variation* 1: 89.

[8] Wyman did not publish on these topics. CD cited the information given in the letter from Jeffries Wyman, [*c*. 15] September 1860, in *Variation* 1: 89 n. 64.

[9] Brown-Séquard 1859–60. In this paper, Charles Édouard Brown-Séquard described his successful attempts to transmit experimentally induced epilepsy through several generations of guinea-pigs. CD discussed this work in *Variation* 2: 24.

[10] CD refers to the burrowing rodent, *Ctenomys*, that he collected in Brazil. In *Journal of researches* 2d ed., p. 51, he stated: 'The man who caught them asserted that very many are invariably found blind. A specimen which I preserved in spirits was in this state; Mr. Reid considers it to be the effect of inflammation in the nictitating membrane.' Mr Reid has not been further identified.

[11] See *Origin*, p. 12, and *Variation* 2: 329.

[12] Wyman's remarks were excised from his letter (letter from Jeffries Wyman, [*c*. 15] September 1860) and sent to Charles Lyell (see preceding letter). Wyman had published several important papers on the anatomy of primates (see Wyman 1849, 1850, and 1855).

[13] See letters to T. H. Huxley, 3 July [1860] and [5 July 1860]. See also Appendix VI.

[14] [R. Owen] 1860a.

From Charles Lyell [after 3 October 1860][1]

wish to understand you can do so— I merely think that the general public would have been rather more carried along with you if you had in such cases as the eye referred to two or more causes at work some unknown giving rise to variation & even occasionally to the great miracle & mystery of mysteries progression & one well known namely Natural Selection.

ever affectly ys | Cha Lyell

My work on antiquity of Man is entirely suspended tho' materials have accumulated but as new Edn. of the Manual has stopped it so may now new or *10th*. Ed. of the Principles tho' each of them will afford opportunity of giving to the world nearly all I can say on the subject[2]

I have just found the passage in Hooker about extinction in St. Helena which astonished me— See—Review of Decandolles Geographie Bot—Hooker's Journal of Botany, vol. 8. p. 54.[3]

In the separate copy given to me it is in page—31 note that J Hooker seems to say that only a few species of plants out of *several hundreds* have survived. As Hooker is absent I cannot ask him but do look at the passage— it will be p 85(?) of Journal—vol. 8.[4]

He rather puzzles me by saying in same note that the island was first botanized half a century ago— It was discovered in 1505 I think, & I suppose the forests burnt down long before "it was first botanized—

Incomplete
DAR 205.9 (Letters)

CD ANNOTATIONS
4.1 I have . . . p. 54. 4.3] *double scored ink*
4.1 I have] '22'[5] *added brown crayon*
5.1 In . . . note] *double scored ink*

[1] Dated by the relationship to the letter to Charles Lyell, 3 October [1860], in which CD asks about Lyell's work on the antiquity of man. See also n. 2, below.
[2] Lyell's work on the antiquity of man was published in C. Lyell 1863. This delayed his plans to prepare new editions of his geological texts, the sixth edition of *Elements of geology* (C. Lyell 1865) and the tenth edition of *Principles of geology* (C. Lyell 1867–8). There are copies of all these works in the Darwin Library–CUL.
[3] [Hooker] 1856.
[4] [Hooker] 1856, p. 156 n. Joseph Dalton Hooker there described Alphonse de Candolle's view that the extinction of plant species was sometimes delayed by the longevity of buried seeds. Hooker commented that the native species of Saint Helena that were rendered extinct many years ago had shown no signs of reappearing through the recent germination of long-buried seeds. See also letter to Charles Lyell, 26 [September 1860].
[5] The number of one of CD's portfolios of notes on palaeontology and extinction.

To H. G. Bronn 5 October [1860]

Down Bromley Kent [Eastbourne]
Oct. 5

Dear & much respected Sir

I ought to apologise for troubling you; but I have at last carefully read your excellent criticisms on my Book.—[1] I agree with much of them, & wholly with your final sentence.[2] The objections & difficulties, which may be urged against my view, are indeed heavy enough almost to break my back; but it is not yet broken! You put very well & very fairly that I can in no one instance explain the course of modification in any particular instance. I could make some sort of answer to your case of the two Rats;[3] & might I not turn round, & ask him, who believes in the separate creation of each species, why one Rat has a longer tail or shorter ears than another? I presume that most people would say that these characters were of some use or stood in some connection with other parts; & if so, natural selection could act on them. But as you put the case, it tells well against me. You argue most justly against my question, whether the many species were created as eggs or as mature &c;[4] I certainly had no right to ask that question.—

I fully agree that there might have been as well 100,000 creations as 8 or 10, or only one. But then on the view of 8 or 10 creations, (i.e. as many as there are distinct types of structure) we can on my view understand the homological & embryological resemblances of all the organisms of each type; & on this ground almost alone I disbelieve in the innumerable acts of creation.— There are only two points on which, I think, you have misunderstood me: I refer only to one

Glacial period as affecting the distribution of organic beings: I did not wish even to allude to the doubtful evidence of Glacial action in the Permian & Carboniferous periods.[5] Secondly, I do not believe that the process of development has always been carried on at the same rate in all different parts of the world.[6] Australia is opposed to such belief.— The nearly contemporaneous equal development in past periods I attribute to the slow migration of the higher & more dominant forms over the whole world; & not to independent acts of development in different parts.— Lastly, permit me to add that I cannot see the force of your objection, that nothing is effected until the origin of life is explained:[7] surely it is worth while to attempt to follow out the action of Electricity, though we know not what electricity is.

If you should at any time do me the favour of writing to me, I should be very much obliged if you would inform me, whether you have yourself examined Brehm's sub-species of Birds;[8] for I have looked through some of his writings, but have never met an Ornithologist who believed in his w⟨orks.⟩ Are these sub-species really characteristic of certain different regions of Germany?

Should you write, I should much like to hear how the German Edition sells.—

With my once more sincerely expressed gratitude & with entire respect | I beg leave to remain | Dear & Honoured Sir | Yours sincerely & respectfully | Charles Darwin

Cleveland Health Sciences Library (Stecher collection)

CD note:[9]

Bronn's Criticisms for New Edit of Origin.

Objects that I cannot precisely say why *two species of Rats [*interl*] Hare & rabbit assumed by selection their *several* [*interl*] present characters— very true I can in no case say this— we know so little of use of parts & laws of correlation.— *But I confess, I thus evade every special difficulty. Why one gets round & another pointed leaves. [*added*]

Objects that in case of two varieties still living in abundance side by side how can intermediate vars *have been exterminated [*above del* 'be living']— But are there such cases, excluding such vars. as albinoe.— Do not they inhabit distinct countries or stations— surely this is the general rule.—

Do you believe in Brehms sub-species. Have you seen them?[10]

Thinks that varieties arising from external conditions are linked together by intermediates—not those produced by n. selection.—

Why of two cells, primordial, one got volition & sensation & one did not.—

Says I ought to answer ['ho' *del*] whether my primordial forms were created as eggs or full-grown &c.— Admits that vegetable-cell w^{d} come first.

I think Schiodte says the eyes not so completely gone, at least from more related to those of external world ['in' *del*] near entrance of cave—[11] ['Cave-Rat regained some sight' *del*][12]

I sh^{d} never suppose with respect to his supposed changes in 2 Rats, that first longer or shorter tail & larger [*interl*] ears were acquired, but that all were modified together.—

*Might I not ask Creationist why tail longer or ear shorter? I could not ask this of him who believes that God makes his creation different for mere variety—like man fashions a pattern for mere variety. [*added pencil*]

Objects there might ⟨have be⟩en 100,000 creations as well as one; I agre⟨e, b⟩ut then these would not have borne signs of common descent in homologies & embryology & rudimentary organs.

Some mistake about my supposing several glacial periods—Permian & Chalk.—

"Why sh[d] the process of development have always carried on same rate in all different parts of world?" I do not believe so—N. Zealand & Australia.

*I cannot see force of your objection because one cannot explain origination of life—*the far grandest problem of any [*interl*] why it w[d] not be gain to explain or account for forms; if this could be done —so we ['co' *del*] do not know what electricity is, but we trace its action.— [*del pencil*]

I fully agree to your final sentence—& I fully admit the many awful difficulties in my view.

Certainly, as he grants 〈 〉 that both yet descended from one common parent.

As I cannot justify my opinions in any one single case, so I need not in any.—is as [*over* 'very'] true *as it is [*above del* '& very'] severe.—

Though I can in no single instance, (except by conjecture, as longer legs of Hare for fleetness & not for *[*form*]* —longer ears to hear with) explain changes ['with' *del*] yet the structure &c led me to conclusion.— Laws of Variation will hereafter be made out far clearer.

[1] CD had received a copy of the first part of Bronn's German translation of *Origin* in April 1860 (see letter to H. G. Bronn, 10 April [1860]); he received the third and final part in July (see letter to H. G. Bronn, 14 July [1860]). An incomplete copy of the book (Bronn trans. 1860) is in the Darwin Library–CUL. At CD's suggestion, Bronn had added to the translation a final chapter giving his own comments on CD's views (see letter to H. G. Bronn, 4 February [1860]). CD refers to reading this final chapter. The chapter, however, was removed from CD's copy. It has not been located in the Darwin Archive, and nor has the translation made for CD by Camilla Ludwig, the Darwins' governess. Notes relating to Bronn's criticism are tipped into CD's copy (see CD note, above). It seems probable that CD did not read much of the German text since all but a few of the pages in his copy are uncut. He addressed Bronn's final chapter of criticisms at length, however, in the third edition of *Origin*.

[2] Bronn trans. 1860, p. 520: 'Nur aus dem Widerstreite der Meinungen wird die Wahrheit hervorgehen und der Urheber dieser Theorie selbst zweifelsohne noch die grosse Befriedigung erleben, der Naturforschung einen neuen Weg geöffnet zu haben.' ['Only out of the conflict of opinions will the truth be forthcoming, and the originator of this theory will himself no doubt experience the great satisfaction of having opened a new path for scientific research.']

[3] Bronn trans. 1860, pp. 504–5. Bronn discussed the case of the two species of rat, the black or house-rat (*Rattus rattus*) and the brown or Norway rat (*R. norvegicus*). Bronn asked how CD's theory could account for their derivation—either one from the other or both from some ancestral progenitor. Differences between the two—colour and the length of ears and tail—did not appear to provide any selective advantage to either. Moreover, no forms intermediate between the two had ever been identified. CD mentioned the case in the third edition of *Origin*, admitting that he could 'give no definite answer to such questions' (*Origin* 3d ed., pp. 140–1; Peckham ed. 1959, p. 232). See also CD note, above.

[4] In *Origin*, p. 483, CD asked: 'Were all the infinitely numerous kinds of animals and plants created as eggs or seed, or as full grown?' Bronn explored this question further, asking whether these original forms were created simultaneously or successively in time. 'Obviously either an entire nature-system must have been created at once or they must have developed gradually from a point in the far distant past.' (translated from Bronn trans. 1860, p. 515). Perhaps because of Bronn's criticisms, the passage in *Origin* was amended in the third edition to include the sentence: 'Undoubtedly these same questions cannot be answered by those who, under the present state of science, believe in the creation of a few aboriginal forms, or of some one form of life.' (*Origin* 3d ed., p. 517).

[5] Bronn referred to 'viele Eiszeiten' (Bronn trans. 1860, p. 517).

[6] Bronn trans. 1860, p. 517. See also *Origin*, p. 313.

[7] Bronn trans. 1860, p. 519. Bronn had made the same point in his review of *Origin* (Bronn 1860a). CD had been careful to omit from the tenets of his theory of species change any reference to the origin of life itself.

[8] CD had studied Brehm 1831 closely when he was writing the chapter on 'Variation under nature' for his 'big book' on species (*Natural selection*). See *Correspondence* vol. 6, letters to E. W. V. Harcourt, 19 August [1856] and 23 August [1856]. He mentioned Christian Ludwig Brehm's reputation for

describing as species birds that other naturalists only recognised as varieties, yet he also noted that 'Brehm was a laborious observer' (*Natural selection*, p. 114).

[9] The note is tipped into CD's copy of Bronn trans. 1860 (Darwin Library–CUL). The final page was marked in pencil by CD: '168'. The meaning of this number is unclear.

[10] See n. 8, above.

[11] CD refers to Jörgen Matthias Christian Schiödte's study of blind cave insects. CD cited Schiödte [1849] in *Origin*, p. 138.

[12] The reference is to Benjamin Silliman Jr's study of the blind rats of the Mammoth Cave in Kentucky (Silliman 1851). See letter from Benjamin Silliman Jr, 27 October 1860.

To Charles Lyell 5 [October 1860]

15 Marine Parade | Eastbourne
Friday 5th

My dear Lyell

I have two notes to thank you for, & I return Wollaston's.[1] It has always seemed to me rather strange that Forbes, Wollaston & Co, should argue from presence of *allied* & not identical species in islands for former continuity of land.[2] They argue I suppose from the species being allied in different regions of the same continent, though specifically distinct. But I think one might on the creation doctrine argue with equal force in directly reverse manner; & say that as species are so often markedly distinct yet allied on islands, all our continents existed as islands first, & their inhabitants were first created on these islands, & since became mingled together so as not to be so distinct as they now generally are on Islands.

The subject of Bats is washed out of my head: I got Mr Mason[3] to collect in Madeira & the species turned out to be European: I cannot remember about Palma; but I gave all to Mr Tomes of Welford[4] & he could tell you: I know not where to look amongst my notes; it would be an awful hunt for me. I failed to get the Azores species.—

I had heard nothing of the sales of "Origin" for months, & am much pleased to hear that the sale continues; this surprises me.—

I have never received back the 2d. or August Atlantic Article;[5] perhaps you sent it to Down. I had three copies; one gone to Annals; one at home marked; & the third was meant for Pictet according to promise to A. Gray;[6] but it cannot be helped if lost; I think pamphets are sometimes lost by Post.—

You have, also, the Theological Dialogue out of Silliman's Journal;[7] I shd. be sorry if that is lost, as I have no other copy: it was sent with the October Atlantic number.—[8]

Falconer is hard upon Isidore Geoffroy, as I do not suppose *anyone*, except F. himself has the facts correct.—[9] When I get home I will look about St. Helena: it was this passage that I believe I remembered. I cannot conceive how he can conjecture about number of plants exterminated. The island was woody in comparatively late periods: see my Journal for some facts.—[10]

My fun with Drosera is getting rather poor fun; for my experiments are getting all perverse & crooked; & yet I cannot stop & am wasting time shamefully.—

Ever yours affect | C. Darwin

Etty still, we think, gains a little strength.

Do you know that Bronn has Chapt. of objections at end of his Translation;[11] Miss Ludwig has translated it for me;[12] & I could lend Translation, if you cared to see it.—

Endorsement: '1860'
American Philosophical Society (231)

[1] The letter from Thomas Vernon Wollaston to Lyell has not been found. They had corresponded regularly about Wollaston's study of the insect and mollusc fauna of Madeira and the Canary Islands (see Wilson ed. 1970).

[2] Edward Forbes and Wollaston supported the idea that former land-bridges could account for the modern distribution patterns of animals and plants. For CD's discussions with Lyell about this theory, see *Correspondence* vols. 3 and 6.

[3] Probably Nathaniel Haslope Mason, who collected plants on Madeira between 1855 and 1857 (R. Desmond 1977).

[4] Robert Fisher Tomes was an authority on bats.

[5] CD had sent Lyell his copy of the second of Asa Gray's articles on *Origin* in the *Atlantic Monthly* ([Gray] 1860b, pp. 229–39). See letter to Charles Lyell, 23 [September 1860].

[6] CD told Gray of his wish to send a copy of the second part of [Gray] 1860b to François Jules Pictet de la Rive in the letter to Asa Gray, 10 September [1860].

[7] [Gray] 1860c. See letter to Charles Lyell, 26 [September 1860].

[8] The third part of [Gray] 1860b. See letter to Charles Lyell, 26 [September 1860].

[9] Hugh Falconer and Isidore Geoffroy Saint-Hilaire. See letter to Charles Lyell, 3 October [1860].

[10] See letter from Charles Lyell, [after 3 October 1860]. CD discussed the history of changes in the vegetation on St Helena in *Journal of researches* 2d ed., pp. 488–9.

[11] Bronn trans. 1860, chap. 15. See preceding letter.

[12] Camilla Ludwig was the Darwin children's governess from 1860 to 1865. The translation she made of Bronn's criticism of *Origin* has not been found.

To Daniel Oliver 5 October [1860][1]

15 Marine Parade | Eastbourne
Oct. 5th

My dear Sir

I thank you for two notes.[2] It was a capital thought your sending me the gum itself (so ingeniously bottled or quilled); I tried it on 3 leaves, & it produced no effect; & I tried thicker gum on 5 other leaves with no effect, & I subsequently proved that these leaves were good ones.[3] If a leaf is feeble & does not secrete copiously, the gum dries & draws together all the hairs which it has touched; & this, I imagine, must have been cause of the apparent inflection in your case. At last I have come to a puzzler, for I find Carb. of Soda causes inflection;[4] but I have sent to London for pure C. of soda, & as that sold by Druggists is not pure. But I strongly suspect I have come across a poser. I am, however, trying my experiments in another fashion, which may throw light on subject.— I find the glands at end of Hairs are absorbers as well as secreters. The change which takes place in the

Hairs after inflection is very curious. The currents & movements in the cells strike me in my ignorance as marvellous.—

You are very kind in your second note to say that I must not apologise for all the trouble which I have caused; but pray thank Mr Croker for enquiring about the Dionæa.—[5] Also please give my best thanks to Sir William for his wish to oblige me;[6] I shall be intensely curious to examine the leaves; if I cannot get a plant. If the Dionæa commonly catches only small fry, I sh^d^ not be surprised at a fat fly being too much for its digestion; at least I have found bits of raw meat *often*, indeed I think generally, kill the leaf of the Drosera. I sh^d^ like to hear whether Croker is pretty certain of this fact.—

I am very glad to hear that you are experimentising on the leaves in water. I have long thought that naturalists make far too few experiments.—

My dear Sir | Yours very sincerely | C. Darwin

Do you possess a copy of my "Origin of Species"; if you do not, I should much like to have the pleasure of sending you one.—[7]

Down House (MS 10: 14)

[1] Dated by the relationship to the letters from Daniel Oliver, 19 September 1860 and 25 September 1860.

[2] The notes from Oliver have not been found.

[3] See letter from Daniel Oliver, 25 September 1860, and letter to Daniel Oliver, 27 [September 1860]. CD's notes on his experiments are in DAR 60.1: 110.

[4] CD's first experiments on *Drosera* had led him to expect to find that only nitrogenous substances would stimulate inflection of the leaf-hairs.

[5] Charles William Crocker was foreman of the propagating department at the Royal Botanic Gardens, Kew. See letter to Daniel Oliver, 15 [September 1860].

[6] William Jackson Hooker was the director of the Royal Botanic Gardens, Kew. See letter to Daniel Oliver, [29 September 1860].

[7] According to a note by Francis Wall Oliver (Daniel Oliver's son) added to the typed copy he made of this letter, CD subsequently sent Oliver a copy of the first edition of *Origin* with a 'letter of dedication'. However, this would appear to be in error: see second letter to Daniel Oliver, 12 [October 1860], and Appendix III. The letter of dedication recalled by F. W. Oliver may relate to the copy of *Journal of researches* that CD sent Oliver later in October.

From Charles Lyell[1] 6 October 1860

The geograph^l^. vars. of some species on the American & European sides of the Atlantic, Vanessa Atalanta, Pteris aquilina, remarkable. See p. 700, 9th. Ed. *Principles*.[2] The time required by $\frac{9}{10}$^th^ of the living species to accommodate themselves to the new circums. will never be granted. Not that a red Indian might not be developed into something as good or better than a White Man but as it w^d^ take 10,000 or more years & he is not allowed as many centuries he must be improved off the face of ye earth & he is therefore not transmutable for any practical purpose of salvation in this world.

In no other way can any theory of develop[t] or transmut[n] be reconciled with the stationary condit[n] of species in general, or the fact that they are dying out instead of becoming altered when their existence is at stake.

Letter to Darwin Oct 6. 1860

Incomplete
Copy
Kinnordy MS, Charles Lyell's journal VII, pp. 21–2

1 The text of the letter has been taken from a copy in Lyell's scientific journal. It is also printed in Wilson ed. 1970, p. 499, though not in the form of a letter. The phrase 'Letter to Darwin' has been associated with a following passage discussing *Apteryx*. Although Lyell had corresponded with CD about *Apteryx* (letter from Charles Lyell, 30 September 1860, and letter to Charles Lyell, 3 October [1860]), the layout of the text seems to indicate that it was the discussion about 'Varieties' rather than that on 'Apteryx' that formed the substance of a letter to CD. Moreover, in his letter to Lyell, 8 October [1860], CD refers to Lyell's 'case of Red Indian'.

2 Lyell refers to the section in C. Lyell 1853 in which he discussed the effects of changes of climate on the distribution of species. Species, he stated, would probably be exterminated before they could adapt to changing circumstances.

From James Drummond 8 October 1860

Hawthornden Farm
October 8th | *1860*

Dear Sir,

Since I wrote you by the last mail for England[1] I have met with Brunoniæ,[2] in flower I ⟨wa⟩s anxious to see it as I perceive that Dr Lindley, has made of it a ⟨se⟩parate *Order*,[3] whatever other ⟨diffe⟩rences may exist between it ⟨and Goo⟩deniaceæ, it agrees with it ⟨in the⟩ remarkable character of ⟨the an⟩thers discharging their pollen some days before the flower expands, in *Brunoniæ* the stigma surrounded by the open indusium, passes up between the anthers when discharging ⟨the⟩ pollen, and when the flower opens ⟨the⟩ stigma is seen with the lips of the *indusium* closed over it, but in this genus there are no overlapping Ciliæ, and it is easily opened, I observed a small bee bussily ⟨em⟩ployed in extracting the pollen, but I suppose there is always e⟨nough⟩ left to fertilize the *seed*, b⟨ees⟩ cannot reach the anthers ⟨of⟩ this Genus they remain *incl⟨osed* in⟩ the undivided part of ⟨ ⟩ of the flower,

I have ⟨observed⟩ several Goodeniaceus plants in which the lips of the indusium never closes, in one a large blue flowered plant I have by ⟨me⟩ at present, the pollen is ⟨shed⟩ as ⟨ ⟩ into the indusial cup when it is not more than a tenth part of an inch high, but by the time the flower is fully expanded the stigma crowned with the indusium has grown to be full half an inch high, it occupies the upper part ⟨of⟩ the flower while the five ⟨divis⟩ions of the corolla occupy ⟨ ⟩, the lips of the open ⟨indusium⟩ is well protected from insects or wet by a mass of long white hairs,

I inclose you a few seeds of the curious Genus *distyllus*[4] if they grow with you I think ⟨they⟩ will clearly explain the ⟨wh⟩ole curious process of fertilization ⟨in⟩ this family, what appear as two very questionable styles when the flower opens, are seen when examined early enough to be two funnel shaped membranes beautifully adaped for the use for which they evidently intend⟨ed⟩ to convey the pollen to the st⟨igma⟩ in this genus and the two sp⟨ecies⟩ of another which I describ⟨ed in⟩ my last letter, the Peta⟨ls, stamens,⟩ indusial cup, and St⟨igma are⟩ all deciduous leaving only the calyx, and fertilized seed-vessels, I have observed a curious property in the seeds of *Disty*⟨*lus*⟩ wherever they happen to lie ⟨when⟩ the first shower of rain f⟨alls⟩ on them, earth, stones, Gum leaves, &c, they adhere firmly by a sort of gummy matter which surrounds the seed, a simillar property exists in the seeds of several ⟨ ⟩ Compositae of the tribe ⟨An⟩gianthæ particularly in the ⟨ ⟩ plant which Steets in the ⟨Plantae⟩ Preissiana, refers to *Styloncerus* ⟨humifusus⟩ of Labillardiere,[5] the little acheniæ of this plant are elegant cup-shaped little things, very beautiful when examined with a high power of the Microscope, these fall off and are blown about by the wind ⟨un⟩till a shower of rain falls when they attach themselves to the soil by their lower end at the same time setting themselves perfectly upright, they ornament many a barren spot, in this country throughout the ⟨dry season⟩ and they are not easily re⟨moved⟩ floods of several hours dur⟨ation⟩ from thunder showers ⟨ ⟩ sometimes happen, cannot ⟨remove⟩ them unless it moves the ⟨soil⟩ to which they are attached,[6] I inclose you some of the seeds of this curious little plant, it was when trying some experiments with this plant that I first noticed the extraordinary movements that take place in the hairs of plants

I am Sir Your | obedient humble Servant | James Drummond

P.S. We naturally associate in ⟨our⟩ minds the discharge of the ⟨pollen⟩ from the anthers with ⟨the fu⟩ll blown flower, but in the Order Goodeniaceæ it seems to have little connection with it, and in a curious *Malvaceus*, *Sida* like, plant we have (I cannot say in flower) for it never flowers but when the flower bud is examined in a very young state it is found to have a large plicate Calyx, very like in shape and size the Calyx of Physalis Peruviana, when that is removed, it is found to have a c⟨orolla of⟩ a pale straw colour clos⟨ely en⟩velloping the anthers and Stigma, ⟨at⟩ this time the anthers ⟨appear⟩ to have shed their po⟨llen and⟩ effectually fertilized the many celled, many seeded Capsule nothing is ever seen of the Corolla, anthers, or stigmas, of this plant untill they are carried up in ⟨a⟩ withered state on the top of the seed vessel when they are grown so large ⟨as⟩ to burs open the calyx

DAR 162.2: 242

CD ANNOTATIONS
3.8 I have observed . . . Servant 4.1] *crossed pencil*
5.4 flower bud] 'Good' *added brown crayon, circled brown crayon*
5.5 very like . . . large 5.11] *scored brown crayon*

CD note:[7]
Drummond: Oct 8./60/
Bee removing pollen from indusium of Brunonia[8] | singular indusium of Distylis | Case of Malvaceous flower which never opens. *& is self fertilised [*interl*]—like Drosera(?) & Chlora perfoliata?

[1] See letter from James Drummond, 17 September 1860. Drummond lived near Perth, Western Australia.
[2] CD had asked Drummond to observe the fertilisation of various species of Goodeniaceae, plants native to Western Australia. The pollen-cup of *Brunonia* is very similar to that of the Goodeniaceae.
[3] John Lindley classified *Brunonia* as the only genus in the order (family, in modern terms) Brunoniaceae (Lindley 1853, p. 657–8).
[4] *Distylis* is native to Indochina and Australia.
[5] Joachim Steetz contributed the section on Compositae to Lehmann ed. 1844–7, 1: 417–90. In this work, Steetz described three species of *Styloncerus*, including *S. humifusus* Spreng., which was also known, Steetz noted, as *Siloxerus humifusus*, the name assigned to the species by Jacques Julien Houton de La Billardière (*ibid.*, 1: 435). See also letter to Daniel Oliver, 20 December [1860].
[6] CD discussed the nature of gummy seeds in a paper published in the *Gardeners' Chronicle and Agricultural Gazette*, 5 January 1861, pp. 4–5. Although CD primarily discussed the species *Pumilio argyrolepis*, sent to him by Drummond, he quoted information from this section of the letter in his paper. See *Correspondence* vol. 9, letter to *Gardeners' Chronicle*, [before 5 January 1861], and *Collected papers* 2: 36.
[7] The note is in DAR 162.2: 242/2; it was written on what seems to be the accompanying envelope.
[8] This sentence was marked by CD in ink.

From W. H. Harvey 8 October 1860

40 Trin. Coll. Dublin
Oct. 8. 1860

My dear Sir

I have to thank you for the patience & good nature with which you have listened to my crudities, & the trouble you have taken to meet several of the points I ventured to bring to your notice—[1] I do not want to lead you into a controversial correspondence, or to be drawn into one myself, but yet there are a few matters in your letter that I would like to reply to.—

1st.— I never supposed you to say that Natural Selection could act without previous Variability. On the contrary, throughout your book Natural Selection is represented as dependent on "*favourable*" variations & conditions, ready *to take advantage of*, to perpetuate & accumulate any *profitable* item of differentiation;—but, in strict language, to *originate* nothing. Hence in my letter I have called her a Wetnurse, rather than a Mistress, & hence I can see how we may accept her as the manufacturer of multitudes of "*species*" (so called), such as they exist in our arrangements; and yet reject her as explaining *unlimited* divarications. The first impression of your book on my mind was that it too boldly assumed *unlimited* variability as the cornerstone of the argument; and this was the first stumbling block in my way, for I am strongly impressed with the notion (perhaps wholly wrong) that there is no law of organic or inorganic nature *unlimited* in its operation. And so, however widely species may vary, I suspect an oscillation in every case. In the case of your divaricating pigeons for instance, I should anticipate, after endless variations from type, either a return to type or extinction; but not the passage into a new type.

Among plants, I can believe in very wide limits of *transmutation*; but not equally among all plants. Orchids, Amaryllideæ, Irideæ, & perhaps most Endogens, & all Cryptogamic plants, are *centrifugal* in a very high degree:[2]—so much so that whole genera or even suborders or Orders of our arrangement may be really no more than Natural Species. Thus, for instance granting Natural Selection to be *partially* true, I can see how all *Gramineæ may* have sprung from one common parentage, one aboriginal grass (a *veritable* Aira *præcox*),—and how all *Cyperaceæ*, in like manner, *may* have sprung from one original Sedge. But here I stop, perhaps illogically, because between Gramineæ & Cyperaceæ I think there is a "*saltus*", or vacant space in which it is difficult to conceive a *strictly* intermediate type. The difference in structure of the seed—the relation of embryo to albumen—implies *opposition* (or *polarity*) to that degree that a half & half structure must be proved to exist before I can believe in it. Now *Gramineæ* & *Cyperaceæ* are unquestionably very ancient types, if the widest distribution & great diversity of specific form be proofs of antiquity; both also must have been dominant in a high degree for myriads of years or ages,—and yet, in embryological structure, in floral arrangement, in foliation & in stem,—we have no reason for supposing that the earliest *grasses* or *sedges* respectively, were materially different from those of today. Both Orders differ from each other by an exact, definite character of seed; and in neither Order do the *genera* rise in organization one above the other, but in both orders the genera, so to say, stand on a dead-level or common platform of organisation. This case (and several similar might be cited) appears to me to support the *notion*, that natural variability has *strict limits*; & to bear, so far as it goes, against the notion of *unlimited* derivation from a Protozoon.

2d.— You regret having used the term Natural *Selection*, as tending to mislead, & propose Nat. *Preservation* as a better phrase. If you will not think it impertinent (understanding your theory so imperfectly as I do), I would suggest the term "Natural *Evolution*" as expressing still more exactly the *idea* conveyed to my mind by your theory. Do you not teach us that all organisms, past & present, have been *evolved*, through the action of secondary or *natural* laws, from one *created* or primæval form? If so, Natural Evolution would well express the combination of all the powers of nature in the production of species.

3d. A word or two respecting the Ostrich's wings.—You suppose he has lost wings, "because the whole structure of a Bird is essentially formed for flight, & the Ostrich is essentially a Bird".[3] But—if all *organs* (*wings* included) have been *gradually* formed, through Nat. Sel. accumulating useful variations in successive generations; —unless the *primal organism* had wings & was essentially a flying animal;—then the *earliest* essential bird, or his remote non-bird ancestor, may have had only rudimentary wings. Now the Ostrich may be a *slightly modified* descendent of one of the earlier types, for his affinity to the Dinornis & other wingless birds suggests it.— If indeed the *primal-bird* were formed "per saltum" or by an act of creation, the case would be different;—we should then suppose his organisation to be perfectly adapted for flight, completely furnished as a *typical* bird should be. But if the primal-bird *ascended* from a lower non-bird form, it is not at all probable that the earlier types would have been so fully furnished either with wings or with feathers as those more recently evolved. Hence, I do not see that it is "more complex" to suppose in this case an *evolution*, than a *suppression*;—when, according to your theory, the evolution *must* have once occurred, whereas the suppression, at best, only *may* have occurred. The only complexity I am guilty of is in supposing the Ostrich to represent an *early* type of bird, not a *late* one. Quere, what degree of carination of the sternum is requisite to constitute an *"essential"* or *typical* bird?— . . . . After all, I only alluded to the Ostrich to show that the *facts* of nature may be interpreted, by the help of your theory, in *opposite* ways;—that if you take *one* explanation, another may take the *opposite*, and both be *equally* plausible. It is a matter to me of indifference, whether the Ostrich be going up the hill or down again.

4th.— You say that I beg the question of "Protococcus being doomed to perennial similarity".— I own to the *begging*;—but pray let me state how little I here beg. Truly we know nothing of the imaginary *Protozoon*; but as you say *you* "have less difficulty in imagining the creation of an asexual cell, encreasing by simple division" than an animal of higher organisation, I hope I am not incorrect in inferring that you suppose the Protozoon to have been such a cell, so encreasing.— Now, *Protococcus nivalis* is simply a spherical cell, asexual, encreasing by simple division. The sphere is also the simplest & most self-balanced form of a cell. Grant it to vary in *outline* however, & in *size*; for it does grow & has an average size when mature.— But however variable in such minor points, the *structure* is essentially the same in all individuals;—it is *a simple, nucleated cell & no more.* The question to be solved then is, *how long has it remained in this condition?*— Of course, the epoch at which it was called into being is wholly undiscoverable, but for the following (theoretical) reasons I think it probable that *Protococcus* is very ancient indeed.— If, as you suppose, *all* organisms were successively evolved from one primordial germ or Protozoon, by accumulations of useful variations from generation to generation, then it is reasonable to infer that organisms which depart least from the type of the Primordial are its nearest "blood relations". There was an age of Protozoa before the world contained anything else, & it was probably a long period, because it must (by theory) have continued till Variability, the Struggle for Life &c &c had evolved something better. There was time for protozoic forms to *diversify* indefinitely, if not to exhaust all possible differentiations to which unicellular bodies could be stretched. In our times the world teems with such forms; their genera & species may be counted by thousands & tens of thousands,—& yet, these are only remnants of the innumerable lost protozoic races that preceded them. Few, comparatively, are capable of fossilisation;—but the Diatomes are, and several of the fossil Diatomes are said to be *specifically* identical with forms now living.— Now, it does not matter, for my argument, whether the Pre-silurian *Protococcus* were *spherical*, or *oval*, or *square*, or *stellæform*; or whether it were *red*, *green* or *brown*; it may have gone through indefinite variations of this kind, and yet I should say it was doomed to perennial similarity, if, through all changes, it remained *a simple, asexual cell.*— Now, as it could not have been *less* complex than a cell & have retained organisation;—and as, at the present day, it is not *more* complex than a cell,—all I can say is that, since it was called into being, it has not materially improved in organisation. So far it has inherited "perennial similarity"; & I see no reason why it should not inherit the same to the end of the world.— But, you may say, why should *Protococcus* & similar organisms, be regarded as relics of a Protozoic, pre-silurian world & not *modern organisations?*— I reply:—either they must have come down to us as undivaricated protozoa, from your Protozoon by uninterrupted succession of generation;—or, they must represent higher animals & plants which once flourished, but which through gradual loss of organs from disuse, & a continual struggle with adverse circumstances, have dwindled down to a monadic condition;—or, they must have been *spontaneously generated*, from age to age, that is to say, the creative act of calling a living cell from the dust, must have been *repeated* an infinite number of times from the dawn of life to the present day.— I can find no other way of accounting for their existence; & I suppose your theory would prefer the *first* supposition, as the simplest—that they have lived, as Protozoa, from the dawn of life to the present day.—

And here I would observe, that if you maintain that Protozoic organisms now living are lineal descendents of the aboriginal Protozoon;—and if you also maintain that all higher animals & plants have descended (or rather *ascended*) from the same aboriginal Protozoon, through the agency of natural laws alone, without the personal interference of an Intelligent First Cause,—you are bound to show theoretically how such widely different results have been arrived at. At starting we have similar elementary organisms, under similar conditions of life, all exposed to the same struggle, all subjected to similar "laws acting round them". Why have some remained Protozoa to the present day, while others have developed into Whales or into Man?— Have the "perennially similar" races had no struggle for life? I suppose not, for they have remained perennially similar. But what kept them out of the struggle while all around them, of their own kind & in their own condition as respects outward circumstances were wriggling upward into higher beings? And what set these latter in motion? A struggle for life. Side by side, in the same sea, with the same water to swim in, the same food to eat, the same sun above them; one Protozoon remained in statu quo *unexterminated*, another *advanced* in organisation & developed nerves, sinews & bones; one perhaps was drawn into an *eddy* & so swam round & round forever; the other floated on a *current* & gradually drifted into a new phase of being.—

5th— You dissent from my requiring myriads of years to overstock the Earth with the imaginary Protozoon & refer to Ehrenberg's calculation of the time it would take for an infusorium to make a cube of rock.[4] The *time required* will depend on the nature of the protozoon. If it secreted *lime*, like one of the Foraminifera, or *silex*, like a Diatome, its dead shells would rapidly form strata; but if it were like a Protococcus or an Amoeba no such result could follow for infinite ages. But my argument had reference to something more than a question of *time*;—I wished to show the difficulty of eliminating *a struggle for life*, of inducing *variation*, & leading, through Nat. Selection, to the evolution of higher animals from a protozoon, so placed in a universal sea.—

Granting however that it secreted lime, like a Foraminifer, & rapidly propagated itself: what would result?— I see nothing at first but a thin film of chalk spread out over the bottom of the ocean, & gradually encreasing in thickness till the sea became shallower & shallower, until (supposing *lime* enough) it was filled up at last;—all which filling up must have occurred before the struggle for life commenced, for so long as clear water & available lime remained the Protozoon (if analogous to his modern kindred) had all he required. And, as the available lime diminished, though the newer generations might have thinner & thinner shells, & might at last come to be membrane coated or gelatinous, they need not have made a single *onward* step;—& the fact of their contriving to do without lime would still further retard the *struggle for life*, for they would have fewer wants to supply, their debris would more slowly accumulate & the clear water would be less rapidly filled up. Suppose however that the lime diminished unequally, & that, in some favoured localities the lime-secreting species continued to propagate;—though you might then have *two* species "naturally selected" where but one had been before, yet the *newer* might be the *lower & feebler type*, for the power of secreting lime would be lost, *by disuse*, & with it whatever apparatus was fitted to that purpose. We may indeed suppose that the power of secreting might remain & be diverted to the secretion of silex or any other earth; and so we should get a third, or more species.— To this I have nothing to object;—all these & similar "specific" changes are compassable by Natural Selection. But it is yet to be shown whether all such *adaptations* to circumstances have not, in every case, their natural limits. We can conceive a Protozoic world so originating & becoming greatly varied in "species" without one of them rising sensibly in organisation above another. Like snow-crystals their *patterns* may be innumerable, but their organisation all of a piece. The modern Protococcus may be very unlike the imaginary Protozoon in outward form & in habits of life, and yet, in simplicity of organisation, it may be identical;—each a nucleated, asexual, spontaneously dividing cell.

The arguments above applied to an imaginary protozoon appear to me to fit as well in the case of a reefbuilding coral;—and if Agassiz observations on the reefs of Florida may be depended upon, namely, that there is no difference in species between the bottom & top of a long persisting reef,[5] it further illustrates the "perennial similarity" of low forms for indefinite periods.

6th.— A word or two more about *Variability*.[6] I fear, by your referring me back to passages in the Origin where Variability is insisted on, you have misunderstood what I meant, when, in my letter, speaking of *Unknown* Laws of Variation, I say:—"If therefore these are necessary, at starting, to set Natural Selection in motion, we are surely calling up a wholly different Agency to any set forth in your Theory" &c. I did not mean that you had *ignored* such Laws, for at the opening of Chap. 5, in speaking of them, you pointedly say that they are not due to *chance*, but to an unknown cause. What I meant was that, to variation from this *Unknown* Cause, & not to "Variability from the indirect & direct action of the external conditions of life, from use & disuse" (p 490) we must, in the present state of science, attribute all the *Major* divarications of the organic world, even on the theory of evolution from a primordial. We must, I think, draw a broad line between Variations arising from allied conditions of life, from use & disuse, and those Variations that introduce *new stages* of being, molding an animal body with symmetry of limbs & definite position of internal organs; that, early in the history of life, introduced a *definite* number of typical formulæ (radiate, articulate, molluscoid, vertebrate, &c), & have *retained* these same types, through *every change* of outward circumstances, to the present day.— Variations arising from altered conditions &c, being *impersonal*, may fairly be referred to secondary causes; —but the other & infinitely the larger class of unresolved Variations, including Correlation of Organs or modelling after a preconceived pattern, imply *personality*, & therefore (to avoid pantheism), I ascribe them to the Creator. Natural Evolution, no doubt, takes advantage of every such organic change & commences a new set of modifications upon it, but (as I think) is insufficient to explain or *originate* it. These *steps* (however originated), and however small in amount each may be, are the *saltations* for which I have all along been contending,—the non-recognition of which, in the theory of Natural Selection, appears to me like leaving out the keystone of the arch.

But it is useless to carry on the discussion. I must again apologise for my prolixity, & remain,

My dear Sir | very truly yours | W. H. Harvey

Charles Darwin Esq, FRS. | &c &c

DAR 98 (ser. 2): 54–7

CD ANNOTATIONS

2.2 On the . . . conditions, 2.3] *cross added pencil*
2.8 The first . . . argument; 2.10] *two crosses added pencil*
3.8 But . . . space 3.9] *cross added pencil*
3.13 Now . . . Protozoon. 3.23] *scored pencil*
4.5 Do you . . . form? 4.6] *cross added pencil*
5.7 Now . . . it.—5.8] *cross added pencil*
5.17 The only . . . one. 5.18] *cross added pencil*
7.4 without . . . at. 7.6] *cross added pencil*
8.1 You . . . rock. 8.3] *cross added pencil*
11.8 What . . . world, 11.11] *scored pencil*; '6 fingered man— hard to say if indirect cause of conditions' *added pencil*
Verso of last page: 'Harvey' *brown crayon*; 'Dominant Forms | Struggle for Existence | Antiquity of World &c' *pencil*

[1] See letter from W. H. Harvey, 24 August 1860, and letter to W. H. Harvey, [20–4 September 1860].

[2] The expression 'centrifugal' variation was coined by Joseph Dalton Hooker to represent changes that made the organism diverge from the parental type. See Hooker 1844–7, p. 315, and *Correspondence* vol. 7, letter from Charles Lyell, 28 October 1859 and n. 4. Harvey and Hooker were close friends.

[3] Harvey quotes from the letter to W. H. Harvey, [20–4 September 1860].

[4] Ehrenberg 1854–6, pp. viii–ix. See letter to W. H. Harvey, [20–4 September 1860].

[5] Agassiz 1851b.

[6] Harvey apparently enclosed with the letter a pamphlet that addressed some of the same points about variability. The pamphlet gave the text of a lecture delivered at the Dublin University Zoological and Botanical Association on 17 February 1860 (Harvey 1860). It is in the Darwin Pamphlet Collection –CUL, marked 'With the writers *repentance.* Oct. 1860.' In it Harvey stated: 'It strikes me that there is fallacy at the very base of Mr. Darwin's argument; and that his whole superstructure rests on the *assumption* that Variability acts indefinitely and continuously, without check or hindrance.' An annotation by CD on the back cover relating to this passage states: 'Preface—good to admit that continued variation is an assumption'.

To Charles Lyell 8 October [1860]

15 Marine Parade | Eastbourne
Oct 8.

My Dear Lyell

I send the Translation of Bronn:[1] the first part of Ch. with generalities & praise is not translated. There are some good hits. He makes an apparently, & in part truly, telling case against me, that I cannot explain why one Rat has longer tail & another longer ears &c. But he seems to muddle in assuming that these parts did not all vary together or one part so insensibly before the other as to be in fact contemporaneous.[2] I might ask the Creationist whether he thinks these differences in the two Rats of any use, or as standing in some relation from laws of growth; & if he admits this, Selection might come into play. He who thinks that God created animals unlike for mere sport or variety, like man fashions his clothes will not admit any force in my argumentum ad hominem.—

Bronn blunders about my supposing several Glacial Periods,—whether or no such ever did occur. He blunders about my supposing that development goes on at same rate in all parts of world. I presume that he has misunderstood this from the supposed migration into all regions of the more dominant forms.—

I have ordered D^r. Bree & will lend it you, if you like, & if it turns out good.[3]

I have written to Down about the 2 missing Reviews.[4]

I do strongly believe in SE. & S.W. corners of Australia having been Islands; & the S.W corner the oldest & most typical. Hooker has speculated on this in his Introduction.—[5] So has Jukes.[6] (By the way capital answer by Jukes in Saturday's Athenæum to Sir —James' wild speculations on change of earth's axis).[7] So did I many long years ago in little Review in Annals & Mag. of N. History on Waterhouse's Mammalia.—[8]

I am very glad that I misunderstood you about species not having the capacity to vary, though in fact few do give birth to new species. It seems that I am very apt to misunderstand you; I suppose I am always fancying objections.— Your case of Red Indian shows me that we agree entirely.—[9]

Miss L. says Bronn is **very** difficult German.—[10]

I had not heard of the Bovey Coal Plants.— I hope Bunbury will undertake them.[11] Will Bunbury's new position interfere much with his Science?[12]

Farewell | My dear Lyell | Ever yours | C. Darwin

I do not know who Bree is. I fancy he is son of old Rev[d]. Bree, a good miscellaneous observer of habits of all creatures.—& Botanist.[13] I had letter yesterday from Thwaites of Ceylon who was much opposed to me.[14] He now says "I find that the more familiar one becomes to your views in connexion with the various phenomena of nature, the more they commend themselves to my mind".—

Endorsement: '1860'
American Philosophical Society (232)

[1] The final chapter of Bronn trans. 1860, containing Heinrich Georg Bronn's criticisms of *Origin*, had been translated for CD by Camilla Ludwig. See letter to Charles Lyell, 5 [October 1860].

[2] See letter to H. G. Bronn, 5 October [1860].

[3] Enclosed with the letter is a printed advertisement for Bree 1860. Charles Robert Bree was a physician at the Essex and Colchester Hospital and a strong opponent of CD's views. There is an annotated copy of Bree 1860 in the Darwin Library–CUL.

[4] [Gray] 1860c and the second part of [Gray] 1860b. See letter to Charles Lyell, 5 [October 1860].

[5] Hooker 1859, p. xxviii. CD had discussed this point with Joseph Dalton Hooker (see *Correspondence* vol. 7, letter to J. D. Hooker, 24–5 November [1858]).

[6] Jukes 1850. See *Correspondence* vol. 7, letter to J. D. Hooker, 27 [November 1858], and letter from J. D. Hooker, 22 December 1858.

[7] Henry James, director-general of the Ordnance Survey of Great Britain, had written a letter to the *Athenæum*, 25 August 1860, pp. 256–7, suggesting that displacement in the earth's mass resulting from the elevation of mountain ranges may in the past have caused shifts in the earth's axis of rotation. He suggested that these displacements produced 'great and extraordinary changes' in climate. Joseph Beete Jukes replied in the *Athenæum*, 8 September 1860, pp. 322–3, pointing out that the mass of the largest mountain ranges was minimal compared with that of the equatorial bulge of the earth. The controversy over this point continued through September, drawing in other physical geologists. Jukes published a second letter in the *Athenæum*, 6 October 1860, p. 451, criticising the catastrophist implications of James's view.

[8] CD's review of the first volume of George Robert Waterhouse's book (Waterhouse 1846–8) appeared in the *Annals and Magazine of Natural History* 19 (1847): 53–6 (*Collected papers* 1: 214–17).

[9] See letter from Charles Lyell, 6 October 1860.

[10] Camilla Ludwig, the Darwin family governess. See n. 1, above.

[11] A large collection of fossil plants was discovered at Bovey Tracey, Devonshire, by William Pengelly early in 1860. Charles James Fox Bunbury, Lyell's brother-in-law and a specialist on fossil flora, may have considered undertaking their examination. In November 1860, he wrote that he was studying a German paper on the 'fossil plants of the brown coal which Charles Lyell lent me' (Bunbury ed. 1891–3, *Middle life* 3: 231). In the event, Oswald Heer, the Swiss palaeobotanist, undertook the description of the Bovey Tracey plants (Heer 1861). See *Correspondence* vol. 9.

[12] Bunbury's father, Henry Edward Bunbury, died in April 1860. On his death Charles Bunbury became 8th baronet and inherited two large estates in Cambridgeshire. See Bunbury ed. 1891–3, *Middle life* 3: 219–20.

[13] CD refers to Robert Francis Bree, canon of Saint Giles's, Camberwell, who collected plants and was

a fellow of the Linnean Society (R. Desmond 1977). Charles Robert Bree was not a relation. See n. 3, above.

[14] The letter from George Henry Kendrick Thwaites has not been found.

To Daniel Oliver [10 October 1860][1]

☞ 15 Marine Parade | Eastbourne | per Brighton Railway
Wednesday

My dear Sir

The morning post goes out immediately— one line to give my address for Plants Shall be **delighted** to try one or two experiments on D. spathulata— Plant c[d] come with Dionæa.—[2] Will write again more leisurely.—

Ever yours | C. Darwin

Down House (MS 10: 13)

[1] The Wednesday before the letter to Daniel Oliver, 12 [October 1860], was written.

[2] CD had asked Oliver to send him a living specimen of *Dionaea* (see letter to Daniel Oliver, [29 September 1860]). He inquired about the Australian species of *Drosera* in the letters to Daniel Oliver, 11 September [1860] and [22–3 September 1860].

To J. S. Henslow 11 October [1860]

15 Marine Parade | Eastbourne
Oct 11.

My dear Henslow

Very many thanks for your pleasant letter just received from Down, with many things which interested me.[1] I am particularly obliged for answer about contracting & expanding red thick fluid substance. I have been observing it with great care. & it really is a most curious phenomenon.— It all takes place within the same cell. & is, I now find certainly not from endosmose or exosmose from any *external* fluid;[2] for a hair cut off & put *dry* between two slips of glass, shows the phenomenon splendidly. I have outline of a whole series of changes; which are infinitely varied & never cease. The movement can never be seen until after the hairs have contracted over a Fly or other substance.[3] But I shall write a paper on the subject: so will not trouble you any more.—

You have told me what I wanted to know, whether it was some well phenomenon.—[4]

Yours affect | C. Darwin

D[r] Bree "pitches" into me handsomely.—[5]

Endorsement: '1860'
DAR 93: 79–80

[1] Henslow's letter has not been found. It was evidently a reply to the letter to J. S. Henslow, 28 [September 1860].

[2] CD discussed the phenomenon of 'aggregation' in *Insectivorous plants*, chapter 3. He stated that at first he believed the 'movements of the masses might be due to the absorption of water' (*ibid.*, p. 41).
[3] CD later found that the process of aggregation was independent of the inflexion of the hairs (*ibid.*, pp. 43, 54–5).
[4] Henslow interlined 'known' after 'well'.
[5] Charles Robert Bree. See previous letter.

To John Murray 12 October [1860][1]

15 Marine Parade | Eastbourne
Oct 12th

My dear Sir

I enclose a draft for the amount of your account with many thanks for all your kindness—[2] Please send receipt to above address.—

Sir C. Lyell tells me that he hears that the Origin sells yet pretty well which pleases & surprises me.—[3] If ever a new Edit. is called for give me some good notice, as I shall have a good many corrections & an Historical Preface.—[4]

My dear Sir | Yours very sincerely | C. Darwin

John Murray (publishers)

[1] The year is given by CD's wish to add a historical preface to the next English edition of *Origin* (see n. 4, below).
[2] A payment of £42 18*s.* 8*d.*, made out to John Murray, is recorded in CD's Account book (Down House MS) on 12 October 1860. The entry reads: 'Murray for copies of Origin (subtract from the last Bill & do not class)'. It appears that CD was paying Murray for presentation copies of the second edition of *Origin* (see Appendix III). He had already received from Murray half of the profits owed to him on the sale of the second edition: his Account book records that £318 6*s.* 8*d.* was paid to him on 9 July 1860. His note indicates that his net profit at that time was £275 8*s.* A further payment from Murray brought the total earnings from sales of the second edition to £636 13*s.* 4*d.* (see 'Journal'; Appendix II).
[3] Three thousand copies of the second edition of *Origin* had been printed. The book was issued in January 1860.
[4] The third edition of *Origin* was published in April 1861. It includes a short historical sketch similar to the one that had already been published in both the American and German editions in 1860 (see letters to Asa Gray, 28 January [1860], and to H. G. Bronn, 14 February [1860]).

To Daniel Oliver 12 [October 1860][1]

15 Marine Parade | Eastbourne
12th

My dear Sir

The Drosera & leaves of Dionæa. arrived quite safe. I am very much interested by them.— Pray give my best thanks to Sir William, & accept them yourself.—[2]

I am very busy & get my constant enemy headach; so will not write any more, but I have got a good deal to say first day I can.

Will you try one easy experiment.— Drop on sensitive hair of Dionæa one drop gently of pure water; & get solution of Carbonate of Ammonia 6 six gr. to 2 two oz of distilled water, & drop this & see if it immediately or within hour or two causes movement. This will not injure leaf.— Perhaps safer to get 4 gr to 2 oz water[3] Also one drop of the C. of Ammonia on disc of leaf.— I want so much to know whether movement analogous of Dionæa & Drosera.—[4]

Yours very truly | C. Darwin

Down House (MS 10: 16)

[1] This and the following letter are dated by the relationship to the letter to Daniel Oliver, 14 October [1860].

[2] William Jackson Hooker had agreed to sending CD a living specimen of *Drosera spathulata* and some detached leaves of *Dionaea.* See letter to Daniel Oliver, [10 October 1860]. There is a note dated 12 October 1860 (headed 'Dionæa from Kew') in DAR 54: 2–3.

[3] Oliver wrote on the letter: '(ordered 2 grs. to 1^{oz} pure in dis^{d} water)'.

[4] CD had hoped to obtain a living specimen of *Dionaea* so that he could conduct his own comparative experiments (see letter to Daniel Oliver, [29 September 1860]).

To Daniel Oliver 12 [October 1860][1]

15. M. P. | Eastbourne
12th

My dear Sir

One line to amend my request, if you try the C. of Ammonia experiment.— please drop the third drop on midrib of leaf on inside or upper side of leaf.—

Yours | C. Darwin

I am sorry that you have my Origin;[2] I hope you won't think me bothersome, but would you like to have my Journal of Researches or Travels?[3] I sh^{d} *like* to send them to you.—

If Dionæa is allied in its moving apparatus & whole structure to Drosera, as I suspect from what little I have yet seen; then the absorbing surface, wherever that may lie, would be shown by C. of Ammonia causing contraction; & that is why I want you to try, middle parts of lateral disc & midrib & sensitive Hairs (& *marginal Hairs?*)

University College London Library, Smith Woodward/DAR

[1] See preceding letter, n. 1.

[2] See letter to Daniel Oliver, 5 October [1860] and n. 6.

[3] The letter is tipped into a presentation copy of *Journal of researches* (1860).

From H. G. Bronn[1] 13 [or 15][2] October 1860

Heidelberg
den 13 [or 15] Oct. 1860

Hochverehrter Herr!

Ich habe mich sehr gefreut wieder einen Brief von ⟨Ihnen⟩ zu erhalten, und zu sehen wie Sie noch immer ⟨un⟩ermüdlich mit der weiteren Ausbildung Ihrer Theorie beschäftigt sind.[3] Sie sagen mir, dass Sie mit mehren meiner Bemerkungen über Ihre Theorie vollkommen einverstanden seyen und nur in zwei Dingen sich meinem Urtheil nicht anzuschließen vermögen, indem es scheine als habe ich Sie missverstanden,—hinsichtlich einer wiederholten EisZeit, und hinsichtlich der Annahme dass die Entwickelung der Schöpfung auf der *ganzen* Erd-Oberfläche gleichen Schritt gehalten. Allein ich erinnere mich nicht in diesen Beziehungen Einreden gegen Ihre Theorie erhoben zu haben; und Sie drücken sich in Ihrem Briefe so unbestimmt über das Abweichende in unseren beiderseitigen Ansichten aus, daß ich nicht im Stande bin etwas bestimmteres ⟨Ihnen zu⟩ antworten. Sollte der Gegensatz zwischen unseren ⟨beide⟩rseitigen Ansichten zu einer weiteren Besprechung erheblich genug erscheinen, so bitte ich mir die Stellen in meinem Buche anzugeben, *wo* in der bezeichneten Hinsicht meine abweichende Meinung ausgedrückt ist.

Sie sagen mir ferner, daß wir viele elektrische Erscheinungen aus einer Theorie über Elektrizität erklären, ohne zu wißen, was Elektrizität seye,—u[d]. daß wir mit gleichem Rechte eine Theorie über den Ursprung des Lebens aufstellen zu können (ohne diesen Ursprung beweisen zu können?). Indeßen beide Fälle sind weit entfernt einander analog zu seyn. Wir *wißen*, dass alle Materie gewisse Eigenschaften besitzt, und dass jeder Elementar Stoffe wieder seine besondre Eigenschaften hat. Nun nehmen wir an daß die Elektrizität *zugleich* mit gewissen Elementarstoffen gegeben seye, wie ihre übrigen Eigenschaften. Denn wir erklären auch viele Erscheinungen aus der Schwere, indem wir annehmen, daß Schwere eine den meisten Körpern zukommende Eigenschaft seye, obwohl wir nicht wißen, wie die Schwere mit dem Stoffe zusammenhänge. Wir begnügen uns daher zu sagen, die Schwere ist eine Eigenschaft der meisten Materien. Wir können dieß aber nicht vom Leben sagen. Das Leben ist keine Eigenschaft, welche der Materie inhärirt. *Leben setzt Leben voraus*, u. *Leben erzeugt Leben*: wo aber noch kein ⟨Leb⟩en ist, da entsteht kein Leben; die Materie ist also unabhängig von dem Leben u. das Leben nicht *nothwendig* an irgend eine Materie geknüpft: überdieß ist es eigenthümlich dass das *ganze Leben* einer jeden Species ein andres Leben als das Leben einer jeden andern Species ist! und sich *so* fortpflanzt.

Wir denken uns Elektrizität und Schwere schon *mit der Materie gegeben*; die Lebenskraft aber ist nicht mit der Materie gegeben und *wir bedürfen für Sie noch einer weiteren Erklärung, welche dort nicht nöthig ist.*[4]

Was Ihre Fragen nach Brehms Subspecies betrifft,[5] so nimmt er an, daß jede Subspecies einen etwas abweichenden Wohnort, eine etwas abweichende Nahrung, eine abweichende Heimath habe. Viele sind, was man klimatische Varietäten nennt (Corvus corone: Corvus frugilegus); die andern könnte man Stations-Varietäten, Futter-Varietäten u. dgl. nennen.

Ich habe gesehen, daß Brehm viele Dutzend Exemplare von jeder Subspecies verschiedener Arten mit der Behauptung vorzeigte, daß *alle* Individuen *einer* Subspecies aus gleichem Vaterlande oder von gleicher Station herstammen. Aber Brehm gibt zu, daß diese Subspecies sich zuweilen miteinander paaren, wenn ein

Männchen kein Weibchen von seiner ⟨ei⟩gnen Subspecies mehr finde, oder umgekehrt. Dieß ist in der That bei Corvus corone u. C. frugilegus zuweilen in Thüringen der Fall, weil in Thüringen der Sommer-Aufenthalt der nordöstlichen C. frugilegus mit dem südwestlichen Corvus corone zusammenstößt. Im Winter leben beide im Südwesten beisammen.

Ich habe als junger Mensch von 15 Jahren zwei Subspecies von Certhia unterschieden, lange zuvor als Brehm sie unterschied, lange zuvor ehe Brehms Namen nur bekannt war Certhia familiaris *Brehm* lebt in den Wäldern, Certhia brachydactyla *Brehm* lebt in den Gärten. Nach Brehm hat sie kürzere Zehen als die erste. Ich habe *diesen* Unterschied nie deutlich finden können. Aber ich habe gewußt, ehe Brehm es sagte, daß die in den Gärten lebende Varietät einen ganz andern Gesang hat, anders gefärbte Eyer legt, und auch ihr Nest etwas anders anlegt! Es ist bekannt, dass Individuen einer Varietät sich lieber mit einander paaren, als mit solchen von andern Varietäten; daß sie daher im Stande sind ihre Varietät fortzupflanzen u. somit sind Brehms Subspecies nichts anders als klimatische u. Stations Varietäten, die sich **in** *sich* fortpflanzen.

Man erkennt in und außerhalb Deutschland klimatische Varietäten an; warum nicht auch Stations-Varietäten? und warum sollten diese Stations-Varietäten sich *in der Regel* durch **In***zucht* fortpflanzen wie es die klimatischen Varietäten thun? Inzwischen will ich nicht behaupten, daß *alle* Subspecies, welche Brehm aufgestellt hat, wirklich so weit ausgebildet sind, um sich in der Regel durch **In***zucht* fortpflanzen.

Sie wünschen zu wißen, wie die Deutsche Übersetzu⟨ng⟩ von Origin of Species verkauft werde? Ich erse⟨he⟩ aus der Anzeige, daß der Preiss 4 Gulden oder etwas weniger als 7 Shilling ist, welche sich nach Abzug des Rabattes auf 6 Shilling reduziren. Schweizerbart schrieb mir neulich, daß der erste Absatz der Übersetzung nicht so stark gewesen seye, als er erwartet habe, daß derselbe aber noch immer fortdauere.[6]

Kann ich Ihnen durch sonstige Auskunft dienen, so werde ich es stets mit Vergnügen thun. Inzwischen verharre ich mit ausgezeichneter Hochschätzung Ihr | ergebenster Diener | H. G. Bronn

DAR 160.3: 317

CD ANNOTATIONS

Verso of last page: '⟨ ⟩ee about Glacial Periods & equal Development over world | Bronn seems to believe largely in Brehm— says he distinguished the two Certhias himself before he had heard of Brehm'[7] *ink*; 'Ch. IV'[8] *brown crayon*

[1] For a translation of this letter, see Appendix I.

[2] The manuscript has deteriorated, and only the bottom half of the numeral is now visible.

[3] See letter to H. G. Bronn, 5 October [1860].

[4] Bronn had discussed this subject in some detail in *Untersuchungen über die Entwickelungs-Gesetze der organischen Welt* (Bronn 1858b, pp. 75–82).

[5] In the letter to H. G. Bronn, 5 October [1860], CD had asked Bronn whether he was familiar with

Christian Ludwig Brehm's work on the birds of Germany (Brehm 1831) and whether he recognised the large number of subspecies that Brehm identified in this work.

[6] The firm of E. Schweizerbart, based in Stuttgart, published the German translation of *Origin* (Bronn trans. 1860). Christian Friedrich Schweizerbart was the head of the firm.

[7] See n. 5, above.

[8] CD's annotation refers to the chapter of *Origin* that discussed natural selection. In the third edition of *Origin*, published in 1861, CD added a section to chapter 4 in which he responded to Bronn's criticism.

From August Wilhelm von Hofmann to Edward Cresy[1] 13 October 1860

R. College of Chemistry
Oct. 13. 1860.

My dear Sir

In reply to your note I hasten to state that I have not myself made many special experiments on the delicacy of tests.[2] I must therefore quote to you such results as are known to me to have been obtained by others, stating at the same time the authority.

1 part of arsenic is detected in 64.800 parts of solution by Marsh's test. (Taylor)[3]

1 part of lead in 158.400 parts of solutions by Sulphide of ammonium. (Taylor)[4]

1 part of iodine in 4.000.000 parts of solution by starch (Price)[5]

Hoping that these illustrations will suit you and Mr. Darwin, whose experiments are really of very great interest.

I remain | My dear Sir | Yours very sincerely | A W. Hofmann

E. Creasy Esq.

DAR 58.1: 4

[1] The letter was forwarded to CD by Edward Cresy. The renowned German organic chemist August von Hofmann was director of the Royal College of Chemistry, part of the Government School of Mines. The school at one time came under the administration of the Metropolitan Board of Works, of which Cresy was principal assistant clerk. Cresy had visited CD in September 1860 (see 'Journal'; Appendix II), at which time CD was engaged in experiments to determine the sensitivity of *Drosera* leaves to certain chemical substances.

[2] Hofmann's great reputation in organic chemistry primarily rested on his work on the nitrogenous bases. The laboratory of the Royal College of Chemistry was well known for its experimental accuracy.

[3] James Marsh, Michael Faraday's assistant at the Royal Institution of Great Britain and a practical chemist at the Royal Military Academy, London, invented the test for arsenic that bears his name. The test involved passing hydrogen gas through arsenical compounds. In recognition of this valuable toxocological achievement, Marsh was awarded the gold medal of the Society of Arts in 1836 (*DNB*). The reference is to Alfred Swaine Taylor's authoritative textbook on poisons, in which the procedure for Marsh's test is described (Taylor 1848, pp. 345–6). Taylor remarked that Marsh's test was 'undoubtedly one of great delicacy' (Taylor 1848, pp. 350–1). See also Coley 1991.

[4] In Taylor 1848, pp. 153–4, the method of detecting lead in solution with 'hydrosulpuret of ammonia' is described. Taylor noted that treating lead acetate solution in this way produced a deep black or brown precipitate 'even when less than the 100,000th part of the salt is dissolved' (*ibid*., p. 436).

[5] Astley Paston Price was Hofmann's assistant at the Royal College of Chemistry and a consulting chemist from 1857. There is no record of his particular test for iodine. Iodine is readily detected by the characteristic blue colour that it immediately produces upon contact with starch paste. In Taylor 1848, p. 305, it is stated that this test is a very delicate one.

To Edward Cresy 14 October [1860][1]

15 Marine Parade, Eastbourne
Oct: 14th.

My dear Sir

I have received Hofman's letter which is extremely interesting to me.[2] It is a great shame to trouble you, but I should be very much obliged for a line here, of explanation, if you can give it.

Does Hofman mean that for instance $\frac{1}{64,00}$ part[3] of a grain of arsenic can be detected;—or that $\frac{1}{4,000,000}$ part of a grain of iodine can be detected; surely that is quite incredible. Or does it mean that in a large body of water containing one part of iodine to 4,000,000 of water, that the starch in the course of time will out of this very weak solution absorb iodine so as to colour it. If this is meant these facts do not, in fact, show how little of these substances produce a visible chemical result.[4]

Pray forgive me, and believe me | Yours very truly | C. Darwin

At present I fully believe after endless experiments that $\frac{1}{64,000}$ of a grain of C. of Ammonia produces on an absorbent gland of the Drosera a visible effect.[5]

Copy
DAR 143

[1] Dated by the relationship to the preceding letter.
[2] See the preceding letter.
[3] The copyist originally wrote '$\frac{1}{64,800}$', which has been corrected in an unidentified hand.
[4] CD only resolved the point to his own satisfaction by performing further experiments on the sensitivity of *Drosera* leaves to chemical substances. His notes on these experiments are in DAR 60.1. His results were given in *Insectivorous plants*, chapters 7–9. With respect to ordinary chemical tests for salts, he reported that about $\frac{1}{4000}$ of a grain of arsenic could be detected, 'but the power of detection depends much on the solutions under trial not being extremely weak' (*Insectivorous plants*, p. 170 n.).
[5] The figure given by CD is repeated in *Insectivorous plants*, p. 146, where CD described immersing four *Drosera* leaves in a weak solution of carbonate of ammonia for some hours. To find the amount of salt that each gland received, he divided the absolute quantity of ammonia in solution by the total number of glands. See also the first letter to Edward Cresy, 2 November [1860].

To Daniel Oliver 14 October [1860][1]

15 Marine Parade | Eastbourne
Oct. 14th

My dear Sir

I am much obliged to you for so kindly telling me about the Australian Orchids, (a subject which interests me greatly, & I have now examined nearly all the British kinds);[2] but I cannot quite understand the description, & without examining the *live* plants, with reference to visits of insects, I believe their means of fertilisation can never be understood. Even Hooker was led into considerable error, not of facts, but of purpose in his curious description of Listera.[3]

I want to consult you on one point, as you have so much knowledge: I have consulted Prof. Henslow, but he knows nothing like it.—[4] The elongated cells in the hairs of Drosera are filled, when expanded, with homogeneous pink fluid: after inflection this fluid *always* separates into colourless fluid & into thick, viscid, *dark* red fluid. This latter fluid undergoes a slow, never ceasing, endless changes of form. The enclosed diagrams (1, 2, 3 &c)[5] represent the same cell full of colourless fluid, with the shaded dark red viscid matter: the outlines were done at the interval of about $1\frac{1}{2}$ minute; & you will see the changes of form & position of the two original masses; they divide, coalesce, leave threads behind them; these threads change into necklaces, & one or more of the beads swell & enlarge, then often divide, coalesce & so on ad infinitum— I have never seen the same changes twice, they are infinitely varied. It is not due to endosmose, for it takes place in Hair cut off & placed dry between two slips of glass.— Has anything like this been described? A very weak solution of C. of Ammonia instantly sets the process at work.—

I hope that you will like to hear a little about Dionæa,—the examination of which I have *much* enjoyed owing to your kindness.— I have made out but very little. The sensitive Hairs, have no gland, or spiral vessel or stomata, & so differ greatly from the Hairs of Drosera; but I have hardly any doubt that the spirally arranged cells are before the contraction of the leaf filled with homogeneous red fluid; for many cells were in the contracted specimens sent by you thus filled And I saw the same, but very slow, movements, as in Drosera, & the red viscid matter was much excited by C. of Ammonia. This seems an important point of agreement with Drosera.

On upper or inside of leaf, the whole surface is studded with minute projections on footstalks, just like the back of a tortoise; & in the cells or compartments, there is pink fluid, which with gf/C. of Ammonia acts as the matter in cells of Drosera. From analogy of Drosera I infer that these tortoises are the secreting & absorbing glands.—[6] One thing surprises me, viz under each footstalk of the tortoise-shells there is what appears a stoma.

These tortoises do not occur on insides of marginal spikes. On back of spikes & on back of leaf, there are many petaloid or star-like little projections on footstalks;

& beneath these there are apparently stomata.[7] The brown matter within these stars is not acted on by C. of Ammonia. The marginal spikes have spiral vessels; & I suspect may be looked at as the extreme marginal hairs of Drosera in a functionless condition.

Now these few observations would lead me to try the following experiments, if I had living plants; but I *beg* you not to try, unless you feel some little interest on the subject.—[8]

Firstly & **chiefly**, to gather one more leaf very quietly, (or look at growing leaf) put it under simple microscope & observe whether the sensitive Hairs are uniformly coloured; then touch Hair with needle, & observe whether the colour does not become *very soon* (how soon?) mottled from the action of segregation, as described, of red viscid matter from red fluid.

Drop *weak* C. of Ammonia on *inside* of marginal spikes,—on disc of leaf,, on back of leaf & on sensitive Hairs, but in in latter case so small a drop as not to run down on to the disc, if this be possible.— These experiments could tell much.—

In how many seconds or minutes does leaf close?

Lastly can you tell me, whether the stomata are ever covered by reticulated cells?

Have stomata & spiral vessels any connection? or relation

First try disc & sensitive Hair; if these do not act, other trials would be useless.—

I have as yet tried only 2 experiments on D. spathulata.—[9]

at 9° A.M put on one leaf $\frac{1}{2}$ minim of solution of Nitrate of Ammonia (3 gr to 1 oz, so that I gave the leaf $\frac{1}{320}$ of a grain which is much too strong), in 6 hours, hairs incurved; in 7 hours hairs met; in 9 hours edges of the leaf itself **much** incurved; so that leaf became semi-cylindrical.—

Gum has produced no effect—

I am in complete muddle about effect of various salts; & I am deeply indebted to you (in main part) for having made me try more experiments.— God forgive me, for having written you so long a letter.— You will soon find out that I can be troublesome with a vengeance.—

With cordial thanks, Yours very truly | C. Darwin

(It has been ridiculous in me, from my ignorance, taking up this subject; but it was purely accidental.)

Do please touch the sensitive Hair; though my fear is whether with simple microscope you could perceive change from uniform to broken colour: yet I felt sure that colour was broken when I first saw through simple lens the Hairs.—I expect the change to be almost instantaneous. It would be very curious if disc alone was sensitive to Ammonia; & the Hairs only to touch, but Heaven knows how it would be.

Perhaps action of C. of Ammonia on the disc would not be rapid (as from a touch on Hair).—

Down House (MS 10: 17)

[1] The year is given by CD's stay in Eastbourne (see 'Journal'; Appendix II).
[2] In *Orchids*, CD thanked Oliver, along with many other botanists and gardeners, for his help in supplying information about foreign orchids (*Orchids*, p. 158 n.).
[3] Hooker 1854b. Joseph Dalton Hooker did not regard the curious modifications of the sexual organs of *Listera* as adaptations for insect pollination.
[4] See letters to J. S. Henslow, 28 [September 1860] and 11 October [1860].
[5] The drawings have not been found. For later drawings illustrating the phenomenon of aggregation, see *Insectivorous plants*, pp. 40–1.
[6] The structure of the glands of *Dionaea muscipula* is described in *Insectivorous plants*, pp. 287–90.
[7] Stomata are openings in the epidermis through which gases can pass into the internal tissue of plants and water and oxygen pass to the outside.
[8] Oliver had only sent CD some leaves of *Dionaea*, not a complete plant. See letters to Daniel Oliver, 11 September [1860], [29 September 1860], and 12 [October 1860].
[9] Oliver had sent a specimen of *Drosera spathulata* from Kew at the end of September. See first letter to Daniel Oliver, 12 [October 1860].

To Johann Nicolaus Trübner?[1] 14 October [1860][2]

15 Marine Parade | Eastbourne
Oct. 14th

Dear Sir

A friend tells me that there is an Article, which I wish to have, in Sillimans N. American Journal of Science "in the September number which will appear in October".—[3] What this means I do not know.— Will you see whether you can get the Sept. or Oct. number in London; & if so send it here, as I much want it immediately; if not, please send it, when procured to Down, Bromley, Kent, whither I shall return in 13 days.—[4]

Dear Sir | Yours very faithfully | C. Darwin

Fitzwilliam Museum Library

[1] The recipient is inferred from CD's wish to acquire a copy of [Gray] 1860c. Trübner's London-based publishing firm specialised in American works and was used by Asa Gray to distribute his works in England (see letter to Charles Lyell, 26 [September 1860]).
[2] The date is based on the reference to [Gray] 1860c.
[3] [Gray] 1860c. Although Asa Gray had sent CD a copy (letter to Asa Gray, 26 September [1860]), CD had sent it on to Lyell to read and had not yet had it back; it may have been lost in the post on its return to Down (see letter to Charles Lyell, 5 [October 1860]).
[4] The family's return to Down was delayed until 10 November 1860 owing to Henrietta Emma Darwin's illness ('Journal'; Appendix II).

From Daniel Oliver [15–16 October 1860][1]

From 'Botanische Zeitung' (Regensburg) 1833. *p. 96* Notice of Prof. Burnett on irritable plants & the insects wh. they secure.[2]

"Ich habe einen Versuch mit 2 Exemplaren der *Dionæa muscipula* gemacht, und zu dem Ende 2 von beinahe gleicher Grösse & Gesundheit gewählt. Die eine davon bekam weder Würmer noch Fliegen u.s.w., während die andere täglich mit kleinen Ueberbleibseln von Thieren gefüttert wurde: das Ergebniss dieser Versuchs war, dass die erste pflanze, bei ihrer strengen Diät, allmählig hinwelkte, die andere aber, bei ihrer Fleischkost(?) sich vortrefflich befand."—

Nuttal, in Gen.ª No. American Plants, p. 277, (note)— "This sensibility is said to reside principally in the 4 capillary processes situated upon the disk of the lamina. While engaged in collecting this plant in its singularly insulated situation near Wilmington, in North Carolina, I had occasion to observe that a detached leaf would make repeated efforts towards disclosing itself to the influence of the Sun, these attempts consisted in an undulating motion of the marginal ciliæ, accompanied by a partial opening & succeeding collapse of the lamina, which at length terminated in a complete expansion & in the destruction of sensibility."—[3]

Note fr. Drummond. Swan River Bot.ʸ Kew J.ˡ Bot.ʸ v. 345.[4]

Speakg. of a sp. of Elatine.—"I have ascertained that in the minute submerged concealed flowers, the anthers burst internally, & the seeds arrive at maturity in a much shorter time than they require when the inflorescence is exposed to the air." x x x x x x

Incomplete
DAR 58.2: 53

CD ANNOTATIONS

1.1 From . . . befand."— 2.6] 'comparison of 2 Plants with flies & without not enough.' *added pencil*
2.3 täglich] *underl pencil*; '!!' *added pencil*
3.1 "This . . . Carolina, 3.4] *double scored pencil*
3.4 I had occasion] *paragraph divided before* 'I had', *brown crayon*
3.6 these attempts . . . sensibility".—3.8] *scored brown crayon*; *line added after* 'sensibility', *ink*; 'Does this refer to young leaf first opening' *added pencil*
4.1 Note . . . x x x x 5.4] *scored ink over pencil*; *one word illeg added ink*
Top of first page: 'Opening' *brown crayon*; '*Dionæa* & **Intercrossing**' *ink*
End of letter: 'Intercrossing' *ink*

[1] Dated by the relationship to the preceding and following letters.

[2] A note in the botanical information section in *Flora, oder allgemeine botanische Zeitung* pt 1 (1833): 96 describes the digestive activity of *Dionaea* and other species. The information appears to have been taken from a work by the medical botanist Gilbert Thomas Burnett, professor of botany at King's College, London. The passage is translated as follows:

> I have conducted an experiment with 2 specimens of *Dionaea muscipula* and for this purpose chose 2 of almost the same size and condition. One got neither worms nor flies, etc. while the other was fed daily with small remnants of animals: the result of this experiment was that the first plant, because of its strict diet, gradually withered away; the other, however, fared excellently on its meat regime.

[3] Nuttall 1818, 1: 277–8. CD cited this information in *Insectivorous plants*, p. 318 n., thanking Oliver for the reference.

[4] Drummond 1853, p. 345. For CD's correspondence about the fertilisation of Australian plants, see the letter to James Drummond, 16 May 1860, and the letters from James Drummond, 17 September 1860 and 8 October 1860.

To Daniel Oliver 17 October [1860]

15 Marine Parade | Eastbourne
Oct 17th

My dear Sir

I am infinitely obliged for your mass of interesting facts & extracts.—[1] The only at all full account which I have seen of Dionæa is by Mr Curtus, quoted by author (Lindley?) in Penny Encyclopædia.[2] Lindley also refers to Dionæa having been fed & flourished on chopped meat.— I think you could see with lens on growing plant whether colour in Sensitive Hairs becomes broken after being touched; that experiment & effect of C. of Ammonia on disc could be very interesting. A good dissector, I daresay could follow the cells with red matter from base of sensitive Hair, & thus follow its *nerves*!—

Many thanks about stomata; it was uncommonly stupid in me not to think of your explanation, which I have hardly a doubt is right;[3] for it was after removing the tortoise-like glands, which are mounted on *very* short footstalks, that I saw the appearance of stomata: I have hardly a doubt that my so-called stomata are 2 cells of the footstalk.—

Thanks for your various analogies & comparisons about the moving red matter: it is beyond me; I shall just publish what I saw.—[4] I think that you have misunderstood me in supposing that this appearance follows only from C. of Ammonia.— it is *better* seen in the Hairs which have naturally contracted over a fly or other object; & this it is, which seems to me to make the case curious.— I have been ascertaining this morning how quickly C. of Ammonia acts; & certainly 13 seconds suffice for the absorption & for a marked change in structure in the glands; in one minute the action reaches the upper part of footstalk.— No other substance (such as Acetate, Oxalate, nitrate of Ammonia &c &c) acts nearly so quickly, though they do act after some time; the acetate of Ammonia is next in quickness.— I *suspect* that milk, urine &c & these salts do not act until they have become decomposed, & yield C. of Ammonia.— I cannot avoid suspecting that we see in the action of these substances on the leaves of Drosera what chemists believe takes place with organic manures on the roots of other plants.—[5] I am going to try tomorrow C. of Ammonia on the root of Drosera & on some other common plant.—[6]

Thanks about Elatine—& for Goodenia Plants.— Pray thank Croker[7] for all great trouble which he has taken.—

I have told Murray to send you copy of my Journal,[8] which I am very glad you did not possess.—

Yours very truly | C. Darwin

The day after tomorrow I shall begin & draw up my paper on Drosera; for I have wasted a shameful lot of time on it—& **yours** also.[9]

The leaves of D. spathulata will not open.—

Down House (MS 10: 18)

[1] See preceding letter. The letter that apparently accompanied the note has not been found.

[2] Moses Ashley Curtis, a resident of North Carolina where *Dionaea* is native, was quoted by John Lindley in [Lindley] 1837, p. 508. CD cited Curtis 1834 in *Insectivorous plants*, p. 301 n.

[3] The stomata of *Dionaea muscipula* were apparently discussed in the missing portion of the preceding letter. For CD's views on the fine structure of the leaves, see the letter to Daniel Oliver, 14 October [1860].

[4] CD described the phenomenon of aggregation in *Insectivorous plants*, chapter 3.

[5] CD was familiar with the prevailing view of agricultural chemists that the roots of plants absorbed nitrogen derived from the decay of animal matter. A copy of Liebig 1840, which discusses this point, is in the Darwin Library–CUL. For CD's reading of the works of Justus von Liebig, see *Correspondence* vol. 4, Appendix IV.

[6] CD's experiments on the absorption of carbonate of ammonia by the roots of *Drosera rotundifolia* were given in *Insectivorous plants*, pp. 141–2. Further experiments on the roots of other plants were ultimately given in his paper on 'The action of carbonate of ammonia on the roots of certain plants' (1882) (*Collected papers* 2: 236–56). By that time, CD had decided that aggregation was the result of an accumulation of waste products in the cells at the tip of the root.

[7] Charles William Crocker. See letter to Daniel Oliver, 5 October [1860].

[8] *Journal of researches* (1860). See second letter to Daniel Oliver, 12 [October 1860].

[9] Although CD did not publish the results of his study of *Drosera* until 1875 (*Insectivorous plants*), he delivered a paper on the subject on 21 February 1861 at a meeting of the Philosophical Club of the Royal Society (Bonney 1919, p. 154). The index to his 'Eastbourne paper' is in DAR 54: 1.

To W. D. Fox 18 October [1860]

15 Marine Parade | Eastbourne
Oct 18th

My dear Fox

Many thanks to you for calling my attention to the Hybrids.[1] But I have seen them & read about them in French— I at first quite disbelieved the case; but I suppose it is true; though by chance I had a note from Lyell this morning, who reports that some French naturalists believe whole story is Cock & a Bull.[2] It is the most curious case that ever was if true; for so many have failed to get Hybrids from Hare & Rabbit.—

We have been here for 4 weeks & return home in a weeks time.— We came here for chance of the sea doing Etty good; & it has certainly to certain extent succeeded.[3] At one time this summer I gave up all hopes of her recovery, but she does gain strength at a snail's pace, & now suffers only from indigestion & weakness.[4] We, all the others, are well; though I cannot boast very much of myself.— I have done little of my regular work this summer; chiefly owing to incessant anxiety & movement on account of Etty. My correspondence about the "Origin" has also been gigantic, & a day hardly passes without one, two or three letters on the subject.

I have amused myself with a little natural History of other kinds; & have lately worked hard at the power of the Drosera or sun-dew in catching flies; & tremendous slaughter the plant makes.—

I shall be very glad to be at home again; for the weather is most dismal.— I hope that the world goes fairly well with you & yours.—

My dear old friend | Yours affectionately | C. Darwin

Endorsement: 'Oct 18/60'[5]
Christ's College Library, Cambridge (Fox 130)

[1] CD refers to the hybrid offspring of hares and rabbits discussed in the letters to Charles Lyell, 18 May [1860] and 5 [July 1860].

[2] The letter from Charles Lyell has not been found. Lyell may have told CD about the work of the French naturalist Pierre Paul Broca. There is an undated abstract of Broca's paper on hybridity (Broca 1858–9), which discusses the hare–rabbit cross, in DAR 205.7(2): 154. CD cited the work in *Variation* 1: 105 n. 7.

[3] The Darwins took Henrietta Emma Darwin to Eastbourne on 22 September 1860 to recuperate following a long illness ('Journal'; Appendix II).

[4] In a letter from Emma Darwin to William Erasmus Darwin probably written in September 1860 (DAR 210.6), Emma said of Henrietta:

> It is very disheartening but I don't think poor Etty makes a bit of progress . . . Mr Williams says we cannot give her too much wine & brandy & all the food we can get her to take He has always said that it would be most lingering & that the only thing was to keep up her strength. He thinks her going on quite favorably & that we must have patience. Poor soul she has a great deal of patience I think & is got in a manner used to it. She is as thin as possible.

Edward Augustus Williams, who practised in Bromley, was the Darwin family physician.

[5] The endorsement is on the accompanying envelope.

From Robert Patterson 18 October 1860

6 College Square North. Belfast
Oct 18. 1860

My dear Sir

I copy for you a note, that I hope may possess some interest.[1] Should you ever think of publishing it, I think it would be only right to the gentleman who is my authority for the fact, to send a copy of the note to him for revision so as to avoid all chance of error.

I am Very truly Yours | Robert Patterson.

Chas Darwin Esq

PS. Your old friend Ogilby,[2] and Dr E. P. Wright of Dublin,[3] spent last Tuesday evening in my house.

[Enclosure]

Copy

Augt 27 1860. When visiting the Copeland-Islands at the entrance to Belfast Lough this day, Captn Nesbitt RN.[4] one of the Elder Brethren of the Trinity House,[5] told me the following fact. In 1845, when the Trinity House bought up all the private Lighthouses round the Coast, the Skerries near Holyhead,[6] had immense numbers of rabbits, which were extensively used by the lighthouse-keepers. A few years ago an American vessel was wrecked there, on board of which there were a number of rats. These swam to the shore. They have now increased prodigiously in numbers, have destroyed the wild rabbits, and are obliged to live in a great degree on the shellfish of the shore. The light-house men not only have lost the wild rabbits, but find a difficulty in rearing any tame ones, because of the attacks of the rats.

DAR 46.1: 89–90

CD ANNOTATIONS
1.3 to send . . . revision] *double scored ink*
Top of first page: 'Struggle for Existence' *ink*; 'Ch 5.'[7] *brown crayon*

[1] Robert Patterson, a Belfast naturalist, had corresponded with CD since 1847. See especially *Correspondence* vol. 5.

[2] William Ogilby, also an Irish naturalist, had in 1842 served with CD on a committee of the British Association for the Advancement of Science charged with drawing up rules for standardising zoological nomenclature. See *Correspondence* vol. 2, letter to H. E. Strickland, 17 February [1842]. Ogilby had been secretary of the Zoological Society from 1840 to 1847.

[3] Edward Perceval Wright was a lecturer in zoology at Trinity College, Dublin. He was editor of the *Natural History Review* until it changed hands in 1860. Afterwards he remained one of a team of editors that included Thomas Henry Huxley and John Lubbock. See letter to T. H. Huxley, 20 July [1860].

[4] Edward Parry Nisbet.

[5] The Corporation of Trinity House is an association of English mariners, chartered in 1514, whose duties include the erection and maintenance of lighthouses and beacons and the supervision of pilots. The Elder Brethren were initially a select group of thirteen charged with the management of the corporation's affairs, two of whom were elected from the Royal Navy and eleven from the merchant service; as a mark of honour, persons of rank and eminence were also sometimes admitted as honorary Elder Brethren (*EB*).

[6] Skerries Lighthouse is located on the highest of a group of rocky islands off the north-west coast of Anglesey, North Wales.

[7] CD refers to chapter 5 of his 'big book' on species (*Natural selection*, pp. 173–212). The information does not seem to have been included in later editions of *Origin*.

To Edward Cresy [19 October 1860][1]

15, Marine Parade, | Eastbourne.

My dear Sir,

I am extremely much obliged for your note of the 16th,[2] forwarded to me here, where we stay one more week.[3] But I grieve to say your letter contained *no* enclosure, i.e. Hofman's letter.[4] I hope that you will ere this have discovered that it was *not* enclosed. On my return home, I shall be very glad to read Dr. Taylor's work,[5] which shall be returned in about a fortnight's time. I am particularly obliged to you for remembering my wish, and I have become more than ever interested in the subject. Thanks for news about B. Museum,[6] which I have not read.

The Sea has done my daughter some good, but she is still very weak. The weather has been utterly dismal here.

With many thanks, | Yours very sincerely, | C. Darwin.

Copy
DAR 143

[1] Dated by the relationship to the letter to Edward Cresy, 14 October [1860], and by the reference to the Darwins remaining in Eastbourne 'one more week' (see n. 3, below).

[2] The letter has not been found. It was a reply to questions posed by CD in his letter to Edward Cresy, 14 October [1860].

[3] Although the Darwins intended to return to Down on 26 October, they had to revise their plans when Henrietta Emma Darwin fell ill again unexpectedly. They did not return to Down until 10 November 1860 ('Journal'; Appendix II).

[4] Cresy apparently intended to forward to CD a letter from the chemist August Wilhelm von Hofmann answering questions CD had posed in his letter to Edward Cresy, 14 October [1860].

[5] Taylor 1848.

[6] Cresy may have related in his letter the final opinion of the Parliamentary select committee appointed to consider the relocation of the natural history collections of the British Museum. The committee, which published its report in August 1860, rejected the proposed move in favour of expanding the existing facilities at the Bloomsbury site ('Report from the select committee on the British Museum', *Reports from Committees: Public Institutions; British Museum; Kensington Museum*, session 24 January – 28 August 1860, 16: 173–525). The Metropolitan Board of Works, of which Cresy was assistant clerk, was asked by the committee to evaluate the building costs of various architects' plans. CD had signed two memorials concerning the proposed move in 1858 (see *Correspondence* vol. 7 and Appendix VI).

To Asa Gray 19 October [1860][1]

Down Bromley Kent [Eastbourne]
Oct. 19th

My dear Gray

Your pleasant letter of Oct. 1st has been forwarded to me to sea-side, & I have been meditating on it.—[2] First let me say how pleased I am to hear that Prof. Henry goes so far with me.—[3] When I sent the Aug. Atlantic to Annals,[4] I said that I myself took the responsibility of adding your name as author; so that I do not think it will be made to appear as if from you.—[5] Nor do I at all know whether it will be therein printed; but the Editor is not very courteous & I do [not] much like writing to him again—[6] I enclosed envelope, begging for it to be returned if not used; & I have not received it. But this does not convince me that it will appear; but it is either now rejected or *printed*, so much time has elapsed, so no time for 2 or 3 corrections. But I have been thinking of a larger scheme, viz to get the whole three published with (as you permit) your name. But unfortunately I am the worst person in world, from mixing so little with people, to judge; & Hooker is away & Huxley in distress.[7] But I intend writing to day to Lyell, who admires your reviews as much as I do (especially I think No 3) & has written me several letters chiefly about them.[8] Now he is a real good judge in all publishing affairs & if he thinks there is any chance of even a small sale, I will consult Murray.[9] I shall say to Murray that as Dr Bree has just published a Book against the Origin[10] & two other Brochures have appeared (all poor), a pamphlet on our side (to a certain extent) might sell. It would not be worth while to *lose* money, not for the money sake, but on account of small circulation & consequently not much good being done. It would be indispensable to have your name & title on Title-page; & very adviseable to have some remark on Title, showing its bearing on Natural Theology or Design.— I daresay it is 10 to 1 against my succeeding; but, as it will cost very little trouble, send me a Title, (subject to Murray's approval). I fear there is no chance of your having time to add some criticisms on Hopkins.—[11] I will let you know as soon as I can see my way. I must wait till I get home & see how long all the Articles are together. They seem to me much too good to be almost lost in a Periodical.— Murray says the Origin goes on selling well.—

I had a letter a week ago from Thwaites of Ceylon:[12] he was at first much opposed to me. he now says "I find that the more familiar one becomes to your views in connexion with the various phenomena of nature, the more they commend themselves to my mind."— Even Harvey, I find, is not nearly so savage against me as he was when he published his foolish pamphlet.[13] Such cases give me much confidence that Natural Selection is not very far from truth.—

Believe me | My dear Gray | Ever yours | C. Darwin

Perhaps it would be worth while for you to write & send me Title, & say whether you would add any criticisms; as in that case I would delay publication, if it is ever to take place.—

Gray Herbarium of Harvard University (32)

[1] The year is given by CD's reference to sending Gray's article to the *Annals and Magazine of Natural History* (see n. 4, below).
[2] Gray's letter has not been found.
[3] Joseph Henry was the secretary of the Smithsonian Institution, Washington, D. C. Previously he had been professor of natural philosophy at Princeton University.
[4] CD sent the second part of Gray's three-part article on *Origin* ([Gray] 1860b, pp. 229–39) to *Annals and Magazine of Natural History* in September 1860. It was published in the November issue (*Annals and Magazine of Natural History* 3d ser. 6 (1860): 373–86). See letter to Asa Gray, 10 September [1860].
[5] The extract from Gray's article was preceded by the following editorial remark (*ibid*., p. 373):

> [In our Number for September last we placed before our readers an extract from the forthcoming volume of Prof. Agassiz's 'Contributions to the Natural History of the United States', relating to the interesting question as to the origin of species, newly raised by Mr. Darwin's well-known book. We now give a notice on the opposite side of the question to that taken by Prof. Agassiz, from the pen of another able naturalist of the United States, for the communication of which we are indebted to Mr. Darwin.—Eds.]

[6] CD presumably refers to Prideaux John Selby, the primary editor of the journal.
[7] Joseph Dalton Hooker was travelling in Syria. Thomas Henry Huxley had suffered the death of his oldest child in September 1860 (see letter to T. H. Huxley, 18 September [1860]).
[8] See letter to Charles Lyell, 23 [September 1860], and letter from Charles Lyell, 25 September 1860.
[9] For Lyell's and John Murray's responses to CD's proposal, see the letter to Asa Gray, 24 October [1860].
[10] Bree 1860.
[11] Hopkins 1860.
[12] George Henry Kendrick Thwaites's letter has not been found, but see the letter to G. H. K. Thwaites, 20 October [1860].
[13] Harvey 1860. See letters from W. H. Harvey, 24 August 1860 and 8 October 1860.

To Daniel Oliver 20 October [1860]

15 Marine Parade | Eastbourne
Oct. 20th

My dear Sir

We return to "Down Bromley Kent" on 26th.—[1]

I have heard from Huxley about the Nat. Hist. Review[2] & it has my good wishes & I shall certainly take it in.— I really know nothing whatever about vegetable irritability (it is quite beyond my scope) except in case of Orchids; I have a large mass of notes with many new facts, but I resolutely, *against my inclination*, put them away a month ago with the determination to work them up & get drawings made at some distant period;[3] for I am convinced that I ought to work on Variation & not amuse myself with interludes.— Drosera was an accident owing to my being from home & having nothing to do.— My paper on Drosera, from containing minute particulars of numerous experiments, would not, I think, be fitted for the Review; & if I do not deceive myself some of the results are sufficiently curious to be published in some old standing Transactions. Where I shall send them I know not yet; as I cannot judge of evidence till all my experiments are tabulated & that will be a long job.—[4] I may add that I told Huxley some months ago that I really could not assist by writing; I may add that my health is so weak that I cannot work above 3 hour a day; & I am at all times a most slow worker.—

I am so much obliged to you for going to try C. of Ammonia on Dionæa.[5] I sh[d] expect (but expectations are oftener wrong than right) that of C. of Ammonia on disc of leaf would cause it to shut slowly; but would produce no effect on back of leaf or on marginal spikes.— What it will do on the sensitive lamina, Heaven only knows.— I sh[d] expect *after* leaf has closed that each of the little tortoise-shell glands on the disc of leaf would secrete a minute bead of liquor. I must have expressed myself badly,—it is the red fluid being broken up or segregated with-in the sensitive lamina of a *closed* leaf, as in Drosera, which seems to me the important point of accordance; & not so much the action of C. of Ammonia on the red-fluid in the sensitive Hairs & tortoise-shell-like glands.— I will try, & sh[d] not be in the least surprised, at C. of Ammonia acting on the fluid in cells of common leaves.—

My dear Sir | With many thanks | Yours very sincerely | C. Darwin

Down House (MS 10: 20)

[1] The Darwins in fact did not return home until 10 November 1860 ('Journal'; Appendix II). See letter to Daniel Oliver, 23 [October 1860].

[2] Thomas Henry Huxley and John Lubbock were planning to take over the editing, in conjunction with several other naturalists, of the *Natural History Review*, a scientific quarterly until then published by naturalists in Dublin. Oliver was also one of the new editors. See letters to T. H. Huxley, 20 July [1860], and to John Lubbock, 20 July [1860].

[3] *Orchids* was published in 1862.

[4] CD eventually published his work on *Drosera* and other insectivorous plants in *Insectivorous plants*, which appeared in 1875.

[5] See letter to Daniel Oliver, 14 October [1860].

To G. H. K. Thwaites 20 October [1860][1]

Down, Bromley, Kent. [Eastbourne]
Oct. 20

My dear M[r] Thwaites

I thank you sincerely for your kind note and very curious little fact about the Ducks.[2] I much hope that you will attend a little more to them. Is the breed in any way different externally from the other ducks of Ceylon which do frequent water?

Has anybody else observed the same facts with these Ducks? Is there any danger in the water to which the Ducks are exposed? How curious it would be if you were to get some eggs from the aquatic breed of some other district and compare their habits with your own ducks. It really is worth a little trouble. I have seen ducks in England dabbling in long grass.

Your sentence about my views more commending themselves to your mind, has pleased me extremely.[3] I care much for the opinion of naturalists and very little for that of the public. The avalanche of Reviews against me has hardly yet ceased; but they have hurt me very little; and I hear from Murray that my Book goes on selling well. It will be a long battle, before it is known how much truth and how much error there is in my Book. But I find that all those who went at first an inch with me go a foot or so now. I know of no renegades as yet. I have no news for I have not seen a naturalist for several months.

With every good wish and thanks | Pray believe me | Yours very sincerely | Charles Darwin.

Copy
DAR 148

[1] The year is given by the reference to reviews of *Origin*. CD mentioned having received a letter from Thwaites in the letter to Asa Gray, 19 October [1860].

[2] Thwaites's letter has not been found. He apparently told CD about a breed of ducks that did not regularly frequent water. CD had discussed similar cases in *Origin*, p. 185. CD did not include any information on the Ceylonese breed in *Variation*.

[3] Thwaites's comment was quoted by CD in his letter to Asa Gray, 19 October [1860].

To Robert Patterson 21 October [1860][1]

Down, Bromley, Kent [Eastbourne]
Oct. 21

My dear Sir

I [would] like to thank you for so kindly taking the trouble of communicating the Rat V. Rabbit case.[2] I am very glad to have all such facts, but it is doubtful whether I would require to use it, so I will not give trouble to your informant of inquiring.

With many thanks, pray[3] believe me my dear Sir | Yours sincerely | Ch. Darwin

Praeger 1935, p. 715

[1] The date as given in Praeger 1935, p. 715.

[2] See letter from Robert Patterson, 18 October 1860.

[3] The printed text from which the transcript is taken reads 'prey'.

From Daniel Oliver [before 23 October 1860][1]

In Bot: Zeitung. 1852. p. 540.

D[r] Milde über d. Reizbarkeit d. Blätter von Drosera rotundifolia[2] (refer[d] to by *Nitschke.*)[3]

Gegen ende des Juni setzte ich mitten auf die Fläche eines recht kräftig vegetirenden Blattes der *Drosera rotundif.*, welche ich seit kurzer Zeit in einem kleinen Napfe mit Sphagnum im Zimmer hatte, vier kleine Fliegen, jede von der Grösse eines Stecknadelknopfes. Die Thierchen blieben fast regungslos daselbst sitzen, & ihre Versuche sich aus dem Schleime herauszuarbeiten waren vergeblich.

Nach ungefähr 5 Minuten betrachtete ich wieder das Blatt & sah zu meinem Erstaunen, dass sich die vorher horizontal abstehenden Drüsenhaare des vorderen Randes nach der Blattfläche zu gebogen & die Fliegen zum Theil bedeckt hatten. Erst am folgenden Tage hatte ich Zeit, das Blatt wieder genauer zu betrachten. Nun hatte sich aber sogar der vordere Rand & die Seiten des Blattes selbst nach der Mitte zu gewölbt & so die Fliegen völlig eingeschlossen. Erst nach Verlauf von 5 Tagen schlugen sich Blattränder & Haare zurück, so dass die natürlich getödteten Fliegen frei auf der Blattfläche dalagen.[4]

DAR 58.2: 55

[1] Dated by the relationship to the following letter.

[2] The extract was copied from Milde 1852. For a translation, see n. 4, below.

[3] Oliver presumably came across the reference to Julius Milde's notice on *Drosera* in a paper on the same subject by Theodor Rudolf Joseph Nitschke published in the 29 June 1860 issue of the *Botanische Zeitung* (Nitschke 1860b, p. 230).

[4] There is a translation of the passage in DAR 58.2: 56. It was evidently made at CD's request, possibly by Camilla Ludwig.

> Towards the end of June, I put into the middle of the surface of a leaf of Drosera, which since a short time ago I was keeping in my room, in a bason of [intentional gap left] four small animals, each fly being of the size of the head of a pin. The little creatures remained fastened to the spot, all their labouring to get out of the slime being useless. After about 5 minutes, I again looked at the leaf, and saw to my astonishment that the gland hairs of the front edge, which before had been standing horizontally had turned towards the surface of the leaf and had already partly covered the flies. When the next day, I again examined the leaf, even the front edge and the sides of the leaf had curved towards the middle, and had completely shut in the flies. 5 Days afterwards the hairs and the edge of the leaf opened again and the flies, which of course were dead, were lying free on the surface of the leaf.

To Daniel Oliver 23 [October 1860][1]

15 Marine Parade | Eastbourne
23d

My dear Mr Oliver.

What wealth of knowledge you have!— But I have hardly read the extract about Drosera;[2] as we are in much distress, my poor eldest daughter has had a relapse.[3] We cannot return on 26th to Down.—

Excuse me at present | Yours very sincerely | C. Darwin

Down House (MS 10: 21)

[1] Dated by the reference to Henrietta Emma Darwin's relapse (see n. 3, below).
[2] See preceding letter.
[3] From entries in Emma Darwin's diary, it appears that Henrietta Emma Darwin became ill with a stomach complaint late on Sunday night, 21 October 1860, and experienced intermittent attacks of vomiting for over a week.

To Asa Gray 24 October [1860]

15 Marine Parade | Eastbourne
Oct 24

My dear Gray

We are in much distress as my poor girl has just had relapse, but is rallying again.—[1] To save you labour in vain, I write hurriedly to say that Lyell, like a good & kind man, has been consulting with Murray about publishing your Reviews as pamphlet;[2] but they say, though it is against their rule to publish pamplets they would break it though this time) that they do not advise it. That it would be absolutely necessary to spend more in advertisement than the cost of publication; but that this would not suffice, that in short, they are well convinced that it is impossible to get a pamphlet circulated. They say they will do it "but it will be a dead loss". Under these circumstances it would be no use to attempt it. I am much vexed.— Lyell personally concurs with Murray's view.—[3]

Lyell has highest opinion of the "talent & science" shown in these 3 Reviews— In short agrees quite with me & he cannot well go further. He adds "it would be well worth while if a little Book could be got up by Asa Gray for the theological part is so admirable & would surely have many readers".—

I now hope Annals of Nat. Hist will put in 2^d Part;[4] & if they do I will try it on for 3^d Part; & I will offer to pay whole expence of paper & printing if they will condescend to accept it. I cannot bear that such admirable essay sh^d. not be preserved in this country.—

I have no idea whether Sat. R. will insert dialogue.—[5]

God Bless you | C. Darwin

Endorsement: '1860'
Gray Herbarium of Harvard University (33)

[1] Henrietta Emma Darwin fell ill on 21 October (Emma Darwin's diary).
[2] See letter to Asa Gray, 19 October [1860].
[3] The letter from Charles Lyell has not been found. Gray's review ([Gray] 1860b) was eventually published privately in 1861 as a separate pamphlet. See *Correspondence* vol. 9.
[4] See letter to Asa Gray, 19 October [1860] and n. 4.
[5] CD refers to another article by Gray ([Gray] 1860c) written in the form of a dialogue about *Origin*. In the hope of having it reprinted in England, CD had sent it to the *Saturday Review*. See letter to Asa Gray, 26 September [1860].

To W. E. Darwin 26 [October 1860][1]

15 Marine Parade | Eastbourne
Friday 26th

My dear William

We have been in great distress about poor dear Etty. Severe vomiting began on Monday & was dreadful for 48 hours; then it mitigated & we thought the attack over, but she had more last night & is very ill.—[2] I do not yet see actual much danger, though great suffering. When we shall be able to get home it is quite impossible to say.— Poor dear Mamma has had many sleepless nights & is very low.— God knows what the end will be.— But I still have great hopes.—

Thank you much for your long pleasant letter received on Monday or Tuesday.—

I am too tired to write any more | My dear son | Your affect. Father | C. Darwin

DAR 210.6

[1] Dated by the reference to Henrietta Emma Darwin's relapse (see n. 2, below).

[2] Emma Darwin's diary chronicles Henrietta's attack, which began on 21 October. See also letter to Daniel Oliver, 23 [October 1860] and n. 3.

To J. S. Henslow 26 October [1860]

15 Marine Parade | Eastbourne
Oct 26th

My dear Henslow

Many thanks for your note & for all the trouble about the seeds, which will be most useful to me next spring.—[1] On my return home I will send the shillings.—[2]

I concluded that Dr Bree had blundered about the Celts.[3] I care not for his dull unvarying abuse of me & singular misrepresentation But at p. 244 he in fact doubts my deliberate word, & that is the act of a man who has not the soul of a gentleman in him. Kingsley is "the celebrated Author & Divine" whose striking sentence I give in 2d Edition with his permission:[4] I did not chose to ask him to let me use his name, & as he did not volunteer, I had of course no choice.

I read with interest your letter in Athenæum.[5] Lyell seems to consider the deposits ordinary fluviatile beds, & not as showing signs of a debacle.[6] It is the most interesting subject which Geology has turned up for many a long year.—

Dr Freke has sent me his paper,[7]—which is far beyond my scope,—something like the capital quiz in the Anti-Jacobins on my Grandfather, which was quoted in the Quarterly Rw.[8]

My poor girl improved during the first four weeks here, but has had this last week a fearful attack, & is much exhausted, & we are much dispirited about her.— When we shall be able to take her home, I cannot conjecture.

My dear old master | Yours affect. | C. Darwin

Endorsement: '*1860*'
DAR 93: 81–2

[1] See letters to J. S. Henslow, 14 May [1860], 17 May [1860], and 28 September [1860].

[2] CD usually paid a small sum to the girls in Henslow's parish who collected seeds for him.

[3] CD refers to Charles Robert Bree's denial of the antiquity of the so-called celts or flint implements found in drift deposits (Bree 1860, p. 220):

> They have never yet been found to occupy any stratum, in which they could not have been buried by the ancient people, who used them as spear heads. Professor Henslow has several which I gave him, and which were found in the gravel-beds in Suffolk, mingled with bones of the mammoth and rhinoceros. Surely we are not called upon to found an argument against special creation, upon such facts as these?

Henslow discussed the subject in a letter to the *Athenæum*, 11 February 1860, pp. 206–7.

[4] Bree 1860, p. 244. Bree insinuated that CD might have fabricated the remark by a 'celebrated author and divine' included in *Origin* 2d ed., p. 481. Bree wrote: 'I think we ought to have had the name of this Divine, given with this remarkable statement. I confess that I have not yet fully made up my mind, that any Divine could have ever penned lines, so fatal to the truths he is called upon to teach.' Charles Kingsley had made the remark in a letter to CD (see *Correspondence* vol. 7, letter from Charles Kingsley, 18 November 1859).

[5] A letter from Henslow discussing his recent examination of celts in drift deposits in France was published in the *Athenæum*, 20 October 1860, p. 516. See also letter to J. S. Henslow, 28 September [1860].

[6] CD refers to a point discussed in Henslow's letter in the *Athenæum* (see n. 5, above). Henslow concluded that at least a few sites showed signs of the gravel having been disturbed, and hence the evidence was insufficient to conclude decisively that the implements were contemporaneous with the gravel deposits. He suggested that localised 'cataclysmic action' of water might have washed out and intermingled deposits from the Chalk with those in the gravel (*Athenæum*, 20 October 1860, p. 516).

[7] Freke 1860. There is a copy of the review of *Origin* by Henry Freke in the Darwin Pamphlet Collection –CUL.

[8] CD refers to a satirical attack on Erasmus Darwin's evolutionary poems composed by John Hookham Frere and George Canning and published in the political journal they edited, the *Anti-Jacobin* ([Frere and Canning] 1798). Samuel Wilberforce had quoted a long passage on the generation of 'filaments of organization' from the satire in his anonymous review of *Origin* ([Wilberforce] 1860, p. 255). Henry Freke also discussed the generation of living matter in his review (Freke 1860).

To John Innes 26 October [1860]

15 Marine Parade | Eastbourne
Oct 26th

Dear Innes

Very many thanks for your kind note & all the trouble you have taken for me.—[1]

Etty progressed very well here for 4 weeks, but I am sure that you will sympathise with us, when you hear that she had a terrible attack of sickness on Monday & which has not yet ceased. She is much prostrated & what the end will be, we know not.— My poor wife is much knocked up.— I am very glad you give a good of Johnny.—[2]

Yours most truly | C. Darwin

Postmark: OC 26 1860
Cleveland Health Sciences Library (Stecher collection)

[1] See letter to John Innes, 11 September [1860].

[2] John William Brodie Innes.

To David Thomas Ansted 27 October [1860][1]

15, Marine Parade, Eastbourne,
Oct. 27.

My dear Ansted,

As I am away from home in consequence of my daughter's health, I do not know your address, and fly this at random, and it is of very little consequence if it never reaches you.[2]

I have just been reading the greater part of your 'Geological Gossip,' and have found part very interesting; but I want to express my admiration at the clear and correct manner in which you have given a sketch of natural selection.[3] You will think this very slight praise; but I declare that the majority of readers seem utterly incapable of comprehending my long argument. Some of the reviewers, who have servilely stuck to my illustrations and almost to my words, have been correct, but extraordinarily few others have succeeded. I can see plainly, by your new illustrations and manner and order of putting the case, that you thoroughly comprehend the subject. I assure you this is most gratifying to me, and it is the sole way in which the public can be indoctrinated. I am often in despair in making the generality of naturalists even comprehend me. Intelligent men who are not naturalists and have not a bigoted idea of the term species, show more clearness of mind. I think that you have done the subject a real service, and I sincerely thank you. No doubt there will be much error found in my book, but I have great confidence that the main view will be, in time, found correct; for I find, without exception, that those naturalists who went at first one inch with me now go a foot or yard with me.

This note obviously requires no answer.

Pray believe me, | Yours sincerely, | C. Darwin.

Copy
DAR 143

[1] Dated by the reference to Ansted 1860.

[2] Ansted was lecturer on geology at the East India Company's military college in Addiscombe, Surrey.

[3] In his popular textbook on geology, Ansted discussed CD's concept of natural selection: in a long chapter entitled 'The battle of life', he gave an account of how, over long periods of geological history, natural selection could engender change from one species to another (Ansted 1860, pp. 155–80).

From A. W. von Hofmann to Edward Cresy 27 October 1860

R. College of Chemistry
Oct. 27. 1860.

My dear Sir

I enclose you the book on poison from which my statement about arsenic was taken. Taylor explains his opinion so well that I need not add a word.[1]

I have made a few experiments with Iodine and find that 0·001 milligramme of iodine dissolved in 1 Gramme of water is scarcely indicated by the starch test,—but that 0·002 Milligrammes dissolved in 2 Grammes of Water is easily distinguished by that test.

0·001 Milligramme	=	0·000015 Grains
0·002 do	=	0·00003 do

Hoping that this information will prove useful | Yours very truly | A W Hofmann

E. Cresy Esq

DAR 58.1: 5

[1] See letter from A. W. von Hofmann to Edward Cresy, 13 October 1860. CD had questioned the system of measurement used by Hofmann (letter to Edward Cresy, 14 October [1860]). Hofmann refers to Taylor 1848. Hofmann's letter was sent to CD by Cresy enclosed with a letter of his own (see letter from Edward Cresy, 30 October 1860).

From Benjamin Silliman Jr 27 October 1860

Office of Silliman's Journal, New Haven, Ct.,
Oct 27 *1860*

My dear Sir—

Our mutual friend D[r] Wyman has informed me of your desire to learn something more of the Rats found in Mammoth Cave—[1] I kept a pair of these rats for some weeks in Louisville after my visit to the Cave in 1850—[2] We captured them in a trap at a place more than half a mile from the entrance to the Cave—[3] The surface of dry dust on the ledges of rock bore abundant record of their tracks but we never could surprise them or obtain a sight of them until by setting a trap we caught them with the flavor of an apple—

We made a great number of experiments to satisfy ourselves that they were blind being at first unwilling to believe that their large lustrous eyes—as large as Squirrels eyes—could be sightless— Our conclusion was clearly to the point of their blindness

Mr Reginald Mantell son of D[r] G A Mantell[4] was there with me in Ky and being at leisure he devoted himself to the study of these animals until an accident (from dogs) destroyed them— He became fully satisfied that after about a month of exposure to a graduated light they came to a dim perception of objects—as they would blink or dodge on the approach of bodies toward the eyes while at first they took no heed of such unless their long silky lashes were touched— So far as I know—no naturalist has described this animal which is no rat in the ordinary sense and it is much to be desired that Someone would spend the time there required for a thorough study of their habits—

I send you a copy of a letter I wrote at the time to Prof Guyot in which some facts are mentioned which may perhaps interest you[5]

Dana is improving steadily[6]—is in excellent spirits—but says he is like Nebedchnazzer turned out to grass—[7] I confidently expect his perfect restoration in another year—

With high regard | I am Dr Sir | Yours truly | B. Silliman Jun

Chas Darwin Esq

Darwin Pamphlet Collection–CUL (bound with Silliman 1851)

CD ANNOTATIONS
1.4 a place . . . Cave—] *double scored brown crayon*
3.3 month] *underl brown crayon*

[1] See letter to Jeffries Wyman, 3 October [1860].
[2] Silliman had formerly been professor of chemistry in the medical school of the University of Louisville, Kentucky (*DAB*).
[3] CD quoted this information in the third edition of *Origin*, pp. 154–7. See also letter to Benjamin Silliman Jr, 4 December [1860].
[4] Reginald Neville Mantell, the son of the geologist Gideon Algernon Mantell, was a family friend of Silliman's.
[5] Silliman refers to a letter to Arnold Henri Guyot published in the *American Journal of Science and Arts* in 1851 (Silliman 1851). Guyot was professor of geology and geography at Princeton University. There is a copy of the paper in the Darwin Pamphlet Collection–CUL.
[6] James Dwight Dana was Silliman's brother-in-law.
[7] 'They shall make thee [Nebuchadnezzar] to eat grass as oxen' (Dan. 4:25).

To William Sharpey 28 October [1860][1]

15 Marine Parade | Eastbourne
Oct. 28

My dear Dr Sharpey

I hear that Mr Oliver is candidate for Professorship of Botany.[2] Therefore I trouble you with a few lines to be laid, if you think fit, before the Committee.

I have lately had much correspondence with Mr Oliver on certain physiological points, which I have been investigating; & I have been quite impressed at the range of his knowledge on facts buried in all sorts of foreign publications. I have been even more impressed at the philosophical caution he has shown in sifting the evidence on certain points laid before him, & in suggesting new experiments. In my judgment, whatever that may be worth, his mind is of a high philosophical order—[3] Pray excuse brevity as my daughter is most dangerously ill—

My dear Sir | Yours very sincerely | C. Darwin
(of Down, Bromly, Kent)

Copy
Down House (MS 10: 70)

[1] Dated by the reference to the professorship of botany at University College London (see n. 2, below).
[2] Daniel Oliver was a candidate for the chair of botany at University College London, which had fallen vacant through John Lindley's appointment as emeritus professor. Sharpey was professor of anatomy and physiology at University College and a member of the senate of the University of London, the governing body in charge of appointments.
[3] Oliver's appointment to the chair was announced in the *Athenæum*, 10 November 1860, p. 633.

From Edward Cresy 30 October 1860

1 Greek St
30 Oct 60.

My dear Sir,

From Taylor on Poisons, the book Dr Hofman refers to I have extracted the passages which I think will be useful or interesting to you—and which answer your question—[1] the smallest portion of arsenic detectable seems $\frac{1}{4000}$ grain and of iodine $\frac{1}{1400}$ grain and $\frac{1}{4400}$ grain of prussic acid, but as you will see much depends on the amount of dilution— Your Drosera beats the chemists hollow, when you have time, I should like to know the details of your experiments Are you quite sure of the amount of dilution and equally so of your *negative* experiments?— Have you eliminated all mechanical causes & effects due to breathing, perspiration soiling your glass rod in previous experiments— I presume also your observation is made under the microscope.[2] Still it is very curious— I hope if I can do anything else for you, you will let me know— I have addressed you at Down fearing you might have left Eastbourne—[3] I only got Hofman's answer yesterday—so he probably had not time till Sunday—

Yours very truly | E Cresy

Charles Darwin Esq—

[Enclosure][4]

From Taylor on Poisons—| Churchill. *1848.*
Iodine—$\frac{1}{14,00}$ grain diffused through 77,000 parts of water gave with starch a light *pink* colour and the colour became *blue* when the proportion amounted to $\frac{1}{470}$ grain in 28,000 parts of distilled water—[5]

Arsenic—"Dr Traill[6] has lately asserted that $\frac{1}{16000}$ part of a grain is precipitated by the silver test and that with $\frac{1}{10,000}$ grain the precipitate is visible to the eye. I have found that $\frac{1}{8000}$ grain dissolved in one drop of water gave a pale yellow film, but the result materially depended on the quantity of water present— Thus $\frac{1}{4000}$ grain of arsenic in ten drops of water was not perceptibly affected by the test but $\frac{1}{2000}$ grain dissolved in four drops of water gave a decidedly yellow precipitate"—[7]

"This test (sulphuretted hydrogen) is extremely delicate in its reaction.— It begins to give a yellow tinge when the liquid contains only $\frac{1}{4000}$ grain of arsenious acid in ten drops of water the arsenic therefore forming about $\frac{1}{40,000}$ part of the solution"—[8]

"There is no doubt that considerably less than the millionth part of a grain of arsenic may by Marsh's test be rendered *visible* on a glass plate;[9] it is possible to distinguish by the eye a piece of leaf gold which would weigh less than the ten-millionth part of a grain but the real question is whether the test will discover arsenic in a single drop of solution made by dissolving one grain of the poison in a million grains or sixteen gallons of water! If not the statement amounts to nothing for it is clear that if more than one drop of such an extremely diluted solution be taken the test is acting upon a larger quantity of arsenic than the above form of expression would indicate— I have generally found that the fractional quantity stated to be detected referred rather to the degree of dilution than to the absolute quantity of poison present, whereas a test may fail to act as we have already seen either from the smallness of the quantity present or from the very large quantity of water in which it is diffused— The results of my own experiments are that where the arsenic is mixed with the acid liquid in a tube capable of holding two fluid ounces very faint and scarcely perceptible deposits begin to be formed on a glass plate with a quantity equal to the $\frac{1}{2160}$ of a grain the diffusion here being equal to two million times the weight of the poison— With $\frac{1}{1080}$ grain in the same quantity of water the arsenic forming therefore one millionth part slight brown annular stains were procured. x x x x x x x x x— With $\frac{1}{720}$ grain the arsenic being in the proportion of about $\frac{1}{800,000}$ of the liquid the stains were much more decided but quite imponderable— With $\frac{1}{100}$ grain in one fluid oz of water, $\frac{1}{48,000}$ part, and $\frac{1}{67}$ grain in 2 fluid ounces, $\frac{1}{64,800}$, the deposits on glass were decided and characteristic—[10]

"Delicacy of Reinsch's process—[11] This test failed to detect $\frac{1}{4000}$ grain in 30 drops of water the dilution being equal to 120,000 times the weight of arsenic— The deposit on copper commenced with a violet coloured film—when the quantity of arsenious acid was equal to $\frac{1}{3000}$ grain in 30 drops of water or under a dilution of 90,000 times its own weight It was also very decided with $\frac{1}{2000}$ grain in the same quantity of water—x x x x x x— the copper was coated in a few seconds when boiled in a solution containing $\frac{1}{4000}$ grain in ten drops of water though the test had failed to detect the same weight of arsenic in three times that quantity of water—[12]

"Corrosive sublimate— The protochloride of tin added to $\frac{1}{8640}$ grain of corrosive sublimate dissolved in one drop of water in a minute tube produced a dark grey discoloration from which in twenty four hours a black substance separated— This however was in such small quantity that it was impossible to determine whether it did or did not contain any mercury"[13]

Prussic acid—"A standard solution was made in which each grain of liquid contained $\frac{1}{4400}$ grain of anhydrous prussic acid— One grain of this very diluted liquid placed on a plate of glass—gave a decided milkiness with nitrate of silver but when the same quantity was mixed with sixty grains of water nitrate of silver had no perceptible effect shewing the powerful influence of dilution on the action of a test $\frac{1}{4400}$ grain of prussic acid being in this instance diffused through 264000 parts of water— $\frac{1}{1460}$ grain in sixty grains of water, dilution 87,600, gave with nitrate of silver a milkiness but no precipitate—With $\frac{1}{440}$ grain, dilution 26400, the milkiness produced by the test increased—& the white cyanide became flocculent—& with $\frac{1}{220}$ grain, dilution 13200, the test gave a decided precipitate"—[14]

DAR 58.1: 6, 58.2: 49–52

[1] Taylor 1848. Cresy refers to the remark made by August Wilhelm von Hofmann in the letter from

A. W. von Hofmann to Edward Cresy, 27 October 1860, which Cresy forwarded to CD. See also the enclosure. In his letter to Edward Cresy, 14 October [1860], CD had inquired about the units of measurement used by Hofmann in the information he gave about the sensitivity of certain chemical tests.

[2] See letter to Edward Cresy, 2 November [1860]. From the notes CD made on his experiments in Eastbourne (DAR 54 and 60.1), it is evident that he was using a microscope.

[3] CD remained in Eastbourne until 10 November ('Journal'; Appendix II).

[4] The enclosure is in DAR 58.2: 49–52. For Alfred Swaine Taylor's work on the detection of poisons, see Coley 1991.

[5] Taylor 1848, p. 305.

[6] Thomas Stewart Traill was professor of medical jurisprudence at Edinburgh University.

[7] Taylor 1848, p. 339.

[8] Taylor 1848, p. 341.

[9] For James Marsh's test for arsenic, see the letter from A. W. von Hofmann to Edward Cresy, 13 October 1860, n. 3.

[10] Taylor 1848, p. 351.

[11] Edgar Hugo Emil Reinsch was a teacher of chemistry in Zweibrücken, Germany. He outlined his test for the presence of arsenic in Reinsch 1841 and discussed further details of it in Reinsch 1842. The test involved boiling the substance with muriatic (hydrochloric) acid: if arsenic was present, a piece of copper wire would be discoloured by the mixture. See Taylor 1848, p. 352.

[12] Taylor 1848, p. 354.

[13] Taylor 1848, p. 409. Corrosive sublimate is mercuric chloride.

[14] Taylor 1848, p. 685.

To Asa Gray 31 October [1860]

15 Marine Parade | Eastbourne
Oct 31

My dear Gray

Since I wrote about a week or 10 days ago, we have had such a week of misery as I did not know man could suffer. My daughter grew worse & worse, with pitiable suffering, so that all the Doctors thought we should lose her.[1] But the stoppage is over, & she has rallied surprisingly; but whether there is much organic mischief & what the final result will be cannot be known, till the miserable issue is decided. But she is quite easy now, & one comes at last to care only for that;; & we have managed to conceal from her, her extreme danger.— You are so kind & sympathetic that I have not resisted telling you our unhappiness.— We shall not be able to remove her home for several weeks, even if the case is not worse than the Doctors now hope & believe.—

I received this morning your letter of 16th Oct, & forwarded that to Decaisne.—[2] Thanks for Atlantic Oct. nor. which I shall find at home; but I have ordered 2 copies for self, so I shall be able to give away.[3]

How I wish you had time to write on affinities in relation to descent with modification; how well you would do it.— Could you not some time bring it in by side wind in discussing affinities of some group in some of your papers? You have in truth far more, over & over again, done more than you promised in getting my views a fair hearing.—

I have been reflecting about getting, as you suggest, if it can be done, 200 or 250 copies of the 3 articles of the Atlantic reprinted from the Plates in America & sent here.[4] Could it be done for 4 or 5 Pounds; I would gladly pay that.— I do not think it would be very material about the paying. But what I do consider *all* important; is that you would prefix title-page with *your name* & titles. If you did not object to this, I would post a copy to all the Scientific men I could get addresses of from the Societies. Unless your name was appended, the articles would not, *I am sure*, be read in England & it would be time & money thrown away. Their merit would not be discovered. But would it not cost you a good deal of trouble to get it done? They might be sent to care of Murray Albemarle St or Williams & Norgate, marked "to be forwarded".—

I am profoundly contented with your sentence "I think it grows more & more likely that species are derived." I am grieved to hear about Dana.—[5] I enclose sketch about Spiranthes; if you will observe your species, I sh^d be *infinitely* obliged. If a very distinct species, the contrivance will probably differ; as the contrivances are endlessly diversified.— I am not sure that you will understand my rough description, but you will if your species is like ours.— I shall publish on Drosera.—

Yours ever my dear Gray | Most truly | C. Darwin

As You speak of determinate movements for an end in plants I am tempted to give case of Orchis pyramidalis which seems to me pretty.[6] The pollinia are attached to *connate* glands of shape of a saddle, & are suspended in cup of liquid with a moveable lip.—

Insert a needle into the nectary, & the lip of cup is moved, & the sticky under surface of saddle sticks to needle & the pollinia are removed. **Immediately** the saddle is exposed to air (& *not* whilst under water) the flaps of the saddle curl in & embrace the needle. I had a moth sent me with 3 pollen-masses attached to proboscis & the sender said it was extraordinary how the moth had *bored* through the sticky glands of some orchis! The movement of the flaps is very conspicuous. At first pollinia occupy this position

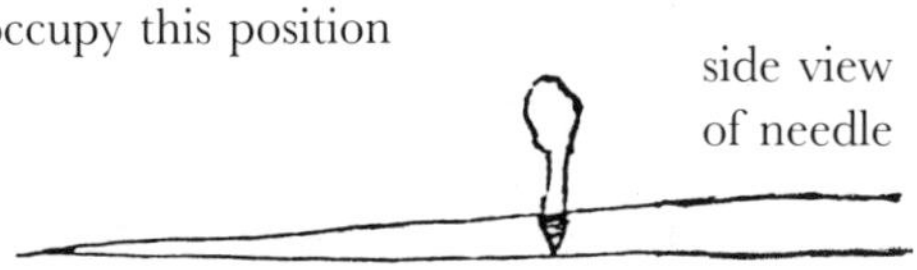

(viewed laterally, one pollen-mass behind the other); & if you push needle into nectary of another flower, the pollen-masses do *not* touch stigmas, which are lateral & double, one on each side of orifice of nectary. But immediately after the flaps of saddle have seized round the needle, another movement commences, & the pollinia move downward, nearly parallel to needle & the two diverge from each other.

Now if if you push needle into nectary, the pollen-masses exactly hit the lateral stigmas & leave pollen-grains on them. The seat of this latter movement, which is always in one direction, lies at the junction of the footstalk of pollinia & the saddle.

[Enclosure]

Spiranthes autumnalis[7]

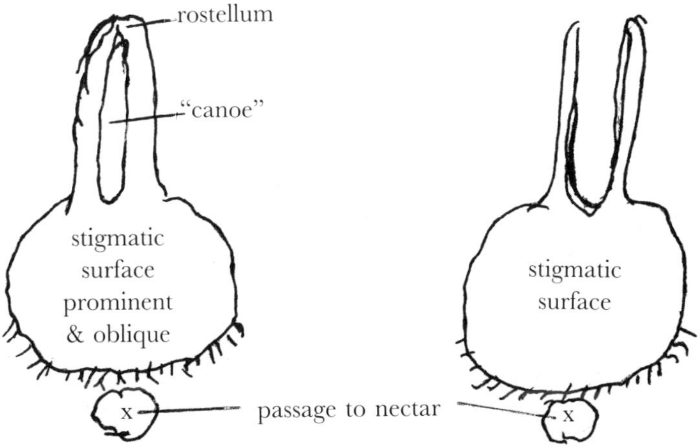

fig 2.

(1) In rather early bud the anthers open & press pollinia against back of rostellum

(2) Back of rostellum opens in early bud & a viscid substance glues the pollinia to the back of the rostellum or strictly to the bottom of what I call the "canoe".[8]

(3) When the flower first opens the rostellum is not ruptured on the *front* surface, (ie on surface above the stigma) Within the substance of the rostellum a brownish hard object can be seen, which is of the shape of a "canoe", pointed at both ends, & standing vertically up, parallel to longer axis of rostellum. The bottom of canoe, as stated, is already glued to the pollinia.— The front & concave or hollow side of the canoe is filled or loaded with very viscid substance, which sets hard in less than a minute when exposed to the air. The concave side of the canoe is, as stated, covered, when flower opens, by the membrane of the rostellum.—

(4) Now if front surface of rostellum be touched longitudinally most delicately by a needle, (or exposed to vapour of chloroform) it instantly splits longitudinally its whole length, & the needle comes into contact with the viscid matter within the canoe. The canoe embedded in the rostellum becomes *firmly* glued to the needle; & the bottom of the canoe is already long since glued to the pollinia. Hence when the

needle is withdrawn the canoe & pollinia are all withdrawn together. The rostellum is then left in the state rudely represented by fig 2. All the flowers which have been visited by insects are in this condition. Change the needle into the proboscis of insect, & you will see how fertilisation is effected. The pollinia become attached nearly parallel to proboscis; & the insect visiting a **second** flower is sure to strike the apices of the pollinia against the obliquely **projecting** viscid stigma.— If you poke thin culm of a grass down to nectar & withdraw it without bending, you will not withdraw the pollinia; but if before you withdraw it, you bend the culm so that its bowed or convex surface touches the rostellum the pollinia will generally be withdrawn. Moths in curling up their probosces *after* sucking perform this action; & I go so far as to believe, whole structure of flower is made in relation to this mode of withdrawing proboscis!!!

Endorsement: '1860'
Gray Herbarium of Harvard University (45 and 124A)

[1] Henrietta Emma Darwin fell ill on 21 October 1860. Emma Darwin recorded in her diary that on 29 October the Darwins' physician Henry Holland was called to Eastbourne. See also letter to T. H. Huxley, 1 November [1860].

[2] Gray's letter has not been found. CD refers to the French botanist Joseph Decaisne, a correspondent of Gray's.

[3] The third and final part of Gray's review of *Origin* published in the *Atlantic Monthly* ([Gray] 1860b, pp. 406–25).

[4] See letters to Asa Gray, 19 October [1860] and 24 October [1860]. CD agreed to Gray's suggestion, and the pamphlet was published in 1861 (Gray 1861). See *Correspondence* vol. 9.

[5] Although James Dwight Dana had returned to the United States in August 1860, he was still too unwell to resume his duties as professor of geology at Yale University and co-editor of the *American Journal of Science and Arts*.

[6] The movements of the pollinia of *Orchis pyramidalis* are described in *Orchids*, pp. 23–8. CD stated that no other adaptations in plants were 'more perfect than those presented by this Orchis' (*Orchids*, p. 28).

[7] CD discussed the pollination mechanism of this orchid in *Orchids*, pp. 116–29. He added a note stating that Gray had examined specimens of *Spiranthes gracilis* and *S. cernua* for him. 'He [Gray] found the same general structure as in our S. autumnalis, and he was struck with the narrowness of the passage into the flower.' (*ibid*., p. 123 n.).

[8] The rostellum of *S. autumnalis* is characterised by a narrow, membranous ridge that CD called the 'boat-formed disc' (*ibid*., p. 116).

To Daniel Oliver [31? October 1860][1]

15. M. P. | Eastbourne
Wednesday

My dear Sir

Just received your letter. which I have hardly read—[2] I write before morning post, to say, that I did not suppose you would care to look at the movement in Drosera;[3] (I am very glad), else I would have told best way, viz to take Hairs *lately* inflected over a fly, cut off, & put either dry or in water between glass, under high power ($\frac{1}{4}$ object) make a sketch of shape of red matter, & look again in 1 or 2

minutes.— There is, **also, sometimes** a strong circulation in the cells.— I refer to cells of *footstalk* of *round*-headed marginal Hairs—
Yours [*illeg*] | C. Darwin

Down House (MS 10: 19)

[1] Dated by the relationship to the letters to Daniel Oliver, 14 October [1860], 17 October [1860], and 20 October [1860], in which CD suggested experiments on *Dionaea* that Oliver might perform. The letter from Daniel Oliver, [before 23 October 1860], and the letter to Daniel Oliver, 23 [October] 1860, indicate that Oliver had not yet mentioned to CD that he intended to undertake such experiments. By the time that CD wrote the letter to Daniel Oliver, 3 November [1860], CD knew that Oliver also intended to look for the 'moving red matter' in the cells of *Drosera rotundifolia* (see n. 3, below).

[2] Oliver's letter has not been found.

[3] CD had asked Oliver to examine actively functioning cells of *Dionaea muscipula* in order to substantiate his descriptions of a process that appeared to be connected with the plant's metabolism. CD told Oliver about a similar process that he had first observed in leaves of *Drosera rotundifolia*. See letters to Daniel Oliver, 14 October [1860], 17 October [1860], and 20 October [1860].

From John Medows Rodwell[1] 31 October 1860

Belle Vue House | Brox Bourne.
Oct. 31. 1860

Dear Sir/

I have purposely postponed writing to you again until I had an opportunity of making further observations upon my white cat. This I have now done and am able to state positively that although both her eyes are undoubtedly blue she is not in the slightest degree deaf.[2] She has been several times put into a room apart with merely the door left ajar, and has always run immediately, at the summons of myself or of any of the servants, from an adjoining room in which it was impossible for her to see us. My friend Dr Hooper of Hoddesdon[3] also assures me that though he has often known cats with both eyes blue to be stone deaf he has also known them to be as fully endowed with the faculty of hearing as any others. But I hope in the course of a short time to able to report to you upon the state of a large number of white cats, as a friend of his, a lady residing not far hence, is a fancier of white cats and has a large number—of which I could only learn yesterday that some with eyes more or less tinged with pink are entirely deaf.

Now that I am speaking of cats I may mention that my late uncle the Rev. W. Kirby of Barham[4] was a great Cat-fancier all his life, and that he was very careful to breed none but Tabbies, which under his care, I am afraid I must not use the word *selection*, as I do not know that he ever eliminated the smaller cats, though he certainly never would allow a white or Tortoiseshell on his premises, attained to a size I never saw elsewhere. He was extremely proud of his Cats, and had petted this one particular Breed for the nearly sixty years that he was Curate and Rector of Barham. From my childish days, till I became his Curate, I always recollect him with a cat on his Table, Knee or Shoulder. And I have no doubt but that careful Breeding had produced their unusual size.

It may perhaps be worth while to mention that about 1843 when I was Incumbent of S. Peter's Saffron Hill, a large portion of the old Fleet Sewer (or River), said never to have been before opened since the days of Q. Elizabeth, was exposed to view. I then saw several enormous Rats which had been taken thence by the workmen, and upon examination they all proved to be blind and almost entirely devoid of hair, and so ferocious were they that the workmen assured me they were deterred from entering the old parts of the Sewer as the rats would unquestionably fly at them. The rats which I saw were taken out at Holborn Bridge, and as there are three arches still remaining there of an old roman Bridge some sixteen or more feet below the present surface, it is *possible* that those rats may have been breeding there for ages, and if like the blind cave animals you mention in C. V. of the Origin of Sp.s—their progenitors lost the power of sight a 1000 years since, and lost as they would, I suppose, at the same time any great ability for migration, this would be a curious illustration of a part of your theory.[5] All I can say is that some eight or ten Rats which I saw

AL incomplete
DAR 47: 167–8

CD ANNOTATIONS
1.1 I have . . . size. 2.10] *crossed pencil*
Top of first page: '*Blind Rats*—' *ink*

[1] Rodwell had been a friend of CD's at Cambridge University. He graduated from Caius College in 1830 and became a clergyman and orientalist.

[2] CD discussed the correlation between blue eyes and deafness in cats in *Origin*, pp. 12 and 144, where he stated that the connection was invariable. He again discussed the topic in *Variation* 2: 329. In *Variation* CD mentioned, without citing Rodwell's case, that he had since heard of a few exceptions to the rule.

[3] John Hooper, MD, of Hoddesdon, Hertfordshire, lived close to the village of Broxbourne where Rodwell resided.

[4] William Kirby, the late rector of Barham, had been a leading entomologist.

[5] CD mentioned in *Origin*, pp. 137–8, the blind cave-rats found in the Mammoth Cave in Kentucky as an illustration of the action of natural selection on useless organs.

From David Forbes [November? 1860][1]

and is about finished—

The question you ask me about the glacial action I can only answer as follows—[2]

From Lat. 13° to lat 30° there is a plateau or rather indistinct remains of one at about 12000 feet elevation composed of debris such as would generally be ascribed to ice action— the rocks are in many parts especially lat 15° 16°. 17°. 18° S. strongly furrowed & grooved similar to those of Norway.[3] the grooves have a general direction and the debris from the plateau in question contains rolled and marked or grooved & scratched pebbles— down where you observed the mass of debris in question,[4] the remains of the plateau is only distinct in spots evidently from the river action, but only some 15 leagues north of same I found it grandly developed—and could not explain it at all untill my later examination of the chain of the Andes northward up to Lat 13. showed me at spots (barometrical measurements) a similar development connecting same with the great Eastern Andean plateau of Silurian debris, described in section— I was informed that a little south of Cauquenes, there exist real glaciers in the Andes—but I found none north of same— my information was tolerably reliable

As I believe fully in your theory of the origin of species I shall be glad to do anything for you in advancing same and shall attend to the questions you send me and send you answers as soon as possible If you have more points I shall attend to them if you drop me a line— The question you ask me about the Chilean breed of sheep—I can in part answer now—[5]

I was also struck with this and at 2 haciendas I staid at endeavoured to come to the bottom of this but not satisfactorily.— The prevalent opinion is that the race was *originally* derived from an intermixture of sheep & goat but as far as I could discover, it is nowhere now produced from this source—and the flocks at all the haciendas I saw are only ofspring of similar sheep and not artificially produced as the mules— One Haciendado at Mendoza stated that, altho if I asked every other Haciendado—in the province they would tell me it was a cross he never knew such a case to have actually occurred & stated his belief that they were a distinct race of sheep

One would think it had been easy to have sifted such a simple question but I did not succed— would an examination (microscopic) of the hair be of service— I could send you some as I have a saddle cover here in Birmingham of this article.—[6]

There is one point I would direct your attention to it is the horse— The horse of S America was originally Spanish or Andalusian—but now presents 3 wonderfully different variety

1. Chilian—exactly same as true Andalusian
2. Pampa horse you know
3. Puno horse—a most strange sort of small horse, slender narrow chested & very peculiar altogether—evidently owing to the rarified atmosphere—[7] I believe I can prove a great change from same cause in the Puno Indians (Aymara) whom I found to be structurally different. for example femur longer than tibia, &c.[8]

Yours very truly | David Forbes

Incomplete
DAR 164: 150

CD ANNOTATIONS
3.1 From . . . reliable 3.14] 'Glacial Action' *added brown crayon*
4.1 As I . . . Forbes 8.1] *crossed pencil*

5.1 I was . . . sheep 5.9] 'Hybrid Sheep—' *added brown crayon*
7.1 There is . . . atmosphere—7.7] '*Horses*' *added brown crayon*

[1] The date is based on the reading of Forbes's paper on the geology of Bolivia and Peru (D. Forbes 1861) at a meeting of the Geological Society on 21 November 1860. Forbes, who had travelled widely in the western regions of South America, visited England in 1860. He returned to Bolivia in 1861 (*DNB*).

[2] In his paper on the geology of the Andes (D. Forbes 1861), Forbes did not discuss the possible effects of glacier action. His predominant concern was rather the identification of the primary and secondary rocks and their mineral veins. CD included Forbes's information in *Origin* 3d ed., p. 403.

[3] Forbes had geologised in Norway during his employment as superintendent of the mining and metallurgical works at Espedal, 1847–57. See *Quarterly Journal of the Geological Society of London* 33 (1877): 42.

[4] The reference is to CD's description of the western coast of South America in *South America*.

[5] CD cited Forbes's information on the sheep of Chile in *Variation* 1: 95. This chapter was completed by 20 March 1861 (*Correspondence* vol. 9, 'Journal'; Appendix II).

[6] Forbes had entered into partnership with a Birmingham nickel-smelting firm before leaving for South America in 1857 (*DNB*).

[7] CD cited Forbes on this point in *Variation* 1: 52.

[8] Forbes read a paper on the Aymara Indians of Bolivia and Peru before the Ethnological Society in 1870 (D. Forbes 1870).

To T. H. Huxley 1 November [1860][1]

15 Marine Parade | Eastbourne
Nov. 1st

My dear Huxley

Your note has been wonderfully interesting. Your term "pithecoid man" is a whole paper & theory in itself.[2] How I hope the skull of the new Macrauchenia has come:[3] it is grand.—

I return Hooker's letter, with very many thanks. The glacial action on Lebanon is particularly interesting considering its position between Europe & Himalaya:[4] I get more & more convinced that my doctrine of *mundane* glacial period is correct, & that it is the most important of all late phenomena with respect to distribution of plants & animals.—[5]

I hope your Review progresses favourably.[6]

I am exhausted & not well so write briefly; for we have had 9 days of as much misery as man can endure.[7] My poor daughter has suffered pitiably, & night & day required three persons to support her. The crisis of extreme danger is over & she is rallying surprisingly, but the Doctors are yet doubtful of ultimate issue. But the suffering was so pitiable I almost got to wish to see her die. She is easy now. When she will be fit to travel home we know not.—

I most sincerely hope that Mrs. Huxley keeps up pretty well.[8] The work which most men have to do is a blessing to them in such cases as yours.—

God Bless you | C. Darwin

Sir H. Holland came here to see her & was wonderfully kind.—[9]

Imperial College of Science, Technology and Medicine Archives (Huxley 5: 141)

[1] Dated by the reference to Henrietta Emma Darwin's illness (see n. 7, below).
[2] Pithecoid: 'resembling in form or pertaining to the apes, esp. the higher or anthropoid apes' (*OED*). Huxley was apparently writing his essay on the zoological relations of man with the lower animals (T. H. Huxley 1861b), at the beginning of which he touched on the implications of demonstrating that humankind had a 'pithecoid pedigree' (T. H. Huxley 1861b, p. 67). Huxley's essay attacked Richard Owen's view that there were great anatomical differences between the higher apes and human beings. Huxley concluded that 'the Quadrumana differ less from man than they do from one another; and that, hence, the separation of *Homo* and *Pithecus* in distinct sub-classes, while *Pithecus* and *Cynocephalus* are retained in one order, is utterly inconsistent with the principle of any classification of the Mammalia by cerebral characters.' (T. H. Huxley 1861b, p. 84).
[3] Huxley read a paper at a meeting of the Geological Society on 21 November 1860 on some vertebrate remains discovered in Bolivia by David Forbes (T. H. Huxley 1861a). Huxley described a new species of *Macrauchenia* that was distinct from that described by Owen in *Fossil Mammalia*, pp. 35–56, on the basis of bones collected by CD during the *Beagle* voyage (see *Correspondence* vol. 2). The remains found by Forbes included a portion of the skull and fragments of teeth; no part of *Macrauchenia* skull had previously been discovered. Owen had found the *Beagle* specimens to have a combination of perissodactyl and artiodactyl characters, but assigned the genus to the perissodactyl Pachydermata. Huxley found characteristics of both divisions in Forbes's skull and tooth specimens and remarked that *Macrauchenia* had 'strong claims to be considered an example of what has been termed "a generalized type"' (T. H. Huxley 1861a, p. 84). For a discussion of the history of the classification of *Macrauchenia*, see Rachootin 1985.
[4] Joseph Dalton Hooker was travelling in the Middle East. He and his party ascended the mountains of Lebanon before visiting Syria and Palestine (L. Huxley ed. 1918, 1: 528–33). In a letter to his brother-in-law Charles James Fox Bunbury apparently written in January 1861, Charles Lyell wrote (K. M. Lyell ed. 1881, 2: 329):

> Did I tell you of Hooker's ascent of Lebanon, 11,000 feet high, and his finding, though now there is no perpetual snow there, old glacier moraines descending 4,000 feet down from the summit, of which he has made some very graphic sketches? All the cedars of Lebanon grow exclusively on these moraines.

[5] See *Origin*, pp. 365–82. CD had discussed the theory at length with Hooker over a number of years. See *Correspondence* vols. 5 and 6.
[6] CD refers to the *Natural History Review*, of which Huxley had recently become senior editor.
[7] Emma Darwin recorded in her diary on 29 October 1860 that Henrietta Darwin's 'Sickness ended.'
[8] Henrietta Anne Huxley, who was pregnant with her fourth child, was still in deep grief over the death of their oldest child Noel in September (L. Huxley ed. 1900, 1: 216, 222–4). See letter to T. H. Huxley, 18 September [1860].
[9] The London physician Henry Holland was called to Eastbourne to examine Henrietta Darwin on 29 October 1860 (Emma Darwin's diary).

To Edward Cresy 2 November [1860][1]

15 Marine Parade, Eastbourne
Nov: 2nd.

Dear Cresy

I have received in a parcel from Down your letter and most interesting enclosures and Taylor's pamphet.[2] Good heavens, what trouble you have taken for me! I am sure I am very much obliged; and the facts are most interesting to me, but I have not yet digested them.— Since writing last we have had a terrible week with my poor girl on the point of death; but she has rallied surprisingly. When we shall be able to move her home we know not as yet at all.— You ask how my experiments were tried; but they have been tried in so many manners that I hardly know how to answer. I will before very long (but all my work has been utterly stopped for a fortnight by this miserable illness) publish an account.—

When I saw you I had tried only putting a minute drop on the leaves on growing plants and observing whether or not they contracted as if over a fly.[3] From many reasons I inferred that the leaves absorbed some nitrogenous element, probably some form of Ammonia; so I thought I would try *under the microscope* the effect of C. of Ammonia; and several other salts and substances, and it is truly wonderful how quickly a minute dose acts and produces marvellous changes in the absorbing glands and in the adjoining cells. I have tried plain water over and over again with no effect. I have over and over again kept a leaf for 1, 2, 3, 4, and 5 hours in water with no effect, and then put these *same* leaves in a few measured drops of very weak solutions of C. of Ammonia (all made, and **re**-made by myself) and the same *peculiar* effects were produced in one hour or $1\frac{1}{2}$ hour as is produced instantaneously by a stronger solution. Generally I have scrutinized every gland and and hair on leaf before experimenting; but it occurred to me that I might in some way affect the leaf; though this is almost impossible, as I scrutinized with equal care those that I put into distilled water (the same water being used for dissolving the C. of Ammonia). I then cut off 4 leaves (not touching them with fingers) and put in plain water and 4 other leaves into the weak solution and after leaving them for $1\frac{1}{2}$ hours I examined every hair on all 8 leaves; no changes on the 4 in water every gland and hair affected in those of Ammonia.

I had measured the quantity of weak solution and I counted the glands which had absorbed the ammonia and were plainly affected; the result convinced me that each gland could not have absorbed more than $\frac{1}{64,000}$ or $\frac{1}{65000}$ of a grain.— I have tried numbers of other experiments all pointing to the same result. Some experiments lead me to believe that very sensitive leaves are acted on by much smaller doses. Reflect how little Ammonia a plant can get growing on poor soil —yet it is nourished. The really surprising part seems to me that the effect should be visible and not under very high power; for after trying high power, I thought it would be safer not to consider any effect which was not plainly visible under $\frac{2}{3}$ object glass and middle eye-piece. The effect which the C. of Ammonia produces is the segregation of the homogeneous fluid in the cells into (first) a cloud of granules and colourless fluid; and subsequently the granules coalesce into larger masses and for hours have the oddest movements coalescing, dividing, coalescing ad infinitum.[4] I do not know whether you will care for these ill-written details; but as you asked I am sure I am bound to comply after all the very kind and **great** trouble which you have taken.

Yours very truly | C. Darwin

Copy
DAR 143

[1] Dated by the relationship to the letter from Edward Cresy, 30 October 1860.

[2] See letter from Edward Cresy, 30 October 1860 and enclosure. Cresy also forwarded to CD the letter from A. W. von Hofmann to Edward Cresy, 27 October 1860. The pamphlet by Alfred Swaine Taylor was probably Taylor 1860, which discussed chemical tests for arsenic and antimony. There is a note on this work in DAR 60.1: 69.

[3] Cresy and his wife visited Down on 18 September 1860, shortly before the Darwins left for Eastbourne (Emma Darwin's diary). CD's notes on his observations on *Drosera* and *Dionaea*, begun in July 1860, are in DAR 54, 60.1, and 60.2.

[4] CD's initial observations on the cytological phenomenon of the 'aggregation of protoplasm' are in DAR 60.1: 77–81.

To Edward Cresy 2 November [1860][1]

15 Marine Parade, Eastbourne
Nov: 2nd.

Dear Cresy

I return by this post Dr. Taylor's pamphlet which is *very* interesting.—[2]

This morning I vowed to myself that I would give you no more trouble; this evening I break my vow! Do read over Hofman's note,[3] it seems to me clear that he asserts that he has easily distinguished 0·002 of milligram of iodine and this equals, as I make out, $\frac{3}{100,000}$ of a grain! Whereas Taylor, in your copy, speaks of only $\frac{1}{1400}$ of a grain.[4] Can you understand this? Pray do not trouble Hofman again, for I shall see him in course of early winter. Now I am on weights, can you assist me in another point. I have used ordinary apothecary measures (which I shall state in my paper), viz, fluid oz. divided into 8 drachms and each drachm into 60 minims,—so that each oz. contains 480 minims.[5] But when I looked into such books as I possessed I found the most ridiculous discrepancies and misprints of weight of one such oz. of distilled water. I applied to good and clever druggist and he says positively one such oz. weighs 437·5 grains and this agreed with *one* of my printed authorities; so I have *provisionally* used it in calculating proportions. Can you anyhow aid me in telling me whether this is correct?[6]

God forgive me for being so troublesome, and believe me, | Yours very sincerely | C. Darwin

Please return Hofman's note.

Copy
DAR 143

[1] The letter was written in the evening, after the preceding letter was posted.

[2] Taylor 1860. See preceding letter.

[3] Letter from A. W. von Hofmann to Edward Cresy, 27 October 1860. See preceding letter.

[4] Taylor 1860. There is a note on this pamphlet in DAR 60.1: 69 listing the amounts of arsenic and antimony that could be detected.

[5] A note giving these calculations and apparently subsequent emendations is in DAR 60.1: 70.
[6] See letter from Edward Cresy, 10 November 1860.

To Daniel Oliver 3 November [1860][1]

15 Marine Parade | Eastbourne
Nov. 3d

My dear Mr Oliver

My daughter has escaped the imminent danger she was in for a week & our minds are now tolerably at ease; & we hope to get home in a week or 10 days, if she progresses favourably.—[2]

I shall be anxious to hear the result of the election for the Professorship.—[3]

I presume your time will have been fully occupied in this important matter, & you will not have had time for any experiments on Dionæa with the C. of Ammonia; nor had time to look at the moving red matter in the Hairs of Drosera.—[4] I am so greatly obliged to you for taking so much trouble to copy the extract from Bot. Zeitung.[5] I presume there is a paper in extenso.— Is this so? The author has missed the point which seems to me most interesting, viz that though the Hairs close equally (or nearly so) over animal, vegetable & inorganic dry substances; they remain closed *much* longer over animal than over other substances. This fact led me to try the nitrogenous & non-nitrogenous fluids; for somehow the plant knows well when it has caught good food.— You were right in anticipating (as I believe you did) that C. of Ammonia acts on the fluid in cells of other plants, as in case of Drosera; I find it so with Lemna & almost as quickly with the roots of a little Euphorbia, as with the hairs of Drosera.[6] I have asked Daubeny whether the action of C. of Amm. on the sap of plants was known; & he says he does not remember having heard of it.[7]

Yours very sincerely | C. Darwin

P.S.[8] There is something which looks odd to me in the roots of the Duck-weed;[9] a new root seems formed within the old one; at least an old outer cylinder is cast off in pieces; but it is out of my line.

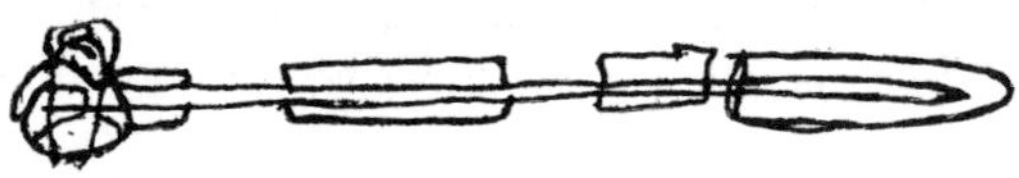

Down House (MS 10: 24)

[1] The year is given by the reference to Henrietta Emma Darwin's illness.
[2] CD and his family returned to Down on 10 November 1860 ('Journal'; Appendix II).
[3] Oliver was a candidate for the chair of botany at University College London. See letters to William Sharpey, 28 October [1860], and to Daniel Oliver, 7 November [1860].
[4] See letter to Daniel Oliver, [31? October 1860].
[5] See letter from Daniel Oliver, [before 23 October 1860]. The extract that Oliver copied from Milde 1852 has been found, but not Oliver's accompanying letter. See also letter to Daniel Oliver, 23 [October 1860].

[6] CD described the effect of a weak solution of carbonate of ammonia on the root cells of various plants in *Insectivorous plants*, pp. 63–5. He stated that he 'dug up in the latter part of October the first weed which I met with, viz. *Euphorbia peplus*' (*ibid.*, p. 63).

[7] The correspondence on this subject between CD and Charles Giles Bridle Daubeny has not been found. Daubeny, professor of botany at Oxford University, was renowned for his studies on the chemistry of plants.

[8] The postscript is on a separate slip of paper, which from the content appears to belong with this letter.

[9] CD refers to the roots of *Lemna*, the common duckweed, which he was studying in relation to the effect of carbonate of ammonia on plants. He had been led to experiment on the roots of plants from the analogy between their nutritive processes and those of the leaves of insectivorous plants. There are notes on such experiments, carried out by CD in December 1860 and January 1861, in DAR 54: 8–9.

To J. M. Rodwell 5 November [1860][1]

15 Marine Parade. | Eastbourne.
Nov.r 5.

My dear Sir.

I am extremely much obliged for your letter; which I can compare only to a plum pudding, so full it is of good things.— I have been rash about the Cats; yet I spoke on what seemed to me good authority: the Revd. W. D. Fox gave me a list of cases of various foreign breeds in which he had observed the correlation & for years he had vainly sought an exception.[2] A French paper also gives numerous cases, & one very curious one of a kitten which *gradually* lost the blue colour in its eyes & as gradually acquired its power of hearing—[3] I had not heard of your uncle Mr Kirby's (whom I, for as long as I can remember have venerated) care in breeding cats—[4] I do not know whether Mr Kirby was your uncle by marriage; but your letters show me that you ought to have Kirby-blood in your veins, that if you had not taken to languages you would have been a first rate naturalist.

Your Rat case is most wonderful— It is hardly fair to ask you to take so much trouble; but I should be very much obliged for a little more information— Did the workmen believe that they were blind? Might it not have been that they were temporarily blinded by being brought suddenly into bright light? Do you remember whether there was any difference in the eyes besides the paleness? Was the appearance of the eyes sufficiently remarkable to excite attention in others? Lastly might I quote your name as my authority—[5] I have been so attacked that I am sure you will not wonder at my requiring to give authorities & all evidence that I can. Dr. Bree, for instance, who has just published a Book against me, speaks with utter contempt of my statement about colour & the action of poison; & logically engages to poison a white & black sheep or pig with the same poison![6] Probably he would select arsenic!

Your case about the Rats interests me especially, because in some French Transactions, there is a long account of some eels taken out of a deep & dark artificial well, which were stone blind & had singularly shaped heads;[7] but these fish were pronounced by Agassiz to be the common species.[8]

Very many thanks for information about horses;[9] My memory has returned so far that I believe I can trace in Heusingers German Pamphlet[10] where I have read an account in some English Book (certainly not in my old friend L. Jenyns books)[11] I feel almost sure the result followed from eating "lousy tares" but whether Horses or Cattle I cannot remember.— Your case will be a most valuable confirmation— I sincerely hope that you will be able to carry out your intention of writing on the "Birth Life & Death of Words". Anyhow you have a capital Title, & some think this the most difficult part of a book.— I remember years ago at the C. of Good Hope; Sir J. Herschel saying to me—I wish some one would treat language as Lyell has treated Geology—[12] What a Linguist you must be to translate the Koran![13] Having a vilely bad head for languages—I feel an awful respect for Linguists.—

I do not know whether my Brother in Law Hensleigh Wedgwoods, Etymological Dict[y] would be at all in your line; but he treats briefly on the genesis of words, & as it seems to me very ingeniously—[14]

You kindly say that you would communicate any facts which might occur to you; & I am sure that I should be most grateful— Of the multitude of letters which I receive not one in a thousand is like yours in value—

With my cordial thanks & apologies for this untidy letter written in haste | Pray believe me, My dear Sir. Yours sincerely obliged. | Ch. Darwin.

We are delayed here by most suffering illness of my daughter but we hope to return to "Down, Bromley, Kent" in about a weeks time.

Copy
DAR 147

[1] Dated by the relationship to the letter from J. M. Rodwell, 31 October 1860.

[2] William Darwin Fox had supplied CD with this information at some point before 1856. See *Correspondence* vol. 6, letters to W. D. Fox, 8 [June 1856] and 14 June [1856].

[3] Sichel 1847. CD cited this paper in *Variation* 2: 329.

[4] See letter from J. M. Rodwell, 31 October 1860. CD had often consulted William Kirby's entomological works during his youthful days as an insect collector.

[5] See letter from J. M. Rodwell, 6 December 1860. CD did not cite Rodwell's information on this point in his later works.

[6] Bree 1860, p. 29. CD had stated that 'white sheep and pigs are differently affected from coloured individuals by certain vegetable poisons.' (*Origin*, p. 12).

[7] Eudes-Deslongchamps 1835 and 1842. CD mentioned the case in his 'big book' on species (*Natural selection*, p. 297).

[8] Louis Agassiz's identification is cited in Eudes-Deslongchamps 1835, p. 50.

[9] Rodwell apparently included this information in the portion of his letter of 31 October 1860 that is now missing. CD cited Rodwell's information in *Variation* 2: 337:

> The Rev. J. Rodwell informs me that his father turned out about fifteen cart-horses into a field of tares which in parts swarmed with black aphides, and which no doubt were honeydewed, and probably mildewed; the horses, with two exceptions, were chesnuts and bays with white marks on their faces and pasterns, and the white parts alone swelled and became angry scabs. The two bay horses with no white marks entirely escaped all injury.

[10] CD mentioned Karl Friedrich Heusinger's paper (Heusinger 1846) on the differential poisoning of animals in *Origin*, p. 12. There is a copy of the paper in the Darwin Pamphlet Collection–CUL.

[11] Leonard Jenyns and CD had been friends since CD's undergraduate days (see *Correspondence* vol. 1). Rodwell, who also attended Cambridge University, may have also known Jenyns. CD refers to Jenyns 1846, an annotated copy of which is in the Darwin Library–CUL.

[12] CD visited John Frederick William Herschel in Cape Town in 1836 before the *Beagle* returned to England. Details of the visit are recorded in Keynes ed. 1988 and in *Correspondence* vol. 1. CD had told Charles Lyell about Herschel's remark in a letter written in 1850 (see *Correspondence* vol. 4, letter to Charles Lyell, [8 March 1850]).

[13] Rodwell published a translation of the Koran in 1861.

[14] Wedgwood 1859–65. See letter from Hensleigh Wedgwood, [January? 1860], and *Correspondence* vol. 6, letter from Hensleigh Wedgwood, [before 29 September 1857].

To Daniel Oliver 7 November [1860][1]

15 Marine Parade | Eastbourne
Nov 7th
(We leave, I hope & think on Saturday.)[2]

My dear Mr Oliver

I am heartily glad to hear that the Professorship is properly decided.—[3]

I have been glad to see the slide, but I am sorry to say the glass *arrived* cracked. I do not think that the structure is so plain as when I made fresh slices. I shall be some time curious to know what you see so remarkable in the sensitive Hairs.— I hope some day you will study the Dionæa & publish on it. I shall make only a few remarks in comparison with Drosera.[4] I can see nothing in Sensitive Hairs more than I saw before, viz elongated cells arranged spirally with a bifid point, with no spiral vessel or stomata.— There is a short footstalk making change of structure in cells & that is all I can see! I conjecture that the sensitive Hairs are homologous with the tortoise-shells or buttons, in a much elongated condition.— I cannot help fancying that by holding footstalk of living leaf & retarding the movement by your thumb with good lens, you could see whether the sensitive Hair became mottled in colour by touching it with needle. (I can tell with Lens whether Hairs of Drosera are mottled.) It **ought** to do so!!— I shall be intensely curious to hear about C. of Ammonia & Dionæa.—[5]

I *have* tried several salts on the Hairs & Roots of Drosera & on other plants.[6] But I know well I have not tried enough; nor intend to do so for the work is quite out of my line; & though I have enjoyed it very much, I have been exceedingly foolish to attempt it. I still think the action of the Carbonate & some other salts of Ammonia remarkable. & worth any one's following out.— I shall briefly give the facts in my paper.— I put all the roots of a little plant of Euphorbia in water with $\frac{1}{8}$th of grain of C. of Ammonia, & it was curious how it wholly altered the appearance of every cell in all the roots.—[7] The action is so rapid, that you cannot put the root under the microscope, before clouds of granules have shot some way up the roots.—

Now I am going to beg a little favour of you— Does not Mr Fitch work at Kew?[8] And will he draw for anyone for payment? If he will be so good as to make 2 or 3 drawings for me, I shd be **very much obliged** & I could repay by P. order.

First, a leaf of Drosera with all the Hairs fully expanded. It is for *woodcut* for any reader who is not a Botanist. It ought to be 3 or 4 times natural size. M[r] F. could judge this. It ought to be done neatly.— The *extreme* marginal Hairs are long-headed; whether on this scale the difference could be represented, I know not.—

Secondly would you put little fly on exact middle of leaf, & after Hairs are well clasped get a drawing made on same scale.

Thirdly, perhaps, sketch of leaf of Dionæa, just to interest Reders.[9] I shall have copied from Trécul the minute structure of Hairs.—[10]

My dear Sir | Yours very sincerely | C. Darwin

Down House (MS 10: 25)

[1] The year is given by the reference to Oliver's appointment as professor at University College London (see n. 3, below).

[2] The Darwins left Eastbourne for Down on Saturday, 10 November 1860 ('Journal'; Appendix II).

[3] Oliver had applied for the professorship of botany at University College London previously held by John Lindley. CD wrote a letter of recommendation for him (see letter to William Sharpey, 28 October [1860]). Oliver's appointment was announced in the *Athenæum*, 10 November 1860, p. 633.

[4] CD refers to the paper on *Drosera* he was preparing. He read it at a meeting of the Philosophical Club of the Royal Society in February 1861 (Bonney 1919, p. 154). He compared the structure of *Dionaea* and *Drosera* in *Insectivorous plants*, pp. 286–320.

[5] See letter to Daniel Oliver, 3 November [1860].

[6] There is a sheet summarising the results of CD's experiments on the effects of various salts on plant cells in DAR 60.2: 89.

[7] See letter to Daniel Oliver, 3 November [1860]. The results of CD's experiments are described in *Insectivorous plants*, pp. 63–5.

[8] Walter Hood Fitch was employed as botanical draughtsman at the Royal Botanic Gardens, Kew, where Oliver also worked.

[9] In fact, Fitch did not supply the drawings for *Insectivorous plants*. All of the figures of *Dionaea* and *Drosera* were drawn by George Howard Darwin, and those of *Aldrovanda* and *Utricularia* by Francis Darwin (*Insectivorous plants*, p. 3 n.). Fitch may have illustrated the manuscript CD was preparing on *Drosera* (see n. 4, above); he prepared the drawings for CD's 1862 paper on *Primula* (see *Correspondence* vol. 9, letter to Daniel Oliver, 23 March [1861]).

[10] Although CD discussed Trécul 1855 in *Insectivorous plants*, he did not use any illustrations from this work.

To W. E. Darwin [9 November 1860]

15 Marine P.
Friday night

My dear William

I write a line to say that Etty has been going on very well, but the Doctors, I can see still think her health very precarious. However we have taken a Bed-carriage & fully expect to be at Down by 4 oclock tomorrow.[1] We shall be a grand procession through the village,—six horses altogether including a cart.—. We all so long to be at home— this place with its miserable associations has been most dreary.

Farewell my dear old man. When shall you come to Down?[2]

Your affect Father | C. Darwin
How does your money hold out? Be sure tell me if you get too poor.—
Farewell my dear William.—

Postmark: NO 10 1860
DAR 210.6

[1] The Darwins left Eastbourne for Down on Saturday, 10 November 1860 ('Journal'; Appendix II).
[2] William was in his third year at Cambridge University.

From Edward Cresy 10 November 1860

1 Greek St—
10 Nov '60—

My dear Sir,

The discrepancy in regard to the weights & measures I think admits of explanation—[1] Previously to 1836 the old Wine measure was in use and served as the basis of the fluid apothecaries measure in this wise— The gallon or congius was the unit this was divided into 8 parts octarii or pints—each of which was considered roughly to weigh a pound—in point of fact it weighed 7289·7 grains a pound avoirdupois is 7000 grains— well then this old wine pint thus roughly assimilated to a pound was divided into 16 *fluid* ounces—&c thus—

20	Minims	=	1 fluid scruple	the same subdivisions
3	scruples	=	1 " drachm	being used as in ordinary
8	drachms	=	1 " ounce—	apothecaries weight only the
16	ounces	=	1 pint—	minim takes the place of the
8	pints	=	1 gallon—	grain—

Subsequently to 1836 the new measure came into operation by which the pint in lieu of containing 28·875 cubic inches was made to hold 34·659 cubic inches to meet this change the apothecaries subdivided the new pint into *20* fluid ounces instead of 16—*still* retaining the old subdivision into drams scruples & minims but their *weights* became materially altered—the old fluid ounce contained 455·6 grains of distilled water the new contains 437·5— the old fluid drachm contained 56·9 grains, the new 54·7 grains— I dont know whether you have the subjoined table but it is a very useful one—

C Congius. Gallon—		O Octarii Pints—		℥ Fluid ounces.		ʒ Fluid Drachms—		♏ Minims		Grains of water—
1	=	8	=	160	=	1280	=	76800	=	76000
		1	=	20	=	160	=	9600	=	8750
				1	=	8	=	480	=	437·5
						1	=	60	=	54·7
								1	=	0·9

It is convenient if you have not a measuring glass to weigh out your measure with a pair of scales in fact it is more accurate for the glasses are often very indifferently graduated only of course you must not consider a very heavy or very light liquid weighs the same per gallon or per pint as water— a gallon of ether weighs $7\frac{1}{2}$ lbs. a gallon of nitric acid 15 lbs—

I have called on Hofman[2] and had from the explanation he gave me written to you that his note was quite correct but having penned the passage I checked the figures from the data he gave me and found they did not work out rightly so I cancelled a page and called again this morning at the College of Chemistry and saw his Secretary,[3] as it was merely a matter of figures I would not trouble the D.r and the Sec confirmed my summing— The original solution was made by dissolving 10 milligrammes of iodine in 1000 grammes of water—that is one to one million— the Dr then took one gramme of the solution containing therefore ·01 milligramme that is $\frac{1}{100}$ of a milligramme, & he found no effect, he then experimented on two grammes of the same solution containing two hundredths of a milligramme with a sensible result— The *strength* of the solution remained the same in both experiments but the *absolute quantity* of iodine present was double in the second— He was kind enough to call in his assistant who made the solution that I might question him and both he and the Dr agreed as to the quantities originally employed, but these quantities when I got home I found did not give the same decimals as the Dr's letter—[4] he gives 0·001 or one thousandth of a milligramme per gramme. I find 10 milligrammes per 1000 grammes = 1 kilogramme = 1 litre—gives ·01 per gramme and in this the Sec says I am right well then.

Milligramme		Grains		
1·	=	·0154	—	$\frac{154}{10,000}$
·1	=	·00154	—	$\frac{154}{100,000}$
·01	=	·000154	—	$\frac{154}{1,000,000}$
·02	=	·000308	—	$\frac{308}{1,000,000}$

So that in the Dr's second experiment he discovered by the starch test 308 millionths of a grain or 3 ten thousandths or $\frac{1}{3333}$ which is credible enough as compared with Alfred Taylor's statement of the visibility of $\frac{1}{4000}$ grain of arsenic—[5] the Dr's error is in having one place of decimals too many—a very likely one and the plague of the decimal system you know you ought to write a milligramme ·001 then the Drs ·01 becomes ·00001 a very clumsy figure to work with—[6] This explanation will I trust prove satisfactory notwithstanding its interminable length which I have run into mindful of the maxim brevis esse laboro, obscurus fio—

While writing Woodward Brit Mus—came in and I always like to stir him up about the Origin,[7] "Ah he says it is a very sad book, it unsettles all one's religious principles and the worst of it is *so much* of it is true— It is just like all those Geologists they say they dont know whether God made the world or the Devil and it is my opinion they dont care!" He says Owen is a worse materialist than you are—[8] What a dreadful man he must be!—

I delayed answering till I could see Hofman— pray write if I can do anything more for you—

Yours most truly | E Cresy.

I sincerely trust from your not mentioning your daughter that she is rapidly amending— Pray present our best respects to Mrs Darwin and expression of our very earnest sympathy for the anxious time you must have passed—

C Darwin Esq—

DAR 58.1: 7, 9

CD ANNOTATIONS
2.6 437·5] '436' *underl red crayon*
2.7 54·7] *underl red crayon*
2.16 0·9] *underl red crayon*
5.1 Milligramme . . . $\frac{1}{3333}$ 5.7] '$\frac{1}{3333}$' *underl red crayon*; 'This is very important point' *added red crayon*
6.1 While . . . Cresy. 8.1] *crossed red crayon*
9.1 I . . . passed—9.3] *crossed pencil*

First CD note:[9]
Look out for weight of fluids.

$\frac{1}{28,800}$ of ^{gr}437.50?
$\frac{1}{60}$ of $\frac{9}{10}$ of ['grain' *del*] 1 grain

['Money' *del*] *Weight & Measures* | *[illeg]* Notes & Book. Look to references in other works about segregation. One of my minute drops from Pins Head is $\frac{1}{28,800}$ of oz in bulk & ∴ $\frac{1}{28,800}$ of ^{gr}437·5, which is weight of 1 oz of water. ['& *[illeg]*f oz' *del*] One minim *of water [*interl*] weighs $\frac{9}{10}$ of gr. (according to Cresy) & $\frac{1}{60}$ (or my minute drop weighs $\frac{1}{60}$ of $\frac{9}{10}$ of gr. i.e. $\frac{15}{1000}$ or $\frac{3}{200}$ of a grain. This is the result gained by taking $\frac{1}{28,800}$ of 2 oz, or quite accurately ·015191.

Second CD note:[10]
The original solution contained 10 milligrammes of iodine in 1 kilogramme of water. i.e. ·01 milligramme per gramme.

Milligrammes		Grains				
1·	=	·0154				
·1	=	·00154				
·01	=	·000154				
·02	=	·000308	=	$\frac{308}{100000}$	=	$\frac{3}{10,000}$

The original solution was 1 part of iodine in 1,000,000 parts of water, and in this $\frac{154}{1000000}$ grains were not discernible, but $\frac{308}{1000000}$ were.

1 See the two letters to Edward Cresy, 2 November [1860].

2 August Wilhelm von Hofmann had sent Cresy information for CD on the sensitivity of chemical tests. See letters from A. W. von Hofmann to Edward Cresy, 13 October 1860 and 27 October 1860.

3 Cresy may be referring to Herbert McLeod, who was assistant chemist to Hofmann at the Royal College of Chemistry (see James ed. 1987).

4 Cresy forwarded the letter from A. W. von Hofmann to Edward Cresy, 27 October 1860, enclosed in the letter from Edward Cresy, 30 October 1860.

[5] Taylor 1860, p. 12. See the enclosure with the letter from Edward Cresy, 30 October 1860.

[6] See second CD note.

[7] Samuel Pickworth Woodward was first-class assistant in the department of geology and mineralogy of the British Museum. He was known to be strongly against the views CD put forward in *Origin* (see *Correspondence* vol. 7, letters from J. D. Hooker, [21 November 1859]).

[8] Richard Owen was superintendent of the natural history departments of the British Museum.

[9] The note is associated with the letter in DAR 58.1: 8. CD crossed in pencil the first page of the note, from 'Look' to 'segregation'.

[10] The note is in DAR 143 with the copies of correspondence between CD and Cresy. Although there is no indication when CD wrote the note, it seems to be related to the information Cresy gives in this letter.

To J. S. Henslow 10 November [1860]

Eastbourne
Nov 10.

My dear Henslow.

We return home this afternoon, as my poor dear girl is now just strong enough to bear removal.— I received your letter forwarded from Down this morning, & very much obliged I am to you for having taken so much trouble about Dr. Bree.[1] I had thought I would have inserted a letter in some Journal on so unprovoked attack on my veracity; but I am glad I did not. No one would ever have dreamed of his interpretations. Again I sincerely thank you.—

I never heard of such a muddle about the stone Hatchets.[2] If you are sure that you can spare & know no one else who would make better use of B. de Perthes Book, I shd like to have a copy.[3] His course of investigation has been a strange one.—

On my return home I will settle about the seeds.—

I am still at work on Drosera. I asked you about the moving red matter in the cells, & now for the chance of your knowing, I want to ask one other question; but please *observe* if I get no answer I shall understand that you do not know.— My question is whether, observations have been made on the action of weak solutions of Carb. of Ammonia (or of other salts) on the fluid contents of the cells of living plants.— I find that C. of Ammonia has a remarkable & rapid action when absorbed by the roots.

My dear Henslow | Ever yours gratefully | C. Darwin

Endorsement: '1860'
DAR 93: 83–4

[1] The physician Charles Robert Bree had attacked *Origin* in a recently published book (Bree 1860). See letter to J. S. Henslow, 26 October [1860].

[2] See letter to J. S. Henslow, 26 October [1860], in which CD discussed Henslow's letter published in the *Athenæum*, 20 October 1860, p. 516. Henslow had written a further letter, published in the *Athenæum*, 3 November 1860, pp. 592–3. He again discussed the confusion about whether the gravel beds had been disturbed or not and showed evidence of the action of water.

[3] CD probably refers to the second volume of Boucher de Crèvecoeur de Perthes 1847–64. CD had discussed the implications of Jacques Boucher de Crèvecoeur de Perthes's work with Joseph Dalton Hooker, and indicated that he had already looked at one of his volumes. See *Correspondence* vol. 7, letter to J. D. Hooker, 22 [June 1859].

To T. H. Huxley 10 November [1860][1]

[Eastbourne]
Nov 10th

My dear Huxley

I have been very glad to receive the circular today about the Review—a grand set of names as Editors!—[2] I have sent in my name, of course, as subscriber.—[3] But I write today merely just to say (as I know full well that we have had your sympathy & that your own great sorrow has not shut up your heart for others' sorrows)[4] that my girl is well enough to be carried home today. Nevertheless I fear the Doctors think her health very precarious, there is great torpor in the mesentery.

Reading over the Prospectus it has occurred to me as just worth your consideration whether you would not offer to give your correspondents information, as far as lies in your power on any point which they may wish to investigate.—[5] Many a country man has no power of ascertaining whether any subject has been investigated.— I *feel* this from being in this predicament myself & having had to write to lots of people to know whether a certain point (action of C. of Ammonia on sap of living plants) has been observed.—

Ever | My dear Huxley | Yours most truly | C. Darwin

Imperial College of Science, Technology and Medicine Archives (Huxley 5: 143)

[1] The year is given by the reference to the new editorial board of the *Natural History Review*.
[2] The editors included, in addition to Huxley, George Busk, William Benjamin Carpenter, Frederick Currey, Joseph Reay Greene, John Lubbock, Robert M'Donnell, Daniel Oliver, Philip Lutley Sclater, Charles Wyville Thomson, and Edward Perceval Wright.
[3] CD's copies of the *Natural History Review* (1861–5) are in the Darwin Library–CUL.
[4] Huxley's oldest son had died in September (see letter to T. H. Huxley, 18 September [1860]).
[5] This service was not offered by the new series of the *Natural History Review*.

To Edward Cresy [12 November 1860][1]

Down, Bromley, Kent
Monday night

Dear Cresy

Upon my life I think you must be the most good-natured man in the world. How capitally you have answered the question about the weight of water; you are as good, indeed a great deal better, than an Encyclopedia. How fortunate for me that I asked you about Hofman's note. What a horrid blunder I should have made! But what trouble you have taken for me.— About the measure and weights I daresay your explanation is quite right, for I looked into old and new books; but there were also most stupid misprints.— I shall permanently keep your letter.[2]

As you seem a little interested about Drosera, I may mention that today I have been testing how light a weight will set a single hair moving, and I find $\frac{1}{50}$ of an inch of finest cotton thread suffices and rather shorter length (measured by micrometer) of my wife's hair suffices; and I have sent today several inches of both to be weighed by good balance in Jermyn St.;[3] and the result will show a very small fraction of a grain; so that the hairs are as sensitive to mechanical as to chemical action.— All the wild Droseras, I find, are dead for this year; so I must now write my paper and cannot try a few more experiments, as I much wished to do.

I have been much amused by your account of your conversation with Woodward and I should much like to hear you take him off. He is a very good man in his way, and his generalisations on Shells are really capital.[4]

We returned home on Saturday evening; and my girl stood the journey well; but she is very weak.

With hearty thanks | Yours very sincerely | C. Darwin

Copy
DAR 143

[1] Dated by the relationship to the letter from Edward Cresy, 10 November 1860, and to CD's return to Down on Saturday, 10 November ('Journal'; Appendix II).

[2] Letter from Edward Cresy, 10 November 1860.

[3] See letter from Trenham Reeks, 15 November 1860. Reeks worked at the Museum of Practical Geology in Jermyn Street, London.

[4] See letter from Edward Cresy, 10 November 1860. CD refers to Samuel Pickworth Woodward's treatise on the Mollusca (S. P. Woodward 1851–6), which he had read and praised in 1856 (see *Correspondence* vol. 4, Appendix IV, 128: 18; and vol. 6, letter to S. P. Woodward, 15 May [1856]). There is an annotated copy of the work in the Darwin Library–CUL.

To G. V. Reed 12 November [1860][1]

Down, Bromley Kent.—
Nov. 12th

My dear Sir

We returned on Saturday night from Eastbourne where we were detained by another severe attack of illness in my poor girl.[2] She has rallied surprisingly & stood the journey very well; but after six months illness she is very weak & I fear in rather precarious state, though thank God there is no immediate danger.— I tell you all this, because we know we have your sympathy.—

If I do **not** hear from you, I will send Leonard on Thursday *morning*;[3] I say morning for the days are now so short, he would hardly have time to return by daylight, if he started after our midday dinner.— In the Spring, if then not

Robert FitzRoy
(Courtesy of the Royal Society of London Archive.)

John Phillips
(Courtesy of the Library of the New York Botanical Gardens, Bronx, New York.)

inconvenient to you, we will change to the afternoon.— Franky seems getting on capitally, thanks to your kindness.[4]

Pray believe me | My dear Sir | Yours very sincerely | Charles Darwin

Buckinghamshire Record Office (D.22.39.5)

[1] Dated by the reference to Eastbourne (see n. 2, below).

[2] The Darwins returned to Down, following a prolonged stay in Eastbourne, on 10 November 1860 ('Journal'; Appendix II). Henrietta Emma Darwin had been very ill during their visit.

[3] CD had arranged for Reed to tutor Leonard Darwin twice a week. See letter to G. V. Reed, 15 September [1860].

[4] Francis Darwin had entered Clapham Grammar School in August 1860. Before going to school he had been tutored by Reed.

To John Phillips 14 November [1860][1]

Down Bromley Kent
Nov. 14th

Dear Phillips,

Many thanks for your kind present of your work Life on the Earth.[2]

I have glanced at the part about the Origin & am sorry, but not surprised, to see that you are dead against me.—

Pray believe me Dear Phillips | Yours very sincerely | Ch. Darwin

Oxford University Museum (Geological collections)

[1] Dated by the reference to Phillips 1860.

[2] There is a copy of Phillips 1860 in the Darwin Library–CUL. The work gives the text of Phillips's Rede lecture at Cambridge University, in which he addressed the question of the origin of species. CD had presented Phillips with a copy of *Origin* (see Appendix III and *Correspondence* vol. 7, letter to John Phillips, 26 November [1859]).

From Trenham Reeks 15 November 1860

Museum, Jermyn St
15th. Novr. 1860.

My dear Mr. Darwin

It really gives me much pleasure to be able to render you any assistance in your interesting experiments and I hope you will never hesitate to let me know where or when I can be of use.[1]

On the other side you have the weights of the three objects.[2] I first weighed them myself in our own balance, and then I asked my friend Mr. Tookey,[3] who is known for his carefulness and accuracy, to do the same in his own assay-balance, which he considers to be about as accurate as can be made.

I may say that in both these balances "the beam is kicked" by the $\frac{1}{1000}$ of a grain or ·001 and practically I do not think Chemists attempt anything more delicate.

The difference in the weights I think may be explained by hydroscopic moisture as they were weighed on different days.

	Reeks	Tookey	Mean
Blotting paper 11 inches long	·564	·563	·5635
Thread 30 inches	·185	·182	·1835
Hair 6 inches long	·009	·010	·0095

Believe me, Yours very sincerely | Trenham Reeks.

DAR 58.1: 10–11

[1] Reeks was curator and librarian of the Museum of Practical Geology, London. He had previously helped CD with his research.

[2] Reeks refers to the verso of the letter, on which he gave the weights of the materials sent to him by CD.

[3] Charles Tookey was employed in the Metallurgical Laboratory of the Royal School of Mines.

To T. H. Huxley 16 November [1860]

Down Bromley Kent
Nov 16

My dear Huxley

Many thanks for your kind note & Lecture.[1] The latter seems to me excellent —the best expose & classification of the higher objects of Natural History that I have ever read. It is really admirable. I like much your bit about the Roman school-boy.—[2] I think you are a little too emphatic against reading, for how the deuce is a poor schoolmaster to learn anything except by reading & comparing with what he sees.[3] Your remarks on absolute necessity of observation are capital. When my son was at Rugby[4] there was a Botanical prize, which he won, by merely getting up Henslow's Botany,[5] & he never actually looked at a single flower!!

My daughter keeps in much the same state.

I had a letter today from R. McDonnell of Dubline[6] (I wonder whether he is the bearded man one sees at B. Assoc^{n} if so I fear he is rash & wild) & he says owing to passage in my Book on Electric fishes he has been dissecting Rays, & believes he finds in same fish the homologues of *both* the anterior & posterior proper electric organs of fish: which, if true, seems to me an interesting fact.—[7]

By the way I hear that Agassiz is coming out in next Part of his Contributions, with heavy thunder against the Origin.[8] On other hand, I hear from L. Horner[9] that Dubois-Reymond[10] expresses strong approbation. I suppose you have heard nothing more from Von Siebold;[11] what a trump-card he would be on our side.

Owing to all the illness of my poor child & constant change of place I make no progress with my work.—

I shall be very curious to see the 1st no. of the Review.— Long may you live as "a buttered angel".

Ever yours most truly | C. Darwin

I suppose we shall have dear old Hooker back soon.—[12]

Imperial College of Science, Technology and Medicine Archives (Huxley 5: 145)

[1] Huxley apparently sent CD an offprint of a lecture he had delivered at the South Kensington Museum on 14 May 1860. The lecture was entitled 'On the study of zoology' and was the fourth in the series 'Lectures addressed to teachers on preparation for obtaining science certificates and the method of teaching a science class' (1860–1). Neither CD's copy of the lecture nor the note CD mentions has been found in the Darwin Archive.

[2] Huxley maintained that a fourth-century Roman boy, 'transplanted into one of our public schools . . . would not meet with a single unfamiliar line of thought; amidst all the new facts he would have to learn, not one would suggest a different mode of regarding the universe from that current in his own time.' (T. H. Huxley 1860c; reprinted in T. H. Huxley 1870, p. 129). CD had long felt that schools focused too much on classical learning (see *Correspondence* vol. 5, letter to W. D. Fox, 7 March [1852], and J. R. Moore 1977).

[3] Huxley stated that 'real knowledge in science means personal acquaintance with the facts, be they few or many' (T. H. Huxley 1860c; reprinted in T. H. Huxley 1870, p. 131). When Huxley edited this essay for re-publication in 1870, he added a footnote addressing the point raised by CD (T. H. Huxley 1870, p. 131 n. 1): 'It has been suggested to me that these words may be taken to imply a discouragement on my part of any sort of scientific instruction which does not give an acquaintance with the facts at first hand. But this is not my meaning . . . The system which I repudiate is that which allows teachers who have not come into direct contact with the leading facts of a science to pass their second-hand information on.'

[4] William Erasmus Darwin attended Rugby School from 1852 to 1858.

[5] Henslow 1837. William began to study botany seriously in 1858 before leaving Rugby (see *Correspondence* vol. 7, letter to W. E. Darwin, [3 May 1858]).

[6] The letter has not been found. Robert M'Donnell was an Irish surgeon and naturalist and co-editor with Huxley of the *Natural History Review*. His name was variously spelled M'Donnell or McDonnell.

[7] In *Origin*, p. 192, CD referred to the origin of electric organs in fish as a 'special difficulty'. The rays, he stated, possessed organs closely analogous to the electric apparatus that did not discharge electricity (*ibid.*, p. 193). M'Donnell's information about a second organ that was the real homologue of the electric battery in the torpedo, published in M'Donnell 1861, was included in the fourth edition of *Origin* (Peckham ed. 1959, p. 350).

[8] Louis Agassiz's critique of *Origin* was published towards the end of 1860 in Agassiz 1857–62, 2 (pt 1): 89–99. The section was entitled 'On Acalephs in general' and discussed CD's views in relation to the definition of species and their classification. The same text had already been printed in the July number of the *American Journal of Science and Arts* and in the September issue of *Annals and Magazine of Natural History* (Agassiz 1860).

[9] Leonard Horner was acquainted with many German naturalists and scholars. His daughter Leonora was married to the noted historian Georg Heinrich Pertz, chief librarian and privy councillor in Berlin. Horner was at the time president of the Geological Society of London.

[10] Emil Heinrich Du Bois-Reymond was professor of physiology at Berlin University. Famous for wide-ranging researches into animal electricity, Du Bois-Reymond was also a noted free-thinker, greatly interested in questions relating to the origin of life. In the 1870s, he delivered a series of lectures on Darwinism, the manuscripts of which are in the Staatsbibliothek Preussischer Kulturbesitz in Berlin.

[11] Huxley corresponded with Karl Theodor Ernst von Siebold, professor of zoology and comparative anatomy in Munich, whose work on parthenogenesis CD had studied closely (see *Correspondence*

vol. 6). However, CD confused Siebold with Karl Ernst von Baer (see letter from T. H. Huxley, 6 August 1860, and letter to T. H. Huxley, 8 August [1860]).

[12] Joseph Dalton Hooker left Beirut en route for London on 5 November 1860 (L. Huxley ed. 1918, 1: 528).

To Daniel Oliver 16 November [1860][1]

Down Bromley Kent
Nov 16th

My dear Mr Oliver

In writing out my paper yesterday I was so astounded at my results that I have got fairly frightened, & have determined to finish my paper, but to publish nothing until next summer[2] I shall have retested my results & tried some of them in another way: so that there is no hurry about the Drawings.[3] If there are good specimens, I shd be glad for Mr Fitch to complete them; otherwise will you ask him to wait till next spring or summer.—

Most heartily do I thank you for your most kind & valuable assistance.— If I could get one or two plants of Dionæa I would experimentise on them; but I shall not of course attempt the anatomy; & if you thought you would undertake the subject, I would not interfere in any way, except by bare allusions to what I have seen; but it is so important to me about the mottling or segregation of colour, that I shd. extremely much wish to ascertain that point, if in the Spring, by the aid of Mr Croker I could purchase (at almost any price) one or two or three of the young plants which he saw.—[4]

I have been corresponding with some of the chemical physiologists & as far as I can find out the curious action of the C. of Ammonia on the roots of plants has not been observed.—

With cordial thanks for all your kindness | My dear Mr Oliver | Yours very sincerely | C. Darwin

P.S. | Can you tell me name of Plant, which grew out of doors in my Fathers garden, 2 to 4 ft high—considerably branched, died down (I think) in winter, bore many *minute* almost white or very pale pink flowers, & which *flowers caught a multitude of flies by their probosces.*—[5] I shd. like to get a plant to try to make out final cause of the catching.—

P.S. 2d. | Perhaps you would like to hear following extraordinary fact.— I have placed over & over again minute atoms of paper, stick, cinders meat flies &c on glands of *single* hairs of Drosera & they always became inflected; so I thought I would test how minute an atom would cause movement: I measured with micrometer several bits of woman's Hair (& thread) which caused movement & found that *under* $\frac{1}{50}$th of inch **amply** sufficed. Of *same* hair I sent **6** inches to be weighed in London by best balance, & it weighed *under* $\frac{1}{100}$ of grain, which shows that $\frac{1}{30,000}$ of a grain suffices!![6]

Prolonged pressure alone causes movement.

Down House (MS 10: 26)

[1] Dated by the relationship to the letter to Daniel Oliver, 7 November [1860].

[2] CD's paper on *Drosera* was delivered at a meeting of the Philosophical Club of the Royal Society on 21 February 1861 (Bonney 1919, p. 154). It was not published, but CD ultimately included his earlier material in *Insectivorous plants*. His index for the paper is in DAR 54: 1.

[3] CD had asked Oliver to find out whether Walter Hood Fitch would make drawings of *Drosera* for him (see letter to Daniel Oliver, 7 November [1860]).

[4] Charles William Crocker was the foreman in the propagating department at the Royal Botanic Gardens, Kew.

[5] Robert Waring Darwin, CD's father, had been a keen gardener who cultivated a variety of rare and unusual plants in the gardens of The Mount, the Darwin residence in Shrewsbury (see *Correspondence* vol. 1).

[6] See letter from Trenham Reeks, 15 November 1860.

To John Lubbock [18 November 1860][1]

Down
Sunday night

My dear Lubbock

I received this morning Ld Cranworths[2] Subscription & now enclose 5 · 5 · 0 for the Nat. School.—[3]

We returned home yesterday week: Etty did the journey pretty well;[4] she was desperately ill at Eastbourne & we gave up all hope; but she has rallied considerably. God knows what the ultimate result will be. I never passed a more miserable time than at Eastbourne.

Pray give our kindest thanks to Mrs Lubbock for her note received this morning.[5] There is no possibility of my wife leaving Etty at present. As for myself I have had some thoughts of going up on Wednesday & Thursday, but I very much doubt whether I shall have the heart. I thank you both most sincerely for so kindly wishing to see me; but when working in London, I am so poor a wretch & so tired that my own room in my Brother's house is the sole refuge for which I am fit.—[6]

I have done hardly anything in science for many weeks; but have lately been employed in drawing up a paper on the movements of Drosera; the result has been that I have got actually frightened at my results, & will not publish till I have retested my work next summer.— If you are at all at leisure, I sh^d^ like to hear a little news of what you have been doing & whom you have seen most of.—

You take in Annals & Mag. of N. Hist; will you tell me whether in last number (or if in any future number) there is reprinted an Article on the "Origin" by Asa Gray from Atlantic Monthly Review.—[7] I wish particularly to know.

I hope you have enjoyed yourself & are not working too hard.

Dear Lubbock | Yours most sincerely | C. Darwin

Endorsement: '1860'
Down House

[1] Dated by an entry in CD's Account book (Down House MS) relating to the sums discussed in the letter (see n. 3, below).

[2] Robert Monsey Rolfe, Baron Cranworth, was a neighbour of CD's and a contributor to several of the Down charities.

[3] The sum of £5 5*s* made out to 'Lubbock for school for Lord Cranworth' was entered in CD's Account book on 19 November 1860. Lubbock was one of the trustees of the Down national school established in 1855 by the Lubbock family.

[4] Henrietta Emma Darwin.

[5] Ellen Frances Lubbock.

[6] Erasmus Alvey Darwin lived in Queen Anne Street, London.

[7] CD had sent the second part of Asa Gray's review of *Origin* ([Gray] 1860b, pp. 229–39) to the *Annals and Magazine of Natural History* in September (see letter to Asa Gray, 26 September [1860]). It was reprinted in the November issue (*Annals and Magazine of Natural History* 3d ser. 6 (1860): 373–86).

From Charles Lyell [before 20 November 1860][1]

was ever returned.

I suspect that the Scandinavian flora was the first, then the Germanic & that the other subfloras got in by various geographical opportunities after the era of submergence or floating ice & when the larger part or the whole area within the 100 fathom line was land dry—

If you refer to DelaBeche's map Geological Researches p. 190[2] or Trimmer Quart. Journ. G. S. p. 293 vol. 9.[3] see maps you will see that a low plain of land skirting the bay of Biscay & France & joining Ireland would when all within the 100 fathom line was dry land allow the Asturian plants to reach Ireland & if this happened when the emeral isle was nearly or quite separated from England it might explain why said southern species never got into England.[4]

Have you read Jamieson of Ellon in our last Quarterly Journal. 1860 p 349 &c.[5] His 5 periods at p 370 can be made to square pretty well with Trimmers sketch maps above referred to.[6] I shall give I think DelaBeches 100 fathom map, & Trimmer's submergence map— But I am not prepared to give a third map but shall say that things were nearly restored to the status quo ante glacier after which came minor oscillations of level.[7] What immense periods these within the glacial episode so brief a portion of the pleiocene!

Has Hooker or any one ever criticized E. Forbes' Botanical migrations of his 5 floras in British Isles?[8] I have not heard whether J. Hooker is returned from Syria.[9]

Phillips does not appear to me in glancing at his chapters to have taken hold strongly of any subject such as the age of the Amiens man or Natural Selection.[10] He gives Sedgwick & others as authorities.[11] In regard to progressive development he seems more cautious than formerly. Has he discovered that that theory tends in favour of transmutation!

Incomplete
DAR 170.2: 80

CD ANNOTATIONS
1.1 was . . . dry—2.4] *crossed ink*
3.1 If you] *opening square bracket added ink*
4.3 I shall . . . transmutation! 6.5] *crossed ink*
Top of first page: '(Glacial)' *brown crayon*

[1] Dated by the relationship to the letter to Charles Lyell, 20 November [1860].
[2] De la Beche 1834, p. 190. The map indicates the position of the 100-fathom line in the seas surrounding Britain. CD's annotated copy of the work is in the Darwin Library–CUL.
[3] Trimmer 1853. The plate facing p. 293 includes sketch maps 'illustrating the changes in the distribution of land and sea during the Pleistocene epoch'.
[4] Lyell refers to the plants of the ancient province of Asturias in northern Spain. The presence of a few Spanish plants, chiefly heaths and saxifrages, in Ireland was an anomaly in the explanation of geographical distribution. See nn. 6 and 7, below.
[5] Jamieson 1860.
[6] Jamieson 1860, pp. 370–1. The text gives a synopsis of five possible stages in the 'Pleistocene history of Scotland', accounting for the accumulation of drift and gravel deposits.
[7] Lyell was studying the relationship between changes in the flora and fauna of Britain and geological changes in the level of the land during the Pleistocene period for his work on the antiquity of man. The maps and the possible configuration of Europe since the Pliocene were subsequently discussed in C. Lyell 1863, pp. 273–80. Lyell also published a map of his own illustrating the effects of 'a downward movement of a hundred fathoms, or 600 English feet, supposed to have been uniform over the whole of the British Isles' (C. Lyell 1863, pp. 275, 278). Lyell was attempting to explain the non-uniform distribution of European species across the British Isles: he accepted the validity of Edward Forbes's hypothesis that organisms had migrated from Spain to Ireland over former land-bridges (C. Lyell 1863, pp. 283–4).
[8] CD had often discussed Forbes's proposal of successive waves of plants and animals spreading throughout the British Isles from other parts of Europe during the Pleistocene period (see E. Forbes 1846) with Lyell, Joseph Dalton Hooker, and Hewett Cottrell Watson. See *Correspondence* vols. 4, 5, and 6. For CD's reply, see the following letter.
[9] See letter to J. D. Hooker, 21 November [1860].
[10] Phillips 1860, in which John Phillips addressed the question of the origin of life on earth. Phillips discussed the human artefacts found at Amiens, France, on pp. 48–50, and natural selection on pp. 200–4. CD's copy of the book is in the Darwin Library–CUL.
[11] Phillips cites Adam Sedgwick in Phillips 1860, pp. 191 and 203 n. 1.

To Charles Lyell 20 November [1860]

Down Bromley Kent
Nov. 20

My dear Lyell

I quite agree in admiration of Forbe's Essay,[1] yet, on my life, I think, it has done in some respects as much mischief as good.[2] Those who believe in vast continental extensions will never investigate means of distribution. Good Heavens look at Heers map of Atlantis!![3] I thought his division & lines of travel of the British Plants very wild & with hardly any foundation. I quite agree with what you say of almost certainty of Glacial epoch having destroyed the Spanish Saxifrages &c in Ireland.— I remember well discussing this with Hooker;[4] & I suggested that a slightly different or more equable & humid climate might have allowed (with perhaps some extension of land) the plants in question to have grown along the *entire western* shores, between Spain & Ireland, & that subsequently they became extinct except at the present points under an oceanic climate. The point of Devonshire now has a touch of the same character.—

I demur in this particular case to Forbe's transportal by ice: the subject has rather gone out of my mind, & it is not worth looking to my M.S. discussion on migration during the Glacial period,[5] but I remember that the distribution of Mammalia & the very regular relation of the alpine plants to points *due* north (alluded to in Origin)[6] seemed to indicate continuous land at close of Glacial period.—

Your chapters, to which you allude, must be very difficult, but I shd think well worth much labour.—[7] I fear your volume on Geolog. Hist. of Man will slip through your fingers.—[8] (I do not think Hooker has criticised Forbes; H. C. Watson has abused him in the Cybele, but I do not remember that it was at all well done.—)[9]

I have not had heart to read Phillips yet,[10] or a tremendous long hostile Review by Prof Bowen in 4to Mem. of American Acad. of Sciences.[11] (By the way I hear Agassiz is going to thunder against me in the next part of Contributions).—[12] Thank you for telling me of the sale of the Origin, of which I had not heard.—[13] There will be sometime,, I presume, a new Edition; & I especially want your advice on one point; & you know I think you the wisest of men, & I shall be *absolutely guided by your advice.*— It has occurred to me that it would *perhaps* be good plan to put a set of notes (some 20 to 40 or 50) to Origin, which now has none, exclusively devoted to errors of my Reviewers.— It has occurred to me that where a Reviewer has erred a common reader might err. Secondly it will show reader that he must not trust implicitly to Reviewers. Thirdly when any special fact has been attacked, I shd like to defend it. I would show no sort of anger. I enclose a mere *rough* specimen, done without any care or accuracy, done from memory alone (to be torn up) just to show the sort of thing which has occurred to me.— *Will you do me the great kindness to consider this well.* It seems to me it would have good effect & give some confidence to reader. It wd horrid bore going through all the Reviews.—

Yours affect | C. Darwin

(Etty goes on pretty well, but is terribly weak.)

[Enclosure]

* Dr Bree (p.) asserts that I explain the structure of the cells of the Hive-Bee by "the exploded doctrine of pressure".—[14] But I do not say one word, which directly or indirectly can be interpreted into any reference to pressure.—

* The Edinburgh Reviewer (vol. — p.) quotes my work as saying that the "dorsal vertebræ" of Pigeons vary in number. & disputes the fact.—[15] I nowhere even allude to the dorsal vertebræ, only to the sacral & caudal vertebræ.—

* The Edinburgh Reviewer throws doubt on these organs being the Branchiæ of Cirripedes. But Prof. Owen in 1854 admits without hesitation that they are Branchiæ, as did John Hunter long ago.—[16]

* The confounded Wealden calculation, to be struck out. & a note to be inserted

to effect that I am convinced of its inaccuracy from Review in Saturday R.[17] & from Phillips, as I see in Table of Contents that he attacks it.[18]

* Mr. Hopkins (Fraser vol. p.) states that I argue in favour of my views from the extreme imperfection of the Geological Record, & says this is the first time in the History of Science he has ever heard of ignorance being adduced as an argument.—[19] I am quoting only from *vague* memory But I repeatedly admit in the most emphatic language which I can use, that the imperfect evidence, which Geology offers in regard to transitional forms, is most strongly opposed to my views.— Surely there is a wide difference in fully admitting an objection & in then endeavouring to show that it is not so strong as it at first appears, & in Mr Hopkins assertion that I found my argument on the Objection.—

* I would also put a note to "Natural Selection" & show how variously it has been misunderstood.

* A writer in Eding. Phil. Journal denies my statement that the woodpecker of La Plata never frequents trees:[20] I observed its habits during two years, but what is more to the purpose, Azara, whose accuracy all admit, is more emphatic than I am in regards to its never frequenting trees.[21] Mr A. Murray denies that it ought to be called a Woodpecker:[22] it has two toes in front & two behind, pointed tail-feathers, a long pointed tongue & the same general form of body the same manner of flight, colouring & voice. It was classed until recently in the same genus, Picus, with all other woodpeckers;[23] but now has been ranked as a distinct genus amongst the Picidæ.— It differs from the typical Picus only in the beak not being quite so strong & in the upper mandible being slightly arched. I think these facts fully justify my statement that it is "in all essential parts of its organisation" a Woodpecker.—

Endorsement: '1860'
American Philosophical Society (233)

[1] E. Forbes 1846. See preceding letter.

[2] In 1856, when he was composing his species manuscript, CD wrote Lyell a long letter expostulating about the 'ill-service' done to science by theories of continental extensions such as Edward Forbes's. See *Correspondence* vol. 6, letter to Charles Lyell, 25 June [1856].

[3] Heer 1855. Lyell was more disposed toward Oswald Heer's hypothesis of a former land-bridge connecting Europe, Africa, and America than was CD. See *Correspondence* vol. 6, letter from Charles Lyell, 1–2 May 1856.

[4] See, for example, *Correspondence* vol. 6, letter from J. D. Hooker, 4 August 1856, and letter to J. D. Hooker, 5 August [1856].

[5] *Natural selection*, pp. 534–66. CD discussed this chapter with Joseph Dalton Hooker in 1856 (see *Correspondence* vol. 6).

[6] *Origin*, p. 367.

[7] Lyell was working on new editions of his *Elements of geology* (C. Lyell 1865) and *Principles of geology* (C. Lyell 1867–8).

[8] Lyell published his work on the geological evidence for the antiquity of man in 1863 (C. Lyell 1863).

[9] Hewett Cottrell Watson accused Forbes of plagiarising his subdivisions of the British flora (Watson 1847–59, 1: 465–72). For Hooker's and CD's views on the controversy between Watson and Forbes, see *Correspondence* vol. 3, letter from J. D. Hooker, [before 3 September 1846], and letter to J. D. Hooker, [3 September 1846].

[10] Phillips 1860.
[11] Bowen 1860b. There is an annotated copy of the review in the Darwin Pamphlet Collection–CUL.
[12] Agassiz 1857–62, 2 (pt 1): 89–99. See letter to T. H. Huxley, 16 November [1860] and n. 8.
[13] See following letter.
[14] Charles Robert Bree made this point, in somewhat different words, in Bree 1860, pp. 165–6. According to CD's theory of the construction of bees' cells, it was a combination of the bee's instinct for beginning cells at the proper relative distance and for building up intermediate planes between the intersection of two spheres, rather than simply pressure, that accounted for the hexagonal shape of the cell. See *Origin*, pp. 224–35.
[15] [R. Owen] 1860a, p. 522. There is an annotated copy of the review in the Darwin Pamphlet Collection–CUL.
[16] [R. Owen] 1860a, p. 489. For CD's reaction to Richard Owen's remarks on this point in the *Edinburgh Review*, see especially the letter to T. H. Huxley, 9 April [1860]. CD refers to R. Owen 1843 (not 1854) and to the anatomist John Hunter, some of whose writings Owen edited.
[17] An anonymous review of *Origin* was published in the *Saturday Review of Politics, Literature, Science, and Art*, 24 December 1859, pp. 775–6, which criticised CD's estimate of the time required to denude the Weald in southern England. CD had already corrected his estimate for the revised American edition of *Origin* and the German translation. See the enclosure to the letter to Asa Gray, 1 February [1860] and n. 16.
[18] See Phillips 1860, pp. 128–30.
[19] Hopkins 1860, 62: 84–5. The article was published in *Fraser's Magazine* in two parts, the first in vol. 61 and the second, referred to here, in vol. 62.
[20] *Edinburgh New Philosophical Journal* n.s. 11 (1860): 289.
[21] Azara 1809, 4: 9–11. CD's copy of this work is in the Darwin Library–CUL.
[22] Murray 1860a, p. 289. CD kept two copies of this review; in the separately paginated reprint (Darwin Pamphlet Collection–CUL), the section on woodpeckers (pp. 18–19) is heavily annotated.
[23] CD originally wrote 'from the the time of Lichenstein' after 'until recently' in this sentence, but then deleted the phrase (see Manuscript alterations and comments). He refers to Martin Heinrich Karl Lichtenstein, a German naturalist who had collected in South America and had classified several species of woodpecker and piculet.

To John Murray 20 November [1860]

Down Bromley Kent
Nov. 20

My dear Sir

Will you be so good as to have sent (charging to my account) a copy of my **Journal**, addressed to

"Ch. Darwin Es
care of Miss Darwin
6 Queen Anne St
Cavendish Sq^{e}"[1]

I hear from Sir C. Lyell that all the Origin was sold at your sale;[2] therefore I suppose sometime there will be new Edition, please observe that I must somehow have clean sheets to correct, as there will be a good many alterations & I wish to keep a copy of each edition for myself.—

Lastly I want to beg a favour; can you procure me a copy of the **second & corrected** edition of the Origin from N. America?[3] I am told I cannot get it through my Bookseller.— I wish much to have it, to see how some long corrections read in type.

I know how busy you are at this time of year, so pray do not write.—

My dear Sir | Yours very sincerely | Ch. Darwin

John Murray (publishers)

[1] Susan Elizabeth Darwin, CD's older sister, was presumably staying with Erasmus Alvey Darwin in London. She lived at The Mount, Shrewsbury.

[2] See preceding letter. John Murray's publishing firm held annual sales for booksellers in November. According to CD's 'Journal' (Appendix II), 3000 copies of the second edition of *Origin* were printed in January 1860. In fact, Murray sold more copies of *Origin* than he had and had to call for a new edition (see letter to T. H. Huxley, 22 November [1860]).

[3] CD refers to the revised American edition of *Origin*, to which he had added a historical preface and supplementary material (*Origin* US ed., pp. v–xi, 116*–21*, 426–32). See Appendix IV.

To J. D. Hooker 21 November [1860]

Down Bromley Kent
Nov. 21.

My dear Hooker

I heard yesterday & today of your safe arrival & that you are rosy, jolly & fat; & most heartily do I rejoice at it. I wish I were a despot, for I would punish you most severely if ever you again worked so hard as you did for months previous to your Expedition.—[1] In a week or two's time you must be a good man & write to me; for some time you will of course have much to do.— Will you give enclosed scrap to Oliver:[2] He has been extraordinarily kind & has helped me in all manner of ways.—

Since your absence we have had a miserable time & poor Etty has been fearfully ill; but she is now rallying very slowly. I have nothing to say & I need not have written; only I could not resist saying how heartily glad I am of your safe & prosperous return.—

Yours affectionately | C. Darwin

I have been working like a madman at Drosera.

Here is a fact for you, which is as certain as you stand where you are, though you wont believe it, that a bit of Hair $\frac{1}{78,000}$ of one grain in weight placed on gland will cause *one* of the gland-bearing hairs of Drosera to curve inwards, & will alter the condition of contents of every cell in the footstalk of the gland.[3]

P.S. I have just received your most kind note. I send the enclosed note though not worth sending, as I am very busy this morning. Hearty thanks about Donkeys.[4] Etty had bad night. My wife very poorly, so not up to write another word. It is somehow a real pleasure to me to know you are at Kew.—

God Bless you.—

Endorsement: '/60'
DAR 115.1: 75

[1] Hooker had just returned from a botanical expedition to Syria.

[2] See following letter.

[3] See letter from Trenham Reeks, 15 November 1860. In *Insectivorous plants*, p. 28, the weight is given as $\frac{1}{78740}$ of a grain.

[4] CD had asked Hooker to report on any striped donkeys that he might encounter in Syria (see letters to J. D. Hooker, 2 September [1860] and 6 September [1860]). Hooker's note to CD on this topic has not been located, but Hooker's diary records two observations: in Syria, he saw '4 asses with banded legs both fore and hind down nearly to hoof', and during the journey back to Beirut, 'Saw two asses with forked end to shoulder stripe' (L. Huxley ed. 1918, 1: 530).

To Daniel Oliver [21 November 1860][1]

[Down]

My dear Mr Oliver.—

Thank you for your note—[2] I do not think the plant can be an Asclepias:[3] as far as my memory serves the flowers were very simple & I think the proboscis was caught between stamens & pistil. The minute white or very pale pink flowers were not in trusses or umbels. I have seen the plant fairly studded with captured flies. My Father called it the Fly-catcher. It died down in winter. The leaves were narrow.—stem thin, much branched smooth & I think slightly succulent. You did send me the curious account of Asclepias, which surprises me much, considering R. Browns paper on importance of insects in its fertilisation.—[4] The conferva seems a good suggestion.—

Yours very sincerely | C. Darwin

In the Fly-catcher the orifice of corolla was small.— It is odd that such a plant shd. not be in Kew.— It lived 30 years ago for very many years in my Father's flower-garden.— Would you ask Sir William if he can recognise my vague account?[5]

Down House (MS 10: 27)

[1] The note seems to be the 'scrap' referred to in the preceding letter.

[2] Oliver's letter has not been found.

[3] CD had asked Oliver whether he could identify a plant that had grown in Robert Waring Darwin's garden (letter to Daniel Oliver, 16 November [1860]).

[4] Brown 1833. There is a lightly annotated copy of the paper in the Darwin Pamphlet Collection–CUL.

[5] William Jackson Hooker. The 'Fly-catcher' was a species of dogbane, *Apocynum androsaemifolium*. See letter from Daniel Oliver, 23 November 1860.

To Henry Walter Bates 22 November [1860]

Down Bromley Kent
Nov. 22

My dear Sir

I thank you sincerely for writing to me & for your very interesting letter. Your name has for very long been familiar to me, & I have heard of your zealous exertions in the cause of Natural History.[1] But I did not know that you had worked with high philosophical questions before your mind.[2] I have an old belief that a good observer really means a good theorist & I fully expect to find your observations most valuable. I am very sorry to hear that your health is shattered; but I trust under a healthy climate it may be restored. I can sympathise with you fully on this score, for I have had bad health for many years & fear I shall ever remain a confirmed invalid.—

I am delighted to hear that you, with all your large practical knowledge of Nat. History, anticipated me in many respects & concur with me.— As you say I have been thoroughily well attacked & reviled, (especially by entomologists, Westwood, Wollaston & A. Murray have all reviewed & sneered at me to their hearts' content)[3] but I care nothing about their attacks; several really good judges go a long way with me, & I observe that all those who go some little way tend to go somewhat further. What a fine philosophical mind your friend, Mr Wallace has, & he has acted in relation to me, like a true man with a noble spirit.—[4] I see by your letter that you have grappled with several of the most difficult problems, as it seems to me, in natural History—such as the distinctions between the different kinds of varieties, representative species &c.[5]

Perhaps I shall find some facts in your paper on intermediate varieties in intermediate regions,—on which subject I have found remarkably little information.— I cannot tell you how glad I am to hear that you have attended to the curious point of Equatorial refrigeration.[6] I quite agree that it must have been small; yet the more I go into that question the more convinced I feel that there was during the Glacial period some migration from N. to S.— The sketch in the Origin gives a very meagre account of my fuller M.S. Essay on this subject.—[7]

I shall be particularly obliged for a copy of your paper when published; & if any suggestions occur to me (not that you require any) or questions I will write & ask.—

Pray believe me, with respect & good wishes | My dear Sir | Yours sincerely | C. Darwin

I have at once to prepare a new Edit of the Origin, & I will do myself the pleasure of sending you a copy; but it will be only very slightly altered.—

Cases of neuter ants, divided into castes, with *intermediate gradations*. (which I imagine are rare) interest me much.[8] V. Origin on the Driver Ants p.—241.— (please look at the passage)

Endorsement: '*1860*'
Cleveland Health Sciences Library (Stecher collection)

[1] The letter has not been found. Bates had recently returned to England, having spent ten years collecting in the Amazon basin. He had first travelled to Brazil with his friend Alfred Russel Wallace in 1848 (Marchant 1916, 1: 24–6). Announcements of his identification of new insect fauna had been regularly published in the *Transactions of the Entomological Society* since 1850.

[2] Bates seems to have told CD about his long-standing belief in transmutation, a view that was well established in the early years of his travels with Wallace. He had been particularly interested in 'tracing the laws which connect together the modifications of forms and colour with the *local* circumstances

of a province or station—tracing as far as possible the actual *affiliation* of the species.' (Marchant 1916, 1: 65). This work eventually culminated in Bates's identification of the phenomenon of mimicry among butterflies and other animals, which he explained on the basis of natural selection (Bates 1861). The missing letter may also have mentioned that Bates was preparing a paper on the variation of Amazonian butterflies (see nn. 5 and 6, below).

[3] CD mistakenly attributed the hostile review that appeared in the *Athenæum* in 1859 to John Obadiah Westwood (see *Correspondence* vol. 7, letter from J. D. Hooker, [21 November 1859], and letter to J. D. Hooker, [22 November 1859]). The author was, in fact, John R. Leifchild. The other reviews to which he refers were by Thomas Vernon Wollaston ([Wollaston] 1860) and Andrew Murray (Murray 1860a).

[4] For Wallace's reaction to the publication of Darwin and Wallace 1858 and to *Origin*, see *Correspondence* vol. 7.

[5] In a paper read at a meeting of the Entomological Society on 5 March 1860, Bates discussed the 'different degree of variability of different species' of butterflies (Bates 1860, pp. 226–7). The second part of the paper, read on 24 November 1860, discussed the origin and dispersal of the different species.

[6] In his paper, Bates referred to CD's theory of the probable cooling of the tropics during the Pleistocene era (Bates 1860, p. 352). Although concurring with CD that there was evidence for the migration of Arctic species through the tropics, he nonetheless concluded that the 'present distribution of the species of *Papilio* does not support the hypothesis of such a degree of refrigeration in the equatorial zone of America, or at least does not countenance the supposition of any considerable amount of extinction.' (Bates 1860, pp. 352–3).

[7] *Origin*, pp. 365–82, and *Natural selection*, pp. 534–66.

[8] Bates had studied Brazilian ants and published descriptions of several species in conjunction with Frederick Smith of the British Museum (Bates and Smith 1854–6).

To James Drummond 22 November [1860][1]

Down Bromley Kent
Nov. 22[d]

Dear Sir

I am extremely much obliged to you for your second interesting letter of Sept. 17th.—[2] What you state about the Leschenaultias not seeding is surprising; but from my experiments I cannot get it out of my head that the agency of insects, though not perhaps indispensable, to their fertilisation, yet is useful to them. You do not mention, but I should much like to know the simple fact whether the flowers of Leschenaultia (or other Goodeniaceæ) are visited by Bees or other insects.—[3]

Your offer to send me seeds of Distylis & the other genus, will be of greatest possible interest to me.—

With my very sincere thanks for all your great kindness, I beg leave to remain | Dear Sir | Yours sincerely obliged | Charles Darwin

Library Board of Western Australia, State Archives, Accession 2275A

[1] Dated by the relationship to the letter from James Drummond, 17 September 1860.

[2] Only one letter from James Drummond with this date has been located in the Darwin Archive (see letter from James Drummond, 17 September 1860).

[3] See letter to James Drummond, 16 May 1860.

To T. H. Huxley 22 November [1860][1]

Down Bromley Kent
Nov. 22

My dear Huxley

I am so much obliged to you for telling me about Mac donnell & Mac donald:[2] Since writing to you I have had a second, such a capital, letter from him.[3] He will in time come round to our view on Species as I believe.— For Heaven sake don't write an anti-Darwinian article; you would do it so confoundedly well.— I have sometimes amused myself with thinking how I could best pitch into myself, & I believe I could give two or three good digs; but I will see you — first, before I will try.— I shall be very impatient to see the Review.[4] If it succeeds, it may really do much, very much good.—

I cannot imagine what name in one of your notes I mistook (what an odd thing that I sh^{d} not read your handwriting) for Von Siebold.[5] I remember you said that the man had written to enquire where one of your Reviews was;—that he was interested in subject;—that in some work on Ethnography he had already expressed similar views, founded, chiefly, I think on geographical considerations; & lastly that perhaps he would take opportunity publickly to say something more on subject. This latter remembrance made me ask whether he had written.— I now remember you began your note by saying he was equal to Owen + Agassiz or some such expression. I cannot have dreamed all this.[6] I suppose I shall some time come to London & shall see you & you must tell me who it is.—

Ever my dear Huxley | Yours most truly | C. Darwin

I heard today from Murray that I must set to work at once on new Edit of Origin. He sold at his sale 700 copies & has not $\frac{1}{2}$ the number. Says the Reviews have not injured sale.— I shall always think those early Reviews, almost entirely yours,—did the subject an *enormous* service.— If you have any important suggestion or criticism to make on any part of Origin I sh^{d} of course be grateful for it. For I mean to correct as far as I can, but not enlarge. How you must be wearied with & hate the subject & it is God's Blessing if you do not get to hate me.—

Adios

Imperial College of Science, Technology and Medicine Archives (Huxley 5: 147)

[1] Dated by the reference to a new edition of *Origin*. According to an entry in his 'Journal' in March 1861, CD began preparing the third English edition in December 1860 (de Beer ed. 1959).

[2] In his letter to T. H. Huxley, 16 November [1860], CD asked about the Irish naturalist with whom he was corresponding. Huxley presumably explained that CD was confusing Robert M'Donnell, the Dublin co-editor of the *Natural History Review*, with John Denis MacDonald, the naval surgeon and microscopist.

[3] The letter has not been found; it is further discussed in the letter to Charles Lyell, 24 November [1860].

[4] The first volume in the new series of the *Natural History Review*.

[5] See letter to T. H. Huxley, 16 November [1860] and n. 11.

[6] CD was confusing Karl Theodor Ernst von Siebold with Karl Ernst von Baer. Huxley had told CD about Baer's general approval of *Origin* in the letter from T. H. Huxley, 6 August 1860.

To John Murray 22 [November 1860][1]

Down Bromley Kent
22^{d}

My dear Mr Murray

Thank you for your note. You do not say that you can send a copy of sheets of Origin for corrections: if you have not please get 2^{d}-hand copy from Mudie[2] & send me & charge to me.—[3]

I hate troubling you to write, so please ask one of your Clerks to write & tell me how soon you wish me to begin to send proofs to Printers, as I have about 10 days or fortnight's work, which I should much like to get finished before I begin.—[4]

All my corrections I *conjecture* would not take above 3 weeks & I Could send some sheets to Printers after I have done half.— But what will be longest job will be to glance over all the endless Reviews to guide me in my corrections—

Yours very sincerely | C. Darwin

Endorsement: '3 Decr 1860'
John Murray (publishers)

[1] The endorsement seems to refer to the date of Murray's response (see n. 3, below). The month is otherwise confirmed by the relationship to the preceding letter and to the letter to John Murray, 3 December [1860].

[2] Mudie's Select Library, a subscription lending library, had purchased 500 copies of the first edition of *Origin* on publication (see *Correspondence* vol. 7, letter to John Murray, 24 November [1859], n. 1). The library was owned by Charles Edward Mudie.

[3] CD evidently received from Murray a copy of the volume on which to mark up his corrections. See letter to John Murray, 3 December [1860].

[4] CD refers to his study of insectivorous plants, the results of which he was writing up. According to his journal (de Beer ed. 1959), CD began to prepare the third edition of *Origin* in December 1860.

From W. H. Harvey to J. D. Hooker[1] 23 November [1860]

Trin. Coll.
Nov. 23.

My dear Hooker

I am glad to find you are safe home again & have not been eaten up by the Druses:[2] Your wife promised me the reading of your letters—but I never received any tell her so. Tell her also that it was *in the City* "Maull & Polyblank"[3] that I gave the Order—but also, that I have never yet received any copies myself—& have written to enquire why delay—& have got no answer. If *she* will use her influence with the wretches, perhaps then we may all get our own.

I dont want to bother you with further writing.— I am going through Lyall's N.W. Algæ,—meaning to print a list in Linn. Journal—but there are *very few* species —& of some, as many as 100 or more specimens.[4]

Cape Flora goes on slowly. Doing Crassulaceæ now—of which Sonder has *a superb set*—far and away better than yours & mine put together.[5]

In working out *Cliffortia* you will be glad to hear that I found some *funny* evidences of *transmutation*. I thought of preparing a short paper thereon—to show gradual passage from very unlike to same. Here is one. You know the *genus* has typically *trefoil* leaves.

= 

Well— C. crenata & pulchella
2 lateral only.

C. obcordata—

Now I find *vars* of C. obcordata going thus—

the *larger* & more uniform or orbicular the side leaves grow, the *smaller* becomes the middle leaf.— Carry on the dwarfing far enough & you get *C. crenata*![6]

I have also found a *divarication* in the fruit; from which I could draw a productive inference—as to *genera* arising from *more than one* species of old genus—whence diff^{t} subtypes under one type.

Yours affy. | *Quid Nunc.*

DAR 47: 218–19

CD ANNOTATIONS
1.1 I am . . . together. 3.2] *crossed pencil*
4.1 In . . . *transmutation.* 4.2] 'Rosaceæ of Jussieu'[7] *ink*
Top of first page: 'Compensation | Analogous Variation' *brown crayon*; 'Harvey' *ink*

1 This letter was sent to CD by Hooker, enclosed with the letter from J. D. Hooker, [26 November – 4 December 1860]. See letter to J. D. Hooker, 4 December [1860].

2 The Druses formed a significant proportion of the population of Syria, where Hooker had recently been travelling. Harvey's allusion is to the Damascus massacre of July 1860, in which thousands of Maronite Christians were killed by Druses. The privileged province of Lebanon was ultimately constituted in 1864 as a response to the long-continued conflict.

3 The firm of Maull and Polyblank in London had taken a photograph of Hooker in 1855 for their published series entitled *Literary and scientific portrait club*.

4 Harvey described the algae collected by David Lyall during an expedition to British Columbia (Harvey 1862).

5 Harvey's study of the plants of the Cape Colony (Harvey and Sonder 1859–65), undertaken with the German botanist Otto Wilhelm Sonder, was not completed until 1865.

6 See letter to J. D. Hooker, 4 December [1860]. *Cliffortia*, a genus of the Rosaceae, is described in Harvey and Sonder 1859–65, 2: 292–304.

7 *Cliffortia* belongs to the family Rosaceae, first described by Antoine Laurent de Jussieu.

From Daniel Oliver 23 November 1860

Kew
23. XI. 60 | Friday morn[g]

My dear Sir

Dr. Hooker has given me your memorandum about the Fly-Catcher.[1] Looking again at the notice of *Asclepias* in Konig & Sims' Annals of Botany.[2] I find reference there to an account of the *Fly-catching Dogbane* (*Apocynum androsaemifolium*) by Curtis in Bot. Mag. 280.[3] This is doubtless the plant. Dr. Hooker recollects it I think at Glasgow. & I believe quite thinks yours must have been the same. Curtis says "The flowers of this *Apocynum* have a sweet honey-like fragrance, which perfumes the air to a considerable distance x x x x when a plant x x is fully blown one may always find flies caught in its blossoms, usually by the trunk, very rarely by the leg x x x —. x x x x x x the sweet viscid substance x secreted by the stigma within the antherae x x the fly endeavours to obtain, & to this end insinuates its trunk first into the lowermost & widest part of the slit, betwixt each of the antherae x x pushing it of necessity upwards: when gratified, not having the sense to place itself in the same position, as that in wh. it stood when it inserted its trunk & to draw it out in the same direction downwards, unfortunately for it, it varies its position, & pulling its trunk upwards, draws it into the narrow part of the slit, where it becomes closely wedged in, & the more it pulls the more securely it is caught," &c &c

See also Darwin (!) *Botanic Garden*, Supplement. *p*—?.[4]

Shall I copy you *all* Curtis' account? We must make a point of looking at this *Apocynum*.

Yours very faithfully | Dan! Oliver

DAR 157a

[1] Letter to Daniel Oliver, [21 November 1860].

[2] The reference was written in the margin of the letter. Oliver refers to an item in the 'Miscellaneous articles' section of *Annals of Botany* 1 (1805): 398–9, edited by Karl Dietrich Eberhard Koenig and John Sims. The article is an abstract of Barton 1806, in which *Asclepias syrica*, the fly-catcher, is described. The abstract also mentions *Apocynum androsaemifolium*.

[3] *A. androsaemifolium* was figured in William Curtis's *Botanical Magazine* 8 (1794): tab. 280.

[4] CD's grandfather, Erasmus Darwin, provided a plate of *A. androsaemifolium* in *The loves of the plants* (E. Darwin 1789–91, pt 2, facing p. 182). He described plants of this species growing in the garden of his brother, Robert Waring Darwin of Elston, Nottinghamshire. Oliver's reference to the 'Supplement' is explained by the plate and description being in a section of the book entitled 'Additional notes'.

To Charles Lyell 24 November [1860]

Down Bromley Kent
Nov. 24th

My dear Lyell

I thank you much for your letter.[1] I had got to take pleasure in thinking how I could best snub my Reviewers; but I was determined in any case to follow your advice, & before I had got to the end of your letter I was convinced of the wisdom of your advice.[2] What an advantage it is to me to have such friend as you.— I shall follow every hint in your letter exactly.—

I have just heard from Murray; he says he sold 700 copies at his sale, & that he has not half the number to supply; so that I must begin at once.[3] But I will & must finish my Drosera M.S. which will take me a week, for at this present moment I care more about Drosera than the origin of all the species in the world. But I will not publish on Drosera till next year, for I am frightened & astounded at my results.—[4] I declare it is a certain fact, that one organ is so sensitive to touch that a weight of $\frac{1}{78,000}$ of a grain (ie seventy-eight times less weight than that, viz $\frac{1}{1000}$ of a grain, which will move the best chemical balance) suffices to cause conspicuous movement.— Is it not curious that a plant sh^{d} be far more sensitive to a touch than any nerve in the human body! Yet I am perfectly sure that this is true.— When I am on my hobby-horse, I never can resist telling my friends, how well my hobby goes, so you must forgive the rider.—

Farewell my wisest & best of Lord Chancellors.

Yours most truly obliged | C. Darwin

Etty goes on pretty well. All the Doctors say any rapid progress is impossible.—

All this dreadful illness for last six months (& that wicked dear little Drosera) has made any progress in my larger Book almost nothing—

P.S. I must tell you one little fact which has pleased me. You may remember that I adduce Electrical Organs of Fish, as one of the greatest difficulties which had occurred to me, & Owen notices the passage in a singularly disingenous spirit.[5] Well McDonnell of Dublin (first rate man) writes to me that he felt the difficulty of whole case as overwhelming against me.[6] Not only are the fishes which have electric organs very remote in scale; but the organ is near Head in some & near tail in others & supplied by wholly different nerves.— It seems impossible that there could be any transition.

Some friend who is much opposed to me seems to have crowed over McDonnell, who reports that he said to himself that if Darwin is right there must be homologous organs both near the Head & Tail in other non-electric fish. He set to work & by Jove he has found them. So that some of difficulty is removed, & is it not satisfactory that my hypothetical notions sh^{d} have lead to pretty discovery. McDonnell seems very cautious; he says years must pass before he will venture to call himself a believer in my doctrine; but that on the subjects which he knows well viz morphology & embryology my views accord well & throw light on whole subject.—[7]

Endorsement: '1860'
American Philosophical Society (234)

[1] The letter has not been found.

[2] See letter to Charles Lyell, 20 November [1860].

[3] CD refers to the new, third edition of *Origin*, which John Murray had called for. See letters to T. H. Huxley, 22 November [1860], and to John Murray, 22 [November 1860].

[4] In fact, CD delayed publication until 1875, when *Insectivorous plants* appeared. He did, however, read

a paper on the subject at a meeting of the Philosophical Club of the Royal Society in February 1861 (Bonney 1919, p. 154). See also letter to Daniel Oliver, 16 November [1860].

[5] The passage was rather in Samuel Wilberforce's anonymous review of *Origin* ([Wilberforce] 1860, p. 246). Wilberforce had been primed on the scientific objections to CD's views by Richard Owen. The passage reads, in part: 'We see no possible solution on the Darwinian theory for the presence at once so marked and so exceptional of these organs.' Wilberforce went on to state that CD's confession of ignorance was 'a solution which could of course equally make the scheme it is intended to serve compatible with any other contradiction.' There is an annotated copy of [Wilberforce] 1860 in the Darwin Pamphlet Collection–CUL.

[6] The letter from Robert M'Donnell has not been found, but see the letters to T. H. Huxley, 16 November [1860] and 22 November [1860].

[7] M'Donnell described his discoveries in M'Donnell 1860 and 1861.

From Charles Lyell 24 November 1860

Nov. 24 1860.

My dear Darwin,

In former editions in speculating on age of Uddevalla & still more elevated beds of glacial period in Norway & Sweden, I have assumed an average rate of rise as $2\frac{1}{2}$ feet in a century. see Manual. p. 119.—[1] 125 feet take 5000 years to rise or to sink. I wish to keep to my old standard or estimate unless shown to be objectionable as applied to countries like Greenland & Sweden in full movement. a mere guess at an average rate. at p. 120 I give 27,500 years for 700 ft of upheaval in recent times. But I expressly guard myself in same page by saying I make no allowance for pauses or for oscillations of level (of the minor kind.)

What I want to do in the new edition is to make another conjecture & to allow for the pauses.[2] When the amount is reversed as if (which I believe with Trimmer Ramsay Jamieson & others) Scotland & Wales have moved first down & then up again 2200 feet or more,[3] then I conceive the chief pauses would be before the downward was converted into an upward movement. Successive sea-beaches & terraces & inland cliffs mark long pauses. Smith of Jordanhill thinks he has a cliff which took more than 100,000 years to cut, subsequently to the glacial re-elevation in Scotland.[4]

Now I propose to conjecture keeping on the safe side & not exposing myself to the charge of exaggerating the probable time, that 4 parts of Europe are stationary for one which moves at the rate of $2\frac{1}{2}$ feet in a hundred years, or that the stationary area exceeds that in motion as 4 to 1. & 4 expresses the period of rest or the pauses where 1 in a given region expresses the movement. Therefore if Scotland has first gone down 2250 feet after the period of land glaciers, as I believe this would take 90,000 years & the re-elevation rather more because it went up higher than it's present level, say 100,000 years, but then I must give 400,000 additional (in round numbers) for the intervening periods of rest.

Thus the last oscillation is about 50 feet in Scotland in each direction, since the 2 vast movements of 2250 down & up. But when we try to estimate the time required for the 2 movements we encounter, as Smith of Jordanhill observes, 2000 years since the Romans built the Pictish wall[5] & find that we do not get back to the close of the last oscillation.

Then I regard this 2000 years (and we know not how many 1000 before) as a part of the great excess of stationary condition.

I also consider the successive beaches cut at Glasgow in the canoe-bearing sands all before the era of the Roman Wall as portions of the same excess of 4 to 1.[6]

I am aware that the North Cape moves, or is said to move, 5 or 6 feet in a century & D Forbes thinks that Chili has gone up from 40 to 60 feet in 350 years, but then north of Arica, he thinks there has been a subsiding area, & about Arica, a stationary one as proved by Indian tumuli & mummies buried near the old shore level & elevated since.[7]

But I look on S. America as rather exceptional. It would be balanced by other more inactive regions of oscillation. The average in Scandinavia would not I think exceed $2\frac{1}{2}$ feet.[8] I am however much more in doubt as to the comparative areas of rest, contrasted with those of movement. I once made them as 9 to 1. If I could stimulate the geographers to make objections it would do good.

Perhaps the contemporaneous inland oscillations, especially in mountain chains, may be greater than the sea-coast ones. But we may leave those speculations out for the present.

There may be scarcely any areas in an absolute state of rest but we cannot yet take account of minute quantities. I wish you to reflect on my principle, & if possible say if 4. to 1. is a reasonable conjecture.

Large plateaux of denudation & inland sea cliffs are monuments of immense pauses.

1st.	period of rest of unknown duration before the extreme cold & glaciers of Scotland	0–? years
2dly.	Period of gradual submergence 2250 feet.	90,000
3dly	— of reelevation 2500 feet	100 000
4thly	Intercalated pauses up to the close of period of reelevation	400,000
		590,000

During this period the whole of the glacial period & the present establishment of provinces of species has occurred—the mammalian fauna greatly changed, but the shells very little[9]

The last oscillation of Scotland of about 50 feet in each direction wd take 4000 years & for pauses 16.000 more which would give 20,000 years. The Glasgow canoes & polished celts come into this brief era, but the Somme valley flint implements probably into some part of the great period of reelevation—during which in Scotland the erratic boulder clay was getting denuded & my Forfar gravel beds manufactured.

Chas Lyell

Copy
Kinnordy MS, Charles Lyell's journal VII, pp. 40–8

[1] C. Lyell 1855, p. 119.

[2] The new edition of Lyell's *Manual of elementary geology*, retitled *Elements of geology*, was published in 1865 (C. Lyell 1865). CD's copy of the work is in the Darwin Library–Down. Although Lyell alluded to these points in his new edition, they were in fact discussed in greater detail in his book on the antiquity of man (C. Lyell 1863, chaps. 3, 13, and 14).

[3] Trimmer 1853, Ramsay 1852, and Jamieson 1860. See letter from Charles Lyell, [before 20 November 1860].

[4] James Smith, commonly known as Smith of Jordanhill (Scotland), wrote to Lyell about this point (Wilson ed. 1970, pp. 502–5). Lyell copied the information into his scientific journal immediately preceding the draft of this letter to CD. Smith had established himself as an acknowledged expert on the elevation of Scottish coastlines with the publication of an influential paper on the subject in 1836 (J. Smith 1836).

[5] The 'Picts' wall' (*OED*) was the common name of the great wall built by the Romans between AD 120 and 130 (*EB* 1970); it is now commonly known as Hadrian's Wall. Running through Cumberland and Northumberland, it was constructed to help defend Roman Britain from the Picts and the Scots.

[6] The discovery of ancient canoes buried in the silt around Glasgow was described by John Buchanan in 1855 and discussed by Lyell in C. Lyell 1863, pp. 48–9. Lyell thought some of them might date from the Bronze Age.

[7] David Forbes, recently returned from a geological expedition to Bolivia and Peru, read a paper on the geology of the area at a meeting of the Geological Society on 21 November 1860. The figures given by Lyell in the letter were probably ascertained in conversation with Forbes. Forbes's results were published in D. Forbes 1861. The geology of the area near the town of Arica, in Peru, is discussed in D. Forbes 1861, p. 11.

[8] This figure is repeated in C. Lyell 1863, p. 58.

[9] Lyell discussed the point in C. Lyell 1865, pp. 158–60.

To Charles Lyell 25 November [1860]

Down Bromley Kent
Nov. 25th

My dear Lyell

I have endeavoured to think over your discussion, but not with much success.— I, of course, see that if you can make out your argument satisfactorily to yourself & others it would be *most* valuable.— You will have to lay down, I think, very clearly on what foundation you argue from 4 parts (which seems to me exceedingly moderate on your part) of Europe being now at rest, with 1 part undergoing movement, how it is that from this that you can argue that the one part which is now moving will have rested since the commencement of Glacial period in the proportion of four to one— I do not pretend to see with any clearness, but does not your argument rest on the assumption, that within a given period, say two or 3 million years, that the whole of Europe necessarily has to undergo movement.—

This may be probable or not so, but it seems to me that you must explain the foundation of your argument from space to time,—which at first to me was very far from obvious.

I can imagine some one saying that it is not fair to argue that the great plains of Europe & the mountainous districts of Scotland & Wales have been at all subjected to same laws of movement.

Looking to the whole world it has been my opinion, from the very size of the continents & oceans, & especially from the enormous ranges of so many mountain-chains (resulting from cracks which follow from vast areas of elevation as Hopkins argues)[1] & from other reasons, it has been my opinion that as a general rule *very large* portions of the world have been simultaneously affected by elevation or subsidence.— I can see that this does not apply so strongly to broken Europe any more that to the Malay Archipelago. Yet had I been asked, I sh^{d} have said that probably nearly the whole of Europe was subjected, during the Glacial period, to periods of elevation & of subsidence. It does not seem to me so certain that the kinds of *partial* movements which we now see going on show us the kind of movement which Europe has been subjected to since the commencement of Glacial period.—

These notions are at least possible, & would they not vitiate your argument? Do you not rest on belief that as Scandinavia & some few other parts are now rising, & a few others sinking, & the remainder at rest; that so it has been since the commencement of the Glacial period.— With my notions, I sh^{d} require this to be made pretty probable before I could put much confidence on your calculations— You have probably thought this all over; but I give you the reflexions which come across me, supposing for the moment, that you took the proportion of space at rest & in movement, as as plainly applicable to time.— I have no doubt that you have sufficient evidence that at the commencement of the Glacial period the land in Scotland, Wales &c stood as high or higher than at present; but I forget the proofs.—

Having burnt my own fingers so consumedly with the Wealden,[2] I am fearful for you; but I well know how infinitely more cautious, prudent & far-seeing you are than I am.— But for Heaven-sake take care of your fingers; to burn them severely, as I have done, is very unpleasant.—

Your $2\frac{1}{2}$ feet per century of elevation seems a *very* handsome allowance. Can D. Forbes really show that great elevation of Chile: I am astounded at it, & I took some pains on point.—[3]

I do not pretend to say that you may not be right to judge of the past movements of Europe by those now & recently going on, yet it somehow grates against my judgment,—perhaps only against my prejudices.—

My dear Lyell | Yours affec^{t} | C. Darwin

P.S. | I do not know whether I have made clear, what I think probable or at least possible; viz that the greater part of Europe has at times been elevated in some degree equably; at other times it has all subsided equally; & at other times might all have been stationary; & at other times it has been subjected to various unequal movements, up & down, as at present.— As a change from elevation to subsidence implies some great subterranean or cosmical change, one may surely calculate on long interval of rest between. Though if cause of change be ever proved to be astronomical even this might be doubtful.

Endorsement: '1860'
American Philosophical Society (235)

[1] William Hopkins published several important memoirs on physical geology that included discussions of this topic. CD probably refers to Hopkins 1838 and 1847. The point was also mentioned in Hopkins's presidential address to the Geological Society in 1852 (Hopkins 1852, p. lxviii).

[2] See letter to Charles Lyell, 20 November [1860] and n. 17.

[3] See preceding letter and n. 7. In his paper (D. Forbes 1861), David Forbes did not give a precise figure for the suggested elevation of Chile. CD refers to his own geological study of the coastline of Chile, 'Observations of proofs of recent elevation on the coast of Chili' (*Collected papers* 1: 41–3).

To Asa Gray 26 November [1860][1]

Down Bromley Kent.
Nov. 26th

My dear Gray.

I have to thank you for two letters.[2] The latter with corrections, written before you received my letter asking for an American Reprint & saying that it was hopeless to print your Reviews as a pamphlet, owing to impossibility of getting pamphlets known.—[3] I am very glad to say that the August or second Atlantic Article has been reprinted in Annals & Mag of N. History;[4] but I have not yet seen it there.

Yesterday I read over with care the third Article;[5] & it seems to me, as before, *admirable*. But I grieve to say that I cannot honestly go as far as you do about Design. I am conscious that I am in an utterly hopeless muddle. I cannot think that the world, as we see it, is the result of chance; & yet I cannot look at each separate thing as the result of Design.— To take a crucial example, you lead me to infer (p. 414) that you believe "that variation has been led along certain beneficial lines".—[6] I cannot believe this; & I think you would have to believe, that the tail of the Fan-tail was led to vary in the number & direction of its feathers in order to gratify the caprice of a few men. Yet if the fan-tail had been a wild bird & had used its abnormal tail for some special end, as to sail before the wind, unlike other birds, everyone would have said what beautiful & designed adaptation. Again I say I am, & shall ever remain, in a hopeless muddle.—

Thank you much for Bowen's 4to Review.[7] The coolness with which he makes all animals to be destitute of reason is simply absurd. It is monstrous at p. 103, that he should argue against the possibility of accumulative variation & actually leaves out entirely Selection![8] The chance that an improved Short-Horn, or improved Pouter-pigeon should be produced by accumulative variation, without Man's selection is as almost infinity to nothing; so with natural species without natural selection. How capitally in the Atlantic, you show that Geology & Astronomy are according to Bowen Metaphysics;[9] but he leaves out this rubbish in the 4to. Memoir.—

I have not much to tell you about my Book.— I have just heard that Dubois Reymond agrees with me.[10] The sale of my Book goes on well, & the multitude of Reviews has not stopped the sale. Murray sold a few days ago at his sale 700 copies; & he has not half; so I must begin at once on new corrected Edition.— I will send you a copy; for chance of your ever rereading; but good Heavens how sick you must be of it.—

Hooker has returned "rosy fat & jolly", I am glad to say; but I have not seen him & not heard much news; except that he found traces of Glacial action on Lebanon.—[11] I have gone on working at Drosera, but shall not publish till next summer, as I am frightened at my results & must retest them; (By the way I have been rereading in consequence some part of your Lesson in Botany,[12] & have been so much pleased with the extremely clear way you put things), but you may rely on the truth of the fact that the *prolonged* weight of an innutritious atom, placed with all care on *one* of the glands, though it weighed only $\frac{1}{78,000}$ of one grain caused conspicuous movement. I got the weight by weighing a length of fine hair & cutting off atoms & measuring them with micrometer. This weight is 78 times less than that with which best balance will turn with; & yet you may rely on it, this suffices to start the movement. Moreover it produces such changes within the cells of the glandular hairs; that an hour after weight had been put on, I could distinguish which hair had carried this fairy weight from all the other 100 & more hairs on the leaf.—

I suppose in summer you take walks in the country: I see in your Flora, you say that Apocynum androsæmifolium is common & another species.[13] Will you observe whether the flowers of both species catch numbers of flies by their probosces, as the former does in England. *And whether Bees visit the flowers.* I mean to get this plant, if I can, & observe it; as a Boy I was surprised at number of flies captured.[14] Please make a memorandum about this plant & **the Spiranthes**.—[15]

My daughter improves very slowly & I have now the heart to work again nearly as hard (strictly as lightly) as I ever can.—

My dear Gray | Ever yours most truly | C. Darwin

Have you read Olmsted Journey in the Back Country; what a remarkably interesting Book.—[16]

Gray Herbarium of Harvard University (27)

[1] Dated by the reference to reprinting the second part of [Gray] 1860b (see n. 4, below).

[2] Neither of Gray's letters has been found.

[3] Letter to Asa Gray, 31 October [1860].

[4] CD refers to the second part of Gray's review of *Origin* published in the *Atlantic Monthly* ([Gray] 1860b, pp. 229–39). On CD's recommendation, it was published in the *Annals and Magazine of Natural History* 3d ser. 6 (1860): 373–86. See letter to Asa Gray, 26 September [1860].

[5] CD received the third part of the article ([Gray] 1860b, pp. 406–25) in September. See letter to Asa Gray, 26 September [1860].

[6] CD annotated this passage ([Gray] 1860b, p. 414) in his copy of the paper (Darwin Pamphlet Collection–CUL): 'Look at variation in domestic productions leading to all sorts of mutations. Fantail Pouter'.

[7] Bowen 1860b. There is an annotated copy in the Darwin Pamphlet Collection–CUL.

[8] Francis Bowen stated that CD's theory was unphilosophic in its dependence on accidental variations. In his copy of the paper, CD wrote (Bowen 1860b, p. 103): '*Without selection* it is fatal.' At the top of the page, he added: 'The chance of an improved Short-Horn being produced by accumulative variation, without man's selection, is as nothing, so with species *not* acted on by natural selection'.

[9] The third part of [Gray] 1860b, pp. 406–25, included a critique of an earlier review of *Origin* by Bowen ([Bowen] 1860a). The point mentioned by CD is in [Gray] 1860b, p. 419.

[10] Emil Heinrich Du Bois-Reymond. According to the letter to T. H. Huxley, 16 November [1860], CD had learned about Du Bois-Reymond's favourable opinion from Leonard Horner.

[11] CD learned of the find in a letter from Joseph Dalton Hooker passed on to CD by Thomas Henry Huxley (see letter to T. H. Huxley, 1 November [1860]). CD added a sentence to the fourth edition of *Origin* describing Hooker's assessment of evidence of glaciation in the Lebanon mountains (Peckham ed. 1959, pp. 591–2). See also letter to T. H. Huxley, 1 November [1860] and n. 4.

[12] Gray 1857. CD's copy is in the Darwin Library–CUL.

[13] Gray 1856, p. 350.

[14] CD remembered the plant growing in the garden of his father, Robert Waring Darwin; he had ascertained its name by questioning Daniel Oliver. See letters to Daniel Oliver, 16 November [1860] and [21 November 1860], and letter from Daniel Oliver, 23 November 1860.

[15] CD had sent Gray a memorandum outlining observations he hoped Gray would make on the American species of the orchid genus *Spiranthes*. See letter to Asa Gray, 31 October [1860] and the enclosure.

[16] CD had read Frederick Law Olmsted's two previous books about his tours through the slave states of America with great interest, remarking that Olmsted 1856 was 'excellent' (*Correspondence* vol. 4, Appendix IV, 128: 23, 25). Olmsted's third book, *A journey in the back country* (Olmsted 1860), chronicled Olmsted's travels from New Orleans to Richmond, Virginia. Many believed it to be an unbiased and accurate picture of conditions in the South. Gray knew Olmsted, having met him in England in 1850 (Dupree 1959, p. 192).

To J. D. Hooker 26 November [1860]

Down Bromley Kent
Nov. 26th

My dear Hooker

Murray writes to me that he sold at his sale some 300 or 400 more copies of the Origin than he has; & that I must *at once* prepare a new Edition, which I intend to correct as far as I can. In beginning to look over my notes I find one (ominous to you) "Hooker has some criticisms for me".— Now I know perfectly well that you have no time to spare; but perhaps you have scored or scribbled notes on margin of your copy; if so would you lend it me, & send it by enclosed address.— If not, would you just glance over the pages to see if you can remember any *important* corrections & not of course trouble yourself about trifles.—

I fear that even this will be bothersome to you at present; but I thought it a pity not just to mention it.—

Will you thank Oliver much for having hunted up so capitally the case of Apocynum androsæmifolium:[1] & ask him whether it is possible to buy plants anywhere;

for I presume you have it not at Kew.— I think you should have it, for I remember my Fathers plants surprised beholders.—

Yours affect | C. Darwin

P.S. I have sent to Linn. Soc. to know if Meyen's Ueber die Secretions organe die Pflanzen: is in Library;[2] if it be not, & you have it, will you lend it me for a very short time, as I want much to consult it on one point about my beloved Drosera.—

I have just heard from Mr Kippist that they have not Meyen's work.[3] Have you?

Endorsement: '/1860'
DAR 115.1: 76

[1] See letter from Daniel Oliver, 23 November 1860.
[2] Meyen 1837.
[3] Richard Kippist was the librarian of the Linnean Society of London.

From J. D. Hooker [26 November – 4 December 1860][1]

Royal Gardens Kew

in solution will excite electric &c currents & these again would affect vitality— be this as it may, I cannot but suppose, that if your observations are good they will lead to most important discoveries in Veg. physiology. I saw at Paris[2] a sensitive grass! of which they promised me a live plant— it's leaves are short & distichous & these close on one another at night; as do those of the sensitive plants. Have you chloroformed your *Drosera* & then tried it?

I inspected a collection of New Caledonian plants at the Jardin de Plantes that altogether confirms my view of the very close affinity of that Flora with the Australian & diversity from that of the Pacific Islands to the East of it.— Also more Australian types are turning up in the Phillippines & China.

Another most curious fact of sporadic distribution fell in my way at Paris; when inspecting a small collection of Aden plants I found one that is *identical* with a Namaqua-land[3] plant! & is found no where else hitherto, it is not even found in the Eastern districts of South Africa— What is still more remarkable is that it belongs to a Natural Order *Loaseæ* which is otherwise wholly & exclusively American. Another & parallel case is that of *Cacteæ* , the only Extra American species of which is found only in Tropical Africa & Ceylon. Another is the *Papyrus* which is found abundantly in Syria, in one spot in Sicily & nowhere else north of the Tropics!

I paid particular attention to your query about the sudden appearance of plants on ascending Lebanon[4] & made a good many observations to the effect that the more remarkable **forms** especially generic do appear very suddenly in great quantities, & am inclined to believe that the lower limit of these is far better defined than the upper limit of those that disappeared.— This applies to *Astragalus*, *Acantholimon*, *Vicia*, & several other plants which are characteristic of the dry soil & climate that prevails above 7000 ft. but not to other plants which are equally peculiar to the elevation but which depend on some little moisture—as *Potentilla*, &c. The Vegetation above 8000 ft was extremely scanty, & I found but one Alpine or Arctic plant (*Oxyria reniformis*) & that was close to the tip-top & very rare. This absence of Alpine plants on the Mts of Asia-minor is a very characteristic feature, & is shared, I am assured, by the Mts of Algeria— Under these Circumstances the presence of so very marked an artic plant as Oxyria is very interesting— it seems to say that an expulsion of other Arctics must have taken place, & the drought would effet this well enough. The Cedars are going owing to the same causes— Every seedling dies, there are no trees under 40–50 years old. from which age up to 500 (perhaps the oldest) there are trees of all (or many) ages.[5]

I have taken up the Genera Plantarum with Bentham in earnest[6] & am going to mend my ways now & for ever—giving up all Societies but the Linnæan, & every unnecessary excitement, keep early hours, cut off all correspondents (except those I love), with short letters—eat well, walk a good deal in the garden & avoid all occasions of sin— I have not had a headache for 3 months now!

Incomplete
DAR 100: 158–60

CD ANNOTATIONS
1.1 in solution . . . it? 1.6] *crossed pencil*
2.1 I . . . ages. 4.17] 'Limitation of Form on Mountain | Arctic Plants | Sporadic Distrib' *brown crayon*
5.1 I . . . now! 5.5] *crossed pencil*

[1] The part of the letter that is now missing apparently included a response to points raised by CD in the preceding letter. See CD's reply, 4 December [1860]. The letter is dated 'December 2 [?], 1860' in L. Huxley ed. 1918, 1: 533.
[2] Hooker passed through France on his way to and from Syria.
[3] Namaqualand was the name given to a region on the west coast of South Africa.
[4] See letter to J. D. Hooker, 6 September [1860].
[5] Hooker composed a memoir on the cedars of Lebanon that was published in 1862 (Hooker 1862b).
[6] The proposal to publish a new and complete *Genera plantarum* had been entertained by Hooker and the botanist George Bentham since 1859. They began work in March 1860 (see letter to J. D. Hooker, 12 March [1860]). The first part of the first volume was published in 1862 (Bentham and Hooker 1862–83).

To John Lubbock 28 November [1860][1]

Down Bromley Kent
Nov. 28

My dear Lubbock

Your remarks seem to me admirably good, & I am sure I sh^{d} have thought so, if I had never heard your name & whether or no they concurred with my notions.—[2] I thank you heartily & am pleased at the high compliment you pay the Origin.—[3] Such remarks as yours, (I do not mean the compliment!) I feel sure are the true way of advancing our general belief on Species; viz by showing that several subordinate questions are illustrated by our view. Such remarks as yours just make the difference whether a memoir does or does not advance Biology. And how few naturalists deal in them!

Now for a few details. Your remarks on links in classification seem to me excellent.[4] I am especially glad to see your observations on variability of secondary sexual characters:[5] this is important likewise in the practical classificatory work of naturalists.

There is misprint & omission of word in one paragraph. p. 168.—[6]

I have omitted one remark on p. 167 (7 l. from top): would it not be well to change "although" into "and"? for when one first reads sentence it seems as if you thought that it was a rule that aberrant forms sh^{d} be rich in genera & species,—which would be almost a contradiction in terms.[7] Your following *good* remarks show that this is not meant.—

With respect to domestication & variability;[8] I do not think your sentences clear; in fact I can hardly understand the last.— They want clearing, altering & amplifying.— Possibly it may aid you just to say to what conclusion I arrived on this subject.— Changes in the conditions of life (at least abrupt changes) certainly tend to induce some degree of sterility: they likewise certainly tend to cause variability in successive generations; but I could never satisfy myself, whether at same time they caused both; whether they caused *either* sterility *or* variability; or whether variability ensued after the sterility ceased. Many of our domestic productions are more fertile than the aboriginal forms in the wild state & are more variable; so that there is no necessary relation between variability & lessened fertility. I am rather inclined to view sterility & variability, as alternatives produced by changes of conditions.* But this again relates only to when a being is first subjected to a considerably change from its natural to domestic conditions.— It is an obscure question, & you had better be cautious.— Thanks for Annals & Owen[9]

I cannot find after search Millepora conplanata; but you have I think the other Millepora; & they differ only in growing in great masses.—

Ever yours | C. Darwin

P.S.* | As far as I remember all that I have said in Origin is that changed conditions of life in some inexplicable way strongly affect the generative system, as shown by the induced tendency to sterility; & therefore one need not be surprised at other & lesser changes of conditions so far affecting the reproductive system, that the offspring are produced not quite like their parents.—

In your last sentence I cannot see why species which are partially variable are most likely to lose their fertility.—

Heaven knows whether you will be able to decipher or understand my scribble—

Down House

[1] The year is given by the reference to Lubbock's paper on the Entomostraca (Lubbock 1862), which was read in June 1860 (see n. 2, below).

[2] Lubbock read a paper on the oceanic Entomostraca collected by T. Toynbee in 1858 and 1859 at a meeting of the Linnean Society of London on 7 June 1860. The paper was published in 1862 (Lubbock 1862). Lubbock appears to have sent CD a proof copy of his article. There is an annotated offprint of the final version in the Darwin Pamphlet Collection–CUL.

[3] Lubbock referred to *Origin* as an 'admirable work' (Lubbock 1862, p. 176).

[4] Lubbock discussed the classification of species that were intermediate in form. He reiterated CD's view that the absence of 'links' was not an argument for the permanence of species (Lubbock 1862, p. 174). The passage in CD's copy of the paper is heavily scored in pencil. In the third edition of *Origin*, CD quoted Lubbock on this point, stating: 'Every species is a link between other allied forms.' (*Origin* 3d ed., p. 323; Peckham ed. 1959, p. 502).

[5] Lubbock 1862, p. 176, where Lubbock stated that the Entomostraca presented 'remarkable examples' of CD's observation that secondary sexual characteristics are very variable. The passage is heavily scored in CD's copy of the paper. CD cited Lubbock on this point in the third edition of *Origin* (*Origin* 3d ed., p. 175; Peckham ed. 1959, p. 306).

[6] The final version of the paper does not have the same page numbers as the proof-sheets seen by CD.

[7] The sentence in the final version reads: 'This latter group [the Cyproidea] is at present poor both in genera and in existing species, and it is in many respects very aberrant.' (Lubbock 1862, p. 175). The passage is marked in CD's copy.

[8] The topic of domestication and variability is not discussed in the final version of the paper.

[9] CD had asked Lubbock to check the contents of the most recent issue of *Annals and Magazine of Natural History* for him (see letter to John Lubbock, [18 November 1860]). The work by Richard Owen has not been identified.

From Charles Lyell 30 November 1860

30. Nov. 1860

You must have thought so much, when on your coral-reef-areas, of subsidence & volcanic areas of elevation,[1] that I think you high authority in speculations of that kind & if you think $2\frac{1}{2}$ feet per century a good allowance (as I believe it to be & above an average)[2] & if you agree with me that between two great movements in an opposite direction there would generally be a long pause, I am satisfied; even if we have no means even of guessing at the probable duration of such pauses. Suppose the passage of a great subterranean lake of liquid lava into a crystalline rock to be the cause of secular upheaval—the heat gradually diminishing & the rock e.g. granite, expanding, it is likely that the rate of upheaval tho' it may be equable & persistent for tens of thousands of years, like the heat of thermal springs, will when on the decline, decrease very gradually & one cannot expect the reheating of the same mass or of one in the same part of the earth's crust & the melting thereof to take place just when the cooling had ceased.

The chances are that before the reversal of the movement there would be a very long interval. But you speak of astronomical causes. Did you ever read of any as yet suggested, which seemed in the least degree feasible, & one cannot doubt the connection in S. America of volcanicity & movements of upheaval & depression?

I understood, however, the force of your objection to my reasoning as to movements from space to time If I were to judge by sea-coast evidence alone I should think a tenth rather than a fifth was all that is in motion. But I cannot help suspecting that the Alps & other mountain chains move more than lower countries. The

Alps seem so often to have risen & sunk considerably. This perhaps strengthens your objection as showing that you cannot apply the rule of one area to another as certain spaces are repeatedly moved while others may continue at rest for one geological period after another. Spratt's proof of Candia or Crete nearly 100 miles long having gone down at one end—Greek cities being submerged & up at the other the old docks being upraised high & dry & in the intermediate space no movement, agrees with many facts to indicate that the Mediterranean is one of the areas of unequal movements in the last 2000 years.[3] Smith of Jordanhill shows great oscillation of rock of Gibraltar since existing Mediterranean shells[4]

On the other hand there are parts of Norway stationary where beaches with recent shells have been upheaved 700 feet. I doubt there having been periods of general-upward or general-downward movement in Europe in Glacial Period.

As to Scotland having been as high above the sea as now before the great submergence, all that I know of the Grampians, which are covered by unstratified till & old moraines, under these more modern gravels which Jamieson has correctly described at all heights lead me to the belief that Jamieson is right, p.p. 368–369, in believing that glaciers or land-ice in Scotland & Ireland acted powerfully before the land sank.[5] You may say, however, that the land need not have been as high as now, but I think it more likely that it was, as the Scotch mountains are not high enough to have had glaciers on them, unless there was a great volume of land & that would almost imply height.

I shall take your hint & be satisfied in claiming, as a minimum, 100,000 years for the downward, & as much for the upward, movement, & then give as my reason for inferring a very long stationary intervening period, that so much of Europe, even where there has been pleistocene & post-pliocene upheaval & depression, as e.g. Scotland, part of Norway & parts of Sicily, have been at rest for last 2000 years.

You will have no objection I presume to my going so far without pretending to define by centuries the length of said interval.

My idea is that subterranean upheaval, like volcanos, shifts it's theatre of action & by degrees visits every country, & if there are Silurian regions of horizontal strata, they too can be proved to have been submerged & denuded, depressed & upheaved. As active volcanos occupy a small area so did extinct ones & so do deep-seated volcanic agencies or those causing upward & downward movements. Rest, therefore, always has been as it is now, the character of the surface generally, & change of level exceptional & inland sea-cliffs, like those of Greece or the Morea, show very long pauses. The quantity of pleistocene & post-pliocene movement makes me rather doubt the continents & oceanic basins remaining as constant as you suppose through several successive geological periods— From the Miocene to our era I can imagine it But not from the Cretaceous to our times.

Copy
Kinnordy MS, Charles Lyell's journal VII, pp. 49–57.

[1] CD's *Coral reefs* was published in 1842. For his discussion with Lyell of the growth of coral islands, see *Correspondence* vol. 2.
[2] CD expressed his approval of Lyell's estimate of an elevation in the land of $2\frac{1}{2}$ feet per century in his letter to Charles Lyell, 25 November [1860]. Lyell cited CD on this point in C. Lyell 1863, p. 286.
[3] Spratt 1854. Lyell discussed Thomas Abel Brimage Spratt's observations on the geology of Crete in C. Lyell 1863, p. 178. Crete was formerly known as the island of Candia.
[4] J. Smith 1844.
[5] Jamieson 1860.

From Alfred Russel Wallace [December? 1860][1]

From *"Von Buch on the Flora of the Canaries"*.[2] "On continents the individuals of one kind of plant disperse themselves very far, and by the difference of stations of nourishment & of soil produce *varieties* which at such a distance not being crossed by other varieties and so brought back to the primitive type, become at length permanent and distinct *species*. Then if by chance in other directions they meet with an other *variety* equally changed in its march, the two are become *very distinct species* and are no longer susceptible of intermixture."

P.S. "Natural Selection" explains *almost* everything in Nature, but there is one class of phenomena I cannot bring under it,—the repetition of the forms & colours of animals in distinct groups, but the two always occurring in the same country & generally on the *very same spot*.[3] These are most striking in insects, & I am constantly meeting with fresh instances. Moths resemble butterflies of the same country— *Papilios* in the east resemble *Euplœas*, in America *Heliconias*. At Amboyna I took on the *same tree* at *the same time* two *longicorns* of distinct genera, but so alike in colour & markings that I only separated them after some days— Here also & at Macassar occurs together a *Malacoderm* & an *Elater* of exactly the same tints of metallic blue & soft orange & also similarly striate,—yet there is no affinity between them. A few days ago only I took a new & curious little *Cicindela* which so closely resembles in size & markings a *Therates* occurring with it, that I never know which it is till I take it out of my net;—yet there is no sign of a change in the structural characters which separate these *genera*. It seems to shew that colour, markings & texture of surface depend strictly on local conditions— Home Entomologists might do something in experiments on breeding insects, varying conditions of food light heat &c as much as they will bear. In domestic var.^s^ have you discovered what tends to produce white, black, or particular coloured variations? or what tends to produce spots rather than stripes.[4]

ARW.

Incomplete
DAR 45: 1

CD ANNOTATIONS
1.1 From . . . intermixture." 1.7] 'Ch 3'[5] *added brown crayon*
2.1 P.S. . . . stripes. 2.19] 'Ch IV'[6] *added brown crayon*

End of letter: 'It is like Electric organs in orders of Fishes[7] | Hemp-seed on Birds[8] | Toads-poison on Parrots feather | Colour correlated with other peculiarities' *ink*

[1] The date is suggested by Wallace's itinerary during the final years of his expedition in Malaysia. The letter clearly precedes the formulation of either Wallace's or Henry Walter Bates's views on insect mimicry and probably falls in the period between Wallace's receiving a copy of *Origin* in February 1860 and the autumn of 1861, when Bates read a paper in which he explained mimicry in terms of CD's views (Bates 1861). Wallace had collected insects from areas around Macassar on the island of Celebes in 1856 (see Wallace 1869, 1: 334 *et seq.*) and from Amboina in 1857 and 1859 (see Wallace 1869, 1: 460 *et seq.*). During 1860, having left Amboina in February, Wallace visited the islands of Ceram, Goram, and Waigiou, any of which might have yielded the beetles mentioned. When he returned to Ternate in December 1860, he was 'stupefied' by his 'year's letters, accounts, papers, magazines, and books' (Wallace 1905, 1: 373). In 1861, Wallace visited Timor and its neighbouring islands, all of which were very poor in beetles (Wallace 1869, 1: 296 *et seq.*).

[2] Buch 1825, later translated into French as Buch 1836. The passage Wallace translates is given in Buch 1825, pp. 132–3, and in Buch 1836, pp. 147–8. From the extract given here, Wallace may be responding to CD's comments in the letter to A. R. Wallace, 18 May 1860, about people who had anticipated his theory. Christian Leopold von Buch was not, however, cited in the historical sketch that CD added to the third edition of *Origin* published in April 1861.

[3] In his published account of his travels in Malaysia, Wallace described how he had noticed that some species of insects, predominantly butterflies and beetles, imitated each other in their shape and colouring. It was while he was on the island of Bourn, in the Moluccas, during May and June 1861, that he first noted mimicry among birds (Wallace 1869, 2: 150–3). At the time this letter was written, Wallace evidently did not believe that the phenomenon was explicable by natural selection. Later, after considering the case of birds and the work of Bates (Bates 1861), he came to believe that mimicry supplied powerful support for natural selection operating in nature. See Wallace 1866 and also Wallace 1905, 1: 401–2.

[4] Coloration of animals had been mentioned in previous correspondence between Wallace and CD. See *Correspondence* vol. 6, letter to A. R. Wallace, 1 May 1857, and vol. 7, letter to A. R. Wallace, 25 January [1859].

[5] The annotation refers to the third chapter in CD's 'big book' on species, 'On the possibility of all organic beings occasionally crossing' (*Natural selection*, pp. 33–91).

[6] 'Ch IV' refers to the chapter in *Origin* entitled 'Natural selection' (*Origin*, pp. 80–130).

[7] CD's comment relates to his regarding the origin of the electric organs of fishes as a particular difficulty for the theory of natural selection (*Origin*, pp. 192–3). Not only was it hard to envisage an intermediate form of electric organ, but it was also difficult to account for the organs occurring in species widely remote in their affinities, for then the organs could not be explained as a product of inheritance from a common ancestor. In *Origin*, pp. 193–4, CD suggested that the case could perhaps be explained by independent adaptations.

[8] CD cited the independently developed adaptations of some plant species to disseminate their seeds by plumes and hooks in the fourth edition of *Origin* (see Peckham ed. 1959, p. 357).

To W. H. Miller 1 December [1860][1]

Down Bromley Kent
Dec 1

Dear Miller

Some months ago you were so kind as to say you would measure the thickness of the walls of the basal and side plates of the cell of Bee.[2] Could you find time to do so soon? Why I want it soon, is that I have lately heard from Murray that he sold at his sale far more copies than he has of the Origin of Species, and that I must immediately prepare a new Edition, which I am now correcting. By the way I hear from Murray that all the attacks heaped on my book do not seem to have at all injured the sale, which will make poor dear old Sedgwick groan.[3] If the basal plates and walls do differ considerably in thickness, as they certainly did in the one or two cells which I measured without particular care (as I never thought the point of any importance) will you tell me the bearing of the fact as simply as you can, for the chance of one so stupid as I am in geometry being able to understand.

Would the greater thickness of the basal plates and of the rims of the hexagons be a good adaptation to carry the vertical weight of the cells filled with honey and supporting clusters of living Bees?

Will you endeavour to screw out time and grant me this favour.

Believe me | Dear Miller | Yours sincerely | Ch Darwin

P.S | If the result of your measurement of the thickness of the walls turns out at all what I have asserted; would it not be worth while to write a little bit of a paper on the subject of your former note; and "pluck" the Bees if they deserve this degradation.[4] Many mathematicians seem to have thought the subject worthy of attention. When the cells are full of honey and hang vertically they have to support a great weight, can the thicker basal plates be a contrivance to give strength to the whole comb, with less consumption of wax, than if all the sides of the hexagons were thickened?

This crude notion formerly crossed my mind; but of course it is beyond me even to conjecture how the case would be.—[5]

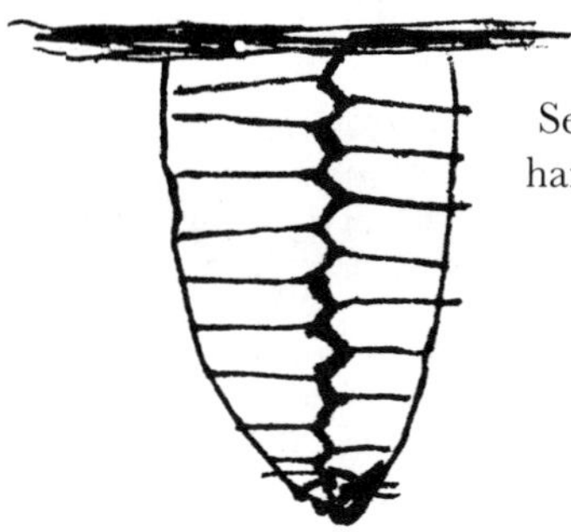

Section of comb hanging vertically

A mathematician Mr Wright has been writing on the geometry of Bees-cells in the U. States in consequence of my Book; but I can hardly understand his paper[6]

Copy
DAR 146

[1] Dated by the reference to Chauncey Wright's paper on bees' cells (see n. 6, below).

[2] Miller had written to CD about the measurements of the cells of bees earlier in the year. See letters to W. H. Miller, 5 June [1860] and [after 5 June 1860].

[3] Adam Sedgwick, a colleague of Miller's at Cambridge University, had written to CD in 1859 of his strong antipathy to *Origin*. See *Correspondence* vol. 7, letter from Adam Sedgwick, 24 November 1859.

[4] Miller's note has not been found. To 'pluck' is to give a failing grade in an examination (*OED*).

[5] In *Origin*, p. 231, CD suggested that the basal plates were about twice as thick as the cell walls. In the third edition of *Origin*, he gave more precise measurements, citing Miller as his authority. He also added (*Origin* 3d ed., p. 252): 'By the above singular manner of building, strength is continually given to the comb, with the utmost economy of wax.' See letter to W. H. Miller, 27 December [1860].
[6] Wright 1860. CD's annotated copy is in DAR 48 (2d ser.): 53–61.

To T. H. Huxley 2 December [1860]

Down Bromley Kent
Dec 2^d

My dear Huxley

I humbly beg pardon. I am getting old & dotty I suppose, for I was distinctly thinking of Von Baer & in both notes wrote von Siebold![1] The Lord have mercy on poor old noddle.— I have not read Phillips though it lies staring me in the face;[2] for I have got fairly sick of hostile Reviews. Nevertheless they have been of use in showing me where to expatiate a little & to introduce a few new discussions. *Of course*, I will send you copy of new Edition.—

I entirely agree with you that the difficulties on my notions are terrific, yet having seen what all the Reviews have to say against me, I have far more confidence in the *general* truth of the doctrine that I formerly had.— Another thing gives me confidence viz. that some who went $\frac{1}{2}$ an inch with me now go further, & some who were bitterly opposed are now less bitterly opposed. And this makes me feel a little disappointed that you are not inclined to think the general view in some slight degree more probable than you did at first. This I consider rather ominous. Otherwise I sh^d be more than contented with your degree of belief.— I can pretty plainly see that if my view is ever to be generally adopted, it will be by young men growing up & replacing the old workers, & these young ones finding that they can group facts & search out new lines of investigation better on the notion of descent, that on that of creation.—

But forgive me for running on so egotistically. Living so solitary as I do one gets to think in silly manner of one's own work.—

Ever yours very sincerely | C. Darwin

Imperial College of Science, Technology and Medicine Archives (Huxley 5: 149)

[1] See letters to T. H. Huxley, 16 November [1860] and 22 November [1860].
[2] Phillips 1860.

To John Murray 3 December [1860]

Down Bromley Kent
Dec. 3^d

My dear Mr Murray

In the copy of Origin sent me for correction p. 31, 32, 41, & 42 are missing.— if you can find them please send them, as it will be a bore copying them in M.S.; but it will not, of course, be worth cutting up a new copy.—

You did not answer my question in how great a hurry you are for the new Edition?—[1] Shall I send the first **half** of sheets when ready. The corrections will enlarge volume by a dozen or so pages; there will also be a short Historical Sketch.[2] I will give list of principle additions & corrections—[3] I sh^{d} like words "with Corrections" on Title.—[4]

It will perhaps save you another note; if you will tell me to what printers I shall send sheets? Also may I direct printers to take *especial* & extra care of merely typical errors so that I may look only to sense; it would save me much time, & I am much pressed for time.

The Printers managed new paging of Index for last Reprint, & I must request same assistance this time.

My dear Mr Murray | Yours very sincerely | C. Darwin

I have just found an old Proof Sheet which will serve.—[5]

John Murray (publishers)
Endorsement: 'Dec 3rd 1860'

[1] See letter to John Murray, 22 [November 1860].

[2] The sketch had been composed for the revised American edition of *Origin* and also included in the German translation (Bronn trans. 1860). See Appendix IV.

[3] The third English edition gives a list of 'Additions and corrections, to the second and third editions' (*Origin* 3d ed., pp. xi–xii; Peckham ed. 1959, pp. 51–7).

[4] The title page of the new edition included the words: 'Third edition, with additions and corrections.'

[5] The final sentence was written in the margin next to the first paragraph, which was then deleted by CD.

To Jeffries Wyman 3 December [1860]

Down Bromley Kent
Dec 3^{d}

My dear Sir

I am very much obliged to you for your kindness in writing to me & for sending me the pamphlets.[1] If you can at any time find out about the relative size of the eyes of the Cave Rat, it would be a very great favour.—[2]

You cannot tell how much your paper on Gestation has interested me.[3] How facts do beat the wildest imagination! The case of the frogs has interested me particularly, because some time since I read up what I could find on subject, thinking it a good case against my notions, from absence of transitional states; but now your case is something on the road. But the unfortunate male fish with the load of eggs in their mouths exceeds everything;[4] & what a curious fact about the foreign eggs in the mouths of some.[5] It is quite a pity that there are not fish of the same group with cuckoo-like habits; your fact would so well have explained how the habit might have arisen.— Speaking of transitions Mr MacDonnell of Dublin writes to me that he has made some curious discoveries on the electrical organs of the Rays,[6] being led thereto by trying to make the case of the Electrical organs, already so very difficult to me, still more unpleasant; but as I understand him his new facts help my views considerably.—

I am very glad to hear that you are collecting facts on the "Bull-dog" fish: I suppose & hope you will bring in about the Nâta cattle.[7] I am now at work in bringing out a Corrected Edition of the "Origin" & I will do myself the great pleasure of sending you a copy when it is published: I could not resist giving briefly your Hog case.[8] By the way I have received another analogous fact in case of Horses.—[9]

I once saw several years ago Dr Ackland, & was charmed with him: I am pleased to hear that he defended me against so redoubtable an opposer as Agassiz.[10] I was rather surprised that the latter did not attack me with more skill.—

Will you kindly forward the enclosed to Prof. Silliman *by any opportunity*: the note is merely to thank him for so kindly writing & giving me information.—[11]

Pray believe me, My dear Sir, with sincere thanks & respect | Yours very truly. | C. Darwin

If you ever write again to me, can you refer me to any paper on the Rattle of Rattle-snake; I want to see some account of the tip of tail in the young before the skin has ever been moulted; & the relation of the rattle in the old to the act of moulting. I want to see what relation there is to the tail ⟨of⟩ Trigonocephalus, the habits of which I mention in my Journal as Vibrating its tail.—[12]

Endorsement: '[1860]'
Jeffries Wyman Jr

[1] Wyman may have sent CD copies of his articles on the chimpanzee (Wyman 1849) and on unusual modes of gestation (Wyman 1859). Annotated copies of these, inscribed by Wyman, are in the Darwin Pamphlet Collection–CUL.

[2] See letter to Jeffries Wyman, 3 October [1860].

[3] In Wyman 1859 Wyman discussed the different ways in which some species of frogs and fish brood their eggs.

[4] Wyman 1859, pp. 11–13. Here, Wyman described how in some species of the family Siluridae in Guiana the males stored eggs and young in their mouths. The passage discussing the point mentioned by CD is heavily scored in CD's copy.

[5] Wyman 1859, p. 13. CD marked the passage in his copy and added the comment: 'What an odd parasitic Habit this might give rise to—'.

[6] The letter from Robert M'Donnell has not been found. See letter to T. H. Huxley, 22 November [1860].

[7] See letter to Jeffries Wyman, 3 October [1860].

[8] See letter to Asa Gray, 3 April [1860], and letter from Jeffries Wyman, [*c.* 15] September 1860.

[9] See letter to J. M. Rodwell, 5 November [1860] and n. 9.

[10] Henry Wentworth Acland, Regius professor of medicine at Oxford, accompanied Prince Edward as his physician during a royal visit to Canada and the United States in the summer of 1860. The prince and his entourage visited Harvard University in October. It was presumably during an occasion on this visit that the criticism of Louis Agassiz was offered.

[11] See letter to Benjamin Silliman Jr, 4 December [1860]. See also letter from Jeffries Wyman, [*c.* 15] September 1860.

[12] See *Journal of researches*, p. 114. Wyman responded to CD's inquiry in the letter from Jeffries Wyman, 8 January [1861] (*Correspondence* vol. 9).

To J. D. Hooker 4 December [1860]

Down Bromley Kent
Dec. 4th

My dear Hooker

What a King of good men you are to write me so long a letter so brimfull of most interesting matter.[1] Thank you heartily for your sympathy (which my wife read) about poor Etty, who keeps in same state; but the Doctors do not despair; I almost do.—

Thank you much for remarks about sudden changes: I have not been cautious enough; but I formerly worked the subject, & in animal kingdom could get go[2] sufficient evidence nor could I get any in Aug. Geoffroy, or Moquin-Tandon which satisfied me.[3] I have already put a sentence or two in the new Edition, & I will now make them still more cautious.—[4] I had thought of the cases of plants which *habitually* bear two kinds of flowers, but I had no means of conjecturing how they were acquired.[5] I **entirely** agree in what you say about convergence, & have already put few sentences on this subject.—[6] I am thus adding little discussions on almost every point which has been handled by Reviewers. I will of course send you a copy, & I shall publish table of additions, so anyone who cares can easily see the more important additions.—

Thanks for Harvey's letter:[7] I will keep it, as *hereafter* it may be very useful, unless I hear you want it back again. What truly wonderful, most uncomfortably wonderful, facts you tell me on Distribution. I am glad you attended to manner in which plants appeared & disappeared; it seems to me in many ways important. Many thanks about Apocynum & Meyen.—[8] The latter I want about some strange movements in cells of Drosera, which Meyen alone seems to have observed. It is very curious but Trècul disbelieves that Drosera really clasps flies![9] I sh^{d} very much wish to talk over Drosera with you. I did chloroform it & the leaves which were already expanded did not recover 30" of exposure for 3 days.—

I used expression weight, for the bit of Hair which caused movement & weighed $\frac{1}{78,000}$ of a grain; but I do not believe it is *weight*, & what it is I cannot, after many experiments, conjecture. The movement in this case does **not** depend on chemical nature of substance. Latterly I have tried experiments on single glands, & a microscopical atom of raw meat causes such rapid movement that I could see it move, like hand of clock. In this case it is nature of object. It is wonderful, the rapidity of the absorption; in 10" weak solution of C. of Ammia changes not colour, but state of contents within the glands. In 2′ 30′ juice of meat has been absorbed by gland & passed from cell to cell all down the pedicel (or hair) of the gland, & caused the sap to pass from the cells on the upper side of the pedicel to the lower side & this causes the curvature of the pedicel. I shall work away next summer, when Drosera opens again for I am much interested in subject.— *After* the glandular Hairs have curved, the oddest changes take place, viz a segregation of the homogeneous pink fluid, & never-ceasing slow movements in the thicker matter. By Jove I sometimes think Drosera is a disguised animal! You know that

I always so like telling you what I do, that you must forgive me scribbling on my beloved Drosera.— Farewell.—

I am so very glad that you are going to reform your ways. I am sure that you would have injured your health seriously. There is poor Dana has done actually nothing, cannot even write a letter, for a year, & it is hoped that in another *year* he may quite recover.—[10]

After this homily good night My dear friend.— Good Heavens I ought not to thank you, but scold you, for writing me so long & interesting a letter

Farewell | C. Darwin

I believe I shall have to go soon for some Water-cure, I cannot sleep & my heart is almost always palpitating.—

P.S. I have been looking at Harvey's case again: it is really very pretty, as case of compensation & analogous variation. When you write urge him, *as an honest man*, to publish it with figures— During your absence I had immense correspondence with him.[11] He is not nearly so savage as he was. If he works in this way he will be a convert to us. Many of his objections were *very* ingenious, but several of them, I think, not well founded.— He sent me his foolish pamphet marked "with repentance". Was this not nice of him.?—[12]

If you have any suggestion on first half of Origin, please let me hear soon, as I must go to press immediately. Poor wretch you will soon remember that I did write a Book on Species.—

P.S. I have just heard from Murray that I must get on with proofs as quickly as I can[13]

Endorsement: '/1860.'
DAR 115.1: 78

[1] Letter from J. D. Hooker, [26 November – 4 December 1860].

[2] CD obviously meant to write 'no'.

[3] CD refers to Saint-Hilaire 1841 and Moquin-Tandon 1841. He read these works on teratology in 1846 and discussed them with Hooker. See *Correspondence* vol. 3, letter to J. D. Hooker, [13 March 1846].

[4] 'It may perhaps be doubted whether monstrosities, or such sudden and great deviations of structure as we occasionally see in our domestic productions, more especially with plants, are ever permanently propagated in a state of nature.' (*Origin* 3d ed., p. 46; Peckham ed. 1959, p. 121).

[5] CD discussed this point in *Origin* 3d ed., p. 46.

[6] CD discussed the problem of 'convergence' with the botanist Hewett Cottrell Watson at the beginning of the year (see letter from H. C. Watson, [3? January 1860], and letter to H. C. Watson, [5–12 January 1860]. He added a section on the topic to the revised American edition of *Origin* (see Appendix IV) and to the third edition of *Origin* (*Origin* 3d ed., pp. 141–3; Peckham ed. 1959, pp. 267–70).

[7] Letter from W. H. Harvey to J. D. Hooker, 23 November [1860].

[8] Meyen 1837. See letter to J. D. Hooker, 26 November [1860].

[9] CD made the same statement in *Insectivorous plants*, p. 1 n.: 'M. Trécul went so far as to doubt whether they [the leaves] possessed any power of movement.' He refers to Trécul 1855.

[10] See letter to J. D. Dana, 30 July [1860], letter from Benjamin Silliman Jr, 27 October 1860, and *Correspondence* vol. 7, letter to J. D. Dana, 30 December [1859].

[11] See letter from W. H. Harvey, 24 August 1860, and letter to W. H. Harvey, [20–4 September 1860].
[12] Harvey 1860. See letter from W. H. Harvey, 8 October 1860.
[13] The second postscript was added to the first in pencil.

To Charles Lyell 4 December [1860][1]

Down Bromley Kent
Decr 4th

My dear Lyell

It certainly seems to me safer to rely solely on slowness of ascertained up & down movement. But you could argue length of probable time before the movement became reversed as in your letter.[2] And might you not add that over whole world it would probably be admitted that a larger area is *now* at rest than in movement? & this I think would be tolerably good reason for supposing long intervals of rest. You might even adduce Europe, only guarding yourself by saying that *possibly* (I will not say probably, though my prejudices would lead me to say so) Europe may at times have gone up & down all together.— I forget whether in former letter you made a strong point of upward movements being always interrupted by long periods of rest.— After writing to you, out of curiosity I glanced at the early Chapts in my Geology of S. America; & the areas of elevation on E. & W. coasts are so vast, & proofs of many successive periods of rest so striking that evidence becomes to my mind striking.

With respect to Astronomical causes of change, in ancient days in the Beagle, when I reflected on the repeated great oscillations of level on very same area; & when I looked at symmetry of mountain-chains over such vast space, I used to conclude that the day would come when slow change of form in the semi-fluid matter beneath the crust would be found cause of volcanic action & of all changes of level. And the late discussion in Athenæum by Sir H. James (though his letter seemed to me mighty poor, & what Jukes wrote good) reminded me of this notion.[3] In case astronomical agencies shd ever be proved or rendered probable, I imagine, as in nutation or precession, that an upward movement or protrusion of fluidified matter below might be *immediately* followed by movement of opposite nature. This is all that I meant.— I have not read Jamieson, or yet got the number.—[4]

I was very much struck with Forbe's explanation of N. of Soda beds & the saliferous crust, which I saw & examined at Iquique.—[5] I often speculated on greater rise inland of Cordillera, & cd never satisfy myself.

How far to lump & split species is indeed a hopeless problem.— It must in the end, I think, be determined by mere convenience.—

I have not read Stur,[6] & am awfully behind hand in many things.

I am *very* glad to hear that you continue to stir them up at Zoolog. Soc.—[7]

I get on slowly with my new Edition; I find that your advice was *excellent*; I can answer all Reviews, without any direct notice of them, by a little enlargement here & there, with here & there a new paragraph.— Bronn alone I shall treat with the respect of giving his objections with his name.—[8] I think I shall improve my Book a good deal, & add only some 20 pages.—

I hear that there is article in Macmillan on the Origin—[9]

Ever yours | C. Darwin

I have not yet read Phillips.—[10]

American Philosophical Society (236)

[1] The year is given by the reference to the new edition of *Origin*, the third edition, which appeared in March 1861.

[2] Letter from Charles Lyell, 30 November 1860. See also letter from Charles Lyell, 24 November 1860.

[3] Henry James, the director-general of the Ordnance Survey, was involved in a controversy over the possible effect of periods of mountain-building on the orbit of the earth. Joseph Beete Jukes had attacked several of James's points in a letter to the *Athenæum*, 6 October 1860, p. 451. CD had already discussed the issues involved with Lyell (see letter to Charles Lyell, 8 October [1860]).

[4] Jamieson 1860 was published in the concluding number of the 1860 volume of the *Quarterly Journal of the Geological Society of London*.

[5] D. Forbes 1861, pp. 13–17. CD described the soda beds near Iquique in Peru in *South America*, pp. 69–72. CD may have seen a copy of David Forbes's paper on the geology of the Andes before it was read at a meeting of the Geological Society on 21 November 1860. It is possible, however, that the information about salinas was in the missing portion of the letter from David Forbes, [November? 1860].

[6] Stur 1860. There is an annotated copy of this monograph on the geographical distribution of the plant genus *Astrantia* in the Darwin Pamphlet Collection–CUL. Dionyz Štúr was a Hungarian botanist and palaeontologist whom Lyell had met during a continental tour in 1856. Lyell had been impressed with his analysis of the geology of the Alps (K. M. Lyell ed. 1881, 2: 228).

[7] Lyell was a fellow of the Zoological Society, but it is not clear in what way he was stirring it up. The society experienced severe financial difficulties in 1860 in spite of charging a fee for admission to the zoological gardens: a member of the public paid 1*s.* on most weekdays and 6*d.* on Mondays. In an attempt to attract more visitors and thus raise more revenue, the society tried reducing the entrance fee to 6*d.* on Saturdays as well during August, September, and October 1860; however, the practice was stopped after fellows of the society protested. It was also proposed that no money should be spent on new buildings and that the society should hold its meetings in the rooms of another society. See Scherren 1905.

[8] CD discussed Heinrich Georg Bronn's criticism of *Origin* (Bronn trans. 1860, pp. 495–520) in *Origin* 3d ed., pp. 139–41.

[9] Fawcett 1860. There is an annotated copy in the Darwin Pamphlet Collection–CUL.

[10] Phillips 1860.

To Benjamin Silliman Jr 4 December [1860][1]

Down Bromley Kent
Decr. 4th

My dear Sir

I must trouble you with a few lines to thank you very sincerely for your note with some additional information on the Cave-Rat[2] & for your printed letter, which I well remember reading some years ago in your Journal.[3] Several of your statements in your letter have interested me much, & the whole subject of these cave blind animals seems to me eminently curious.—

I am much obliged to you for telling me about Prof. Dana;[4] pray give him my very kind remembrances; I believe no one, except his personal friends, will more rejoice at his perfect recovery than I shall.—

I am preparing a new corrected Edition of my "Origin" & shall take the liberty to quote a few words from your letter on the Cave-Rat.—[5]

With sincere thanks & respect, pray believe me | My dear Sir | Yours sincerely | C. Darwin

New York Botanical Garden Library (Charles Finney Cox collection)

[1] Dated by the relationship to the letter from Benjamin Silliman Jr, 27 October 1860. CD's letter to Silliman was enclosed with the letter to Jeffries Wyman, 3 December [1860], for Wyman to forward.

[2] Letter from Benjamin Silliman Jr, 27 October 1860.

[3] Silliman 1851 was published in the *American Journal of Science and Arts*, of which Silliman was principal editor. There is an annotated copy in the Darwin Pamphlet Collection–CUL.

[4] James Dwight Dana.

[5] *Origin* 3d ed., p. 154; Peckham ed. 1959, p. 253.

To T. H. Huxley 5 [December 1860][1]

Down Bromley Kent
5th

My dear Huxley

I must thank you for your note, which made me laugh heartily.[2] On my honour bright I did not look at you as the venerable figure which you pourtray.[3] What I meant was that a long course of work, which the young are only likely to do, will be necessary to confirm my views, if they are ever to be confirmed. I did however fancy that you had got to think the probability less, which did not surprise me, as I used formerly myself to fluctuate in extreme degree.— You may guess, therefore, how your note has pleased me.— I am too sensitive of a *shade* of opinion in such men as you, but I find I can now honestly say that I care nothing about common reviewers. You will be a real good Christian if you do not hate me, as you are so "pounded" on opposite sides.—

I have read few first pages of the Macmillans article,[4] & it pleases me that he quotes Mill's Logic & declares that I have philosophised in right spirit, & that all the talk about Baconian induction is cant & rubbish.—[5] I well remember that you took nearly same line & cited undulatory theory of light.—[6]

Here I am again running on about my wearisome self.— Farewell my best of defenders & friends. I hope that Mrs. Huxley is going on pretty well.—[7] Do not work too hard; I am always fearful that you will break down with so many irons in the fire.—

Adios | C. Darwin

P.S. I have just reread your note. What a fool old Gray is (not but what I like him);[8] he understands my book no more than a pig does. He told me the last time I saw him "you see natural selection cannot possibly apply to plants".— As far as accurate & extended knowledge & acute philosophy go, he would be a precious deal nearer the truth, if he were to say "old Huxley" & "young Darwin".

I see Fawett in Macmillan bring in you & the B. of Oxfords Ape! You must see how well he praises your rejoinder.[9]

Imperial College of Science, Technology and Medicine Archives (Huxley 5:130)

[1] Dated by the reference to Henry Fawcett's article on *Origin*. See n. 4, below.

[2] Huxley's letter has not been found.

[3] In the letter to T. H. Huxley, 2 December [1860], CD lamented that Huxley had not accepted his views more fully and remarked that if they were ever to be adopted, it would be by 'young men growing up & replacing the old workers'.

[4] Henry Fawcett's article, entitled 'A popular exposition of Mr. Darwin on the origin of species', was published in the December 1860 issue of *Macmillan's Magazine* (Fawcett 1860).

[5] Stating that all hostile critics had repudiated CD's theory by claiming that it was not based on 'a rigorous induction', Fawcett pointed out that CD's method accorded well with John Stuart Mill's exposition of the deductive method consisting of direct induction, ratiocination, and then verification (Fawcett 1860, pp. 83–4).

[6] During his confrontation with Samuel Wilberforce at the 1860 meeting of the British Association for the Advancement of Science, Huxley is reported to have said that CD's theory was an explanation of the phenomena of natural history in the same way as the undulatory theory was an explanation of the phenomena of light. See L. Huxley ed. 1900, 1: 184 n. 1, and Appendix VI.

[7] Henrietta Anne Huxley gave birth to Leonard Huxley on 11 December 1860. She was still grieving over the death of their oldest child, Noel Huxley, in September.

[8] John Edward Gray.

[9] Expressing his surprise that opponents had not discussed the 'most formidable argument' against CD's theory, namely geology, Fawcett stated: 'It was sad, indeed, to think that the opponents of the theory sought to supply this omission by summoning to their aid a species of oratory which could deem it an argument to ask a professor if he should object to discover that he had been developed out of an ape.' (Fawcett 1860, pp. 88). For a fuller account of the discussion at the Oxford meeting of the British Association for the Advancement of Science, see Appendix VI.

To John Murray 5 December [1860][1]

Down Bromley Kent
Dec. 5

My dear Mr Murray

One line to thank you for your note & to remind you (in case you sh^d^ forget) that copies of the Diagram must be Printed off.—[2] May I *earnestly* beg you to have the Plate *carefully* inspected; for if it prints the small letters & figures badly, I assure you that parts of my Book will be as unintelligible as Hebrew.—

I shall be glad to hear, when you have decided how many copies you will print off—the more the better for me in all ways, as far as compatible with safety;[3] for I hope never again to have to make so many corrections or rather additions which I have made in hopes of making my many rather stupid reviewers at least understand what is meant.— I hope & think I shall improve the Book considerably.—

Yours very sincerely | C. Darwin

There is very good article in Macmillan on the "Origin".—[4]

John Murray (publishers)

[1] Dated by the reference to the new edition of *Origin* and to the article in *Macmillan's Magazine* (see n. 4, below).

[2] The diagram in *Origin* illustrating divergence was inserted between pages 116 and 117. It had been separately printed for the first edition (see *Correspondence* vol. 7).

[3] According to CD's journal (de Beer ed. 1959), 2000 copies of the third edition of *Origin* were initially printed. See also Peckham ed. 1959, p. 21.

[4] Fawcett 1860.

To J. D. Hooker 6 December [1860][1]

[Down]
Dec 6th.

My dear Hooker.

Get hold of Macmillan for Fawcetts article on the Origin, for he quotes your splendid speech at Oxford with admirable effect.—[2] He alludes to Huxley, the Bishop & the Ape.—[3] He evidently thinks the discussion at Oxford was first rate

C. D

DAR 115.1: 79

[1] Dated by the reference to Henry Fawcett's article on *Origin* (Fawcett 1860).

[2] The final paragraph of Fawcett's essay quoted a passage said to be taken from Hooker's speech at the Oxford meeting of the British Association for the Advancement of Science defending CD and his views (Fawcett 1860, p. 92):

> I knew of this theory fifteen years ago; I was then entirely opposed to it; I argued against it again and again; but since then I have devoted myself unremittingly to natural history; in its pursuit I have travelled round the world. Facts in this science which before were inexplicable to me became one by one explained by this theory, and conviction has been thus gradually forced upon an unwilling convert.

[3] See letter to T. H. Huxley, 5 [December 1860] and n. 9.

From J. D. Hooker [6–11 December 1860][1]

Royal Gardens Kew

Ed I.[2]

P. 9. Is the term *Sport* confined to buds & offsets. If so, they must necessarily be rare under nature.[3]

p. 18. Many people have remarked that the *Quarterly* so disposed of Horner's pottery dating that they wonder you should quote it.—[4] I know nothing.

p. 19. Bottom of page. I suppose you know that a Greyhound will *run* a Scotch Terrier.
p. 29. I think the "explanation" of the Gardeners disbelief, is more due to want of reversion, & the fact that all future varieties wander still further from the wild apple.
p. 58. Bottom. "Finally x x x &c"— this is very unclear, though I see the meaning
p. 109. Middle of page. "Each area is already fully stocked"— 110 at top. "No region is as yet fully stocked"—
p. 132. Forbes' & Gould's statements are repeated in next page.
p. 193. replace *h* by *c* & read '*Matteucci*
p. 197. Is not the Bamboo here alluded to a Palm?
219 "Slaves carry masters" 221 "Masters carry slaves."
353. Perhaps better to say that most Natural Genera "almost invariably continuous in area"—for many of *most* natural are cosmopolitan.
369 line 17. for "Some few" read "*many*"
379. An equally curious case is that of a group of several N. Chili species reappearing in N. Mexico & California[5]
416 It is the position of the rudimental florets of grapes that Brown found so important, not their characters.
459. line 7 from bottom The introducing "*which we may consider*" is very confusing— ? take out comma before it—

DAR 104: 218

CD ANNOTATIONS
1.1 P. 9. . . . nature. 1.2] '?' *added brown crayon*
2.1 p. 18. . . . nothing. 2.2] *crossed pencil*
3.1 p. 19. . . . Terrier. 3.2] '?' *added brown crayon*
4.1 p. 29. . . . apple. 4.3] *crossed brown crayon*
5.1 p. 58. . . . meaning 5.1] *crossed pencil*; *cross added brown crayon*
6.1 p. 109. . . . '*Matteucci* 8.1] *crossed ink*
10.1 219 . . . slaves." 10.1] *crossed ink*
11.1 353. . . . "*many*" 12.1] *crossed pencil*
13.1 379. . . . California 13.2] '[Glacial]' *added brown crayon*
14.1 416 . . . before it— 15.2] *crossed pencil*

[1] The letter had not been received by CD when he wrote to Hooker on 6 December (see preceding letter). CD replied to it on 11 December (letter to J. D. Hooker, 11 December [1860]).
[2] CD had asked Hooker to send him his criticisms of *Origin* to assist with the corrections and additions to the third edition (see letters to J. D. Hooker, 26 November [1860] and 4 December [1860]).
[3] Hooker's point is that buds and offsets in plants are almost always garden phenomena. CD did not change the passage in the third edition of *Origin*.
[4] In the first edition of *Origin*, CD cited the results of Leonard Horner's geological excavation of alluvial deposits in Egypt (Horner 1858) as support for the statement that 'civilized men may have lived in the valley of the Nile thirteen or fourteen thousand years ago' (*Origin*, p. 18). The empirical basis for Horner's chronology, however, was challenged by the classicist William Smith in an unsigned article published in the *Quarterly Review* in April 1859 ([W. Smith] 1859; see *Wellesley index* 1: 742). In fact, CD

had already replaced this sentence in the revised American edition of *Origin* (see letter to Asa Gray, 1 February [1860] and n. 7).

5 In the first edition of *Origin*, p. 379, CD discussed how only a few southern organisms seemed to have migrated north, giving as an example the case of 'a few southern vegetable forms on the mountains of Borneo and Abyssinia.' He did not add Hooker's example to the third edition.

From J. M. Rodwell 6 December 1860

Belle Vue House | Broxbourne.
Dec. 6. 1860

My dear Sir/

I ought ere this to have replied to your last[1]—especially as more than a fortnight has elapsed since I completed, so far as this neighbourhood is concerned, my enquiries about the white Cats. My neighbour M^rs. Ellis of Hoddesdon[2] has three white cats all of which are, and always have been stone deaf and thus far your original statement is confirmed. But—per contra—M^r. Bedingfield the proprietor of a well known horticultural & fishing rendezvous for Cockneys in the adjacent parish of Nazing[3] has five white cats and a kitten all, like my cat, fully endowed with the faculty of hearing. When I tell you that in answer to his call they came rushing from all parts of the premises (with 4 black cats) more like a flight of tame pigeons than any thing else, and clustered around his legs, mewing for food, there could not remain a shadow of a doubt, than these $9\frac{1}{2}$ pussies were in every sense, wide awake.[4]

I will now answer your questions in regard to my Saffron Hill quondam Parishioners the rats.[5] My attention was drawn to the blindness of the rats by my Sexton, and I presume that he could only have heard it from the workmen, as his profession lay rather with dead Xtians than dead animals, but as the poor man is himself dead, I cannot clear this point up. It is quite possible that the rats may have been temporarily blinded, as you suggest, by sudden exposure to the light, and if this be the usual effect upon cave-animals suddenly brought into the glare of day, it w^d. I suppose be very unsafe to assert that this had not been the result in the present case. All that I now recollect about the appearance of the eyes is that they were of that peculiar white or ashy hue which one commonly sees in the blind. But I will make further enquiry, whenever an opportunity offers, from some of the contractors for our city Sewers who *may* possibly be able to give reliable & valuable information. For any thing that I have said, you are most welcome to use my name, but I should think that my information about the rats is scarcely matter-of fact enough for your purpose.

I am almost afraid that in what I am about to write I shall get out of my depth and betray my ignorance. However I will venture to state what has occurred to me viz—that your whole theory of the origin of Species might possibly find its best confirmation in observations of Microscopists upon Infusorial Animalcules and especially upon the rudimentary spores such as Euglena Protococcus &c. And for this reason especially that within a few hours even, one has the opportunity of watching the developement of generation after generation; and because observations upon such forms of life patiently prolonged over a series of years would most likely detect some minute structural change of sufficient amount & regularity to form a law. I should very much like to hear what such observers as the authors of the recent Microgr. Dictionary, Mr Griffiths & Henfrey would think of this idea.[6] It would be easy to calculate what the proportion would be between 5 years observation of the Generations of an infusorium or a spore, as compared with any of the longest recorded observations of the developement of any of the larger animals or plants— and what would be the chance of detecting an organic change.—

I hope I am making my meaning clear. At all events I think that I understand myself— And it was your book which put the idea into my mind in connection with observations I have made upon Volvox Globator,—of the following nature. I never yet found V^{x} in any pools in which under altered circumstances of temperature & moisture I did not find the individual animalcules or rather green spores of wh. V^{x} is composed. In the Pools on Clapham Common & Norwood Hills in 1857 I found large q^{s} of the V^{x} well developed—but in 1858 I found nothing in the same pools but the Separate spores

AL incomplete
DAR 47: 169–70

CD ANNOTATIONS
1.1 I ought . . . awake. 1.12] *crossed pencil*
2.13 For . . . purpose. 2.15] *scored pencil*
3.1 I am . . . spores 4.8] *crossed pencil*
Top of first page: 'Cats Deaf' *pencil, del pencil*; 'Rats Blind (Rodwell)' *pencil*

1 Letter to J. M. Rodwell, 5 November [1860].
2 There is a Mrs Ellis listed as living at Rose Hill, Hoddesdon, Hertfordshire, in the *Post Office directory of the six home counties* for 1859.
3 Probably John Benningfield of the Crown Hotel, Nazing, Essex (*Post Office directory of the six home counties* 1859).
4 In the first edition of *Origin*, pp. 12 and 144, CD stated that deafness in cats was 'invariably' correlated with blue eyes. He changed the wording in the third edition to read 'generally'.
5 See letter from J. M. Rodwell, 31 October 1860, and letter to J. M. Rodwell, 5 November [1860].
6 Griffith and Henfrey 1860.

To W. E. Darwin [8 December 1860]

Down
Saturday night.—

My dear William

I write one line to remind you to bring back Babington.—[1]

Secondly, my curiosity is excited about the blind Fawcett by his capital article in Macmillan on The Origin.[2] Can you find out anything about him. Did he take high degree? What county does he come from? Is he rich?[3]

How jolly it will be to have you home again; I am almost forgetting how to play at Billiards.—[4]

The Boys got out your drawings this evening & we were all looking at them & voted it a horrid shame if you do not go on practising a little.[5] I am hard at work at a new & corrected Edit. of the Origin; Murray having sold 400 more copies that he has in hand.—

Aunt Catherine came to day, but she returns on Tuesday[6] & Miss Pugh goes on Monday.—[7]

Farewell | My dear old man | Yours affect^{y} | C. Darwin

Postmark: DE 9 60
DAR 210.6

[1] Probably C. C. Babington 1851, a copy of which is in the Darwin Library–CUL.

[2] Fawcett 1860.

[3] Henry Fawcett was a fellow of Trinity Hall, Cambridge University, and a close friend of the geologist and mathematician William Hopkins. In 1863 he was appointed professor of political economy at Cambridge. He had been blinded in a shooting accident in 1858.

[4] William went home for the Christmas holidays on 13 December (Emma Darwin's diary).

[5] William had begun to paint in 1857 (see *Correspondence* vol. 6, letter to W. E. Darwin, [November 1857]).

[6] According to Emma Darwin's diary, Emily Catherine Darwin visited Down from 8 to 11 December 1860.

[7] Miss Pugh had been governess to the Darwin children during the years 1856 and 1857 (Freeman 1978). Afterwards, she maintained close contact with the family and was a frequent visitor to Down.

From Alfred Swaine Taylor to Edward Cresy[1] 10 December 1860

St James's Terrace | Regents Park
Dec 10 1860

Dear Mr Cresy

I saw a passage in the Chemical News of Nov 24th,[2] which I think will interest your friend Mr Darwin as to the recognition of saline matter in an almost infinitesimal state of division

At p 281 of that number (published by Mitchell Red Lion Court Fleet Street)[3] you will find it stated that by a peculiar optical property $\frac{1}{3000,000}$ of a milligramme ($\frac{1}{65}$th grain) of chloride of sodium or common table-salt may be discovered! This is equal to $\frac{1}{195,000,000}$th of a grain!

A dilution to the extent of $\frac{1}{500,000}$th grain admits of detection by the ordinary test,—nitrate of silver. The facts which you mention in your note regarding the Drosera serving to detect nitrogen by its sensitiveness, are very remarkable.[4] The nitrogen in a salt of ammonia and the nitrogen in gun-cotton[5] gutta percha—India rubber or Indigo are in very different states. It would be curious to try these substances—also ferrocyanide of potassium and ivory shavings

I am glad you think that Herapath has undergone a proper rasping in my paper[6]

I am | Your's very truly | Alfred S Taylor

E Cresy Esqr

DAR 58.1: 14–15

CD ANNOTATIONS
3.3 The nitrogen . . . potassium 3.6] *scored pencil*
End of letter: 'Does not eating flies make Drosera so animalised— | Try whether Drosera plant contains much nitrogen. | Try with bit of zinc & silver or gold leaf, whether volume cited is potent?'[7] *ink*

[1] Taylor was the author of a book on poisons (Taylor 1848) about which Cresy and CD had been corresponding. See particularly the letter from Edward Cresy, 30 October 1860. Cresy subsequently sent Taylor's letter to CD (see letter to Edward Cresy, 12 December [1860]).

[2] The *Chemical News* 2 (1860): 281 reported the discovery of a new alkali metal (later called caesium or cesium) by Robert Wilhelm Eberhard Bunsen and Gustav Robert Kirchhoff with the aid of the new technique of spectrum analysis (see Bunsen and Kirchhoff 1860). The great advantage of this method, the report noted, was its extraordinary delicacy.

[3] C. Mitchell and Co., founded by Charles Mitchell, published the two 1860 volumes of *Chemical News*; thereafter the journal was published by other companies.

[4] CD had been experimenting on the effects of various salts of ammonia on *Drosera* leaves. His results were given in *Insectivorous plants*, chapter 7.

[5] CD found that gun-cotton, a highly explosive, manufactured compound that consists of a series of nitrates of cellulose, was not digested by *Drosera* (*Insectivorous plants*, p. 125).

[6] William Herapath, the Bristol medical chemist, was frequently on the opposing side of Taylor in medical jurisprudence cases, most notably in the notorious 1856 murder trial of William Palmer (the Rugeley poisoner). In 1860 Taylor issued a detailed paper demonstrating that the test for arsenic established by Edgar Hugo Emil Reinsch and much favoured by Herapath was inaccurate: the copper wire used in the test itself contained arsenic compounds. Taylor used this article to take issue with criticisms levelled at him by Herapath in a letter to the *Lancet* (1859 (2)): 248. (Taylor 1860, pp. 219–20 and 220 n. 1). See also Coley 1991.

[7] CD's note refers to the statement in the letter that a dilution of common salt to the extent of $\frac{1}{500,000}$ of a grain can be detected by nitrate of silver. In *Insectivorous plants*, pp. 174–88, he gave the results of testing the response of *Drosera* leaves to very dilute solutions of various salts. His manuscript notes are in DAR 54, 60.1, and 60.2.

To Asa Gray 11 December [1860][1]

Down Bromley Kent
Dec 11th

My dear Gray

I have just got your letter of Nov. 26th.—[2] Not to lose the first chance, I write to say that I will **gladly** pay half of the (about) 20£, & will distribute some copies & leave others in the agents or Trübners hands:[3] I will put a few advertisements in newspapers & see what I can do to get it noticed in a few Periodicals.— But I am a bad man for that, as I live so retired.—

I must hear somehow to whom the consignment of the copies are sent & you can tell me exact sum to repay you.—[4] I am heartily glad of this. But I fear it will be impossible to get it known in England so as to sell many.—

If upon reflection you would not like to go to expence do not consider, of course yourself bound.—

Just to exemplify use of your pamphlet. Bishop of London[5] was asking Lyell what he thought of the Review in Quarterly, & Lyell's answer was read "Asa Gray in Atlantic".[6]

I will write to Huxley about Mr. Wright's article[7] but I have no hope, as H. said a good strong Anti-Darwinian Article was what he wanted.— I know nothing about payment, but will enquire & let you hear.—

In Haste | Ever yours | C. Darwin

I heartily acquiesce either in *better* or cheaper Edition—

Hooker will I am sure help in getting pamplet known.—

Gray Herbarium of Harvard University (38)

[1] Dated by the reference to the forthcoming publication of Gray 1861.

[2] The letter has not been found. Charles Lyell, when CD consulted him on the matter, had asked John Murray whether he would reprint Gray's review of *Origin* ([Gray] 1860b) as a separate pamphlet, but Murray had advised against it on the grounds of cost (see letter to Asa Gray, 24 October [1860]). See also letters to Asa Gray, 31 October [1860] and 26 November [1860]. The review was eventually published privately by Gray and CD in March 1861 (Gray 1861). See *Correspondence* vol. 9.

[3] The London-based publisher Johann Nicolaus Trübner had established many American connections; he had acted as the British distributor of some of Gray's earlier works.

[4] An entry in CD's Account book (Down House MS) dated 25 February 1861 indicates that he paid £7 10*s.* to 'Asa Gray for Pamphlet'.

[5] Archibald Campbell Tait, university reformer and liberal theologian, was appointed bishop of London in 1856.

[6] The review of *Origin* in the *Quarterly Review* was by Samuel Wilberforce, bishop of Oxford ([Wilberforce] 1860). For Charles Lyell's favourable opinion of Gray's articles ([Gray] 1860b), see letter to Charles Lyell, 23 [September 1860], and letter from Charles Lyell, 25 September 1860. See also letters to Asa Gray, 19 October [1860] and 24 October [1860].

[7] Gray apparently asked CD whether he thought Thomas Henry Huxley would be interested in publishing an article on CD's theory by his friend Chauncey Wright. Huxley had recently taken over the principal editorship of the *Natural History Review*. See letter to T. H. Huxley, 11 December [1860], and *Correspondence* vol. 9.

To J. D. Hooker 11 December [1860]

Down Bromley Kent
Dec. 11th

My dear Hooker

Sincere thanks for suggestions; some of which had occurred to me.—[1] I will profit by all that I can.—& two or three are excellent hits.

Please tell me **soon** is it Bamboo or Palm: it grows in your great Hot-house & is used in Java to catch thieves.—[2]

I do not understand what you mean by "a greyhound will run a Scotch Terrier".— Nor do I see, why sports in buds or offsets "must *necessarily* be rare under nature"— But these cannot be important points.— I understand everything else.—

I heard from A. Gray this morning;[3] at my suggestion he is going to reprint the 3 Atlantic articles as Pamplet & send 250 copies to England, for which I intend to pay half cost of whole Edition, & shall give away & try to sell by getting a few advertisements put in, & if possible Notices in Periodicals.— A. Gray said he shd. ask you to help, but I have told him that I would **gladly** pay half cost.—

Ever my dear Hooker | Yours | C. Darwin

David Forbes has been carefully working the Geology of Chile, & as I value praise for accurate observation far higher than for any other quality, forgive (if you can) the *insufferable* vanity of my copying the last sentence in his note. "I regard your monograph on Chile as without exception one of the finest specimens of geological enquiry".—[4] I feel inclined to strut like a Turkey-cock!

[Enclosure][5]

The Crustacea on shores of N. Chile & California offer parallel case with your plants: but I shall keep details for my larger work if it ever can be done.—[6]

Endorsement: '1860'
DAR 115.1: 80, 78E

[1] Letter from J. D. Hooker, [6–11 December 1860].
[2] In his letter, Hooker stated that he thought the trailing 'bamboo' of the Malay Archipelago, referred to by CD in *Origin*, p. 197, was in fact a palm. The rattan or cane palms are armed with formidable hooked prickles. CD changed the text of the third edition of *Origin* to read 'palm'.
[3] Asa Gray's letter has not been found, but see the preceding letter.
[4] The letter has not been found, but see the letter from David Forbes, [November? 1860]. Forbes had recently read a paper on the geology of Bolivia and Peru at a meeting of the Geological Society of London (D. Forbes 1861).
[5] The note is in DAR 115.1: 78E. Although the paper and the ink of the note are the same as those of the letter, it is not clear whether it was enclosed with this letter; since it continues the discussion of points raised in Hooker's letter of [6–11 December 1860], it would appear to have been written at about this time.
[6] The larger work CD envisioned as providing the empirical evidence summarised in *Origin* was never published in its entirety. CD abandoned the plan after the appearance in 1868 of *Variation*.

To T. H. Huxley 11 December [1860]

Down Bromley Kent
Dec. 11th

My dear Huxley

Please read enclosed, as I did not like to refuse.[1] But I know well from what you said, what your answer will be.— Still less than simply insert would you (after seeing it) pay for article.— Please return the enclosed, & let me have answer for Asa Gray.— The part in question is on second Page of Gray's letter— Wright, I believe, is excellent mathematician & I *think* self taught & educated & so I suppose poor.—

Ever yours | In Haste | C. Darwin

Imperial College of Science, Technology and Medicine Archives (Huxley 5: 166)

[1] The enclosed item was presumably the letter from Asa Gray, dated 26 November, that is now missing (see letter to Asa Gray, 11 December [1860]). It apparently included an account of an article by Chauncey Wright that Gray hoped Huxley would publish in the *Natural History Review*.

To Richard Kippist?[1] 11 December [1860][2]

Down Bromley Kent
Dec. 11th.

Dear Sir

I will return the two vols. in Hand on Thursday morning, & will you be so kind as to have the following Books ready for me.—

Mém du Mus. d'Hist. Nat. *Tom 8.*[3]

Edinburg. Phil. Journal XIV (published *about 35* years ago)[4]

Reports of Brit. Assoc. for the year, when Prof. Owen was President;[5] it is the volume containing his address which I wish to see.— Will you kindly look for volume.—

Dear Sir | Yours faithfully | C. Darwin

University of Virginia Library, Manuscripts, Special Collections, Darwin-Evolution Collection (#3314)

[1] The recipient is suggested by the reference in the letter to J. D. Hooker, 26 November [1860], to Kippist's writing to CD about a book that CD wished to borrow from the library of the Linnean Society. Kippist was librarian to the society.

[2] The year is given by the reference to articles CD needed for the preparation of the third edition of *Origin* (see nn. 4 and 5, below).

[3] It is not clear which paper in this volume of the *Mémoires du Muséum d'Histoire Naturelle* CD wished to read. None of them is cited in the third edition of *Origin* or in *Variation*. CD may possibly have consulted a paper by Adolphe Théodore Brongniart on the classification and distribution of fossil plants (Brongniart 1822) or one by René Joachim Henri Dutrochet on the crossing of plants (Dutrochet 1822).

[4] This volume of the *Edinburgh Philosophical Journal* contained an article by the zoologist Robert Edmond Grant, who had befriended CD during his time at Edinburgh University, that referred to Grant's belief in transmutation (Grant 1826). The paper is cited in the 'Historical sketch' included in the third edition of *Origin*, p. xiv:

> In 1826, Professor Grant, in the concluding paragraph in his well-known paper (Edinburgh Philosophical Journal, vol. xiv. p. 283) on the Spongilla, clearly declares his belief that species are descended from other species, and that they become improved in the course of modification.

This passage had not appeared in the first version of the historical sketch published in the revised American and German editions of *Origin*. For CD's recollection of learning about Grant's transformist views, see *Autobiography*, p. 49.

[5] Richard Owen was president of the British Association for the Advancement of Science in 1858. CD cited his presidential address (R. Owen 1858) in the historical sketch in the third edition of *Origin*, pp. xvi–xvii, stating that Owen had addressed the question of species change in speaking about 'the axiom of the continuous operation of creative power, or of the ordained becoming of living things.'

To Edward Cresy 12 December [1860][1]

Down, Bromley, Kent
Dec: 12th.

Dear Cresy

Thank you much for your most interesting note. I have ordered a Chemical News.— I particularly beg you to give Dr. A. Taylor my cordial thanks for his note full of valuable suggestions.[2] After writing out the greater part of my paper on

Drosera, I thought of so many points to try and I wished to re-test the basis of one large set of experiments, namely to feel still more sure than I am, that drop of plain water *never* produces any effect, that I have resolved to publish nothing this year.— For I found in record of my daily experiments one suspicious case.—[3] I must wait till next summer. It will be difficult to try any solid substance containing nitrogen, such as ivory; for two quite distinct causes excite the movement, namely mechanical irritation and presence of nitrogen. When a solid substance is placed on leaf it becomes clasped, but is released sooner than when a nitrogenous solid is clasped; yet it is difficult (except with raw meat and flies) to be sure of the result owing to differences in vigour of different plants. The last experiments which I tried before my plants became too languid are very curious, and were tried by putting microscopical atoms on the gland itself of single hairs; and it is perfectly evident that an atom of human hair $\frac{1}{76,000}$ of a grain (as ascertained by weighing a length of hair)[4] in weight, causes conspicuous movement. I do not believe (for atoms of cotton thread acted) it is the chemical nature; and some reasons make me doubt whether it is actual weight; it is not the shadow and I am at present, after many experiments, confounded to know what the cause is. That these atoms did really act and alter the state of the contents of all the cells in the glandular hair, which moved, was *perfectly* clear. But I hope next summer to make out a good deal more.—

I have been much interested by your account of your discussion, and I thank you as a bold and true Defender of the Faith.[5] It is very interesting to hear of such discussions going on in London on scientific questions out of the regular Societies. I am much pleased at all that you tell me. The article in National is by Carpenter.[6] I am pleased that you were struck by the "Snakes" in Huxley's article; it seemed capital to me.—[7] Dr Bree is nobody scientifically. His book is a mass of misrepresentations and blunders.[8] By far, in my opinion, the best Review is by Asa Gray in Atlantic Monthly in U. States.[9] If you cared to see it, I could lend it you. I am very hard at work in preparing a corrected Edit. of the Origin (for Murray sold at sale 400 more copies than he has). I shall not allude to any Reviews, but so far profit as to expand where I have been most misunderstood, and add several considerable discussions on points which I formerly ought to have noticed. I will send you a copy when it is out.

Thanks for advice about transitional cases. Excuse this rather hurried note, and with cordial thanks for all your great kindness

Yours very sincerely | C. Darwin

My daughter *very slowly*, but steadily improves in health.

Copy
DAR 143

[1] The year is given by the reference to CD's work on *Drosera*.

[2] Cresy's letter has not been found. It apparently accompanied the letter from Alfred Swaine Taylor that Cresy forwarded to CD (letter from A. S. Taylor to Edward Cresy, 10 December 1860).

[3] CD had composed a '*Careful* Resume of effect of Plain water' on *Drosera*, dated 'Nov. 26th. 1860.' (DAR 54: 7). The notes he made at the time of performing such experiments are in DAR 60.1.
[4] See letter from Trenham Reeks, 15 November 1860.
[5] Cresy was a supporter of CD's views.
[6] [Carpenter] 1860a.
[7] In his anonymous review of *Origin* in the *Westminster Review* ([T. H. Huxley] 1860b, p. 556), Thomas Henry Huxley made the statement: 'Extinguished theologians lie about the cradle of every science as the strangled snakes beside that of Hercules'.
[8] Bree 1860. See letters to J. S. Henslow, 11 October [1860] and 26 October [1860].
[9] [Gray] 1860b.

To George Charles Wallich 12 December 1860[1]

Down. Bromley. Kent.
Dec. 12. 1860.

Dear Sir.

I beg permission to thank you sincerely for sending me your "Notes on animal Life at vast Depths".[2] It has interested me extremely.

You have indeed made a grand beginning at an admirable field of research.

If you would not think me very unreasonable, you would do me a great favour, if you would inform me on one point not noticed in your Notes. In the account given in the Times, it is stated that the Machine or Borer, *either often or sometimes penetrated through the Foraminiferous deposit into different underlying matter.*[3] This would show that the Foraminiferous deposit was sometimes or often thin; and this is the point on which I am anxious for information. It bears on the decay of the Exuviæ of organisms at the bottom of the sea; & is important for me in relation to some few passages in my Book on the Origin of Species, of which I am now preparing a corrected Edition.—[4]

You allude, also, to bare rock at the bottom of the Sea. Have you any reason to believe that extensive areas are bare?— About the Borer I had with hesitation thought of quoting the Times; but if the facts are true, I should of course infinitely prefer just alluding to the case on your authority.[5]

What a wonderful fact about the Ophiuræ, & what a capital proof of the Foraminifera having been alive is their discovery in their stomachs.[6]

Do you not think that you are rather bold in inferring that the basaltic pebbles were rounded at such great depths? Are you sure they were not dropped by icebergs either recently or at the close of the Glacial period?

With my best thanks & apologies for troubling you, I remain, | Dear Sir, | Yours very faithfully | C. Darwin

P.S. Is it not a most curious fact that the Water at such profound depths, & under such a vast pressure, should retain Oxygen for the respiration of the animals mentioned by you?

Copy
British Museum (Natural History) (General Library MSS/DAR: 6)

[1] The letter, which is in Wallich's hand, was probably copied at Francis Darwin's request. The date is as given on the copy.

[2] Wallich 1860. There is an annotated copy in the Darwin Pamphlet Collection–CUL. Wallich, who was an army surgeon in India, had been seconded in 1860 to HMS *Bulldog*, which was conducting an extensive survey of the Atlantic seabed for the proposed submarine cable link between England and America. The Admiralty, at the request of the promoters of the telegraph scheme, undertook to obtain the deep-sea soundings along a route running from Scotland to Labrador via the Faroe Islands and Greenland. Francis Leopold McClintock was the commander. The survey, which proved to be very arduous, lasted from June to October 1860. The *Bulldog* returned to England in poor condition in November (Markham 1909, pp. 252–62). Ultimately the plan to lay a cable along this route was abandoned.

[3] *The Times*, 5 December 1860, p. 5, reported the preliminary results of the North Atlantic telegraph expedition.

[4] In *Origin*, p. 288, CD discussed how 'Shells and bones will decay and disappear when left on the bottom of the sea'. He further stated that the sediment, contrary to popular opinion, was probably not deep, owing to the constantly changing conditions of the seabed.

[5] See letter from G. C. Wallich, 14 December [1860].

[6] As proof that *Globigerina*, a genus of Foraminifera, lived at much greater depths of the sea than had previously been supposed, Wallich stated that he had found specimens in the digestive cavity of starfish living in nearly 1300 fathoms of water (Wallich 1860, pp. 11, 22–3). CD marked these passages in his copy of the paper.

From T. H. Huxley [before 14 December 1860][1]

P.S. I have just received your note & the inclosure from Asa Gray—[2] How can you imagine there can be a doubt about the reply? Of course I shall be exceedingly glad to have it if it is good— I wish to see articles both *for* & *against*, so as to have it clearly understood that the Review is a 'Champ libre'—[3]

Paying for it however is what I unfortunately do not see my way to at present— Until the circulation reaches 6 or 7 hundred we shall hardly be able to pay anything & even then I am afraid not much— But get us the article if you can & if your friend is the most rabid advocate you ever had so long as he argues the question scientifically & is'n't abusive (beyond reason) I am quite sure Busk will agree with me on the propriety of inserting it[4]

My desire to have an anti-Darwinian article—is not so much because I want to see you criticised & because I want all the world to understand that it is not a party journal.

I return A. Grays note. Will he write for us?

AL incomplete
Gray Herbarium of Harvard University (39)

[1] Dated by the relationship to the following letter. CD forwarded the postscript to Asa Gray.

[2] See letter to T. H. Huxley, 11 December [1860].

[3] Huxley was one of the editors of the new series of the *Natural History Review*.

[4] George Busk, a close friend of Huxley's, was also one of the editors of the *Natural History Review*.

To Asa Gray 14 December [1860][1]

Down Bromley Kent
Dec 14th

My dear Gray,

The enclosed note of Huxleys you will see shows that he will be glad of Mr Wrights article, if approved of; but he cannot pay for it.[2] I really do not believe that there is any other Review which would accept it, for all have treated of the Origin.— If Mr Wright cannot better dispose of it; it had better be sent to Messrs Williams & Norgate Henrietta St., Covent Garden; as they publish the Review.— I feel sure that the Review will have great merit, & so far be worthy of Mr Wright's article; & it would be a great pity if it were not published, as you say it is so good.— You will see that Huxley asks whether you would ever contribute to his Review; but you must, I fear, be pledged to Silliman's.[3]

I shall be anxious to hear your final decision on manner of publishing your Three Articles.[4] I still think Nor 2 best, though I believe most would prefer Nor 3.[5]

I shd. rather like to see sub rosâ Harvey's letter.—[6] I have had much correspondence with him; & he wrote me some very nice letters. I return Ticknor & Fields letter,[7] as you might want to refer to it.—

My dear Gray | Yours most truly | C. Darwin

There is a nice, but too flattering article on the Origin by a young man, whom many think most highly of.—[8]

Gray Herbarium of Harvard University (39)

[1] Dated by the reference to the forthcoming publication of Gray 1861 (see n. 4, below).

[2] See preceding letter. Gray had asked CD whether he could get an article by his friend Chauncey Wright published in England. See letter to Asa Gray, 11 December [1860].

[3] CD refers to the *American Journal of Science and Arts*, edited by James Dwight Dana and Benjamin Silliman Jr. Gray was one of three contributing editors.

[4] At CD's request, Gray was arranging for copies of his review of *Origin* ([Gray] 1860b) to be separately printed in America. A number of these were intended for distribution in Britain. The pamphlet was published early in 1861 (Gray 1861). See letter to Asa Gray, 11 December [1860].

[5] The review had been published in three parts. For CD's comments on the second part, see the letter to Asa Gray, 26 September [1860].

[6] William Henry Harvey's letter to Asa Gray, dated 3 November 1860, is in the Gray Herbarium, Harvard University. Harvey commented on Gray's review of *Origin* ([Gray] 1860b): 'Almost thou persuadest me to be a Darwinite—not *quite* so, but thou persuadest me to be a Grayite.'

[7] The Boston publishing firm Ticknor and Fields had agreed to print Gray's pamphlet, for distribution by Trübner and Company in London (Dupree 1959, p. 299).

[8] Probably Fawcett 1860.

From G. C. Wallich 14 December [1860][1]

17 Campden Hill Road | Kensington
Decr. 14th. 6.pm—

Dear Sir—

Your kind note has but now been delivered to me, otherwise I should made a point of replying before—[2]

The assertions made in the Article in the 'Times' to which you refer are erroneous—[3] so erroneous indeed, both on the matter of the deposits and the sounding gear, that I thought it advisable to address the Editors giving a correct but very brief statement of facts, with my name—for insertion. For reasons best known to themselves no notice whatsoever was taken of my communication—

I had pointed out whilst at sea that the *surface layer* of the Foraminiferous deposit *sometimes* differed in certain respects from the substratum. The difference being due to the fact that whilst live Globigerinæ occurred in the former, retaining the brownish-yellow color of their Sarcodic contents in the latter they only occurred dead the color of the sarcode being thus rendered dusky at the same time that a larger proportion of decayed animal & vegetable matter contributed to the difference— This statement was twisted into a declaration that "by these contrivances not only were soundings rendered more certain but *very often* the understratum of the bottom was found to be composed of *entirely* different *material* from that which lay upon the surface, & which was ordinarily brought up by the rod or lead in ordinary deep sea soundings"! The Italics are mine—

I have no doubt that the Foraminferous as well as all other deep sea deposits vary greatly in thickness— Until soundings & surveys are conducted in a much more systematic and searching manner, it is almost impossible to arrive at anything like positive results— For instance where the intervals between Soundings extend from 20 to *70* miles, I maintain that we cannot pronounce upon the general character of the bottom—or do much more than suggest what may be the character of the intervals. But there is I think quite sufficient evidence to shew that in many portions of the deep sea bed, the deposits are of no great thickness. At times we come across deposits of great thickness (several inches, for that is the limit to which any form of sounding apparatus hitherto employed has been able to penetrate) but almost entirely deficient in either minute calcareous or silicious organic remains. On the other hand we meet with deposits in which the entire mass is composed of nearly pure Globigerina shells, alive & dead, as in the cases I just referred to. I do not see how to account for this remarkable fact except on the supposition that in the former instance the deposits are the result of some sediment-laden current, flowing along slowly at great depths in all probability—whereas, in the latter, currents charged with nutritive matter necessary for the increase and perpetuation of the Rhizopodous population, cause it to flourish only within given limits— In the course traversed by the Bulldog I observed that wheresoever it was possible for the Gulf Stream to send off a ramification, Globigerina was to be met with. Where a main branch of that stream was known to flow, these organisms were in greatest profusion & in the most flourishing condition. And wheresoever the influence of the Gulf stream was either absent or imperceptible (as for instance between Greenland & Labrador) the Globigerinæ and indeed all other minute organisms were almost altogether unrepresented—

As regards the occurrence of bare rock I would observe that at 682 Fathoms between the Farœ Islands & Iceland I obtained a living Serpula, the stout calcareous tube of which was broken—together with several pieces of stone & but a faint trace of mud— From this I cannot help inferring that the bottom was comparatively bare of deposit— The Serpula could not have lived imbedded— Its size necessitated its being attached to some object of considerable size. One Stone weighed about an ounce. No ice or glacial drift now passes along this part of the sea— I cannot believe that large tracts are bare—but strange to say the sounding in which the Serpula & stones occurred with so little admixture of mud is the deepest we encountered between the points named—the last place in short at which we should a priori have inferred the existence of bare rock or gravel! The interval between this sounding and those on either side of it was very great— Alas I had no voice in determining where to sound.

Sir Chas Lyell has convinced me that I rushed at a far too hasty conclusion on the matter of the deep currents as producing abrasion & rounding of gravel—The presence of littoral Foraminifera tempted me somewhat too forcibly, I suppose— Although no Glacial drift occurs now at the points referred to, it is, as he and yourself suggest, far more likely that the gravel is an ancient glacial deposit—and of course I bow reverently and with implicit confidence under such authority— Hereafter the error shall be duly corrected—

The occurrence of the Ophiocomæ *13* in number is indeed a wonderful fact—[4] I cannot concieve the presence of any *free* gas possible at the depth at

Incomplete
Darwin Pamphlet Collection–CUL (bound with Wallich 1860)

CD ANNOTATIONS
1.1 Your . . . communication—2.5] *crossed pencil*
4.8 the deposits . . . penetrate) 4.10] *scored pencil*
6.1 Sir Chas . . . depth at 7.2] *crossed pencil*
Top of first page: 'Dr Wallich' *pencil*

[1] Dated by the relationship to the letter to G. C. Wallich, 12 December 1860, and by the reference to an article in *The Times* (see n. 3, below).

[2] Letter to G. C. Wallich, 12 December 1860.

[3] *The Times*, 5 December 1860, p. 5.

[4] During his researches Wallich had found a number of specimens of *Ophiocoma* (then classified as starfish but now as brittle-stars) in almost 1300 fathoms of water. In his article, Wallich pointed out that his find showed that a highly organised species was able to withstand far greater water pressure than was previously believed possible. See Wallich 1860, pp. 22–4.

To W. D. Fox 17 December [1860]

Down Bromley Kent
Dec 17th

My dear Fox

Your remarks on William pleased us much: He is now at home & gives a good account of your appearance.—[1]

I shall be proud to send you the Photograph of myself by Post. & will order a copy for you.—[2]

I have no idea whether an engraving of my Grandfather could be purchased; there are Photographs of his picture to be had at Cambridge & I will tell William to get (if he can) one & send it you.[3] The engraving of my Father is hideous, but I presume my sisters could get it at Shrewsbury, if you wished for it.—[4]

Thank for your enquiries about poor Etty who has had a terrible succession of illnesses; she is now very slowly, but steadily improving.

You also ask after Caroline: she also was very ill this summer; but is now fairly well. She & her three daughters live at Leith Hill Place, 6 miles from Dorking.[5]

As for myself I have been of late rather below my low mark; & so end my long Bulletin. I am at present busy with a new & somewhat enlarged & corrected Edit. of the well abused Origin, which the Public like, whatever the Reviewers may do.— I have done very little in Nat. Hist. this summer owing to Etty's illness.—

You say not a word about your innumerable progeny;[6] so that I hope that they are all well. Farewell | My dear Fox | Yours very truly | C. Darwin

I forgot.— you offer me a good inkstand. This is by far too good an offer not to be gratefully accepted.

You might send it to "6, Queen Anne St. | Cavendish Sq^{e}".—[7]

Postmark: DE 18 60
Christ's College Library, Cambridge (Fox 130a)

[1] William Erasmus Darwin had apparently called on Fox before arriving at Down for the Christmas holidays. According to Emma Darwin's diary, 'William came' on 13 December 1860.

[2] CD's comments in the following letter suggest that he intended to send Fox a copy of a photograph taken by Maull and Fox (see the frontispiece). It is not known precisely when the photograph was taken. Francis Darwin dated it as '1854?' (*LL* 1: frontispiece); Nora Barlow believed it was taken when CD was 'aged 51', or in 1860 (*Autobiography*, frontispiece). An entry in CD's Account book (Down House MS) in February 1858 suggests that CD may have sat for the photograph in the summer of 1857.

[3] Erasmus Darwin, CD's grandfather, was the brother of Fox's grandfather William Alvey Darwin. CD probably refers to photographs of the portrait by Joseph Wright of Derby, *circa* 1780. The original is in the Darwin Museum–Down. Another portrait of Erasmus Darwin by Wright, dated 1770, is in the National Portrait Gallery, London.

[4] CD refers to the mezzotint of Robert Waring Darwin made by Thomas Lupton in 1839 after the portrait in oil by James Pardon. See *Correspondence* vol. 4, facing p. 188. Both the engraving and the portrait are in the Darwin Museum–Down.

[5] Caroline Sarah Wedgwood, aged sixty, was CD's oldest sister. Her three daughters were Sophy Marianne, Margaret Susan, and Lucy Caroline Wedgwood (*Darwin pedigree*).

[6] By 1860 Fox had 15 children.

[7] The address of CD's brother, Erasmus Alvey Darwin.

To J. D. Hooker 17 December [1860][1]

Down Bromley Kent
Dec. 17th

My dear Hooker

Thanks about the Hooked Palm,[2] & for your jolly note written by a gentleman lolling in his arm-chair.

I have just been ordering a Photograph of self for a friend; & have ordered one for you & for Heaven-sake oblige me & burn that now hanging up in your room. It makes me look atrociously wicked.—[3]

When will it do to send me the Apocynum androsæfolium (I enclose address):[4] can you find out for me whether it seeds in open air, or had I better keep it in Greenhouse *to seed.* In the Spring I must get you to look for long pistils & short pistils in the rarer species of Primula & in some allied Genera.— It holds with P. Sinensis You remember all the fuss I made on this subject last Spring:[5] well the other day at last I had time to weigh the seeds, & by Jove the Plants of Primrose & Cowslip with short pistils & large-grained pollen are rather *more* fertile than those with long pistils & small grained pollen. I find that they require the action of insects to set them, & I never will believe that these differences are without some meaning.

Some of my experiments lead me to suspect that the large-grained pollen suits the long pistils & the small-grained pollen suits the short pistils. But I am determined to see, if I cannot make out the mystery next Spring.[6]

How does your Book on Plants brew in your mind?[7] Have you begun it? Farewell.— Since receiving your note, I have read Phillips & such weak, washy stilted stuff I never read—[8]

Farewell | C. Darwin

Remember me most kindly to Oliver.[9] He must be astonished at not having a string of questions, I fear he will get out of practice!

P.S.—| If I can keep pretty well I shall be at Club on Thursday.[10] How I wish there was any chance of seeing you there.— If you do come, do let me sit by you.—

DAR 115.1: 81

1 The year is given by the reference to Phillips 1860.

2 See letter from J. D. Hooker, [6–11 December 1860], and letter to J. D. Hooker, 11 December [1860].

3 See preceding letter. The photograph that CD refers to as making him look 'atrociously wicked' is that taken by Maull and Polyblank *circa* 1855. See *Correspondence* vol. 5, facing p. 448 and letter to J. D. Hooker, 27 May [1855].

4 CD had been discussing the identity of this plant, which he remembered as growing in the garden of The Mount in Shrewsbury, with Daniel Oliver, a colleague of Hooker's at the Royal Botanic Gardens, Kew. See letter to Daniel Oliver, 16 November [1860], and letter from Daniel Oliver, 23 November 1860.

5 See letters to J. D. Hooker, 26 April [1860], 11 May [1860], and 12 July [1860].

6 CD read a paper on the two forms of flowers, or dimorphic condition, of *Primula* at a meeting of the Linnean Society in November 1861. See *Correspondence* vol. 9 and *Collected papers* 2: 45–63. The paper was eventually published in the *Journal of the Proceedings of the Linnean Society (Botany)* 6 (1862): 77–96.

7 CD had urged Hooker to compile his published materials into a new volume. See letters to J. D. Hooker, 8 February [1860], 12 March [1860], and 18 [March 1860].

8 Phillips 1860.

9 See n. 4, above. Oliver was also assisting CD with his study of insectivorous plants.

10 There was a meeting of the Philosophical Club of the Royal Society on Thursday, 20 December 1860, at which William Benjamin Carpenter reported on George Charles Wallich's findings from

the North Atlantic telegraph survey. CD's name does not appear in the list of members present (Philosophical Club minutes, Royal Society).

To James Drummond 20 December [1860][1]

Down Bromley Kent
Decr 20th

Dear Sir

I received your extremely interesting letter of Oct 8th this morning.[2] I thank you much for the Distylis seeds which I will plant in greenhouse in the Spring.[3] I will also have the pleasure of exaning the seed of the Compos. plant, which you describe as so remarkable.[4] It seems a trifling point, but I have been *extremely* much interested by what you tell me of the small Bee extracting the pollen from within the indusium of the Brunonia. If you have any opportunity of observing, I should so much like to hear whether Bees do the same for Leschenaultia; or whether, as I suspect, in sucking the nectar of the flowers, their hairy abdomens do not brush over & partly open the slightly protuberant upper lip of the indusium—[5]

Another fact which you mention has interested me particularly, namely the Malvaceous plant setting seed with its flower never opening. I have observed something of this in Drosera rotundifolia & in Chlora perfoliata. Have you observed the Malvaceous plant during more than one year? And are you sure that it has not *two* periods of flowering (like some Campanulas) at one period opening its flowers & another period not opening them.—

Pray accept my cordial thanks for your very great kindness & believe me | Dear Sir | Yours truly obliged | Charles Darwin

P.S. I have kept back this note for a couple of days just to say that I have been interested by watching the achenia of the Styloncerus placed on damp paper. The vesicles which contain the dried mucus seem to me a pretty & curious adaptation; & the whole object very pretty under the microscope. I will communicate with Dr Hooker, & if he thinks fit, I will insert an extract out of your note together with some account of the structure of the achenia, in some Periodical.[6]

Library Board of Western Australia, State Archives, Accession 2275A

[1] Dated by the relationship to the letter from James Drummond, 8 October 1860.

[2] Letter from James Drummond, 8 October 1860.

[3] Drummond sent CD some seeds of *Distylis*, a genus of Goodeniaceae, in his letter of 8 October 1860.

[4] Drummond also enclosed in his letter of 8 October 1860 seeds of the plant *Styloncerus humifusus*. The genus is confined to Australia.

[5] CD evidently received an answer from Drummond on this point in a letter that is now missing. In CD's paper on the fertilisation of *Leschenaultia formosa*, published in the *Cottage Gardener*, 28 May 1861, p. 151, CD stated that Drummond had seen a bee 'cleverly opening the indusium and extracting pollen' (see *Collected papers* 2: 42–3).

[6] See letter to J. D. Hooker, 26 December [1860].

To Daniel Oliver 20 December [1860][1]

Down Bromley Kent
Dec. 20th

My dear Mr Oliver

If you have Rafinesque's New Flora of N. America, will you kindly look & tell me the date of Part First.[2]

Drumond of Swan River has sent me seeds of a Compos. Plant.—Styloncerus Humifusus of Labillard.[3] As, after looking at them, I shall not want them, I have thought I would just mention that I had them, in case they shd. of any use at Kew; not that I suppose they would.— He has sent me seeds of Distylis which I shall plant.— You sent me a Goodenia, also, so I shall have plenty of this order now.— I have been pleased at a prophecy which I made to myself coming true, viz that Bees would open the indusium & get out the pollen & thus accidentally carry it from flower to flower, for Drummond writes that he watches a small Bee busily employed in extracting the pollen out of the indusium of a Brunonia.— This feat, would I expect puzzle our European Bees.—

Yours very truly | C. Darwin

Down House (MS 10: 28)

[1] The year is given by the reference to the letter from James Drummond, 8 October 1860.

[2] Rafinesque 1836, which was published in four parts. CD wished to consult the work in connection with the historical sketch he was including in the third edition of *Origin*. At some point, Oliver wrote the bibliographical information on the back of the letter. See letter to J. D. Hooker, 29 December [1860].

[3] See preceding letter and letter from James Drummond, 8 October 1860 and n. 5.

To Asa Gray 21 December [1860][1]

Down Bromley Kent
Dec. 21

My dear Gray

One word to ask you as soon as ever you have definitely fixed on title of the Reviews[2] (if, as I hope, you publish) to send it me & the London Agent or Publisher (it would be best if a regular publisher like Trübner); for if I can hear before my new Edit of Origin is bound (& Printers have not yet sent me 1st sheet) I would ask Murray to permit me to bind in every copy a slip, as advertisement.—[3] If possible price had better be fixed likewise.—

In Haste | Ever yours | C. Darwin

I have only just thought of the manner of advertisement

by
Asa Gray
Professor of Botany, &c. &c. ?
Trubner & Co: London. Price —?—

Gray Herbarium of Harvard University (J. L. Gray autograph collection 50)

[1] Dated by the reference to the forthcoming publication of Gray 1861 (see n. 2, below).
[2] CD hoped that Gray would be able to republish his review of *Origin* ([Gray] 1860b) as a separate pamphlet that could be distributed in England. See letter to Asa Gray, 11 December [1860].
[3] CD added a 'Postscript' to the preliminary pages of the third edition of *Origin* to advertise the publication of Gray's reprint. It reads (*Origin* 3d ed., p. xii):

> An admirable, and, to a certain extent, favourable Review of this work, including an able discussion on the Theological bearing of the belief in the descent of species, has now been separately published by Professor Asa Gray as a pamphlet, about 60 pages in length. It is entitled, 'Natural Selection not inconsistent with Natural Theology. A Free Examination of Darwin's Treatise on the Origin of Species, and of its American Reviewers. By Asa Gray, M.D., Fisher Professor of Natural History in Harvard University. Reprinted from the Atlantic Monthly for July, August, and October, 1860. London: Trübner and Co., 60, Paternoster-row. Boston: Ticknor and Fields. 1861.' (Price 1*s.* 6*d.*)

When first published in the *Atlantic Monthly*, the review was unsigned. The leaf of advertisements bound at the end of this edition of *Origin* lists only works published by John Murray (see Freeman 1977, pp. 78–9).

From Sarah Elizabeth Wedgwood 25 December [1860?][1]

Christmas day

Dear Charles

Charlotte[2] reports from Mr Wallis[3] that he only once saw sundew open— he mentioned the time & place, but was sorry he could not recollect the time of day— he took so⟨me⟩ home & kept ⟨*two words missing*⟩ he never s⟨aw it⟩ open again.—It was a long time ago— he said he understood they did open some time in the day, but did not know at what hour. Smith[4] says nothing about ⟨it. He s⟩eems to know ⟨very lit⟩tle, for he says

Incomplete
DAR 181

[1] The year is inferred from the reference to William Wallis, whom CD first met in 1860 while visiting Sarah Elizabeth Wedgwood, Emma Darwin's sister, in Sussex. See 'Journal' (Appendix II).
[2] Charlotte Langton, Emma's other sister, also lived in Hartfield, Sussex.
[3] William Wallis was the surgeon of Hartfield and an orchid collector. He had assisted CD in his study of orchids and of the sundew (*Drosera rotundifolia*) when the Darwins visited Hartfield in July 1860. See letters to J. D. Hooker, 29 July [1860], and to W. E. Darwin, [30 July 1860].
[4] Probably Frederick Smith, a farmer residing at Thornhill farm, Hartfield, Sussex (*Post Office directory of the six home counties* 1859).

To J. D. Hooker 26 December [1860][1]

Down Bromley Kent
Dec. 26th

My dear Hooker

Will you have kindness to read enclosed.[2] Is the little fact new? Is it too trifling (on principle that "Lex de minimis non curat")[3] to be worth printing? *Do not let me publish rubbish.*— If worth publishing, shall I send it to Linn. Soc. or to Gardeners Chronicle?[4] I looked at the seeds for simple amusement, & then the case seemed a **little** curious, & I thought Mr Drummond would perhaps like to see his fact published.—[5] To save you trouble I send envelope directed & stamped.—

Please do not let me publish absurdly small fact,—perhaps already well known.—

Ever yours | C. Darwin

I call plant Siloxerus because Steudel[6] gives it as prior to Styloncerus.—[7] I enclose seeds of no use to me or to anyone, I suppose.—

P.S. | I wrote to Oliver[8] to ask him for date of publication in Pritzel[9] or elsewhere, of "Rafinesques New Flora of N. America Part I."[10] Can you tell me? I hear that Oliver will not return till Jan 5th.— If you can answer me, please tell him.—

DAR 115.1: 82

[1] The year is provided by the reference to the letter from James Drummond, 8 October 1860.
[2] CD refers to a manuscript on the adaptations of some Australian seeds for dispersal. At the time, he believed that the parent plant was called *Styloncerus humifusus* (see letter to Daniel Oliver, 20 December [1860]). The plant was later identified as *Pumilo argyrolepis*.
[3] 'The law does not concern itself with trifles'.
[4] The paper was published in the *Gardeners' Chronicle and Agricultural Gazette*, 5 January 1861, pp. 4–5 (see *Collected papers* 2: 36–8).
[5] James Drummond, who lived in Swan River, Western Australia, had sent the seeds to CD. See letter from James Drummond, 8 October 1860, and letter to James Drummond, 20 December [1860].
[6] Steudel 1841 was the most authoritative source of the period for botanical names. CD's copy is in the Darwin Library–CUL.
[7] Neither of these names was used in CD's published paper (see n. 2, above).
[8] See letter to Daniel Oliver, 20 December [1860].
[9] Pritzel 1851 was a frequently consulted bibliography of botanical books.
[10] Rafinesque 1836.

To W. H. Miller 27 December [1860][1]

Down Bromley Kent.
Dec 27

My dear Miller,

I hope this will come in time to save you any more trouble.[2] I have just measured some walls in the same manner as I did before and I find the same thickness in both basal plates and hexagonal prism, viz about $\frac{5}{2000}$ which is rather thicker than you made them. I can only account for my strange blunder by my having two micrometers, one ruled with lines exactly twice as far apart. I will take average of your six measurements and give that as *about* the thickness of walls.[3] With cordial thanks for all your kindness, and grief at causing you so much trouble by my scandalous blunder which I will correct.

Believe me yours very sincerely | C Darwin

P.S. | Unless you happen to have made more measurements, I will give the thickness not on your authority; but if you do give me accurate thickness I will then say that you measured thickness for me.

Copy
DAR 146

[1] The year is given by the reference to the preparation of the third edition of *Origin* (see n. 3, below).
[2] CD probably refers to work carried out by Miller in response to the letter to W. H. Miller, 1 December [1860]. No letter from Miller detailing his results has been found.
[3] The 'blunder' to which CD alludes is his statement in the first edition of *Origin*, p. 231, that the walls of bees' cells were 'about one four-hundredth of an inch in thickness; the plates of the pyramidal basis being about twice as thick.' The third edition was corrected to read (*Origin* 3d ed., p. 252):

> These walls, as Professor Miller has kindly ascertained for me, vary greatly in thickness; being, on an average of twelve measurements made near the border of the comb, $\frac{1}{353}$ of an inch in thickness; whereas the basal rhomboidal plates are thicker nearly in the proportion of three to two, having a mean thickness, from twenty-one measurements, of $\frac{1}{229}$ of an inch.

From J. D. Hooker 28 December 1860

Kew
Dec 28 /60

Dear Darwin

Your notes about the little Compos. are very interesting indeed—[1] I would certainly send them to G. C. which will woodcut the illustration—[2] The case is a great extension of the well-known fact, that many Compositæ (the common groundsel is one)[3] have a layer of vesicles full of mucus on the achænium, which swell & burst when wet—& so hold down the seed to ground, but I know of no case like this, of the mucus cells being developed only on the lower sides of the *oblique* Achænium—[4] This side is the *outer* as respects the position of the floret in the capitulum, & it is a curious fact that various other Compos: with *erect* (or but slightly inclined) achænia have a similar glandular development on the outer surface only—& especially such as inhabit S. Africa— The Selection principle comes in here capitally— The laws of growth render all the achænia in most capitula of comp. lop sided, especially such comps. as have the outermost flowers female (the case with all these). in falling it is the lop-side that touches the ground, & the most glandular that adhere best & hence grow &c &c &c.[5]

The same phenomena occur in **seeds** (not seed vessels as in Compos.) of Cruciferæ, & it is curious that that genus of Crucif. in which the viscid coat is thickest is Australian. I have followed the subject a little in Crucif.— Some Crucif. you know have dehiscent fruit: others indehiscent: I find the mucous coat general in the dehiscent—never in the indehiscent.—but some genera have upper half of capsule indehiscent, lower dehiscent. in these **all** the seeds are mucous, or **all** not mucous.— this is only inteligeable on Nat. Select. principle & shews that the simple physiolog. peculiarity of testa is more permanent than structure of Capsule.— this further falls in with my principle of classification of the Order— What is more odd is that the mucous or dry testa is a more constant character than form & direction of Embryo! I did want to write a good paper on whole subject in reference to classification & origin of *genera* & species too in order Cruciferæ, but have not strength of mind for it.[6]

You ask how my Book on plants brews?[7]—as flat as beer with bad malt.— I have no *Torula cerevisiæ*[8] in me any more.— I have arrived at a hale old age & shall take care of it. & am seriously thinking of taking out a commission of Lunacy against my one hard working friend Huxley— I feel all sorts of inward monitors advising me against over excitement or [9]

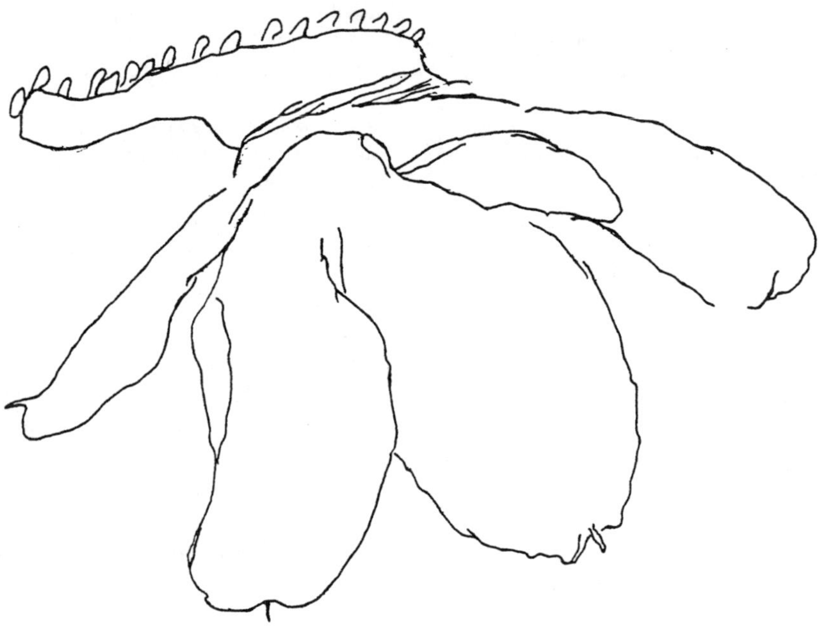

AL incomplete
DAR 100: 143–4, 146–8

CD ANNOTATIONS
3.1 You . . . or 3.5] *crossed ink*
Top of first page: 'Pumilio argyrolepis V. G. Chronicle for my paper p. 5. Jan 5. 1860'[10] *ink*

CD note:[11]
I think I might describe seeds of Pumilio & say I wondered by what gradation produced— D Groundsel, I suppose no Bladders then how is this?— As achenia becomes more oblique. bladders more developed??—

[1] See letter to J. D. Hooker, 26 December [1860].

[2] The paper, a note on the achenes of *Pumilo argyrolepis*, was sent to the *Gardeners' Chronicle and Agricultural Gazette*. It was published in the issue of 5 January 1861. See *Correspondence* vol. 9, letter to *Gardeners' Chronicle*, [before 5 January 1861]. The paper includes two woodcuts of the underside of the mature flower.

[3] Groundsel (*Senecio vulgaris*) is a common yellow-flowered weed of waste ground.

[4] CD's paper described the way in which the underside of the fruit (the achenium) is covered by numerous little bladders; when the bladders become damp, they burst and make the seed case sticky, allowing the achenium to rest upright and become firmly attached to an object.

[5] In his paper, CD stated: 'Whether the tendency of the achenia to place themselves upright be of any service to the plant would be hard to determine; but it seems possible that the salver-shaped calyx, which, as we have seen, so rapidly absorbs moisture and carries it down to the lower surface of the achenia, might aid in utilising dew or showers of fine rain.' (*Collected papers* 2: 38).

[6] The plant family Cruciferae was, in Hooker's time, believed to be a natural family characterised by the pod-like capsules that enclose the seeds and by the shape of the embryo. Nevertheless, the grouping of the smaller divisions and the definition of the genera gave rise to much dispute. Hooker had been interested in the family since his early work on the Kerguelen Land cabbage, a plant that he considered the simplest of all Cruciferae. See *Correspondence* vol. 4, letter from J. D. Hooker, [*c.* 4 March 1847].

[7] See letter to J. D. Hooker, 17 December [1860] and n. 7.

[8] *Torula cerevisiae* is the yeast micro-organism that acts in the fermentation of beer.

[9] The diagrams were drawn by CD and enclosed with the letter to Hooker, 26 December [1860]. They are similar to those published in the *Gardeners' Chronicle* (*Collected papers* 2: 37).

[10] Hooker probably supplied CD with the name *Pumilo argyrolepis* in the missing portion of the letter. The annotation refers to the species whose seeds CD had just described; in the manuscript sent to Hooker, he had called the plant *Siloxerus* (see letter to J. D. Hooker, 26 December [1860]). CD erred in writing '1860' rather than '1861'.

[11] The note is in DAR 100: 145. It was marked 'CH VIII' in brown crayon, a reference to chapter 8 of CD's 'big book' on species. In that chapter, entitled 'Difficulties on the theory of natural selection in relation to passages from form to form' (*Natural selection*, pp. 339–86), CD discussed the difficulties associated with supposing gradual transitions in the functional change of organs or parts.

To John Innes 28 December [1860][1]

Down Bromley Kent
Dec. 28th

Dear Innes

Many thanks for your kind & pleasant letter. I cannot give a very good account of poor Etty, who had a relapse a week ago, which threw her a good deal back &

she has hardly yet got up to her standard a fortnight ago & that was not high.—[2] It is enough to make one despair.—

I am very sorry that you have such poor success in building: I am sure you must miss a home of your own with your various animals & pets. I had heard that Ainslie wants £4000, which it is impossible to believe he will ever get.[3] The rogue must be cleaned out some day & have to sell at fair price.—

As for news we have none. I daresay you heard that Christopher Osborne is dead[4]—the third death this year in the Friendly Club![5] You will have heard of the new Doctor, next door to where Mr Edwards is now staying.[6] I have not seen him yet.—

What you say about my Book pleases me & I *do* look at it as high compliment. I never expected to convert people under 20 year, though firmly convinced *now* that I am in the main right.— For a week hardly passes without my hearing of some good judge coming some little way with me. And those who go an inch will surely have to go a yard with me. By far the greater part of the opposition is just the same as that made when the sun was first said to stand still & the world to go round. I am now passing through the press a new Edition.— Did you see the Quarterly Review, the B. of Oxford[7] made really splendid fun of me & my grandfather.[8]

On account of Etty we lead a more retired life than ever, though this seems hardly possible. But to night Emma & some relations in the House are going to a Ball at the Lubbocks. I see John[9] occasionally but have not seen any other member of the Family for an age.—

I am heartily glad that you can give a pretty good account of your son & M^rs. Innes,[10] to whom pray give our very kind remembrances & Believe me Dear Innes | Yours very truly | C. Darwin

Cleveland Health Sciences Library (Stecher collection)

1 Dated by the reference to the review of *Origin* by the 'B. of Oxford' ([Wilberforce] 1860).

2 Henrietta Emma Darwin had been ill for some time. She had been recuperating from severe attacks suffered in July and again in October 1860.

3 Robert Ainslie, formerly of Tromer Lodge, Down, was a methodist minister, apparently at that time residing in Great Grimsby, Lincolnshire (see *Crockford's*). CD probably refers to the asking price on Tromer Lodge, which Ainslie had vacated in 1858. At the time, CD wrote to William Erasmus Darwin that 'the beast is ruined & soon will be clear of the village.—' See *Correspondence* vol. 7, letter to W. E. Darwin, 14 [May 1858]. Innes, who was hoping to buy a house in Down, had considered purchasing Tromer Lodge.

4 Christopher Osborne had formerly been a bricklayer in Down (*Post Office directory of the six home counties* 1845).

5 CD was the treasurer of the Down Friendly Society.

6 Mr Edwards has not been identified. The 'new Doctor' may have been the surgeon Stephen Paul Engleheart (*Post Office directory of the six home counties* 1862).

7 Samuel Wilberforce was bishop of Oxford. CD refers to Wilberforce's anonymous review of *Origin* ([Wilberforce] 1860).

8 [Wilberforce] 1860, pp. 254–5. See also letter to J. D. Hooker, [20? July 1860].

[9] John Lubbock.
[10] John William Brodie Innes and Eliza Mary Innes.

To J. D. Hooker 29 December [1860][1]

Down Bromley Kent
Dec. 29th.

My dear Hooker

Thank you for your most interesting letter, which I have not had time to consider yet.—[2] The case seems as if it would do hereafter as good illustration of gradation.— Thanks, also, for A. Gray letters.[3] I answered him some time ago about the proposed Review.[4] Huxley will take it in if it seems *good.*—[5] I write now, with many apologies you poor wretch for troubling you, to say that you forgot to answer my question of date of Rafinesque Flora of N. America Part I.—[6] Poor Naturalist as he was, he has good sentence about species & vars. which I must quote in my Historical Sketch & I sadly want the date at once.—[7]

Poor Etty has had another Relapse, but has rallied again to nearly standard a week ago.—

I am profoundly glad that you will take care of yourself.— I was always certain no one's mind could stand your incessant work.— I feel sure in a year or two you will be as brisk as ever.—

I know well that my head would have failed years ago, had not my stomach always saved me from a minute's over work.— I am indeed so glad that you are resolved to work more quietly.— I think I have quoted to you poor Dana's case, & my prediction to him a few years ago.—[8]

My dear old friend | C. D.

DAR 115.1: 83

[1] Dated by the relationship to the letter from J. D. Hooker, 28 December 1860.
[2] Letter from J. D. Hooker, 28 December 1860.
[3] In a letter dated 10 December 1860, Gray wrote: 'Send this note to dear Darwin after perusal' (Archive, Royal Botanic Gardens, Kew).
[4] Asa Gray had asked CD whether he could find a publisher for a review of *Origin* written by Chauncey Wright. See letters to Asa Gray, 11 December [1860] and 14 December [1860].
[5] See letter to T. H. Huxley, 11 December [1860], and letter from T. H. Huxley, [before 14 December 1860].
[6] See letters to Daniel Oliver, 20 December [1860], and to J. D. Hooker, 26 December [1860].
[7] CD included Constantine Samuel Rafinesque among those naturalists who had discussed the transmutation of species in the historical sketch prefixed to the third edition of *Origin*. Referring to Rafinesque 1836, he wrote (*Origin* 3d ed., p. xv):

> Rafinesque, in his 'New Flora of North America,' published in 1836, wrote(p. 6) as follows:— All species might have been varieties once, and many varieties are gradually becoming species by assuming constant and peculiar characters:" but further on (p. 18), he adds, "except the original types or ancestors of the genus."

[8] James Dwight Dana was recovering from a breakdown in his health. For the expressions of concern CD made to him, see *Correspondence* vol. 7, Supplement, letter to J. D. Dana, 15 June [1851].

From Gideon Lincecum 29 December 1860

Long Point. Texas.
29. Decr 1860.

Charles Darwin. M.A.,
Down, Bromley, Kent. England.

Dear Sir.

While in my little, quiet office this evening, carefully examining your valuable work "on the origin of species" &^{c} the fact that I had received, hot from the press, and read with so much pleasure, the "Temple of Nature", which, as far as I know, was the last published work written by your Grandfather, Erasmus Darwin.[1] And now, with emotions of approving interest, find myself profitably perusing the highly useful labours of his truly investigating Grandson, who, himself alludes favorably, to having a grown up son; it occurred to me, and I was forcibly impressed with the thought that I have been here long enough, all the time in possession of good eyes and ears, long ago set in motion on the plane of investigation, from the suggestions of your worthy Grandfathers inimitable works, to enable me to state some facts relative to your favourite theory of natural selection, that may not be offensive to you. With that thought, by natural selection—thoughts breeds, and I am a natural—came the desire to communicate,—not much, for writing has become irksome to me—a few facts of striking character, in natural movements, which are, as I conceive, in harmony with your theory. You record many cases of equal force, mine will be a few more, and at the worst will not damage your Labors, as you can, very conveniently, avoid their consideration.

In my journal of observations I find many cases, applicable to your theory of natural selection, but in my present state of mind, I feel more inclined to state some of my observations on the agricultural successes of one of our many species of Texas ants.[2] It may interest you some. It cannot injure. Consider it. If you like it, and want more, write.[3]

The species of Formica,[4] which I have named Agricultural; is a large brownish red ant, dwells in paved cities, is a farmer, thrifty and healthy; is delligent and thoughtful, making suitable and timely arrangements for the changing seasons; in short, he is endowed with capacities sufficient to contend with much skill and ingenuity; and untiring patience, with the varying exigencies with which he may encounter in the life conflict.

When selects a situation upon which to locate a city, if it is on ordinarily dry land, he bores a hole, around which he elevates the surface three, sometimes six inches; forming a low, circular mound; with a very gentle inclination from the center to its outer limits; which on an average is three to four feet from the entrance. But if the location is made on a low, or flat wet land, liable to inundation, though the ground may be perfectly dry at the time he does the work, he nevertheless elevates his mound in the form of a pretty sharp cone, to the hight of fifteen to twenty inches, sometimes even more, having the entrance near the apex. Around this, and its the same case with the upland cities, he clears the ground of all obstructions, levels and smoothes the surface to the distance of three or four feet from the gate of the city, giving it the appearance of a handsom pavement, as it really is. Upon this pavement not a spine of any green thing is permitted to grow, except a single species of grain bearing grass. Having planted it in a circle around, and two or three feet from the center of the mound, he nurses and cultivates it with constant care cutting away all other grasses and weeds that may spring up amongst it, and all round outside of the farm circle to the extent of one or two feet. The cultivated grass grows luxuriently, producing a heavy crop of small, white, flinty seeds, which under the microscope very much resembles the rice of commerce. When it gets ripe it is carefully harvested, and carried by the workers, chaff and all into the grainery cells, where it is divested of the chaff and packed away; the chaff is taken out and thrown beyond the limits of the pavement.

During protracted spells of wet weather, it sometimes happens that their provision stores become damp, and liable, as they are invariably seeds of some kind, to sprout and spoil. If this has occurred, the first fair day after the rain, they bring out the damp and damaged stores expose them to the sun till they are dry, when they carry back and pack away all the sound seeds, leaving all that are sprouted to waste.

In a peach orchard, not far from my residence, is a considerable elevation, on the top of which, there is an extensive bed of rock. In the sand beds overlying portions of this rock, are five cities of the agricultural ants. They are evidently quite ancient cities, and may have occupied this elevated rock for thousands of years. My term of observations on their manners and customs, has been limited to the time the cattle, by the orchard enclosure, have been kept away from their Rice farms 12 years. Those cities which are outside of the inclosures, as well as the protected cities, are, at the proper season, invariably planted with the ant Rice. And we accordingly see it sprining up, in the farm circle about the first of November, every year. Of late years however, since the numbers of farms and cattle have greatly increased, and the cows are eating out the grass much closer than formerly, preventing the ant farms and every where else from maturing seeds, I notice that the agricultural ants—but no other species of ant—are locating their cities along the turn rows in the fields, walks in the gardens, inside about the gates &c. where they can cultivate their farms without molestation from the cattle. And here, if it was not for the dread I feel of offending that host of investigators, who know all things, and who are satisfied there is no truth beyond the range of their own observations, I should be almost wicked enough to state it as my opinion, that these granivorous agriculturists, had intentionally gone in, and unobtrusively settled themselves in the unplowed portions of our inclosures for the purpose of escaping from the depredations of the cattle. This species of ant do no damage to the farms, however numerous.

There can be no doubt of the fact, that the peculiar grain bearing grass, mention above, is intentionally planted; in farmer like manner, the ground upon which it stands, is carefully divested of all other grasses and weeds during the time of its growth, and that when it is ripe and the grain taken care of, they cut away the dry stubble, and carrying it out of the way, leave the pavement unincumbered until the insuing autumn, when the same ant rice, and in the same circle appears again, receiving the same agricultural attention as did the previous crop, and so on, year after year, as *I know* to be the case, in all situations where they are protected from graminivorous animals.

I have no theological clique to curb the play of my mind such as it is, nor is it fettered to the cosmogony of any priestly doccuments, ancient or modern; and, so, I may here undisturbed, suggest that it may not exceed a few short millions of years, since this now, bold, healthy, thrifty, city paving, agricultural species of emmet, belonged perhaps, to a feeble variety of hunters and herders, who, in the struggle for existance, were entirely dependent on the natural droppings of the seeds of vegetation, such insects and worms as their hunters could master, and the Aphis for subsistence. Then the "struggle for existence" was difficult, alike to all the communities of that variety. At a certain period however, and near the gates of a particular city, a seed of the favourite grain bearing grass by carelessness had been left,—or if you like, and it does not overreach the bounds of their intelligence in my estimation, some thoughtful ant planted it there for experiment.— Be this as it may, the seed sprouted and grew up to a flourishing tuft, producing a bountiful crop of the favourite Rice; which in due time was, by the inhabitants of the city, carefully harvested. This, in addition to what had been collected from the country around, smartly enlarged their stores of food, whereby greater numbers would survive the winter, in greater strength they would be able to compete in the *struggle for life*, during the coming season; again a few plants of the rice would be tried; the results of course, would be encouraging; and thus by little and little the agricultural habit gradually obtained. By the good effects of less labour and more food, the population of the city were steadily increasing in form, size, vigor, and numbers. And so, the slow, but sure modefying principles and powers of natural selection, rolled on, gathering as it progressed through an unimaginable series of ages until they found themselves occupying a great paved city, surrounded by a stately farm and crowded with a lusty, warlike population. colony after colony had gone out, and, carrying with them the agricultural habit, they had dotted the far off regions with their robust, dominant, farm cities; all so changed in strength, habits and general appearance, that they bore no similarity to the type of the original variety from which they had branched.—But there were none of the parent variety to be found for comparrison. they were extinct. Had been exterminated by the forces, under favorable conditions, of Natural Selection. The no-cornmaking, pastoral *Abel*, had been overcome,—rooted out by the vigorous, prolific, Agricultural *Cain*. Thus, the transitional link between this, now distinct and wonderfully developed species, and probably, the *well* digging Species, who are about the same size and colour, subsisting on the foilage of vegetation, are, to Science forever lost.

I might ennumerat twenty odd Species of Texas ants. The two species I have alluded to are however, the most interesting. But I must abstain from any attempt at giving you even a glimpse of the history of their stupendeous public works, governmental systems; laws, civil and military; slaves, prisoners, distructive wars &c. &c. It would require a year or two, to perform the work in the stile it merits

Old men, like myself, can do but little of anything, and ought not to promise themselves much; but I think I will, sometime when I feel in a proper condition for it, write out a short account of what I have seen during the past twelve years, amongst our abundant Animal and vegetable fossils; our insect kingdom and our blooming prairies.

I am a native American, born in the smoke of our first revolution; was raised and have always been a dweller in the wild border countries, and now, I think the chances are pretty good for me to make my exit in the turmoil and smoke of another revolution.[5]

Most Respectfully. | Gideon Lincecum.

Linnean Society of London

CD ANNOTATIONS
0.1 Long . . . write. 2.5] *crossed pencil*
6.15 And here, . . . cattle. 6.21] *crossed pencil*; *enclosed in square brackets, pencil*
8.1 I have . . . Lincecum. 12.1] *crossed pencil*; *opening square bracket before* 'I', *pencil*

[1] Lincecum was an American physician and self-taught naturalist who practised medicine in Texas. He refers to E. Darwin 1803, a work published posthumously. American editions were issued in 1804 and 1808.

[2] Lincecum had been studying several local species of ants. Apparently unaware of their scientific classification, he catalogued and named over fifty species himself (Burkhalter 1965, p. 208).

[3] No letters from CD to Lincecum have been located, but see the letter from Gideon Lincecum, 4 March 1861 (*Correspondence* vol. 9).

[4] CD subsequently sent the letter to George Busk, zoological secretary of the Linnean Society of London (see *Correspondence* vol. 9, letter to George Busk, 5 April [1861]). A section of the letter, from the beginning of the third paragraph to the end of the sixth, was read at a meeting of the society on 18 April 1861 (*Proceedings and Journal of the Linnean Society* 6 (1861): 29). The letter was corrected and edited, possibly by Busk.

[5] Lincecum was born in 1793. He refers to the political turmoil in the United States over the question of slavery and states' rights that threatened to dissolve the union.

To W. H. Miller 31 December [1860][1]

Down Bromley Kent
Dec. 31.

My dear Miller—

How infinitely kind you have been to do the work so completely.[2] But I am in truth almost ashamed to think how much of your valuable time you have given to this work.—

At some future time when I give all my experiments on the combs, I shall give your results in rather fuller detail. Now, I will give on your authority only the general result.—[3]

From several circumstances (which I did not previously think of) the measurements at the upper thick end of the comb are, I think, most trustworthy; for the Bees might have failed in building from want of wax, & their combs were taken from Hives which naturally perished; or they might have subsequently gnawed away wax from the border-cells which they did not mean to fill with honey & I know they often economise every particle of wax.

I am extremely much pleased that I turn out not so *grossly* wrong, as I feared, for it has annoyed me more than you would suppose. My having been not grossly wrong has, however, evidently been mere chance.

My original rough measurements were made, I remember, in middle of a great piece of comb;[4] & the two or three measurements made the other day & when I wrote to you, giving up the ghost, were made on border.

Again let me thank you most sincerely for your very great kindness, & believe | My dear Miller | Yours very truly | C. Darwin

American Philosophical Society (189)

[1] The year is given by the inclusion of Miller's information in the third edition of *Origin*, published in April 1861.

[2] Miller's letter has not been found. See letters to W. H. Miller, 1 December [1860] and 27 December [1860].

[3] See *Origin* 3d ed., p. 252. See also letter to W. H. Miller, 27 December [1860].

[4] CD's notes on bees' cells, primarily dating from 1858, are in DAR 48 (ser. 2): 5–75.

APPENDIX I
Translations of letters

From Charles Théophile Gaudin[1] [15 August 1860][2]

In recent years we have had a new race of bees, which have a longer proboscis and can suck red clover. A hive of honey weighs up to 90 lb., whereas a good hive of the ordinary bee weighs only 60 lb. This bee does not sting; it likes humans. It is called the Ligurian bee It is raised in the Grisons,[3] and a Swiss has this summer taken a shipment of them to the United States.

I shall be delighted to provide Mr Darwin with any information he would like on this subject[4]

Charles Th Gaudin

Eglantine, Lausanne

Incomplete
DAR 47: 164

[1] For the transcription of this letter in its original French, see pp. 321–2. This incomplete note, written in pencil, was given to CD by the original recipient of the letter, probably Hugh Falconer (see nn. 2 and 3, below).

[2] The date is written at the top of the letter in Hugh Falconer's hand. The letter may have been given to CD when he visited Falconer in London (see letter to Charles Lyell, 28 August [1860]). According to Emma Darwin's diary, CD went to London on 21 August 1860.

[3] Grisons is a canton in south-east Switzerland. At this point, the letter was torn into two parts. The parts were subsequently glued together and attached to another piece of paper, on which notes about the sequence of fossil elephants were written in Falconer's hand. Gaudin, a Swiss palaeobotanist, studied the fossils of the Arno River valley and incorporated some of Falconer's findings in Gaudin 1859. Gaudin refers to comments made about bees and clover in *Origin*, pp. 94–5. The information given in the note was used, without attribution, in the fourth edition of *Origin* (see Peckham ed. 1959, p. 184).

[4] The information in the note related to CD's researches on the role of insects in the cross-fertilisation of leguminous plants such as clover; it also includes information about a new kind of bee, called the Ligurian bee, that had recently been introduced into Britain. On the first point, see the letter to the *Entomologist's Weekly Intelligencer*, [20 June 1860]; on the second, see the letters to the *Cottage Gardener*, [after 8 May 1860], and to W. B. Tegetmeier, 20 July [1860] and n. 8.

From H. G. Bronn[1] 13 [or 15][2] October 1860

Heidelberg
13 [or 15] Oct. 1860

Honoured Sir:

I was pleased to have received another letter from you and to see how you are still untiringly occupied with the further development of your theory.[3] You say that you are in full agreement with most of my remarks about your theory, and that only in two things are you unable to agree with my opinion in which it seems that I have misunderstood you,—with regard to a recurring glacial period, and with regard to the assumption that the development of creation on the entire earth's surface proceeded at the same rate. But I do not remember having raised objections to your theory in these respects; and you express yourself in your letter so vaguely about the deviation in our mutual ideas that I am not in a position to answer you more specifically. If the contrast between our mutual views seems great enough to warrant further discussion, then please point out to me the passages in my book *where* in this regard my different opinion is expressed.

You say further that we explain many electrical phenomena on the basis of a theory of electricity without knowing what electricity really is,—and that we are able with as much justification to formulate a theory about the origin of life (without being able to prove that origin?). But, the two cases are far from being analogous. We *know* that all matter possesses certain properties and that each particular element also has its own special properties. Now we assume that electricity is *at the same time* associated with certain elements, along with their remaining properties. For we also explain many phenomena on the basis of gravity by assuming that gravity is a property pertaining to most bodies, although we do not know how gravity is connected to matter. We satisfy ourselves in saying that gravity is a property of most matter. However, we cannot say this of life. Life is not a property that is inherent in matter. *Life presupposes life*, and *life generates life*: but where there is no life, then no life can arise. Matter is therefore independent of life and life is not necessarily bound to any kind of matter. Moreover, it is a curious fact that the *entire life* of each species is another life apart from the life of any other species! and propagates itself in that way.

We think that electricity and gravity are *associated with matter*; the Lebenskraft, however, is not associated with matter, and *we require for it a further explanation which is not needed for them.*[4]

As regards your question about Brehm's subspecies,[5] he assumes that each subspecies has a somewhat different station, different nourishment, and a different habitat. Many are what one calls climatic varieties (Corvus corone; Corvus frugilegus); others could be designated 'station-varieties', 'food-varieties', etc.

I have seen that Brehm presented many dozens of examples of each subspecies of various species claiming that *all* individuals of *a* subspecies derive from the same country or the same station. However, Brehm admits that these subspecies occasionally will pair with each other if a male finds no female of his own subspecies, or the reverse. This is in fact the case with Corvus corone and C. frugilegus now and then in Thüringen, because in Thüringen the summer residence of the north-eastern C. frugilegus adjoins that of the south-western Corvus corone. In winter both live together in the south-west.

When I was a boy of 15 I distinguished 2 subspecies of Certhia long before Brehm distinguished them, in fact long before Brehm's name was known. Certhia familiaris

Brehm lives in the woods; C. brachydactyla *Brehm* lives in gardens. According to Brehm the latter has shorter toes than the former. I have never been able to find *this* distinction clearly. However, I knew before Brehm said so that the variety living in gardens has an entirely different song, lays eggs that are different in colour, and also constructs her nest somewhat differently. It is known that individuals of one variety prefer to breed with each other rather than with those of another variety and that they are therefore in the position to propagate their variety, and thus Brehm's subspecies are nothing other than climatic- and station-varieties that propagate within their own group.

Climatic varieties are known inside and outside Germany; why not also station-varieties and why should not these station-varieties propagate themselves *as a rule* through **in***breeding* as the climatic-varieties do? Still, I do not want to maintain that *all* subspecies that Brehm has adduced are actually so widely developed that they generally reproduce through **in***breeding*.

You would like to know how the German translation of Origin of Species is selling. I see from the announcement that the price is 4 guilders or somewhat less than 7 shillings, which after deducting the trade discount reduces to 6 shillings. Schweizerbart wrote to me recently that the first sale of the translation was not as strong as he had expected, but that it yet continues steadily.[6]

If I can provide you with some additional information, I would be pleased to do so. In the meantime I remain with the highest esteem your devoted servant | H. G. Bronn

DAR 160.3: 317

CD ANNOTATIONS

Verso of last page: '⟨ ⟩ee about Glacial Periods & equal Development over world | Bronn seems to believe largely in Brehm—says he distinguished the two Certhias himself before he had heard of Brehm'[7] *ink*; 'Ch. IV'[8] *brown crayon*

[1] For the transcription of this letter in its original German, see pp. 425–7.

[2] The manuscript has deteriorated, and only the bottom half of the numeral is now visible.

[3] See letter to H. G. Bronn, 5 October [1860].

[4] Bronn had discussed this subject in some detail in *Untersuchungen über die Entwickelungs-Gesetze der organischen Welt* (Bronn 1858b, pp. 75–82).

[5] In the letter to H. G. Bronn, 5 October [1860], CD had asked Bronn whether he was familiar with Christian Ludwig Brehm's work on the birds of Germany (Brehm 1831) and whether he recognised the large number of subspecies that Brehm identified in this work.

[6] The firm of E. Schweizerbart, based in Stuttgart, published the German translation of *Origin* (Bronn trans. 1860). Christian Friedrich Schweizerbart was the head of the firm.

[7] See n. 5, above.

[8] CD's annotation refers to the chapter of *Origin* that discusses natural selection. In the third edition of *Origin*, published in 1861, CD added a section to chapter 4 in which he responded to Bronn's criticism.

APPENDIX II
Chronology 1860

This appendix contains a transcription of Darwin's 'Journal' for the year 1860. Darwin commenced his 'Journal' in August 1838 and continued to maintain it until December 1881. In this small notebook, measuring 3 inches by $4\frac{1}{2}$ inches, Darwin recorded the periods he was away from home, the progress and publication of his work, and important events in his family life.

The version published by Sir Gavin de Beer as 'Darwin's journal' (de Beer ed. 1959) was edited before the original 'Journal' had been found and relied upon a transcription made by an unknown copyist. The original, now in the Darwin Archive, Cambridge University Library (DAR 158), reveals that the copyist did not clearly distinguish between the various types of entries it contains.

From 1845 onward, Darwin recorded all that pertained to his work (including his illnesses, since these accounted for time lost from work) on the left-hand pages of the 'Journal', while the periods he was away from home and various family events were noted on the right-hand pages. In order to show clearly Darwin's deliberate separation of the types of entries he made in his 'Journal', the transcription has the left- and right-hand pages labelled, with the left-hand page of an opening preceding the right-hand page.

All alterations, interlineations, additions, and the use of a different ink or pencil have been noted. In addition, the editors have inserted relevant information throughout this transcription of the 'Journal' for 1860. These interpolations are enclosed in square brackets to distinguish them from Darwin's own entries.

[Left]

1860

Jan. 9th. Began looking over M.S. for Work on Variation[1] (with many interruptions)

Letters & preparations for foreign Editions[2]

March 24th Began Introduction to Vol. *on Variation [*interl*]

June 10th Finished **2d** Ch. on Pigeons.[3] arranged papers for next Ch. & Experiments.[4]

Aug 11th. Began Ch. III[5]

During July at Hartfield & afterwards & at Eastbourne, worked at Drosera[6]

[Right]

1860

Jan. Murray says 6000 copies of my Journal [7] [*interl*]

Jany 7th The 2d Edition, *3000 copies [*interl*] of Origin was published

[24–7 January. Visited London. Attended meeting of the Philosophical Club on 26 January.][8]
[10 February. Attended lecture by Thomas Henry Huxley at the Royal Institution of Great Britain.][9]
[14–16 February. John Stevens Henslow visited Down.][10]
[27–8 February. Visited London. Called on Andrew Crombie Ramsay. Came home unwell with an attack of fever and pleurisy.][11]
[5–6 March. Went to London with Henrietta and Leonard. Saw Dr Headland.][12]
[9–12 March. Charles and Mary Elizabeth Lyell visited.][13]
[10–14 March. The Tollets and Emily Catherine Darwin visited.][14]
[28 March. Went to London.][15]
[5 April. Joseph Dalton Hooker came to Down.][16]
[7 April. Thomas Henry and Henrietta Anne Huxley came.][17]
[21 April. Went to London. Attended meeting of the Philosophical Club on 23 April.][18]
[28 April. Robert Mackintosh and children came. Etty poorly.][19]
May 22. The 1st Edit of Origin in U. States was 2500 copies.—[20]
[3 June. Henry Holland came to examine Henrietta.][21]
June 28th to July 7th Sudbrook Park[22]
[7 July. Visited Hooker at Kew on return to Down.][23]
July 10 to Aug 2d Hartfield (Etty long ill)[24]
[26 July. Henry Holland visited.] [25]
[21 August. Went to London.][26]
[8–14 September. Hensleigh and Fanny Mackintosh Wedgwood, Erasmus Alvey Darwin, and Sarah Elizabeth Wedgwood visited.][27]
[18 September. Edward Cresy and his wife visited.][28]
Sept. 22d to Eastbourne. Returned Nov. 10th seven weeks, owing to Etty illness.[29]
[29 October. Henry Holland visited.][30]
[23–8 November. Susan Elizabeth Darwin visited.][31]
[8–11 December. Emily Catherine Darwin visited.][32]
[15 December. Thomas Henry Huxley visited.][33]

	£[34]
May 28 1st Edit	180
July & Oct. 2d Edit	636.13.4
May. American Edit	21.17.6
— (61) do 1861.	10.10
1861 April. 30 3d Ed	372[35]

1 *Variation*, published in two volumes in 1868, was the first of a three-part work CD intended to publish to explain in greater detail the operation of natural selection. He described the project in a letter to his publisher John Murray of 22 December [1859] (*Correspondence* vol. 7): 'I think I have got good scheme for larger work of three separate volumes with separate titles & indices for each but in addition with one general Title.—' His plans changed even as he began to prepare the work. In a letter to Charles Lyell, 6 June [1860], CD told Lyell:

> If my M.S. spreads out, I think I shall publish one volume exclusively on "Variation of animals & plants under domestication". I want to show that I have not been quite so rash as many suppose.—

[2] CD was attempting to arrange the publication of an American edition of *Origin* and for translations into French and German. Through the efforts of Asa Gray and Heinrich Georg Bronn, editions of *Origin* were published in the United States and in Germany in 1860. A French edition was not available until 1862.

[3] Chapters 5 and 6 of *Variation* were devoted to pigeons. CD had begun writing the first chapter on pigeons in June 1858, but his work was interrupted by the receipt of Alfred Russel Wallace's manuscript setting out his species theory. CD finished his 'skeleton & Pigeon M.S.' before beginning to write his 'Abstract' (*Origin*). See *Correspondence* vol. 7, 'Journal' (Appendix II).

[4] The 'Experiments' CD refers to were probably a series of botanical experiments he was planning to make on various topics. See especially the letters to J. D. Hooker, 7 June [1860] and 12 [June 1860], and to Asa Gray, 8 June [1860].

[5] In the letter to Charles Lyell, 11 August [1860], CD states that he has 'recommenced work & am at dogs'. This material eventually formed part of chapter 1 of *Variation.*

[6] See nn. 24 and 29, below. When CD came to publish the results of his study of *Drosera* fifteen years later, he opened the work with the autobiographical remarks (*Insectivorous plants*, pp. 1, 2–3):

> During the summer of 1860, I was surprised by finding how large a number of insects were caught by the leaves of the common sun-dew (*Drosera rotundifolia*) on a heath in Sussex . . . but it was soon evident that Drosera was excellently adapted for the special purpose of catching insects, so that the subject seemed well worthy of investigation.

[7] In 1860, John Murray issued a new printing of CD's journal from the Beagle voyage, *Journal of researches.* The text was identical to the second edition published in 1845 and reprinted in 1852, but it included a 'Postscript' (p. vii), in which CD corrected three errors in the text. See Freeman 1977, pp. 36, 40.

[8] Emma Darwin's diary; Philosophical Club Minute book, Royal Society of London. During this trip to London, CD consulted the London physician Edward Headland (see letter to J. D. Hooker, [22 January 1860]). It is likely that CD also visited John Murray (see letters to John Murray, 23 [January 1860] and [25 January 1860]).

[9] Letters to T. H. Huxley, [26 January 1860] and 2 [February 1860]; to Charles Lyell, 12 [February 1860]; and to J. D. Hooker, 14 February [1860]. Huxley's lecture was entitled 'On species and races, and their origin' (T. H. Huxley 1860a).

[10] Emma Darwin's diary; letter to J. S. Henslow, 9 February [1860].

[11] Emma Darwin's diary; letters to A. C. Ramsay, 23 February [1860], and to J. D. Hooker, 3 March [1860].

[12] Emma Darwin's diary; letter to W. E. Darwin, [4 March 1860]. See n. 8, above.

[13] Emma Darwin's diary; letter to J. D. Hooker, 12 March [1860].

[14] Emma Darwin's diary. The members of the Tollet family who visited were probably Emma's childhood friends, Georgina and Ellen Harriet Tollet.

[15] Emma Darwin's diary. CD may have visited the British Museum. See letter to G. R. Waterhouse, 1 April [1860], and letter from Frederick Smith, 3 April 1860.

[16] Emma Darwin's diary; letter to J. D. Hooker, 18 [March 1860].

[17] Emma Darwin's diary; letter to T. H. Huxley, 20 March [1860]. On 10 April 1860, Emma recorded in her diary: 'Ball at High Elms'. High Elms, Kent, was the home of the Lubbock family.

[18] Philosophical Club Minute book, Royal Society of London. See also the letter to Charles Lyell, 15 April [1860], for CD's intention to attend the soirée given by Benjamin Collins Brodie, president of the Royal Society, on 24 April and to breakfast with Lyell the next morning.

[19] Emma Darwin's diary. Robert Mackintosh, the brother of Frances Mackintosh Wedgwood, visited until 1 May and his children, Ronald and Eva, until 7 May. Henrietta Emma Darwin fell ill with

a 'remittent Fever' on 28 April (Emma Darwin's diary; letter to J. D. Hooker, 7 May [1860]). The fever was later diagnosed as 'a form of Typhus fever' (letter to J. D. Hooker, 11 May [1860]).

20 See letter to Asa Gray, 22 May [1860] and the first enclosure. CD received £21 17*s.* 6*d.* as his share of the profits on the sale of the first American edition.

21 Emma Darwin's diary. Henry Holland, physician in ordinary to Queen Victoria and a distant relation of the Darwin family, was called in owing to Henrietta Emma Darwin's continuing illness. He was consulted again on 26 July, when the family was in Hartfield, Sussex.

22 CD visited Edward Wickstead Lane's hydropathic establishment at Sudbrook Park, near Richmond, Surrey.

23 Letters to J. D. Hooker, [2 July 1860] and 12 July [1860].

24 Hartfield, a village in Sussex, was the home of Emma's sisters Sarah Elizabeth Wedgwood and Charlotte Langton, the wife of Charles Langton.

25 See n. 21, above. For Henry Holland's diagnosis of Henrietta's illness and prospects for recovery, see the letter to J. D. Hooker, 29 July [1860].

26 Emma Darwin's diary. CD visited Hugh Falconer (letter to Charles Lyell, 28 August [1860]).

27 Emma Darwin's diary.

28 Emma Darwin's diary. At the time of Cresy's visit, CD was experimenting on the effect of various substances in stimulating a response in the leaves of *Drosera* (see letters to Daniel Oliver, 11 September [1860] and 15 [September 1860]). Cresy subsequently assisted CD's research by putting him into contact with the organic chemist August Wilhelm von Hofmann (see letters from A. W. von Hofmann to Edward Cresy, 13 October 1860 and 27 October 1860, and letter from Edward Cresy, 30 October 1860).

29 See n. 6, above. The Darwins, who hoped that a visit to the seaside would improve Henrietta Darwin's health, stayed at 15 Marine Parade, Eastbourne, on the Sussex coast. Emma went ahead with Henrietta after a short stay at Hartfield (see n. 24, above). The Darwins' stay in Eastbourne was prolonged owing to Henrietta's 'terrible attack of sickness' on 22 October (letter to John Innes, 26 October [1860]). She was violently ill for over a week, so much so that the doctors thought her life was in danger (letter to Asa Gray, 31 October [1860]).

30 Emma Darwin's diary.

31 Emma Darwin's diary.

32 Emma Darwin's diary.

33 Letter from T. H. Huxley to J. D. Hooker, 19 December 1860 (L. Huxley ed. 1900, 1: 222–3). Huxley wrote: 'I was at Down on Saturday and saw Darwin. He seems fairly well, and his daughter was up and looks better than I expected to see her.'

34 Written in pencil. The list is a record of CD's share of the profits for the first three English editions of *Origin* and the first and second American editions. For a transcript of John Murray's records on *Origin*, see Peckham ed. 1959, pp. 775–6.

35 Written beneath deleted '480'. John Murray had made an accounting error on CD's share of the profits for *Origin* 3d ed. See *Correspondence* vol. 9, letters to John Murray, 30 April [1861] and 3 May [1861].

APPENDIX III
Presentation copies of *Origin*

Volume 7 of *The Correspondence of Charles Darwin* (1858–1859) included an appendix entitled 'Presentation copies of *Origin*' (Appendix VIII). The list of those persons to whom Darwin sent presentation copies of *Origin* was compiled, in the absence of CD's manuscript list, on the basis of evidence provided by the extant correspondence. Using the correspondence, the editors identified 45 recipients of either the first or second edition of *Origin*.

Since the publication of volume 7, however, the editors have discovered Darwin's own manuscript lists of the presentation copies of all of the various editions of *Origin*, as well as his lists for other published works, in DAR 210.18.[1] The list of the presentation copies of *Origin* reveals that Darwin actually presented a total of ninety copies of the first edition, twenty-two of the second edition, and three of the German translation (retaining one copy of each for himself).

In comparing the list based on correspondence (*Correspondence* vol. 7, Appendix VIII) with the manuscript list, it is interesting to note that at least one name included in the first list, that of William Jardine, is not present on Darwin's own list. A possible explanation for this apparent discrepancy is that the publisher of *Origin*, John Murray, himself sent out forty-one review copies of the first edition (Peckham ed. 1959, p. 775). As one of the editors of the *Edinburgh New Philosophical Journal*, Jardine most likely received a copy for review in the journal, for which he subsequently thanked Darwin (see *Correspondence* vol. 7, letter from William Jardine, 20 December 1859). Michael Faraday is also known to have received a presentation copy of the first edition of *Origin*, which was sent to him by Murray rather than by Darwin himself.[2]

Darwin's 'List of copies' for the first edition of *Origin* is reproduced below. Darwin apparently drafted the list between August and October 1859 (see *Correspondence* vol. 7, letters to John Murray, 31 August [1859] and 15 October [1859]). On the first page of this two-page list, Darwin subsequently added, in the blank space of the right-hand side of the page, the names of the three people to whom he planned to send copies of the German translation of *Origin*, which he received in July 1860 (letter to H. G. Bronn, 14 July [1860]).

Also included here is the shorter list Darwin drew up in January 1860 of those to whom he presented a copy of the (slightly altered) second edition of *Origin*. The names that appear on this list fall into several different categories: naturalists like Charles Lyell or Thomas Henry Huxley who were interested in the changes introduced in the new edition; naturalists like George Robert Gray who might have a scientific interest in the volume; others, like John Baily, who had in some way assisted Darwin's studies; and several more family members or friends. It also includes German naturalists and individuals connected with possible foreign editions.

In drawing up the list of foreign naturalists who should receive a copy, Darwin relied on the advice of Joseph Dalton Hooker for botanists and Thomas Henry Huxley for zoologists. For the correspondence relating to presentation copies for European naturalists, see *Correspondence* vol. 7, letters to J. D. Hooker, 15 October [1859], and to T. H. Huxley, 15 October [1859].

The two lists are reproduced below in the same order that Darwin wrote them, but for stylistic purposes the format has been slightly altered. The first page of the original 'List of copies' for the first edition is written with only one column, with the exception of two names of German zoologists added 'by Huxleys advice' and those who were to receive copies of the German translation. On the second page of the manuscript list, Darwin used two columns. Rather than following Darwin's format, we have arranged both pages in two columns, indicating in notes any pertinent information about the original format. Since the names of those receiving the German translation were obviously added to the list at a later date, they have been appended to the bottom of the first list. The second list was originally one column which, through subsequent additions, eventually spilled over into a second column. Again, pertinent information about the arrangement of the original list is given in footnotes.

The two lists are followed by a biographical register, arranged alphabetically, identifying the recipients and giving some indication as to why Darwin would have sent them a presentation copy of *Origin*. Individuals not otherwise mentioned in the volume are only identified here and are not included in the main biographical register.

List of copies

Pictet Geneva
A. De Candolle. do
H. G. Bronn Heidelberg
Quatrefages
I. G. St. Hilaire
D' Archiac
de Verneuil
Keyserling
Barrande
Decaisne
Milne Edwards
Soc. Geologique[3]
Steenstrup

J. D. Dana
Agassiz
Asa Gray
James Hall
Siebold, Munich
Dr Andersson Acad. of Svense Stockholm[4] }
Agardh Lund, Sweden } per Hooker
Grisebach Gœttingen Hanover }
Oswald Heer Zurich
Caspary Kœnigsberg
Dr. Weddell Bagneres de Bigorres Haute Pyrenees

Hon. W. Elliot. Madras
E. Blyth Calcutta
Dr Thomson do
G. H. Thwaites Ceylon
Dr Buist Bombay
C. Moore. Sydney
Dr F. Müller Adelaide
Richard Hill Spanish Town
R. Swinhoe Amoy
A. R. Wallace, care of Hamilton Gray & Co. Singapore
Leuckhart[5] } by Huxley advice
Victor Carus? } by Huxley advice
W. B Tegetmeier
G. Waterhouse
S. P. Woodward
Prof. Owen
Fred Smith
Dr Gray[6]
H. C. Watson
Sir H. Holland
Dr Lindley
Royal Soc
Linnean Soc
Geolog. Soc.
J. Crawfurd 4 Eaton Place
J. Gould
Dr Falconer
Dr Hooker
J. Prestwich
Herbert Spencer
Dr Carpenter
Prof. Ramsay
Prof. Huxley
L. Horner
J. G. Jeffreys Esq
J. Lubbock
G. Norman
Rev J. Innes
Earl Stanhope
Hensleigh W.
Harry W.
Jos. W.
F. W.
Miss. W[7]
E. Darwin
Admiral FitzRoy
Hon. Charles Murray
Self.
Miss Darwin[8]
— G. Tollet
Rev W. D. Fox
Eyton
L. Jenyns
Henslow
Buckman
Sedgwick
Phillips
Brent
Dr Whewell
Herschel
Dr Lane
Bunbury
E. Vernon Harcourt
Rev C. Kingsley
Prof. Miller
J. M. Herbert
T. V. Wollaston
R. Chambers.
Sir C. Lyell
Miss Butler

German Edition 4 copies[9]
Self
Huxley
Van der Hoeven
Prof. Moleschott

Second Edition

(1) Mr H. Hemmings Baldwins Gate | Whitmore, Newcastle Staffordshire
(2) Self.[10]
3 —self for John Morris[11]
(3)[12] William
4. Henry Norman[13]
5 Sir C. Lyell

6 G. R. Gray
7 Adam White
8 Sulivan
9 Lonsdale
10 Tomes
11. E. Cresy
13. Asa Gray for new Edit.
14. H. Wedgwood
15. Mr J. Baily
16 Miss Allen
17. Huxley—
18. Talandier[14]
19 German Bookseller[15]
20. F. Moore, India H
21. Petermann.
22 Kölliker per Huxley.
23 Liebig.

[1] This oversight was a result of a peculiarity in the previous archival organisation of DAR 210, housed in a box containing 34 different folders. DAR 210.18 itself was previously divided into four different folders, only one of which was ever given to readers who ordered this material. The presentation lists for the various editions of *Origin*, as well as lists for *Coral reefs* and for subsequent publications, were discovered in the remaining 3 folders.

[2] Letter from Michael Faraday to John Murray, 2 December 1859 (John Murray (publishers) Archive).

[3] The Société Géologique de France was founded in 1830.

[4] CD wrote 'Acad. of Svense' above the entry.

[5] CD obviously added this and the following name at a later date: they are written in the blank space in the middle of the page preceding the presentation list for the German edition.

[6] John Edward Gray.

[7] Sarah Elizabeth (Elizabeth) Wedgwood.

[8] Emily Catherine (Catherine) Darwin.

[9] The names of those who were to receive the German translation of *Origin* were written in the blank space on the right-hand side of the page. They were added at a later date, probably in July 1860.

[10] CD deleted this entry in pencil.

[11] CD wrote 'for John Morris' in pencil above the deleted 'for anyone'.

[12] From this point down to number 13, CD renumbered the list in pencil, increasing the units by one.

[13] It has not been possible to identify Henry Norman.

[14] It appears from the arrangement of the names that CD added new names to the original list of thirteen on several different occasions. The second column on the manuscript page has '21', '22', and '23' written above the numbers 18 through 20. Above the entire second column, CD wrote, apparently at an earlier time, in pencil and circled in pencil: 'Perhaps compl to M. Talandier'. Pierre Theodore Alfred Talandier was attempting to secure a publisher for a French translation of *Origin*.

[15] The German translation of *Origin* was published by E. Schweizerbart'sche Verlagshandlung of Stuttgart, directed by Christian Friedrich Schweizerbart.

Biographical register of recipients of *Origin*

Agardh, Jacob Georg (1813–1901). Professor of botany, University of Lund, Sweden. (*DSB*.) CD apparently sent Agardh a presentation copy of *Origin* at Joseph Dalton Hooker's suggestion.

Agassiz, Jean Louis Rodolphe (Louis) (1807–73). Professor of natural history, Harvard University. (*DAB*, *DSB*.) CD had corresponded with Agassiz intermittently since 1841. In 1858, Agassiz presented CD with a copy of his *Essay on classification* (the first part of Agassiz 1857–62), in which he described his own

Agassiz, J. L. R., cont.
view of species that was very much opposed to the theory CD proposed (see *Correspondence* vol. 7, letter to Louis Agassiz, 21 February [1858]).

Allen, Emma (1780–1866). Emma Darwin's aunt who lived together with her sister Frances Allen (1781–1871) at Heywood Lodge, Tenby, Wales. (Freeman 1978.)

Andersson, Nils Johan (1821–80). Swedish botanist who explored the Americas on the *Eugenie* expedition. (*SBL.*) Andersson had worked at the Royal Botanic Gardens, Kew, in 1857 (see *Correspondence* vol. 6, letters to J. D. Hooker, 22 August [1857] and 11 September [1857]). During this time, CD had directed some queries to him concerning the colouring of horses in Norway. Hooker apparently recommended that CD send Andersson a presentation copy of *Origin* (see *Correspondence* vol. 7, letter to J. D. Hooker, 15 October [1859]).

Archiac, Étienne Jules Adolphe Desmier de Saint-Simon, Vicomte d' (1802–68). French army officer, geologist, palaeontologist, and historian of geology. An active member of the Société Géologique de France. (*DSB*, Sarjeant 1980.) In 1849 CD had read the first two volumes of Archiac's work on the progress of geology in France (Archiac 1847–60) (see *Correspondence* vol. 4, letter to Charles Lyell, [2 September 1849] and n. 10).

Baily, John. Poulterer and dealer in live birds. (Freeman 1978.) Baily had supplied CD with live pigeons and fowls for his study of variation in domestic breeds (see *Correspondence* vols. 5 and 6).

Barrande, Joachim (1799–1883). Palaeontologist and stratigrapher who spent many years collecting and describing the fossils of the Bohemian basin. (*DSB.*) CD wanted to send *Origin* to a number of 'good & *speculative*' foreigners, and included Barrande among their number (see *Correspondence* vol. 7, letter to T. H. Huxley, 15 October [1859]). Barrande's work on the palaeontology of Bohemia (Barrande 1852–1911) was on CD's list of books to be read (see *Correspondence* vol. 4, Appendix IV, 128: 179). CD cited Barrande's views on Silurian fossils and the affinities of ancient species in *Origin.*

Blyth, Edward (1810–73). Curator of the museum of the Asiatic Society of Bengal. (*DNB*, *DSB.*) Blyth provided CD with much information about variation in domesticated animals. He also sent CD specimens of fowl skins (see *Correspondence* vols. 5 and 6).

Brent, Bernard Philip (d. 1867). Bird fancier and author. (*CDEL.*) CD had obtained information from Brent about the breeding of pigeons and other fancy birds (see *Correspondence* vol. 6, letter from B. P. Brent, 23 October 1857, and vol. 7, Supplement, letter from B. P. Brent, [after August 1856]). He was cited as an authority in *Origin* and in *Natural selection.*

Bronn, Heinrich Georg (1800–62). Professor of natural and technical sciences and director of the zoological collection at Heidelberg. (*DSB*, *NDB.*) In 1846, CD had studied the second volume of Bronn's *Handbuch einer Geschichte der Natur* (Bronn 1841–9), which contains a discussion of variation in organic life (*Correspondence* vol. 4, Appendix IV, 119: 16a). In February 1860, Bronn offered to superintend the translation of *Origin* into German. In the end, he prepared the translation himself and added to it a final chapter of criticism.

Buckman, James (1814–84). Professor of geology and botany at the Royal Agricultural College, Cirencester. (*DNB*, Sarjeant 1980.) Buckman had assisted CD in the preparation of his cirripede monograph (see *Correspondence* vol. 4, letter to Robert Fitch, 3 December [1849]). CD also made use of Buckman's study of variation in agrarian plants (see *Correspondence* vol. 6, letters to James Buckman, 4 October [1857], and to the *Gardeners' Chronicle*, [before 12 November 1857] and n. 6).

Buist, George (1805–60). Anglo-Indian journalist and scientific observer. Editor of the *Bombay Times* until 1859. Author of numerous papers on the geology, geography, and agriculture of India. (*DNB*.) In 1857, CD asked Hugh Falconer to direct some queries about the markings of horses to contacts in India, and Falconer had sent them to Buist (see *Correspondence* vol. 6, letter to Hugh Falconer, 23 November 1857 and n. 2). However, there is no evidence that CD ever received any information from Buist on this point: he is not acknowledged in *Origin* or in *Variation*.

Bunbury, Charles James Fox, 8th Baronet (1809–86). Botanist and palaeobotanist. (R. Desmond 1977, Sarjeant 1980.) CD had mentioned his belief in the transmutation of species to Bunbury in 1845 (see *Correspondence* vol. 3, letter from Charles Lyell, [after 2 August 1845], n. 5). Bunbury had provided CD with information about the flora of Cape Colony (see *Correspondence* vol. 6), which CD used in his 'big book' on species (*Natural selection*).

Butler, Mary (d. 1865). Patient at Moor Park and Ilkley Wells hydropathic establishments who became CD's friend and visited the family at Down (see *Correspondence* vol. 7). CD stayed at Moor Park while he was preparing *Origin* and discussed the work with Butler. He later wrote to ask her to visit the Ilkley establishment while he and his family were there recuperating from his ordeal of seeing *Origin* through the press (see *Correspondence* vol. 7, letter to Mary Butler, 11 September [1859]).

Candolle, Alphonse de (1806–93). Professor of botany and director of the botanic gardens, Geneva, 1835–50. After 1850, concentrated on his own research. (*DSB*.) CD had been greatly interested by Candolle's work on geographical distribution (A. de Candolle 1855). See *Correspondence* vol. 7, letter to Alphonse de Candolle, 11 November [1859].

Carpenter, William Benjamin (1813–85). Physician and naturalist. Registrar of the University of London. (*DNB*, *DSB*.) There were several discussions in Carpenter's *Principles of comparative physiology* (Carpenter 1854) that were of interest to CD in relation to his species work. In particular, Carpenter discussed hybridity, monsters, and the origin of variation (Carpenter 1854, pp. 632–40); in CD's copy (Darwin Library–CUL), the section discussing these topics is heavily scored.

Carus, Julius Viktor (1823–1903). Extraordinary professor of comparative anatomy and director of the zoological museum, University of Leipzig. (*DSB*, *NDB*.)

Carus, J. V., cont.

The list indicates that Thomas Henry Huxley recommended that CD send Carus a presentation copy of *Origin.*

Caspary, Johann Xaver Robert (1818–87). German botanist. Ordinary professor of botany and director of the botanic garden in Königsberg, 1858. (*ADB*, *NDB*.) CD may have sent Caspary a presentation copy of *Origin* on Joseph Dalton Hooker's recommendation.

Chambers, Robert (1802–71). Partner in the publishing firm W. & R. Chambers. Joint editor of *Chambers's Journal.* (*DNB*, *DSB*.) CD and Chambers shared an interest in geology, and CD believed, correctly, that Chambers was the author of the evolutionary treatise *Vestiges of the natural history of creation* (1844) (see *Correspondence* vol. 4).

Crawfurd, John (1783–1868) Orientalist and linguist who held several civil and political posts in Java, India, Siam, and Cochin China. After his return to England in 1827, he devoted his time to promoting studies connected with Indo-China. (*DNB*.) CD was interested in Crawfurd's ethnographical works on India and China.

Cresy, Edward (1823–70). Principal assistant clerk at the Metropolitan Board of Works and architect to the fire brigade. (*DNB*.) Cresy was the son of Edward Cresy Sr (1792–1858), the architect and engineer who advised CD about the purchase of Down House (see *Correspondence* vol. 2). The son, also an architect and engineer, lived near Down village. CD and Cresy had corresponded for some years on diverse subjects including natural history (see *Correspondence* vols. 4 and 5).

Dana, James Dwight (1813–95). Professor of geology, Yale University. (*DAB*, *DSB*.) CD had corresponded with Dana since 1849 about geology and invertebrate zoology (see particularly *Correspondence* vols. 5 and 6). Dana had recently published his belief in the fixity of species (Dana 1857), a copy of which is in the Darwin Pamphlet Collection–CUL.

Darwin, Emily Catherine (1810–66). CD's sister. Resided at The Mount, Shrewsbury until 1863. (*Darwin pedigree.*)

Darwin, Erasmus Alvey (1804–81). CD's brother. He and CD shared an interest in various scientific studies from their youth. Became involved in mathematical problems connected with the construction of bees' cells during the preparation of CD's 'big book' on species (see *Correspondence* vol. 7).

Darwin, William Erasmus (1839–1914). CD's oldest child. Student at Christ's College, Cambridge, 1859–62. (*Alum. Cantab.*) William assisted CD's study of reversion by sending him observations on the coloration of horses (see *Correspondence* vol. 7).

Decaisne, Joseph (1807–82). French botanist. Professor of plant cultivation, Muséum d'histoire naturelle, Paris. (*DBF*, *NBU*.) Decaisne was a friend of Joseph Dalton Hooker's who had come to agree with his views on the great variability of species. CD had asked Hooker's opinion about whether to send *Origin* to Decaisne (*Correspondence* vol. 7, letter to J. D. Hooker, 15 October [1859]).

Elliot, Walter (1803–87). Indian civil servant and archaeologist who wrote articles on Indian natural history and culture. Member of the council of the governor of Madras. (*DNB.*) At CD's request, Elliot supplied him with skins of Indian pigeons and poultry and a translation of a Persian tract on domestic pigeons (see *Correspondence* vols. 6 and 7).

Eyton, Thomas Campbell (1809–80). Shropshire naturalist. Inherited the family estate in Eyton, Shropshire, in 1855 and built a museum that included a fine collection of skins and skeletons of European birds. (*DNB.*) A friend of CD's at Cambridge, Eyton and CD corresponded during the preparation of CD's species book (see *Correspondence* vols. 6 and 7).

Falconer, Hugh (1808–65). Palaeontologist and botanist. Superintendent of the Calcutta botanic garden and professor of botany at the Calcutta Medical College, 1848–55. (*DNB, DSB.*) After Falconer settled in London following his retirement in 1855, CD applied to him for information about various topics (see *Correspondence* vols. 6 and 7). Falconer was interested in reading *Origin* as soon as it was published (see *Correspondence* vol. 7, letter from Hugh Falconer, 25 October and 12 November [1859]).

FitzRoy, Robert (1805–65). Naval officer, hydrographer, and meteorologist. Commander of the *Beagle*, 1828–36. Chief of the meteorological department of the Board of Trade. (*DNB, DSB.*) CD had been invited onto the *Beagle* not only as naturalist but also as companion to FitzRoy (see *Correspondence* vol. 1). The two maintained a good relationship with one another during the voyage and following their return until after the publication of *Origin.*

Fox, William Darwin (1805–80). Rector of Delamere, Cheshire. (*Alum. Cantab.*) CD's second cousin and friend who introduced him to entomology while they were students at Cambridge University. Fox was interested in CD's work on species, and the two corresponded on many topics relating to CD's research (see *Correspondence* vols. 6 and 7). CD cited Fox on several points in his 'big book' on species (*Natural selection*).

Geoffroy Saint-Hilaire, Isidore (1805–61). Professor of zoology, Muséum d'histoire naturelle, Paris. (*DBF, DSB.*) CD read a number of Geoffroy Saint-Hilaire's works (see *Correspondence* vol. 4, Appendix IV) and cited him as an authority several times in *Origin.*

Gould, John (1804–81). Taxidermist to the Zoological Society of London, 1826. (*DNB, DSB.*) Described the birds collected on the *Beagle* voyage in *Birds*, which was completed by George Robert Gray following Gould's departure for Australia in 1838. Influenced CD's interpretation of the Galápagos finches (see *Correspondence* vol. 2).

Gray, Asa (1810–88). American botanist. Fisher professor of natural history, Harvard University. (*DAB, DSB.*) CD first wrote to Gray in 1855 and corresponded extensively with him thereafter about subjects discussed in his 'big book' on species, especially the geographical distribution of species and varieties and the naturalisation of plants. CD sent Gray a sketch of his theory in 1857. See *Correspondence* vols. 5–7.

Gray, George Robert (1808–72). Assistant in the zoological department of the British Museum. (*DNB*, Gilbert 1977.) Gray completed *Birds* following John Gould's departure for Australia (see *Correspondence* vol. 2).

Gray, John Edward (1800–75). Keeper of the zoological collections of the British Museum. (R. Desmond 1977, *DNB*.) Gray encouraged CD to undertake a monograph on the Cirripedia and arranged for him to borrow the British Museum's cirripede collection (see *Correspondence* vol. 4).

Grisebach, August Heinrich Rudolf (1814–79). Professor of botany, Göttingen University. (*DSB*, *NDB*.) CD made extensive use of Grisebach's flora of the Balkans (Grisebach 1843–4) when he was tabulating the number of species and varieties in large and small genera for his 'big book' on species (see *Correspondence* vol. 6).

Hall, James (1811–98). Palaeontologist to the New York State Geological Survey. Expert in the field of North American stratigraphy and invertebrate palaeontology. A prolific author. (*DAB*, *DSB*.) Charles Lyell may have recommended that CD send *Origin* to Hall: he had worked with Hall during a trip to the United States in 1852 (K. M. Lyell ed. 1881, 2: 184).

Harcourt, Edward William Vernon (1825–91). Authority on the ornithology of Madeira. (*Modern English biography*.) Harcourt provided CD with information on migratory birds and occasional visitors to the Madeiran islands (see *Correspondence* vol. 6). CD cited Harcourt in *Origin* and in *Natural selection*.

Heer, Oswald (1809–83). Professor of botany and entomology, Zürich University, and teacher of taxonomic botany at the Technische Hochschule. Specialist on the Tertiary flora of Europe. (*DSB*, *NDB*.) CD had read Heer's paper on the fossil flora of Madeira (Heer 1855) in 1856 (see *Correspondence* vol. 4, Appendix IV, 128: 20).

Hemmings, Henry. A servant of Sarah Elizabeth Wedgwood (1778–1856), who resided in Down village until her death. Hemmings then returned to Staffordshire, remaining in touch with the Darwin family.

Henslow, John Stevens (1796–1861). Professor of botany, Cambridge University. Rector of Hitcham, Suffolk. (*DNB*, *DSB*.) Henslow was CD's tutor at Cambridge University and the person who recommended him for the *Beagle* voyage. He described some of the *Beagle* plant specimens (see *Correspondence* vol. 1). CD corresponded with Henslow regularly and Henslow often visited the Darwins in Down. CD considered him his 'dear old master in natural history' (*Correspondence* vol. 7, letter to J. S. Henslow, 11 November 1859).

Herbert, John Maurice (1808–82). County court judge in South Wales. (*Alum. Cantab.*) Herbert was one of CD's closest friends at Cambridge University (see *Correspondence* vol. 1).

Herschel, John Frederick William (1792–1871). Astronomer, chemist, mathematician, and philosopher. Resigned from his position as master of the mint in 1855 owing to ill health and lived a retired life in Hawkshurst, Kent. (*DNB*,

DSB.) Herschel's *Preliminary discourse on the study of natural philosophy* (1831) had a formative influence on CD's view of scientific methodology (*Autobiography*, pp. 67–8). CD first met Herschel when the *Beagle* called at the Cape of Good Hope (see *Correspondence* vol. 1).

Hill, Richard (1795–1872). Jamaican-born magistrate and planter. Assisted Philip Henry Gosse in the preparation of works on the natural history of Jamaica. (Cundall 1920.) Hill supplied CD with information about the birds and natural history of Jamaica (see *Correspondence* vols. 6 and 7).

Hoeven, Jan van der (1801–68). Lecturer of zoology, University of Leiden. Author of an influential textbook on zoology and comparative anatomy (Hoeven 1828–33) that was translated into English. (*BWN*, *DSB*.) There is no clue in the extant correspondence to indicate why CD sent Hoeven a copy of the German translation of *Origin*.

Holland, Henry (1788–1873). Physician. Physician in ordinary to Queen Victoria, 1852. President of the Royal Institution of Great Britain. (*DNB*, *Physicians*.) Distant relative of the Darwins and Wedgwoods; he often treated members of the Darwin family.

Hooker, Joseph Dalton (1817–1911). Botanist. Assistant director of the Royal Botanic Gardens, Kew. (*DNB*, *DSB*.) Friend and confidant of CD. In 1847, Hooker read CD's 1844 manuscript essay outlining his species theory, and later he read and commented on chapters of CD's 'big book' on species (*Natural selection*) and *Origin*. In his 'Introductory essay' to the *Flora Tasmaniæ* (Hooker 1859), he applied CD's theory to explain the geographical distribution of plants. See *Correspondence* vols. 4–7.

Horner, Leonard (1785–1864). Geologist and educationist. Inspector of factories, 1833–56. President of the Geological Society of London, 1860–2. Father-in-law of Charles Lyell and Charles James Fox Bunbury. (*DNB*, *DSB*.) CD had known Horner for many years and had sought his opinion about many of his published works. Horner had recently become interested in the study of the antiquity of the human race (see *Correspondence* vol. 7).

Huxley, Thomas Henry (1825–95). Lecturer on natural history at the Royal School of Mines and naturalist to the Geological Survey of Great Britain. (*DNB*, *DSB*.) CD had sought Huxley's opinion about his discussion of several zoological issues during the preparation of his 'big book' on species. Huxley's approval of and support for his species theory was important to CD (see *Correspondence* vols. 6 and 7). CD presented Huxley with both the first and the second editions of *Origin* and a copy of the German translation prepared by Heinrich Georg Bronn.

Innes, John (1817–94). Perpetual curate, later vicar, of Down, 1846–69. (*Crockford's*.) Innes was a close friend of the Darwin family. In 1858, he assisted CD in his study of the construction of bees' cells (*Correspondence* vol. 7).

Jeffreys, John Gwyn (1809–85). Conchologist, zoologist, and solicitor. (*DNB*, *DSB*.) In 1851, Jeffreys sent CD his cirripede collection, many specimens of which CD described in his monograph on the group (see *Correspondence* vol. 5).

Jenyns, Leonard (1800–93). Clergyman and naturalist. Settled in Bath in 1860. Founded the Bath Natural History and Antiquarian Field Club. Brother-in-law of John Stevens Henslow. (*DNB*.) Jenyns was one of the few naturalists to whom CD revealed his views on the transmutation of species before the publication of *Origin* (see *Correspondence* vol. 3, letter to Leonard Jenyns, 12 October [1844]). He shared CD's enthusiasm for entomology in Cambridge, and subsequently described the *Beagle* fish collection. CD later sought information from him about variation and varieties.

Keyserling, Alexandr Andreevich (1815–91). Russian geologist, palaeontologist, and botanist. Accompanied Roderick Impey Murchison on his expedition to Russia in 1841. (*DSB*, Sarjeant 1980.) CD considered Keyserling to be one of the 'good & *speculative* foreigners to whom it would be worth while to send copies' of *Origin* (see *Correspondence* vol. 7, letter to T. H. Huxley, 15 October [1859]).

Kingsley, Charles (1819–75). Author and clergyman. Rector of Eversley, Hampshire, 1844–75. Chaplain to the Queen, 1859–75. Appointed professor of modern history at Cambridge in 1860. (*Alum. Cantab.*, *DNB*.) Although best known for his historical novels and sermons, in which he preached what came to be known as Christian Socialism, Kingsley also took an active interest in natural history.

Kölliker, Rudolf Albert von (Albert) (1817–1905). Professor of physiology and comparative anatomy, and of anatomy, University of Würzburg. Co-editor of the *Zeitschrift für wissenschaftliche Zoologie*. (*DSB*.) On Thomas Henry Huxley's recommendation, CD had sent Kölliker a presentation copy of *Living Cirripedia* (1854) (see *Correspondence* vol. 5). CD may have sent Kölliker a copy of the second edition of *Origin* with a view to his supervising the German translation (see *Correspondence* vol. 7, letter to T. H. Huxley, [9 December 1859]).

Lane, Edward Wickstead (1823–89). Physician. Proprietor until 1859 of a hydropathic establishment at Moor Park, near Farnham, Surrey; from 1860, at Sudbrook Park, near Ham, Surrey. (Freeman 1978, *Medical directory* 1860.) CD became friends with Lane and his family during stays at Moor Park beginning in 1857 (see *Correspondence* vols. 6 and 7).

Leuckart, Karl Georg Friedrich Rudolf (Rudolf) (1822–98). Professor of zoology, University of Giessen, 1850–69. (*DSB*, *NDB*.) CD had sent Leuckart a presentation copy of *Living Cirripedia* (1854), apparently at Thomas Henry Huxley's suggestion. See *Correspondence* vol. 5, letters to T. H. Huxley, 2 September [1854] and 8 September [1854].

Liebig, Justus von (1803–73). Professor of chemistry at Munich University, 1852. Published influential works on the application of chemistry to agricultural improvement. (*DSB*, *NDB*.) CD had read several of Liebig's books (see *Correspondence* vol. 4, Appendix IV).

Lindley, John (1799–1865). Professor of botany at University College London. Editor of the *Gardeners' Chronicle*. (*DNB*, *DSB*.) Lindley was a good friend of

Joseph Dalton Hooker's. CD gathered much material for his species work from the *Gardeners' Chronicle* and frequently sent to Lindley for publication short articles on various topics and notes requesting information.

Lonsdale, William (1794–1871). Geologist. Retired from his post as assistant secretary and librarian to the Geological Society in 1842. Subsequently pursued his geological studies in various parts of the west of England. (*DNB*, *DSB*.) Lonsdale had described some of CD's coral specimens for *Volcanic islands*. See *Correspondence* vol. 1. CD considered Lonsdale as a possible editor of his species sketch of 1844 in the event of his death (see *Correspondence* vol. 3, letter to Emma Darwin, 5 July 1844).

Lubbock, John (1834–1913). Banker, politician, and naturalist. (*DNB*, *DSB*.) Resided with his family at High Elms, near Down House. CD had encouraged Lubbock's interest in natural history, particularly microscopic studies of insects and Crustacea (see *Correspondence* vols. 4–6).

Lyell, Charles (1797–1875). Uniformitarian geologist. (*DNB*, *DSB*.) Lyell had long been a friend and mentor of CD. Lyell had encouraged CD to begin writing his species book, urging this more strongly after the receipt of Alfred Russel Wallace's manuscript on natural selection. Lyell spoke to John Murray about publishing *Origin*. He read and commented on the manuscript prior to publication. See *Correspondence* vols. 1–7.

Miller, William Hallowes (1801–80). Professor of mineralogy, Cambridge University. (*DNB*, *DSB*.) Miller had identified some of the minerals in CD's collection from South America (see *Correspondence* vol. 3). He also assisted CD with his study of the geometry of bees' cells (see *Correspondence* vol. 7).

Milne-Edwards, Henri (1800–85). French zoologist. Professor of entomology, Muséum d'histoire naturelle, Paris. (*DSB*.) CD had dedicated *Living Cirripedia* (1854) to Milne-Edwards, having been greatly influenced by his work on the development and embryology of Crustacea (see *Correspondence* vol. 4, Appendix II). CD had a 'profound' respect for Milne-Edwards's opinion (see letter to J. L. A. de Quatrefages de Bréau, 21 January [1860]).

Moleschott, Jakob (1822–93). Dutch physician and physiologist. Professor of physiology in Zürich, 1856. (*ADB*, *DSB*.) Author of influential works on nutrition, including *Der Kreislauf des Lebens* (1852), which related the functioning of the brain to its structure and nutritive composition. This book became one of the key works of German materialism.

Moore, Charles (1820–1905). Botanist. Director of the botanic garden, Sydney, Australia. (*Aust. dict. biog.*, R. Desmond 1977.) In 1858, Moore had answered CD's queries about the naturalisation of British plants in Australia (see *Correspondence* vol. 7).

Moore, Frederic. Assistant in the museum of the East India Company, Leadenhall Street, London. CD may have sent him a copy of the second edition of *Origin* owing to their common interest in the natural history of India and mutual friendship with Edward Blyth. See *Correspondence* vol. 6.

Morris, John (1810–86). Professor of geology, University College London. (*DNB*, Sarjeant 1980.) Morris had loaned CD specimens of fossil cirripedes for use in the preparation of his monograph on the group (see *Correspondence* vol. 4). In 1856, Morris informed CD about the geographical distribution of molluscs (see *Correspondence* vol. 6).

Mueller, Ferdinand Jakob Heinrich von (1825–96). German-born botanist and explorer who emigrated to Australia in 1847. Director of the botanic garden, Melbourne. (*Aust. dict. biog.*, R. Desmond 1977.) CD corresponded with Mueller about the naturalisation of British plants in Australia (see *Correspondence* vols. 6 and 7).

Murray, Charles Augustus (1806–95). Diplomat and author. Envoy and minister plenipotentiary to the court of Persia until 1859, when he was appointed to the court of Saxony. (*DNB*.) Murray assisted CD's study of variation in fowls and other birds by sending him skins from Persia (see *Correspondence* vols. 5 and 6).

Norman, George Warde (1793–1882). Writer on finance. Resided at The Rookery in Bromley, Kent. (Freeman 1978, *Post Office directory of the six home counties* 1859.) Norman was a good friend of the Darwin family.

Owen, Richard (1804–92). Comparative anatomist. Superintendent of the natural history departments of the British Museum, 1856. (*DNB, DSB*.) Owen described the fossil mammals CD collected during the *Beagle* voyage. He influenced CD's views on classification. Until the publication of *Origin*, he and CD maintained an amicable relationship and corresponded about various subjects.

Petermann, August Heinrich (1822–78). German geographer and cartographer. Assisted Alexander Keith Johnston in the production of an English edition (1848) of Heinrich Karl Wilhelm Berghaus's *Physikalischer Atlas*. Established a lithographic firm in London in 1847 that produced maps for the Royal Geographical Society. Promoted exploration of Africa. Returned to Germany in 1854. Professor of geography at Gotha. (*ADB*.) Petermann published papers on the geographical distribution of animals and plants.

Phillips, John (1800–74). Geologist. Deputy reader of geology at Oxford University, 1853; professor, 1860. Assistant general secretary of the British Association for the Advancement of Science, 1832–62. (*DNB, DSB*.) Phillips was president of the Geological Society in 1859 when the society awarded CD its Wollaston Medal. CD had corresponded with Phillips about various geological points over a number of years (see *Correspondence* vols. 4–7).

Pictet de la Rive, François Jules (1809–72). Swiss zoologist and invertebrate palaeontologist. Professor of zoology, Geneva. (Gilbert 1977, Sarjeant 1980.) Having been struck by Pictet de la Rive's discussion of the species question in Pictet de la Rive 1853–7, CD believed he might be swayed by the geological evidence presented in *Origin* (see *Correspondence* vol. 7).

Prestwich, Joseph (1812–96). Geologist. Expert on the Tertiary geology of Europe and prominent in the study of early human remains. Proprietor of the family wine business. (*DNB*, *DSB*.) CD told Thomas Henry Huxley that he sent Prestwich a copy of *Origin* 'not as a friend' but because he thought he was 'doubting' (see *Correspondence* vol. 7, letter to T. H. Huxley, 27 November [1859]).

Quatrefages de Bréau, Jean Louis Armand de (1810–92). French zoologist and anthropologist. Professor of the natural history of man, Muséum d'histoire naturelle, Paris. (*DSB*.) CD had sent Quatrefages de Bréau a presentation copy of *Living Cirripedia* (1854), presumably on the advice of Thomas Henry Huxley (see *Correspondence* vol. 5, letter to T. H. Huxley, 2 September [1854]). CD was familiar with several of his scientific works, and the two corresponded on several occasions (see *Correspondence* vols. 5–7).

Ramsay, Andrew Crombie (1814–91). Geologist with the Geological Survey of Great Britain. Lecturer at the Government School of Mines, 1852–71. (*DNB*, *DSB*.) CD had sent Ramsay a presentation copy of *South America* (see *Correspondence* vol. 3). The two had frequently corresponded about various geological questions, and CD based several of his statements in *Origin* concerning denudation and fossil remains on information Ramsay provided (see *Correspondence* vol. 7).

Schweizerbart, Christian Friedrich (1805–79). Director of the E. Schweizerbart'sche Verlagshandlung of Stuttgart, 1841–67. (Nägele 1926.) On the recommendation of Heinrich Georg Bronn, undertook the publication of the German translation of *Origin*.

Sedgwick, Adam (1785–1873). Woodwardian professor of geology, Cambridge University, 1818–73. Canon of Norwich. (*DNB*, *DSB*.) CD had accompanied Sedgwick on a geological tour of North Wales when an undergraduate at Cambridge. He had a high opinion of Sedgwick's geological work (see *Correspondence* vols. 1, 6, and 7).

Siebold, Karl Theodor Ernst von (1804–85). Professor of zoology and comparative anatomy, University of Munich, 1853. Co-editor of the *Zeitschrift für wissenschaftliche Zoologie*. (*DSB*, *DNB*.) CD was familiar with several of Siebold's works. At Thomas Henry Huxley's suggestion, CD sent him a presentation copy of *Living Cirripedia* (1854). See *Correspondence* vols. 5–7.

Smith, Frederick (1805–79). Entomologist in the zoological department of the British Museum, 1849. Expert on Hymenoptera. (Gilbert 1977.) CD corresponded with Smith and visited him on a number of occasions to gain information on various topics, including the social habits of insects and cell formation among bees and wasps. See *Correspondence* vol. 7.

Spencer, Herbert (1820–1903). Writer of papers on evolution and numerous works on philosophy and social science. (*DNB*, *DSB*.) In 1856, Spencer sent CD a presentation copy of his book on the principles of psychology (Spencer 1855) and in 1858 sent him the first volume of his collected essays (Spencer 1858–63), which included the evolutionary tract 'The development hypothesis'. See *Correspondence* vols. 6 and 7.

Stanhope, Philip Henry, 5th Earl (1805–75). Historian. President of the Society of Arts, 1846–75. FRS 1827. (*DNB*.) CD had been a guest at Stanhope's seat

Stanhope, P. H., cont.
in Chevening, Kent, and visited his London home on several occasions to join 'one of his parties of historians and other literary men' (*Autobiography*, p. 111; *Correspondence* vol. 4).

Steenstrup, Johannes Japetus Smith (Japetus) (1813–97). Danish zoologist. Professor of zoology and director of the Zoology Museum, University of Copenhagen, 1846–85. (*DBL*, *DSB*.) Steenstrup sent CD the collection of fossil cirripedes from the Copenhagen Zoology Museum for use in the preparation of *Fossil Cirripedia* (1851). They also corresponded about Steenstrup's excavations of ancient sites of human habitation along the coasts of Denmark (see *Correspondence* vols. 4 and 7).

Sulivan, Bartholomew James (1810–90). Naval officer of the marine department of the Board of Trade. Lieutenant in HMS *Beagle*, 1831–6. (*DNB*.) CD and Sulivan, who shared an interest in natural history, became friends during the *Beagle* voyage and maintained contact thereafter. After Sulivan's return to England in 1856, the two met occasionally.

Swinhoe, Robert (1836–77). Consular official in Amoy, China. (R. Desmond 1977, *Modern English biography*.) Swinhoe supplied CD with information on dogs, horses, and pigeons in China, and may have sent him skins of pigeons and poultry. See *Correspondence* vol. 5.

Talandier, Pierre Theodore Alfred (1822–90). Master of French at the Royal Military College, Sandhurst, 1859. (Royal Military Academy Archives.) CD sent Talandier a copy of the second edition of *Origin* after he expressed an interest in preparing a French translation. In the end, Talandier failed to find a publisher for the work. See letters to Edward Cresy, 15 January [1860] and 20 January [1860], and to John Murray, 23 [January [1860].

Tegetmeier, William Bernhard (1816–1912). Editor, journalist, lecturer, and naturalist. (Richardson 1916.) CD had corresponded extensively with Tegetmeier since 1855 about variation in domestic fowls and pigeons. Tegetmeier introduced him to pigeon fanciers at meetings of London clubs and exhibitions and facilitated CD's acquisition of live birds as well as skins and skeletons. He also contributed to CD's study of the construction of bees' cells. See *Correspondence* vols. 5–7.

Thomson, Thomas (1817–78). Superintendent of the Calcutta botanic garden and professor of botany at the Calcutta Medical College, 1854–61. (R. Desmond 1977, *DNB*.) Travelling companion of Joseph Dalton Hooker's during part of his 1849 botanical expedition in the Himalayas. The two collaborated in producing the *Flora Indica* (Hooker and Thomson 1855). Thomson provided CD with information about the ability of plants to withstand a varying range of temperature (see *Correspondence* vol. 7).

Thwaites, George Henry Kendrick (1811–82). Botanist and entomologist. Superintendent of the Peradeniya botanic gardens, Ceylon. (*DNB*.) CD met Thwaites at the 1847 meeting of the British Association for the Advancement of Science. He facilitated CD's study of domestic pigeons and sent him information about the effect of climate and altitude on plants of Ceylon (see *Correspondence* vol. 6). CD cited Thwaites in his 'big book' on species (*Natural selection*).

Tollet, Georgina. Daughter of George Tollet, of Betley Hall, Staffordshire, who was a close friend of Josiah Wedgwood II. Georgina had been a friend of Emma Darwin's since childhood. (Freeman 1978.) She agreed to read CD's manuscript of *Origin* before it went to press and 'look out for errors' in style (see *Correspondence* vol. 7, letter to John Murray, 5 April [1859]).

Tomes, Robert Fisher (b. 1823). An authority on bats and fellow of the Zoological Society of London. Lived in Welford, Warwickshire. (*Post Office directory of Birmingham, Warwickshire (Staffordshire, and Worcestershire)* 1864.) Tomes had examined for CD specimens of bats from Palma, Canary Islands (letter to Charles Lyell, 5 [October 1860]).

Verneuil, Philippe Édouard Poulletier de (1805–73). French geologist and palaeontologist who specialised in Palaeozoic stratigraphy and fossils. In 1860, he carried out important geological work in Spain. (*DSB*, Sarjeant 1980.) CD had read the work on the geology of Russia of which Verneuil was co-author (Murchison, Verneuil, and Keyserling 1845) (see *Correspondence* vol. 4, Appendix IV, 119: 22a).

Wallace, Alfred Russel (1823–1913). Naturalist and collector in the Amazon and the Malay Archipelago, who in 1858 formulated independently of CD a theory of natural selection. (*DNB*, *DSB*.) CD had read Wallace's 1855 paper on the geographical distribution of species, and the two had corresponded about the topic. At CD's request, Wallace supplied him with domestic poultry and specimens of bees and their combs from the Malay Archipelago. In 1858, Wallace sent an essay on varieties, later published in Darwin and Wallace 1858, that prompted CD to begin preparing an abstract of his 'big book' on species, published as *Origin* (see *Correspondence* vols. 5–7).

Waterhouse, George Robert (1810–88). Naturalist and entomologist. Keeper of the geology department of the British Museum. (*DNB*, Gilbert 1977.) Waterhouse described CD's mammalian specimens collected on the *Beagle* voyage. CD often consulted him about various topics, such as the construction of bees' cells and the principles of classification.

Watson, Hewett Cottrell (1804–81). Botanist with interests in phytogeography and phrenology. (*DNB*, *DSB*.) A supporter of the idea of the progressive development of plant species, Watson supported CD's work on species and corresponded with him over a period of several years about the geographical distribution of plant species and varieties (see particularly *Correspondence* vols. 6 and 7).

Weddell, Hugh Algernon (1819–77). British-born botanist and physician who worked in France. Specialist on South American flora. (R. Desmond 1977.) CD consulted Weddell's monograph on the Urticaceae (Weddell 1856) when he was calculating the number of species and varieties in large and small genera (see *Correspondence* vol. 7).

Wedgwood, Francis (Frank) (1800–88). Master-potter and partner in the Wedgwood pottery works at Etruria, Staffordshire. (*Alum. Cantab.*) CD's cousin and brother-in-law.

Wedgwood, Henry Allen (Harry) (1799–1885). Barrister. CD's cousin and brother-in-law.

Wedgwood, Hensleigh (1803–91). Philologist and barrister. (*DNB.*) CD's cousin and brother-in-law. Hensleigh Wedgwood had long supported CD's scientific work, having encouraged him to publish *Journal of researches*. CD included in *Origin* examples drawn from Hensleigh's study of etymology to show that the natural relationship of languages may possibly be analogous to change in plant and animal species (see *Correspondence* vol. 6).

Wedgwood, Josiah, III (Jos) (1795–1880). Of Leith Hill Place, Surrey. CD's cousin and brother-in-law. (*Burke's peerage* 1980.)

Wedgwood, Sarah Elizabeth (Elizabeth) (1793–1880). Resided at The Ridge in Hartfield, Sussex. CD's cousin and sister-in-law.

Whewell, William (1794–1866). Mathematician and historian and philosopher of science. Master of Trinity College, Cambridge. (*DNB*, *DSB.*) CD had known Whewell since his undergraduate days at Cambridge (see *Autobiography*, pp. 66–7). President of the Geological Society of London in 1837, Whewell was responsible for CD's election as secretary of the society in 1838 (see *Correspondence* vol. 2). Author of an influential work on scientific methodology (Whewell 1840) that greatly influenced CD.

White, Adam (1817–79). Naturalist employed in the zoological department of the British Museum. (*DNB*, Gilbert 1977.) White had described some of the *Beagle* arachnids, and CD had a high opinion of his zoological work (see *Correspondence* vols. 2 and 5).

Wollaston, Thomas Vernon (1822–78). Entomologist and conchologist who collected on Madeira. (*DNB*, Gilbert 1977.) CD and Wollaston had corresponded for years about the natural history of Madeira and the variation and distributional patterns of insular fauna. He had been a guest at Down in April 1856 when CD, along with Thomas Henry Huxley and Joseph Dalton Hooker, 'made light of all species & grew more & more unorthodox—' (*Correspondence* vol. 6, letter from Charles Lyell, 1–2 May 1856). CD knew, however, that Wollaston would not accept the notion of unlimited variation of species.

Woodward, Samuel Pickworth (1821–65). Naturalist. Assistant in the department of geology and mineralogy, British Museum, 1848–65. (*DNB*, Sarjeant 1980.) CD had corresponded with Woodward about cirripedes during the preparation of CD's monograph (see *Correspondence* vols. 4 and 5). CD had studied Woodward's *Manual of the Mollusca* (1851–6) and questioned him about the variability of species in different geological periods (see *Correspondence* vol. 6).

APPENDIX IV
New material added to the American edition of *Origin*

Soon after *Origin* was published, Darwin received a letter from Asa Gray offering to arrange an American reprint of the book and to secure the author a share in possible profits. Darwin responded favourably to Gray's proposal in his letter of 21 December [1859] (*Correspondence* vol. 7):

> I sh^{d} be infinitely obliged if you could aid an American Reprint; & could make, for my sake & Publisher's, any arrangement for any profit.— The new Edit. is only Reprint; yet I have made a **few** important corrections. I will have the clean sheets sent over in a few days . . . I sh^{d} be glad for the new Edit to be reprinted, & not the old.—

Darwin was motivated by more than possible profit, for he further added that he thought it 'of importance that my notions sh^{d} be read by intelligent men, accustomed to scientific argument though *not* naturalists. It may seem absurd but I think such men will drag after them those naturalists, who have too firmly fixed in their heads that a species is an entity.—'

After Gray had contacted the Boston publishing firm of Ticknor and Fields about this prospect, he discovered that two New York publishers were already well on the way toward publishing the book. Indeed, by early in January D. Appleton & Co. had *Origin* in type—taken from the first English edition—and were preparing for distribution. Acting on Darwin's behalf, Gray duly contacted D. Appleton to inquire about author's copyright and the possibility of reprinting the second English edition, transmitting their response to Darwin (see letters from Asa Gray, [10 January 1860], [17 January 1860], and 23 January 1860).

Although D. Appleton was not obliged by United States law to honour foreign copyright, they agreed to grant Darwin a share of the profits from their sales. However, they demurred about changing the text of their edition since their plates had been fixed through the process of stereotyping (see letter from Asa Gray, 23 January [1860] and n. 2). The firm agreed, however, to consider preparing a new edition at some future date and asked Darwin to provide them with any changes he wanted to make.

Darwin immediately set to work. Although at first sceptical of Baden Powell's suggestion that he prefix to *Origin* a list of authors who had maintained the modification of species (two letters to Baden Powell, 18 January 1860), Darwin subsequently changed his mind. On 31 January he told Joseph Dalton Hooker that he was preparing a 'historical sketch' for *Origin* that cited those naturalists who had preceded him in espousing favourable views of the transmutation of species; Darwin sent this off to Gray enclosed in his letter of [8 or 9 February 1860]. He had earlier sent Gray some additions and corrections that he hoped could be included (see letters to Asa Gray, 28 January [1860] and 1 February [1860]).

A month later, in his letter of 8 March [1860], Darwin sent Gray several more substantive corrections to be added to any second American edition. These included a long discussion in response to early criticisms of natural selection (especially that given by Hewett Cottrell Watson in his letter of [3? January 1860]) that Darwin wanted inserted at the conclusion of chapter four. Darwin had informed Charles Lyell about some of the changes he intended to make in the American edition in the letter to Lyell, 18 [and 19 February 1860]. Darwin suggested to Gray that the title page of the new edition should read: '"Reprinted from corrected Second Edition with additional corrections"' (letter to Asa Gray, 1 February [1860]).

By 1 May 1860, D. Appleton had sold the bulk of the 2250 copies resulting from three separate printings of *Origin* (see letter to Asa Gray, 22 May [1860] and enclosure) and were preparing to produce the promised 'revised and augmented' American edition. Most of the emendations that Darwin had sent were incorporated into the new edition. However, at some point, probably early in May, Darwin had sent a list of further changes for the new American edition that were received too late to be incorporated into the text and were printed in a 'Supplement' added to the end of the edition.

This 'revised and augmented' American edition of *Origin* was available in July 1860 (see [Gray] 1860b, p. 116). It is interesting to note that it was in several respects different from the two British editions: it carried the whale–bear case, which had been removed from the second English edition, yet it incorporated many of the other changes Darwin made to the second edition; it also included a number of significant alterations and additions that were only later added, in a somewhat modified form, to the third edition of *Origin*, published in March 1861 (see Freeman 1977, p. 83). As Gray's biographer A. Hunter Dupree has noted, 'Gray's great accomplishment . . . was to provide the American public with an edition of the *Origin* in some respects even ahead of the current English edition.' (Dupree 1959, p. 271). In fact, this edition was the only one available in the United States until 1873, when D. Appleton prepared a new edition taken from the sixth English edition of 1872 (Freeman 1977, pp. 85–6).

The text printed below is divided into three sections: the 'historical sketch' printed as a preface to the American edition (*Origin* US ed., pp. i–xi), the new material added to chapter 4 on 'Natural selection' (*ibid.*, pp. 116*–121*), and the final list of changes added as a Supplement to the volume (*ibid.*, pp. 426–32).

PREFACE

Contributed by the Author to this American Edition.

I will here attempt to give a brief, but I fear imperfect, sketch of the progress of opinion on the Origin of Species. The great majority of naturalists have believed that species were immutable productions and have been separately created: this view has been ably maintained by many authors. A few naturalists, and several who have not particularly studied natural history, believe, on the other hand, that species undergo modification, and that the existing forms of life have descended by true generation from pre-existing forms. Passing over authors of the classical period, and likewise Demaillet and Buffon, with whose writings I am not familiar, Lamarck was the first man, whose view that species undergo change excited much attention. This justly celebrated naturalist published his Philosophie Zoologique in 1809, and his Introduction to Hist. Nat. des animaux sans Vertèbres in 1815, in which works he upholds the doctrine that species are descended from each other. He seems to have been chiefly led to this conclusion by the difficulty of distinguishing species and varieties,—by the almost perfect gradation of the forms in certain groups, and by the analogy of domestic productions. With respect to the means of modification, he attributed something to the action of external conditions, something to the crossing of already existing forms, and much to use and disuse or the effects of habit. To this latter agency he seems to attribute all the beautiful adaptations in nature,—such as the long neck of the giraffe for browsing on the branches of trees.* But he likewise believed in a law of progressive development; and as all the forms of life thus tended to progress, in order to account for the presence of very simple productions at the present day, he maintained that such forms were now spontaneously generated.

Geoffroy Saint Hilaire, as is stated in his Life by his Son, as early as 1795, suspected that what we call species are various degenerations of the same type. It was not until 1828 that he published his conviction that the same forms have not been perpetuated since the origin of all things. Geoffroy seems to have relied chiefly on the conditions of life, or the "monde ambiant," as the cause of change; but he was cautious, and, as his son says, "C'est donc un problème à reserver entièrement a l'avenir, supposé même que l'avenir doive avoir prise sur lui."

In England, the Hon. and Rev. W. Herbert, afterwards Dean of Manchester, in his work on the Amaryllidaceæ (1837, p. 19, 339), declares that "horticultural experiments have established, beyond the possibility of refutation, that botanical species are only a higher and more permanent class of varieties." He extends the same view to animals. The Dean believes that single species of each genus were created in an originally highly plastic condition, and that these by intercrossing and by variation have produced all our existing species.

In 1843–44, Prof. Haldeman (in the Boston (U. S.) Journal of Nat. Hist., vol. iv., p. 468) has ably given the arguments for and against the hypothesis of the development and modification of species: he seems to me to lean towards the side of change.

The Vestiges of Creation appeared in 1844. In the last or tenth and much improved edition (1853, p. 155), the anonymous author says: "The proposition

* It is curious how completely my grandfather, Dr. Erasmus Darwin, anticipated these erroneous views in his Zoonomia (vol. i. pp. 500–510), published in 1794.

determined on after much consideration is, that the several series of animated beings, from the simplest and oldest up to the highest and most recent, are, under the providence of God, the results, *first*, of an impulse which has been imparted to the forms of life, advancing them, in definite times, by generation, through grades of organisation terminating in the highest dicotyledons and vertebrata, these grades being few in number, and generally marked by intervals of organic character which we find to be a practical difficulty in ascertaining affinities; *second*, of another impulse connected with the vital forces, tending in the course of generations to modify organic structures in accordance with external circumstances, as food, the nature of the habitat and the meteoric agencies, these being the 'adaptations' of the natural theologian." The author apparently believed that organisation progresses by sudden leaps; but that the effects produced by the conditions of life are gradual. The author argues with much force on general grounds that species are not immutable productions. But, I cannot see how the two supposed "impulses" account in a scientific sense for the numerous and beautiful co-adaptations, which we see throughout nature;—I cannot see that we thus gain any insight how, for instance, a woodpecker has become adapted to its peculiar habits of life. The work, from its powerful and brilliant style, though displaying in the earlier editions little accurate knowledge and a great want of scientific caution, immediately had a very wide circulation. In my opinion, it has done excellent service in calling in this country attention to the subject, and in removing prejudices.

In 1846, the veteran geologist, M. J. d'Omalius d'Halloz, published in an excellent, though short, paper (Bulletins de l'Acad. Roy. Bruxelles, tom. xiii., p. 581), his opinion that it is more probable that new species have been produced by descent with modification, than that they have been separately created: the author first promulgated this opinion in 1831.

M. Isidore Geoffroy Saint Hilaire, in his Lectures delivered in 1850 (of which a Résumé appeared in the Revue et Mag. de Zoolog., Jan. 1851), briefly gives his reasons for believing that specific characters "sont fixes, pour chaque espèce, tant qu'elle se perpétue au millieu des mêmes circonstances, ils se modifient, si les circonstances ambiantes viennent à changer." "En résumé, *l'observation* des animaux sauvages démontre déjà la variabilité *limitée* des espèces. Les *expériences* sur les animaux sauvages devenus domestiques, et sur les animaux domestiques redevenus sauvages, la démontrent plus clairement encore. Ces même expériences prouvent, de plus, que les differences produites peuvent être de valeur *générique*."

Mr. Herbert Spencer, in an Essay (originally published in the Leader, March, 1852, and republished in his Essays, 1858), has contrasted the theories of the creation and development of organic beings with remarkable skill and force. He argues from the analogy of domestic productions, from the changes which the embryos of many species undergo, from the difficulty of distinguishing species and varieties, and from the principle of general gradation, that species have been modified; and he attributes the modification to the change of circumstances. The author (1855) has treated Psychology on the principle of the necessary acquirement of each mental power and capacity by gradation.

In 1852 (Revue Horticole, p. 102), M. Naudin, a distinguished botanist,* has expressly stated his belief that species are formed in an analogous manner as varieties are under cultivation; and the latter process he attributes to man's power of selection. But he does not show how selection can act under nature. He believes, like Dean Herbert, that species when nascent were more plastic. He lays weight on what he calls the principle of finality, "puissance mystérieuse, indéterminée; fatalité pour les uns; pour les autres, volonté providentielle, dont l'action incessante sur les êtres vivants détermine, à toutes les époques de l'existence du monde, la forme, le volume et la durée de chacun d'eux, en raison de sa destinée dans l'ordre de choses dont il fait partie. C'est cette puissance qui harmonise chaque membre à l'ensemble en l'appropriant à la fonction qu'il doit remplir dans l'organisme général de la nature, fonction qui est pour lui sa raison d'être."

In 1853, a celebrated geologist, Count Keyserling (Bulletin de la Soc. Geolog., 2d ser., tom. x., p. 357) suggested that as new diseases, supposed to have been caused by some miasma, have arisen and spread over the world, so at certain periods the germs of existing species may have been chemically affected by circumambient molecules of a particular nature, and thus have given rise to new forms.

The "Philosophy of Creation" has been treated in an admirable manner by the Rev. Baden Powell, in his Essays on the Unity of Worlds, 1855. Nothing can be more striking than the manner in which he shows that the introduction of new species is "a regular, not a casual, phenomenon," or, as Sir John Herschel expresses it, "a natural in contradistinction to a miraculous process." I think this work can hardly have failed to have produced a great effect in every philosophical mind.

The third volume of the Journal of the Linnean Society (August, 1858) contains papers by Mr. Wallace and myself, in which, as stated in the introductory remarks to this volume, the theory of Natural Selection is promulgated.

In June, 1859, Professor Huxley gave a lecture before the Royal Institution on the Persistent Types of Animal Life. Referring to such cases, he remarks: "It is difficult to comprehend the meaning of such facts as these, if we suppose that each species of animal and plant, or each great type of organisation, was formed and placed upon the surface of the globe, at long intervals, by a distinct act of creative power; and it is well to recollect that such an assumption is as unsupported by tradition or revelation as it is opposed to the general analogy of nature. If, on the other hand, we view 'Persistent Types' in relation to that hypothesis which supposes the species living at any time to be the result of the gradual modification of pre-existing species,—a hypothesis which, though unproven, and sadly damaged by some of its supporters, is yet the only one to which physiology lends any countenance—their existence would seem to show that the amount of modification which living beings have undergone during geological time is but very small in relation to the whole series of changes which they have suffered."

* M. Lecoq, another French botanist, entertains, I believe, analogous views on the modification and descent of species.

In November, 1859, the first edition of this work was published. In December, 1859, Dr. Hooker published his Introduction to the Tasmanian Flora: in the first part of this admirable essay he admits the truth of the descent and modification of species; and supports this doctrine by many original and valuable observations.

Charles Darwin

Down, Bromley, Kent, Feb. 1860.

[Darwin's response to early criticisms of the operation of natural selection].

A distinguished naturalist has objected that the continued action of natural selection and divergence will tend to make an indefinite number of specific forms. As far as mere inorganic conditions are concerned, it seems probable that a sufficient number of species would soon become adapted to all considerable diversities of heat, moisture, &c.; but I fully admit that the mutual relations of organic beings are the most important, and as the number of species in any country goes on increasing, the organic conditions of life will become more and more complex. Consequently there seems at first to be no limit to the amount of profitable diversification of structure, and therefore no limit to the number of species which might be produced. We do not know that even the most prolific area is fully stocked with specific forms; at the Cape of Good Hope, which supports such an astonishing number of species, many European plants have become naturalised. But geology shows us, at least within the whole immense Tertiary period, that the number of species of shells, and probably of mammals, has not increased. What, then, checks an indefinite increase in the number of species? Firstly, the amount of life (I do not mean the number of specific forms) supported on any area must have a limit, depending so largely as it does on physical conditions: therefore where very many species are supported, each, or nearly each, will be few in individuals; and any species with scanty numbers would be liable to extermination from accidental fluctuations in the nature of seasons and in the number of its enemies. The process of extermination would in such cases be rapid, whereas the process of the production of new species would always be slow. Imagine the extreme case of as many species as individuals in England, and the first severe winter or very dry summer would exterminate thousands on thousands of species; and individuals of other species would take their places. Secondly, I suspect that when any species becomes very rare, close interbreeding will tend to exterminate it; at least authors have thought that this comes into play in accounting for the deterioration of aurochs in Lithuania, of red deer in Scotland, and of bears in Norway, &c. Thirdly, as far as animals are concerned, some species are closely adapted to prey on some one other being; but if this other being had been rare, it would not have been any advantage to the animal to have been produced in close relation to its prey; therefore it would not have been produced by natural selection. Fourthly, when any species becomes few in number, the process of modification will be slower, for the chance of favourable variations arising will be lessened;

therefore if we suppose an area to be inhabited by very many species, each, or nearly each, species will be poor in individuals, and consequently the process of modification and of giving birth to new forms will be retarded. Fifthly, and this I am inclined to think is the most important element, a dominant species which has already beaten many competitors in its own home, will tend to spread and supplant many others. Alph. de Candolle has shown that those species which spread widely, tend generally to spread *very* widely; and consequently they will tend to exterminate several species in several areas, and thus check the inordinate increase of specific forms throughout the world. Hooker has recently shown that in the S. E. corner of Australia, where apparently there are many invaders from different quarters of the world, the endemic Australian species have apparently been greatly reduced in number. How much weight to attribute to these several causes, I do not pretend to assign; but conjointly I think they must limit in each country the tendency to an indefinite augmentation of specific forms.

Natural Selection acts, as we have seen, exclusively by the preservation and accumulation of variations, which are beneficial under the organic and inorganic conditions of life to which each creature is at each successive period exposed. The ultimate result will be that each creature will tend to become more and more improved in relation to its conditions of life. This improvement will, I think, inevitably lead to the gradual advancement of the organisation of the greater number of living beings throughout the world. But here we enter on a very intricate subject, for naturalists have not defined to each other's satisfaction what is meant by an advance in organisation. Amongst the Vertebrata, the degree of intellect and an approach in structure to man clearly come into play. It might be thought that the amount of change which the various parts and organs undergo in their development from the embryo to maturity would suffice as a standard of comparison; but there are cases, as with certain parasitic crustaceans, in which several parts of the structure become less perfect and even monstrous, so that the mature animal cannot be called higher than its larva. Von Baer's standard seems the most widely applicable and the best; namely, the amount of differentiation of the different parts (in the adult state, as I should be inclined to add), and their specialisation for different functions; or as Milne Edwards would express it, the completeness of the division of physiological labour. But we shall see how obscure a subject this is, if we look, for instance, to Fish, amongst which some naturalists rank those as highest which, like the sharks, approach nearest to reptiles; whilst other naturalists rank the common bony or teleostean fishes as the highest, inasmuch as they are most strictly fish-like, and differ most from the other vertebrate orders. Still more plainly we see the obscurity of the subject, by turning to plants, where the standard of intellect is of course quite excluded; and here some botanists rank those plants as highest which have every organ, as sepals, petals, stamens and pistils, fully developed in each flower; whereas other botanists, probably with more truth, look at the plants which have their several organs much modified and somewhat reduced in number as being of the highest rank.

If we look at the differentiation and specialisation of the several organs of each being, when adult (and this will include the advancement of the brain for intellectual purposes), as the best standard of highness of organisation, natural selection will clearly lead towards highness; for all physiologists admit that the specialisation of organs, inasmuch as they perform in this state their functions better, is an advantage to each being; and hence the accumulation of variations tending towards specialisation is within the scope of natural selection. On the other hand, we can see, bearing in mind that all organic beings are striving to increase at a high ratio and to seize on every ill-occupied place in the economy of nature, that it is quite possible for natural selection gradually to fit an organic being to a situation in which several organs would be superfluous and useless; and in such cases there might be retrogression in the scale of organisation. Whether organisation on the whole has actually advanced from the remotest geological periods to the present day will be more conveniently discussed in our chapter on geological succession.

But it may be objected, that if all organic beings thus tend to rise in the scale, how is it that throughout the world a multitude of the lowest forms still exist, and how is it that in each great class some forms are far more highly-developed than others? Why have not the more highly-developed forms everywhere supplanted and exterminated the lower? Lamarck, who believed in an innate and inevitable tendency towards perfection in all organic beings, seems to have felt this difficulty so strongly, that he was led to suppose that new and simple forms were continually being produced by spontaneous generation. I need hardly say that science in her progress has forbidden us to believe that living creatures are now ever produced from inorganic matter. On my theory the present existence of lowly organized productions offers no difficulty; for natural selection includes no necessary and universal law of advancement or development; it only takes advantage of such variations as arise and are beneficial to each creature under its complex relations of life. And it may be asked, what advantage, as far as we can see, would it be to an infusorian animalcule—to an intestinal worm—or even to an earth-worm, to be highly organized? If it were no advantage, these forms would be left by natural selection unimproved or but little improved; and might remain for indefinite ages in their present little advanced condition. And geology tells us that some of the lowest forms, as the infusoria and rhizopods, have remained for an enormous period in nearly their present state. But to suppose that most of the many now-existing low forms have not in the least advanced since the first dawn of life, would be rash; for every naturalist who has dissected some of the beings now ranked as very low in the scale, must often have been struck with their really wondrous and beautiful organisation.

Nearly the same remarks are applicable, if we look to the great existing differences in the grades of organisation within almost every class excepting birds; for instance, to the coexistence of mammals and fish in the vertebrata,—or to the coexistence of man and the ornithorhynchus amongst mammalia,—or amongst fish, of the shark and Amphioxus, which latter fish in the extreme simplicity of its structure closely approaches the invertebrate classes. But mammals and fish hardly come into competition with each other; the advancement of certain mammals or of the whole class to the highest grade of organisation would not lead to their taking the place of and thus exterminating fishes. Physiologists believe that the brain must be bathed by warm blood to be highly active, and this requires aërial respiration; so that warm-blooded mammals, when inhabiting the water, live under some disadvantages compared with fishes. In this latter class members of the shark family would not, it is probable, tend to supplant the Amphioxus; the struggle for existence in the case of the Amphioxus must lie with members of the invertebrate classes. The three lowest orders of mammals—namely, marsupials, edentata and rodents—coexist in South America in the same region with numerous monkeys. Although organisation, on the whole, may advance throughout the world, yet the scale of perfection will still present all degrees for the high advancement of certain whole classes, or of certain members of each class, does not at all necessarily lead to the extinction of those groups with which they do not enter into close competition. In some cases, as we shall hereafter see, lowly-organized forms seem to have been preserved to the present day, from having inhabited peculiar or isolated stations where they have been subjected to less severe competition; and where they have not advanced in organization owing to their scanty individual numbers, which, as already explained, retards the chance of favourable variations arising.

Finally, I believe that lowly-organised forms now exist in numbers throughout the world and in nearly every class, from various causes. In some cases favourable variations may never have arisen for natural selection to act on and accumulate. In no case, perhaps, has time sufficed for the utmost possible maximum of development. In some few cases there may have been what we must call retrogression of organisation. But the main cause lies in the circumstance, that under very simple conditions of life, a high organisation would be of no service—possibly would be of actual disservice, as being of a more delicate nature, and more liable to be put out of order and thus injured.

A difficulty, diametrically opposite to this which we have just been considering, might be advanced; namely, looking to the dawn of life, when all organic beings, as we may imagine, presented the simplest structure, how could the first steps in advancement or in the differentiation and specialization of parts arise? I can make no sufficient answer, and can only say that we have no facts to guide us, and therefore that all speculations on this subject would be baseless and useless.

SUPPLEMENT

The following additions and alterations, prepared by the authors expressly for this edition, were received too late to be incorporated in their proper places.

Page 46,[1] eight lines from bottom, after "not generally propagated," insert:

If it could be shown that monstrosities were even propagated for a succession of generations in a state of nature, modification might be effected (with the aid of natural selection) more abruptly than I am inclined to believe they are.

Page 79,[2] six lines from bottom, after word "nature," insert parenthesis—reading the whole sentence thus:

Man can act only on external and visible characters: nature (if I may be allowed for brevity-sake to personify the natural preservation of favoured individuals during the struggle for existence) cares nothing for appearances, except in so far as they may be useful to any being.

Page 168,[3] first line, after "structure of the eye," insert:

(though in the fish Amphioxus), the eye is an extremely simple condition without a lens)

Page 168, sixth line from top, omit all seventeen lines beginning with "In the Articulata," and ending with "living crustaceans, and bearing in mind," and insert as follows:

[In the great kingdom of the Articulata we can start from an optic nerve, simply coated with pigment, which sometimes forms a sort of pupil, but is destitute of a lens or any other optical mechanism. From this rudimentary eye which can distinguish light from darkness, but nothing else, there is an advance towards perfection along two lines of structure, which Müller thought were fundamentally different; namely, firstly, stemmata, or the so-called "simple eyes," which have a lens and cornea; and secondly, "compound eyes," which seem to act mainly or solely by excluding all the rays from each point of the viewed object, except the pencil that comes in a line perpendicular to the convex retina. In the class of compound eyes, besides endless differences in the form, proportion, number and position of the transparent cones coated by the pigment which act by exclusion, we have additions of a more or less perfect concentrating apparatus; thus in the eyes of the meloe the facets of the cornea are "slightly convex both externally and internally; that is, lens-shaped." In many crustaceans there are two cornea, the external smooth and the internal divided into facets, within the substance of which, as Milne Edward says, "renflemens lenticulaires paraissent s'etre développés;" and sometimes these lenses can be detached in a layer distinct from the cornea. The transparent cones, which were supposed by Müller to act solely by excluding the divergent pencils of light, usually adhere to the cornea, but not rarely they are detached from it, and have their free ends convex, and in this case they must act, I presume, as converging lenses. Altogether, so diversified is the structure of the compound eyes, that Müller makes three main classes, with no less than seven subdivisions of structure. He makes a fourth main class—namely, "aggregates" of stemmata; and he adds that "this is the transition-form between the mosaic-like compound eyes, unprovided with a concentrating apparatus and organs of vision with such an apparatus."

With these facts, here too briefly and imperfectly given, which show how much graduated diversity there is in the eye of our existing articulata, and bearing in mind, &c., &c.

Page 169,[4] tenth line from top, after "transparent tissue," add:

with spaces filled with fluid and with a nerve sensitive, &c., &c.

Page 169, sixteenth line from top, after "there is a power," insert in brackets:

[natural selection]

Page 170,[5] twenty-first line from top, after "insensible steps," insert:

Certain plants, as some Leguminosæ, Violaceæ, &c., bear two kinds of flowers; one having the normal structure of the order, the other kind being degraded, though sometimes more fertile than the perfect flowers; if the plant ceased to bear its perfect flowers, and this did occur during several years with an imported specimen of Aspicarpa in France, a great and sudden transition would in fact be effected in the nature of the plant.

Pages 293 and 294.[6] Omit thirty lines, beginning, "*On the state of Development,*" and ending with "class, may have beaten the highest molluscs," and insert as follows:

On the state of Development of ancient compared with living Forms.— We have seen in the Fourth Chapter that the degree of differentiation and specialisation of the parts of all organic beings, when come to maturity, is the best standard as yet suggested of their degree of perfection or highness. We have also seen that, as the specialisation of parts or organs is an advantage to each being, so natural selection will constantly tend thus to render the organisation of each more specialised and perfect, and in this sense higher; not but that it may, and will, leave many creatures fitted for simple conditions of life with simple and unimproved structures. In another and more general manner we can see that, on the theory of natural selection, the more recent forms will tend to be higher than their progenitors; for each new species is formed by having had some advantage in the struggle for life over other and preceding forms. If, under a nearly similar climate, the eocene inhabitants of one quarter of the world were put into competition with the existing inhabitants of the same or some other quarter, the eocene fauna or flora would certainly be beaten and exterminated; as would a secondary fauna by an eocene, and a palæozoic fauna by a secondary fauna. So that by this fundamental test of victory in the battle for life, as well as by the standard of the specialisation of organs, modern forms ought, on the theory of natural selection, to stand higher than ancient forms. Is this the case? A large majority of palaeontologists would certainly answer in the affirmative; but in my imperfect judgment, I can, after having read the discussions on this subject by Lyell and by Hooker in regard to plants, concur only to a certain[7] limited extent. Nevertheless, it may be anticipated that the evidence will be rendered more decisive by future geological research.

The problem is in many ways excessively intricate. The geological record, at all times imperfect, does not extend far enough back, as I believe, to show with unmistakable clearness that within the known history of the world organisation has largely advanced. Even at the present day, looking to members of the same class, naturalists are not unanimous which forms are highest; thus some look at the Selaceans or sharks from their approach in some important point of structure to reptiles as the highest fish; others look at the teleosteans as the highest. The ganoids stand intermediate between the selaceans and teleosteans; the latter, at the present day, are largely preponderant in number, but formerly selaceans and ganoids alone existed; and in this case, according to the standard of highness chosen, so will it be said that fishes have advanced or have retrograded in organisation. To attempt to compare in the scale of highness members of distinct types seems hopeless; who will decide whether cuttle-fish be higher than a bee? that insect which the great Von Baer believed to be "in fact more highly organised than a fish, although upon another type." In the complex struggle for life it is quite credible that crustaceans, for instance, not very high in their own class, might beat the cephalopods or highest molluscs; and such crustaceans, though not highly developed, would stand very high in the scale of invertebrate animals, if judged by the most decisive of all trials,—the law of battle.

Besides this inherent difficulty in deciding which forms are the most advanced in organization, we ought not solely to compare the highest members of a class at any two distant periods,—though undoubtedly this is one and perhaps the most important element in striking a balance—but we ought to compare all the members, high and low, at the two periods. At an ancient epoch the highest and lowest molluscs—namely, cephalopods and brachiopods—swarmed in numbers; at the present time, both these orders have been greatly reduced, whereas other orders, intermediate in grade of organisation, have largely increased; consequently some naturalists have maintained that molluscs were formerly more highly developed than at present; but a stronger case can be made out on the other side by considering the vast reduction at the present day of the lowest molluscs; more especially as the existing cephalopods, though so few in number, are more highly organised than their ancient representatives. We ought, also, to consider the relative proportional numbers of the high and low classes in the population of the world at the two periods; if, for instance, at the present day there were fifty thousand kinds of vertebrate animals, and if we had reason to believe that at some former period there were only ten thousand kinds, we ought to look at this increase in number of the highest class, which implies a great displacement of lower forms, as a decided advance in the organisation of the world, whether or not it were the higher vertebrata which had thus largely increased. We can thus see how hopelessly difficult it will apparently forever be to compare with perfect fairness under such extremely complex relations the standard of organisation of the imperfectly known faunas of successive periods of the earth's history.

We shall appreciate (under one important point of view) this difficulty the more clearly, by looking to the case of certain existing faunas and floras. From the extraordinary manner in which European productions have recently spread over New Zealand, &c., &c.

Page 417,[8] 25th line, after "facts above specified," insert:

It is no valid objection that science as yet throws no light on the origin of Life. Who can explain what is the essence of the Attraction of gravity? Although Leibnitz accused Newton of introducing "occult qualities and miracles into philosophy;" yet this unknown element of attraction is now universally looked at as a *vera causa* perfectly well established.]

[I see no good reason why the views given in this volume should shock the religious feelings of any one. It is satisfactory, as showing how transient such impressions are, to remember that the greatest discovery ever made by man, namely, the law of gravity, was attacked by Leibnitz, "as subversive of natural and inferentially of revealed religion." A celebrated author and divine, &c., &c.

Page 420,[9] fifteen lines from top, after "deceitful guide," omit whole remainder of paragraph, and insert, instead, as follows:

Nevertheless, all living things have much in common; in their chemical composition, their cellular structure, their laws of growth, and their liability to injurious influences. We see this in so trifling a circumstance as that the same poison often similarly affects plants and animals, or that the poison secreted by the gall-fly produces monstrous growths on the wild rose or oak tree. In all organic beings the union of a male and female elemental cell seems occasionally to be necessary for the production of a new being. In all, as far as is at present known, the germinal vesicle is the same. So that every individual organic being starts from a common origin. If we look even to the two main divisions—namely, to the animal and vegetable kingdoms—certain low forms are so far intermediate in character that naturalists have disputed to which kingdom they should be referred; and on the principle of natural selection with divergence of character, it does not seem utterly incredible that from some such intermediate production both animals and plants might possibly have been developed. Therefore I should infer that probably all the organic beings which have ever lived on this earth have descended from some one primordial form, into which life was first breathed by the Creator. But this inference is chiefly grounded on analogy, and it is immaterial whether or not it be accepted. The case is different with the members of each great class, as the Vertebrata or Articulata; for here, as has just been remarked, we have in the laws of homology and embryology, &c., some distinct evidence that all have descended from a single primordial parent.]

[1] The pagination of the American edition of *Origin* does not correspond to that of the English edition. See the first edition of *Origin*, p. 44.

[2] *Origin*, p. 83.

[3] *Origin*, pp. 187–8. This phrase and the following, longer discussion of the modification of the eye were added to the third edition of *Origin* (*Origin* 3d ed., pp. 363–6). See also letter from John Lubbock, [after 28 April 1860?].

[4] *Origin*, p. 188.

[5] *Origin*, p. 190.

[6] *Origin*, pp. 336–7. This material was added to *Origin* 3d ed., pp. 363–6.

[7] The word is misspelled 'cretain' in the original text.

[8] The change was made to the text of the second edition of *Origin* (*Origin* 2d ed., p. 481). It was included in a slightly modified form in *Origin* 3d ed., pp. 514–15.

[9] *Origin* 2d ed., p. 484. The material was included in *Origin* 3d ed., pp. 518–19.

APPENDIX V
Patrick Matthew's letter to the *Gardeners' Chronicle*

Patrick Matthew, the author of *Naval timber and arboriculture* (Matthew 1831), published a letter in the *Gardeners' Chronicle and Agricultural Gazette*, 7 April 1860, pp. 312–13, in which he claimed to have formulated a principle of natural selection prior to the publication of Darwin's *Origin of species*. Darwin responded to Matthew's claim in a letter published in the *Gardeners' Chronicle and Agricultural Gazette*, 21 April 1860, pp. 362–3 (see letter to *Gardeners' Chronicle*, [13 April 1860]). He included Matthew's name among his predecessors in the revised 'Historical sketch' published in the third edition of *Origin*. Although Matthew's letter has recently been reprinted (see Dempster 1983), since it is referred to on several occasions in this volume, it has been included here for the reader's easy reference.

Nature's law of selection.

Trusting to your desire that every man should have his own, I hope you will give place to the following communication.

In your Number of March 3d I observe a long quotation from the *Times*, stating that Mr. Darwin "professes to have discovered the existence and *modus operandi* of the natural law of selection," that is, "the power in nature which takes the place of man and performs a selection, *suâ sponte*," in organic life. This discovery recently published as "the results of 20 years' investigation and reflection" by Mr. Darwin turns out to be what I published very fully and brought to apply practically to forestry in my work "Naval Timber and Arboriculture," published as far back as January 1, 1831, by Adam & Charles Black, Edinburgh, and Longman & Co., London, and reviewed in numerous periodicals, so as to have full publicity in the "Metropolitan Magazine," the "Quarterly Review," the "Gardeners' Magazine," by Loudon, who spoke of it as *the* book, and repeatedly in the "United Service Magazine" for 1831, &c. The following is an extract from this volume, which clearly proves a prior claim. The same volume contains the first proposal of the steam ram (also claimed since by several others, English, French, and Americans,) and a navy of steam gun-boats as requisite in future maritime war, and which, like the organic selection law, are only as yet making way:—

"There is a law universal in nature, tending to render every reproductive being the best possibly suited to its condition that its kind, or that organised matter, is susceptible of, which appears intended to model the physical and mental or instinctive powers, to their highest perfection, and to continue them so. This law sustains the lion in his strength, the hare in her swiftness, and the fox in his wiles. As nature, in all her modifications of life, has a power of increase far beyond what is needed to supply the place of what falls by Time's decay, those individuals who possess not the requisite strength, swiftness, hardihood, or cunning, fall prematurely without reproducing—either a prey to their natural devourers, or sinking under disease, generally induced by want of nourishment, their place being occupied by the more perfect of their own kind, who are pressing on the means of subsistence."

"Throughout this volume, we have felt considerable inconvenience, from the adopted dogmatical classification of plants, and have all along been floundering between species and variety, which certainly under culture soften into each other. A particular conformity, each after its own kind, when in a state of nature, termed species, no doubt exists to a considerable degree. This conformity has existed during the last 40 centuries. Geologists discover a like particular conformity—fossil species—through the deep deposition of each great epoch, but they also discover an almost complete difference to exist between the species or stamp of life on one epoch from that of every other. We are therefore led to admit, either of a repeated miraculous creation; or of a power of change, under a change of circumstances, to belong to living organised matter, or rather to the congeries of inferior life, which appears to form superior. The derangements and changes in organised existence, induced by a change of circumstance from the interference of man, affording us proof of the plastic quality of superior life, and the likelihood that circumstances have been very different in the different epochs, though steady in each, tend strongly to heighten the probability of the latter theory.

"When we view the immense calcareous and bituminous formations, principally from the waters and atmosphere, and consider the oxidations and depositions which have taken place, either gradually, or during some of the great convulsions, it appears at least probable, that the liquid elements containing life have varied considerably at different times in composition and weight; that our atmosphere has contained a much greater proportion of carbonic acid or oxygen; and our waters aided by excess of carbonic acid, and greater heat resulting from greater density of atmosphere, have contained a greater quantity of lime and other mineral solutions. Is the inference then unphilosophic, that living things which are proved to have a circumstance-suiting power—a very slight change of circumstance by culture inducing a corresponding change of character—may have gradually accommodated themselves to the variations of the elements containing them, and, without new creation, have presented the diverging changeable phenomena of past and present organised existence?

"The destructive liquid currents, before which the hardest mountains have been swept and comminuted into gravel, sand, and mud, which intervened between and divided these epochs, probably extending over the whole surface of the globe, and destroying nearly all living things, must have reduced existence so much, that an unoccupied field would be formed for new diverging ramifications of life, which, from the connected sexual system of vegetables, and the natural instincts of animals to herd and combine with their own kind, would fall into specific groups, these remnants, in the course of time, moulding and accommodating their being anew to the change of circumstances, and to every possible means of subsistence, and the millions of ages of regularity which appear to have followed between the epochs, probably after this accommodation was completed, affording fossil deposit of regular specific character.

"There are only two probable ways of change—the above, and the still wider deviation from present occurrence—of indestructible or molecular life (which seems to resolve itself into powers of attraction and repulsion under mathematical figure and regulation, bearing a slight systematic similitude to the great aggregations of matter), gradually uniting and developing itself into new circumstance-suited living aggregates, without the presence of any mould or germ of former aggregates, but this scarcely differs from new creation, only it forms a portion of a continued scheme or system.

"In endeavouring to trace, in the former way, the principle of these changes of fashion which have taken place in the domiciles of life, the following questions occur:—Do they arise from admixture of species nearly allied producing intermediate species? Are they the diverging ramifications of the living principle under modification of circumstance? Or have they resulted from the combined agency of both? Is there only one living principle? Does organised existence, and perhaps all material existence, consist of one Proteus principle of life capable of gradual circumstance-suited modifications and aggregations, without bound under the solvent or motion-giving principle, heat or light? There is more beauty and unity of design in this continual balancing of life to circumstance, and greater conformity to those dispositions of nature which are manifest to us, than in total destruction and new creation. It is improbable that much of this diversification is owing to commixture of species nearly allied, all change by this appears very limited, and confined within the bounds of what is called species; the progeny of the same parents, under great difference of circumstance, might, in several generations, even become distinct species, incapable of co-reproduction.

"The self-regulating adaptive disposition of organised life, may, in part, be traced to the extreme fecundity of Nature, who, as before stated, has, in all the varieties of her offspring, a prolific power much beyond (in many cases a thousandfold) what is necessary to fill up the vacancies caused by senile decay. As the field of existence is limited and pre-occupied, it is only the hardier, more robust, better suited to circumstance individuals, who are able to struggle forward to maturity, these inhabiting only the situations to which they have superior adaptation and greater power of occupancy than any other kind; the weaker, less circumstance-suited, being prematurely destroyed. This principle is in constant action, it regulates the colour, the figure, the capacities, and instincts; those individuals of each species, whose colour and covering are best suited to concealment or protection from enemies, or defence from vicissitude and inclemencies of climate, whose figure is best accommodated to health, strength, defence, and support; whose capacities and instincts can best regulate the physical energies to self-advantage according to circumstances—in such immense waste of primary and youthful life, those only come forward to maturity from the strict ordeal by which Nature tests their adaptation to her standard of perfection and fitness to continue their kind by reproduction.

"From the unremitting operation of this law acting in concert with the tendency which the progeny have to take the more particular qualities of the parents, together with the connected sexual system in vegetables, and instinctive limitation to its own kind in animals, a considerable uniformity of figure, colour, and character, is induced, constituting species; the breed gradually acquiring the very best possible adaptation of these to its condition which it is susceptible of, and when alteration of circumstance occurs, thus changing in character to suit these as far as its nature is susceptible of change.

"This circumstance-adaptive law, operating upon the slight but continued natural disposition to sport in the progeny (seedling variety), does not preclude the supposed influence which volition or sensation may have over the configuration of the body. To examine into the disposition to sport in the progeny, even when there is only one parent, as in many vegetables, and to investigate how much variation is modified by the mind or nervous sensation of the parents, or of the living thing itself during its progress to maturity; how far it depends upon external circumstance, and how far on the will, irritability, and muscular exertion, is open to examination and experiment. In the first place, we ought to investigate its dependency upon the preceding links of the particular chain of life, variety being often merely types of approximations of former parentage; thence the variation of the family, as well as of the individual, must be embraced by our experiments.

"This continuation of family type, not broken by casual particular aberration, is mental as well as corporeal, and is exemplified in many of the dispositions or instincts of particular races of men. These innate or continuous ideas or habits seem proportionally greater in the insect tribes, those especially of shorter revolution; and forming an abiding memory, may resolve much of the enigma of instinct, and the foreknowledge which these tribes have of what is necessary to completing their round of life, reducing this to knowledge, or impressions and habits, acquired by a long experience. This greater continuity of existence, or rather continuity of perceptions and impressions, in insects, is highly probable; it is even difficult in some to ascertain the particular stops when each individuality commences, under the different phases of egg, larva, pupa, or if much consciousness of individuality exists. The continuation of reproduction for several generations by the females alone in some of these tribes, tends to the probability of the greater continuity of existence, and the subdivisions of life by cuttings (even in animal life) at any rate must stagger the advocate of individuality.

"Among the millions of specific varieties of living things which occupy the humid portion of the surface of our planet, as far back as can be traced, there does not appear, with the exception of man, to have been any particular engrossing race, but a pretty fair balance of powers of occupancy—or rather, most wonderful variation of circumstance parallel to the nature of every species, as if circumstance and species had grown up together. There are indeed several races which have threatened ascendancy in some particular regions, but it is man alone from whom any general imminent danger to the existence of his brethren is to be dreaded.

"As far back as history reaches, man had already had considerable influence, and had made encroachments upon his fellow denizens, probably occasioning the destruction of many species, and the production and continuation of a number of varieties or even species, which he found more suited to supply his wants, but which, from the infirmity of their condition—not having undergone selection by the law of nature, of which we have spoken, cannot maintain their ground without its culture and protection.

"It is, however, only in the present age that man has begun to reap the fruits of his tedious education, and has proven how much 'knowledge is power.' He has now acquired a dominion over the material world, and a consequent power of increase, so as to render it probable that the whole surface of the earth may soon be overrun by this engrossing anomaly, to the annihilation of every wonderful and beautiful variety of animated existence, which does not administer to his wants principally as laboratories of preparation to befit cruder elemental matter for assimiliation by his organs."

"Much of the luxuriance and size of timber depending upon the particular variety of the species, upon the treatment of the seed before sowing, and upon the treatment of the young plant, and as this fundamental subject is neither much attended to nor generally understood, we shall take it up *ab initio.*

"The consequences are now being developed of our deplorable ignorance of, or inattention to, one of the most evident traits of natural history, that vegetables as well as animals are generally liable to an almost unlimited diversification, regulated by climate, soil, nourishment, and new commixture of already formed varieties. In those with which man is most intimate, and where his agency in throwing them from their natural locality and dispositions has brought out this power of diversification in stronger shades, it has been forced upon his notice, as in man himself, in the dog, horse, cow, sheep, poultry—in the Apple, Pear, Plum, Gooseberry, Potato, Pea, which sport in infinite varieties, differing considerably in size, colour, taste, firmness of texture, period of growth, almost in every recognisable quality. In all these kinds man is influential in preventing deterioration, by careful selection of the largest or most valuable as breeders; but in timber trees the opposite course has been pursued. The large growing varieties being so long of coming to produce seed, that many plantations are cut down before they reach this maturity, the small growing and weakly varieties, known by early and extreme seeding, have been continually selected as reproductive stock, from the ease and conveniency with which their seed could be procured; and the husks of several kinds of these invariably kiln-dried, in order that the seeds might be the more easily extracted. May we, then, wonder that our plantations are occupied by a sickly short-lived puny race, incapable of supporting existence in situations where their own kind had formerly flourished—

particularly evinced in the genus Pinus, more particularly in the species Scots Fir; so much inferior to those of Nature's own rearing, where only the stronger, more hardy, soil-suited varieties can struggle forward to maturity and reproduction?

"We say that the rural economist should pay as much regard to the breed or particular variety of his forest trees, as he does to that of his live stock of horses, cows, and sheep. That nurserymen should attest the variety of their timber plants, sowing no seeds but those gathered from the largest, most healthy, and luxuriant growing trees, abstaining from the seed of the prematurely productive, and also from that of the very aged and over-mature; as they, from animal analogy, may be expected to give an infirm progeny, subject to premature decay."— See "Naval Timber and Arboriculture," pages 364 and 365, 381 to 388; also 106 to 108. *Patrick Matthew, Gourdee Hill, Errol N.B., March 7.*

APPENDIX VI
Report of the British Association meeting in Oxford, 26 June – 3 July 1860

Several letters in this volume contain references to events that took place at the 1860 meeting of the British Association for the Advancement of Science held in Oxford from 26 June to 3 July 1860. Darwin had planned to attend the meeting, having arranged his accommodation in Oxford, but in the end was unable to attend because of his own poor health and that of his daughter, Henrietta Emma Darwin. Indeed, Darwin spent the week of the meeting at Edward Wickstead Lane's hydropathic establishment at Sudbrook Park in Richmond, Surrey, undergoing treatment for a stomach that had 'utterly broken down' (letter to Charles Lyell, 25 [June 1860]).

Undoubtedly the most famous incident of the meeting was the verbal encounter between Samuel Wilberforce, bishop of Oxford, and Thomas Henry Huxley that occurred during the Saturday, 30 June, meeting of Section D (zoology and botany, including physiology). Although details concerning the 'debate' between Huxley and Wilberforce over Darwin's theory were not well reported in the contemporary press (see Ellegård 1990), a relatively consistent narrative can be gleaned from personal recollections. Various accounts of the involvement and perspective of particular participants can be found in the respective *Lives and letters* of: Thomas Henry Huxley (L. Huxley ed. 1900, 1: 179–89); Joseph Dalton Hooker (L. Huxley ed. 1918, 1: 521–4); Charles Lyell (K. M. Lyell ed. 1881, 2: 335–6); John Lubbock (Hutchinson ed. 1914, 1: 50); John William Draper (Fleming 1950); and also Darwin (*LL* 2: 320–3, F. Darwin ed. 1892, pp. 236–9). (For a summary of the various accounts of the Huxley–Wilberforce debate, see Jensen 1988.) Discussions of Darwin's theory that occurred in other sections of the meeting are less well known.

The following account of the 1860 meeting of the British Association in Oxford has been drawn from the *Athenæum*, which provided the most complete contemporary report of the meeting and which Darwin himself read. Only those passages in the coverage of the proceedings of the various sections that directly relate to discussions of Darwin's theory have been transcribed below, preceded by their precise attribution.

Athenæum, 7 July 1860, p. 19: Introduction to the reports

. . . Yet the main interest of the week has unquestionably centred in the Sections, where the intellectual activities have sometimes breathed over the courtesies of life like a sou'-wester, creating the waves of conversation with white and brilliant foam. The flash, and play, and collisions in these Sections have been as interesting and amusing to the audiences as the Battle at Farnborough or the Volunteer Review to the general British public. The Bishop of Oxford has been famous in these intellectual contests, but Dr. Whewell, Lord Talbot de Malahide, Prof. Sedgwick, Mr. Crawford, and Prof. Huxley have each found foemen worthy of their steel, and made their charges and countercharges very much to their own satisfaction and the delight of their respective friends. The chief cause of contention has been the new theory of the Development of Species by Natural Selection—a theory open—like the Zoological Gardens (from a particular cage in which it draws so many laughable illustrations)—to a good deal of personal quizzing, without, however, seriously crippling the usefulness of the physiological investigations on which it rests. The Bishop of Oxford came out strongly against a theory which holds it possible that man may be descended from an ape,—in which protest he is sustained by Prof. Owen, Sir Benjamin Brodie, Dr. Daubeny, and the most eminent naturalists assembled at Oxford. But others—conspicuous among these, Prof. Huxley—have expressed their willingness to accept, for themselves, as well as for their friends and enemies, all actual truths, even the last humiliating truth of a pedigree not registered in the Herald's College. The dispute has at least made Oxford uncommonly lively during the week.

Athenæum, 7 July 1860, pp. 25–6: Thursday session of Section D.—Zoology and botany, including physiology. President: John Stevens Henslow

'On the Final Causes of the Sexuality of Plants, with particular Reference to Mr. Darwin's Work "On the Origin of Species by Natural Selection,"' by Dr. DAUBENY.— Dr. Daubeny began by pointing out the identity between the two modes by which the multiplication of plants is brought about, the very same properties being imparted to the bud or to the graft as to the seed produced by the ordinary process of fecundation, and a new individual being in either instance equally produced. We are, therefore, led to speculate as to the final cause of the existence of sexual organs in plants, as well as in those lower animals which can be propagated by cuttings. One use, no doubt, may be the dissemination of the species; for many plants, if propagated by buds alone, would be in a manner confined to a single spot. Another secondary use is the production of fruits which afford nourishment to animals. A third may be to minister to the gratification of the senses of man by the beauty of their forms and colours. But as these ends are only answered in a small proportion of cases, we must seek further for the uses of the organs in question; and hence the author suggested that they might have been provided, in order to prevent that uniformity in the aspect of Nature, which would have prevailed if plants had been multiplied exclusively by buds. It is well known that a bud is a mere counterpart of the stock from whence it springs, so that we are always sure of obtaining the very same description of fruit by merely grafting a bud or cutting of a pear or apple tree upon another plant of the same species. On the other hand, the seed never produces an individual exactly like the plant from which it sprang; and hence, by the union of the sexes in plants, some variation from the primitive type is sure to result. Dr. Daubeny remarked that if we adopt in any degree the views of Mr. Darwin with respect to the origin of species by natural selection, the creation of sexual organs in plants might be regarded as intended to promote this specific object. Whilst, however, he gave his assent to the Darwinian hypothesis, as likely to aid us in reducing the number of existing species, he wished not to be considered as advocating it to the extent to which the author seems disposed to carry it. He rather desired to recommend to naturalists the necessity of further inquiries, in order to fix the limits within which the doctrine proposed by Mr. Darwin may assist us in distinguishing varieties from species.

Prof. HUXLEY, having been called on by the Chairman, deprecated any discussion of the general question of the truth of Mr. Darwin's theory. He felt that a general audience, in which sentiment would unduly interfere with intellect, was not the public before which such a discussion should be carried on. Dr. Daubeny had brought forth nothing new to demand or require remark.— Mr. R. DOWDEN, of Cork, mentioned, first, two instances in which plants had been disseminated by seeds, which could not be effected by buds; first, in the introduction of *Senecio squalida*, by the late Rev. W. Hincks; and, second, in the diffusion of chicory, in the vicinity of Cork, by the agency of its winged seeds. He related several anecdotes of a monkey, to show that however highly organized the Quadrumana might be, they were very inferior in intellectual qualities to the dog, the elephant and other animals. He particularly referred to his monkey being fond of playing with a hammer; but although he liked oysters as food, he never could teach him to break the oysters with his hammer as a means of indulging his appetite.— Dr. WRIGHT stated that a friend of his, who had gone out to report on the habits of the gorilla— the highest form of monkey—had observed that the female gorilla took its young to the sea-shore for the purpose of feeding them on oysters, which they broke with great facility.— Prof. OWEN wished to approach this subject in the spirit of the philosopher, and expressed his conviction that there were facts by which the public could come to some conclusion with regard to the probabilities of the truth of Mr. Darwin's theory. Whilst giving all praise to Mr. Darwin for the courage with which he had put forth his theory, he felt it must be tested by facts. As a contribution to the facts by which the theory must be tested, he would refer to the structure of the highest Quadrumana as compared with man. Taking the brain of the gorilla, it presented more differences, as compared with the brain of man, than it did when compared with the brains of the very lowest and most problematical form of the Quadrumana. The deficiencies in cerebral structure between the gorilla and man were immense. The posterior lobes of the cerebrum in man presented parts which were wholly absent in the gorilla. The same remarkable differences of structure were seen in other parts of the body; yet he would especially refer to the structure of the great toe in man, which was constructed to enable him to assume the upright position; whilst in the lower monkeys it was impossible, from the structure of their feet, that they should do so. He concluded by urging on the physiologist the necessity of experiment. The chemist, when in doubt, decided his questions by experiment; and this was what was needed by the physiologist.— Prof. HUXLEY begged to be permitted to reply to Prof. Owen. He denied altogether that the difference between the brain of the gorilla and man was so great as represented by Prof. Owen, and appealed to the published dissections of Tiedemann and others. From the study of the structure of the brain of the Quadrumana, he maintained that the difference between man and the highest monkey was not so great as between the highest and the lowest monkey. He maintained also, with regard to the limbs, that there was more difference between the toeless monkeys and the gorilla than between the latter and man. He believed that the great feature which distinguished man from the monkey was the gift of speech.

Athenæum, 14 July 1860, pp. 64–5: Saturday session, Section D.—Zoology and botany, including physiology

'On the Intellectual Development of Europe, considered with Reference to the Views of Mr. Darwin and others, that the Progression of Organisms is determined by Law,' by Prof. DRAPER, M.D., of New York.— The object of this paper was to show that the advancement of man in civilization does not occur accidentally or in a fortuitous manner, but is determined by immutable law. The author introduced his subject by recalling proofs of the dominion of law in the three great lines of the manifestation of life. First, in the successive stages of development of every individual, from the earliest rudiment to maturity; secondly, in the numberless organic forms now living contemporaneously with us, and constituting the animal series; thirdly, in the orderly appearance of that grand succession which in the slow lapse of geological time has emerged, constituting the life of the Earth, showing therefrom not only the evidences, but also proofs of the dominion of law over the world of life. In those three lines of life he established that the general principle is, to differentiate instinct from automatism, and then to differentiate intelligence from instinct. In man himself three distinct instrumental nervous mechanisms exist, and three distinct modes of life are perceptible, the automatic, the instinctive, the intelligent. They occur in an epochal order, from infancy through childhood to the more perfect state. Such holding good for the individual, it was then affirmed that it is physiologically impossible to separate the individual from the race, and that what holds good for the one holds good for the other too; and hence that man is the archetype of society, and individual development the model of social progress, and that both are under the control of immutable law: that a parallel exists between individual and national life in this, that the production, life, and death of an organic particle in the person, answers to the production, life, and death of a person in the nation. Turning from these purely physiological considerations to historical proof, and selecting the only European nation which thus far has offered a complete and completed intellectual life, Prof. Draper showed, that the characteristics of Greek mental development answer perfectly to those of individual life, presenting philosophically five well-marked ages or periods,—the first being closed by the opening of Egypt to the Ionians; the second, including the Ionian, Pythagorean, and Eleatic philosophies, was ended by the criticisms of the Sophists; the third, embracing the Socratic and Platonic philosophies, was ended by the doubts of the Sceptics; the fourth, ushered in by the Macedonian expedition and adorned by the splendid achievements of the Alexandrian school, degenerated into Neoplatonism and imbecility in the fifth, to which the hand of Rome put an end. From the solutions of the four great problems of Greek philosophy, given in each of these five stages of its life, he showed that it is possible to determine the law of the variation of Greek opinion, and to establish its analogy with that of the variations of opinion in individual life. Next, passing to the consideration of Europe in the aggregate, Prof. Draper showed that it has already in part repeated these phases in its intellectual life. Its first period closes with the spread of the power of Republican Rome, the second with the foundation of Constantinople, the third with the Turkish invasion of Europe: we are living in the fourth. Detailed proofs of the correspondence of these periods to those of Greek life, and through them to those of individual life, are given in a work now printing on this subject, by the author, in America. Having established this conclusion, Prof. Draper next briefly alluded to many collateral problems or inquiries. He showed that the advances of men are due to external and not to interior influences, and that in this respect a nation is like a seed, which can only develope when the conditions are favourable, and then only in a definite way; that the time for psychical change corresponds with that for physical, and that a nation cannot advance except its material condition be touched,—this having been the case throughout all Europe, as is manifested by the diminution of the blue-eyed races thereof; that all organisms and even man are dependent for their characteristics, continuance, and life on the physical conditions under which they live; that the existing apparent invariability presented by the world of organization is the direct consequence of the physical equilibrium, but that if that should suffer modification, in an instant the fanciful doctrine of the immutability of species would be brought to its proper value. The organic world appears to be in repose because natural influences have reached an equilibrium. A marble may remain motionless for ever on a level table, but let the table be a little inclined, and the marble will quickly run off; and so it is with organisms in the world. From his work on Physiology, published in 1856, he gave his views in support of the doctrine of the transmutation of species; the transitional forms of the animal and also the human type; the production of new ethnical elements, or nations; and the laws of their origin, duration, and death.

The announcement of this paper attracted an immense audience to the Section, which met this morning in the Library of the New Museum. The discussion was commenced by the REV. MR. CRESSWELL, who denied that any parallel could be drawn between the intellectual progress of man and the physical development of the lower animals. So far from the author being correct with regard to the history of Greece, its masterpieces in literature—the Illiad and Odyssey—were produced during its national infancy. The theory of intellectual development proposed was directly opposed to the known facts of the history of man.— SIR B. BRODIE stated, he could not subscribe to the hypothesis of Mr. Darwin. His primordial germ had not been demonstrated to have existed. Man had a power of self-consciousness —a principle differing from anything found in the material world, and he did not see how this could originate in lower organisms. This power of man was identical with the Divine Intelligence; and to suppose that this could originate with matter, involved the absurdity of supposing the source of Divine power dependent on the arrangement of matter.— The BISHOP OF OXFORD stated that the Darwinian theory, when tried by the principles of inductive science, broke down. The facts brought forward did not warrant the theory. The permanence of specific forms was a fact confirmed by all observation. The remains of animals, plants, and man found in those earliest records of the human race—the Egyptian catacombs, all spoke of their identity with existing forms, and of the irresistible tendency of organized beings to assume an unalterable character. The line between man and the lower animals was distinct: there was no tendency on the part of the lower animals to become the self-conscious intelligent being, man; or in man to degenerate and lose the high characteristics of his mind and intelligence. All experiments had failed to show any tendency in one animal to assume the form of the other. In the great case of the pigeons quoted by Mr. Darwin, he admitted that no sooner were these animals set free than they returned to their primitive type. Everywhere sterility attended hybridism, as was seen in the closely-allied forms of the horse and the ass. Mr. Darwin's conclusions were an hypothesis, raised most unphilosophically to the dignity of a causal theory. He was glad to know that the greatest names in science were oppposed to this theory, which he believed to be oppposed to the interests of science and humanity.— Prof. HUXLEY defended Mr. Darwin's theory from the charge of its being merely an hypothesis. He said, it was an explanation of phenomena in Natural History, as the undulating theory was of the phenomena of light. No one objected to that theory because an undulation of light had never been arrested and measured. Darwin's theory was an explanation of facts; and his book was full of new facts, all bearing on his theory. Without asserting that every part of the theory had been confirmed, he maintained that it was the best explanation of the origin of species which had yet been offered. With regard to the psychological distinction between man and animals; man himself was once a monad—a mere atom, and nobody could say at what moment in the history of his development he became consciously intelligent. The question was not so much one of a transmutation or transition of species, as of the production of forms which became permanent. Thus the short-legged sheep of America were not produced gradually, but originated in the birth of an original parent of the whole stock, which had been kept up by a rigid system of artificial selection.— Admiral FITZROY regretted the publication of Mr. Darwin's book, and denied Prof. Huxley's statement, that it was a logical arrangement of facts.— Dr. BEALE pointed out some of the difficulties with which the Darwinian theory had to deal, more especially those vital tendencies of allied species which seemed independent of all external agents.— Mr. LUBBOCK expressed his willingness to accept the Darwinian hypothesis in the absence of any better. He would, however, express his conviction, that time was not an essential element in these changes. Time alone produced no change.— Dr. HOOKER, being called upon by the President to state his views of the botanical aspect of the question, observed, that the Bishop of Oxford having asserted that all men of science were hostile to Mr. Darwin's hypothesis,—whereas he himself was favourable to it,—he could not presume to address the audience as a scientific authority. As, however, he had been asked for his opinion, he would briefly give it. In the first place, his Lordship, in his eloquent address, had, as it appeared to him, completely misunderstood Mr. Darwin's hypothesis: his Lordship intimated that this maintained the doctrine of the transmutation of existing species one into another, and had confounded this with that of the successive development of species by variation and natural selection. The first of these doctrines was so wholly opposed to the facts, reasonings, and results of Mr. Darwin's work, that he could not conceive how any one who had read it could make such a mistake,— the whole book, indeed, being a protest against that doctrine. Then, again, with regard to the general phenomena of species, he understood his Lordship to affirm that these did not present characters that should lead careful and philosophical naturalists to favour Mr. Darwin's views. To this assertion Dr. Hooker's experience of the Vegetable Kingdom was diametrically opposed. He considered that at least one half of the known kinds of plants were disposable in groups, of which the species were connected by varying characters common to all in that group, and sensibly differing in some individuals only of each species; so much so that, if each group be likened to a cobweb, and one species be supposed to stand in the center of that web, its varying characters might be compared to the radiating and concentric threads, when the other species would be represented by the points of union of these; in short, that the general characteristics of orders, genera, and species amongst plants differed in degrees only from those of varieties, and afforded the strongest countenance to Mr. Darwin's hypothesis. As regarded his own acceptation of Mr. Darwin's views, he expressly disavowed having adopted them as a creed. He knew no creeds in scientific matters. He had early begun the study of natural science under the idea that species were original creations; and it should be steadily kept in view that this was merely another hypothesis, which in the abstract was neither more nor less entitled to acceptance than Mr. Darwin's: neither was, in the present state of science, capable of demonstration, and each must be tested by its power of explaining the mutual dependence of the phenomena of life. For many years he had held to the old hypothesis, having no better established one to adopt, though the progress of botany had, in the interim, developed no new facts that favoured it, but a host of most suggestive objections to it. On the other hand, having fifteen years ago been privately made acquainted with Mr. Darwin's views, he had during that period applied these to botanical investigations of all kinds in the most distant parts of the globe, as well as to the study of some of the largest and most different Floras at home. Now, then, that Mr. Darwin had published it, he had no hesitation in publicly adopting his hypothesis, as that which offers by far the most probable explanation of all the phenomena presented by the classification, distribution, structure, and development of plants in a state of nature and under cultivation; and he should, therefore continue to use his hypothesis as the best weapon for future research, holding himself ready to lay it down should a better be forthcoming, or should the now abandoned doctrine of original creations regain all it had lost in his experience.

APPENDIX VII
Reviews of *Origin*, 1859–60

This brief list identifies the reviews of *Origin* mentioned in Darwin's correspondence of November and December 1859 (see *Correspondence* vol. 7) and in letters written during 1860. The list excludes articles and comments about Darwin's theories that are not, strictly speaking, reviews, for example, addresses at meetings of learned societies and editorials in periodicals; also excluded are reviews not mentioned in the correspondence unless referred to in Darwin's own manuscript list of reviews (see below). For a more complete list of the reviews of *Origin*, see Ellegård 1990; other studies of the early reception in Britain of Darwin's evolutionary views include Hull 1973, Burkhardt 1974, and Bowler 1985.

Several of the reviews listed below also include notices of other scientific books, such as Joseph Dalton Hooker's introductory essay to the Tasmanian flora (Hooker 1859) and Baden Powell's *Essays on the spirit of the inductive philosophy* (Powell 1855). One ([R. Owen] 1860a) also discusses Darwin's and Alfred Russel Wallace's joint paper delivered at the Linnean Society in 1858 (Darwin and Wallace 1858).

Many reviews were sent as presentation copies to Darwin; others he obtained himself. Darwin frequently wrote on his copies the name of the author, when known, and the date of publication. In many, he also made marginal annotations. Most of Darwin's copies of the reviews of *Origin* are in the Darwin Pamphlet Collection–CUL.

A manuscript headed 'List Reviews of Origin of Sp & of C. Darwins Books' is held at Down House (Down House MS 8). This list served as Darwin's index to his collection of reviews, each being assigned a number. It includes several reviews of *Origin* that are not mentioned in the correspondence; their bibliographic details have been transcribed here at the end of the list since these reviews were obviously known to Darwin. This manuscript is prefaced by the following note in Darwin's hand: 'Some of the most absurd or unjust articles were by Harvey *& Westwood [*interl*] in G. Chronicle 1860[1] Wollaston in Annals & Mag. of Nat. Hist[2] *& Haughton[3] [*interl*] & Owen[4] see remarks on latter in Saturday Review.—[5] Hopkins in Fraser[6] says my view differs in nothing from Lamarck.'

The following list is arranged chronologically according to the month of publication. For those reviews that were unsigned, the author's name, if known, is given in square brackets with the source of the identification supplied in a note. If the review is in the Darwin Pamphlet Collection, and if it contains annotations, this information follows the entry.

November 1859

[John R. Leifchild].[7] On the origin of species. *Athenæum*, 19 November 1859, pp. 659–60.

December 1859

Thomas Henry Huxley. Time and life: Mr. Darwin's "Origin of species." *Macmillan's Magazine* 1 (1859–60): 142–8. [Darwin Pamphlet Collection–CUL.]

[John Crawfurd].[8] *Examiner*, 3 December 1859, pp. 772–3.

[Anon.]. Darwin's Origin of species. *Saturday Review*, 24 December 1859, pp. 775–6.

[Thomas Henry Huxley].[9] The Darwinian hypothesis. *The Times*, 26 December 1859, p. 8.

[Richard Chambers].[10] Charles Darwin on the origin of species. *Chambers's Journal* 2 (1859): 388–91. [Darwin Pamphlet Collection–CUL.]

[Joseph Dalton Hooker].[11] *Gardeners' Chronicle and Agricultural Gazette*, 31 December 1859, pp. 1051–2.

January 1860

[Heinrich Georg Bronn].[12] *Neues Jahrbuch für Mineralogie, Geognosie, Geologie und Petrefakten-Kunde* (1860), pp. 112–16. [Darwin Pamphlet Collection–CUL.]

[William Benjamin Carpenter].[13] Darwin on the origin of species. *National Review* 10 (1860): 188–214.

Andrew Murray. On Mr Darwin's theory of the origin of species. *Proceedings of the Royal Society of Edinburgh* 4 (1857–62): 274–91. [Darwin Pamphlet Collection–CUL; annotated.]

February 1860

[Thomas Vernon Wollaston].[14] *Annals and Magazine of Natural History* 3d ser. 5 (1860): 132–43. [Darwin Pamphlet Collection–CUL; annotated.]

March 1860

[Asa Gray].[15] Review of Darwin's theory on the origin of species by means of natural selection. *American Journal of Science and Arts* 2d ser. 29 (1860): 153–84. [Darwin Pamphlet Collection–CUL; annotated.]

François Jules Pictet de la Rive. Sur l'Origine de l'espèce. *Bibliothèque Universelle. Revue Suisse et Étrangère. Archives des Sciences Physique et Naturelles* n.s. 7 (1860): 233–55. [Darwin Pamphlet Collection–CUL; annotated.]

[Richard Simpson].[16] Darwin on the origin of species. *Rambler* n.s. 2 (1860): 361–76.

[Adam Sedgwick].[17] Objections to Mr. Darwin's theory of the origin of species. *Spectator*, 24 March 1860, 285–6. [Reprinted, with revisions, in the *Spectator*, 7 April 1860, pp. 334–5.]

April 1860

[Thomas Henry Huxley].[18] Darwin on the origin of species. *Westminster Review* n.s. 17 (1860): 541–70. [Darwin Pamphlet Collection–CUL; annotated.]

[Francis Bowen].[19] *North American Review* 90 (1860): 474–506. [Darwin Pamphlet Collection–CUL; annotated.]

[Richard Owen].[20] Review of Origin of species and other works. *Edinburgh Review, or Critical Journal* 111 (1860): 487–532. [Darwin Pamphlet Collection–CUL; annotated.]

[William Benjamin Carpenter].[21] *British and Foreign Medico-Chirurgical Review* 25 (1860): 367–404. [Darwin Pamphlet Collection–CUL; annotated.]

John William Dawson. Review of "Darwin on the origin of species by means of natural selection". *Canadian Naturalist* 5 (1860): 100–20. [Darwin Pamphlet Collection–CUL; annotated.]

[Samuel Haughton].[22] Biogenesis. *Natural History Review* 7 (1860): 23–32. [Darwin Pamphlet Collection–CUL; annotated.]

Antoine Auguste Laugel. Nouvelle théorie d'histoire naturelle. L'Origine des espèces. *Revue des Deux Mondes* 2d ser. 26 (1860): 644–71. [Darwin Pamphlet Collection–CUL; annotated.]

[Luke Berke].[23] Darwin on the origin of species. *The Future: a Monthly Journal of Research and Criticism in the Physical and Historical Sciences* 1 (1860): 10–15, 27–30. [Darwin Pamphlet Collection–CUL; annotated.]

May 1860

Adam Sedgwick. On Darwin's theory. *Cambridge Chronicle*, 19 May 1860, pp. 4–5.

[John Duns].[24] *North British Review* 32 (1860): 455–86. [Darwin Pamphlet Collection –CUL; annotated.]

[Anon.]. Professor Owen on the Origin of species. *Saturday Review*, 5 May 1860, pp. 573–4.

[John Amory Lowell].[25] *Christian Examiner and Theological Review* 5th ser. 6 (1860): 449–64.

Francis Bowen. Remarks on the latest form of the development theory. *Memoirs of the American Academy of Arts and Sciences* n.s. 8 (1861): 97–122. [Darwin Pamphlet Collection–CUL; annotated.]

June 1860

Rudolph Wagner. Louis Agassiz's Prinzipien der Classifikation der organischen Körper insbesondere der Thiere mit Rücksicht auf Darwin's Ansichten.

Göttingische gelehrte Anzeigen pt 2 (1860): 761–800. [Darwin Pamphlet Collection–CUL; annotated.]

[Anon.]. Species. *All the Year Round*, 2 June 1860, pp. 174–8; 7 July 1860, pp. 293–9. [Darwin Pamphlet Collection–CUL.]

William Hopkins. Physical theories and the phenomena of life. *Fraser's Magazine* 61 (1860): 739–52; 62 (1860): 74–90.

July 1860

Theophilus Parsons. On the Origin of species. *American Journal of Science and Arts* 2d ser. 30 (1860): 1–13. [Darwin Pamphlet Collection–CUL; annotated.]

Louis Agassiz. On the Origin of species. *American Journal of Science and Arts* 2d ser. 30 (1860): 142–54. [Darwin Pamphlet Collection–CUL; annotated.]

[Asa Gray].[26] Darwin on the origin of species. *Atlantic Monthly* 6 (1860): 109–16, 229–39. Darwin and his reviewers. *Atlantic Monthly* 6 (1860): 406–25. [Darwin Pamphlet Collection–CUL; annotated.]

[Samuel Wilberforce].[27] Darwin's Origin of species. *Quarterly Review* 108 (1860): 225–64. [Darwin Pamphlet Collection–CUL; annotated.]

[Anon.]. Species in our domesticated animals and cultivated plants. *Journal of Agriculture, and the Transactions of the Highland and Agricultural Society of Scotland* n.s. 69 (1860): 333–53. [Darwin Pamphlet Collection–CUL; annotated.]

[Anon.]. *London Review and Weekly Journal of Politics, Literature, Art and Society* 1 (1860): 11–12, 32–3, 58–9.

October 1860

Henry Freke. Observations upon Mr. Darwin's recently published work—'On the Origin of species by means of natural selection'. Dublin: printed for the author. [Darwin Pamphlet Collection–CUL; annotated.]

December 1860

Henry Fawcett. A popular exposition of Mr. Darwin on the origin of species. *Macmillan's Magazine* 3 (1860–1): 81–92. [Darwin Pamphlet Collection–CUL; annotated.]

[The following reviews are not mentioned in the correspondence but are included in Darwin's manuscript list of reviews.]

April 1860

[William Jardine].[28] *Edinburgh New Philosophical Journal* n.s. 11 (1860): 280–9.

[George Henry Lewes].[29] Studies in animal life. Chapter 4. *Cornhill Magazine* (1860), pp. 438–47. [Darwin Pamphlet Collection–CUL.]

August 1860

[Anon.]. The Origin of species controversy. *Weldon's Register of Facts and Occurrences Relating to Literature, the Sciences, and the Arts* (1860), pp. 1–7. [Darwin Pamphlet Collection–CUL.]

September 1860

[Anon.]. *Calcutta Review* (1860): 64–88. [Darwin Pamphlet Collection–CUL.]

December 1860

[Anon.]. *The Geologist* 3 (1860): 464–72. [Darwin Pamphlet Collection–CUL; annotated.]

1 The references are to a series of letters printed in the *Gardeners' Chronicle and Agricultural Gazette* early in 1860 that criticised CD's theory. For William Henry Harvey's criticism, see the letter to Charles Lyell, 18 [and 19 February 1860]. For that of John Obadiah Westwood, see CD's letter to the *Gardeners' Chronicle*, [14–19 January 1860].

2 [Wollaston] 1860.

3 [Haughton] 1860b.

4 [R. Owen] 1860a.

5 The reference is to the second article on CD's theory published in the *Saturday Review*, 5 May 1860.

6 Hopkins 1860.

7 The publisher's marked copy of the *Athenæum* cites John R. Leifchild as the author (City University Library, London). See *Correspondence* vol. 7, letter to J. D. Hooker, [20 November 1859]. The attribution is confirmed in Marchand 1941, p. 226.

8 CD received a letter from John Crawfurd admitting the authorship of the review prior to its publication. See *Correspondence* vol. 7, letters to Charles Lyell, 2 December [1859], and to T. H. Huxley, [5 December 1859].

9 See letter to T. H. Huxley, 1 January [1860].

10 CD wrote Chambers's name on his copy.

11 See letter to J. D. Hooker, 14 January [1860].

12 CD wrote Bronn's name on his copy. Bronn was a co-editor of the journal in which the review appeared. Footnotes to the article are signed 'BR'. See also letter to T. H. Huxley, 2 [February 1860], and Hull 1973, p. 118.

13 *Wellesley index* 3: 153. See also Hull 1973, p. 88.

14 CD wrote Wollaston's name on his copy. See also Hull 1973, p. 127.

15 CD wrote Gray's name on his copy.

16 *Wellesley index* 2: 779.

17 See letter to Charles Lyell, [24 March – 3 April 1860]. See also Hull 1973, p. 159.

18 *Wellesley index* 3: 630. CD wrote Huxley's name at the top of his copy.

19 CD wrote Bowen's name on his copy.

20 *Wellesley index* 1: 509. CD wrote Owen's name on his copy. See also Hull 1973, p. 175.

21 CD wrote Carpenter's name on his copy.

22 CD wrote Haughton's name on his copy. See also Hull 1973, p. 217.

23 CD wrote 'By a Mr. Berke' at the top of his copy. Luke Berke was the editor of the *Future.*

24 *Wellesley index* 1: 684. CD wrote Duns's name on his copy, apparently on the authority of Robert

[24] *Wellesley index* 1: 684. CD wrote Duns's name on his copy, apparently on the authority of Robert Chambers. See *LL* 2: 311 n.

[25] See letter to Charles Lyell, 14 [June 1860].

[26] CD wrote Gray's name on his copy. The three-part article was reprinted in 1861 under Gray's name (Gray 1861).

[27] *Wellesley index* 1: 743 and Wilberforce 1874. CD wrote at the top of his copy: 'Bishop of Oxford aided by Owen & Murchison(?)'. See letter to Charles Lyell, 30 July [1860].

[28] CD wrote Jardine's name on his copy.

[29] *Wellesley index* 1: 325. CD wrote Lewes's name on his copy.

MANUSCRIPT ALTERATIONS AND COMMENTS

The alteration notes and comments are keyed to the letter texts by paragraph and line numbers. The precise section of the letter text to which the note applies precedes the square bracket. The changes recorded are those made to the manuscript by CD or his amanuensis; readers should consult the Note on editorial policy in the front matter for details of editorial practice and intent. The following terms are used in the notes as here defined:

del	deleted
illeg	illegible
interl	interlined, i.e., inserted between existing text lines
omitted	omitted by the editors to clarify the transcription
over	written over, i.e., superimposed

To Thomas Henry Huxley 1 January [1860]
1.5 (do not repeat this)] *square brackets in MS*
1.5 repeat] *before del* 'you'
1.6 Father & Son] *interl*

To Joseph Dalton Hooker 3 January [1860]
1.2 a] *added*
1.7 How] 'H' *over illeg*
1.8 (here . . . you).—] *square brackets in MS*; 'from you).—' *added*
2.2 (sneer no^{r} 2.)] *square brackets in MS*
2.4 $a\times x$] *after del* '$a\times$'
3.1 Orders] *after del* 'Several'; 'O' *over* 'o'
3.1 next] *after del* 'being'
3.2 S] *over* 'E'
3.3 affinity of] 'of' *over* 'to'
4.3 vegetation] *interl*
4.3 & level] *interl*
4.5 as of Glacial . . . productions] *interl*
4.6 productions of greater] *after del* 'pl'
4.6 those] *above del* 'that'
4.7 seizing] *after del* 'gettin'
4.10 (p. civ)] *interl*
4.11 viz p. civ] *interl*
5.1 is] *interl*
5.1 cases] *above del* 'facts'
5.2 (do you . . . deserts?)] *square brackets in MS*
6.1 invaders] *above del* 'natives'
6.3 which was] *interl*
6.4 *important.*] *before del* 'Does it not also apply to any invaded land as'
7.3 Australian orders] *interl before del* 'genera'
7.5 (together . . . N. Zealand)] *interl*
7.5 (bitter sneer no^{r} 3.)] *square brackets in MS*
8.1 am] *interl*
8.5 *Leguminosæ*] *before del caret*
8.6 migrate] *after del illeg*
8.6 to] *over* '&'
8.8 ; especially . . . sea-water. 8.9] *added over* '.—'
9.3 world,] *comma over full stop*
9.5 the list] 'the' *interl*
9.5 (N.B . . . does not lie.) 9.7] *square brackets in MS*
10.4 Should] *after del* 'Were'
10.9 introduced by man] *interl*
11.1 form] *after del* 'an'
14.3 lapse] *above del* 'lapse'

To Charles Lyell 4 [January 1860]
1.2 of species] *interl*
2.4 strictly] *interl*

To Hewett Cottrell Watson [5–11 January 1860]
3.5 that] *interl*
3.11 (*falsely* in my opinion)] *parentheses and underline added*; 'in my opinion)' *interl*
3.11 require] *above del* 'wish for'

To Thomas Bridges 6 January 1860
15.1 wild] *interl*
15.1 *White*] *underline added*

To William Benjamin Carpenter 6 January [1860]
1.9 specifics] 'c' *added*
1.14 loss] *above del* 'changes'

To Asa Gray 7 January [1860]
3.5 to] *after del* 'for'
3.6 species] *over* 'speecies'

To Leonard Jenyns 7 January [1860]
1.7 of these classes of facts] *interl*
1.8 of organic beings] *interl*
1.10 During] 'D' *over* 'd'
2.1 certainly] *interl*
4.4 People] *above del* 'they'

To T. H. Huxley 9 January [1860]
2.2 504] *interl*

To Charles Lyell 10 January [1860]
1.2 nearly] *interl*
1.5 by inadvertence] *interl*
2.6 by] *before del* 'a concluding'
2.7 (p. 489 new Edit)] *interl*
2.7 with other animals.] *interl*
2.12 Unless] *after del* 'Ulless'
4.1 write] *altered from* 'writes'
5.1 in Address . . . Edinburgh 5.2] *interl*
5.4 more &] *interl*
5.5 that] *interl*
5.5 caves] *after del* 'same genus in'
5.6 the genus] 'the' *over* 'this'
7.1 had a] 'a' *over* 'an'

To T. H. Huxley 11 January [1860]
1.3 to what is there said,] *interl*
1.6 same two] *interl*
1.8 & shall . . . larger book] *added*
1.9 hybridity)] ')' *over comma*
1.11 in pure species.] *interl*
2.1 might] *above del* 'can'
2.2 adaptation to] *above del illeg*
2.8 improbable] *after del* 'impossible'
2.9 on] *after del* 'of'
3.3 this] *interl*
3.4 rest] *after del* 'make'
3.5 it] *added*
4.3 in M.S.] *interl*
4.5 attack] 'k' *over* 'ks'
4.6 papers] *above del* 'answer'

To J. D. Hooker 14 [January 1860]
3.9 another] *altered from* 'a'

To Charles Lyell 14 January [1860]
2.3 another] *interl*
2.5 ; I remember . . . to Forbes. 2.6] *added*
2.7 I find] 'I' *over illeg*
2.8 unless . . . migration 2.9] *interl*
2.9 littoral shells of] *interl*
2.11 for . . . littoral shells,] *interl*
2.12 the period] 'the' *interl*
2.13 followers] 's' *added*
5.1 known to have been] *interl*
5.4 (p. 268 new Edit.)] *interl*
5.4 the **vars** of] *interl*
5.5 is a] 'is' *interl*
5.5 of sterility] *interl*
5.8 ; & . . . discussion.] *added*
6.2 a rather] 'a' *interl*
6.3 a firm] 'a' *interl*
6.4 to show this,] *interl*
7.2 Europe] 'Eu' *over illeg*
7.5 countries] *over* '—'
9.3 a man] 'a' *interl*
9.4 that] *after del* 'all'
9.5 The Bookseller . . . book!!! 9.6] *added*
10.1 a] *interl*
10.2 (Since . . . to translate.) 10.4] *added*; *square brackets in MS*

To Edward Cresy 15 January [1860]
1.2 Royal] *interl*
1.5 of Park . . . *[illeg]*] *added*; '?' *interl*

To Williams and Norgate 16 January [1860]
0.2 16th] '6' *over* '5'

To Baden Powell 18 January [1860]
2.17 I believe . . . Germans.] *added*
3.4 scientific] *interl*
4.3 most fully] *above del illeg*

To Baden Powell 18 January [1860]
1.1 addressed . . . Club] *interl*

To William Bernhard Tegetmeier 20 January [1860]
2.6 has] *over* 'have'

To T. H. Huxley 21 [January 1860]
3.1 on Wednesday at Museum] *interl*

6.1 In Haste] *added*

To Jean Louis Armand de Quatrefages de Brèau 21 January [1860]
4.1 has read . . . whether he 4.2] *interl*

To J. D. Hooker [22 January 1860]
1.2 sheet] *above del* 'note'
7.1 A capital . . . you.—] *pencil*

To John Murray 23 [January 1860]
3.2 Wednesday] *after del* 'Tues'

To John Murray [25 January 1860]
1.1 in note addressed to Down] *added*
2.2 also, hope] 'h' *over illeg*
3.2 Edition of] 'of' *over* '.—'
3.2 my] *over* 'the'
6.2 it was] *interl*
9.2 already] *interl*

To T. H. Huxley [26 January 1860]
2.2 at] *interl*

To Asa Gray 28 January [1860]
2.5 had I] *after del illeg*
3.1 strong] *interl*
3.2 (excepting] *parenthesis over comma*
3.7 received] *interl*
3.9 brief] *interl*
3.16 publisher,] *interl*
4.4 & so excellent a geologist] *interl*

To John Murray 30 [January 1860]
1.4 in text] *interl*
1.5 present] *interl*
1.5 if] *after del* 'other'
2.1 small] 'll' *over* 'ty'
3.1 My dear Sir] *added*

To J. D. Hooker 31 [January 1860]
2.1 Naudins] *above del* 'his'
4.2 ?] *over* '—'

To Asa Gray 1 February [1860]
2.2 few] *added*
2.2 , all that I have.] *interl*
5.1 (The Article in Times by Huxley.—)] *added*

To Richard Kippist? 1 February [1860]
1.1 & bring some Books] *added*

To T. H. Huxley 2 [February 1860]
1.4 what do you think?? 1.5] *interl*
5.1 tickets] *after del illeg*
5.1 us for] *interl*
6.2 himself] *interl*

To Herbert Spencer 2 February [1860]
5.3 on Development] *interl*
5.5 having no strength to spare 5.6] *interl*
5.8 in inverted commas 5.9] *interl*

To John Murray 3 February [1860]
2.1 Now] *added*
3.1 Talandier] 'd' *over* 'i'

To John Murray 4 February [1860]
2.1 but proof . . . day. 2.2] *interl*
2.2 corrections] *after del* 'corre'

To Philip Lutley Sclater 4 February [1860]
1.2 my Journal] *above del* 'Book'

To John Murray 5 February [1860]
3.1 page] *interl*
3.1 Journal?] '?' *over del full stop*

To J. D. Hooker 8 February [1860]
1.7 cash] *after del* 'or'
2.1 night before] *interl*
2.1 last] *before del* 'night'
2.4 classification] 'f' *over illeg*
5.1 last] *before del interl* 'Phil.' *and del* 'Club'
5.1 Royal Soc. meeting] *interl*
5.2 *all*] *interl*
6.1 talk] *after del* 'tak'
6.1 sure] *added*
6.1 then] *interl*

To Asa Gray [8 or 9 February 1860]
1.3 ; some . . . authority.] *interl*
1.3 Agassiz] 'A' *over* '&'
1.5 have] *after del* 'we'
1.5 yet] *interl*
1.7 (& before)] *interl*
2.4 in America,] *interl*
2.5 several] *interl*
2.6 likewise] *interl*
2.6 corrections] *interl*
4.5 the strongest] 'the' *interl*
4.5 marked] *interl*
5.3 health;] *interl*

To Charles James Fox Bunbury 9 February [1860]
2.3 to turn] 'to' *interl*
2.7 what] *after del* 'st'
3.4 in accounting . . . Light. 3.5] *interl*
3.7 to some extent.] *interl*
3.9 large] *interl*
4.1 Ch.] *after del* 'the'
4.2 of Character] *interl*
4.3 almost] *interl*
5.4 &] *over full stop*
5.4 & in giving . . . known facts 5.5] *added*; '&' *over full stop*
7.1 that] *interl*

To John Stevens Henslow 9 February [1860]
1.1 (N. Kent Station)] *interl*
1.2 (**not** Shortland Station)] *square brackets in MS*

To Charles Lyell 12 [February 1860]
2.4 is very] 'is' *over illeg*
2.4 Bunbury's . . . well written. 2.5] *added*
4.1 German] *interl*

To J. D. Hooker 14 February [1860]
1.5 the] *over* 'a'
1.6 He said . . . time failed.] *added*
1.7 an Exposition] 'an' *interl*
2.1 no] *over illeg*
2.1 doctrine of Nat. Selection 2.2] *interl*
2.8 shown to be] *interl*
3.1 of a] *interl*
3.2 on marble.] *interl*
5.2 on] *over* 'of'

To Charles Lyell 15 and 16 [February 1860]
1.3 (I think)] *parentheses over commas*
1.9 other formation] *interl*
2.2 have used] 'have' *interl*
2.6 clever] *after del* 'also'
3.3 in] *above del* 'of'
3.4 new species] 'new' *interl*
3.5 seems to] *above del* 'struck'
3.5 that about] 'that' *interl*
3.5 the Celebes] 'the' *interl*
3.6 aberrant] *interl*
4.1 Man] *above del* 'man'
4.2 Remember . . . Tertiaries. 4.3] *added*
4.6 Eye!] '!' *over full stop*
5.3 candid] *above del* 'honest'
5.3 any opposer] *above del* 'anyone'
5.6 geology,] *before del* 'that'
7.2 on] *over illeg*

To Smith, Elder, and Company? 17 February [1860]
1.2 the charges] 'the' *interl*
1.2 for me] *interl*

To Asa Gray 18 February [1860]
1.2 returned &] *interl*
1.3 Lyell] *after del* 'Hooker—'
2.2 both] *after del* 'abo'
3.3 are] *interl*

To Charles Lyell 18 [and 19 February 1860]
1.8 the Planets] 'the' *added*
2.1 lebensweise"] *before omitted closing quotes*
4.2 My] 'M' *over illeg*
5.8 groups] *after del* 'cases'
6.1 with flowers of two kinds] *interl*
6.3 of species . . . flowers.] *interl*
6.4 (as he supposes)] *interl*
7.1 is] *interl*
7.2 I will] *after del* 'per'
8.4 agree] *after del* 'qui'
8.6 an intestinal] 'an' *interl*
8.6 worm] *above del* 'work'
8.7 highly developed,—] *above del* 'so;—'
8.11 (Perhaps . . . so)] *interl*
8.12 any future] *above del* '2^{d}'
8.13 such] *interl*
8.13 H.] *over illeg*
8.14 about infinite . . . species] *interl*
8.17 ?] *over full stop*
9.3 & less . . . Nat. Selection, 9.4] *interl*
9.5 & where . . . be slower.—9.6] *interl*

To J. D. Hooker [20 February 1860]
1.2 together] *after del* 'with me'
2.1 I suppose . . . wild plant? 2.2] *added*
2.2 monsters] 'mo' *over illeg*
2.3 likewise] *after del* 'no'
2.4 that all] 'that' *interl*
2.5 Natural] *below del* 'that'
3.7 alone] *interl*
5.2 attacks] *interl*

To J. D. Hooker [23 February 1860]
1.2 It] *over* 'I'
1.3 very] *interl*
2.2 remainder of Gray's] *interl*

To Charles Lyell 23 February [1860]
2.1 forms] *interl*
3.6 easily] *interl*
3.7 abrupt] *interl*
3.10 monstrous] *interl*
3.10 with . . . state of nature 3.11] *interl*
3.14 more than very rarely] *above del* 'often'
4.5 As it] *after del* 'As it'
4.7 the movements] 'the' *interl*
4.7 a] *added*
4.10 curious?] '?' *over colon*
8.2 (as] *parenthesis over comma*

To François Jules Pictet de la Rive 23 February [1860]
1.7 offering to] *interl*
1.7 send] *altered from* 'sending'
2.4 of Heidelberg] *interl*
2.6 the] *over illeg*

To Herbert Spencer 23 [February 1860]
1.2 has] *altered from* 'have'
1.3 it] *interl*
1.3 in a] *above del illeg*
2.3 very] *interl*

To Asa Gray 24 February [1860]
3.6 latest] *altered from* 'later'
5.2 to me] *after del* 'clear'
5.2 & omniscient] *interl*
5.6 by Brewster] *interl*

To W. B. Tegetmeier 24 [February? 1860]
3.6 **one**] *interl*
5.3 of other kinds] *interl*

To Heinrich Georg Bronn [*c.* 25 February 1860]
5.1 an animal] *above del* 'it'
5.2 Habit may] 'Habit' *interl*
5.5 optic nerve] *above del* 'eyes'

To Charles Lyell 25 February [1860]
3.4 some time ago] *interl*
3.5 2^{d}] *interl*
3.8 English] *interl*
4.1 today] *interl*

To J. D. Hooker 26 [February 1860]
1.6 monstrous] *interl*

To J. D. Hooker [26 February 1860]
1.3 does] *after del* 'is'
2.1 some] *before omitted point*
2.2 a bunch] *after del* 'so'
2.3 Australian] *interl*
2.3 perhaps also Eucalyptus] *added*
3.1 pollen] *after del illeg*

To J. D. Hooker 3 March [1860]
2.1 note.] *full stop altered from semicolon*
2.1 That] 'T' *over* 't'
2.5 nearly] *interl*
2.5 actual] *interl*
4.4 It . . . duty.—] *added*
8.1 Huxley's] 'H' *over illeg*
9.1 his] *over illeg*
9.2 noticed] 'd' *over* 's'

To a Bookseller 4 March [1860]
0.2 March] *before del* 'Feb'
2.1 *2*,] *over illeg*
2.2 published . . . Str.] *added*

To William Erasmus Darwin [4 March 1860]
2.3 (] *over comma*
2.8 certainly] *altered from* 'certainy'
4.9 now] *interl*

To T. H. Huxley 4 March [1860]
2.1 I never . . . before] *added*

To Albrecht Carl Ludwig Gotthilf Günther 6 March [1860]
1.5 no . . . snakes] *added before del* 'of my Journal'

To Asa Gray 8 March [1860]
1.3 for p. 126] *after del* 'at'
1.11 share] *altered from* 'have shared'
2.7 run] *altered from* 'runs'
2.8 very slight] *interl*
5.1 P.S.] *before omitted point*
7.2 & a bird . . . the Bush.—] *interl*
7.2 two] *over* 'in'

To Charles Lyell 12 [March 1860]
1.2 with the little . . . improvement] *interl*
1.3 the case] *above del* 'in'
2.5 after] *after del illeg*
4.1 subject] 'b' *over* 'j'

To J. D. Hooker 18 [March 1860]
1.3 Emma] *over* '&'

To George Henry Kendrick Thwaites 21 March [1860]
1.7 (& I quite . . . retrograding)] *interl*
1.9 for instance Diatomaceæ] *interl before del* 'they'
3.1 how far] *after del* 'what I mean by the manner'
3.9 ones opinion;] *interl*

To William Darwin Fox 22 [March 1860]
1.1 showed in its] *above del* 'exhibited in'
1.3 than] *over illeg*

To Charles Lyell [24 March – 3 April 1860]
1.1 show!] '!' *over* '.—'
1.2 not to] 'n' *over* 't'
2.1 I will . . . afternoon.—] *added*
3.3 breaks] *above del* 'intervals'

To Miss Holland [April 1860]
1.3 made] *interl*
1.4 before the] 'the' *interl*

To F. J. Pictet de la Rive 1 April [1860]
1.3 yours] *over illeg*

To George Robert Waterhouse 1 April [1860]
1.3 a selected] 'a' *interl*
2.5 instinctively] *interl*
2.5 & Bees] *interl*
2.12 each Bee] *above del* 'them'
3.3 or side] *interl*

To J. S. Henslow 2 April [1860]
2.3 by Babington] *interl; caret over semicolon*
2.4 at first introduction.] *interl*
3.3 selling] *before omitted point*

To Asa Gray 3 April [1860]
1.5 (3[d] time)] *interl*
2.3 of structure] *interl*
2.5 Everglades] *after del* 'ev'
5.9 denudation] *interl*
5.18 ever] *after del* 'in'

To Frederick Smith 4 April [1860]
1.4 as] *over illeg*

To W. B. Carpenter 6 April [1860]
1.3 not,] *before omitted comma*
1.6 is] *over illeg*
1.9 It is . . . balanced.—] *added*

To Maxwell Tylden Masters 7 April [1860]
1.7 of Sweet Pea] *interl*
1.7 close together] *interl*
1.8 other] *before omitted point*
1.8 he] 'h' *over* 'th'
2.3 perhaps] *interl*

To Albert Way 7 April [1860]
1.7 some] *interl*
2.3 ?] *over comma*

To T. H. Huxley 9 April [1860]
1.3 so-called] *interl*
1.3 almost] *interl*
3.1 young] *over illeg*
3.3 naturalists] *above del* 'they'
3.4 &c,] *added*
3.4 it] *interl*
3.4 that I] 'that' *interl*
4.1 imply] *after del illeg*

To John Murray 9 April [1860]
1.1 marked from Author 1.2] *added*
4.2 Upon] *over illeg*
4.3 well.—] *before del* 'as yet.—'

To Charles Lyell 10 April [1860]
1.1 are] *interl*
1.4 "The] *over illeg*
5.5 all the] 'all' *interl*
5.11 organs] *interl*

To T. H. Huxley 11 April [1860]
2.3 to him] *interl*

To J. D. Hooker 13 [April 1860]
1.5 are] 'a' *over illeg*
1.9 a friend] *after illeg*
3.2 partly] *interl*
3.2 remove] *altered from* 'removes'

To T. H. Huxley 14 April [1860]
1.7 of the calcareous organisms] *interl*
1.11 soundings.—] *above del* 'deep sea'
4.1 laying] 'l' *over* 'y'

To Charles Lyell 15 April [1860]
2.4 have] *after del* 'are'
2.7 preposterous] *altered from* 'perposterous'
2.10 interposition . . . beings,] *above del* 'natural cases,'

5.1 is that] 'is' *interl*
5.2 question] 'n' *over* 'ns'

To W. B. Tegetmeier 17 April [1860]
1.2 (& the answers . . . "no" 1.3] *interl*
4.4 further . . . published] *interl*

To J. D. Hooker 18 [April 1860]
2.4 doctrine] 'r' *over* 'i'
9.1 (Leschenaultia formosa.—)] *square brackets in MS*
10.1 (In a flower . . . closed.—)] *square brackets in MS*; 'the' *interl*

To Asa Gray 25 April [1860]
1.1 on] *over* 'of'
3.3 (ie yours & this) 3.4] *interl*
3.6 April 1st] *interl*
6.6 as not] *interl*

To J. D. Hooker 26 April [1860]
1.5 the humid surface] *above del* 'it'
2.3 more spitefully] 'more' *above del* 'so'
2.3 than] *over illeg*
2.4 the Review] *above del* 'it'
3.5 in] *after del* 'flying'
3.5 &] *over* 'or'
3.5 Dictamnus???] *interl*
3.7 then] *after del* ', &'
4.1 Corydalis] 'C' *over* 'c'

To J. D. Hooker 27 April [1860]
1.1 tear] 't' *over* 'k'
3.3 this] *interl*
4.1 Mem. . . . Fraxinella—] *added*
5.1 kind] *interl*
6.1 side] *after del* 'basal'
6.1 flower,] *above del* 'flower'
8.1 into gangway] 'gangway' *interl*
11.1 Polygala] *altered from* 'Polyagla'
11.1 has pistil] *above del* 'is'
12.1 or] *over* 'of'

To Charles Lyell 27 and 28 April [1860]
2.1 President of] *interl*
3.9 dog or] *interl*
3.10 wild] *interl*
10.2 particulars] *below del* 'authorit'
10.4 it] *interl*
10.5 was] *after del* 'is'
10.7 colchicus &] *interl*
12.2 would] *above del* 'is'
12.2 be an] *interl*
12.2 infinite] ^{1}'i' *over* 'a'
12.5 of] *over* '—'
12.5 But] *above del* 'And'

To Andrew Murray 28 April [1860]
1.3 have no] *after del illeg*
4.1 to call . . . passages] *added*
6.2 being] *after del* 'keep'
7.3 selection] *interl*
7.4 female;] *semicolon over full stop*
7.4 or conversely] *interl*
8.3 but] *interl*
8.4 facts?] *before del* 'The case of'
9.3 some good] *interl*
9.5 from experiments] *interl*
9.5 on] *over illeg*
10.4 &c] *interl*
10.7 without any selection] *interl*
10.7 under] *above del illeg*
11.3 knowing] *before del* 'when'
11.4 impossibility] *above del* 'certainty'
11.5 distinguishing a linking] *above del* 'knowing a'
11.6 or very little] *interl*
11.6 absence] *after del* 'at'
11.8 Mr Salter . . . Survey] *interl*
11.12 geological] *interl*
11.13 of shells] *interl*
12.3 might] *above del* 'would'
12.6 dreamed] *above del* 'dreamed'
12.6 gaining] *after del* 'or'
13.1 rank] *above del* 'call'
14.1 346] *interl*
14.2 **close**] *after del* 'near'
15.6 than I.] *above del* '.—'
15.11 &] *above del* 'but'
15.12 specified] *after del* 'found'
16.1 the blind] *above del* 'the'
17.2 in] *after del* 't'
18.1 of the habits] *interl*
18.1 La Plata] *interl*
18.2 (p. 184)] *interl*
18.3 "*every*] *after del* 'all'
18.5 that] *after del illeg*

To Andrew Murray 28 [April 1860]
1.2 males] *before del* 'or females'
1.2 of a rare] 'of' *over* 'in'
2.2 without facts in support.] *interl*

4.3 , as the theory . . . to do.—] *added*
4.5 combined . . . Nat. Selection. 4.6] *interl*
5.1 do;] *semicolon altered from colon*

To J. D. Hooker 30 April [1860]
1.5 to ask you] *interl*
1.7 is] *interl*
2.2 of growth] *interl*
2.4 special] *interl*
2.5 kind] *interl*
3.2 to depend] 'to' *added*
3.2 depend on the] *above del* 'for'
3.2 the flower] 'the' *interl*
4.2 he remarked] 'he' *added*

To Charles Lyell 4 May [1860]
1.2 peat in] *interl*
1.3 at] *over illeg*
1.5 flint] *over illeg*
1.5 in such] *after del* 'by boxes full'
1.6 **in Peat**] *interl*
3.1 as] *over illeg*
3.2 supposed] *interl*
6.3 or other Reviewers] *interl*
9.3 Tradesman] *after del illeg*
9.3 is also] 'is' *interl*

To Andrew Murray 5 May [1860]
1.6 hermaphrodite] *interl*
2.1 the case] 'the' *over illeg*
2.2 has been] *above del* 'is'

To J. D. Hooker 7 May [1860]
1.1 (whatever . . . may be) 1.2] *interl*
1.4 I know . . . for pollen. 1.5] *interl*
2.1 my experimental] *interl*
2.2 which . . . "male plants"] *interl*
2.5 sets of] *interl*
2.5 stigma] *above del* 'pistil'
2.5 slightly] *interl*
2.6 & certainly . . . roughness] *interl*
2.7 very] *above del* 'pretty'
2.7 more] *after del* 'much'
2.12 an] 'a' *over illeg*
2.13 Likewise . . . Oxlips.—2.14] *added*
6.2 plants] *interl before del* '& flowers'
6.3 plants] *interl before del* '& flowers'
7.1 I measured . . . Micrometer.—] *pencil*

To T. H. Huxley 7 May [1860]
3.2 but] *interl*
4.1 proportion] *altered from* 'proportions'
5.1 increased] *above del* 'greater'
5.3 or slower] *interl*

To J. S. Henslow 8 May [1860]
2.1 alters] *above del* 'alters'
2.2 viz] *interl*
3.3 at the bottom . . . soul] *interl*
4.6 from] *above del* 'about'
4.8 of Light] *interl*
4.14 &c.—] *over* '.—'

To Charles Lyell 8 [May 1860]
1.2 should be] *above del* 'is'
3.8 varieties or] *interl*
3.9 two] *interl*
4.2 from old . . . conversely] *interl*
4.5 flora] *after del* 'dry'
4.5 of dry land] *interl*
5.1 Pigeon] *altered from* 'Pigeons'
8.1 extinct] *added*

To W. B. Tegetmeier 8 May [1860]
2.6 till today] *interl*

To J. D. Hooker 13 [May 1860]
1.6 liveth.] *before omitted point*

To J. S. Henslow 14 May [1860]
4.8 any] *interl*

To J. D. Hooker 14 May [1860]
1.2 do] *after del* 'I'
1.3 a] *above del* 'the point of small'
1.3 holding handle nearly] *interl*
1.3 the pistil &] *interl*
1.4 direction &] *interl*
1.5 the copious] *interl*
2.5 clearer] *before omitted point*
3.1 (honour bright)] *parentheses over commas*
3.1 for a long time] *interl*
5.2 the] *over* 'She'
5.2 Fever] *interl*
7.1 the] *interl*
7.3 gathered by the children] *interl*

To J. D. Hooker 15 [May 1860]
1.4 Wolf] *altered from* 'Woff'
2.1 de facto] *interl*
2.2 period.] *full stop over comma*
2.2 But] *interl*

2.2 his] *interl*
2.4 claim] *after del illeg*
2.5 the reason] 'the' *interl*
2.14 in answer . . . assigned me.—2.15] *interl*
3.2 Candidates] 'C' *over* 'c'
3.5 M.P.] *interl*
4.3 of this size &] *interl*
5.4 at their attacks,] *above del* 'at'
5.9 so] *before omitted point*
7.1 on the] *after del illeg*
7.1 to day] *interl*

To James Drummond 16 May 1860
2.1 creeping in] 'in' *interl*
2.3 I find . . . brush.] *interl*
2.10 the] *interl*
3.2 Botanists] *after del* 'naturalists'
4.2 Charles Darwin] *in CD's hand*

To J. S. Henslow 17 May [1860]
7.1 Stamens] *after del illeg*
7.1 in water] *interl*
7.3 smoother] *after del illeg*
8.2 in water] *interl*

To W. D. Fox 18 May [1860]
1.1 curious] 'cu' *over illeg*
1.2 & origin] *interl*
2.3 of its internal organs] *interl*
5.1 various] *interl*
5.2 now] *interl*
5.3 me] 'm' *over* 'l'

To Asa Gray 18 May [1860]
1.8 saving] 'sa' *over illeg*
1.10 the whole] 'the' *interl*
1.13 with] *after del* 'for'
1.15 the Geologist] *interl*
1.16 the Entomologist] *interl*
1.16 of Edinburgh] *interl*
1.18 Geolog.] *after del illeg*

To Charles Lyell 18 May [1860]
2.5 an old] *interl*
2.5 A] *over illeg*
2.9 when you have done with it] *interl*
4.9 & be] *interl*
5.1 or Sunday] *interl*

To Alfred Russel Wallace 18 May 1860
1.6 of pursuers] *after del* 'of'
2.6 leisure] *above del* 'pleasure'
3.1 envious &] *interl*
4.3 in Lecture] 'in' *over illeg*
5.2 this] *above del* 'do'
5.3 but very . . . paragraphs] *interl*
5.8 yet] *interl*

To J. D. Hooker 20 May [1860]
2.3 He succeeded . . . cowslip] *added*
2.6 703] *below del* '30'
3.3 by] *over illeg*
3.5 no] *over illeg*
3.11 rather] *interl*
3.11 adaption to different] *interl*
3.15 on . . . intelligible] *added*
5.2 the Origin.] *interl*
5.2 They are] *over* 'It is'

To Asa Gray 22 May [1860]
1.1 (of May 7th)] *interl*
1.5 to the Appletons] *interl*
1.11 in Silliman] *interl*
1.12 Origin] *above del* 'Journal'
1.13 from remarks] 'from' *interl*
1.15 article] *above del* 'Review'
1.16 if] 'f' *over* 't'
1.18 publickly that] *interl*
2.1 in North] *after del* 'by'
2.6 point] *before omitted point*
2.6 slip] *altered from* 'slipp'
2.8 & the latter] *above del* 'whom'
9.3 also] *interl*
13.1 512. . . . quotation. 13.3] *added*

To J. D. Hooker 22 [May 1860]
1.2 in my part] *interl*
2.4 this plant,] *interl*
4.2 I can] *after del* 'for'
5.2 Origin] 'O' *over* 'o'
8.2 us] *after del* 'me'
10.2 wished] *above del* 'begun'
10.4 with *pollen shedding*] *interl*
10.6 little] *interl*
10.6 the] *interl*
10.9 if you . . . enough] *interl*
10.10 only] *above del* 'only'
10.10 7 or] *interl*
10.14 & curious] *interl*
10.15 The pollen . . . water.—10.16] *added*
11.7 scales on the buds] *interl*
11.7 by the] *over illeg*
11.7 all uselessly] *after del* ', &'

To Charles Lyell 22 May [1860]
2.2 afterwards] *interl*
3.1 from Williams & Norgate] *interl*
6.3 grew] *above del* 'grey'

To John Lubbock 25 May [1860]
1.5 & that M[r] S. refused] *interl*
1.8 (from manner alone)] *interl*
1.9 committee] *before del comma*
1.11 & I informed . . . this;] *added*

To J. D. Hooker 29 [May 1860]
1.5 clear,] *interl*
1.8 though] *before abandoned caret*
1.8 you] *interl*
2.2 almost] *interl*
5.1 I have looked . . . to me.—5.2] *added pencil*

To John Lubbock 29 [May 1860]
1.5 Or you might say this.—] *interl*

To J. D. Hooker 30 May [1860]
4.1 then] *interl*

To Charles Lyell 1 [June 1860]
2.1 review] *altered from* 'reviews'
3.1 Phillips' . . . published.] *added*
5.3 (& he . . . judge) 5.4] *parentheses over commas*
7.1 of dog] *interl*
7.2 many] *above del* 'may'
7.4 (according as 1st day reckoned) 7.5] *interl*
7.5 days.] *before del* 'according'; *full stop over comma*
8.2 general] *interl*
8.2 in these respects] *interl*

To George Rolleston 1 June [1860]
1.2 which] 'w' *over illeg*

To J. D. Hooker 5 June [1860]
2.3 our] 'o' *over* 't'
3.1 letters] 's' *added*
3.2 from (received today)] *interl*
3.4 & me] *interl*
5.3 it] *interl*
7.1 though present] *interl*
7.4 & sent] *interl*
7.8 as I] 'as' *over* 'I'
9.1 Fraser] 'F' *over illeg*
9.3 of proof] *interl*
9.4 there is] *interl*
10.6 so-called] *interl*
11.1 (I] '(' *over comma*

To Charles Lyell 6 June [1860]
2.2 in N. British] *interl*
2.2 else] *interl*
2.4 Melipona;] *semicolon over colon*
2.7 during] *interl*
3.2 as] *interl*
3.2 vars. of] *interl*
4.1 exclusively] *interl*
4.2 animals &] *interl*
5.2 Schaafhausen] *before del* '—'
5.2 on natural selection.] *interl*
5.5 i.e. whether . . . structure 5.6] *interl*
5.6 first] *interl*
5.6 steps] *before del* 'first'
5.8 has] *interl*
5.8 changed.] *altered from* 'changes insensibly'
6.1 (N.B . . . weight!) 6.3] *square brackets in MS*
7.1 in Origin] *interl*
7.1 animals] *above del* 'insects'
7.2 *exclusively*] *interl*
7.3 (but overlooked bearing of fact)] *interl*
7.4 (a Carabidous Beetle) 7.5] *interl*
7.6 in dark] *after del* 'by'
7.6 & would] '&' *over* '.—'
7.7 tenant.] *interl*
9.2 confidently] *interl*
9.4 oblivion of fact] *interl*
9.7 productions of] *interl*

To George Rolleston 6 June [1860]
2.2 in a] 'a' *over* 'an'
2.2 near] *above del* 'adjoining'

To J. D. Hooker 7 June [1860]
1.4 afterwards] *interl*
1.6 lies] *interl*
1.6 of indusium;] *interl*
1.6 after flower sometime expanded 1.7] *interl*
1.11 style] *above del* 'pistil'
1.12 (at the . . . pushed out) 1.13] *parentheses over commas*
2.2 but] *interl*
2.3 it] *after del* 'ready'
2.4 of hairs] *interl*
2.6 with] *before del* 'not one'
2.6 outer] *interl*
2.7 the $1\frac{1}{2}$ hour] *interl*
2.7 outside] *after del full stop*
2.7 & no trace . . . horns. 2.8] *added*
2.11 pot] 't' *over* 'ts'
2.11 it] *after del* 'they'

4.1 today] *interl*
5.1 little] *interl*
5.1 at] *over illeg*
5.3 each] *after del* 'the'
5.6 brushing] *before del* 'out'
5.6 out of] 'out' *interl*

To Asa Gray 8 June [1860]
1.5 last] *interl*
1.5 Review] *above del* 'Mag.', *pencil*
1.8 He] *before del* 'in'
1.8 discusses] *altered from* 'discussing'
1.9 wholly] *interl*
1.11 for] *after del* 'that'
2.2 as he admits] *interl*
2.5 very] *interl*
2.5 Huxley.] *above del* 'Owen.'
2.7 as] *interl*
3.1 or] *over illeg*
7.1 (by me)] *interl*
9.2 this] *added*
9.3 the size] 'the' *interl*
9.3 which] *before omitted caret*
9.4 in the . . . plants. 9.5] *interl*
9.4 "female"] [2]'e' *over* 'es'

To Emma Gärtner 9 June [1860]
1.7 as] *interl*

To Henry Tibbats Stainton 11 June [1860]
3.1 to me] *interl*

To J. D. Hooker 12 [June 1860]
1.1 herewith] *interl*
1.2 of species;] *interl*
2.1 observations] *after del* 'much'
3.2 some one] *above del* 'your'
3.3 This would be invaluable for me—3.4] *added*
3.5 of Bee-orchis] *added*
9.2 Squared] 'S' *over illeg*
9.2 bricks] *above del* 'bricks'

To John Higgins 13 June [1860]
4.1 My dear Sir] *added*
5.3 probably] *interl*
5.3 Alford] 'Al' *over* 'Af'

To Charles Lyell 14 [June 1860]
2.6 forms] *before del* 'in dis'
3.2 it to] *interl*
3.3 He] *over illeg*
5.2 than on] *interl*

To W. B. Carpenter 17 June [1860]
2.3 , as I often do,] *interl*

To J. D. Hooker 17 June [1860]
1.3 touch from a] *interl*
1.3 not held close to point 1.4] *above del* 'held at 2 inches off'
1.4 cause] *above del* 'make'
1.5 or even less] *interl*
1.8 (] *over comma*
1.11 a] *interl*
1.11 only] *interl*
2.4 the rostellum] 'the' *interl*
2.5 indefinite ie] *interl*
2.6 as in Listera] *interl*
2.9 the opposite] 'the' *over illeg*
2.10 is the case] *above del* 'takes place'
3.1 Cephalanthera grandiflora] *above del* 'Epipactis'
3.2 in two friable column close] *interl*
3.3 rostellum)] ')' *over comma*
3.4 with the glue] *interl*
5.1 It requires no sort of answer.] *added pencil*
6.1 *externally*] '*ex*' *over illeg*
7.1 thank Heaven] *interl*
7.2 of **Bee Orchis**] *interl*
7.3 of some orchis] *added*

To Charles Lyell 17 June [1860]
1.2 i.e.] *above del* 'or'
1.4 very] *interl*
3.1 not] *interl*

To Thomas Campbell Eyton 18 June [1860]
1.3 Pray] *before omitted point*

To W. D. Fox 18 June [1860]
2.1 with exact dates] *interl*

To Charles Lyell 20 [June 1860]
4.2 from] *over* 'his'
5.4 effected] [1]'e' *over illeg*

To John Higgins 21 June [1860]
2.3 Anwick] *after del* 'Aln'

To Charles Lyell 25 [June 1860]
4.2 have been] 'have' *interl*

To John Obadiah Westwood 25 June [1860]
1.1 Hive] *interl*

To J. D. Hooker 26 [June 1860]
3.6 so] *interl*
5.4 very minute] *interl*
5.5 to attempt to walk] *interl*
6.1 as far . . . concerned] *interl*

To J. D. Hooker [2 July 1860]
1.2 a] *interl*
1.5 I know] *interl*
2.1 that I] 'that' *over illeg*
2.4 (] *over comma*
3.2 much] *interl*
4.2 Thursday] *above del* 'Thurday'
6.2 with] *before omitted point*

To Asa Gray 3 July [1860]
1.2 try to] *interl*
1.11 Nevertheless] 'I' *over illeg*
2.2 a savage] 'a' *interl*
2.2 on my Book] *interl*
2.4 my] *altered from* 'I'
4.2 on] *altered from* 'in'
6.1 strong] *below del* 'strong'
7.3 over the] 'the' *interl*
8.4 probably] *interl*
9.3 in] *interl*
9.4 a] *above del* 'the'
9.6 into which] *interl*
9.7 crawled] *before del* 'into'

To T. H. Huxley 3 July [1860]
2.2 some] *above del* 'an'
2.4 (& . . . others) 2.5] *parentheses over commas*
2.7 (how . . . I know not.)] *square brackets in MS*
2.8 would.] *above del* 'would.'
2.12 rather] *interl*
2.14 utterly] *before omitted point*
3.4 in letter] *interl*

To John Lubbock [4 July 1860]
2.4 or Wednesday] *interl*

To T. H. Huxley [5 July 1860]
3.3 except . . . told him.—] *added*

To Charles Lyell 5 [July 1860]
1.3 glad to] *before del illeg*
1.4 flint] *interl*
2.3 that] *added*
2.6 on the question] *interl*
4.4 could] *altered from* 'c^{d}'

To J. O. Westwood 9 July [1860]
1.4 pin] *above del* 'Bee'
2.2 on] *after del* 'of'
2.2 the (I think)] *interl*
3.1 accident] *before omitted point*
3.1 is] 'is' *over illeg*
4.1 for the] 'for' *over* 'the'

To Samuel Pickworth Woodward 9 [July 1860]
1.3 the best] 'the' *interl*
1.6 of which] 'of' *interl*
2.2 volcanic] *interl*
2.3 *sub-aerial*] *interl*
2.3 at] *over illeg*
5.1 In Haste] *added*
6.3 flowed] *after del* 'water'
7.1 Iceland] *after del* 'Eo'

To J. D. Hooker 12 July [1860]
3.1 greatly] *after omitted point*
4.1 it is . . . in Houses] *interl, pencil*
4.2 & impregnation by Diptera] *added*
5.1 bent] *after del* 'bu'
5.2 This . . . nectaries? 6.7] *del*
5.7 gangway into] *interl*
7.1 (N.B . . . above points) 7.2] *square brackets in MS*
7.1 Lily] *interl*
8.3 seizes] *after del* 'cease'
9.2 flowers of] *interl*
9.2 the pollen in indusium] *interl*
10.5 by gemmation] *after del* 'between'

To Emma Gärtner 14 July [1860]
1.3 (which I . . . my children) 1.4] *square brackets in MS*

To J. D. Hooker [17 July 1860]
1.1 416] *over* '415'
1.3 as I much wish to see all.—] *added*
3.1 last] *interl*

To J. D. Hooker 19 [July 1860]
2.3 no-care-devil] *interl*
4.1 parts of fructification] *interl*
4.3 junction] *above del* 'part'

To J. D. Hooker [20? July 1860]
3.1 the] *interl*
3.1 for] *interl*

To T. H. Huxley 20 July [1860]
1.4 that] *interl*
1.5 Is] 'I' *over illeg*
1.8 are] *above del* 'were'
2.6 thorough] *above del* 'thorough'
4.4 may] *above del* 'will'
4.6 than] *over* 'that'
5.1 scheme] *above del* 'plan'
6.3 now] *above del illeg*

To John Lubbock 20 July [1860]
5.1 R.?] '?' *over* '—'

To W. B. Tegetmeier 20 July [1860]
1.2 young] *interl*
2.3 health;] *semicolon over full stop*
2.3 so that . . . some time.—2.4] *added*
2.5 Worker] 'r' *over* 'rs'
2.6 the Italian] *after omitted point*
4.3 exactly] *interl*
4.3 of Pigeons] *interl*

To Asa Gray 22 July [1860]
2.2 republish] *altered from* 'republished'
3.2 Monthly] *interl*
3.4 as] *after del* 'of'
3.7 living] *interl*
4.1 (Wilberforce] *parenthesis over comma*
4.8 his] *interl*
5.5 his former views] *interl*
5.6 this] *interl*

To Charles Hardy 27 July [1860]
0.1 (Down Bromley Kent)] *square brackets in MS*
0.1 (now . . . Tonbridge Wells.) 0.2] *square brackets in MS*
1.4 also] *interl*
2.1 single] *interl*

To J. D. Hooker 29 July [1860]
1.5 how] *after del* 'yo'
4.2 insect-catching power of] *interl*

To W. E. Darwin [30 July 1860]
1.6 but we] 'b' *over* '&'
5.1 last] *interl*
6.1 dear] 'r' *over* 'rd'

To Charles Lyell 30 July [1860]
1.6 (except] *parenthesis over comma*
1.12 has been] *above del* 'is'
1.13 (by the way . . . Oxford)] *square brackets in MS*
2.4 of the Origin] *interl*

To J. D. Hooker 7 August [1860]
3.2 Rudolph] *before omitted point*
5.6 (at least greater part)] *interl*; *parentheses over commas*
5.8 in one direction] *interl*
5.11 when] *after del* 'if'
5.13 (who] '(' *over comma*
5.14 & not ask this favour] *interl*
5.16 : it] *over* '.—'
5.16 it would . . . movements.—] *added*
7.1 say] *interl*

To T. H. Huxley 8 August [1860]
1.2 Owen] 'O' *over* 'o'
1.5 write] *after del illeg*
1.6 & in Silliman . . . Agassiz] *added*
2.7 He says . . . Report. 2.8] *added*

To Asa Gray 11 August [1860]
2.4 distinction] *above del* 'difference'
2.4 no] *after del* 'now'
2.16 us] *over full stop*
3.3 (ie great & sudden variations)] *interl*; *square brackets in MS*
3.6 such] *above del* 'their'
3.6 occurrences] 's' *added*
3.13 Sir R.] *interl*
3.15 true Fish] 'true' *interl*
4.2 (I fear] *parenthesis over full stop*

To Charles Lyell 11 August [1860]
2.5 R.] *added*
5.4 Review] *above del* 'Essay'
6.4 that there] 'there' *over illeg*
7.5 3] *over* '2'
9.1 makes] *after del illeg*
9.2 as] *over* '—'
9.2 in] *after del* 'I'

To Edward Cresy 25 August [1860]
1.6 that] *over illeg*

To Charles Lyell 28 August [1860]
3.3 abused] 'bu' *over illeg*

To J. D. Hooker 31 [August 1860]
1.3 It is] *after del* 'I want'
1.4 an] *altered from* 'a'
1.4 imported] *interl*

To Charles Lyell 1 September [1860]
1.2 lately] *interl*
2.4 certain] *interl*
2.4 (viz . . . species) 2.5] *interl*
2.5 in other] 'in' *interl*
2.5 of other islands 2.6] *interl*
3.1 even] *above del* 'even'
3.3 the extremely] 'the' *interl*
3.4 the very] *interl*
3.6 existence] *altered from* 'existences'
3.6 as atolls.—] *added*
4.1 existing] *after del* 'r'
4.4 the] *interl*
4.4 animals] *interl*
4.5 characters] *interl*
4.5 Bird] *after del* ')'
4.6 as is . . . Fish, 4.7] *interl*
4.8 hand.] *interl*
4.14 true] *interl*
5.3 Is^{d} is] *interl*
6.1 (N.B . . . Islands) 6.2] *square brackets in MS*
6.1 2 or] *interl*
6.3 of Bats] *interl*
6.4 in N. Zealand] *above del* 'there'
6.4 they] *after del illeg*
6.7 an] *interl*
7.1 polar] *interl*
7.5 wanderers] *above del* 'they'
7.6 individuals . . . an island] *above del* 'new unaltered immigrants'
7.7 Bermuda] *above del* 'New Zealand'
8.1 infer] *over illeg*
8.3 a] *interl*
8.3 at the Galapagos;] *above del* 'there; when'
8.4 country it is] *interl*
8.4 (I believe)] *interl*
8.5 probably] *interl*
8.6 variation] *after del* 'gradati'
8.7 (in itself . . . circumstance) 8.8] *interl*
9.4 in question] *interl*
9.5 particular] *interl*
9.7 a subject] 'a' *interl*

To W. B. Tegetmeier 1 September [1860]
1.1 obliged] *interl*
1.2 have] *over* 'has'
1.6 to think] 'to' *interl*
1.6 to be] *interl*

To J. D. Hooker 2 September [1860]
3.2 is very] 'is' *interl*
3.4 it] *interl*
4.1 or Asses] *interl*

To J. D. Hooker 6 September [1860]
1.1 all] *interl*
1.5 on shoulder] *interl*
1.6 legs] *after del* 'stri'
1.6 is an] 'is' *added*; 'an' *interl*
1.6 absurdity] 'ity' *added*
1.7)] *over comma*
3.2 from its gripe] *interl*
3.5 **thin**] *interl*
3.5 fluid] *interl*
3.8 raw] *interl*
3.13 certainly] *after del* 'cet'
3.13 one] *interl*
3.14 experiment on] 'on' *above del* 'of'
3.14 (] *over semicolon*
3.16 of solution] *interl*
4.1 action] *above del* 'structure'
4.2 (for . . . subject) 4.4] *square brackets in MS*
4.8 permanently] *interl*
4.9 now] *interl*
4.10 most delicately] *interl*
4.10 ie] *above del* 'or'
4.10 surface] *interl*
4.11 down its] *interl*
4.16 now split] *interl*
4.17 two] *interl*

To Asa Gray 10 September [1860]
1.2 stupid] *interl*
1.7 this] *interl*
1.13 the wisdom] 'the' *over* 'its'
1.14 of whole number 1.15] *interl*
3.1 a] *interl*
3.3 it] *interl*
3.18 age,] *interl*
3.20 in Zoology] 'in' *interl*

To T. H. Huxley 10 September [1860]
1.4 (although . . . circulation) 1.5] *interl*; *square brackets in MS*
2.1 in addition] *interl*
4.4 on] *over* 'of'

To George Gordon 11 September [1860]
1.1 &] *over full stop*
5.2 9 or] *interl*

To Daniel Oliver 11 September [1860]
2.2 on the] *after del* 'in manner'

5.2 Will . . . Pods.] *added*
7.3 $\frac{1}{2400}$th] *after del* 'less'

To Charles Lyell 12 September [1860]
1.6 (do not . . . strong) 1.7] *square brackets in MS*
1.7 yet it] *above del* '&'
1.10 that] *over* 'so'
1.11 with] *above del* 'in'
1.11 the same] 'the' *added*
1.11 structure] *above del* 'condition'
1.13 the difficulty] 'the' *interl*
1.14 geological] *interl*
2.1 I see I misunderstand you] *over del* 'One point I vehemently object to, ie your supposing that with any length of time an *existing* marsupial animal would be developed into placental; I quite'
3.3 weight] *before del* 'on'
4.1 becoming modified 4.2] *above del* 'varying'
4.3 a discussion] 'a' *interl*
4.3 point.] *before del* 'Not only this, but'
4.3 It] 'I' *over* 'i'
4.4 one] *over* 'or'
4.5 , ie . . . species.] *interl*
4.7 in each genus] *interl*
4.8 as I often have] *interl*
4.10 so . . . modified] *interl*
4.11 engraved] *interl*
7.2 only] *interl*
7.4 well as] *interl*
7.5 (I see . . . p. 321) 7.6] *interl*; *square brackets in MS*
7.8 seem to] *above del* 'must'
7.8 Tertiary] *after del* 'called'
8.1 you on] *interl*
8.2 & Australian Trigonia,] *interl*
8.3 hardly] *above del* 'do not'
9.1 do] *interl*
9.1 a difficulty] 'a' *interl*
9.3 at] 't' *over* 's'
9.3 as] *interl*
9.5 there] 'th' *over illeg*
9.5 much] *before del* 'more'
9.9 or homological] *interl*
11.1 a wish] 'a' *interl*

To Daniel Oliver 15 [September 1860]
2.4 that] *interl*
4.1 in Veg. . . . bottom 4.2] *interl*
4.3 the names] *after del* 'the'
5.1 Croker] *interl*
5.1 that] *after del* 'the'
6.2 tantalising] *over illeg*
8.1 if you like the job] *interl*
8.1 *small*] *interl*

To George Varenne Reed 15 September [1860]
1.5 owing] *interl*
1.7 (until . . . short,)] *parentheses over commas*
2.3 to] *interl*

To George Gordon 17 September [1860]
5.1 in] *before del* 'by'
5.1 by] *over* '&'

To Daniel Oliver 21 [September 1860]
1.2 (& . . .wine)] *interl*
1.3 case of the] *interl*
1.4 weaker & weaker] *above del* 'less & less'; '&' *not del in error*
1.4 Sulphate] *after del* 'With'
1.5 none or] *interl*
4.2 of white sugar] *interl*
4.2 Nor . . . starch.—] *added*
5.1 I have . . . was produced. 5.2] *added*

To Daniel Oliver [22–3 September 1860]
4.1 of] *before del* 'the'
4.1 your] *added*
4.3 .—] *added, pencil*
5.1 Any corrosive Sublimate??] *added, pencil*

To Charles Lyell 23 [September 1860]
1.4 Species!] 'S' *over* 's'
2.3 of N. H.] 'of' *before omitted point*
4.1 the] *interl*
5.6 a similar or nearly similar 5.7] *interl before del* 'the same'
6.3 argument] *above del* 'organ'
7.3 (please read the paragraph.)] *interl*
7.4 two] *interl*
7.4 if] *before omitted point and del* 'Placenta'
7.4 Marsupials] *above del* 'placentals'
7.7 being] *after del* 'f'
7.7 would] *interl*
7.8 depend] *altered from* 'depends'
7.9 Owen . . . out. 7.10] *added*
8.4 before the] 'the' *interl*
8.8 belonging to] *above del* 'of'
9.2 the] *interl*
9.3 by man] *interl*
9.5 aboriginal] *after del* 'endemic'

9.7 primæus] *under del line*
10.1 the stumps] 'the' *interl*
10.2 the paper] 'the' *interl*
10.3 Coniferæ] *before del* 'It seems rather doubtful whether it can be of much use to the *species*; one tree thus robbing another of its sap.—'
10.4 (which . . . spring up again) 10.5] *parentheses over commas*
10.5 in the fir-tribe] *interl*
10.5 up] *interl*
10.5 but] *after del* 'for'
11.1 (Agassiz's . . . apes.) 11.3] *square brackets in MS*
11.2 in Nott & Gliddon??] *interl*
11.3 does not go] *above del* 'goes'
11.3 much] *before del* ']'
11.4 descent of] *interl*
11.5 from one progenitor] *interl*
11.5 all] *after del* 'them'
11.6 certainly] *interl*
11.11 domestic] *interl*
11.15 rather] *interl*
13.2 (N.B . . . by man) 13.3] *square brackets in MS*
13.3 the] *after del* 'other'
13.4 *much*] *interl*
13.4 neither . . . elsewhere 13.5] *interl*
13.7 throughout the world] *interl*
13.8 could] *interl*
15.1 mammals] *after del* 'reptiles'
15.2 between Fish & Batrachians.—] *interl*
15.2 This . . . more 15.3] *interl after del* 'probably more'

To Daniel Oliver 24 [September 1860]
0.2 24th] *over* '23d'
2.2 those] *after del* 'only'
2.2 alone] *interl*
2.6 This . . . Chile.—] *added*
3.5 unisexual] *after del* 'do have'
3.6 of fertilisation] *interl*
4.2 I am . . . "indefinite".—] *added*
4.8 forms] 's' *added*
4.10 Violets] *before del* ')'
4.11 for its self-fertilisation] *interl*
4.12 an occasional] *interl*
5.2 or . . . me,] *interl*

To Asa Gray 26 September [1860]
1.7 the Hybrids] 'the' *interl*
1.9 unite] *above del* 'cross'
2.1 (just . . . Hooker) 2.2] *interl*
2.5 (I don't know though)] *interl*
2.9 *far*] *interl*
2.13 on those] 'on' *added*
3.1 Annals &] *interl*
3.6 October] *after del* 'Augus'
4.5 use an] *interl*
4.12 great] *interl*
4.13 away.] *before omitted point*
7.1 sea-side] '-side' *added*
8.4 consequence of] 'of' *interl*
8.5 of Ammonia] *interl*
9.1 on Drosera] *interl*
9.1 throwing] *before omitted point*
10.2 next summer] *interl*
10.4 decked] *interl*
10.5 But . . . Spiranthes.—] *added*
11.3 now] *after del* 'know'
11.6 during] *after del* 'more'
11.6 &] *over illeg*

To Charles Lyell 26 [September 1860]
1.5 insect] *interl*
1.9 grossly] *after del* 'b'
1.10 extinct] *after del* 'fos'
1.11 the] *interl*
3.1 almost] *after del comma*
3.3 of C. livia] *interl*
4.2 when] *above del* 'what'
4.4 systems] *interl*
6.1 or dozen] *interl*
6.3 You] 'Y' *over* 'y'
6.4 about Dogs] *interl*
7.1 of man] *interl*
7.3 great] *above del* 'much'
7.3 differences] 's' *added*
9.1 one . . . corrected) 9.2] *added*

To Daniel Oliver 27 [September 1860]
0.2 Thursday] *below del* 'Wednesday'
1.1 of 25th] *interl*
1.3 you I] *interl*
1.5 saline solutions &] *interl*
1.5 solutions] *below del* 'mixtures'
1.7 would . . . but it] *added*
1.8 solution] *interl*
1.8 to 1 oz] *interl*
1.9 growing] *interl*
1.11 the hairs . . . leaf] *above del* 'all'
2.3 the pretty] 'the' *interl*
3.5 Vapour of] *interl*
4.2 Neither] *after del* 'Now'
4.2 or D. longifolia 4.3] *interl*

7.1 of Drosera] *interl*
8.1 **itself**] *interl*
8.2 In] *before omitted point*

To J. S. Henslow 28 [September 1860]
1.1 (I . . . sea-air) 1.2] *square brackets in MS*
5.2 pink] *interl*
5.3 becomes] 'be' *over illeg*
5.6 or substance] *added*
5.6 an] *interl*
5.9 in short . . . form.] *interl*

To Charles Lyell 28 [September 1860]
2.1 breaks] *above del* 'breaks'
2.4 high] *above del* 'low'
4.5 such] 's' *over* 'S'
4.6 that is so wrong; 4.7] *interl*
5.1 A. Gray where] *interl*
6.1 Did . . . attention.—6.2] *added*
7.4 see] *above del* 'go'
7.4 him] *after del* 'there'
9.2 accumulated & arrived at 9.3] *interl before del* 'got'

To Daniel Oliver [29 September 1860]
1.2 structure of] *interl*

To Charles Lyell 3 October [1860]
2.1 is] *interl*
8.2 out of very] *above del* 'in'
8.2 ever] *interl*
8.6 when . . . in connection; 8.7] *interl*
8.7 countries.] *over full stop*
8.8 nearly all] *interl*
8.8 varies] *after del illeg*
8.9 new specific] 'new' *interl*
8.10 organic & inorganic] *interl*
8.10 whenever . . . of nature. 8.11] *interl*
8.11 to large] 'to' *interl*
8.12 one] *over* 'a'
8.12 each] 'e' *over illeg*
8.14 in] *altered from* 'into'
9.2 genealogy] 'a' *added*
10.2 do] *after del* 'are'
10.4 these] 'se' *added*
10.5 all] *interl*
10.6 same bones in the] *interl*
10.7 reduced length] *after del* 'already'
12.2 the case] *interl*
12.3 the Indian Brothers] *interl*
14.3 the term] 'the' *interl*
17.1 work.] *interl*

To Jeffries Wyman 3 October [1860]
3.1 that it] 'that' *interl*
3.6 the Rat was blind &] *interl*
4.2 *hind*] *above del 'hind'*
4.5 If you do will you inform me.—] *interl*
5.4 I have . . . disease. 5.5] *added*
5.9 not] *after del illeg*
5.9 when] *above del* 'with'
5.10 deaf on] *interl* 'deaf'
5.13 that the] 'the' *altered from* 'this'
5.13 blueness] *interl*
5.14 or arrest of development] *interl*
5.14 the deafness] 't' *over semicolon*
5.14 deafness continues; 5.15] *interl*
6.3 colts] *above del* 'than'
6.3 than] added
7.5 from some conversation with him] *interl*

To H. G. Bronn 5 October [1860]
2.5 almost] *interl*
2.11 in past periods 2.12] *interl*
3.5 of Germany] *interl*

To Charles Lyell 5 [October 1860]
1.1 I return] 'I' *interl*
1.3 & not identical] *interl*
1.3 in islands] *interl*
1.4 I suppose] *interl*
1.5 distinct.] *interl*
1.5 on the creation doctrine] *interl*
1.8 mingled] *after del* 'blended'
1.9 together so] 'so' *over* 'in'
1.9 generally] *interl*
2.1 M^{r}] *added*
2.3 of Welford] *interl*
3.1 of the] 'the' *interl*
4.1 Atlantic] *interl*
4.1 you sent] *above del* 'it is'
4.1 it] *altered from* 'at'
4.2 to Down.] 'to' *interl*
5.1 Theological] *interl*
5.2 Atlantic] *interl*

To Charles Lyell 8 October [1860]
1.2 is] *interl*
1.3 against me,] *above del* 'why'
1.6 differences] *above del* 'parts'
1.7 in the] *after del* 'of'
1.7 or] *over* '&'
5.3 (By . . . earth's axis) 5.4] *square brackets in MS*

10.1 Rev^{d}] *interl after del interl* 'Reverend'
10.2 I had . . . mind".—10.5] *added*

To Daniel Oliver [10 October 1860]
1.2 Plant] 't' *over* 'ts'

To J. S. Henslow 11 October [1860]
1.3 fluid] *interl*
1.4 the same] *above del* 'one'
1.8 The movement] *above del* 'They'
1.8 be] *above del* 'have'

To Daniel Oliver 12 [October 1860]
3.2 get] *after del* 'then'
3.2 Carbonate] *interl above del* 'C'
3.2 six] *above* '6'
3.2 two] *above* '2'
3.3 immediately] *after del* 'soon'
3.4 This . . . water] *added*
3.4 get] *before del* '4 oz'
3.4 4 gr to 2 oz] *above del* '2 gr to 4 oz of'

To Daniel Oliver 12 [October 1860]
4.1 & whole structure] *interl*
4.4 parts] 'p' *over illeg*
4.4 sensitive] 'si' *over illeg*

To Daniel Oliver 14 October [1860]
1.1 Australian] *interl*
1.2 (] *over* '—'
2.10 then] *after del* 'di'
3.4 hardly any] *above del* 'no'
3.6 thus filled] *interl*
3.7 , but very slow,] *interl*
3.7 Drosera, & the] 'the' *over* 'they'
3.7 red viscid matter was 3.8] *above del* 'were'
4.1 upper or] *interl*
4.3 the matter] 'the' *interl*
4.4 of Drosera] *interl*
4.4 tortoises] *interl*
4.5 viz] *interl*
6.1 (or look at growing leaf) 6.2] *interl*
6.5 described, of] ', of' *over* '.—'
7.1 , on back of leaf 7.2] *interl*
7.2 in latter case] *interl*
10.1 or relation] *added*
11.1 First try . . . useless.—] *added*
13.2 much] *interl*
13.2 strong] *above del* 'much'
13.4 became] *above del* 'is'
15.1 deeply] ^{2}'e' *over* 'p'
15.2 (in main part)] *after del* 'for having'
17.1 (It . . . accidental.) 17.2] *added*; *square brackets in MS*
18.1 the] 't' *over* 'a'
18.3 through simple lens] *interl*
19.1 on the disc] *interl*

To Daniel Oliver 17 October [1860]
1.1 extracts.—] *before del* 'I'
1.1 The] *over* 'the'
1.1 only] *added*
1.3 Dionæa] *above del* 'Drosera'
1.4 on growing plant 1.5] *interl*
1.5 that] *over* '&'
1.7 red] *before omitted point*
3.6 this morning] *interl*
3.6 acts] 's' *added*
3.7 suffice] 'ce' *over* 'ces'
3.8 such] *over* 'as'
3.10 after some time] *interl*
3.10 is] *interl*
3.13 on the leaves of Drosera] *interl*
3.13 with] *after del* 'in'
3.15 on some] 'on' *interl*

To W. D. Fox 18 October [1860]
1.4 the] *interl*
2.7 also] *interl*

To Asa Gray 19 October [1860]
1.4 myself] *interl*
1.9 so no time . . . corrections. 1.10] *interl*
1.11 (as you permit)] *parentheses over commas*
1.14 chiefly] *interl*
1.26 the Articles] 'the' *altered from* 'thre'

To Daniel Oliver 20 October [1860]
2.14 writing; I may add] 'I' *below del* 'you'; 'may add' *above del* 'must know'
3.2 on disc of leaf 3.3] *after del* 'would'
3.6 of leaf] *interl*
3.7 with-in] 'with-' *added*
3.8 Drosera,] *comma over semicolon*

To Asa Gray 24 October [1860]
1.4 though] *after del* '('
1.4 to publish pamplets] *interl*
3.1 of Nat. Hist] *interl*

To J. S. Henslow 26 October [1860]
2.2 & singular misrepresentation] *interl*

2.2 in fact] *after del* 'says'
2.5 in 2^{d} Edition] *interl*
3.2 the most] 'the' *over illeg*
5.2 about] 'ab' *over illeg*

To John Innes 26 October [1860]
0.2 26] '6' *over* '7'

To Asa Gray 31 October [1860]
1.2 as] *interl before del* 'that'
1.3 so that] 'so' *interl*
1.8 our] *after del* 'how'
1.9 now] *interl*
2.3 I shall] 'I' *interl*
4.1 200] *after del* '150 or'
4.1 or 250 4.2] *interl*
4.4 the paying] 'the' *interl*
4.8 & it . . . away.] *interl above del semicolon*
5.1 I am] *after del* 'Your sent'
5.3 Spiranthes;] *semicolon after omitted comma*
7.1 As] *interl*
7.4 a moveable] 'a' *interl*
7.5 a needle] 'a' *interl*
7.5 the nectary] 'the' *interl*
7.5 & the lip] '&' *interl*
7.5 the sticky] 'the' *interl*
7.7 air] *after del* 'dry'
7.8 embrace the] 'the' *interl*
7.8 to proboscis 7.9] *interl*
7.10 is] *interl*
7.11 one pollen-mass] *after del* '('
7.11 pollen-mass] *interl*
7.11 and if] 'if' *interl*
7.13 another] 'a' *over illeg*
7.14 & double] '&' *over* 'one'
7.15 seized] *after del illeg*
7.15 round] *interl*
7.16 the two] *interl after del interl* '&'
7.17 lateral] *interl*
7.19 lies at] *above del* 'is'
7.19 the footstalk] 'the' *interl*
9.1 in early bud] *interl*
9.2 of the rostellum . . . to the] *above del* 'or'
10.2 surface] *after del* 'sti'
10.3 hard] *interl*
10.6 or loaded] *interl*
10.6 less] *after del* 'about'
10.7 concave side of the] *interl*
10.8 by] *after del* 'with'
10.8 membrane of] 'of' *over illeg*
11.1 longitudinally] *interl*
11.2 (or . . . chloroform)] *interl*
11.3 within the canoe. 11.4] *before del* ', is firml'
11.4 embedded in the rostellum] *interl*
11.7 then] *interl*
11.7 the state] 'the' *interl*
11.8 the needle] 'the' *interl*
11.8 of insect, 11.9] *interl*
11.10 a **second**] 'a' *over* 'is'
11.15 *after* sucking] *interl*

To Daniel Oliver [31? October 1860]
1.1 which] *after del* 'I w'
1.3 Hairs *lately*] *interl*
1.5 1 or 2] *interl*
1.6 minutes.—] '—' *over* '&'

To T. H. Huxley 1 November [1860]
2.1 on] *over* 'of'

To Daniel Oliver 3 November [1860]
2.1 for] 'f' *over* 'of'
3.7 dry] *interl*
3.8 over other] 'over' *interl*
3.8 fact] *interl*
3.12 the roots] 'the' *interl*
3.13 the hairs of] *interl*

To Daniel Oliver 7 November [1860]
0.3 (We leave, . . . Saturday.)] *square brackets in MS*
0.3 & think] '&' *over comma*
2.7 There is a] 'is a' *interl*
2.9 with the] 'the' *above del* 'one of'
2.9 in a much] 'in a' *interl*
2.10 & retarding . . . thumb 2.11] *interl*
2.12 (I can . . . mottled.) 2.13] *interl; square brackets in MS*
3.1 several] *above del* 'many'
3.5 remarkable] *after del* 'extremely'
3.6 the facts] 'e' *over* 'em'; 'facts' *interl*
3.6 all the roots of] *interl*
3.6 Euphorbia] *interl*
3.9 of granules] *interl*
6.2 a drawing] 'a' *interl*

To J. S. Henslow 10 November [1860]
1.4 Journal] *after del* 'letter'
2.3 has been] *above del* 'is'

To T. H. Huxley 10 November [1860]
1.3 (] *over comma*
2.1 just] *interl*

To T. H. Huxley 16 November [1860]
1.4 a little] 'a' *over* 'i'
1.4 how] *added in margin*
3.4 in same fish] *interl*
3.4 proper] *after del* 'ele'
4.2 from L. Horner that 4.3] *interl*

To Daniel Oliver 16 November [1860]
1.3 shall] *interl*
6.3 glands of] *interl*
6.5 of woman's Hair (& thread)] *interl*
6.5 that *under*] '*under*' *interl*

To John Lubbock [18 November 1860]
4.2 result] 't' *over* 'ts'
5.1 & Mag. of N. Hist] *interl*
5.3 I wish particularly to know.] *interl*

To Charles Lyell 20 November [1860]
1.3 Good Heavens . . . Atlantis!! 1.4] *interl*
1.8 & humid] *interl*
1.11 now] *interl*
2.1 in this particular case] *interl*
3.3 (I do not . . . done.—) 3.4] *square brackets in MS*
4.8 or 50] *interl*
4.13 done without] 'done' *interl*
4.13 (] *over comma*
6.1 (Etty . . . weak.)] *square brackets in MS*
7.1 (p.)] *interl*
7.2 But I do not say] *interl above del* 'I say not'
8.1 (vol. —p.)] *interl*
8.2 & disputes the fact.] *interl*
8.3 caudal vertebræ] 'vertebræ' *over* '.—'
9.3 Branchiæ,] *comma over full stop*
11.1 (Fraser vol. p.)] *interl*
11.3 in the History of Science] *interl*
11.3 he] 'e' *over* 'as'
11.4 I am . . . memory] *added*
11.7 fully] *interl*
11.9 Objection] *after del* 'Imperfection.—'
13.3 admit] *above del* 'praise'
13.6 tongue] *above del* 'tail'
13.6 the same . . . voice. 13.7] *interl*
13.7 in the same genus] *after del* 'from the the time of Lichenstein'
13.8 but now] *after del* '& no one has for a moment doubted that it is'
13.9 only] *interl*
13.10 facts] *interl*
13.11 "in all] 'in' *interl after del interl* 'in'

To J. D. Hooker 21 November [1860]
1.6 He] 'H' *over illeg*
5.2 placed on gland] *interl*
5.4 contents of] *interl*

To Daniel Oliver [21 November 1860]
1.3 white or very pale pink] *interl*
1.6 stem thin, . . . succulent.] *interl*
3.2 30 years ago] *interl*

To Henry Walter Bates 22 November [1860]
1.5 theorist] [2]'t' *over* 'ts'
3.6 period] *interl*
6.2 very] *interl*
7.2 on the Driver Ants] *interl*

To T. H. Huxley 22 November [1860]
1.2 a second,] *interl*
1.3 on Species] *interl*
1.3 don't write 1.4] *after del illeg*
2.1 (what . . . handwriting) 2.2] *square brackets in MS*
2.6 more] *interl*
4.1 today] *interl*

To John Murray 22 [November 1860]
1.1 copy of] 'of' *over illeg*
3.2 to Printers] *interl*

To Charles Lyell 24 November [1860]
2.4 the origin] 'the' *over illeg*
2.6 a weight 2.7] *after del* 'the'
7.6 near tail 7.7] 'near' *interl*

To Charles Lyell 25 November [1860]
1.2 I, . . . valuable.—1.3] *added*
1.5 at rest] *after del* ', & having been lately,'
1.5 with] *above del* 'whereas'
1.7 now moving] 'now' *interl*
1.7 since . . . period] *interl*
1.12 to me] *after del* 'seemed'
1.12 was] *interl*
3.1 from] *after del* 'that'
3.5 *very large*] *after del* 'the world'
3.10 *partial*] *interl*
3.11 the commencement] 'the' *interl*
4.5 before . . . calculations] *interl*
4.5 put] *after del* 'lay mu'
4.7 supposing] *after del* 'assuming'
5.1 fingers] *after del* 'fig'
5.1 I am] *after del* 'that'

5.3 But for] 'for' *interl*
5.4 done,] *over comma*
5.4 unpleasant] 'un' *over* 'pl'
7.2 now] *after del illeg*
7.3 prejudices] 'pre' *over illeg*
9.2 viz] *interl*
9.2 has] 's' *over* 'd'
9.3 equably] *above del* 'of equally'
9.3 it has] *above del* 'had'
9.4 it has been] *interl*
9.5 As . . . doubtful. 9.8] *added*

To Asa Gray 26 November [1860]
1.2 asking . . . Reprint &] *interl*
1.4 very] *interl*
2.1 over] *above del* 'over'
3.4 The chance] *after del* 'It would be'
4.4 corrected] *interl*
4.4 I will . . . of it.—4.6] *added*
5.8 the glands] 'the' *interl*
5.8 of one grain] *interl*
5.11 that with which] *interl*
5.14 100 & more hairs] 'hairs' *interl*
6.3 the flowers of] *interl*
6.5 it;] *semicolon altered from colon*
7.2 (strictly as lightly)] *parentheses over commas*

To J. D. Hooker 26 November [1860]
1.8 not] *over illeg*
5.2 die Pflanzen] *interl*

To John Lubbock 28 November [1860]
1.4 (I do not mean the compliment!)] *square brackets in MS*
1.5 general] *interl*
3.1 is] *above del* 'are'
3.1 of word] *interl*
4.4 Your] *interl after del* 'Fol'
5.4 (at least abrupt changes)] *interl*; '(' *over comma*
5.6 in successive generations] *interl*
5.6 at] *after del interl* 'these two results'
5.7 whether they] *after del* 'or'
5.7 *either*] *interl*
5.8 our] *after del* 'the'
5.9 the aboriginal forms] *interl*
5.10 necessary] *interl*
6.1 after search] *interl*
6.2 in great] 'in' *interl*
8.1 in Origin] *interl*
8.2 strongly] *interl*
8.3 induced] *interl*

To T. H. Huxley 2 December [1860]
1.1 humbly] *after del* 'am'
1.4 hostile] *interl*
2.3 the doctrine] 'the' *added*
2.4 viz.] *after del* 'th'
2.11 better] *interl*
2.11 on the] 'on' *over* 'in'

To John Murray 3 December [1860]
1.2 them,] *over illeg*
3.2 & extra] *interl*

To Jeffries Wyman 3 December [1860]
1.2 the relative] 'the' *interl*
1.3 it] *before omitted point*
2.8 might have 2.9] *before omitted point*
3.4 when it is published] *interl*
7.2 in the young] *interl*
7.4 to the] *interl before del* 'to the'

To J. D. Hooker 4 December [1860]
2.8 already] *interl*
3.6 in cells of] *above del illeg*
4.1 caused movement &] *interl*
4.7 the rapidity] 'the' *interl*
4.9 (or hair)] *interl*
4.10 upper] 'u' *over* 'p'
4.11 curvature] *after del* 'fluid in'
4.14 pink fluid,] *interl*
6.2 you,] *before omitted comma*
9.1 P.S. . . . I can 11.2] *in pencil*

To Charles Lyell 4 December [1860]
1.3 over] *above del* 'over'
1.10 to you,] *added*
1.12 periods of] *interl*
2.1 in the Beagle,] *interl*
2.4 slow] *interl*
2.4 in the semi-fluid] *after del* 'from alteration'
2.5 of all] 'of' *interl*
2.8 ever] [1]'e' *over illeg*
2.8 proved or] 'or' *over comma*
7.2 a little] *interl*

To Benjamin Silliman Jr 4 December [1860]
3.1 a new] 'a' *interl*

To T. H. Huxley 5 [December 1860]
1.6 guess] *after del* 'guess'
1.7 too] *interl*
1.9 a real] 'a' *interl*

2.1 the] *interl*
2.4 nearly] *interl*
5.4 go] 'o' *over* 'oes'

To John Murray 5 December [1860]
1.3 assure you] 'you' *interl*
2.3 again] *interl*
2.3 which I have made 2.4] *interl*

To Asa Gray 11 December [1860]
1.2 copies] *interl*
1.4 newspapers] *interl*
2.1 the consignment of the] *interl*
2.3 so] *over* '.—'
8.1 known.] *before omitted point*

To J. D. Hooker 11 December [1860]
1.1 some] *after del* 'several'
1.2 & two or three are excellent hits.] *added*
4.2 Atlantic] *interl*
6.4 one of] *interl*

To T. H. Huxley 11 December [1860]
1.2 (after seeing it) 1.3] *interl*
1.4 of Gray's letter] *interl*

To W. D. Fox 17 December [1860]
2.1 the] *over* 'a'
3.1 my] *over illeg*
3.3 but] *over* '&'
3.4 for] 'fo' *over illeg*
8.1 not to be 8.2] 'not' *added*

To J. D. Hooker 17 December [1860]
2.2 oblige me &] *interl*
3.4 some] *above del* 'any'
3.4 It . . . Sinensis 3.5] *interl*
4.1 Some] *after del* '&'; 'S' *over* 's'
4.2 small] *above del* 'short'

To Daniel Oliver 20 December [1860]
2.4 Distylis] 'lis' *over illeg*

To J. D. Hooker 26 December [1860]
1.5 like] *after del* 'to'
2.1 already] *interl*
5.3 me,] *interl*

To J. D. Hooker 29 December [1860]
4.1 head] 'a' *over* 'd'
4.2 saved] *above del* 'safed'
4.3 to you] *interl*

To William Hallowes Miller 31 December [1860]
1.2 to this 1.3] 'to' *over illeg*
2.2 authority] *after del* 'ath'
3.4 Hives] 'H' *over illeg*
3.4 naturally] *interl*
3.4 subsequently] *interl*
3.5 with honey] *interl*
5.1 a great piece 5.2] 'a' *interl*

BIBLIOGRAPHY

The following bibliography contains all the books and papers referred to in this volume by author–date reference or by short title. Short titles are used for some standard reference works (e.g., *DNB*, *OED*), for CD's books and papers, and for editions of his letters and manuscripts (e.g., *Descent*, *LL*, *Notebooks*). These works are listed in alphabetical order according to the title. Other references are given in author–date form.

ADB: *Allgemeine deutsche Biographie.* Under the auspices of the Historical Commission of the Royal Academy of Sciences. 56 vols. Leipzig: Duncker and Humblot. 1875–1912.

Afzelius, Adam. 1798. Observations on the genus Pausus, and description of a new species. *Transactions of the Linnean Society of London* 4: 243–75.

Agassiz, Jean Louis Rodolphe. 1842. *Nomenclator zoologicus, continens nomina systematica generum animalium tam viventium quam fossilium.* Solothurn.

——. 1848. *Nomenclatoris zoologici index universalis.* Solothurn.

——. 1851a. Observations on the blind fish of the Mammoth cave. *American Journal of Science and Arts* 2d ser. 11: 127–8.

——. 1851b. Results of an exploration of the coral reefs of Florida. *Proceedings of the American Academy of Arts and Sciences* 2: 262.

——. 1857–62. *Contributions to the natural history of the United States of America.* 2 monographs (8 pts) in 4 vols. Boston and London.

——. 1859. *An essay on classification.* London. Reprint of vol. 1, pt 1, of *Contributions to the natural history of the United States of America* (1857).

——. 1860. On the Origin of species. *American Journal of Science and Arts* 2d ser. 30: 142–54. Reprinted in *Annals and Magazine of Natural History* 3d ser. 6: 219–32.

Allan, Mea. 1967. *The Hookers of Kew, 1785–1911.* London: Michael Joseph.

Alum. Cantab.: *Alumni Cantabrigienses. A biographical list of all known students, graduates and holders of office at the University of Cambridge, from the earliest times to 1900.* Compiled by J. A. Venn. Part 2. From 1752 to 1900. 6 vols. Cambridge: Cambridge University Press. 1940–54.

Alum. Oxon.: *Alumni Oxonienses: the members of the University of Oxford, 1715–1886.* By Joseph Foster. 4 vols. Oxford. 1888.

Anca, Francesco. 1860. Notes on two newly discovered ossiferous caves in Sicily. *Report of the 30th meeting of the British Association for the Advancement of Science held at Oxford,* Transactions of the sections, p. 73.

Anderson, Thomas. 1860. Florula Adenensis. A systematic account, with descriptions, of the flowering plants hitherto found at Aden. *Journal of the Linnean Society (Botany)* 5, suppl. 1: i–xxiv, 1–47.

Ansted, David Thomas. 1860. *Geological gossip: or, stray chapters on earth and ocean.* London.

Appel, Toby A. 1988. Jeffries Wyman, philosophical anatomy, and the scientific reception of Darwin in America. *Journal of the History of Biology* 21: 69–94.

Archiac, Étienne Jules Adolphe Desmier de Saint-Simon, Vicomte d'. 1847–60. *Histoire des progrès de la géologie de 1834 à 1845.* 8 vols. Paris.

Army list: *The army list.* London. 1815–81.

Arnott, Neil. 1861. *A survey of human progress, from the savage state to the highest civilization yet attained.* London.

Audubon, John James and Bachman, John. 1846–54. *The viviparous quadrupeds of North America.* 3 vols. New York.

Aust. dict. biog.: *Australian dictionary of biography: 1788–1850; 1851–1890; 1891–1939.* Edited by Douglas Pike, Bede Nairn, Geoffrey Serle, and John Ritchie. 12 vols. and index. Melbourne: Melbourne University Press; London and New York: Cambridge University Press. 1966–91.

Autobiography: *The autobiography of Charles Darwin 1809–1882. With the original omissions restored.* Edited, with appendix and notes, by Nora Barlow. London: Collins. 1958.

Azara, Felix d'. 1809. *Voyages dans l'Amérique méridionale.* 4 vols. Paris.

Babbage, Charles. 1837. *The ninth Bridgewater treatise. A fragment.* London.

Babington, Anna Maria, ed. 1897. *Memorials, journal and botanical correspondence of Charles Cardale Babington.* Cambridge.

Babington, Charles Cardale. 1848. On *Anacharis alsinastrum*, a supposed new British plant. *Annals and Magazine of Natural History* 2d ser. 1: 81–8.

——. 1851. *Manual of British botany, containing the flowering plants and ferns arranged according to the natural orders.* 3d ed. London.

Baer, Karl Ernst von. 1834. Das allgemeine Gesetz der Entwickelungsgeschichte der Natur. In *Vorträge aus dem Gebiete der Naturwissenschaften und der Oekonomie gehalten vor einem Kreise gebildeter Zuhörer in der physikalisch-ökonomischen Gesellschaft zu Königsberg.* Königsberg.

——. 1859. Über Papuas und Alfuren. Ein Commentar zu den beiden ersten Abschnitten der Abhandlung 'Crania Selecta ex Thesauris Anthropologicus Academiae Imperialis Petropolitanae'. *Mémoires de l'Académie Impériale des Sciences naturelles de Saint-Pétersbourg* 6th ser. 8: 269–346.

Barnhart, John Hendley. 1965. *Biographical notes upon botanists, compiled by J. H. Barnhart and maintained in the New York Botanical Garden Library.* 3 vols. Boston, Mass.

Barrande, Joachim. 1852–1902. *Système silurian du centre de la Bohême.* 8 vols. in 29. Prague and Paris.

Barrett, Charles Raymond Booth. 1893. *The Trinity House of Deptford Strond.* London.

Barton, Benjamin Smith. 1806. Memorandum on a new vegetable muscipula. *Transactions of the American Philosophical Society* 6: 79–82.

Bates, Henry Walter. 1860. Contributions to an insect fauna of the Amazon valley. Diurnal Lepidoptera. *Transactions of the Entomological Society of London* n.s. 5 (1858–61): 223–8, 335–61.

Bates, Henry Walter. 1861. Contributions to an insect fauna of the Amazon valley. Lepidoptera: Heliconidæ. [Read 21 November 1861.] *Transactions of the Linnean Society of London* 23 (1862): 495–566.

Bates, Henry Walter and Smith, Frederick. 1854–6. Descriptions of some species of Brazilian ants belonging to the genera *Pseudomyrma*, *Eciton*, and *Myrmica* (with observations on their economy). *Transactions of the Entomological Society of London* n.s. 3: 156–69.

Bechstein, Johann Mathäus. 1793–1805. *Gemeinnützige Naturgeschichte Deutschlands nach allen drey Reichen.* 4 vols. Leipzig.

Belcher, Edward. 1844. *Zoology of the Voyage of H.M.S. 'Sulphur'.* Edited by Richard Brinsley Hinds. 2 vols. London.

Bentham, George. 1826. *Catalogue des plantes indigènes des Pyrénées et du Bas-Languedoc, avec des notes et observations sur les espèces nouvelles ou peu connues; précédé d'une notice sur un voyage botanique, fait dans les Pyrénées pendant l'été de 1825.* Paris.

Bentham, George and Hooker, Joseph Dalton. 1862–83. *Genera plantarum ad exemplaria imprimis in herbariis kewensibus servata definita.* 3 vols. London.

BHGW: *Biographisch-literarisches Handwörterbuch zur Geschichte der exacten Wissenschaften.* By Johann Christian Poggendorff. Vols. 1–4, Leipzig: Johann Ambrosius Barth. 1863–1904. Vol. 5, Leipzig and Berlin: Verlag Chemie. 1926.

Binney, Edward William. 1847. *On the origin of coal.* Manchester: privately printed. Reprinted in *Memoirs of the Literary and Philosophical Society of Manchester* 2d ser. 8 (1848): 148–194.

——. 1848. Sketch of the Drift deposits of Manchester and its neighbourhood. *Memoirs of the Literary and Philosophical Society of Manchester* 2d ser. 8: 195–234.

Birds: Part 3 of *The zoology of the voyage of H.M.S. Beagle.* By John Gould. Edited and superintended by Charles Darwin. London. 1838–41.

Blainville, Henri Marie Ducrotay de. 1841–55. *Ostéographie, ou description iconographique comparée du squelette et du système dentaire des cinq classes d'animaux vertébrés récents et fossiles.* 3 vols. and atlas, 2 vols. Paris.

BLC: *The British Library general catalogue of printed books to 1975.* 360 vols. and supplement (6 vols.). London. 1979–88.

BLKO: *Biographisches Lexicon des Kaiserthums Oesterreich, enthaltend die Lebensskizzen der denkwürdigen Personen, welche seit 1750 in den österreichischen Kronländern geboren wurden oder darin gelebt und gewirkt haben.* By Constant von Wurzbach. 60 vols. Vienna. 1856–90.

Blyakher, L. Y. 1982. *History of embryology in Russia from the middle of the eighteenth century to the middle of the nineteenth century.* Translated from the Russian by H. I. Youssef and B. A. Malek, with an introduction by Jane Maienschein. Washington, D.C.: Smithsonian Institution and the National Science Foundation.

BNB: *Biographie nationale publiée par l'Académie royale des sciences, des lettres et des beaux-arts de Belgique.* 28 vols. Brussels. 1866–1944.

Bonney, Thomas George. 1919. *Annals of the Philosophical Club of the Royal Society, written from its minute books.* London: Macmillan and Co.

Boucher de Crèvecoeur de Perthes, Jaques. 1847–64. *Antiquités celtiques et antédiluviennes.* 3 vols. Paris.

[Bowen, Francis]. 1860a. Review of Origin of species. *North American Review* 90: 474–506.

——. 1860b. Remarks on the latest form of the development theory. [Read 27 March, 10 April, and 1 May 1860.] *Memoirs of the American Academy of Arts and Sciences* n.s. 8 (1861): 97–122.

Bowler, Peter J. 1985. Scientific attitudes to Darwinism in Britain and America. In Kohn, David, ed., *The Darwinian heritage.* Princeton: Princeton University Press in association with Nova Pacifica.

Bravard, Auguste. 1857. *Observaciones geológicas sobre diferentes terrenos de transporte, en la Hoya del Plata.* Buenos Aires.

——. 1858. *Monografia de los terrenos marinos Terciarios de las Cercanias del Parana.* Parana.

Bree, Charles Robert. 1860. *Species not transmutable, nor the result of secondary causes; being a critical examination of Mr. Darwin's work entitled 'Origin and variation of species'.* London.

Brehm, Christian Ludwig. 1831. *Handbuch der Naturgeschichte aller Vögel Deutschlands.* Ilmenau.

Brewster, David. 1855. *Memoirs of the life, writings, and discoveries of Sir Isaac Newton.* 2 vols. Edinburgh.

Bridges, Thomas. 1933. *Yamana–English. A dictionary of the speech of Tierra del Fuego.* Edited by Ferdinand Hestermann and Martin Gusinde. Mödling, Austria: privately printed.

Bridges, E. Lucas. 1948. *Uttermost part of the earth.* London.

British authors: *British authors of the nineteenth century.* Edited by Stanley J. Kunitz. New York. 1964.

Broca, Paul. 1858–9. Mémoire sur l'hybridité en général, sur la distinction des espèces animales et sur les métis obtenus par le croisement du lièvre et du lapin. *Journal de la Physiologie de l'Homme et des Animaux* 1: 432–71, 684–729; 2: 218–58, 345–96.

Bromfield, William Arnold. 1856. *Flora Vectensis: being a systematic description of the phænogamous or flowering plants and ferns indigenous to the Isle of Wight.* London.

Brongniart, Adolphe. 1822. Sur la classification et la distribution des végétaux fossiles en général, et sur ceux des terrains de sédiment supérieur en particulier. *Mémoires du Muséum d'Histoire Naturelle* 8: 203–40.

Bronn, Heinrich Georg. 1841–9. *Handbuch einer Geschichte der Natur.* 3 vols. Stuttgart.

——. 1858a. *Morphologische Studien über die Gestaltungs-Gesetze der Naturkörper, überhaupt und der organischen insbesondere.* Leipzig and Heidelberg.

——. 1858b. *Untersuchungen über die Entwickelungs-Gesetze der organischen Welt während der Bildungs-Zeit unserer Erd-Oberfläche.* Stuttgart.

——. 1860a. Review of Origin of species. *Neues Jahrbuch für Mineralogie, Geognosie, Geologie und Petrefakten-Kunde*, pp. 112–16. Reprinted in Hull 1973, pp. 120–4.

Bronn, Heinrich Georg. 1860b. *Über den Stufengang des organischen Lebens von den Insel-Felsen des Ozeans bis auf die Festländer. Eine Fest-Rede mit erläuternden Beilagen.* Stuttgart.

——, trans. 1860. *Charles Darwin über die Entstehung der Arten im Thier- und Pflanzen-Reich durch natürliche Züchtung.* Stuttgart.

Brougham, Henry Peter. 1839. *Dissertations on subjects of science connected with natural theology: being the concluding volumes of the new edition of Paley's work.* 2 vols. London.

——. 1858. Recherches analytiques et expérimentales sur les alvéoles des abeilles. *Comptes Rendus Hebdomadaires des Séances de l'Academie des Sceinces* 46: 1024–9.

Brown, Robert. 1814. General remarks, geographical and systematical, on the botany of Terra Australis. Appendix 3, pp. 533–613, in vol. 2 of Flinders, Matthew, *A voyage to Terra Australis.* 2 vols., and atlas. London.

——. 1833. On the organs and mode of fecundation in Orchideæ and Asclepiadeæ. [Read 1 and 15 November 1831.] *Transactions of the Linnean Society of London* 16: 685–745.

Brown-Séquard, Charles Edouard. 1859–60. Hereditary transmission of an epileptiform affection accidentally produced. *Proceedings of the Royal Society of London* 10: 297–8.

Buch, Christian Leopold von. 1825. *Physicalische Beschreibung der Canarischen Inseln.* Berlin.

——. 1836. *Description physique des iles Canaries, suivie d'une indication des principaux volcans du globe.* Translated by C. Boulanger. Paris.

Bunbury, Frances Joanna, ed. 1891–3. *Memorials of Sir C. J. F. Bunbury, Bart. Early life* 1 vol.; *Middle life* vols. 1–3; *Later life* vols. 1–5. Mildenhall.

Bunsen, Robert Wilhelm von and Kirchhoff, Gustav Robert. 1860. Chemische Analyse durch Spectralbeobachtungen. *Annalen der Physik und Chemie* 110: 161–89.

Burke's landed gentry: *Burke's genealogical and heraldic history of the landed gentry.* By John Burke. 1–18 editions. London: Burke's Peerage Ltd. 1836–1972.

Burke's peerage: *Burke's peerage and baronetage.* 1–105 editions. London: Burke's Peerage (Genealogical Books). 1826–1980.

Burkhalter, Lois Wood. 1965. *Gideon Lincecum, 1793–1874. A biography.* Austin and London: University of Texas Press.

Burkhardt, Frederick B. 1974. England and Scotland: the learned societies. In Glick, T. F., ed., *The comparative reception of Darwinism.* Austin: University of Texas Press.

Butler, Joseph. 1736. *The analogy of religion, natural and revealed, to the constitution and course of nature.* London.

Bynum, William F. 1984. Charles Lyell's *Antiquity of man* and its critics. *Journal of the History of Biology* 17: 153–87.

BWN: *Biographisch Woordenboek der Nederlanden, bevattende Levensbeschrijvingen van zoodanige Personen, die zich op eenigerlei wijze in ons Vaderland hebben vermaard gemaakt.* Edited by K. J. R. van Harderwijk and G. D. J. Schotel. 20 vols. Haarlem. 1852–77.

Calendar: *A calendar of the correspondence of Charles Darwin, 1821–1882.* Edited by Frederick Burkhardt and Sydney Smith. New York and London: Garland. 1985.

Candolle, Alphonse de. 1855. *Géographie botanique raisonnée; ou, exposition des faits principaux et des lois concernant la distribution géographique des plantes de l'époque actuelle.* 2 vols. Paris and Geneva.

Candolle, Augustin Pyramus de. 1820. Géographie botanique. In vol. 18 of Cuvier, Frédéric Georges, ed., *Dictionnaire des sciences naturelles*. 61 vols. and 12 vols. of plates. Strasbourg and Paris. 1816–45.

Carpenter, William Benjamin. 1854. *Principles of comparative physiology*. 4th ed. London.

[——]. 1860a. Darwin on the origin of species. *National Review* 10: 188–214.

[——]. 1860b. Review of Origin of species. *British and Foreign Medico-Chirurgical Review* 25: 367–404.

Catalogue of graduates, University of Dublin: *A catalogue of graduates who have proceeded to degrees in the University of Dublin, from the earliest recorded commencements to July, 1866, with supplement to December 16, 1868*. Dublin: Hodges, Smith, and Foster. 1869.

CDEL: *A critical dictionary of English literature and British and American authors, living and deceased.* By S. Austin Allibone. 3 vols. Philadelphia and London. 1859–71. Supplement by John Foster Kirk. 2 vols. Philadelphia and London. 1891.

[Chambers, Robert]. 1844. *Vestiges of the natural history of creation.* London.

[——]. 1847. *Vestiges of the natural history of creation.* 6th ed. London.

[——]. 1853. *Vestiges of the natural history of creation.* 10th ed. London.

Chapman, Alvan Wentworth. 1860. *Flora of the southern United States . . . arranged according to the natural system. The ferns by Daniel C. Eaton.* New York.

Claparède, Edouard. 1860. Zur Morphologie der zusammengesetzten Augen bei den Arthropoden. *Zeitschrift für wissenschaftliche Zoologie* 10: 191–214. Reprinted in *Annals and Magazine of Natural History* 3d ser. 6 (1860): 455–7.

Coley, Noel G. 1991. Alfred Swaine Taylor, MD, FRS (1806–1880): forensic toxicologist. *Medical History* 35: 409–27.

Collected papers: *The collected papers of Charles Darwin.* Edited by Paul H. Barrett. 2 vols. Chicago and London: University of Chicago Press. 1977.

Complete peerage: *The complete peerage of England Scotland Ireland Great Britain and the United Kingdom extant extinct or dormant.* By George Edward Cokayne. New edition, revised and much enlarged. Edited by the Hon. Vicary Gibbs and others. 12 vols. London: The St Catherine Press. 1910–59.

Coral reefs: *The structure and distribution of coral reefs. Being the first part of the geology of the voyage of the Beagle, under the command of Capt. FitzRoy, R.N. during the years 1832 to 1836.* By Charles Darwin. London. 1842.

Correspondence: *The correspondence of Charles Darwin.* Edited by Frederick Burkhardt and Sydney Smith. Cambridge: Cambridge University Press. 1985–.

Corsi, Pietro. 1988. *Science and religion: Baden Powell and the Anglican debate, 1800–1860.* Cambridge: Cambridge University Press.

Crockford's: *The clerical directory, a biographical and statistical book of reference for facts relating to the clergy and the Church.* London. 1858. Continued as *Crockford's clerical directory etc.* London. 1860–.

Cundall, Frank. 1920. Richard Hill. *Journal of Negro History* 5: 37–44.

Cross and self fertilisation: *The effects of cross and self fertilisation in the vegetable kingdom.* By Charles Darwin. London. 1876.

Curtis, Moses Ashley. 1834. Enumeration of plants growing spontaneously around Wilmington, North Carolina, with remarks on some new and obscure species. *Boston Journal of Natural History* 1 (1834–7): 82–140.

Cuvier, Georges. 1804–8. Sur les espèces d'animaux dont proviennent les os fossiles répandus dans la pierre à plâtre des environ de Paris. *Annales du Muséum d'Histoire Naturelle* 3: 275–303, 364–87, 442–72; 4: 66–75; 6: 253–83; 9: 10–15, 16–44, 89–102, 205–15, 272–82; 12: 271–84.

Dalton, B. J. 1974. Sir George Gray and the Keppel affair. *Historical Studies* 16 (1974–5): 192–215.

DAB: *Dictionary of American biography.* Under the auspices of the American Council of Learned Societies. 20 vols., index, and 7 supplements. New York: Charles Scribner's Sons; London: Oxford University Press. 1928–81.

Dana, James Dwight. 1857. Thoughts on species. *American Journal of Science and Arts* 2d ser. 24: 305–16.

Darwin, Charles and Wallace, Alfred Russel. 1858. On the tendency of species to form varieties; and on the perpetuation of varieties and species by natural means of selection . . . communicated by Sir Charles Lyell . . . and J. D. Hooker. [Read 1 July 1858.] *Journal of the Proceedings of the Linnean Society (Zoology)* 3 (1859): 45–62.

Darwin, Erasmus. 1794–6. *Zoonomia; or, the laws of organic life.* 2 vols. London.

——. 1789–91. *The botanic garden; a poem, in two parts.* Pt 1. *Containing the economy of vegetation.* London. 1791. Pt 2. *The loves of the plants. With philosophical notes.* Lichfield. 1789.

——. 1803. *Temple of nature; or, the origin of society: a poem, with philosophical notes.* 2 pts. London.

Darwin, Francis, ed. 1892. *Charles Darwin; his life told in an autobiographical chapter, and in a selected series of his published letters.* London.

——. 1914. William Erasmus Darwin, 1839–1914. *Christ's College Magazine* 29: 16–23.

Darwin pedigree: *Pedigree of the family of Darwin.* Compiled by H. Farnham Burke. Privately printed. 1888. Reprinted in Freeman, Richard Broke, *Darwin pedigrees.* London: privately printed. 1984.

Daubeny, Charles Giles Bridle. 1860. Remarks on the final causes of the sexuality of plants, with particular reference to Mr. Darwin's work 'On the Origin of species by natural selection'. *Report of the 30th meeting of the British Association for the Advancement of Science held at Oxford*, Transactions of the sections, pp. 109–10.

Dawson, John William. 1859. On a terrestrial mollusk, a chilognathous myriapod, and some new species of reptiles, from the coal-formation of Nova Scotia. *Quarterly Journal of the Geological Society of London* 16 (1860): 268–77.

Dawson, John William. 1860a. *Archaia; or, studies of the cosmogony and natural history of the Hebrew Scriptures.* Montreal and London.

——. 1860b. Review of "Darwin on the origin of species by means of natural selection". *Canadian Naturalist* 5: 100–20.

DBF: *Dictionnaire de biographie Française.* Under the direction of J. Balteau, M. Barroux, M. Prevost, and others. 17 vols. (A–Humann). Paris: Librairie Letouzey et Ané. 1933–89.

DBL: *Danskbiografisk leksikon.* Founded by Carl Frederick Bricka, edited by Povl Engelstoft and Svend Dahl. 26 vols. and supplement. Copenhagen: J. H. Sghultz. 1933–44.

de Beer, Gavin, ed. 1959a. Darwin's journal. *Bulletin of the British Museum (Natural History) Historical Series* 2: 3–21.

De la Beche, Henry Thomas. 1834. *Researches in theoretical geology.* London.

Delarouzée, Charles. 1859. Description d'un nouveau genre de carabiques. *Annales de la Société Entomologique de France* 3d ser. 7: 65–9.

Dempster, W. J. 1983. *Patrick Matthew and natural selection: nineteenth century gentleman-farmer, naturalist and writer.* Edinburgh: Paul Harris Publishing.

Desmond, Adrian. 1982. *Archetypes and ancestors: palaeontology in Victorian London, 1850–1875.* London: Blond & Briggs.

Descent: *The descent of man, and selection in relation to sex.* By Charles Darwin. 2 vols. London. 1871.

Desmond, Ray. 1977. *Dictionary of British and Irish botanists and horticulturalists; including plant collectors and botanical artists.* London: Taylor and Francis.

Dictionnaire des parlementaires Français. Edited by Adolphe Robert, Edgar Bourloton, and Gaston Cougny. 5 vols. Paris. 1889–91.

Dictionnaire universel des contemporains: *Dictionnaire universel des contemporains contenant toutes les personnes notables de la France et des pays étrangers.* Edited by G. Vapereau. 1–5 editions. Paris. 1858–80.

Dixon, Edmund Saul. 1848. *Ornamental and domestic poultry; their history and management.* London.

DNB: *Dictionary of national biography.* Edited by Leslie Stephen and Sidney Lee. 63 vols. and 2 supplements (6 vols.). London: Smith, Elder, and Co. 1885–1912. *Dictionary of national biography 1912–85.* Edited by H. W. C. Davis, J. R. H. Weaver, and others. 8 vols. London: Oxford University Press. 1927–90.

Donoso-Barros, R. 1975. Introduction. In *Zoology of the voyage of H.M.S. 'Beagle'. Pt 5: Reptiles, by Thomas Bell.* Facsimile reprint. Laurence, Kansas: Society for the Study of Amphibians and Reptiles.

Doubleday, Henry. 1842. On the Bardfield oxlip. *Annals and Magazine of Natural History* 9: 515.

Draper, John William. 1856. *Human physiology, statical and dynamical; or, the conditions and course of the life of man.* London.

Draper, John William. 1860. On the intellectual development of Europe, considered with reference to the views of Mr. Darwin and others, that the progression of organisms is determined by law. *Report of the 30th meeting of the British Association for the Advancement of Science held at Oxford,* Transactions of the sections, pp. 115–16.

Drummond, James. 1853. Extracts from various letters from Mr James Drummond, relating to the botany of Swan River. *Hooker's Journal of Botany and Kew Gardens Miscellany* 5: 312–15, 344–7, 398–406.

DSAB: *Dictionary of South African biography.* Edited by W. J. de Kock, D. W. Krüger, and C. J. Beyers. Vol. 1, Cape Town: Nasionale Boekhandel Beperk; vols. 2–3, Cape Town: Tafelberg Publishers; vol. 4, Durban and Pretoria: Butterworth & Co. 1968–81.

DSB: *Dictionary of scientific biography.* Edited by Charles Coulston Gillispie. 14 vols., supplement, and index. New York: Charles Scribner's Sons. 1970–80.

Dujardin, Félix. 1841. *Histoire naturelle des zoophytes. Infusoires, comprenant la physiologie et la classification de ces animaux, et la manière de les eťudier a l'aide du microscope.* Paris.

Duméril, André Marie Constant and Bibron, Gabriel. 1834–54. *Erpétologie générale, ou histoire naturelle complète des reptiles.* 9 vols. in 10 and atlas. Paris.

Duncan, David, ed. 1908. *The life and letters of Herbert Spencer.* London: Methuen & Co.

[Duns, John]. 1860. Review of Origin of species. *North British Review* 32: 455–86.

Duperrey, Louis Isidore. 1826–30. *Voyage autour du monde . . . sur la corvette de sa Majesté, La Coquille, pendant les années 1822, 1823, 1824 et 1825. Zoologie, par MM. Lesson et Garnot.* 3 vols. Paris.

Dupree, A. Hunter. 1951. Some letters from Charles Darwin to Jeffries Wyman. *Isis* 42: 104–10.

——. 1959. *Asa Gray, 1810–1888.* Cambridge, Mass.: The Belknap Press of Harvard University.

Dutrochet, René Joachim Henri. 1822. Recherches sur l'accroissement et la reproduction des végétaux. *Mémoires du Muséum d'Histoire Naturelle* 8: 12–46, 241–96.

——. 1828. *Nouvelles recherches sur l'endosmose et l'exosmose, suivies de l'application expérimentale de ces actions physiques á la solution du problème de l'irritabilité végétale.* Paris.

Earl, George Windsor. 1853. *Contributions to the physical geography of south-eastern Asia and Australia . . . reprinted, with additional notes, from 'The Journal of the Indian Archipelago'.* London.

East-India register and army list: *The East-India register and army list.* Compiled from the official returns, by permission of the Honourable the East-India Company. 16 vols. London. 1845–60.

EB: *Encyclopædia Britannica.* 11th ed. 29 vols. Cambridge: Cambridge University Press. 1910–11.

EB (1970): *Encyclopædia Britannica.* 23 vols. and index. Chicago, London, Toronto, Geneva, Sydney, Tokyo, Manila: William Benton for Encyclopædia Britannica. 1970.

Ehrenberg, Christian Gottfried. 1854–6. *Mikrogeologie. Das Erden und Felsen schaffende Wirken des unsichtbar kleinen selbstständigen Lebens auf der Erde.* 1 vol. and 1 vol. plates and supplement. Leipzig.

Eliot, George, *pseud.* (Marian Evans Cross). 1859. *Adam Bede.* 3 vols. Edinburgh and London.

Ellegård, Alvar. 1990. *Darwin and the general reader: the reception of Darwin's theory of evolution in the British periodical press, 1859–1872.* Reprint edition. Chicago: University of Chicago Press.

Elliot, Clark A. 1979. *Biographical dictionary of American science: the seventeenth through the nineteenth centuries.* Westport, Conn. and London: Greenwood Press.

Emma Darwin (1904): *Emma Darwin, wife of Charles Darwin: a century of family letters.* Edited by Henrietta Litchfield. 2 vols. Cambridge: privately printed by Cambridge University Press. 1904.

Emma Darwin: *Emma Darwin: a century of family letters 1792–1896.* Edited by Henrietta Litchfield. Revised edition. 2 vols. London: John Murray. 1915.

Essays and reviews. London 1860.

Eudes-Deslongchamps, Jacques Amand. 1835. Note sur une anquille retirée d'un puits, au mois de Juillet 1831. *Mémoires de la Société Linnéenne de Normandie* 5: 47–51.

——. 1842. Note. *Mémoires de la Société Linnéenne de Normandie* 7: xxix.

Expression: *The expression of the emotions in man and animals.* By Charles Darwin. London. 1872.

Fairmaire, Léon. 1859. Miscellanea entomologica. Troisième partie. *Annales de la Société Entomologique de France* 3d ser. 7: 21–64.

Falconer, Hugh. 1857–8. On the species of *Mastodon* and elephant occurring in the fossil state in Great Britain. Pt 1. *Mastodon. Quarterly Journal of the Geological Society of London* 13: 307–60; Pt 2. *Elephas.* 14: 81–4.

——. 1860. On the ossiferous Grotta di Maccagnone, near Palermo. *Quarterly Journal of the Geological Society of London* 16: 99–106.

——. 1863. On the American fossil elephant of the regions bordering the Gulf of Mexico, (*E. Columbi*, Falc.); with general observations on the living and extinct species. *Natural History Review* 2d ser. 3: 43–114.

Farley, John. 1974. *The spontaneous generation controversy from Descartes to Oparin.* Baltimore and London: The Johns Hopkins University Press.

Fawcett, Henry. 1860. A popular exposition of Mr. Darwin on the origin of species. *Macmillan's Magazine* 3 (1860–1): 81–92.

FitzRoy, Robert. 1860. On British storms, illustrated with diagrams and charts. *Report of the 30th meeting of the British Association for the Advancement of Science held at Oxford*, Transactions of the sections, pp. 39–44.

Fleming, Charles and Tibbins, J. 1841–4. *Royal dictionary, English and French, and French and English.* 2 vols. Paris.

Fleming, Donald. 1950. *John William Draper and the religion of science.* Philadelphia: University of Pennsylvania Press.

Flett, John Smith. 1937. *The first hundred years of the Geological Survey of Great Britain.* London: His Majesty's Stationery Office.

Forbes, David. 1861. On the geology of Bolivia and southern Peru. [Read 21 November 1860.] *Quarterly Journal of the Geological Society of London* 17: 7–62.

——. 1870. On the Aymara Indians of Bolivia and Peru. *Journal of the Ethnological Society of London* n.s. 2 (1869–70): 193–298.

Forbes, Edward. 1846. On the connexion between the distribution of the existing fauna and flora of the British isles, and the geological changes which have affected their area, especially during the epoch of the northern drift. *Memoirs of the Geological Survey of Great Britain and of the Museum of Economic Geology in London* 1: 336–432.

Forms of flowers: *The different forms of flowers on plants of the same species.* By Charles Darwin. London. 1877.

Foskett, Daphne. 1972. *A dictionary of British miniature painters*. 2 vols. London: Faber and Faber.

Fossil Cirripedia (1851): *A monograph on the fossil Lepadidæ, or pedunculated cirripedes of Great Britain.* By Charles Darwin. London. 1851.

Fossil Mammalia: Pt 1 of *The zoology of the voyage of H.M.S. Beagle*. By Richard Owen. Edited and superintended by Charles Darwin. London. 1838–40.

Foster, Michael and Lankester, Edwin Ray, eds. 1898–1903. *The scientific memoirs of Thomas Henry Huxley*. 4 vols. and supplement. London: Macmillan and Co.; New York: D. Appleton.

Freeman, Richard Broke. 1977. *The works of Charles Darwin; an annotated bibliographical handlist.* 2d ed. Folkestone, Kent: William Dawson and Sons; Hamden, Conn.: Archon Books, Shoe String Press.

——. 1978. *Charles Darwin; a companion*. Folkestone, Kent: William Dawson and Sons; Hamden, Conn.: Archon Books, Shoe String Press.

Freeman, Richard Broke and Gautrey, Peter Jack. 1972. Charles Darwin's queries about expression. *Bulletin of the British Museum (Natural History) Historical Series* 4 (1970–5): 205–19.

Freke, Henry. 1860. Observations upon Mr. Darwin's recently published work—'On the Origin of species by means of natural selection'. Dublin: printed for the author.

[Frere, John Hookham and Canning, George]. 1798. The loves of the triangles. *Anti-Jacobin; or, Weekly Examiner* 2: 162–74, 200–5, 274–80.

Fulton, Robert. [1874–6]. *The illustrated book of pigeons. With standards for judging.* London, Paris, and New York.

Gärtner, Karl Friedrich von. 1844. *Versuche und Beobachtungen über die Befruchtungsorgane der vollkommeneren Gewächse und über die natürliche und künstliche Befruchtung durch den eigenen Pollen.* Pt 1 of *Beiträge zur Kenntniss der Befruchtung der vollkommeneren Gewächse.* Stuttgart.

——. 1849. *Versuche und Beobachtungen über die Bastarderzeugung im Pflanzenreich. Mit Hinweisung auf die ähnlichen Erscheinungen im Thierreiche.* Stuttgart.

Gascoigne, Robert Mortimer. 1984. *A historical catalogue of scientists and scientific books: from the earliest times to the close of the nineteenth century.* New York and London: Garland Publishing.

Gaudin, Charles Théophile. 1859. Modifications apportées par M. Falconer à la faune du Val d'Arno. *Bulletin des Séances de la Société Vaudoise des Sciences Naturelles* 6: 130–1.

Geikie, Archibald, ed. 1875. *Life of Sir Roderick I. Murchison Bart . . . based on his journals and letters with notices of his scientific contemporaries and a sketch of the rise and growth of palæozoic geology in Britain.* 2 vols. London.

Geoffroy Saint-Hilaire, Étienne. 1830. *Principes de philosophie zoologique, discutés en Mars 1830, au sein de l'Académie Royale des Sciences.* Paris.

Geoffroy Saint-Hilaire, Isidore. 1847. *Vie, travaux et doctrine scientifique d'E. Geoffroy Saint-Hilaire.* Paris.

——. 1851. Cours de zoologie (mammifères et oiseaux), fait au Muséum d'histoire naturelle, en 1850. *Revue et Magasin de Zoologie Pure et Appliqué* 2d ser. 3: 12–20.

——. 1854–62. *Histoire naturelle générale des règnes organiques, principalement étudiée chez l'homme et les animaux.* 3 vols. Paris.

Gilbert, Pamela. 1977. *A compendium of the biographical literature on deceased entomologists.* London: British Museum (Natural History).

Godron, Dominique Alexandre. 1853. Considérations sur les migrations des végétaux. *Memoires de l'Académie des Sciences et Lettres de Montpellier* 2 (1851–4): 167–97.

——. 1859. *De l'espèce et des races dans les êtres organisés.* 2 vols. Paris.

Göppert, Heinrich Robert. 1842. *Beobachtungen über das sogenannte Ueberwallen der Tannenstöcke, für Botaniker und Forstmänner.* Bonn.

Grant, Robert. 1826. On the structure and nature of the Spongilla friabilis. *Edinburgh Philosophical Journal* 14: 270–84.

Graves, Algernon. 1901. *A dictionary of artists who have exhibited works in the principal London exhibitions from 1760 to 1893.* 3d ed., with additions and corrections. London: Henry Graves and Co.

Gray, Asa. 1854. *Botany. Phanerogamia.* United States Exploring Expedition during the years 1838–42, under the command of Charles Wilkes, U.S.N., 1 vol. (no more published) and atlas. Philadelphia.

——. 1856. *A manual of the botany of the northern United States . . . including Virginia, Kentucky, and all east of the Mississippi: arranged according to the natural system.* 2d ed. New York.

——. 1857. *First lessons in botany and vegetable physiology.* New York.

——. 1859. Diagnostic characters of new species of phænogamous plants, collected in Japan by Charles Wright, botanist of the U.S. North Pacific Exploring Expedition . . . With observations upon the relations of the Japanese flora to that of North America, and of other parts of the northern temperate zone. With an appendix by N. J. Andersson. *Memoirs of the American Academy of Arts and Sciences* n.s. 6 (1857–9): 377–452.

[——]. 1860a. Review of Darwin's theory on the origin of species by means of natural selection. *American Journal of Science and Arts* 2d ser. 29: 153–84.

[——]. 1860b. Darwin on the origin of species. *Atlantic Monthly* 6: 109–16, 229–39; Darwin and his reviewers. 6: 406–25.

[Gray, Asa]. 1860c. Discussion between two readers of Darwin's treatise on the origin of species, upon its natural theology. *American Journal of Science and Arts* 2d ser. 30: 226–39.

——. 1861. *Natural selection not inconsistent with natural theology. A free examination of Darwin's treatise on the origin of species, and of its American reviewers.* Boston and London.

——. 1876. *Darwiniana: essays and reviews pertaining to Darwinism.* New York. Reprinted with an introduction by A. Hunter Dupree. Cambridge, Mass.: Harvard University Press. 1963.

Gregory, J. W. 1916. *Henry Darwin Rogers: an address to the Glasgow University Geological Society 20th January, 1916.* Glasgow: James MacLehose and Sons.

[Griffin, Charles], ed. 1861. *A dictionary of contemporary biography: a handbook of the peerage of rank, worth, and intellect. Containing memoirs of nearly one thousand eminent living individuals.* London and Glasgow.

Griffith, John William and Henfrey, Arthur. 1860. *The micrographic dictionary; a guide to the examination and investigation of the structure and nature of microscopic objects.* 2d ed. 2 vols. London.

Grisebach, August Heinrich Rudolph. 1843–4. *Spicilegium floræ Rumelicae et Bithynicae exhibens synopsin plantarum quas aest. 1839 legit.* 2 vols. Brunswick.

Groenland, Johannes. 1855. Note sur les organes glanduleux du genre *Drosera. Annales des Sciences Naturelles (Botanique)* 4th ser. 3: 297–303.

Grote, Arthur. 1875. Introduction, being a memoir of the late Mr. Ed. Blyth, C.M.Z.S., and Hon. Member Asiatic Soc. of Bengal. *Journal of the Asiatic Society of Bengal* 43, pt 2 (extra number, August 1875): iii–xxiv.

GSE: *Great Soviet encyclopedia* (translation of 3d ed. of *Bol'shaia Sovetskaia entsiklopediia.* Moscow. 1970–9). 31 vols. New York: Macmillan; London: Collier Macmillan. 1973–82.

Günther, Albert Charles Lewis Gotthilf. 1860. On a new snake from the Galapagos Islands. *Proceedings of the Zoological Society of London* pt 28: 97–8.

Gunther, Albert E. 1975. *A century of zoology at the British Museum through the lives of two keepers, 1815–1914.* Folkestone, Kent: William Dawson and Sons.

Haeckel, Ernst. 1860. Ueber die Augen und Nerven der Seesterne. *Zeitschrift für wissenschaftliche Zoologie* 10: 183–90.

Haldeman, Samuel Steman. 1843–4. Enumeration of the recent freshwater Mollusca which are common to North America and Europe; with observations on species and their distribution. *Boston Journal of Natural History* 4: 468–84.

Hansard: *Hansard's parliamentary debates. Third series, commencing with the accession of William IV.* 356 vols. 26 October 1830 to 5 August 1891. London.

Harris, Michael. 1974. *A field guide to the birds of Galapagos.* London: Collins.

Harvard University catalogue: *Harvard University quinquennial catalogue of the officers and graduates 1636–1925.* Cambridge, Mass: Harvard University Press. 1925.

Harvey, William Henry. 1854. *The sea-side book; being an introduction to the natural history of the British coasts.* 3d ed. London.

——. 1860. *An inquiry into the probable origin of the human animal, on the principles of Mr. Darwin's theory of natural selection, and in opposition to the Lamarckian notion of a monkey parentage.* Dublin: privately printed.

——. 1862. Notice of a collection of algae made on the north-west coast of North America, chiefly at Vancouver's Island, by David Lyall. *Journal of the Proceedings of the Linnean Society (Botany)* 6: 157–77.

Harvey, William Henry and Sonder, Otto Wilhelm. 1859–65. *Flora Capensis: being a systematic description of the plants of the Cape Colony, Caffraria, & Port Natal.* 3 vols. Dublin.

Haughton, Samuel. 1860a. Presidential address. *Journal of the Geological Society of Dublin* 8 (1857–60): 137–56.

[——]. 1860b. Biogenesis. *Natural History Review* 7: 23–32.

Headland, Frederick William. 1852. *An essay on the action of medicines in the system.* London.

Hearne, Samuel. 1795. *A journey from Prince of Wales's Fort in Hudson's Bay to the Northern ocean . . . in the years 1769, 1770, 1771 & 1772.* London.

Heer, Oswald. 1855. Ueber die fossilen Pflanzen von St. Jorge in Madeira. *Neue Denkschriften der allgemeinen Schweizerischen Gesellschaft für die gesammten Naturwissenschaften* 15: 1–40.

——. 1861. The fossil flora of Bovey Tracey. *Proceedings of the Royal Society of London* 11 (1860–2): 453–5.

Henslow, John Stevens. 1830. On the specific identity of the primrose, oxlip, cowslip, and polyanthus. *Magazine of Natural History* 3: 406–9.

——. 1837. *Principles of descriptive and physiological botany.* Published in Dionysius Lardner's *Cabinet cyclopædia.* London.

Herbert, William. 1822. On the production of hybrid vegetables; with the result of many experiments made in the investigation of the subject. [Read 21 December 1819.] *Transactions of the Horticultural Society of London* 4: 15–50.

Herschel, John Frederick William. 1831. *A preliminary discourse on the study of natural philosophy.* Published in Dionysius Lardner's *Cabinet cyclopædia.* London.

Heusinger, Karl Friedrich. 1846. Ueber die verschiedenartige Wirkung gewisser äusserer Einflüsse auf verschieden gefärbte Thiere. *Wochenschrift für die gesammte Heilkunde* 18: 277–83.

Hoeven, Jan van der. 1828–33. *Handboek der Dierkunde.* 2 vols. Rotterdam.

Hooker, Joseph Dalton. 1844–7. *Flora Antarctica.* Pt 1 of *The botany of the Antarctic voyage of H.M. Discovery Ships Erebus and Terror in the years 1839–1843, under the command of Captain Sir James Clark Ross.* 2 vols. in 1. London.

——. 1846. Description of *Pleuropetalum*, a new genus of Portulaceæ, from the Galapago Islands. *London Journal of Botany* 5: 108–9.

Hooker, Joseph Dalton. 1847a. An enumeration of the plants of the Galapagos Archipelago; with descriptions of those which are new. [Read 4 March, 6 May, and 16 December 1845.] *Transactions of the Linnean Society of London* 20 (1846–51): 163–233.

——. 1847b. On the vegetation of the Galapagos Archipelago, as compared with that of some other tropical islands and of the continent of America. [Read 1 and 15 December 1846.] *Proceedings of the Linnean Society of London* 1 (1849): 312, 313, 314; *Transactions of the Linnean Society of London* 20 (1846–51): 235–62.

——. 1853. *Introductory essay to the flora of New Zealand.* London.

——. 1853–5. *Flora Novæ-Zelandiæ.* Pt 2 of *The botany of the Antarctic voyage of H.M. Discovery Ships Erebus and Terror, in the years 1839–1843, under the command of Captain Sir James Clark Ross.* 2 vols. London.

——. 1854a. *Himalayan journals; or, notes of a naturalist in Bengal, the Sikkim and Nepal Himalayas, the Khasia mountains, &c.* 2 vols. London.

——. 1854b. On the functions and structure of the rostellum of *Listera ovata. Philosophical Transactions of the Royal Society of London* 144: 259–63.

——. 1855–60. *Flora Tasmaniæ.* Pt 3 of *The botany of the Antarctic voyage of H.M. Discovery Ships Erebus and Terror, in the years 1839–1843, under the command of Captain Sir James Clark Ross.* 2 vols. London.

[——]. 1856. Review of Alphonse de Candolle, *Géographie botanique raisoneé. Hooker's Journal of Botany and Kew Garden Miscellany* 8: 54–64, 82–8, 112–21, 151–7, 181–91, 214–19, 248–56.

——. 1859. *On the flora of Australia, its origin, affinities, and distribution; being an introductory essay to the Flora of Tasmania.* Reprinted from the *Botany of the Antarctic expedition*, pt 3, *Flora of Tasmania*, vol. 1. London.

——. 1861. An account of the plants collected by Dr. Walker in Greenland and Arctic America during the expedition of Sir Francis M'Clintock. *Journal of the Proceedings of the Linnean Society (Botany)* 5: 79–88.

——. 1862a. Outlines of the distribution of Arctic plants. [Read 21 June 1860.] *Transactions of the Linnean Society of London* 23: 251–348.

——. 1862b. On the cedars of Lebanon, Taurus, Algeria, and India. *Natural History Review* n.s. 2: 11–18.

Hooker, Joseph Dalton and Thomson, Thomas. 1855. *Flora Indica: being a systematic account of the plants of British India together with observations on the structure and affinities of their natural orders and genera.* London.

Hopkins, William. 1838. Researches in physical geology. *Transactions of the Cambridge Philosophical Society* 6: 1–84.

——. 1847. Report on the geological theories of elevation and earthquakes. *Report of the 17th meeting of the British Association for the Advancement of Science held at Oxford*, pp. 33–92.

——. 1852. Presidential address. *Quarterly Journal of the Geological Society of London* 8: xxi–lxxx.

——. 1860. Physical theories and the phenomena of life. *Fraser's Magazine* 61: 739–52; 62: 74–90. Reprinted in Hull 1973, pp. 229–72.

Horner, Leonard. 1858. An account of some recent researches near Cairo, with the view of throwing light upon the geological history of the alluvial land of Egypt. Pt 2. *Philosophical Transactions of the Royal Society of London* 148: 53–92.

Hull, David L. 1973. *Darwin and his critics: the reception of Darwin's theory of evolution by the scientific community*. Cambridge, Mass.: Harvard University Press.

Hutchinson, Horace Gordon, ed. 1914. *Life of Sir John Lubbock, Lord Avebury*. 2 vols. London: Macmillan and Co.

Huth, Alfred Henry. 1880. *The life and writings of Henry Thomas Buckle*. 2 vols. London.

Hutton, Frederick Wollaston. 1899. *Darwinism and Lamarckism: old and new*. London.

Huxley, Leonard, ed. 1900. *Life and letters of Thomas Henry Huxley*. 2 vols. London: Macmillan and Co.

——, ed. 1918. *Life and letters of Sir Joseph Dalton Hooker*. 2 vols. London: John Murray.

Huxley, Thomas Henry, trans. 1853. Fragments relating to philosophical zoology. Selected from the works of K. E. von Baer. In Henfrey, Arthur and Huxley, Thomas Henry, eds., *Scientific memoirs, selected from the transactions of foreign academies of science, and from foreign journals. Natural history*. London: Taylor and Francis.

——. 1856–7. Lectures on general natural history. *Medical Times & Gazette* n.s. 12: 429–32, 481–4, 507–11, 563–7, 618–23; 13: 27–30, 131–4, 157–60, 278–81, 383–6, 462–3, 537–8, 586–8, 635–9; 14: 133–5, 181–3, 255–7, 353–5, 505–8, 638–40; 15: 159–62, 186–9, 238–41, 467–71.

——. 1858. Report on the examination of specimens of bottom. In Dayman, Joseph, *Deep sea soundings in the North Atlantic Ocean between Ireland and Newfoundland, made in H.M.S. Cyclops*. London.

[——]. 1859a. The Darwinian hypothesis. *The Times*, 26 December 1859, p. 8.

——. 1859b. *Oceanic Hydrozoa*. London.

——. 1860a. On species and races, and their origin. *Notices of the Proceedings at the Meetings of the Members of the Royal Institution of Great Britain* 3 (1858–62): 195–200. Reprinted in Foster and Lankester, eds. 1898–1903, 2: 388–94.

[——]. 1860b. Darwin on the origin of species. *Westminster Review* n.s. 17: 541–70.

——. 1860c. On the study of zoology. [Read 14 May 1860.] In *Lectures addressed to teachers on preparation for obtaining science certificates and the method of teaching a science class*. 2d ser. London. 1860–1. Reprinted in T. H. Huxley 1870.

——. 1861a. On a new species of Macrauchenia (M. boliviensis). *Quarterly Journal of the Geological Society of London* 17: 73–84.

——. 1861b. On the zoological relations of man with the lower animals. *Natural History Review* n.s. 1: 67–84. Reprinted in Foster and Lankester, eds. 1898–1903, 2: 471–92.

——. 1870. *Lay sermons, addresses, and reviews*. London.

Insectivorous plants: *Insectivorous plants*. By Charles Darwin. London. 1875.

IBN: *Index bio-bibliographicus notorum hominum*. Part C: *Corpus alphabeticum sectio generalis*. Edited by Jean-Pierre Lobies and Otto and Wolfram Zeller. 52 vols. (A–Demópoulos, Demetrios). Osnabrück: Biblio Verlag. 1974–91.

Jäger, Gustav. 1851. *Zum Andenken an den den 1. Mai 1772 geborenen den 1. September 1850 zu Calw gestorbenen Botaniker Dr. Carl Friedrich von Gärtner*. Stuttgart.

James, Frank A. J. L., ed. 1987. *Chemistry and theology in mid-Victorian London: the diary of Herbert McLeod, 1860–1870*. Microfiche edition. London: Mansell Publishing.

Jamieson, Thomas F. 1860. On the drift and rolled gravel of the north of Scotland. *Quarterly Journal of the Geological Society of London* 16: 347–71.

[Jardine, William]. 1860. Review of Origin of species. *Edinburgh New Philosophical Journal* n.s. 11: 280–9.

Jeffreys, John Gwyn. 1856. On the marine Testacea of the Piedmontese coast. *Annals and Magazine of Natural History* 2d ser. 17: 155–88.

Jensen, J. Vernon. 1988. Return to the Wilberforce–Huxley debate. *British Journal for the History of Science* 21: 161–79.

Jenyns, Leonard. 1846. *Observations in natural history: with an introduction on habits of observing, as connected with the study of that science. Also a calendar of periodic phenomena in natural history; with remarks on the importance of such registers*. London.

——. 1856. On the variation of species. *Report of the 26th meeting of the British Association for the Advancement of Science held at Cheltenham*, Transactions of the sections, pp. 101–5.

Johnston, Alexander Keith, ed. 1856. *The physical atlas of natural phenomena*. A new and enlarged edition in 5 pts. Edinburgh and London.

Jordan, Alexis. 1852. De l'origine des diverses variétés ou espèces d'arbres fruitiers et autres végétaux généralement cultivés pour les besoins de l'homme. *Mémoires de l'Académie Impériale des Sciences, Belles-lettres et Arts de Lyon* 2d ser. 2: 65–161.

Journal of researches: *Journal of researches into the geology and natural history of the various countries visited by H.M.S. Beagle, under the command of Captain FitzRoy, R.N., from 1832 to 1836*. By Charles Darwin. London. 1839.

Journal of researches 2d ed.: *Journal of researches into the natural history and geology of the countries visited during the voyage of H.M.S. Beagle round the world, under the command of Captain FitzRoy, R.N.* Second edition, corrected, with additions. By Charles Darwin. London. 1845.

Journal of researches (1860): *Journal of researches into the natural history and geology of the countries visited during the voyage of H.M.S. Beagle around the world, under the command of Capt. FitzRoy R.N.* Reprint edition. By Charles Darwin. London. 1860.

Jukes, Joseph Beete. 1850. *A sketch of the physical structure of Australia, so far as it is present known*. London.

——. 1857. *The student's manual of geology*. Edinburgh.

Junker, Thomas. 1991. Heinrich Georg Bronn und die Entstehung der Arten. *Sudhoffs Archiv für Geschichte der Medizin und der Naturwissenschaften* 75: 180–208.

Kant, Immanuel. 1781. *Critik der reinen Vernunft*. Riga.

Kelaart, Edward Frederick. 1846. *Flora Calpensis. Contributions to the botany and topography of Gibraltar and its neighbourhood: with plan and views of the rock*. London.

Keynes, Richard Darwin, ed. 1988. *Charles Darwin's Beagle diary*. Cambridge: Cambridge University Press.

Knight, Thomas Andrew. 1828. On some circumstances relating to the economy of bees. *Philosophical Transactions of the Royal Society of London* pt 2: 319–23.

Krohn, August David. 1859. Beobachtungen über den Cementapparat und die weiblichen Zeugungsorgane einiger Cirripedien. *Archiv für Naturgeschichte* 25: 355–64.

——. 1860a. Beobachtungen über die Entwickelung der Cirripedien. *Archiv für Naturgeschichte* 26: 1–8. Reprinted in *Annals and Magazine of Natural History* 3d ser. 6 (1860): 423–8.

——. 1860b. *Beiträge zur Entwickelungsgeschichte der Pteropoden und Heteropoden.* Leipzig.

Lamarck, Jean Baptiste Pierre Antoine de Monet de. 1809. *Philosophie zoologique; ou exposition des considérations relatives à l'histoire naturelle des animaux.* 2 vols. Paris.

Lamont, Augusta. 1950. *Records and recollections of Sir James Lamont of Knockdow.* Privately printed.

Larousse XX: *Larousse du XXe siècle.* Under the direction of Paul Augé. 6 vols. Paris: Librairie Larousse. 1928–33.

Lartet, Edouard Amant Isidore Hippolyte. 1860. On the coexistence of man with certain extinct quadrupeds, proved by fossil bones, from various Pleistocene deposits, bearing incisions made by sharp instruments. *Quarterly Journal of the Geological Society of London* 16: 471–5.

Laugel, Antoine Auguste. 1860. Nouvelle théorie d'histoire naturelle. L'Origine des espèces. *Revue des Deux Mondes* 2d ser. 26: 644–71.

Lecoq, Henri. 1845. *De la fécondation naturelle et artificielle des végétaux et de l'hybridation.* Paris.

Leftwich, A. W. 1976. *A dictionary of entomology.* London.

Lehmann, Johann Georg Christian, ed. 1844–7. *Plantae Preissianae sive enumeratio plantarum quas in Australasia Occidentali et Meridionali-Occidentali annis 1838–41 allegit Ludovicus Preiss.* 2 vols. Hamburg.

Lewes, George Henry. 1860. Studies in animal life. *Cornhill Magazine* 1: 61–74, 198–207, 283–95, 438–47, 598–607, 682–90.

Liebig, Justus von. 1840. *Organic chemistry in its applications to agriculture and physiology.* Edited by Lyon Playfair. London.

Linder, Jules. 1859. Description de trois nouvelles espèces d'Anophthalmus trouvées en France. *Annales de la Société Entomologique de France* 3d ser. 7: 71–3.

[Lindley, John]. 1837. Dionæ. In vol. 8, p. 508, of *Penny Encyclopedia.* 27 vols. in 12 and supplement (3 vols.). London. 1833–43.

——. 1846. *The vegetable kingdom.* London.

——. 1853. *The vegetable kingdom.* 3d ed. London.

Link, Heinrich Friedrich. 1821. *Die Urwelt und das Alterthum, erläutert durch die Naturkunde.* 2 vols. Berlin.

Linnaeus (Carl von Linné). 1811. *Lachesis Lapponica; or, a tour in Lapland, now first published from the original manuscript journal of the celebrated Linnæus; by J. E. Smith.* 2 vols. London.

Living Cirripedia (1854): *A monograph of the sub-class Cirripedia, with figures of all the species. The Balanidæ (or sessile cirripedes); the Verrucidæ, etc.* By Charles Darwin. London. 1854.

Livingstone, David. 1857. *Missionary travels and researches in South Africa; including a sketch of sixteen years' residence in the interior of Africa, and a journey from the Cape of Good Hope to Loanda on the west coast; thence across the Continent, down the river Zambesi, to the Eastern Ocean.* London.

LL: *The life and letters of Charles Darwin, including an autobiographical chapter.* Edited by Francis Darwin. 3 vols. London. 1887.

Locke, John. 1690. *An essay concerning human understanding.* London.

London and provincial medical directory: *London and provincial medical directory, and general medical register.* London. 1848–60.

Lowe, Richard Thomas. 1854. Catalogus molluscorum pneumonatorum insularum Maderensium: or a list of all the land and fresh-water shells, recent and fossil, of the Madeiran islands: arranged in groups according to their natural affinities; with diagnoses of the groups, and of the new or hitherto imperfectly defined species. *Proceedings of the Zoological Society of London* 22: 161–218.

[Lowell, John Amory]. 1860. Darwin's Origin of species. *Christian Examiner and Theological Review* 5th ser. 6: 449–64.

Lubbock, John. 1862. On some oceanic *Entomostraca* collected by Captain Toynbee. [Read 7 June 1860.] *Transactions of the Linnean Society of London* 23: 173–91.

Lyell, Charles. 1830–3. *Principles of geology, being an attempt to explain the former changes of the earth's surface, by reference to causes now in operation.* 3 vols. London.

——. 1839. On the origin of the tubular cavities filled with gravel and sand, called 'sandpipes,' in the Chalk near Norwich, with additional facts by J. B. Wigham. *Report of the 9th meeting of the British Association for the Advancement of Science held at Birmingham*, Transactions of the sections, pp. 65–6.

——. 1853. *Principles of geology; or, the modern changes of the earth and its inhabitants considered as illustrative of geology.* 9th ed. London.

——. 1855. *A manual of elementary geology; or, the ancient changes of the earth and its inhabitants as illustrated by geological monuments.* 5th ed. London.

——. 1857. *Supplement to the fifth edition of a 'Manual of elementary geology'.* 2d ed. London.

——. 1863. *The geological evidences of the antiquity of man; with remarks on theories of the origin of species by variation.* London.

——. 1865. *Elements of geology; or, the ancient changes of the earth and its inhabitants as illustrated by geological monuments.* 6th ed. London.

——. 1867–8. *Principles of geology; or, the modern changes of the earth and its inhabitants considered as illustrative of geology.* 10th ed. 2 vols. London.

Lyell, Charles and Dawson, John William. 1853. On the remains of a reptile (*Dendrerpeton acadianum*, Wyman and Owen) and of a land shell discovered in the interior of an erect fossil tree in the coal measures of Nova Scotia. *Quarterly Journal of the Geological Society of London* 9: 58–63.

Lyell, Katherine Murray, ed. 1881. *Life, letters and journals of Sir Charles Lyell, Bart.* 2 vols. London.

Macdonald, Frederick C. 1929. *Bishop Stirling of the Falklands.* London: Seeley, Service & Co.

Marchand, Leslie A. 1941. *The Athenæum: a mirror of Victorian culture.* North Carolina: University of North Carolina Press. Reprint edition. New York: Octagon Books. 1971.

Marchant, James. 1916. *Alfred Russel Wallace; letters and reminiscences.* 2 vols. London: Cassell and Co.

Marcou, Jules. 1896. *Life, letters and works of Louis Agassiz.* 2 vols. London and New York.

Marginalia: *Charles Darwin's marginalia.* Edited by Mario A. di Gregorio, with the assistance of N. W. Gill. Vol. 1. New York: Garland Publishing. 1990.

Markham, Clements. 1909. *Life of Admiral Sir Leopold McClintock.* London: John Murray.

Marshall, William. 1852. Excessive and noxious increase of *Udora canadensis* (*Anacharis alsinastrum*). *Phytologist* 4: 705–15.

——. 1857–8. The American water-weed *Anacharis alsinastrum. Phytologist* n.s. 2: 194–7.

Masters, Maxwell Tylden. 1860. On the relation between the abnormal and normal formations in plants. *Notices of the Proceedings at the Meetings of the Members of the Royal Institution of Great Britain* 3 (1858–62): 223–7.

——. 1869. *Vegetable teratology, an account of the principal deviations from the usual construction of plants.* London.

Matthew, Patrick. 1831. *On naval timber and arboriculture.* London and Edinburgh.

M'Donnell, Robert. 1860. On the formation of sugar and amyloid substances in the animal economy. *Report of the 30th meeting of the British Association for the Advancement of Science held at Oxford*, Transactions of the sections, p. 129.

——. 1861. On an organ in the skate which appears to be the homologue of the electrical organ of the torpedo. *Natural History Review* n.s. 1: 57–60.

MDCB: *The Macmillan dictionary of Canadian biography.* Edited by W. Stewart Wallace. 4th ed., revised, enlarged, and updated by W. A. McKay. Toronto: Macmillan of Canada. 1978.

Meade, Richard Henry. 1855. Monograph on the British species Phalangiidæ or Harvest-men. *Annals and Magazine of Natural History* 2d ser. 15: 393–416.

Medical directory: *The Medical directory . . . including the London and provincial medical directory, the medical directory for Scotland, the medical directory for Ireland, etc.* London. 1870–.

Medical directory for Ireland: *The medical directory and general medical register for Ireland.* London. 1852–60.

Meyen, Franz Julius Ferdinand. 1837. *Ueber die Secretions-organe der Pflanzen.* Berlin.

Meyer, Christian Erich Hermann von. 1861. Pterodactylus spectabilis aus dem lithographischen Schiefer von Eichstätt. *Palæontographica: Beiträge zur Naturgeschichte der Vorwelt* 10 (1861–3): 37–45.

——. 1862. Pterodactylus micronyx aus dem lithographischen Schiefer von Solenhofen. *Palæontographica: Beiträge zur Naturgeschichte der Vorwelt* 10 (1861–3): 47–52.

Milde, Julius. 1852. Ueber de Reizbarkeit der Blätter von *Drosera rotundifolia*. *Botanische Zeitung* 10: 540.

Milne-Edwards, Henri. 1851. *Introduction à la zoologie générale, ou considérations sur les tendances de la nature dans la constitution du regne animal.* Paris.

——. 1867. *Rapport sur les progrès récents des sciences zoologiques en France.* (Recueil de rapports sur les progrès des lettres et des sciences en France.) Paris.

Miner, T. B. 1849. *The American bee keeper's manual; being a practical treatise on the history and domestic economy of the honey bee.* London.

ML: *More letters of Charles Darwin: a record of his work in a series of hitherto unpublished letters.* Edited by Francis Darwin and Albert Charles Seward. 2 vols. London: John Murray. 1903.

Modern English biography: *Modern English biography containing many thousand concise memoirs of persons who have died since the year 1850.* By Frederic Boase. 3 vols. and supplement (3 vols.). Truro: printed for the author. 1892–1921.

Moffat, C. B., ed. 1898. *Life and letters of Alexander Goodman More . . . with a preface by Frances M. More.* Dublin.

Moore, Charles. 1858. On Triassic beds near Frome, and their organic remains. *Report of the 28th meeting of the British Association for the Advancement of Science held at Leeds*, Transactions of the sections, pp. 93–4.

——. 1860. On the contents of three square yards of Triassic drift. *Report of the 30th meeting of the British Association for the Advancement of Science held at Oxford*, Transactions of the sections, pp. 87–8.

Moore, James R. 1977. On the education of Darwin's sons: the correspondence between Charles Darwin and the Reverend G. V. Reed, 1857–1864. *Notes and Records of the Royal Society of London* 32 (1977–8): 51–70.

——. 1985. Darwin of Down: The evolutionist as squarson-naturalist. In Kohn, David, ed., *The Darwinian heritage.* Princeton: Princeton University Press in association with Nova Pacifica.

Moquin-Tandon, Horace Bénédict Alfred. 1841. *Éléments de tératologie végétale, ou histoire abrégée des anomalies de l'organisation dans les végétaux.* Paris.

Morren, Charles François Antoine. 1836. On the agency of insects in causing sterility in flowers by the removal of the masculine organs, observed amongst the Asclepiadae. *Proceedings of the Royal Entomological Society of London* 1: xliv–xlv.

——. 1838. Recherches sur le mouvement et l'anatomie du Stylidium graminifolium. *Nouveaux Mémoires de l'Académie Royale des Sciences et Belles Lettres de Bruxelles* 11: 1–22.

Morris, John. 1854. *A catalogue of British fossils: comprising the genera and species hitherto described; with references to their geological distribution and to the localities in which they have been found.* 2d ed. London.

Müller, Hermann. 1856. Beschreibung eines augenlosen Käfers (Glyptomerus cavicola) aus der Familie der Staphylinen. *Entomologische Zeitung* 17: 308–12.

——. 1857. Ueber die Lebensweise der augenlosen Käfer in den Krainer Höhlen. *Entomologische Zeitung* 18: 65–74.

Müller, Johannes Peter. 1826. *Zur vergleichenden Physiologie des Gesichtssinnes des Menschen und der Thiere, nebst einem Versuch über die Bewegungen der Augen und über den menschlichen Blick.* Leipzig.

——. 1834–7. *Handbuch der Physiologie des Menschen für Vorlesungen.* 2 vols. Coblenz.

——. 1838–42. *Elements of physiology.* Translated from the German by William Baly. 2 vols. London.

Munby, Giles. 1859. *Catalogus plantarum in Algeria sponte nascentium.* Oran.

Murchison, Roderick Impey; Verneuil, Edouard de; and Keyserling, Alexander von. 1845. *The geology of Russia in Europe and the Ural Mountains.* 2 vols. London and Paris.

Murray, Andrew. 1857. On insect-vision and blind insects. *Edinburgh New Philosophical Journal* n.s. 6: 120–38.

——. 1860a. On Mr Darwin's theory of the origin of species. [Read 20 February 1860.] *Proceedings of the Royal Society of Edinburgh* 4 (1857–62): 274–91.

——. 1860b. On the disguises of nature: being an inquiry into the laws which regulate external form and colour in plants and animals. *Edinburgh New Philosophical Journal* n.s. 11: 66–90.

Nägele, Erwin, ed. 1926. *Jubiläums-Katalog der E. Schweizerbart'schen Verlagshandlung.* Stuttgart: E. Schweizerbart.

Natural selection: *Charles Darwin's Natural Selection; being the second part of his big species book written from 1856 to 1858.* Cambridge: Cambridge University Press. 1975.

Naudin, Charles Victor. 1852. Considérations philosophiques sur l'espèce et la variété. *Revue Horticole* 4th ser. 1: 102–9.

Navy list: *The navy list.* London. 1815–70.

NBU: *Nouvelle biographie universelle.* Edited by Jean Chrétien Ferdinand Hoefer. 46 vols. Paris. 1852–66.

NDB: *Neue deutsche Biographie.* Under the auspices of the Historical Commission of the Bavarian Academy of Sciences. 16 vols. (A–Melanchthon). Berlin: Duncker and Humblot. 1953–90.

NDBA: *Nuevo Diccionario Biografico Argentino (1750–1930).* Edited by Vicente Osvaldo Cutolo. 7 vols. Buenos Aires: Editorial Elche. 1968–85.

Newberry, John Strong. 1860. Notes on the ancient vegetation of North America; . . . In a letter to Prof. Dana, dated Santa Fe New Mexico, Oct. 15th, 1859. *American Journal of Science and Arts* 2d ser. 29: 208–18.

Nitschke, Theodor Rudolf Joseph. 1858. *Commentatio anatomico-physiologica de Droserae rotundifoliae irritabilitate.* Doctoral dissertation. Breslau.

——. 1860a. Wachsthumsverhältnisse des rundblättrigen Sonnenthaues. *Botanische Zeitung* 18: 57–61, 65–9.

——. 1860b. Ueber die Reizbarkeit der Blätter von *Drosera rotundifolia* L. *Botanische Zeitung* 18: 229–34, 237–43, 245–50.

——. 1861a. Morphologie des Blattes von *Drosera rotundifolia* L. *Botanische Zeitung* 19: 145–51.

——. 1861b. Anatomie des Sonnenthaublattes (*Drosera rotundifolia* L.). *Botanische Zeitung* 19: 233–5.

Notebooks: *Charles Darwin's notebooks, 1836–1844. Geology, transmutation of species, metaphysical enquiries.* Transcribed and edited by Paul H. Barrett, Peter J. Gautrey, Sandra Herbert, David Kohn, and Sydney Smith. Cambridge: Cambridge University Press for the British Museum (Natural History).

Nott, Josiah Clark and Gliddon, George R. 1854. *Types of mankind: or, ethnological researches . . . illustrated by selections from the inedited papers of S. G. Morton . . . and by additional contributions from L. Agassiz, W. Usher, and . . . H. S. Patterson.* Philadelphia and London.

——, eds. 1857. *Indigenous races of the earth; or, new chapters of ethnological inquiry . . . by Louis Ferdinand Alfred Maury . . . with communications from J. Leidy and L. Agassiz.* London and Philadelphia.

NUC: *The national union catalog.* Pre-1956 imprints. 685 vols. and supplement (vols. 686–754). London and Chicago: Mansell. 1968–81.

Nuttall, Thomas. 1818. *The genera of North American plants and a catalogue of the species to the year 1817.* 2 vols. Philadelphia.

OED: *The Oxford English dictionary. Being a corrected re-issue with an introduction, supplement and bibliography of a new English dictionary.* Edited by James A. H. Murray, Henry Bradley, W. A. Craigie, and C. T. Onions. 12 vols. and supplement. Oxford: Clarendon Press. 1970. Supplement. Edited by R. W. Burchfield. 4 vols. Oxford: Clarendon Press. 1972–86.

Oliver, Daniel. 1859. Observations on the structure of the stem in certain species of the natural orders Caryophylleæ and Plumbagineæ. *Transactions of the Linnean Society of London* 22: 289–94.

——. 1860. On Sycopsis. *Transactions of the Linnean Society of London* 23 (1862): 83–9.

——. 1861. Notes on Aurantiaceae. *Journal of the Proceedings of the Linnean Society* 5, 2d suppl.: 1–44.

Olmsted, Frederick Law. 1856. *A journey in the seaboard slave states, with remarks on their economy.* New York.

——. 1860. *A journey in the back country.* 2 vols. New York.

Onions, C. T., ed. 1966. *The Oxford dictionary of English etymology.* Oxford: Oxford University Press.

Oppenheimer, Jane. 1959. An embryological enigma in the *Origin of species.* In Glass, Bentley *et al.*, eds. *Forerunners of Darwin: 1745–1859.* Baltimore: The Johns Hopkins University Press.

Orchids: *On the various contrivances by which British and foreign orchids are fertilised by insects, and on the good effects of intercrossing.* By Charles Darwin. London. 1862.

Origin: *On the origin of species by means of natural selection, or the preservation of favoured races in the struggle for life.* By Charles Darwin. London. 1859.

Origin 2d ed.: *On the origin of species by means of natural selection, or the preservation of favoured races in the struggle for life.* By Charles Darwin. London. 1860.

Origin 3d ed.: *On the origin of species by means of natural selection, or the preservation of favoured races in the struggle for life.* With additions and corrections. By Charles Darwin. London. 1861.

Origin US ed.: *On the origin of species by natural selection, or the preservation of favoured races in the struggle for life.* A new edition, revised and augmented by the author. By Charles Darwin. New York. 1860.

Ospovat, Dov. 1981. *The development of Darwin's theory: natural history, natural theology, and natural selection, 1839–1859.* Cambridge: Cambridge University Press.

[Owen, Richard]. 1833–40. *Descriptive and illustrated catalogue of the physiological series of comparative anatomy contained in the Museum of the Royal College of Surgeons in London.* 5 vols. London.

——. 1839. Description of an extinct species of serpent found at Sheppey. *Proceedings of the Geological Society of London* 3 (1838–42): 165–6. (Reprinted under the title Description of some Ophidiolites (*Palæophis toliapicus*) from the London Clay at Sheppey, indicative of an extinct species of serpent. *Transactions of the Geological Society of London* 2d ser. 6 (1842): 209–10.)

——. 1843. *Lectures on the comparative anatomy and physiology of the invertebrate animals . . . from notes taken by William White Cooper*. London.

——. 1844. Description of a fossil molar tooth of a *Mastodon* discovered by Count Strzelecki in Australia. *Annals and Magazine of Natural History* 14: 268–71.

——. 1846a. On the geographical distribution of extinct Mammalia. *Annals and Magazine of Natural History* 17: 197–200.

——. 1846b. *A history of British fossil mammals, and birds.* London.

——. 1853. *Descriptive catalogue of the osteological series contained in the Museum of the Royal College of Surgeons of England.* 2 vols. London.

——. 1858. Presidential address. *Report of the 28th meeting of the British Association for the Advancement of Science held at Leeds*, pp. xlix–cx.

——. 1859a. *On the classification and geographical distribution of Mammalia.* London.

——. 1859b. On the orders of fossil and recent Reptilia, and their distribution in time. *Report of the 29th meeting of the British Association for the Advancement of Science held at Aberdeen*, pp. 153–66.

[——]. 1860a. Review of Origin of species and other works. *Edinburgh Review, or Critical Journal* 111: 487–532.

——. 1860b. *Palæontology; or, a systematic summary of extinct animals and their geological relations.* Edinburgh.

Owen, Richard Startin, ed. 1894. *The life of Richard Owen . . . With the scientific portions revised by C. Davies Sherborn, also an essay on Owen's position in anatomical science by the Right Hon. T. H. Huxley.* 2 vols. London.

Pallas, Pyotr Simon. 1780. Mémorie sur la variation des animaux. Pt 1. *Acta Academiæ Scientiarum Imperialis Petropolitanæ* 4, pt 2: 69–102.

Parkyns, Mansfield. 1853. *Life in Abyssinia: being notes collected during three years' residence and travels in that country.* 2 vols. London.

Parsons, Theophilus. 1860. On the Origin of species. *American Journal of Science and Arts* 2d ser. 30: 1–13.

Peckham, Morse, ed. 1959. *The Origin of species by Charles Darwin; a variorum text.* Philadelphia: University of Pennsylvania Press.

Phillips, John. 1860. *Life on the earth. Its origin and succession.* Cambridge and London.

Physicians: The roll of the Royal College of Physicians of London. Vol. 3: 1801–25. By William Munk. 2d ed., revised and enlarged. London. 1878. Vol. 4: *Munk's roll, vol. 4: Lives of the fellows of the Royal College of Physicians of London 1826–1925.* Compiled by G. H. Brown. London: Royal College of Physicians. 1955.

Pictet de la Rive, François Jules. 1853–7. *Traité de paléontologie, ou histoire naturelle des animaux fossiles.* 4 vols. Paris.

——. 1860. Sur l'origine de l'espèce. *Bibliothèque Universelle. Revue Suisse et Étrangère. Archives des Sciences Physique et Naturelles* n.s. 7: 233–55.

Pistor, C. M. Wilhelm. 1831. *Das Ganze der Feld- und Hoftaubenzucht, nebst einer Anleitung zum Abrichten der Brief- oder Posttauben.* Hanau.

Plarr, Victor Gustave. 1930. *Plarr's lives of the fellows of the Royal College of Surgeons of England.* Revised by Sir D'Arcy Power. 2 vols. London: Simpkin Marshall.

Poitevin, Prosper. 1856–60. *Nouveau dictionnaire universel de la langue française.* 2 vols. Paris.

Pope, Alexander. 1733–4. *An essay on man. Address'd to a friend.* 4 pts. London.

Porter, Duncan M. 1985. The *Beagle* collector and his collections. In Kohn, David, ed., *The Darwinian heritage.* Princeton: Princeton University Press in association with Nova Pacifica.

Post Office directory of Birmingham, Warwickshire (Staffordshire, and Worcestershire). London. 1845–.

Post Office directory of Cheshire. London. 1857–.

Post Office directory of Gloucestershire, with Bath and Bristol, Herefordshire, and Shropshire. London. 1856–.

Post Office directory of Monmouthshire and . . . South Wales. London. 1871–.

Post Office directory of the six home counties, viz., Essex, Herts, Kent, Middlesex, Surrey and Sussex. London. 1845–.

Post Office London directory. London. 1802–.

Pouchet, Félix Archimède. 1859. *Hétérogénie, ou traité de la génération spontanée.* Paris.

Powell, Baden. 1855. *Essays on the spirit of the inductive philosophy, the unity of worlds, and the philosophy of creation.* London.

——. 1856. *The unity of worlds and of nature: three essays on the spirit of the inductive philosophy; the plurality of worlds; and the philosophy of creation.* 2d ed. London.

——. 1860. On the study of the evidences of Christianity. In *Essays and reviews.* London.

Praeger, William E. 1935. Six unpublished letters of Charles Darwin. *Papers of the Michigan Academy of Science, Arts and Letters* 20: 711–15.

Priestley, Joseph. 1779–86. *Experiments and observations relating to various branches of natural philosophy; with a continuation of the observations on air.* 3 vols. London.

Pritchard, Michael. 1986. *A directory of London photographers, 1841–1908.* Watford: ALLM Books.

Pritzel, Georg August. 1851. *Thesaurus literaturæ botanicæ omnium gentium.* Leipzig.

Quatrefages de Bréau, Jean Louis Armand de. 1858. Rapport sur la question concernant la reproduction des infusoires. *Comptes rendus hebdomadaires des séances de l'Académie des Sciences* 46: 274–9.

——. 1859. *Études sur les maladies actuelles du ver à soie*. Paris.

——. 1860. *Nouvelles recherches faites en 1859 sur les maladies actuelles du ver à soie*. Paris.

——. 1867–8. Anthropologie. Cours de M. de Quatrefages. *Revue de cours scientifique* 5: 366–9.

Rachootin, Stan Philip. 1985. Owen and Darwin reading a fossil: *Macrauchenia* in a bony light. In Kohn, David, ed., *The Darwinian heritage*. Princeton: Princeton University Press in association with Nova Pacifica.

Raffeneau-Delile, Alire. [1813]. *Flore d'Egypt* . 2 vols. Paris.

——. [1824]. *Flore d'Egypt*. 2d ed. 2 vols. Paris.

Rafinesque, Constantine Samuel. 1836. *New flora and botany of North America*. 4 pts. Philadelphia.

Ramsay, Andrew Crombie. 1852. On the superficial accumulations and surface-markings of North Wales. [Read 26 March 1851.] *Quarterly Journal of the Geological Society of London* 8: 371–6.

Réaumur, René Antoine Ferchault de. 1734–42. *Memoires pour servir à l'histoire des insectes*. 6 vols. Paris.

Reinsch, Edgar Hugo Emil. 1841. Ueber das Verhalten des metallischen Kupfers zu einigen Metalllösungen. *Journal für Praktische Chemie* 24: 244–50.

——. 1842. Ueber die Kupferarsenikprobe. *Repertorium für die Pharmacie* 2d ser. 27: 313–21.

Reissek, Siegfried. 1843. Beitrag zur Teratognosie der Thesienblüte. *Linnaea. Ein Journal für die Botanik in ihrem ganzen Umfange* 17: 641–55.

Reptiles: Part 5 of *The zoology of the voyage of H.M.S. Beagle*. By Thomas Bell. Edited and superintended by Charles Darwin. London. 1842–3.

Richardson, Edmund William. 1916. *A veteran naturalist; being the life and work of W. B. Tegetmeier*. London: Witherby & Co.

Richardson, H. D. 1847. *Pigs; their origin and varieties, and treatment under disease*. Dublin.

Richardson, John. 1852. Eskimos, their geographical distribution. *Edinburgh New Philosophical Journal* 52 (1851–2): 322–3.

Richmond, Charles W. 1902. Note on Pinaroloxias inornata (Gould). *Proceedings of the Biological Society of Washington* 15: 247–8.

Riedel, Wilhelm K. 1824. *Die Taubenzucht in ihrem ganzen Umfange, oder vollständige Anweisung zur Kentniss des Taubenschlags*. Ulm.

Roger, Jacques. 1989. *Buffon, un philosophe au Jardin du Roi*. Paris: Fayard.

Rogers, Emma, ed. 1896. *Life and letters of William Barton Rogers*. 2 vols. Boston and New York.

Roll of graduates of the University of Glasgow: A roll of the graduates of the University of Glasgow from 31st December, 1727 to 31st December, 1897. Compiled by W. Innes Addison. Glasgow: James MacLehose & Sons. 1898.

Royle, John Forbes. 1839. *Illustrations of the botany and other branches of the natural history of the Himalayan Mountains, and of the flora of Cashmere.* 2 vols. London.

Rudwick, Martin John Spencer. 1970. *Living and fossil brachiopods.* London: Hutchinson University Library.

Rugby School register. Vol. 1, from April 1675 to October 1857. Revised and annotated by Godfrey A. Solly. Rugby: George Over. 1933.

Ruse, Michael. 1975. Darwin's debt to philosophy: an examination of the influence of the philosophical ideas of John F. W. Herschel and William Whewell on the development of Charles Darwin's theory of evolution. *Studies in History and Philosophy of Science* 6: 159–81.

Russell-Gebbett, Jean. 1977. *Henslow of Hitcham: botanist, educationalist and clergyman.* Lavenham, Suffolk: Terence Dalton.

Sabine, Edward, ed. 1840. *Narrative of an expedition to the polar sea, in the years 1820, 1821, 1822, & 1823. Commanded by lieutenant, now admiral, Ferdinand von Wrangell, of the Russian imperial navy.* [Translated by Elizabeth Juliana Sabine.] London.

Saint-Hilaire, Augustin François César Prouvençal de. 1841. *Leçons de botanique comprenant principalement la morphologie végétale.* Paris.

Sarjeant, William A. S. 1980. *Geologists and the history of geology: an international bibliography from the origins to 1978.* 5 vols. and suppl. (2 vols.). 1987. London: Macmillan.

SBL: Svenskt biografiskt lexikon. Edited by Bertil Boëthius, Erik Grill, Birgitta Lager-Kromnow, and Göran Nilzén. 26 vols. (A–Nilsson). Stockholm: Albert Bonnier and P. A. Norstedt. 1918–89.

Schaaffhausen, Hermann. 1853. Über beständigkeit und Umwandlung der Arten. *Verhandlungen des Naturhistorischen Vereines der preussischen Rheinlande und Westphalens* 10: 420–51.

Scherren, Henry. 1905. *The Zoological Society of London. A sketch of its foundation and development and the story of its farm, museum, gardens, menagerie and library.* London: Cassell and Co.

Schiödte, Jörgen C. [1849]. Bidrag til den underjordiske Fauna. *Det Kongelige Danske Videnskabernes Selskabs Skrifter. Naturvidenskabelig og Mathematisk* n.s. 2 (1851): 1–39.

Secord, James A. 1986. *Controversy in Victorian geology: the Cambrian–Silurian dispute.* Princeton: Princeton University Press.

——. 1989. Behind the veil: Robert Chambers and 'Vestiges'. In Moore, James R., ed. *History, humanity and evolution: essays for John C. Greene.* Cambridge and New York: Cambridge University Press.

[Sedgwick, Adam]. 1860. Objections to Mr. Darwin's theory of the Origin of species. *Spectator*, 24 March 1860, pp. 285–6. Reprinted with revisions in *Spectator*, 7 April 1860, pp. 334–5.

Selwyn, Alfred Richard Cecil. 1858. On the geology of the gold-fields of Victoria. *Quarterly Journal of the Geological Society of London* 14: 533–8.

——. 1860. Notes on the geology of Victoria. *Quarterly Journal of the Geological Society of London* 16: 145–50.

Sichel, Jules. 1847. Sur un rapport remarquable entre le pigment des poils et de l'iris et la faculté d'ouie chez certain animaux. *Annales des Sciences Naturelles. Zoologie* 3d ser. 8: 239–41.

Sidebotham, Joseph. 1849. Experiments on the specific identity of the cowslip and primrose. *Phytologist* 3: 703–5.

Siebold, Karl Theodor Ernst von. 1856. *Wahre Parthenogenesis bei Schmetterlingen und Bienen. Ein Beitrag zur Fortpflanzungsgeschichte der Thiere*. Leipzig.

Silliman, Benjamin. 1851. On the Mammoth Cave of Kentucky. *American Journal of Science and Arts* 2d ser. 11: 332–9.

Simcock, A. V. 1984. *The Ashmolean Museum and Oxford science, 1683–1983*. Oxford: Museum of the History of Science.

Smith, Audrey Z. 1986. *A history of the Hope entomological collections in the University Museum Oxford; with lists of archives and collections*. Oxford: Clarendon Press.

Smith, James. 1836. On indications of changes in the relative level of sea and land in the west of Scotland. *Proceedings of the Geological Society of London* 2 (1833–8): 427–9.

——. 1844. On the geology of Gibraltar. *Proceedings of the Geological Society of London* 4: 480.

Smith, Joshua Toulmin. 1848. *Ventriculidæ of the Chalk*. London.

[Smith, William]. 1859. Bunsen's Egypt and the chronology of the Bible. *Quarterly Review* 105: 382–421.

Smyth, William Henry. 1824. *Memoir descriptive of the resources, inhabitants, and hydrography of Sicily and its islands*. London.

South America: *Geological observations on South America. Being the third part of the zoology of the Beagle, under the command of Capt. FitzRoy, R.N. during the years 1832 to 1836*. By Charles Darwin. London. 1846.

[Spencer, Herbert]. 1852. A theory of population deduced from the general law of animal fertility. *Westminster Review* n.s. 1: 468–501.

——. 1855. *Principles of psychology*. London.

——. 1858–63. *Essays; scientific, political, and speculative . . . Reprinted chiefly from the quarterly reviews*. 2 vols. London.

——. 1860–2. *First principles*. 6 pts. London.

Spratt, Thomas. 1854. Extract of a letter from Captain Spratt, R.N. on Crete. Communicated by Colonel Leake. *Journal of the Royal Geographical Society* 24: 238–9.

Steinberg, S. H., ed. 1953. *Cassell's encyclopaedia of literature*. 2 vols. London: Cassell and Company.

Stenhouse, John. 1990. Darwin's captain: F. W. Hutton and the nineteenth-century Darwinian debates. *Journal of the History of Biology* 23: 411–42.

Steudel, Ernst Gottlieb. 1841. *Nomenclator botanicus seu: synonymia plantarum universalis, enumerans ordine alphabetico nomina atque synonyma tum generica tum specifica*. 2d ed. 2 pts. Stuttgart and Tübingen.

Stur, Dionyz. 1860. Beiträge zu einer Monographie des Genus *Astrantia*. *Sitzungsberichte der Mathematisch-Naturwissenschaftlichen Classe der Kaiserlichen Akademie der Wissenschaften (Wien)* 40: 469–524.

Sturm, Jakob. 1844. *'Anophthalmus' blindlaufkäfer. Neue Gattung aus der Familie der-Caraben.* Nürnberg.

——. 1847. *Beschreibung einer neuen Art von Anophthalmus blindlaufkäfer.* Nürnberg.

Taxonomic literature: *Taxonomic literature. A selective guide to botanical publications and collections with dates, commentaries and types.* Edited by Frans A. Stafleu and Richard S. Cowan. 2d ed. 7 vols. Utrecht: Bohn, Scheltema & Holkema. The Hague: W. Junk, Publishers. 1976–88.

Taylor, Alfred Swaine. 1848. *On poisons in relation to medical jurisprudence and medicine.* London.

——. 1860. Facts and fallacies connected with the research for arsenic and antimony; with suggestions for a method of separating these poisons from organic matter. *Guy's Hospital Reports* 3d ser. 6: 201–65.

Tegetmeier, William Bernhard, ed. 1856–7. *The poultry book: including pigeons and rabbits; comprising the characteristics, management, breeding, and medical treatment of poultry; being the results of personal observation and practice of the Rev. W. Wingfield, G. W. Johnson, Esq. , and other celebrated breeders of the different varieties.* London.

——. 1860. Natural cross breeding in bees. *Transactions of the Entomological Society of London* n.s. 5 (1858–61): Proceedings, p. 126.

Temple, Frederick. 1860. *The present relations of science to religion: a sermon.* Oxford and London.

Tennent, James Emerson. 1859. *Ceylon, an account of the island, physical, historical and topographical, with notices of its natural history, antiquities and productions.* 2 vols. London.

Torrey, John and Gray, Asa. 1838–43. *A flora of North America: containing abridged descriptions of all the known indigenous and naturalized plants growing north of Mexico: arranged according to the natural system.* 2 vols. New York and London.

Trécul, A. 1855. Organisation des glandes pédicellées des feuilles du *Drosera rotundifolia. Annales des Sciences Naturelles (Botanique)* 4th ser. 3: 303–11.

Trimmer, Joshua. 1853. On the origin of the soils which cover the Chalk of Kent. Pt 3. *Quarterly Journal of the Geological Society of London* 9: 286–96.

Tristram, Henry Baker. 1860. Presidential address. *Transactions of the Tyneside Naturalists' Field Club* 4 (1858–60): 191–228.

Tuckwell, William. 1900. *Reminiscences of Oxford.* London, Paris, and Melbourne: Cassell and Co.

Tyndall, John. 1860. *The glaciers of the Alps.* London.

Unger Franz. 1852. *Versuch einer Geschichte der Pflanzenwelt.* Vienna.

Variation: *The variation of animals and plants under domestication.* By Charles Darwin. 2 vols. London. 1868.

Venables, Edmund. 1860. *The Isle of Wight. A guide.* London.

Volcanic islands: *Geological observations on the volcanic islands visited during the voyage of H.M.S. Beagle, together with some brief notices of the geology of Australia and the Cape of Good Hope. Being the second part of the geology of the voyage of the Beagle, under the command of Capt. FitzRoy, R.N. during the years 1832 to 1836.* By Charles Darwin. London. 1844.

Wagner, Rudolph. 1860a. Bericht über die Arbeiten in der allgemeinen Zoologie und der Naturgeschichte des Menschen im Jahre 1859. *Archiv für Naturgeschichte* 26, pt 2: 1–18.

——. 1860b. Louis Agassiz's Prinzipien der Classifikation der organischen Körper insbesondere der Thiere mit Rücksicht auf Darwin's Ansichten im Auszuge dargestellt und besprochen. *Göttingische gelehrte Anzeigen* pt 2: 761–800.

——. 1861. Bericht über die Arbeiten in der allgemeinen Zoologie und der Naturgeschichte des Menschen im Jahre 1860. *Archiv für Naturgeschichte* 27, pt 2: 1–36.

——. 1862. Nahrforschung und Theologie. *Jahrbücher für deutsche Theologie* 7: 166–9.

Wallace, Alfred Russel. 1860. On the zoological geography of the Malay Archipelago. Communicated by Charles Darwin. *Journal of the Proceedings of the Linnean Society (Zoology)* 4: 172–84.

——. 1866. On the phenomena of variation and geographical distribution as illustrated by the Papilionidæ of the Malayan region. *Transactions of the Linnean Society of London* 25: 1–71.

——. 1869. *The Malay Archipelago; the land of the orang-utan, and the bird of paradise. A narrative of travel, with studies of man and nature.* 2 vols. London.

——. 1905. *My life: a record of events and opinions.* 2 vols. London: Chapman & Hall.

Wallich, George Charles. 1860. *Notes on the presence of animal life at vast depths in the sea; with observations on the nature of the sea bed, as bearing on submarine telegraphy*. London: privately printed.

Waterhouse, George Robert. 1846–8. *A natural history of the Mammalia.* 2 vols. London.

Watson, Hewett Cottrell. 1835–7. *The new botanist's guide to the localities of the rarer plants of Britain.* 2 vols. London.

——. 1845. On the theory of 'progressive development,' applied in explanation of the origin and transmutation of species. *Phytologist* 2: 108–13, 140–7, 161–8, 225–8.

——. 1847. Further experiments bearing upon the specific identity of the cowslip and the primrose. *Phytologist* 2: 852–4.

——. 1847–59. *Cybele Britannica; or, British plants, and their geographical relations.* 4 vols. London.

——. 1860. *Part first of a supplement to the Cybele Britannica.* London.

Weddell, Hugh Algernon. 1856. *Monographie de la famille des Urticés.* Paris.

Wedgwood, Barbara and Wedgwood, Hensleigh. 1980. *The Wedgwood circle, 1730–1897: four generations of a family and their friends*. London: Studio Vista.

Wedgwood, Hensleigh. 1848. *On the development of the understanding.* London.

——. 1859–65. *A dictionary of English etymology.* 3 vols. in 4. London.

Wellesley index: *The Wellesley index to Victorain periodicals 1824–1900.* Edited by Walter E. Houghton and others. 5 vols. Toronto: University of Toronto Press; London: Routledge and Kegan Paul. 1966–89.

Whewell, William. 1840. *The philosophy of the inductive sciences, founded upon their history.* 2 vols. London and Cambridge.

Whitaker's Almanack: *An Almanack.* By Joseph Whitaker. London. 1869–.

White, Gilbert. 1789. *The natural history and antiquities of Selborne, in the county of Southampton.* 2 vols. London.

White, Joseph Blanco. 1845. *The life of the Rev. Joseph Blanco White, written by himself; with portions of his correspondence.* Edited by John Hamilton Thom. 3 vols. London.

[Wilberforce, Samuel]. 1860. On the Origin of species. *Quarterly Review* 108: 225–64.

——. 1874. *Essays contributed to the 'Quarterly Review'.* 2 vols. London.

Wilson, Leonard Gilchrist, ed. 1970. *Sir Charles Lyell's scientific journals on the species question.* New Haven and London: Yale University Press.

Wiltshear, F. G. 1913. The botany of the Antarctic voyage. *Journal of Botany: British and Foreign* 51: 355–8.

Winsor, Mary Pickard. 1979. Louis Agassiz and the species question. *Studies in History of Biology* 3: 89–117.

——. 1991. *Reading the shape of nature: comparative zoology at the Agassiz museum.* Chicago: University of Chicago Press.

Winstanley, D. A. 1947. *Later Victorian Cambridge.* Cambridge: Cambridge University Press.

Wollaston, Thomas Vernon. 1854. *Insecta Maderensia; being an account of the insects of the islands of the Madeiran group.* London.

——. 1857. *Catalogue of the coleopterous insects of Madeira in the collection of the British Museum.* London.

[——]. 1860. Review of Origin of species. *Annals and Magazine of Natural History* 3d ser. 5: 132–43. Reprinted in Hull 1973, pp. 127–40.

——. 1864. *Catalogue of the coleopterous insects of the Canaries in the collection of the British Museum.* London.

Woodward, Horace B. 1882. A memoir of Dr S. P. Woodward with a list of his published papers. *Transactions of the Norfolk and Norwich Naturalists' Society* 3 (1879–84): 279–312.

Woodward, Samuel Pickworth. 1851–6. *A manual of the Mollusca; or, a rudimentary treatise of recent and fossil shells.* 3 pts. London.

——. 1860. Volcanoes. In *Encyclopædia Britannica.* 8th ed. 21 vols. and index. Edinburgh. 1855–61.

Wrangel, Ferdinand Petrovich. 1839. *Reise des Kaiserlich-russischen flotten-lieutenants Ferdinand v. Wrangel längs der Nordküste von Sibirien und auf dem Eismeere, in den Jahren 1820 bis 1824* . Translated from the Russian and edited by G. Engelhardt. Berlin.

Wright, Chauncey. 1860. The economy and symmetry of the honey-bees' cells. *Mathematical Monthly* 2: 304–19.

WWW: *Who was who. A companion to Who's who.* 7 vols. and cumulated index, 1897–1980. London: Adam & Charles Black. 1920–81.

Wyman, Jeffries. 1849. A report on the cranium of the Engéena (*Troglodytes gorilla*). *Proceedings of the Boston Society of Natural History* 3 (1848–51): 179.

Wyman, Jeffries. 1850. A description of two additional crania of the Engéena, (*Troglodytes gorilla, Savage*,) from Gaboon, Africa. *American Journal of Science and Arts* 2d ser. 9: 34–45.

——. 1855. An account of the dissection of a black chimpanzee, (Troglodytes niger). *Proceedings of the Boston Society of Natural History* 5 (1854–6): 274–5.

——. 1859. On some unusual modes of gestation. *American Journal of Science and Arts* 2d ser. 27: 5–13.

Yeo, Richard. 1984. Science and intellectual authority in mid-nineteenth-century Britain: Robert Chambers and *Vestiges of the natural history of creation. Victorian Studies* 28: 1–31.

Zaslow, Morris. 1975. *Reading the rocks . . . the story of the geological survey of Canada, 1842–1972*. Toronto: Macmillan Company of Canada.

BIOGRAPHICAL REGISTER AND INDEX TO CORRESPONDENTS

This list includes all persons mentioned in the letters and notes that the editors have been able to identify and all correspondents. Dates of letters to and from correspondents are given in chronological order. Letters to the correspondents are listed in roman type; letters from correspondents in italic type; third-party letters are in italic type for both correspondents. Following the register, a list of all biographical sources referred to in the entries is given. These works are also listed in the main bibliography.

Abdy, John Thomas (1822–99). Regius professor of civil law at Cambridge University, 1854–73; examiner in civil law for the law tripos, 1858–69. (*Alum. Cantab.*)

Acland, Henry Wentworth (1815–1900). Physician. Lee's reader of anatomy at Christ Church, Oxford, 1845–58. Regius professor of medicine, University of Oxford, 1858–94. Appointed honorary physician to Edward, Prince of Wales, 1860. FRS 1847. (*DNB*, *Physicians.*)

Afzelius, Adam (1750–1837). Swedish botanist. Visited England in 1789. Undertook exploring expeditions to Sierra Leone, acting as botanist for the Sierra Leone Company, 1792–3 and 1794–6. In Africa, he formed natural history and ethnographical collections that were later acquired by Uppsala University. Returning to Sweden, became lecturer in botany, Uppsala, 1803–5; acting professor of materia medica and dietetics, 1812. (*SBL.*)

Agassiz, Jean Louis Rodolphe (Louis) (1807–73). Swiss geologist and zoologist. Professor of natural history, Neuchâtel, 1832–46. Emigrated to the United States in 1846. Professor of natural history, Harvard University, 1847–73. Established the Museum of Comparative Zoology, Harvard, in 1859. Foreign member, Royal Society, 1838. (*DAB*, *DSB.*)

Ainslie, Robert. Methodist minister. Lived at Tromer Lodge, Down, until 1858. (*Post Office directory of the six home counties* 1855.)

Albert Francis Charles Augustus Emmanuel, Prince Consort of England (1819–61). Married Queen Victoria in 1840. A patron of the arts and sciences. Established a model farm at Windsor in 1840. Chancellor of Cambridge University, 1847. Put forward the idea of the Great Exhibition and was actively involved in its organisation, 1849–51. President of the British Association for the Advancement of Science, 1859. (*DNB.*)

Alton, Joseph Wilhelm Eduard d' (1772–1840). German naturalist and art historian. Extraordinary professor, University of Bonn, 1818–26; ordinary professor of archaeology and art history, 1826–40. (*ADB*.)

Anca, Francesco, Baron (fl. 1856). Sicilian writer and naturalist. (*IBN*.)

Anderson, Charles Henry John, 9th Baronet (1804–91). Restored churches in Lincolnshire. Contributed many papers to archaeological and architectural societies. Succeeded to baronetcy in 1846. (*Modern English biography*.)

Anderson, Thomas (1832–70). Botanist. Superintendent, Calcutta botanic gardens, 1861–8. Fellow of the Linnean Society, 1859. (R. Desmond 1977.)

Anderson-Henry, Isaac (1800–84). Horticulturist and hybridiser. Practised as a solicitor in Edinburgh until he acceded to his estates, whereupon he abandoned the profession. Bred new plant varieties and raised rare plants in his garden at Hay Lodge, Edinburgh. Fellow of the Linnean Society, 1865. (R. Desmond 1977, *Proceedings of the Linnean Society of London* (1886–7), pp. 42–4.)

Anne. Nursery maid at Down House.

Ansted, David Thomas (1814–80). Professor of geology, King's College, London, 1840–53. Assistant-secretary of the Geological Society, 1844–7. From 1845, lecturer on geology at the military college at Addiscombe, Surrey, and professor of geology at the College of Civil Engineers, Putney. Consultant geologist and mining engineer from about 1850. FRS 1844. (*DNB*, Sarjeant 1980.)

27 October [1860]

Appleton, Daniel (1785–1849). Publisher. Founded the New York publishing firm D. Appleton & Co. in 1838 in partnership with his son, William Henry Appleton. (*DAB*.)

Appleton, William Henry (1814–99). Publisher. Partner in the New York publishing firm D. Appleton & Co., 1838–48. Became head of the firm in 1848 on the retirement of his father, Daniel Appleton. D. Appleton & Co. was the medium through which the works of British scientists such as CD, Thomas Henry Huxley, and Herbert Spencer reached the American public. (*DAB*.)

Arnott, George Arnott Walker (1799–1868). Botanist. Worked with William Jackson Hooker in Glasgow, 1830–40. Regius professor of botany, Glasgow, 1845–68. (R. Desmond 1977, *DNB*.)

Arnott, Neil (1788–1874). Physician and meteorologist. Surgeon with the East India Company in China, 1807–9, 1810–11. Practised as a physician in London, 1811–55. Gave lectures at the Philomathic Institution published as *Elements of physics* (1827). One of the founders of the University of London, 1836. FRS 1838. (*DNB*.)

16 February [1860?]

Audubon, John James (1785–1851). American ornithologist and illustrator. Used his hunting skill and latent artistic talents to shoot and paint birds. His published illustrations, such as those in *The birds of America* (1827–38) secured him an entry into the scientific community in Britain. (*DAB*, *DSB*.)

Azara, Félix d' (1746–1811). Spanish explorer and army officer. Sent to South America to survey Spanish and Portuguese territories. Published works on the fauna of Paraguay. (*NBU.*)

Babbage, Charles (1791–1871). Mathematician and pioneer in the design of mechanical computers. FRS 1816. (*DNB, DSB.*)

Babington, Charles Cardale (1808–95). Botanist and archaeologist. Greatly involved in the natural history activities at Cambridge for more than forty years. A founding member of the Entomological Society in Cambridge and the Cambridge Antiquarian Society. Professor of botany, Cambridge University, 1861–95. An expert on plant taxonomy. FRS 1851. (*DNB, DSB.*)

Bachman, John (1790–1874). American clergyman and naturalist. A close friend and collaborator of John James Audubon. (*DAB*, Elliott 1979.)

Bacon, Francis (1561–1626). Lord chancellor, 1618–26. Author of important works on philosophy of science. (*DNB, DSB.*)

Baer, Karl Ernst von (1792–1876). Estonian naturalist. Professor of anatomy at Königsberg University, 1817; of zoology, 1826–34. Professor of zoology at the Academy of Sciences, St Petersburg, 1834–67. Carried out embryological researches and, in 1826, discovered the mammalian egg. Propounded the influential view that development proceeds from the general to the specific. Foreign member, Royal Society, 1854. (*DSB, NDB.*)

Baillière, Jean Baptiste Marie (1797–1885). French publisher. Founded the company of J. B. Baillière et fils. (*DBF.*)

Baily, John. Poulterer and dealer in live birds at 113 Mount Street, Berkeley Square, London. Author of works on the management of domestic and game fowl. (Freeman 1978, *Post Office London directory* 1860.)

Baily, John, Sr. Poultry judge. Father of John Baily.

Baily, William Hellier (1819–88). Draughtsman with the Geological Survey, 1844; assistant geologist, 1845; geologist, 1853. Palaeontologist to the Irish branch of the Geological Survey, 1857–88. (*Modern English biography*, Sarjeant 1980.)

Baird, Spencer Fullerton (1823–87). American vertebrate zoologist. Professor of natural history, Dickinson College, Pennsylvania, 1846–50. Assistant secretary, Smithsonian Institution, 1850–78; secretary, 1878–87. Instrumental in augmenting the zoological and anthropological collections of the Smithsonian's National Museum. (*DAB, DSB.*)

Baldwin, A. S. Physician and naturalist. Of Jacksonville, Florida. Member of the Boston Society of Natural History, for which he formed a collection of the freshwater shells of East Florida. Became president of the Florida Medical Association. Studied climate and its sanitary relation to medicine. (*NUC.*)

Balfour, John Hutton (1808–84). Physician and botanist. Professor of botany, Glasgow University, 1841–5. Professor of botany and Regius keeper of the Royal Botanic Garden, Edinburgh, 1845–79. FRS 1856. (*DNB, DSB.*)

Baring, Thomas (1799–1873). Financier. Conservative MP for Great Yarmouth, 1835–7; for Huntingdon, 1844–73. FRS 1860. (*DNB.*)

Barnard, Anne (1825–99). Daughter of John Stevens Henslow. Married R. Cary Barnard, an army officer, in 1859. Botanical artist. (R. Desmond 1977.)

Barrande, Joachim (1799–1883). French-born palaeontologist and stratigrapher. Employed as a tutor by the French royal family until their exile in 1830. Railway engineer in Prague, where he became interested in fossils. Collected and described the fossils of the Bohemian basin, 1840–83. The results of his studies were published in 30 volumes entitled *Systême Silurien du centre de la Bohême* (1852–1902). (*DSB.*)

Bartlett, Abraham Dee (1812–97). Taxidermist in London, 1834–52. Superintendent of the natural history department of the Crystal Palace, 1852–9; of the Zoological Society's gardens, Regent's Park, 1859–97. (*Modern English biography.*)

Barton, Benjamin Smith (1766–1815). American physician and naturalist. Appointed professor of natural history and botany at the College of Pennsylvania in 1790; professor of theory and practice of medicine, 1813–15. (*DAB.*)

Bate, Charles Spence (1819–89). Dentist and scientific writer. Practised dentistry in Swansea, 1841–51, then in Plymouth. Secretary of the Plymouth Institution, 1854–60; president, 1861–2 and 1869–70. Studied the Crustacea. FRS 1861. (*DNB.*)

Bates, Henry Walter (1825–92). Naturalist. Undertook a joint expedition to the Amazon with Alfred Russel Wallace, 1848–50; continued to explore the area alone until 1859. Provided the first comprehensive scientific explanation of the phenomenon subsequently known as Batesian mimicry. Published *The naturalist on the River Amazons* in 1863. Assistant secretary, Royal Geographical Society, 1864–92. FRS 1881. (*DNB, DSB.*)
22 November [1860]

Baxter, William. Ran a chemist's and stamp office in Bromley, Kent. (*Post Office directory of the six home counties* 1859.)

Bechstein, Johann Matthäus (1757–1822). German forestry scientist and ornithologist. Teacher at the Salzmann educational establishment, Schnepfenthal, 1784–94. Director of the Forestry Academy, Dreißigacker bei Meiningen, 1801–22. (*NDB.*)

Belcher, Edward (1799–1877). Naval officer engaged in hydrographic surveys, 1830–47. Commander of an expedition to the Arctic in search of John Franklin, 1852. Admiral, 1872. (*DNB.*)

Bell, Thomas (1792–1880). Dental surgeon at Guy's Hospital, London, 1817–61. Professor of zoology, King's College, London, 1836. President of the Ray Society, 1843–59. One of the secretaries of the Royal Society, 1848–53. President of the Linnean Society, 1853–61. Described the reptiles from the *Beagle* voyage. FRS 1828. (R. Desmond 1977, *DNB.*)

Belloc, Louise Swanton (1796–1881). French writer and translator. Daughter of an Irish officer. Her numerous translations into French included books by Maria Edgeworth and Charles Dickens; in 1860 she translated part of *Journal of researches*. (*DBF.*)

Benningfield, Mr. Proprietor of the Crown Hotel, Nazing, Essex. (*Post Office directory of the six home counties* 1859.)

Bentham, George (1800–84). Botanist. Pursued his botanical studies at the Royal Botanic Gardens, Kew. President of the Linnean Society, 1861–74. Published *Genera plantarum* (1862–83) with Joseph Dalton Hooker. FRS 1862. (*DNB*, *DSB*.)

Berghaus, Heinrich Karl Wilhelm (1797–1884). German geographer and cartographer. Lecturer at the Academy of Architecture in Berlin. (*IBN*.)

Berke, Luke. Editor of *The Future. A monthly journal of research and criticism in the physical and historical sciences*. (*BLC*.)

Berkeley, Miles Joseph (1803–89). Clergyman and botanist. Perpetual curate of Apethorpe and Wood Newton, Northamptonshire, 1833–68. An expert on British fungi. FRS 1879. (*DNB*, *DSB*.)

Bewicke, C. Of Palmeira, Madeira. Collected insects for his friend Thomas Vernon Wollaston.

Bibron, Gabriel (1806–48). French zoologist who described the reptiles collected by Alcide Charles Victor Dessalines d'Orbigny in South America. (*DBF*.)

Binney, Edward William (1812–81). Palaeobotanist. Solicitor in Manchester. A founder of the Manchester Geological Society, 1838; president, 1857–9, 1865–7. FRS 1856. (R. Desmond 1977, *DNB*.)

Blainville, Henri Marie Ducrotay de (1777–1850). French anatomist and zoologist. Professor of comparative anatomy, Muséum d'histoire naturelle, 1832. Foreign member, Royal Society, 1832. (*DBF*, *DSB*.)

Blanc, Louis (1811/13–82). French socialist. Founded the *Journal du progrès* in 1839. Published his first book, *Organisation du travail*, in 1840. Made member of provisional French government in February 1848. Forced into exile, living in London, after the riots in Paris of May 1848. Returned to Paris and re-entered French political life in 1870. (*DBF*, *NBU*.)

Blanco y Crespo, José Maria. *See* White, Joseph Blanco.

Blyth, Edward (1810–73). Owned an unsuccessful druggist's business in Tooting. Wrote zoological articles and edited zoological works before being appointed curator of the museum of the Asiatic Society of Bengal in 1841. He retired from this post in 1862 and returned to England, where he continued to write on zoology and on the question of the origin of species. (*DNB*, *DSB*.)

Bond, Frederick (1811–89). Entomologist and ornithologist. Elected member of Entomological Society, 1841; of Zoological Society in 1854. Collector of British Lepidoptera. (*Entomologist's Monthly Magazine* 25 (1888–9): 384; Gilbert 1977.)
[16? June 1860]; *26 June 1860*

Bonnet, Charles (1720–93). Swiss naturalist and philosopher. Although educated for the law, his main interest was natural history. Discovered parthenogenesis in aphids in 1746. Studied invertebrate regeneration, entomology, and plant physiology. Advocate of preformation. Wrote influential works on theoretical biology. (*DSB*.)

Boott, Francis (1792–1863). Physician and botanist. Had a successful medical practice in London, 1825–32. Lecturer on botany at the Webb Street School of Medicine, London, 1825. Secretary, Linnean Society, 1832–9; treasurer, 1856–61. (R. Desmond 1977, *DNB*.)
29 February 1860

Boucher de Crèvecoeur de Perthes, Jacques (1788–1868). Director of customs at Abbeville, France. Amateur archaeologist who investigated the evidence of what became known as Stone Age cultures discovered at Abbeville. (*DSB*.)

Bowen, Francis (1811–90). Philosopher. Alvord professor of natural religion, moral philosophy, and civil polity, Harvard College, 1853–89. (*DAB*.)

Bravard, Pierre Joseph Auguste (1803–61). French geologist and palaeontologist. Emigrated to Buenos Ayres in 1852. Inspector general of mines in Argentina and director of the national museum in Paraná. (*DBF*, *NDBA*.)

Bree, Charles Robert (1811–86). Physician and zoologist. (Freeman 1978, *IBN*.)

Bree, Robert Francis (1746–1842). Botanist and clergyman. Curate of St Giles's, Camberwell, 1814–19. Collected plants. Fellow of the Linnean Society, 1815. (*Alum. Cantab.*, R. Desmond 1977.)

Brehm, Christian Ludwig (1787–1864). German clergyman and ornithologist. His bird collection formed the basis of the Rothschild museum in Tring, Hertfordshire. (*NDB*.)

Brent, Bernard Philip (d. 1867). Bird fancier and author. Studied pigeon breeding in France and Germany. (*CDEL*.)
[1860?]; *[May–June 1860?]*

Brewster, David (1781–1868). Scottish physicist who specialised in optics. Invented the kaleidoscope, 1816. Edited the *Edinburgh Encyclopædia*, 1807–29. Assisted in the establishment of the British Association for the Advancement of Science in 1831. Knighted, 1832. Principal, University of St Andrews, 1838–59; University of Edinburgh, 1859–68. FRS 1815. (*DNB*, *DSB*.)

Brewster, Jane Kirk. Married David Brewster in 1857. (*DNB*.)

Bridges, Thomas (1842–98). Missionary on Keppel Island, West Falklands, 1856–69; established a mission at Ushuaia in the Beagle Channel. Author of a Fuegian grammar and dictionary (Bridges 1933). (Bridges 1948.)
6 January 1860; *[October 1860 or after]*

Broca, Pierre Paul (1824–80). French surgeon and anthropologist. MD, Paris, 1849. Assistant professor of the medical faculty in Paris and surgeon of the Central bureau, 1853. Pioneer in the field of anthropology. Became professor of pathology at Paris in 1867; of clinical surgery, 1868. Vice-president, French Academy of Medicine. (*DBF*, *DSB*.)

Brodie, Benjamin Collins (1783–1862). Surgeon. Professor of comparative anatomy and physiology, Royal College of Surgeons, 1816. Sergeant-surgeon to Queen Victoria. President of the Royal Society, 1858–61. Created baronet, 1834. FRS 1810. (*DNB*, *DSB*.)

Brodie-Innes, John William. *See* Innes, John William Brodie.

Bromfield, William Arnold (1801–51). Scottish botanist. Qualified as a physician in 1826, but never practised medicine. Resided in Ryde, Isle of Wight, from 1836. Travelled widely; died in Syria. (R. Desmond 1977, *Hooker's Journal of Botany* 3 (1851): 373–82.)

Brongniart, Adolphe Théodore (1801–76). French palaeobotanist and taxonomist. A founder of the *Annales des Sciences Naturelles*, 1824. Professor of botany, Muséum d'histoire naturelle, 1833. Foreign member, Royal Society, 1852. (*DBF*, *DSB*.)

Bronn, Heinrich Georg (1800–62). German palaeontologist. Became ordinary professor of natural science at Heidelberg in 1833. Translated and superintended the first German edition of *Origin* (1860). (*DSB*, *NDB*.)

4 February [1860]; 14 February [1860]; [*c.* 25 February 1860]; 21 March [1860]; 10 April [1860]; 13 April [1860]; 14 July [1860]; 5 October [1860]; *13 [or 15] October 1860*

Brooke, James (1803–68). Raja of Saráwak, Borneo, 1841–63. Appointed British commissioner and consul-general of Borneo in 1847. (*DNB*.)

Brougham, Henry Peter, Baron Brougham and Vaux (1778–1868). Reformer. A founder of the *Edinburgh Review*, 1802. Lord chancellor, 1830–4. Published a number of scientific works, including *A discourse on natural theology* (1835). FRS 1803. (*DNB*.)

Brown, Robert (1773–1858). Botanist. Naturalist on a voyage to Australia commanded by Matthew Flinders (*DNB*), 1801–5; published descriptions of the plants he collected. Librarian to Joseph Banks, 1810–20. Keeper of the botanical collections, British Museum, 1827–58. FRS 1811. (*DNB*, *DSB*.)

Brown-Séquard, Charles-Édouard (1817–94). French physiologist. Practised medicine in France and America before accepting a professorship at Virginia Medical College, Richmond, in 1854. Returned to Paris in 1855, but moved to England in 1858. Physician, National Hospital for the Paralysed and Epileptics, 1860–3. Professor of physiology and pathology, Harvard Medical School, 1864–7. Professor of medicine, Collège de France, 1878–94. Conducted pioneering research in neurology and endocrinology. FRS 1861. (*DBF*, *DSB*.)

Buch, Christian Leopold von (1774–1853). A widely travelled German geologist and geographer. Foreign member, Royal Society, 1828. (*DSB*, *NDB*.)

Buchanan, Francis Hamilton (1762–1829). Physician. Served in the Bengal Medical Service, 1794–1815. Superintendent, Calcutta botanic garden, 1814–15. Took the name Hamilton when he succeeded to his mother's estate. FRS 1806. (R. Desmond 1977, *DNB*.)

Buchanan, John (1802–78). Banker in Glasgow and writer on antiquarian subjects, particularly Roman remains in Scotland. (*Roll of graduates of the University of Glasgow*.)

Buckle, Henry Thomas (1821–62). Historian. (*DNB*.)

Buffon, Georges Louis Leclerc, Comte de (1707–88). French naturalist and scientific administrator. (*DBF*, *DSB*.)

Bunbury, Charles James Fox, 8th Baronet (1809–86). Botanist and palaeobotanist. Collected plants in South America, 1833–4; in South Africa, 1838–9. Accompanied Charles Lyell to Madeira in 1853. Married Frances Joanna Horner, Charles Lyell's sister-in-law, in 1844. FRS 1851. (R. Desmond 1977, Sarjeant 1980.)
30 January 1860; 9 February [1860]

Bunbury, Frances Joanna (1814–94). Daughter of Leonard Horner. Married Charles James Fox Bunbury in 1844. (*Burke's peerage* 1980, Freeman 1978.)

Bunbury, Henry Edward, 7th Baronet (1778–1860). Army officer and politician. Became a major-general in 1814. Succeeded to the baronetcy in 1821. Became MP for Suffolk in 1830. Author of several historical works. (*DNB*.)

Bunsen, Robert Wilhelm Eberhard (1811–99). German chemist. Ordinary professor of chemistry, University of Marburg, 1842; University of Heidelberg, 1852–89. Carried out important work in spectroscopy in the 1860s. Foreign member, Royal Society, 1858. (*DSB*, *NDB*.)

Burdach, Karl Friedrich (1776–1847). German physiologist. Practised medicine, 1799–1811. Professor of anatomy, physiology, and forensic medicine, University of Dorpat, 1811–13. Became professor of anatomy at Königsberg in 1813. Director of the Königliche Anatomische Anstalt in Königsberg, 1817. Influenced by the Romantic philosophy of nature. Made significant contributions to embryology and neuroanatomy. (*ADB*, *DSB*.)

Burke, William (1792–1829). Criminal. He and his accomplice, William Hare, suffocated their victims and sold their bodies to an anatomy school. Found guilty of murder and hanged. (*DNB*.)

Burnett, Gilbert Thomas (1800–35). Medical botanist. First professor of botany at King's College, London, 1831–5. (*DNB*.)

Burns, Robert (1759–96). Scottish poet. (*DNB*.)

Busk, George (1807–86). Naval surgeon and naturalist who served on the hospital ship at Greenwich, 1832–55. Member of several scientific societies and president of the Microscopical Society, 1848–9. Hunterian professor of comparative anatomy, Royal College of Surgeons, 1856–9. Specialised in palaeontology and in the study of Bryozoa. FRS 1850. (*DNB*, *DSB*.)

Butler, Joseph (1692–1752). Clergyman. Rector at Houghton-le-Skerne, 1722–6; at Stanhope, Weardale, 1726–40. Bishop of Bristol, 1740–50; of Durham, 1750–2. Published several works on religious issues. (*DNB*.)

Button, Jemmy. A Fuegian who was brought to England in 1830 by Robert FitzRoy and subsequently returned to Tierra del Fuego on the *Beagle* in 1833. (Freeman 1978.)

Byron, George Gordon, 6th Lord (1788–1824). Poet. (*DNB*.)

Candolle, Alphonse de (1806–93). Swiss botanist whose home was a centre of botanical activity. Professor of botany and director of the botanic gardens, Geneva, 1835–50. After 1850, concentrated on his own research. Son of Augustin Pyramus de Candolle. Foreign member, Royal Society, 1869. (*DSB*.)

Candolle, Augustin Pyramus de (1778–1841). Swiss botanist. Professor of natural history, Academy of Geneva, 1816–35. Foreign member, Royal Society, 1822. (*DSB.*)

Canning, Charles John (1812–62). Governor-general of India, 1855–62; first viceroy, 1858–62. In office during the Indian Mutiny and negotiated the transfer of the government of India from the East India Company to the Crown in 1858. Created earl, 1859. (*Complete peerage*, *DNB.*)

Canning, George (1770–1827). Statesman. Became MP for Newport in 1794. Under-secretary of state for foreign affairs, 1796–1801. Published and contributed to the *Anti-Jacobin*, 1799–1801. Treasurer of the navy, 1804–6. Foreign minister, 1807–9; 1822–6. MP for Liverpool, 1812–22; for Harwich, 1822–7. Chancellor of the Exchequer, 1826–7. (*DNB.*)

Carpenter, William Benjamin (1813–85). Physician and naturalist. Fullerian professor of physiology at the Royal Institution, 1844–56; professor of forensic medicine at University College London; lecturer on physiology at the London Hospital, 1845–56. Registrar of the University of London, 1856–79. FRS 1844. (*DNB*, *DSB.*)

6 January [1860]; 6 April [1860]; 17 June [1860]

Cattell, John. Florist, nurseryman, and seedsman in Westerham, Kent. (R. Desmond 1977, *Post Office directory of the six home counties* 1859.)

12 May 1860

Chambers, Robert (1802–71). Publisher, writer, and geologist. Partner, with his brother William Chambers, in the Edinburgh publishing company W. & R. Chambers. Joint editor of *Chambers's Journal* from 1832. Anonymous author of *Vestiges of the natural history of creation* (1844). (*DNB*, *DSB.*)

Chapman, Alvan Wentworth (1809–99). American botanist and physician. Began medical practice in 1835. Self-taught botanist who published *Flora of the southern United States* in 1860. Friend of Asa Gray. (*DAB*, *DSB.*)

Claparède, Jean Louis René Antoine Édouard (Édouard) (1832–71). Entomologist. Professor of zoology and comparative anatomy, Academy of Geneva. (Gilbert 1977, *IBN.*)

Clark, Henry James (1826–73). Zoologist and botanist. Private assistant to Louis Agassiz, 1854–63. Assistant professor of zoology, Harvard Scientific School, 1860–5. Travelled in Europe for health reasons in 1865. Professor of botany, zoology, and geology, Pennsylvania State College, 1866–9; University of Kentucky, 1869–72. Joined Massachusetts Agricultural College in 1872. (*DAB.*)

Clark, William (1788–1869). Anatomist and clergyman. Professor of anatomy at Cambridge University, 1817–66. FRS 1836. (*DNB.*)

Clowes, William (1807–83). Prominent London printer. Took over the printing firm established by his father, William Clowes (1779–1847). (*DNB.*)

Colburn, Henry (d. 1855). London publisher who published the first edition of CD's *Journal of researches* (1839). (*DNB.*)

Cooke, Josiah Parsons (1827–94). American chemist, teacher, and author. Appointed tutor of mathematics, Harvard University, 1849; elected Erving professor of chemistry and mineralogy, 1850. (*DAB.*)

Cottage Gardener
[after 8 May 1860]

Covington, Syms (1816?–61). Became CD's servant on the *Beagle* in 1833 and remained with him as assistant, secretary, and servant until 1839. Emigrated to Australia in 1839. (Freeman 1978.)

Cranworth, Baron. *See* Rolfe, Robert Monsey.

Cresy, Edward (1823–70). Principal assistant clerk at the Metropolitan Board of Works and architect to the fire brigade. (*DNB*.)
15 January [1860]; 20 January [1860]; 25 August [1860]; *13 October 1860*; 14 October [1860]; [19 October 1860]; *27 October 1860*; *30 October 1860*; 2 November [1860]; 2 November [1860]; *10 November 1860*; [12 November 1860]; *10 December 1860*; 12 December [1860]

Crocker, Charles William (1832–68). Foreman, propagating department, Royal Botanic Gardens, Kew, 1857–63. (R. Desmond 1977.)

Crompton, Charles John (1797–1865). Justice of the Queen's Bench, 1852–65. Knighted, 1852. (*DNB*.)

Crompton, Peter. Medical practitioner in Derby. Father of Charles John Crompton (*DNB*.)

Cross, Mary Ann (neé Evans) (1819–80). Novelist who published under the name George Eliot. (*DNB*.)

Currey, Frederick (1819–81). Mycologist. Called to the bar, 1844. Secretary, Linnean Society, 1860–80; vice-president and treasurer, 1880–1. One of the editors of the *Natural History Review*, 1861. FRS 1858. (*Alum. Cantab.*, *DNB*.)

Curtis, Moses Ashley (1808–72). American botanist and minister. Ordained in the Episcopal Church, 1834. Carried out missionary work in North Carolina for most of his life. Studied and collected North American lichens and fungi. (*DAB*.)

Curtis, William (1746–99). Botanist. Established botanic garden in Lambeth in 1777. Owned a nursery in Brompton, 1789–1811; in Chelsea, 1812–13. Founded the *Botanical Magazine* in 1787. (R. Desmond 1977.)

Cuvier, Georges (1769–1832). French systematist, comparative anatomist, palaeontologist, and administrator. Professor of natural history, Collège de France, 1800–32; professor of comparative anatomy, Muséum d'histoire naturelle, 1802–32. Permanent secretary of the Académie des sciences from 1803. Foreign member, Royal Society, 1806. (*DBF*, *DSB*.)

Dana, James Dwight (1813–95). American geologist and zoologist. Naturalist with the United States Exploring Expedition to the Pacific, 1838–42; described the geology and the zoophytes and Crustacea from the voyage. An editor of the *American Journal of Science and Arts* from 1846. Professor of geology, Yale University, 1856–90. Foreign member, Royal Society, 1884. (*DAB*, *DSB*.)
30 July [1860]

Daniell, William Freeman (1818–65). Surgeon and botanist. Assistant surgeon to the British Army in West Africa, 1841–53. Sent plants to the Royal Botanic Gardens, Kew. (R. Desmond 1977, *DNB*.)

Darwin, Anne Elizabeth (Annie) (1841–51). CD's oldest daughter. (*Darwin pedigree.*)

Darwin, Elizabeth (1847–1926). CD's daughter. (*Darwin pedigree.*)

Darwin, Emily Catherine (1810–66). CD's sister. Resided at The Mount, Shrewsbury, until she married Charles Langton in 1863. (*Darwin pedigree.*)

Darwin, Emma (1808–96). Youngest daughter of Josiah Wedgwood II. Married CD, her cousin, in 1839. (*Emma Darwin.*)

Darwin, Erasmus (1731–1802). Physician, botanist, and poet. Advanced an evolutionary theory similar to that subsequently expounded by Lamarck. CD's grandfather. FRS 1761. (*DNB, DSB.*)

Darwin, Erasmus Alvey (1804–81). CD's brother. Matriculated Christ's College, Cambridge, 1822; MB, 1828. At Edinburgh University, 1825–6. Qualified but never practised as a physician. Lived in London from 1829 to his death. (*Alum. Cantab.*)

Darwin, Francis (Frank) (1848–1925). CD's son. BA, Trinity College, Cambridge, 1870. Collaborated with CD on several botanical projects, 1875–82. Lecturer in botany, Cambridge University, 1884; reader, 1888–1904. Edited CD's letters. FRS 1882. (*DNB, DSB.*)

Darwin, George Howard (1845–1912). CD's son. BA, Trinity College, Cambridge, 1868; fellow, 1868–78. Called to the bar in 1872 but did not practise law. Plumian professor of astronomy and experimental philosophy, Cambridge University, 1883–1912. FRS 1879. (*DNB, DSB.*)

Darwin, Henrietta Emma (Etty) (1843–1927). CD's daughter. Married Richard Buckley Litchfield (*Alum. Cantab.*) in 1871. (*Burke's landed gentry* 1952.)

Darwin, Leonard (1850–1943). CD's son. Attended the Royal Military Academy, Woolwich; entered the Royal Engineers in 1871; major, 1889. Served on several scientific expeditions, including those for the observation of the transit of Venus in 1874 and 1882. Instructor in chemistry, School of Military Engineering, Chatham, 1877–82. Staff, Intelligence Department, War Office, 1885–90. Liberal Unionist MP, Lichfield division of Staffordshire, 1892–5. President, Royal Geographical Society, 1908–11; Eugenics Education Society, 1911–28. Chairman, Bedford College, London University, 1913–20. (Sarjeant 1980, *WWW.*)

Darwin, Robert Waring (1724–1816). Of Elston. Brother of Erasmus Darwin. Had a taste for poetry and botany. Published *Principia botanica* in 1787. (*Burke's landed gentry* 1952, *DNB.*)

Darwin, Robert Waring (1766–1848). Physician. MD, Leiden, 1785. Had a large practice in Shrewsbury and resided at The Mount, which he built *c.* 1796–8. Third son of Erasmus Darwin by his first wife, Mary Howard. Married Susannah, daughter of Josiah Wedgwood I, in 1796. CD's father. FRS 1788. (Freeman 1978.)

Darwin, Susan Elizabeth (1803–66). CD's sister. Lived at The Mount, Shrewsbury, until her death. (*Darwin pedigree.*)

Darwin, William Alvey (1726–83). Brother of Erasmus Darwin. Grandfather of William Darwin Fox. (*Darwin pedigree*.)

Darwin, William Erasmus (1839–1914). CD's oldest child. Attended Rugby School. BA, Christ's College, Cambridge, 1862. Banker in Southampton. (*Alum. Cantab.*)
[4 March 1860]; [30 July 1860]; 26 [October 1860]; [9 November 1860]; [8 December 1860]

Daubeny, Charles Giles Bridle (1795–1867). Professor of chemistry, Oxford, 1822–55; of botany, 1834; of rural economy, 1840. An active supporter of the British Association for the Advancement of Science. FRS 1822. (*DNB*, *DSB*.)

Davidson, Thomas William St Clair (1817–85). Artist and palaeontologist. Expert on fossil brachiopods. FRS 1857. (*DNB*, Sarjeant 1980.)

Dawson, John William (1820–99). Canadian geologist and educationist. Investigated the geology of the maritime provinces with Charles Lyell in 1846 and 1852. Superintendent of education for Nova Scotia, 1850. Appointed principal and professor of geology at McGill University, 1855. FRS 1862. (*DNB*, *DSB*.)

Dayman, Joseph. Commander with the Royal Navy. (*Navy list* 1860.)

Decaisne, Joseph (1807–82). French botanist at the Jardin des plantes, Paris, 1824. Appointed professor of statistical agriculture, Collège de France, 1848, and professor of plant cultivation, Muséum d'histoire naturelle, 1850. Foreign member, Royal Society, 1877. (*DBF*, *NBU*.)

De la Beche, Henry Thomas (1796–1855). Geologist. First director of the Geological Survey of Great Britain, 1835–55. Established the Museum of Economic geology and the School of Mines. FRS 1819. (*DNB*, *DSB*.)

Delarouzée, Charles (1835–60). Entomologist. Travelled in the south of France and in the Middle East, collecting and describing insects, 1856–9. Discovered many new species. Died of consumption. (*Annales de la Société Entomologique de France* 4th ser. 1 (1861): 259–64, Gilbert 1977.)

Demaillet. *See* Maillet, Benoît de.

Despard, George Packenham (1813–81). Clergyman. Secretary of the Patagonian Mission Society, and of the Bristol and Clifton Clerical Association. Missionary in Tierra del Fuego, based on Keppel Island, 1856–61. Emigrated to Australia in the 1860s where he was minister of St Paul's, Sandhurst, until 1881. (*Alum. Cantab.*)

Dickens, Charles John Huffam (1812–70). Novelist. (*DNB*.)

Dixon, William Hepworth (1821–79). Historian and traveller. Editor of the *Athenæum*, 1853–69. (*DNB*.)

Doubleday, Edward (1811–49). Entomologist. Brother of Henry Doubleday. (*DNB*.)

Doubleday, Henry (1808–75). Quaker tradesman in Epping. Naturalist and entomologist noted for his work on the systematics of Lepidoptera. (R. Desmond 1977, *DNB*.)
3 May 1860; *16 May 1860*

Douglas, John William (1814–1905). Official in the Customs House from about 1834 to 1884. Secretary of the Entomological Society, 1849–56; president, 1861–2. Became an editor of the *Entomologist's Monthly Magazine* in 1874. (*Entomologist's Monthly Magazine* 41 (1905): 221–2, Gilbert 1977.)

Draper, John William (1811–82). English-born American chemist. Professor of chemistry and natural philosophy, Hampden-Sidney College, Virginia, 1836–8. Became professor of chemistry, University of the City of New York (New York University), in 1838 and president of its Medical School in 1850. (*DAB*.)

Drummond, James (1784–1863). Botanist. Emigrated to Australia, 1829. Superintendent of the government gardens, Western Australia. (*Aust. dict. biog.*, *DNB*.)
16 May 1860; *17 September 1860*; *8 October 1860*; 22 November [1860]; 20 December [1860]

Drysdale, Lady. Mother-in-law of Edward Wickstead Lane. (*Emma Darwin* (1904).)

Du Bois-Reymond, Emil Heinrich (1818–96). German physiologist who particularly studied animal electricity. A founder of the Physikalische Gesellschaft in Berlin, 1845. Instructor in anatomy, Berlin Academy of Art, 1848–53. Elected member of Prussian Academy of Sciences, 1851; permanent secretary, 1876. Appointed professor of physiology at the University of Berlin, 1858. (*ADB*, *DSB*, *NDB*.)

Dujardin, Félix (1801–60). French protozoologist. Professor of geology and mineralogy, Toulouse, 1839. Became professor of zoology and botany in Rennes in 1840. Made a corresponding member of the Académie des sciences shortly before his death. Published important works on the infusoria. (*DBF*, *DSB*.)

Duméril, André Marie Constant (1774–1860). French entomologist. Professor of anatomy, Paris, 1801–12. Became professor of herpetology and ichthyology at the Jardin des plantes, Paris, in 1813. (*DBF*, Gilbert 1977.)

Duns, John (1820–1909). Free Church minister in Torphichen, West Lothian. Amateur naturalist. (*NUC*.)

Duperrey, Louis Isidore (1786–1865). French naval officer and hydrographer. Carried out observations on the earth's shape and on terrestrial magnetism. (*DBF*, *DSB*.)

Dutrochet, René Joachim Henri (1776–1847). French military medical officer, 1808–9. Abandoned medical practice and devoted himself to natural science. Elected corresponding member of the Académie des sciences, 1819; full member, 1831. Primarily studied plant physiology and is noted for his work on the phenomena of osmosis and diffusion. (*DBF*, *DSB*.)

Dyster, Frederick Daniel. (b. *c.* 1810). Surgeon and naturalist, with a keen interest in marine zoology, who lived in Tenby, Wales. MD, Giessen, 1848. Founder of and medical officer at a twenty-four bed hospital on Madeira, 1843–7. Fellow of the Linnean Society. Friend of Thomas Henry Huxley. (L. Huxley ed. 1900, *London and provincial medical directory* 1860, *Post Office directory of Monmouthshire and . . . South Wales* 1871.)

Earl, George Windsor. Author of works on the Indian Archipelago and Australia. (*NUC.*)

Eden, George, Earl of Auckland (1784–1849). Statesman. Governor-general of India, 1836–42. First lord of the Admiralty, 1846–9. President of the Royal Asiatic Society, vice-president of the Horticultural and Zoological Societies of London, and a trustee of the British Museum. (*DNB.*)

Edward, Prince of Wales (1841–1910). Became King of Great Britain in 1901. (*DNB.*)

Ehrenberg, Christian Gottfried (1795–1876). German naturalist, microscopist, and traveller. Studied the development of coral reefs and worked extensively on infusoria. Professor at Berlin University. Foreign member, Royal Society, 1837. (*DSB*, *NDB.*)

Eliot, George. *See* Cross, Mary Ann.

Ellis, Mrs. Of Rose Hill, Hoddesdon, Hertfordshire. (*Post Office directory of the six home counties* 1859.)

Engleheart, Stephen Paul. Surgeon in Down, Kent. (*Post Office directory of the six home counties* 1862.)

Entomologist's Weekly Intelligencer
[20 June 1860]

Erskine, Henry Knight. Lieutenant-colonel commandant of the Royal Aberdeenshire Highlanders. (*Army list* 1860.)

Esquilant, F. C. Pigeon fancier. Became secretary of the Philoperisteron Society in 1847. Later became president of the National Peristeronic Society. (Fulton [1874–6].)

Eudes-Deslongchamps, Jacques Amand (b. *c.* 1800). French naturalist. Professor of zoology and dean of the science faculty at Caen. (*Dictionnaire universel des contemporains.*)

Eyre, Edward John (1815–1901). Emigrated to Australia, 1833, where he took up sheep farming. Explored the Australian interior. Lieutenant-governor of New Zealand, 1846–53; of St Vincent, West Indies, 1854–60. Acting governor of Jamaica, 1861; governor in chief, 1864. Brutally repressed a native uprising in 1865 and was recalled to England. (*Aust. dict. biog.*, *DNB.*)

Eyton, Thomas Campbell (1809–80). Shropshire naturalist. Friend and Cambridge contemporary of CD. Author of several works on natural history. On coming into possession of the family estate at Eyton, Shropshire, in 1855, he built a museum in which he formed one of the finest collections of skins and skeletons of the birds of Europe. (*DNB.*)
18 June [1860]

Fairmaire, Léon (1820–1906). Studied classics in Paris. A reversal in fortune led him to begin a career in public service administration in 1848. He retired in 1878. Joined the Société Entomologique de France in 1852; president, 1854–81. Studied and collected Coleoptera. (*DBF*, Gilbert 1977.)

Falconer, Hugh (1808–65). Palaeontologist and botanist. Superintendent of the botanic garden, Saharanpur, India, 1832–42. Superintended the arrangement of Indian fossils for the British Museum, 1844. Superintendent of the Calcutta botanic garden and professor of botany, Calcutta Medical College, 1848–55. FRS 1845. (*DNB*, *DSB*.)

9 July [1860]; 12 July [1860]

Faraday, Michael (1791–1867). Apprentice to a bookbinder, 1804. Attended scientific lectures, including those of Humphry Davy (*DNB*). Appealed to Davy for patronage; became his assistant at the Royal Institution in 1812 and director of the laboratory in 1825. Began his own researches in 1816 and contributed extensively to the fields of electrochemistry, magnetism, and electricity. FRS 1824. (*DNB*, *DSB*.)

Fawcett, Henry (1833–84). Statesman. BA, Cambridge, 1856. Blinded in a shooting accident in 1858. Elected professor of political economy, Cambridge University, 1863. MP for Brighton, 1865–74. Became MP for Hackney and post-master general in 1880. FRS 1882. (*DNB*.)

Fitch, Walter Hood (1817–92). Botanical artist at the Royal Botanic Gardens, Kew, from 1841. Produced over 10,000 published drawings. Among the works he illustrated was Joseph Dalton Hooker's *Illustrations of Himalayan Plants* (1855). Fellow of the Linnean Society, 1857. (Gilbert 1977, *Proceedings of the Linnean Society of London* (1891–2), p. 68.)

Fitton, William Henry (1780–1861). Physician and geologist. President of the Geological Society, 1827–9; vice-president, 1831–46. FRS 1815. (*DNB*, *DSB*.)

FitzRoy, Robert (1805–65). Naval officer, hydrographer, and meteorologist. Commander of the *Beagle*, 1828–36. Tory MP for Durham, 1841–3. Governor of New Zealand, 1843–5. Superintendent of the dockyard at Woolwich, 1848–50. Chief of the meteorological department of the Board of Trade, 1854. Vice-admiral, 1863. FRS 1851. (*DNB*, *DSB*.)

Fleming, Charles (1806–75). Professor. Author of an English–French dictionary (Fleming and Tibbins 1841–4). (*NUC*.)

Forbes, David (1828–76). Geologist and philologist. Travelled in South America, 1857–60, in search of ores of nickel and cobalt for the firm of Evans & Askins, nickel-smelters of Birmingham. After more years of travelling, he settled in England and became foreign secretary to the Iron and Steel Institute. FRS 1856. Brother of Edward Forbes. (*DNB*.)

[November? 1860]

Forbes, Edward (1815–54). Zoologist, botanist, and palaeontologist. Naturalist on board HMS *Beacon*, 1841–2. Appointed professor of botany, King's College, London, 1842. Palaeontologist with the Geological Survey, 1844–54. Became professor of natural history, Edinburgh University, in 1854. FRS 1845. (*DNB*, *DSB*.)

Forbes, James David (1809–68). Professor of natural philosophy, Edinburgh University, 1833–60. Secretary of the Royal Society of Edinburgh, 1840–51. Principal, University of St Andrews, 1860–8. FRS 1832. (*DNB*, *DSB*.)

Fox, William Darwin (1805–80). Clergyman. CD's second cousin. A close friend of CD's at Cambridge who shared his enthusiasm for entomology. Maintained an active interest in natural history throughout his life and provided CD with much information. Rector of Delamere, Cheshire, 1838–73. Spent the last years of his life at Sandown, Isle of Wight. (*Alum. Cantab.*)
22 [March 1860]; 18 May [1860]; 18 June [1860]; 18 October [1860]; 17 December [1860]

Francis, William. Printer. Assistant and later partner of Richard Taylor in the printing firm Taylor and Francis. Editor of *Annals and Magazine of Natural History*.

Franklin, John (1786–1847). Naval officer and Arctic explorer. Lieutenant-governor of Van Diemen's Land (Tasmania), 1837–43. Leader of the 1845 expedition in search of a north-west passage during which all hands perished. FRS 1823. (*DNB.*)

Fraser, Louis. Naturalist to the Niger expedition, 1841–2. Curator of the Zoological Society museum, 1843–5. Conservator of Lord Derby's menagerie at Knowsley until 1850. Undertook an expedition to South America, where he collected many rare birds and other animals. Upon his return, he became a dealer in birds for a time. (*DNB.*)

Freke, Henry. Irish physician. Resident medical officer, Dr Steevens's Hospital, Dublin. Physician, Fever Hospital, Kilmainham. Lecturer on clinical medicine, and later, professor of physic, Dr Steevens's Hospital Medical College. (*Catalogue of graduates, University of Dublin*; *Medical directory for Ireland* 1860.)

Frere, John Hookham (1769–1846). Diplomatist and author. MP for West Looe, Cornwall, 1796–1802. Helped to produce the *Anti-Jacobin, or Weekly Examiner*, 1797–8. Succeeded George Canning as under-secretary of state for the foreign office in 1799. Held various diplomatic positions in Spain, Portugal, and Germany. Inherited his father's estate in 1807. Settled in Malta in the 1820s and amused himself with literary work. (*DNB.*)

Galton, Francis (1822–1911). Traveller, statistician, and scientific writer. Explored in south-western Africa, 1850–2. Carried out various hereditarian inquiries. Founder of the eugenics movement. CD's cousin. FRS 1860. (*DNB, DSB.*)
13 May [1860]

Galton, John Howard (1794–1862). Of Hadzor House, Worcestershire. High sheriff of Worcestershire, 1834. Uncle of Francis Galton. (*Burke's landed gentry* 1879.)

Gardeners' Chronicle
[14–19 January 1860]; [13 April 1860]; [4–5 June 1860]; 15 September [1860]

Gärtner, Emma. Daughter of Karl Friedrich von Gärtner.
9 June [1860]; 14 July [1860]

Gärtner, Karl Friedrich von (1772–1850). German physician and botanist. From 1824, he devoted his time solely to the study of plant hybridisation. (*DSB, NDB.*)

Gaudin, Charles Théophile (1822–66). Swiss palaeobotanist. (*NUC.*)
[15 August 1860]

Geoffroy Saint-Hilaire, Étienne (1772–1844). French zoologist. Professor of zoology, Muséum d'histoire naturelle, 1793. Devoted much attention to embryology and teratology. (*DBF*, *DSB.*)

Geoffroy Saint-Hilaire, Isidore (1805–61). French zoologist. Succeeded his father, Étienne Geoffroy Saint-Hilaire, as professor at the Muséum d'histoire naturelle in 1841. Continued his father's work in teratology. Became professor of zoology at Paris in 1850. (*DBF*, *DSB.*)
28 January [1860]

Gliddon, George Robins (1809–57). United States vice-consul at Alexandria for about 20 years. Lectured in Boston, New York, and Philadelphia on Egyptian antiquities. Agent for Honduras interoceanic railway at the time of his death. Author of works on archaeology and ethnology. (*Modern English biography.*)

Godron, Dominique Alexandre (1807–80). French botanist, zoologist, and ethnologist. Became head of the science faculty, Nancy, in 1854, where he established a natural history museum and a botanic garden. (*DBF.*)

Göppert, Heinrich Robert (1800–84). German botanist. Lecturer in medicine and botany at the University of Breslau, 1827–31; extraordinary professor, 1831–9; ordinary professor from 1839. Studied plant physiology and fossil trees. One of the founders of German palaeobotany. (*BHGW*, *NDB.*)

Gordon, George (1801–93). Scottish botanist and clergyman. Minister of Birnie, near Elgin, 1832–89. (Gilbert 1977.)
11 September [1860]; 17 September [1860]

Gould, John (1804–81). Ornithologist and artist. Taxidermist to the Zoological Society of London, 1826–81. Travelled in Australia, 1838–40. Described the birds collected on the *Beagle* and *Sulphur* expeditions. Published forty-one illustrated folio volumes. FRS 1843. (*DNB*, *DSB.*)

Grant, Robert Edmond (1793–1874). Scottish physician and zoologist. An early advocate of the transmutation of species. Befriended CD in Edinburgh. Professor of comparative anatomy and zoology, University College London, 1827–74. FRS 1836. (*DNB*, Sarjeant 1980.)

Gray, Asa (1810–88). American botanist. Fisher professor of natural history, Harvard University, 1842–73. Devoted much time to the Harvard botanic garden and herbarium. Foreign member, Royal Society, 1873. (*DAB*, *DSB.*)
5 January 1860; 7 January [1860]; *[10 January 1860]*; *[17 January 1860]*; *23 January 1860*; 28 January [1860]; 1 February [1860]; [8 or 9 February 1860]; 18 February [1860]; *20 February 1860*; 24 February [1860]; 2 March [1860]; 8 March [1860]; 3 April [1860]; 25 April [1860]; 18 May [1860]; 22 May [1860]; 8 June [1860]; 3 July [1860]; *[10 July 1860]*; 22 July [1860]; 11 August [1860]; 10 September [1860]; 26 September [1860]; 19 October [1860]; 24 October [1860]; 31 October [1860]; 26 November [1860]; 11 December [1860]; 14 December [1860]; 21 December [1860]

Gray, George Robert (1808–72). Zoologist; an expert on insects and birds. Assistant in the zoological department of the British Museum, 1831–72. Brother of John Edward Gray. FRS 1865. (*DNB*, Gilbert 1977.)

Gray, John Edward (1800–75). Naturalist. Assistant keeper of the zoological collections of the British Museum, 1824; keeper, 1840–74. FRS 1832. (R. Desmond 1977, *DNB*.)

Greene, Joseph Reay. Irish naturalist. One of the editors of the *Natural History Review*. Studied Protozoa and coelenterates. (*Catalogue of graduates, University of Dublin*; L. Huxley ed. 1900.)

Grey, Eliza Lucy (née Spencer) (b. 1823). Married George Grey in 1839. (Dalton 1974.)

Grey, George (1812–98). Army officer and Australian explorer. Governor of South Australia, 1841–5; of New Zealand, 1845–53 and 1861–8; of Cape Colony, 1854–61. Settled in New Zealand, 1870–94. Prime minister of New Zealand, 1877–9. Knighted, 1848. (*Aust. dict. biog.*, *DNB*.)

Griffin, Charles (1819–62). Proprietor of Richard Griffin & Co., a bookselling and publishing company. Bookseller in Glasgow, 1853–9; managed the London business, 1859–61. Partner with Henry Bohn in Stationers' Hall, 1861–2. (*Modern English biography*.)
29 January [1860]

Griffith, John William (1819?–1901). Physician. Member of the Royal College of Physicians. Medical officer for Health, Clerkenwell district of London. President of the Clerkenwell Medical Ethics Society. Author of several works on microscopy. (*CDEL*, *London and provincial medical directory* 1860, *NUC*.)

Grisebach, August Heinrich Rudolf (1814–79). German botanist. Travelled through the Balkan Peninsula and north-western Asia Minor, 1839–40, studying the flora of these regions. Professor of botany, Göttingen University, 1847. (*DSB*, *NDB*.)

Groenland, Johannes (1824–91). Botanist. (*NUC*.)

Grote, George (1794–1871). Historian. One of the founders of University College London; vice-chancellor of the college, 1862. FRS 1857. (*DNB*.)

Grove, William Robert (1811–96). Judge and physical scientist. Professor of experimental philosophy, London Institution, 1841–6. An active member of the Royal Society; treasurer and chairman of the executive committee of the Philosophical Club, 1847. Became a member of the Queen's Counsel in 1853. Defended William Palmer, the 'Rugeley poisoner', in 1856. FRS 1840. (*DNB*, *DSB*.)

Günther, Albrecht Carl Ludwig Gotthilf (1830–1914). German-born zoologist. Joined the staff of the British Museum in 1857; assistant in reptiles and fishes, 1862. Keeper of the zoological department from 1875. Edited the *Record of Zoological Literature*, 1865–9. Catalogued several of the museum's collections of snakes, fish, and amphibia. FRS 1867. (*CDEL*, *NDB*.)
6 March [1860]

Guyot, Arnold Henri (1807–84). Swiss geographer and glacial geologist who emigrated to the United States in 1848. Studied the distribution of erratic boulders in Switzerland, 1840–7. Professor of physical geography and geology at Princeton, 1854–84, where he founded a museum. (*DAB*, *DSB*.)

Hachette, Louis Christophe François (1800–64). Founder of the Paris publishing house L. Hachette et Cie, which published literary and scientific works, especially serial publications and journals. (*Dictionnaire universel des contemporains*.)

Hadrian (Publius Aelius Hadrianus) (76–138). Roman emperor, 117–138. Patron of literature and learning. (*EB*.)

Haeckel, Ernst Heinrich Philipp August (Ernst) (1834–1919). German zoologist. MD, Berlin, 1857. Lecturer in comparative anatomy, University of Jena, 1861–2; extraordinary professor of zoology, 1862–5; ordinary professor of zoology and director of the Zoological Institute, 1865–1909. Influenced by *Origin*. His *Generelle Morphologie der Organismen* (1866) linked morphology to the study of the phylogenetic evolution of organisms. Propounded the theory that ontogeny recapitulates phylogeny. (*DSB*, *NDB*.)

Haldeman, Samuel Steman (1812–80). American philologist and zoologist. Professor of natural history, University of Pennsylvania, 1851–5; of comparative philology, 1868–80. (*DAB*, Elliott 1979.)

Hall, James, 4th Baronet (1761–1832). Geologist and chemist. Succeeded to the baronetcy in 1776. Travelled in Europe, 1783–6, where he met several European scientists and became actively interested in geology and chemistry. Was a strong supporter of James Hutton (1726–97), carrying out experimental work to counter criticisms of Hutton's *Theory of the earth* (1795). MP for Michael, Cornwall, 1807–12. Became president of the Royal Society of Edinburgh in 1812. FRS 1806. (*DNB*, *DSB*.)

Hall, James (1811–98). American geologist and palaeontologist. Palaeontologist to the New York State geological survey from 1843. Also state geologist of Iowa, 1855–8; of Wisconsin, 1857–60. Appointed director of New York State Museum in 1866. Made state geologist in 1893. (*DAB*, *DSB*.)

Hampden, Renn Dickson (1793–1868). Clergyman. Made principal of St Mary Hall, Oxford, in 1833. Professor of moral philosophy, University of Oxford, 1834–6; Regius professor of divinity, 1836–47. Bishop of Hereford, 1847–68. (*DNB*.)

Hanbury, Daniel (1825–75). Pharmacist and partner in the London firm Allen and Hanbury. On the board of examiners of the Pharmaceutical Society, 1860–72. FRS 1867. (*DNB*.)

Harborough, Lord. *See* Sherard, Robert.

Harcourt, Edward William Vernon (1825–91). Author of *A sketch of Madeira* (1851). An active member of the artillery volunteer movement from 1862. Sheriff of Oxfordshire, 1875. MP for Oxfordshire, 1878–85; for South Oxfordshire, 1885–6. (*Modern English biography*.)

Hardy, Charles (1803–85). Clergyman. Curate of the Subdeanery Church, Chichester, 1827–32. Vicar of North and South Hayling, Hampshire, 1832–80. (*Alum. Cantab.*, *Crockford's*.)
23 July 1860; 27 July [1860]

Hardy, Charlotte. Wife of Charles Hardy.

Harkness, Robert (1816–78). Professor of geology at Queen's College, Cork, 1853–78. FRS 1856. (*DNB*.)

Harper & Brothers. New York publishing company first established under the name J. & J. Harper in 1818 by James Harper (1795–1869) and his brother John (1797–1875). Two more brothers later joined the firm, Joseph Wesley Harper (1801–70) in 1823, and Fletcher Harper (1806–77) in 1825. The name was changed to Harper & Brothers in 1833. (*DAB*.)

Harrowin, W. Of Anne Harrowin & Son, ironmongers, in Kingsland High Street, London. (*Post Office London directory* 1860.)

Harvey, William Henry (1811–66). Irish botanist. Colonial treasurer in Cape Town, 1836–42. Keeper of the herbarium, Trinity College, Dublin, 1844; professor of botany, 1856–66. (*DNB*, *DSB*.)
24 August 1860; [20–4 September 1860]; *8 October 1860*; *23 November [1860]*

Haughton, Samuel (1821–97). Clergyman and palaeobotanist. Professor of geology at Dublin University, 1851–81. MD, Dublin, 1862; registrar of the medical school. FRS 1858. (R. Desmond 1977, *DNB*, Sarjeant 1980.)

Headland, Edward (1803–69). London physician. Became a member of the Royal College of Surgeons in 1848. President of the Medical Society of London. One of the first physicians to charge for his services rather than for physic. (*Modern English biography*.)

Headland, Frederick William (1830–75). CD's doctor from 1860. Awarded Fothergillian Gold Medal by the Medical Society of London in 1852 for his essay on the action of medicines (Headland 1852). (*Lancet* (1875(2)): 431, *Physicians*.)

Hearne, Samuel (1745–92). Explorer and colonial administrator in Canada. (*DNB*, *MDCB*.)

Heer, Oswald (1809–83). Swiss botanist, palaeontologist, and entomologist. An expert on Tertiary flora. Lecturer in botany, University of Zürich, 1834–5; undertook direction of botanic garden in 1834; associate professor, 1835–52; professor of botany and entomology, 1852–83. Contracted tuberculosis in 1849 and spent the winter of 1850–1 on Madeira. (*DSB*, *NDB*.)

Hemming, George Wirgman (1821–1905). Mathematician and law reporter. Entered Lincoln's Inn in 1844; called to the bar, 1850. Reporter in the Chancery courts, 1859–94. Junior counsel to the treasury, 1871–5. Appointed to Queen's Counsel, 1875. Standing counsel to Cambridge University, 1875–9. (*DNB*.)

Henfrey, Arthur (1819–59). Botanist. An advocate of German physiological botany in Britain. Lecturer in botany, St George's Hospital, London, 1847. Professor of botany, King's College, London, 1854–9. FRS 1852. (*DNB*, *DSB*.)

Henry, Joseph (1797–1878). American physicist. Professor of mathematics and natural philosophy, Albany Academy, New York, 1826–32. Professor of natural philosophy, College of New Jersey at Princeton, 1832–46. Became the first secretary and director of the Smithsonian Institution in 1846. Elected member of the American Philosophical Society in 1835; helped to organise the American Association for the Advancement of Science. (*DAB.*)

Henslow, Frances Harriet. *See* Hooker, Frances Harriet.

Henslow, Leonard Ramsay (1831–1915). Oldest son of John Stevens Henslow. Curate of Hitcham, Suffolk, 1854–6; of Bangor-Monachorum, Flintshire, 1856–60; of Great Chart, Kent, 1860–3. Rector of St Mary Magdalene, Pulham, Norfolk, 1863–70; of Zeals, Wiltshire, 1870–1915. (*Alum. Cantab.*)

Henslow, John Stevens (1796–1861). Clergyman, botanist, and mineralogist. Professor of mineralogy, Cambridge University, 1822–7; professor of botany, 1825–61. Extended and remodelled the Cambridge botanic garden. Curate of Little St Mary's Church, Cambridge, 1824–32; vicar of Cholsey-cum-Moulsford, Berkshire, 1832–7; rector of Hitcham, Suffolk, 1837–61. CD's teacher and friend. (*DNB*, *DSB.*)

3 February [1860]; 9 February [1860]; 2 April [1860]; *7 April 1860*; *5 May 1860*; 8 May [1860]; *10 May 1860*; 14 May [1860]; 17 May [1860]; 16 July [1860]; 28 [September 1860]; 11 October [1860]; 26 October [1860]; 10 November [1860]

Herapath, William (1796–1868). Analytical chemist. One of the founders of the Chemical Society of London and of the Bristol Medical School; professor of chemistry and toxicology, Bristol, 1828. Retained in many criminal trials involving suspected poisoning, where he was frequently in opposition to Alfred Swaine Taylor. (*DNB.*)

Herbert, William (1778–1847). Naturalist, classical scholar, linguist, politician, and clergyman. Noted for his work on plant hybridisation. Rector of Spofforth, Yorkshire, 1814–40. Dean of Manchester, 1840–7. (*DNB*, *DSB.*)

Herschel, John Frederick William (1792–1871). Astronomer, mathematician, chemist, and philosopher. Member of many learned societies. Carried out astronomical observations at the Cape of Good Hope, 1834–8. Master of the mint, 1850–5. Created baronet, 1838. FRS 1813. (*DNB*, *DSB.*)

Herzen, Alexander Ivanovich (1838–70). Russian radical. Left Russia in 1847; settled in London in 1852. Opened a Russian printing press in 1853 which, from 1857 to 1865, printed the radical newspaper 'Kolokol' ('The Bell'), which was smuggled into Russia. Had great influence among Russian radicals and numerous contacts among the progressive thinkers of Europe. (Steinberg ed. 1953.)

Heusinger von Waldegg, Johann Friedrich Christian Karl (Karl Friedrich) (1792–1883). German physician. Entered the Prussian Army as a military doctor in 1813. Appointed extraordinary professor, Jena, 1821. Ordinary professor of anatomy and physiology, Würzburg, 1824. Professor of practical medicine and director of the clinic, Marburg, 1829–83. (*ADB*, *NDB.*)

Higgins, John (1796–1872). Land agent. Born in Shrewsbury but settled in Alford, Lincolnshire, in 1819. Agent for CD's farm at Beesby, Lincolnshire.
13 June [1860]; 21 June [1860]

Hill, Richard (1795–1872). Jamaican-born magistrate and planter who was educated in England. Active in the anti-slavery movement. Naturalist who specialised in the ornithology of Jamaica. Assisted Philip Henry Gosse (1810–88) in the preparation of works on the natural history of the island. (Cundall 1920.)

Hinds, Richard Brinsley (1812–47?). Served as surgeon and naturalist aboard HMS *Sulphur*, 1836–42. (R. Desmond 1977, Plarr 1930.)

Hochstetter, Ferdinand (1829–84). German geologist. Geologist for the Reichsanstalt in Vienna, 1854–6. Lecturer in geology, University of Vienna, 1856–7. Accompanied the *Novara* expedition, 1857–9. Professor of geology and mineralogy at the Polytechnic Institute, Vienna, 1860–9. Undertook further travels, acting as geological advisor on the construction of the Turkish and trans-Siberian railways, 1869–72. From 1877, director of the Hofmuseum, Vienna. (*BHGW*, *NDB*.)

Hoeven, Jan van der (1801–68). Dutch comparative anatomist and anthropologist. Lecturer of zoology, Leiden, 1826–68. An adherent of Cuvier's theory of catastrophism; believed in creation rather than evolution, attempting in his works to show how the infinite wisdom of the Creator is reflected in the harmony of the world's creatures. (*BWN*, *DSB*.)

Hofmann, August Wilhelm von (1818–92). German organic chemist. Student of Justus von Liebig in Giessen, 1836; Liebig's personal assistant, 1843. Director, Royal College of Chemistry, London, 1845–65. Professor of chemistry, Berlin, 1865–92. (*DSB*, *NDB*.)
13 October 1860; *27 October 1860*

Hogan, Arthur Riky. Curate of Corsham, Wiltshire, 1856–8; of Piddleton, Dorset, 1860–3. Vicar of Watlington, Oxfordshire, from 1865. One of the editors of the *Natural History Review*. (*Crockford's*.)

Holland, Elizabeth. Of Knutsford, Cheshire. (*Post Office directory of Cheshire* 1857.)

Holland, Hannah. Of Knutsford, Cheshire. (*Post Office directory of Cheshire* 1857.)

Holland, Henry (1788–1873). Physician. Distant relative of the Darwins and Wedgwoods. Travelled extensively. Physician in ordinary to Queen Victoria, 1852. President of the Royal Institution for many years. Created baronet, 1853. FRS 1815. (*DNB*, *Physicians*.)

Holland, Lucy. Of Knutsford, Cheshire. (*Post Office directory of Cheshire* 1857.)

Holland, Mary. Of Knutsford, Cheshire. (*Post Office directory of Cheshire* 1857.)

Holland, Miss. *See* Holland, Elizabeth, Hannah, Lucy, or Mary.
[April 1860]

Hooker, Brian Harvey Hodgson (1860–1932). Fifth child of Frances Harriet and Joseph Dalton Hooker. (Allan 1967.)

Hooker, Charles Paget (1855–1933). Third child of Frances Harriet and Joseph Dalton Hooker. Became a doctor and practised for a time in Cirencester, then at Coltishall, Norwich, Norfolk. Licentiate in midwifery, 1879. (Allan 1967, *Medical directory* 1885.)

Hooker, Frances Harriet (1825–74). Daughter of John Stevens Henslow. Married Joseph Dalton Hooker in 1851. (R. Desmond 1977.)

Hooker, Joseph Dalton (1817–1911). Botanist. Accompanied James Clark Ross on his Antarctic expedition, 1839–43, and published the botanical results of the voyage. Botanist to the Geological Survey of Great Britain, 1845. Travelled in the Himalayas, 1847–50. Assistant director, Royal Botanic Gardens, Kew, 1855–65; director, 1865–85. Worked chiefly on taxonomy and plant geography. Son of William Jackson Hooker. Friend and confidant of CD. FRS 1847. (*DNB*, *DSB*.)

3 January [1860]; *5 January 1860*; 14 [January 1860]; [22 January 1860]; 31 [January 1860]; 8 February [1860]; 14 February [1860]; [20 February 1860]; [23 February 1860]; 26 [February 1860]; [26 February or 4 March 1860]; 3 March [1860]; 12 March [1860]; 18 [March 1860]; 13 [April 1860]; 18 [April 1860]; *[20 April 1860]*; 26 April [1860]; 27 April [1860]; *[28 April 1860]*; 30 April [1860]; 7 May [1860]; *10 May 1860*; 11 May [1860]; 13 [May 1860]; 14 May [1860]; 15 [May 1860]; 20 May [1860]; 22 [May 1860]; 29 [May 1860]; 30 May [1860]; 5 June [1860]; 7 June [1860]; *8 June 1860*; 12 [June 1860]; 17 June [1860]; 26 [June 1860]; *2 July 1860*; [2 July 1860]; [3 July 1860]; [4 July 1860]; 12 July [1860]; [17 July 1860]; 19 [July 1860]; [20? July 1860]; 29 July [1860]; 7 August [1860]; 31 [August 1860]; 2 September [1860]; 6 September [1860]; 21 November [1860]; *23 November [1860]*; 26 November [1860]; *[26 November – 4 December 1860]*; 4 December [1860]; 6 December [1860]; *[6–11 December 1860]*; 11 December [1860]; 17 December [1860]; 26 December [1860]; *28 December 1860*; 29 December [1860]

Hooker, Maria (1797–1872). Wife of William Jackson Hooker. (Allan 1967, R. Desmond 1977.)

Hooker, William Henslow (1853–1942). Oldest child of Frances Harriet and Joseph Dalton Hooker. Businessman. (Allan 1967.)

Hooker, William Jackson (1785–1865). Botanist. Regius professor of botany, Glasgow University, 1820–41. Director of the Royal Botanic Gardens, Kew, 1841–65. Father of Joseph Dalton Hooker. Knighted, 1836. FRS 1812. (*DNB*, *DSB*.)

Hooper, John. Physician who lived in Hoddesdon, Hertfordshire. (*Post Office directory of the six home counties* 1859.)

Hope, Frederick William (1797–1862). Entomologist and clergyman. Gave his collection of insects, prints, and drawings to Oxford University and founded a professorship of zoology, 1849. FRS 1834. (*DNB*, Gilbert 1977.)

Hopkins, William (1793–1866). Mathematician and geologist. A highly successful mathematics tutor at Cambridge. President of the Geological Society, 1851–3. Specialised in quantitative studies of geological and geophysical questions. FRS 1837. (*DNB*, *DSB*.)

Horace (Quintus Horatius Flaccus) (65 BC – 8 BC). Roman poet. (*EB*.)

Horner, Leonard (1785–1864). Geologist and educationist. Warden of University College London, 1827–31. Inspector of factories, 1833–56. A promoter of science-based education at all social levels. President of the Geological Society, 1845–7, 1860–2. Father-in-law of Charles Lyell and Charles James Fox Bunbury. FRS 1813. (*DNB*, *DSB*.)

Horner, Leonora. *See* Pertz, Leonora.

Huber, Jean Pierre (1777–1840). Swiss entomologist. Son of François Huber (1780–1831) and editor of his father's works. (Gilbert 1977.)

Hughes, Thomas (1822–96). Writer. Author of *Tom Brown's school days* (1857) which describes life at Rugby School. Called to the bar, 1848. Appointed to Queen's Counsel, 1869. County-court judge, 1882–96. (*DNB*.)

Humbert, Alois (1829–87). Swiss naturalist and palaeontologist. Curator of the Geneva museum. (Sarjeant 1980.)

Hunter, John (1728–93). Surgeon and anatomist. His collection of zoological specimens formed the basis of the Hunterian Museum of the Royal College of Surgeons. FRS 1767. (*DNB*, *DSB*.)

Hutton, Frederick Wollaston (1836–1905). Geologist and army officer. Commissioned in Royal Welsh Fusiliers, 1835; captain, 1862. Appointed assistant geologist to the New Zealand geological survey in 1871; provincial geologist of Otago, 1873. Professor of natural science in the University of Otago, 1877–90. Professor of biology and geology, University of New Zealand at Christ Church, 1890–3; curator of the university's museum, 1893–5. (*DNB*.)
8 March 1860

Hutton, Thomas. Captain in the Bengal Army; invalided in 1841. Author of works on natural history and scriptural geology. Fellow of the Geological Society. (*East-India register and army list* 1860.)

Huxley, Henrietta Anne (Nettie) (1825–1915). Lived in Australia, where she met Thomas Henry Huxley in 1847. Came to England and married Huxley in 1855. (Freeman 1978.)

Huxley, Leonard (1860–1933). Son of Thomas Henry Huxley. Biographer. Assistant editor of the *Cornhill Magazine*, 1901; editor, 1916–33. (*DNB*.)

Huxley, Noel (1856–60). Oldest child of Henrietta Anne and Thomas Henry Huxley. Died of scarlet fever. (L. Huxley ed. 1900.)

Huxley, Thomas Henry (1825–95). MB, London, 1845. Assistant-surgeon in HMS *Rattlesnake*, 1846–50, during which time he investigated Hydrozoa and other marine invertebrates. Lecturer on natural history, Royal School of Mines, 1854; professor, 1857. Naturalist to the Geological Survey of Great Britain, 1855. Served on several Royal commissions. Hunterian professor of comparative anatomy, Royal College of Surgeons, 1863–9. Fullerian professor of physiology, Royal Institution, 1863–7. President of the Royal Society, 1883–5. FRS 1851. (*DNB*, *DSB*.)
1 January [1860]; 9 January [1860]; 11 January [1860]; 21 [January 1860];

Huxley, T. H., cont.

[26 January 1860]; 2 [February 1860]; 4 February [1860]; 4 March [1860]; 20 March [1860]; 9 April [1860]; 11 April [1860]; 14 April [1860]; [after 20 April 1860]; 7 May [1860]; 3 July [1860]; [5 July 1860]; 20 July [1860]; [30? July 1860]; *6 August 1860*; 8 August [1860]; 10 September [1860]; 18 September [1860]; 1 November [1860]; 10 November [1860]; 16 November [1860]; 22 November [1860]; 2 December [1860]; 5 [December 1860]; 11 December [1860]; *[before 14 December 1860]*

Innes, Eliza Mary. Married John Innes in 1847. (*Burke's landed gentry* 1879.)

Innes, John (1817–94). Curate of Farnborough, Kent, 1842. Perpetual curate, later vicar, of Down, 1846–69. Left Down in 1862 after inheriting property in Scotland; changed his name to Brodie-Innes *c.* 1860. (*Crockford's*, J. R. Moore 1985.)

18 July [1860]; 6 September [1860]; 11 September [1860]; 26 October [1860]; 28 December [1860]

Innes, John William Brodie (1848–1923). Son of John Innes. Barrister and novelist. Called to the bar, 1876. Advocate of the Scottish bar, 1888. Interested in antiquarian research, romance, demonology, witchcraft, and criminology. (*Alum. Cantab.*, Freeman 1978).

Jäger, Gustav (1832–1917). German physician and zoologist. Professor of zoology and physiology, Polytechnic and Veterinary School, Stuttgart, and Agricultural College, Hohenheim. (Freeman 1978, *NUC*.)

James, Henry (1803–77). Lieutenant, Royal Engineers, 1831; captain, 1846; colonel, 1857. Appointed to the Ordnance Survey, 1827. Local superintendent, Geological Survey of Ireland, 1843–6. Transferred to admiralty employment in 1846, but returned to the Ordnance Survey in 1850; director-general, 1854–75. Director of the topographical and statistical department of the War Office, 1857–70. (*DNB*.)

Jamieson, Thomas Francis (1829–1913). Scottish agriculturalist of Ellon, Aberdeenshire. Factor on the Ellon estate for many years; later took the farm of Mains, Waterton. Appointed Fordyce lecturer on agricultural research, University of Aberdeen, 1862. Carried out notable researches on Scottish Quaternary geology and geomorphology. Became a fellow of the Geological Society in 1862. (*Geological Magazine* 50 (1913): 332–3, Sarjeant 1980.)

Jane. Maid at Down House.

Jardine, William, 7th Baronet (1800–74). Naturalist. A founder of the *Annals and Magazine of Natural History*, 1841. Commissioner on salmon fisheries of England and Wales, 1860. FRS 1860. (*DNB*.)

Jeffreys, John Gwyn (1809–85). Conchologist, zoologist, and solicitor. Conducted extensive deep-sea dredging researches. FRS 1840. (*DNB*, *DSB*.)

Jenner, William, 1st Baronet (1815–98). Appointed professor of pathological anatomy at University College London, 1849. Holme professor of clinical medicine, 1860. Assistant, then full, physician to University College Hospital, 1849–76; elected consulting physician in 1879. Physician extraordinary to Queen Victoria, 1861; physician in ordinary, 1862. Established, through clinical studies, definitive evidence that typhus and typhoid were two different diseases. FRS 1864. (*DNB.*)

Jenyns, Leonard (1800–93). Naturalist and clergyman. Brother-in-law of John Stevens Henslow. Vicar of Swaffham Bulbeck, Cambridgeshire, 1828–49. Moved near Bath in 1850 and settled in Bath itself in 1860. Founded the Bath Natural History and Antiquarian Field Club in 1855. Member of many scientific societies. Described the *Beagle* fish specimens. Adopted the name Blomefield in 1871. (R. Desmond 1977, *DNB.*)
4 January 1860; 7 January [1860]

Johnston, Alexander Keith (1804–71). Geographer. Published numerous atlases and maps. Published the first English atlas of physical geography, 1848. Constructed the first globe of physical geography for the Great Exhibition, 1851. (*DNB.*)

Jones, P. H. Pigeon fancier of Fulham noted as a breeder of barbs. Became secretary of the National Peristeronic Society, a pigeon fanciers' club. (Fulton [1874–6].)

Jordan, Claude Thomas Alexis (Alexis) (1814–97). French botanist. Conducted field research, 1836–46, to complete and correct existing French flora. Assembled an important private herbarium. After giving up his botanical expeditions, worked in his own experimental gardens. (*DSB.*)

Jukes, Joseph Beete (1811–69). On the staff of the Geological Survey of Newfoundland, 1839–40. Naturalist aboard HMS *Fly*, 1842–6. Worked in North Wales for the Geological Survey of Great Britain, 1846–50. Director of the Irish branch of the Geological Survey, 1850–69. FRS 1853. (*DNB, DSB.*)
27 February 1860

Jussieu, Antoine Laurent de (1748–1836). French botanist. Appointed deputy to the professor of botany at the Jardin du roi, Paris, 1770. Studied plant taxonomy. Published his *Genera plantarum* in 1789. Promoted the 'natural system' of classification. After the founding of the Muséum d'histoire naturelle in 1793, became professor of botany; director, 1802–26. (*DSB, NBU, Taxonomic literature.*)

Kant, Immanuel (1724–1804). German philosopher. Lecturer in physics and philosophy, University of Königsberg, 1755–70; professor of logic and metaphysics, 1770–97. (*DSB, NDB.*)

Kelaart, Edward Frederick (*c.* 1818–60). Ceylonese-born physician and botanist. Served in the Ceylon Medical Service. In Gibraltar, 1843–5. An active member of the Ceylon branch of the Royal Asiatic Society. (R. Desmond 1977.)

Keppel, Henry (1809–1904). Naval officer. Became captain in the Royal Navy in 1837 and admiral in 1869. Served in the Crimean War. Groom in waiting to Queen Victoria, 1858–60. Awarded KCB, 1857. (*DNB.*)

Keppel, Katherine Louisa (née Crosbie) (d. 1859). Married Henry Keppel in 1839. (*DNB.*)

Keyserling, Alexandr Andreevich (1815–91). Russian geologist, palaeontologist, and botanist. Accompanied Roderick Impey Murchison on his expedition to Russia in 1841. (*DSB*, Sarjeant 1980.)

Kingsley, Charles (1819–75). Author and clergyman. Lecturer on English literature, Queen's College, London, 1848–9. Professor of modern history at Cambridge University, 1860–9. Rector of Eversley, Hampshire, 1844–75. Chaplain to the Queen, 1859–75. (*Alum. Cantab.*, *DNB.*)

Kippist, Richard (1812–82). Botanist. Librarian of the Linnean Society of London, 1842–81. Specialist on Australian plants. (*DNB*, R. Desmond 1977.)
1 February [1860]; 11 December [1860]

Kirby, William (1759–1850). Clergyman and entomologist. Vicar of Barham, Suffolk, 1797–1850. One of the first fellows of the Linnean Society, 1788. FRS 1818. (*Alum. Cantab.*, *DNB.*)

Kirchhoff, Gustav Robert (1824–87). German physicist. Extraordinary professor, Breslau, 1850–4; Heidelberg, 1854–75. Professor of theoretical physics, Berlin, 1875–86. Made major contributions to the study of electromagnetic radiation and electrical currents. A close friend of Robert Wilhelm Eberhard Bunsen, the two collaborated in laying the foundation of the method of spectral analysis, 1857–63. (*DSB*, *NDB.*)

Knight, Thomas Andrew (1759–1838). Botanist and horticulturist. Correspondent to the Board of Agriculture from 1795. President of the Horticultural Society, 1811–38. Applied scientific principles and techniques to the practical problems of the grower or breeder; carried out important work on the phenomenon now known as geotropism. FRS 1805. (R. Desmond 1977, *DSB.*)

Kölreuter, Joseph Gottlieb (1733–1806). German botanist. Professor of natural history and director of the court gardens at Karlsruhe. Carried out extensive hybridisation experiments on plants. (*DSB*, *NDB.*)

Koenig, Karl Dietrich Eberhard (1774–1851). Mineralogist. Assistant keeper, Department of Natural History, British Museum, 1807–13; keeper, 1813. Later directed the mineralogical department, holding this post until his death. FRS 1810. (R. Desmond 1977, *DNB.*)

Kölliker, Rudolf Albert von (Albert) (1817–1905). Swiss anatomist and physiologist. Professor of physiology and comparative anatomy, University of Würzburg, 1847–64; professor of anatomy, 1849–97. Foreign member, Royal Society, 1860. (*DSB*, *NDB.*)

Krohn, August David (1803–91). Russian invertebrate anatomist and embryologist who travelled extensively. (Blyakher 1982.)

La Billardière, Jacques Julien Houton de (1755–1834). French naturalist and traveller. Botanised in the Middle East, the Canary Islands, the Cape of Good Hope, and Tasmania. (*NBU*, *NUC.*)

Lamarck, Jean Baptiste Pierre Antoine de Monet de (1744–1829). French naturalist. Held various botanical posts at the Jardin du roi, 1788–93. Professor of zoology, Muséum d'histoire naturelle, 1793. Believed in spontaneous generation and in the progressive development of animal types; propounded a theory of transmutation in 1809. (*DSB.*)

Lamont, James (1828–1913). Arctic and African traveller. MP for Buteshire, Scotland, 1865–8. Fellow of the Geological Society of London. (Freeman 1978, *Rugby School register.*)
[23 February 1860]; 5 March [1860]

Lane, Edward Wickstead (1823–89). Physician. Proprietor of a hydropathic establishment at Moor Park, near Farnham, Surrey; from 1860, at Sudbrook Park, near Ham, Surrey. Later moved to Harley Street, London. (Freeman 1978, *London and provincial medical directory* 1860, *Medical directory* 1880.)

Langton, Charles (1801–86). Rector of Onibury, Shropshire, 1832–40. Left the Church of England in 1840. Resided at Maer, Staffordshire, 1840–6; at Hartfield Grove, Sussex, 1847–62. Married Charlotte Wedgwood in 1832. After her death, he married Emily Catherine Darwin in 1863. (*Alum. Oxon.*, *Emma Darwin.*)

Langton, Charlotte (1797–1862). Emma Darwin's sister. Married Charles Langton in 1832. Resided at Maer, Staffordshire, 1840–6; at Hartfield Grove, Sussex, 1847–61. (*Emma Darwin.*)

Lartet, Édouard Amant Isidore Hippolyte (1801–71). French palaeontologist and prehistorian. Trained as a lawyer but never practised. Began serious palaeontological research in the 1830s. Became interested in the question of the antiquity of man in the late 1850s, carrying out investigations in 1860 that provided proof of the contemporaneity of man with extinct animal species. Named professor of palaeontology, Museum d'histoire naturelle, 1869, but became too ill to carry out his duties. (*DSB*, Sarjeant 1980.)

Laugel, Antoine Auguste (Auguste) (1830–1914). French writer on an extensive range of subjects, including politics, history, and psychology. (*NUC.*)

Lecoq, Henri (1802–71). French naturalist and vulcanologist. Professor of natural history, University of Clermont-Ferrand. (Sarjeant 1980.)

Lehmann, Johann Georg Christian (1792–1860). German botanist. Professor of physics and natural history at the Gymnasium Academicum in Hamburg, 1818–60; head librarian, 1818–51. Helped to establish the botanic garden in Hamburg, which he directed for a time, as one of the best in Germany. (*ADB.*)

Leibniz, Gottfried Wilhelm (1646–1716). German mathematician, metaphysician, and philosopher. (*ADB*, *DSB*, *NDB.*)

Leifchild, John R. (b. 1815). Author and reviewer. Commissioner of coalfields. (*NUC*, *Wellesley index.*)

Leonhard, Karl Cäsar von (1779–1862). German geologist. Professor of mineralogy, Heidelberg University, 1818–62. Editor of the *Neues Jahrbuch für Mineralogie, Geognosie, Geologie und Petrefakten-Kunde*, 1807–30; joint editor with H. G. Bronn, 1830–62. (*ADB*, *DSB.*)

Lesson, René Primevère (1794–1849). French naval surgeon and naturalist. Medical officer on board the *Coquille*, 1822–5. Published the results of his zoological investigations made during the voyage. Taught botany at the naval medical school, Rochefort, 1829–31; appointed professor of pharmacy, 1831. Became the top-ranking navy pharmacist in 1835. (*DSB*, *NBU*.)

Lewes, George Henry (1817–78). Writer. Studied physiology from 1855 and published several works on the subject. Common law husband of Mary Ann Evans (George Eliot). (*DNB*.)

Lichtenstein, Martin Karl Heinrich (1780–1857). Naturalist and traveller. MD, Jena, 1802. Tutor to the son of the governor of Cape Colony, and house doctor, 1804–6, during which time he studied the natural history of the Cape. Surgeon-major in Dutch army before returning to Germany, where he wrote up results of his travels. Appointed professor of zoology, Berlin, 1811; director of the zoological museum, 1813–57. Travelled and collected in South America. (*ADB*.)

Liebig, Justus von (1803–73). German organic chemist. Professor of chemistry in Gießen, 1825–51; in Munich, 1852–73. Developed artificial agricultural fertilisers. Foreign member, Royal Society, 1840. (*DSB*, *NDB*.)

Lincecum, Gideon (1793–1874). American physician and naturalist. Studied medicine privately; taught in Georgia. Was for several years an Indian trader in the Choctaw country of central Mississippi. Practised medicine in Cotton Gin Port and Columbus, Mississippi, 1830–48. Settled in Texas in 1848. Studied the behaviour of agricultural ants. (*DAB*.)
29 December 1860

Linder, Jules (1830–69). French entomologist. *Conseiller de préfecture* in Nice. Known for his investigations of French caves. (*Berliner Entomologische Zeitschrift* 113 (1869): iv, Gilbert 1977.)

Lindley, John (1799–1865). Botanist and horticulturist. Assistant secretary to the Horticultural Society, 1822–41; vice-secretary, 1841–58. Professor of botany at London University (later University College London), 1828–60. Editor of the *Gardeners' Chronicle* from 1841. FRS 1828. (*DNB*, *DSB*.)

Link, Heinrich Friedrich (1767–1851). Professor of zoology, botany, and chemistry, University of Rostock, 1792–1811; of chemistry and botany, University of Breslau, 1811–15. In 1815, appointed professor of botany and director of the botanic garden at the University of Berlin. Had a strong interest in the philosophical foundations of the natural sciences. (*DSB*, *NDB*.)

Linnaeus (Carl von Linné) (1707–78). Swedish botanist and zoologist. Proposed a system for the classification of the natural world and reformed scientific nomenclature. (*DSB*.)

Livingstone, David (1813–73). Explorer and missionary in Africa, 1841–56. Published an account of his travels in 1857. Consul for the east coast of Africa and commander of an expedition to explore East and Central Africa, 1858–64. Explored the Nile basin, 1866–73. FRS 1858. (*DNB*.)

Livius, Titus (Livy) (59 BC–AD 17). Roman historian. (*EB.*)

Locke, John (1632–1704). Philosopher. (*DNB, DSB.*)

Logan, William Edmond (1798–1875). Geologist. Produced geological maps of the South Wales coal district, 1832–7. Left Britain for Nova Scotia in 1840, and was appointed head of the Geological Survey of Canada in 1842. (*DNB.*)

Longman, Thomas (1804–79). Became a partner in 1832 in the publishing firm of Longman, founded by his great-great-uncle, Thomas Longman (1699–1755). Became head of the firm in 1842. (*DNB.*)

Louis Napoleon (1808–73). President of the French Republic, 1848–52. Proclaimed Emperor of the French, as Napoleon III, in 1852. Defeated in the war with Prussia in 1870, after which he lived in Chislehurst, England. (*EB.*)

Lowe, Richard Thomas (1802–74). Clergyman and naturalist. Chaplain on Madeira, 1832–54. Rector of Lea, Lincolnshire, 1854–74. (R. Desmond 1977, *DNB.*)

Lowell, John Amory (1798–1881). Boston businessman and philanthropist. Trustee of the Lowell Institute, Boston. Member of the Harvard Corporation. Fellow of the American Academy of Arts and Sciences. A generous supporter of botanical studies. (Dupree 1959, *NUC*, Rossiter 1971.)

Lubbock, Constance Mary (b. 1860). Daughter of Ellen Frances and John Lubbock.

Lubbock, Ellen Frances (d. 1879). Married John Lubbock in 1856. (*Burke's peerage* 1880.)

Lubbock, John (1834–1913). Banker, politician, and naturalist. Son of John William Lubbock and a neighbour of CD in Down. Studied entomology and anthropology. An active supporter of CD's theory of natural selection. Succeeded his father to become 4th baronet in 1865. Created Baron Avebury in 1900. FRS 1858. (*DNB, DSB.*)

[after 28 April 1860?]; 25 May [1860]; 29 [May 1860]; [4 July 1860]; 20 July [1860]; [18 November 1860]; 28 November [1860]

Lubbock, John William, 3d Baronet (1803–65). Astronomer, mathematician, and banker. Treasurer and vice-president of the Royal Society, 1830–5, 1838–47. First vice-chancellor of London University, 1837–42. CD's neighbour in Down. FRS 1829. (*DNB, DSB.*)

Lubbock, Montagu (b. 1842). Younger brother of John Lubbock. (*Burke's peerage* 1880.)

Lucas, Samuel (1818–68). Journalist and author. A frequent contributor to *The Times*. (*DNB.*)

Ludwig, Camilla. Governess to the Darwin children. Joined the household in 1860. (Freeman 1978.)

Lupton, Thomas Goff (1791–1873). Engraver. Mainly responsible for the introduction of steel, to replace less durable copper, in mezzotint engraving. (*DNB.*)

Lyall, David (1817–95). Botanist. Assistant surgeon on HMS *Terror* in James Clark Ross's Antarctic voyage, 1839–42. Surgeon and naturalist on HMS *Acheron* in New Zealand; on HMS *Assistance* in the Arctic, 1852. On British Columbia Boundary Commission, 1858–61. Collected plants in the Antarctic, in Australia, New Zealand, the United States, and Canada. Friend of Joseph Dalton Hooker. (R. Desmond 1977, *Proceedings of the Linnean Society of London* (1894–5): 33–5.)

Lyell, Charles (1797–1875). Uniformitarian geologist whose *Principles of geology* (1830–3) and *Elements of geology* (1838) appeared in many editions. Professor of geology, King's College, London, 1831–3. President of the Geological Society, 1834–6, 1849–50. Travelled widely and published accounts of his trips to the United States. Knighted, 1848; created baronet, 1864. Scientific mentor and friend of CD. FRS 1826. (*DNB*, *DSB*.)
4 [January 1860]; 10 January [1860]; 14 January [1860]; 12 [February 1860]; *[13–14 February 1860]*; 15 and 16 [February 1860]; 18 [and 19 February 1860]; 23 February [1860]; 25 February [1860]; 12 [March 1860]; [24 March – 3 April 1860]; 10 April [1860]; 15 April [1860]; 27 and 28 April [1860]; *2 May 1860*; 4 May [1860]; *7 May 1860*; 8 [May 1860]; 18 May [1860]; 22 May [1860]; 1 [June 1860]; 6 June [1860]; 14 [June 1860]; *15 June 1860*; 17 June [1860]; *19 June 1860*; 20 [June 1860]; 25 [June 1860]; 5 [July 1860]; 30 July [1860]; 11 August [1860]; 28 August [1860]; *28 August 1860*; 1 September [1860]; *8 September 1860*; 12 September [1860]; *18 September 1860*; 23 [September 1860]; *25 September 1860*; 26 [September 1860]; *27 September 1860*; 28 [September 1860]; *30 September 1860*; 3 October [1860]; *[after 3 October 1860]*; 5 [October 1860]; *6 October 1860*; 8 October [1860]; *[before 20 November 1860]*; 20 November [1860]; 24 November [1860]; *24 November 1860*; 25 November [1860]; *30 November 1860*; 4 December [1860]

Lyell, Mary Elizabeth (1808–73). Oldest child of Leonard Horner. Married Charles Lyell in 1832. An accomplished conchologist. (Freeman 1978.)

McClintock, Francis Leopold (1819–1907). Naval commander and Arctic traveller. Given command of HMS *Intrepid* and accompanied Edward Belcher's Arctic expedition, 1852–4. Commanded an expedition in search of the missing Arctic explorer, John Franklin, commissioned by his wife; brought back proof of his death and that of his party in 1859. Knighted, 1860. Achieved rank of admiral in 1884. (*DNB*.)

Macdonald, John Denis (1826–1908). Entered navy as assistant surgeon in 1849. Served on HMS *Herald*, 1852–9. Carried out extensive microscopic studies on material collected at sea. Inspector-general of hospitals and fleets, 1880–6. FRS 1859. (*DNB*.)

M'Donnell, Robert (1828–89). Irish surgeon. His name was variously spelt M'Donnell or McDonnell. Served as a surgeon in the Crimean War, 1855. Demonstrator of anatomy, Carmichael School of Medicine, 1856; afterwards lecturer on anatomy and physiology. Surgeon to Jervis Street Hospital, Dublin, 1863; to Steeven's Hospital, 1866. Medical superintendent of the Mountjoy government prison from 1857 to 1867. Elected president, Royal College of Surgeons of Ireland, 1877; president, Academy of Medicine in Ireland, 1885–8. FRS 1865. (*DNB.*)

Mackintosh, Eva. Daughter of Robert and Mary Mackintosh. (Wedgwood and Wedgwood 1980.)

Mackintosh, Mary (Molly). American-born wife of Robert Mackintosh. (Wedgwood and Wedgwood 1980.)

Mackintosh, Robert (1806–64). Brother of Frances Mackintosh Wedgwood. (Wedgwood and Wedgwood 1980.)

Mackintosh, Ronald. Son of Robert and Mary Mackintosh. (Wedgwood and Wedgwood 1980.)

McLeod, Herbert (1841–1923). Studied under August Wilhelm von Hofmann at the Royal College of Chemistry. Became junior assistant there in 1858; assistant chemist, 1860. Assistant to Edward Frankland, 1865–71. Supplemented his earnings by giving lectures and undertaking translations. Appointed professor of experimental science, Royal Engineering College, Cooper's Hill, 1871. (James ed. 1987.)

Maillot, Benoît de (1656–1738). French diplomatist and traveller. Author of *Telliamed, ou entretiens d'un philosophe indien avec un missionnaire française* (1748), which presented a view of the cosmos as undergoing natural changes at random and a transformist conception of the development of forms of life. (*DSB.*)

Malthus, Thomas Robert (1766–1834). Clergyman and political economist. First professor of history and political economy at the East India Company College, Haileybury, 1805–34. Quantified the relationship between growth in population and food supplies in *An essay on the principle of population* (1798). FRS 1818. (*DNB, DSB.*)

Mantell, Gideon Algernon (1790–1852). Physician, geologist, and palaeontologist. Scientific lecturer and author. Made a noted collection of fossils. Secretary, Geological Society, 1841–2; vice-president, 1848–9. FRS 1825. (*DNB, DSB.*)

Mantell, Reginald Neville. Microscopist. Son of Gideon Algernon Mantell. (Sarjeant 1980.)

Marsh, James (1794–1846). Employed for many years as practical chemist to the Royal Arsenal at Woolwich. Assistant to Michael Faraday, Royal Military Academy, 1829–46. Invented the test for arsenic that bears his name. (*DNB.*)

Marshall, William (1815–90). Solicitor and botanist. (R. Desmond 1977.)
9 April [1860]

Mason, Nathaniel Haslope (*fl.* 1850–70). Collected plants on Madeira. Published sets of Madeira ferns and woods. Fellow of the Linnean Society, 1856. (R. Desmond 1977.)

Masson, David (1822–1907). Biographer and editor. Appointed professor of English literature, University College London, 1853. Edited *Macmillan's Magazine*, 1859–67. Professor of rhetoric and English literature, University of Edinburgh, 1865. (*DNB*.)

Masson, Victor (1807–79). Publisher. Joined the publishing firm of Crochard in 1838; became sole proprietor in 1846. Published elegant, often splendid, editions of scientific works. (*Dictionnaire universel des contemporains*.)

Masters, Maxwell Tylden (1833–1907). Physician. Lecturer, St George's Hospital, 1855–68. Editor, *Gardeners' Chronicle*, 1865–1907. FRS 1870. (R. Desmond 1977.)

7 April [1860]; *[after 7 April 1860]*; 13 April [1860]

Masters, William (1796–1874). Nurseryman in Canterbury. Founded the Canterbury Museum, 1823; honorary curator, 1823–46. (R. Desmond 1977.)

8 May 1860

Matthew, Patrick (1790–1874). Author of works on political and agricultural subjects. Advanced a theory of natural selection in the 1830s. (R. Desmond 1977.)

Maull and Fox. Photographers of Cheapside and Piccadilly, London. A partnership between Henry Maull and possibly John Fox, a miniature portrait painter. (*Post Office London directory* 1855, Pritchard 1986.)

Maull and Polyblank. Photographic artists of Picadilly and Gracechurch Street, London. A partnership between Henry Maull and George Henry Polyblank. (*Post Office London directory* 1860.)

Meade, Richard Henry (1814–99). Fellow, Royal College of Surgeons, 1845. Practised medicine in London and lectured on botany at St Bartholomew's Hospital. Succeeded to a surgical practice at Bradford in 1840. Wrote papers on entomology; studied Diptera and was an authority on the Muscidae. (*Entomologist's Monthly Magazine* 36 (1899): 46–7, Gilbert 1977.)

23 January 1860

Meyen, Franz Julius Ferdinand (1804–40). Prussian botanist and physician. Naturalist on board the *Prinzess Louise*, 1830–2. Published the results of the voyage, 1834–43. Carried out microscopic investigations of plants and animals. (*ADB*, *DSB*.)

Meyer, Christian Erich Hermann von (1801–69). German palaeontologist. Entered financial administration of the Bundestag in 1837, carrying out his palaeontological studies in his spare time; became director of finances in 1863. Offered professorship at Göttingen, but declined it in order to maintain his scientific independence. Studied the fossils of the Solenhofen slate, describing many pterodactyls. (*ADB*, *DSB*.)

Miall, Louis Compton (1843–1921). Became interested in natural history through his brother, a medical student. Received no formal training, though later attended Leeds School of Medicine. Became secretary of Bradford Philosophical Society, *c.* 1853. Appointed curator to the museum of the Leeds Philosophical Society, 1871. Professor of biology, Yorkshire College of Science, University of Leeds, 1876–1907. (*Entomologist's Monthly Magazine* 57 (1921): 93–4, Gilbert 1977.)

Milde, Karl August Julius (Julius) (1824–71). German botanist. Taught at secondary schools in Breslau from 1851, carrying out his botanical research in his spare time. His main interest was in cryptogamic plants. In 1860, ill health forced him to take a leave of absence; returned in 1862 but again forced to leave. Shortly before his death, he was awarded the title of royal professor. (*ADB.*)

Mill, John Stuart (1806–73). Philosopher and political economist. (*DNB, DSB.*)

Miller, William Hallowes (1801–80). Mineralogist and crystallographer. Professor of mineralogy, Cambridge University, 1832–80. FRS 1838. (*DNB, DSB.*)
5 June [1860]; [after 5 June 1860]; 1 December [1860];
27 December [1860]; 31 December [1860]

Milne-Edwards, Henri (1800–85). French zoologist. Professor of hygiene and natural history, École centrale des arts et manufactures, 1832. Professor of entomology, Muséum d'histoire naturelle, 1841, with responsibility for the collections of crustaceans, myriapods, and arachnids as well as insects; professor of mammalogy, 1861. Foreign member, Royal Society, 1848. (*DSB.*)

Milton, John (1608–74). Poet. (*DNB.*)

Miner, T. B. American author of works on the management of bees and poultry. (*NUC.*)

Mitchell, Charles (1807–59). Bookseller and advertisement agent for town and country newspapers in Red Lion Court, Fleet Street, London, from about 1836 until his death. (*Modern English biography.*)

Moir, Mrs. Mother-in-law of Henry Knight Erskine.

Moleschott, Jakob (1822–93). Dutch physician and physiologist. General practitioner, Utrecht, 1845–7. Lecturer in physiology and anthropology, Heidelberg, 1847–54. Appointed professor of physiology, Zürich, 1856; of experimental physiology and physiological chemistry, Turin, 1861. Noted as an advocate of materialist philosophy. (*ADB, BHGW, DSB.*)

Moquin-Tandon, Horace Bénédict Alfred (1804–63). French naturalist. Professor of zoology, Marseilles, 1829; professor of botany, Toulouse, 1833; professor of natural history, faculty of medicine, Paris, 1853. (*Larousse XX, NBU.*)

Moore, Charles (1815–81). Bookseller and geologist from Bath who studied the stratigraphy of Somerset. His collection of fossils formed the basis of the geological museum of the Bath Royal Literary and Scientific Institute. (*DNB*, Sarjeant 1980.)

More, Alexander Goodman (1830–95). Naturalist. Lived on Isle of Wight before settling in Ireland in the 1860s. Studied the flora of Ireland, publishing his *Cybele Hibernica* in 1866. In his zoological work he concentrated mainly on birds. Assistant in the Dublin Natural History Museum, 1867–81; curator, 1881–7. Fellow of the Linnean Society, 1856. (R. Desmond 1977, Gilbert 1977, *Natural Science* 6 (1895): 351.)

24 June 1860; [30 June 1860]; 3 July 1860; 3 August 1860; 5 August 1860; 9 August 1860; 5 September 1860

Morren, Charles François Antoine (1807–58). Belgian botanist and horticultural writer. Taught physics at the industrial school in Gant, 1831–5; studied medicine and qualified in 1835. Extraordinary professor of botany, University of Liège, 1835–7; ordinary professor, 1837–54. (*BNB*, Gilbert 1977.)

Morris, John (1810–86). Geologist. Originally a pharmaceutical chemist in Kensington. Professor of geology, University College London, 1854–77. (*DNB*, Sarjeant 1980.)

Mudie, Charles Edward (1818–90). Founder of Mudie's Lending Library, 1842. (*DNB*.)

Mueller, Ferdinand Jakob Heinrich von (1825–96). German-born explorer and botanist who emigrated to Australia in 1847. Government botanist in Victoria, 1853. Botanist to the North-west Australia Expedition, 1855–7. Director of the botanic garden, Melbourne, 1857–73. FRS 1861. (*Aust. dict. biog.*, R. Desmond 1977.)

Müller, Hermann (1829–83). German botanist and entomologist. School-teacher in Schwerin, 1854–5. Studied blind cave insects in Krain, 1855. Teacher of natural sciences at the Realschule in Lippstadt, 1855–83; became director of the school. After settling in Lippstadt, studied the local flora, in particular the mosses. CD's *Orchids* directed Müller's attention to the pollination and fertilisation of flowers, on which he published several papers and books. (Gilbert 1977, *Science* 2 (1883): 487–8.)

Müller, Johannes Peter (1801–58). German comparative anatomist, physiologist, and zoologist. Became professor of anatomy and physiology at Berlin University in 1833. Foreign member, Royal Society, 1840. (*ADB*, *DSB*.)

Munby, Giles (1813–76). Studied medicine and botany in Edinburgh and Paris. Collected plants in Algeria, 1839–59. (R. Desmond 1977, *DNB*.)

Murchison, Roderick Impey (1792–1871). Geologist noted for his work on the Silurian system. A leading figure in the Geological Society, the British Association for the Advancement of Science, and the Royal Geographical Society. Director-general of the Geological Survey of Great Britain, 1855. Knighted, 1846; created baronet, 1866. FRS 1826. (*DNB*, *DSB*.)

1 May [1860]

Murray, Andrew (1812–78). Entomologist and botanist. Practised in the legal profession in Edinburgh until 1860. Appointed assistant secretary to the Royal Horticultural Society, 1860. Became an expert on insects harmful to crops. In entomology, specialised on the Coleoptera; in botany, on the Coniferae.

(*Entomologist's Monthly Magazine* 14 (1878): 215–16, Gilbert 1977.)
28 April [1860]; 28 [April 1860]; *3 May 1860*; 5 May [1860]

Murray, John (1808–92). Publisher and author of guide-books. CD's publisher from 1845. (*DNB*.)
23 [January 1860]; [25 January 1860]; *27 January 1860*; 29 January [1860]; 30 [January 1860]; 3 February [1860]; 4 February [1860]; 5 February [1860]; 4 April [1860]; 9 April [1860]; 18 May [1860]; 7 June [1860]; 3 August [1860]; 12 October [1860]; 20 November [1860]; 22 [November 1860]; 3 December [1860]; 5 December [1860]

Napoleon III. *See* Louis Napoleon.

Naudin, Charles Victor (1815–99). French botanist. Experimented widely on plants, particularly studying acclimatisation and hybridity. Published a theory of transmutation based on hybridisation. (*Dictionnaire universel des contemporains*, *DSB*.)

Newberry, John Strong (1822–92). American physician, geologist, and stratigrapher. Appointed professor of geology, Columbia University (now George Washington University), 1856. Participated in the Colorado River expedition, 1857–8. During the Civil War, served with the United States Sanitary Commission. Professor of geology and palaeontology, School of Mines of Columbia University, New York, 1866–92. State geologist of Ohio, 1869–74. (*DAB*, Sarjeant 1980.)

Newton, Isaac (1642–1727). Natural philosopher. Advanced the theory of gravitation in *Principia mathematica* (1687) and of light in *Opticks* (1704). President of the Royal Society, 1703–27. FRS 1672. (*DNB*, *DSB*.)

Nisbet, Edward Parry. One of the Elder Brethren of Trinity House. (Barrett 1893, *Whitaker's Almanack* 1869.)

Nitschke, Theodor Rudolf Joseph (1834–83). German botanist. Extraordinary professor of botany, Münster, 1867–75; ordinary professor, 1875–83. Appointed director of the botanic gardens, 1867. (*ADB*.)

Norman, Ebenezer. Schoolmaster in Down whom CD employed as a copyist. (Freeman 1978.)
[24 September 1860]

Norman, George Warde (1793–1882). Writer on finance. Resided at the Rookery, Bromley, Kent. A family friend of the Darwins. (Freeman 1978, *Post Office directory of the six home counties* 1859.)

Norton, Charles Eliot (1827–1908). American editor, author, and teacher. Began his working life in the importing business, but after 1855 entered a literary career. As well as writing and editing books, he contributed to the *Atlantic Monthly* and for a time co-edited the *North American Review*. Taught history of art and literature at Harvard University, 1873–97. (*DAB*.)

Nott, Josiah Clark (1804–73). American physician and ethnologist. Practised medicine in Columbia, South Carolina, 1829–35; in Mobile, Alabama, 1835–57. Professor of anatomy, University of Louisiana, 1857–8. Helped to found the Medical College of Alabama in Mobile in 1858 and was appointed professor of surgery. Wrote extensively on scientific subjects including yellow fever, surgery, hypnotism, and ethnology. (*DAB*.)

Nuttall, Thomas (1786–1859). Botanist. Brought up a printer. Emigrated to the United States in 1807. Travelled in America from about 1811 to 1834; made a number of botanical discoveries. Curator of the botanic gardens and professor of natural history, Harvard University, 1825–34. Returned to England in 1842. Author of many works on North American plants. (R. Desmond 1977, *Modern English biography*.)

Ogilby, William (1808–73). Irish barrister and naturalist. Practised as a barrister in London, 1832–46, before settling in Ireland. Honorary secretary of the Zoological Society, 1839–46. (*Alum. Cantab.*, *Quarterly Journal of the Geological Society* 45 (1888–9): Proceedings, p. 46.)

Oken, Lorenz (1779–1851). German naturalist and leading exponent of *Naturphilosophie.* Professor of physiology, Munich, 1827–32; of natural history, Zürich University, 1832–51. (*ADB*, *DSB*, *EB*.)

Oliver, Daniel (1830–1916). Assistant in the herbarium of the Royal Botanic Gardens, Kew, 1858; keeper, 1864–90. Professor of botany, University College London, 1861–88. (R. Desmond 1977.)

11 September [1860]; 15 [September 1860]; *19 September 1860*; 21 [September 1860]; [22–23 September 1860]; 24 [September 1860]; *25 September 1860*; 27 [September 1860]; [29 September 1860]; 5 October [1860]; [10 October 1860]; 12 [October 1860]; 12 [October 1860]; 14 October [1860]; *[15–16 October 1860]*; 17 October [1860]; 20 October [1860]; *[before 23 October 1860]*; 23 [October 1860]; [31? October 1860]; 3 November [1860]; 7 November [1860]; 16 November [1860]; [21 November 1860]; *23 November 1860*; 20 December [1860]

Oliver, Francis Wall (1864–1951). Palaeobotanist and ecologist. Lecturer in botany, University College London, 1888; professor, 1890–1929. Professor of botany, Cairo University, 1929–35. Son of Daniel Oliver. FRS 1905. (R. Desmond 1977.)

Olmsted, Frederick Law (1822–1903). American landscape architect. Farmed on Staten Island, New York for several years; also pursued literary interests and travelled widely. Wrote extensively on slavery and the economic and social conditions of the American South. Appointed superintendent of Central Park, New York City, 1857; architect in chief, 1858. Superintendent, Frémont Mariposa mining estates, California, 1863–5. Reappointed landscape architect, Central Park, 1865, until his removal in 1878. Thereafter worked independently. (*DAB*.)

Omalius d'Halloy, Jean Baptiste Julien d' (1783–1875). Belgian geologist, ethnologist, and politician. Author of major early studies of the geology of northern France and Belgium. Defended the theory of organic change, believing that species change in response to changes in environment, in *Élements de géologie* (1831). (*BNB*, *DSB*.)

Orbigny, Alcide Charles Victor Dessalines d' (1802–57). French palaeontologist who travelled widely in South America, 1826–34. Professor of palaeontology, Muséum d'histoire naturelle, 1853. (*DSB*.)

Osborne, Christopher (d. 1860). Bricklayer in Down, Kent. (*Post Office directory of the six home counties* 1845.)

Owen, Richard (1804–92). Comparative anatomist. Assistant-conservator of the Hunterian Museum, Royal College of Surgeons, 1827; Hunterian professor, 1836–56. Superintendent of the natural history departments, British Museum, 1856–84. Described the *Beagle* fossil mammal specimens. FRS 1834. (*DNB*, *DSB*.)

Paley, William (1743–1805). Anglican clergyman and philosopher who propounded a popular system of natural theology. (*DNB*, *DSB*.)

Pallas, Pyotr Simon (1741–1811). German naturalist and geographer who travelled extensively in the Russian empire. Foreign member, Royal Society, 1763. (*ADB*, *DSB*.)

Palmer, Roundell, 1st Earl of Selborne (1812–95). Lawyer and politician. Conservative MP for Plymouth, 1847–57; for Richmond, Yorkshire, 1861–72. Attorney-general, 1863–6. Lord chancellor, 1872. Created Baron Selborne, 1872. FRS 1860. (*DNB*.)

Palmer, William (1824–56). The Rugeley poisoner. Practised medicine in Rugeley, Staffordshire, for several years from 1846. Became an owner and breeder of horses, soon getting into pecuniary difficulties. In 1855 charged with the murder by poisoning of John Parsons Cook, a betting man; subsequently it was found that Palmer's wife and brother, who had died previously, had also been poisoned. Palmer was found guilty of Cook's murder and hanged in 1856. (*DNB*.)

Pardon, James (fl. 1811–29). Artist. Of London, Canterbury (1814–25), and Yoxford, Suffolk (1828–9). Had several exhibitions at the Royal Academy. Specialised in portraits but also painted landscapes and miniatures. (Foskett 1972, Graves 1901.)

Parfitt, Edward (1820–93). Naturalist and gardener. His interest in botany and entomology intensified during an enforced stay in the Cape Colony following a shipwreck. Became gardener to John Milford, Exeter, in 1848. Curator to the Archaeological and Natural History Society of Somerset, 1859–61. Librarian to the Devon and Exeter Institute, 1861–93. (R. Desmond 1977, *DNB*.)

Parker, Henry (1828–92). Fellow, Oriel College, Oxford, 1851–85. CD's nephew. (*Alum. Oxon.*, *NUC*.)

Parker, John William (1792–1870). Publisher and printer. Became manager of the printing business of William Clowes the elder (1779–1847). Appointed superintendent of the Cambridge University Press, 1829. Established his own premises in 1832, and was appointed publisher to the Christian Knowledge Society. Printer to the University of Cambridge, 1836–54. His eldest son, John William Parker (b. 1820), entered the business in 1843 but died in 1860. (*DNB*.)

Parkyns, Mansfield (1823–94). Traveller. After visiting Constantinople and Cairo, travelled in Abyssinia, 1843–6, publishing a book describing his experiences. Attaché to the embassy at Constantinople, 1850–2. After settling in England, became official assignee in bankruptcy in 1858, first in Exeter then in London. Later, comptroller of the court of bankruptcy until 1884. (*DNB*.)

Parslow, Joseph (1809/10–98). CD's manservant at 12 Upper Gower Street *c.* 1840–2 and butler at Down House until 1875. (Freeman 1978.)

Parsons, Theophilus (1797–1882). American lawyer. Practised law in Taunton, then in Boston, Massachusetts, 1822–7. Dane professor of law, Harvard Law School, 1848–69. Became a prolific writer on legal subjects. (*DAB*.)

Pascoe, Francis Polkinghorne (1813–93). Entomologist. Surgeon in the Royal Navy until 1843; retired after inheriting property. Resided first in Cornwall, then, after the death of his wife, in London. There he devoted himself to science, forming a large entomological collection later given to the Natural History Museum, South Kensington. Although a believer in evolution, he persistently opposed the theory of natural selection. (R. Desmond 1977, *DNB*, Gilbert 1977.)

Pasteur, Louis (1822–95). French chemist and microbiologist. Professor of chemistry, science faculty, Strasbourg, 1849–54. Professor of chemistry and dean of the science faculty, Lille, 1854–7. Administrator and director of scientific studies, École normale, Paris, 1857–67; director of the laboratory of physiological chemistry, 1867–88. Professor of chemistry at the Sorbonne, 1867–74. Director of the Institut Pasteur, Paris, 1888–95. Famous for his work on fermentation and for experiments providing evidence against the theory of spontaneous generation. (*Dictionnaire universel des contemporains*, *DSB*.)

Patterson, Robert (1802–72). Naturalist and merchant in Belfast. Author of zoological works. A founder of the Natural History Society of Belfast, 1821; served as its president for many years. An active member of the British Association for the Advancement of Science. FRS 1859. (*DNB*.)
18 October 1860; 21 October [1860]

Pengelly, William (1812–94). Geologist. An expert on the geology of Devonshire; explored the plant-bearing deposit at Bovey Tracey and several caves including Brixham Cave and Kent's Hole at Torquay. FRS 1863. (*DNB*, Sarjeant 1980.)

Pertz, Georg Heinrich (1795–1876). German historian who held many public offices in Berlin. Married Leonora Horner in 1853. (*ADB*, *EB*.)

Pertz, Leonora (b. 1818). Daughter of Leonard Horner. Translator, with her sister Joanna B. Horner, of German works. Married Georg Heinrich Pertz in 1853. (*ADB*.)

Phillips, John (1800–74). Geologist. Keeper of the museum of the Yorkshire Philosophical Society, 1825–40. Assistant secretary of the British Association for the Advancement of Science, 1832–62. Professor of geology, King's College, London, 1834–40. Palaeontologist to the Geological Survey of Great Britain, 1840–4. Deputy reader in geology, Oxford University, 1853; professor, 1860–74. FRS 1834. (*DNB*, *DSB*.)
14 November [1860]

Phillips, William Walker. Resided at Down Hall farm. (*Post Office directory of the six home counties* 1859.)

Pictet de la Rive, François Jules (1809–72). Swiss zoologist and invertebrate palaeontologist. Professor of zoology, Geneva, from 1835. (Gilbert 1977, Sarjeant 1980.)
19 February 1860; 23 February [1860]; 1 April [1860]

Pistor, C. M. Wilhelm. German breeder. Author of a manual on pigeon and dove breeding.

Poitevin, Prosper (*c.* 1810–84). French grammarian. (*Dictionnaire universel des contemporains.*)

Pope, Alexander (1688–1744). Poet. (*DNB.*)

Pouchet, Félix Archimède (1800–72). French biologist and naturalist. Qualified in medicine in 1827. Soon afterwards, became director of the Muséum d'histoire naturelle in Rouen, where he also held teaching posts. Prolific author and populariser of science. Chiefly remembered as Louis Pasteur's adversary in the debate over spontaneous generation. (*Dictionnaire universel des contemporains, DSB.*)

Powell, Baden (1796–1860). Writer on theological topics and active in university reform. Savilian professor of geometry at Oxford, 1827–60. FRS 1824. (*DNB, DSB.*)
18 January [1860]; 18 January [1860]

Prado, Casiano de (1797–1866?). Spanish geologist. Undertook a geological survey of Spain from 1848. (Sarjeant 1980.)

Prestwich, Joseph (1812–96). Geologist. Proprietor of the family wine business in London. An expert on the Tertiary geology of Europe. Prominent in studies of early mankind. Professor of geology at Oxford, 1874–88. FRS 1853. (*DNB, DSB.*)
12 March [1860]

Price, Astley Paston (1826–86). Chemist. Became assistant to August Wilhelm Hofmann at the Royal College of Chemistry in 1845. Held an appointment at the School of Mines. Chemist to a Swansea silver works, 1851–7. From 1857, practised as a consulting chemist in London. (*Modern English biography.*)

Priestley, Joseph (1733–1804). Theologian and man of science. Educated for the dissenting ministry and spent most of his adult life as a teacher or preacher. Prolific writer on theology, history, education, metaphysics, and scientific subjects. Became the chief propagandist for Unitarian beliefs in England. Increasing signs of political persecution prompted his emigration to the United States in 1794. (*DNB, DSB.*)

Pritzel, Georg August (1815–74). Librarian and botanical writer. Gained an appointment at the Royal Library, Berlin, in 1851. In 1855, became archivist at the Royal Academy of Sciences, Berlin. His main contribution to botany comprised his bibliographical works. (*ADB.*)

Pugh, Miss. Governess of the Darwin children, 1856–7. Later became insane and was confined to an asylum. (Freeman 1978.)

Purnell, Jane Kirk. *See* Brewster, Jane Kirk.

Quatrefages de Bréau, Jean Louis Armand de (1810–92). French zoologist and anthropologist. Undertook a study of invertebrates and, in 1844, accompanied Henri Milne-Edwards on a zoological excursion off the coast of Sicily. Awarded the chair of natural history at the Lycée Henri IV, Paris, 1850; professor of the natural history of man, Muséum d'histoire naturelle, 1855. Foreign member, Royal Society, 1879. (*DSB*.)
21 January [1860]; 30 March [1860]

Raffeneau-Delile, Alire (1778–1850). French botanist. (*NUC*.)

Rafinesque, Constantine Samuel (1783–1840). Naturalist and archaeologist. Until 1814, called himself Rafinesque-Schmaltz, using his mother's maiden name. Lived in Palermo, Sicily, 1805–15, where for a time he was secretary and chancellor to the American consul. Emigrated to the United States in 1815. Professor of botany, natural history, and modern languages, Transylvania University, Kentucky, 1818–26. Wrote prolifically, but his writings were erratic and his descriptions of plants and fishes often inaccurate. (*DAB*, *DSB*.)

Ramsay, Andrew Crombie (1814–91). Geologist with the Geological Survey of Great Britain, 1841; senior director for England and Wales, 1862; director-general, 1871–81. Professor of geology, University College London, 1847–52; lecturer on geology at the Government School of Mines, 1852–71. Knighted, 1881. FRS 1849. (*DNB*, *DSB*.)
21 February 1860; 23 February [1860]

Réaumur, René Antoine Ferchault de (1683–1757). French mathematician, physicist, and natural historian. Invented the thermometer scale that bears his name. Foreign member, Royal Society, 1738. (*DSB*.)

Rede, Robert (d. 1519). Chief justice of the common pleas, 1509. Endowed the Rede lectures at Cambridge University. (*DNB*.)

Reed, George Varenne (1816–86). Curate of Hayes, Kent, 1837–9; of Tingewick, Buckinghamshire, 1839–54. Rector of Hayes, 1854–86. Tutor to George Howard, Francis, and Leonard Darwin. (*Alum. Cantab.*)
15 September [1860]; 12 November [1860]

Reeks, Trenham (1823/4–79). Mineralogist. Curator and librarian of the Museum of Practical Geology, 1851. (*Modern English biography*.)
15 November 1860

Reeve, Lovell Augustus (1814–65). Conchologist. Set up a shop in King William Street, Strand, for the sale of natural history specimens and the publication of conchological works. Founded the publishing company Reeve Brothers. (*DNB*, Sarjeant 1980.)

Reinsch, Edgar Hugo Emil (1809–84). German chemist. Teacher and headmaster at a technical school in Zweibrücken, 1842. Headmaster, school of business and agriculture, Erlangen, 1851–77. (*BHGW*.)

Reissek, Siegfried (1819–71). Austrian naturalist. Assistant keeper, Royal botanical collection, Vienna, 1845–67; appointed head keeper, 1867. From 1848, corresponding member of the Academy of sciences, Vienna. (*BLKO*.)

Richardson, John (1787–1865). Arctic explorer, physician, and naturalist. Physician to the Royal Hospital at Haslar, 1838. Inspector of hospitals, 1840. Commanded an expedition in search of John Franklin, 1847–9. FRS 1825. (*DNB*, Sarjeant 1980.)

Riedel, Wilhelm K. Pastor and school inspector in Pfuhl, Germany. Published a work on pigeon breeding.

Rodwell, John Medows (1808–1900). Clergyman and orientalist. Friend and contemporary of CD's at Cambridge. Rector of St Peter's, Saffron Hill, London, 1836–43; of St Ethelburga's, Bishopsgate, 1843–1900. Commenced oriental studies as a young man. Published an English version of the Koran in 1861. (*DNB*.)

31 October 1860; 5 November [1860]; *6 December 1860*

Rogers, Henry Darwin (1808–66). American naturalist. Professor of geology and mineralogy, University of Pennsylvania, 1835. Director of the Geological Survey of New Jersey, 1835; of Pennsylvania, 1836. Regius professor of natural history, University of Glasgow, 1855–66. (*DAB*.)

Rogers, William Barton (1804–82). American geologist. Professor of natural philosophy, University of Virginia, 1835–53. Appointed state geologist for the Geological Survey of Virginia, 1835. Established the Massachusetts Institute of Technology, 1861; president, 1862–70 and 1878–81. Brother of Henry Darwin Rogers. (*DAB*, *DSB*.)

Rolfe, Robert Monsey, Baron Cranworth (1790–1868). Lord chancellor, 1853–8, 1865–6. Created Baron Cranworth, 1850. Lived at Holwood House, near Keston, Kent. (*DNB*.)

15 August [1860]

Rolleston, George (1829–81). Physician. Fellow, Royal College of Physicians, 1859. A physician to the British civil hospital at Smyrna, 1855–6. Appointed Regius professor of physic, Oxford, 1857; Linacre professor of anatomy and physiology, 1860–81. (*DNB*.)

1 June [1860]; 6 June [1860]

Ross, James Clark (1800–62). Naval officer and polar explorer. Discovered the north magnetic pole in 1831. Commander of an expedition to the Antarctic, 1839–43; of a search expedition for John Franklin, 1848–9. Rear-admiral, 1856. FRS 1828. (*DNB*, *DSB*.)

Royle, John Forbes (1799–1858). Surgeon and naturalist in the service of the East India Company. Superintendent of the botanic garden in Saharanpur, India, 1823–31. Professor of materia medica, King's College, London, 1837. In charge of the East India House correspondence relating to vegetable productions, 1838. A founder of the Philosophical Club of the Royal Society, 1847. FRS 1837. (R. Desmond 1977, *DNB*.)

Sabine, Elizabeth Juliana (1807–79). Scientific translator. Married Edward Sabine (1788–1883) in 1826. Assisted her husband in his literary work. (*Modern English biography*.)

Saint-Hilaire, Augustin François César Prouvençal (Auguste de) (1779–1853). French naturalist. Surveyed the flora and fauna of Brazil, 1816–22. Appointed professor at the faculty of sciences, Paris, 1830. (*DSB*.)

Salt, Thomas. Partner in the Shrewsbury law firm of Dukes and Salt.

Salter, John William (1820–69). Geologist and palaeontologist. Apprenticed to James de Carle Sowerby (*DNB*), 1835. Assistant to Adam Sedgwick, 1842–6. Assistant to Edward Forbes on the Geological Survey of Great Britain, 1846; palaeontologist to the survey, 1854–63, where his duties included the preparation of specimens for exhibition in the Museum of Practical Geology. (*DNB*.)

Schaaffhausen, Hermann (1816–93). German physician, anthropologist, and vertebrate palaeontologist. Studied Quaternary mammals and fossil man. (*NUC*, Sarjeant 1980.)

Schiödte, Jörgen Matthias Christian (1815–84). Danish zoologist. (Gascoigne 1984.)

Schlagintweit, Adolph (1829–57). German geologist and explorer. Lecturer in geology, University of Munich, 1853. Accompanied his brothers on a scientific expedition to India and Tibet, during which he was put to death by the amir of Kashgar. (*ADB*, *EB*.)

Schlagintweit, Hermann Rudolph Alfred (1826–82). German geographer and explorer. Lecturer in physical geography at the University of Berlin, 1851–4. Undertook with his brothers a scientific expedition to India and Tibet, 1854–7. (*ADB*, *BHGW*.)

Schlagintweit, Robert (1833–85). German traveller and explorer. With his brothers Adolph and Hermann, undertook a scientific expedition, sponsored by the East India Company and supported by the Prussian government, to India and Tibet, 1854–7 (*ADB*, *BHGW*.)

Schwann, Theodor Ambrose Hubert (1810–82). German physiologist. On qualifying MD from Berlin in 1834, became one of Johannes Peter Müller's research assistants. Discovered pepsin and studied alcoholic fermentation. During this time, he was philosophically a Christian rationalist, but from 1839 abandoned rationalism for mysticism. Became professor of anatomy at Louvain, Belgium, in 1839. Professor of anatomy, Liège, 1848–58; of physiology, anatomy, and embryology, 1858–79. (*ADB*, *DSB*.)

Schweizerbart, Christian Friedrich (1805–79). Director of the E. Schweizerbart'sche Verlagshandlung of Stuttgart, 1841–67. Publisher of the German translation of *Origin* (1860) and of CD's later works. (Nägele 1926.)

Schweizerbart, Wilhelm Emanuel (1785–1870). Founded the E. Schweizerbart'sche Verlagshandlung in Stuttgart in 1826. Publisher of many German scientific works. Retired in 1841. (Nägele 1926.)

Sclater, Philip Lutley (1829–1913). Lawyer and ornithologist. Secretary of the Zoological Society, 1859–1903. FRS 1861. (*DSB*, Scherren 1905.)

[3? February 1860]; 4 February [1860]; *11 February 1860*; 14 February [1860]

Sebright, John Saunders, 7th Baronet (1767–1846). Whig politician and agriculturist. Published works on animal breeding. (*DNB*.)

Sedgwick, Adam (1785–1873). Geologist and clergyman. Woodwardian professor of geology, Cambridge University, 1818–73. Canon of Norwich, 1834–73. FRS 1821. (*DNB*, *DSB*.)

Selby, Prideaux John (1788–1867). Naturalist. High sheriff of Northumberland, 1823. Particularly interested in ornithology, forestry, and entomology. In 1837, helped to found the *Magazine of Zoology and Botany*, later known as the *Annals and Magazine of Natural History*, of which he was an editor. (R. Desmond 1977, *DNB*.)

Selwyn, Alfred Richard Cecil (1824–1902). Assistant geologist with the Geological Survey of Great Britain, 1845–52. Director of the Geological Survey of Victoria, Australia, 1853–69; of Canada, 1869–94. FRS 1874. (*DNB*.)

Sharpey, William (1802–80). Physiologist. Professor of anatomy and physiology at University College London, 1836–74. Secretary of the Royal Society, 1853–72. FRS 1839. (*DNB*, *DSB*.)
28 October [1860]

Sherard, Robert, Earl of Harborough (1798–1859). Styled Lord Sherard, 1799–1807. On his death the earldom became extinct. (*Complete peerage*.)

Sheridan, Richard Brinsley (1751–1816). Dramatist and parliamentary orator. (*DNB*.)

Sichel, Jules (1802–68). French ophthalmologist. Founded an ophthalmological clinic in Paris in 1836, which particularly served the poor. Published numerous papers on ophthalmology and was an accomplished linguist. Also studied entomology, his particular interest being the Hymenoptera. (*Annales de la Société de Lyon* 17 (1869): 383–410, *Dictionnaire universel des contemporains*, Gilbert 1977.)

Sidebotham, Joseph (1824–85). Calico printer in Manchester. Botanist and entomologist. One of the founders of the Manchester Field Naturalists' Society. Collected diatoms. (R. Desmond 1977, Gilbert 1977.)

Siebold, Karl Theodor Ernst von (1804–85). German invertebrate zoologist. Professor of zoology and comparative anatomy, University of Munich, 1853–83. Co-editor with Rudolf Albert von Kölliker of the *Zeitschrift für wissenschaftliche Zoologie*. Foreign member, Royal Society, 1858. (*DSB*, *NDB*.)

Silk, George. Lifelong friend of A. R. Wallace.

Silliman, Benjamin (1779–1864). American chemist, geologist, and mineralogist. Professor of chemistry and natural history, Yale University, 1802–53. Founder and first editor of the *American Journal of Science and Arts*, 1818. (*DAB*, *DSB*.)

Silliman, Benjamin, Jr (1816–85). American chemist. Teaching assistant to his father Benjamin Silliman, 1837. An editor of the *American Journal of Science and Arts*, 1838–85. Appointed professor of practical chemistry, Yale, 1846. Succeeded his father as professor of chemistry and natural history at Yale in 1853. From 1849 to 1854, served as professor of chemistry in the medical department, University of Louisville, Kentucky. (*DAB*, *DSB*.)
27 October 1860; 4 December [1860]

Simpson, Richard (1820–76). Biographer and Shakespearian scholar. Vicar, Mitcham, Surrey, 1844–5; then converted to the Roman Catholic faith. After several years of continental travel, returned to England, where he edited *The Rambler*, a liberal Catholic magazine, until 1862. In his later years, chiefly conducted research on Shakespeare. (*British authors.*)

Sims, John (1749–1831). Scottish physician and botanist who settled in London. Appointed physician to the Surrey dispensary and to the Charity for Delivering Married Women at their own Houses. Physician to Princess Charlotte. Edited *Annals of Botany*, 1804–6; *Botanical Magazine*, 1801–26. FRS 1814. (R. Desmond 1977, *Physicians.*)

Sismondi, Jessie de (1777–1853). Married Jean Charles Léonard Simonde de Sismondi in 1819. Emma Darwin's aunt. (*Emma Darwin.*)

Skipworth, Septimus Patrick (1839–1917). A friend of William Erasmus Darwin's at Cambridge. BA, Cambridge, 1861. Sometime surveyor and inspector for Caistor, Lincolnshire, until 1893. Farmed in Lincolnshire. (*Alum. Cantab.*)

Skirving, Robert Scot. Of Edinburgh. Author of several tracts on the labour question in Scotland. (*BLC, NUC.*)

[1860?]; *[1860?]*

Smith, Frederick (1805–79). Entomologist in the zoological department of the British Museum, 1849. Specialised in the Hymenoptera. (*Entomologist* 12 (1879): 89–92, Gilbert 1977.)

Smith, Frederick. Farmer of Thornhill farm, Hartfield, Sussex. (*Post Office directory of the six home counties* 1859.)

3 April 1860; 4 April [1860]

Smith, James (1782–1867). Scottish antiquarian, numismatist, and geologist. Made major early contributions to Quaternary geology; also wrote on the geology of Madeira, Gibraltar, and Portugal. (Sarjeant 1980.)

Smith, James Edward (1759–1828). Botanist. Founded the Linnean Society in 1788 and served as president until his death. FRS 1785. (*DNB, DSB.*)

Smith, Joshua Toulmin (1816–69). Geologist, publicist, and constitutional lawyer. Lived in the United States, 1837–42; lectured in Boston on phrenology and philosophy. Returned to England and was called to the bar, 1849. Studied geology in his spare time. First president of the Geologists' Association, 1859. Studied the fossils of the Chalk. (*DNB*, Sarjeant 1980.)

6 January 1860

Smith, William (1813–93). Lexicographer and classicist. Became a master at University College school in the 1830s. Published many classical grammars and text-books. (*DNB.*)

Smith, Elder, and Company. Partnership of George Smith (1789–1846) and Alexander Elder (1790–1876). Publishers and East India agents. (*DNB, Modern English biography.*)

17 February [1860]

Smyth, William Henry (1788–1865). Naval officer and scientific writer. A founder of the Royal Geographical Society, 1830; president, 1849–50. President of the Royal Astronomical Society, 1845–6. Admiral, 1863. FRS 1826. (*BHGW*, *DNB*.)

Snow, George (d. 1885). Carrier between Down and London. (Down churchyard, *Post Office directory of the six home counties* 1858.)

Sonder, Otto Wilhelm (1812–81). German botanist and pharmacist who specialised on the flora of South Africa. (*DSAB*.)

Sowerby, James (1757–1822). Naturalist and artist. Published several illustrated works on natural history. (*DNB*, *DSB*.)

Spencer, Herbert (1820–1903). Writer. Civil engineer on the railways, 1837–41 and 1844–6. Supported the complete suffrage movement and in 1844 became sub-editor of *The Pilot*, a newspaper devoted to the movement. Sub-editor of *The Economist*, 1848–53. From 1852, author of papers on evolution and numerous works on philosophy and the social sciences. (*DNB*, *DSB*.)
  2 February [1860]; *22 February 1860*; 23 [February 1860]

Spratt, Thomas Abel Brimage (1811–88). Vice-admiral, hydrographer, and author. Entered the navy in 1827, and served on several ships that surveyed the Mediterranean; retired from the navy in 1870. Combined his practical contributions to navigation with archaeological studies of the Greek Archipelago. FRS 1856. (*DNB*.)

Stainton, Henry Tibbats (1822–92). Entomologist. Founder and editor of the *Entomologists' Annual* (1855–74) and of the *Entomologist's Weekly Intelligencer*, 1856–61. FRS 1867. (*DNB*.)
  11 June [1860]; 20 June [1860]

Steetz, Joachim (1804–62). German botanist. (*NUC*.)

Stephens, Thomas Selwood (b. 1825). Clergyman. Curate of Wanstead, Essex, 1853–9; of Down, Kent, 1859–67. Rector of St Erme, Cornwall, 1867–1904. (*Alum. Oxon.*, *Crockford's*.)

Steudel, Ernst Gottlieb (1783–1856). German botanist. Established a medical practice in Esslingen, Württemberg, in 1806. Head physician for the district, 1828–56. Studied botany in his spare time, publishing numerous botanical works. (*ADB*.)

Stewardson, Thomas (1807–68). American physician and naturalist who lived in Philadelphia, Pennsylvania. For a time, physician to the Pennsylvania Hospital. (*CDEL*.)
  8 May 1860

Stirling, Waite Hockin (1829–1923). Assistant missionary on Keppel Island, West Falklands; bishop of Falkland Islands, 1869–1901. Secretary of the Patagonian Mission in London, 1857–62. (*Alum. Oxon.*, *Crockford's*.)

Strickland, Hugh Edwin (1811–53). Geologist and zoologist. An advocate of reform in zoological nomenclature. FRS 1852. (*DNB*, Sarjeant 1980.)

Strzelecki, Paul Edmund de (1797–1873). Polish-born explorer and geologist. Explored the Australian interior, 1839–40. Naturalised as a British subject in about 1850. FRS 1853. (*Aust. dict. biog.*, *DNB.*)

Štúr, Dionyz (1827–93). Hungarian stratigrapher, palaeobotanist, and vertebrate palaeontologist. Worked in Hungary, Rumania, and Yugoslavia. (*NUC*, Sarjeant 1980.)

Sturm, Jakob (1771–1848). Publisher and editor of illustrated works in natural history. Trained in the art of copperplate, and combined this skill with an interest in botany and entomology. Amassed a large entomological collection. (*ADB.*)

Sulivan, Bartholomew James (1810–90). Naval officer and hydrographer. Lieutenant in HMS *Beagle*, 1831–6. Surveyed the Falkland Islands in HMS *Arrow*, 1838–9. Commander of HMS *Philomel*, 1842–6. Resided in the Falkland Islands, 1848–51. Commanded HMS *Lightning* in the Baltic, 1854–5. Naval officer of the marine department of the Board of Trade, 1856–65. Admiral, 1877. (*DNB.*)

Swainson, William (1789–1855). Naturalist and illustrator who collected in Sicily and in Brazil. Follower of the quinary system of classification. Emigrated to New Zealand in 1837. FRS 1820. (*DNB*, *DSB.*)

Sykes, William Henry (1790–1872). Naturalist and military officer in the East India Company. Statistical reporter to the Bombay government, 1824–31. Royal commissioner on lunacy, 1835–45. Elected to the board of directors of the East India Company, 1840; chairman, 1856. FRS 1834. (R. Desmond 1977, *DNB.*)

Symonds, William Samuel (1818–87). Geologist, author, and clergyman. Curate of Offenham, near Evesham, 1843–5. Became rector of Pendock, Worcestershire, in 1845. Devoted his spare time to studying local archaeology and geology and that of Wales. In the 1870s, visited Auvergne and the Ardèche to search for traces of ancient glaciers. (*DNB*, Sarjeant 1980.)

Tait, Archibald Campbell (1811–82). Clergyman and educationist. Headmaster, Rugby School, 1842–50. Dean of Carlisle, 1850–6. Known as a university reformer, he was nominated a member of Oxford University commission in 1850. Bishop of London, 1856–69. Archbishop of Canterbury, 1869–82. (*DNB.*)

Talandier, Pierre Theodore Alfred (1822–90). Political exile from France, 1848. Master of French at the Royal Military College, Sandhurst, 1859; professor, 1868. Returned to France in 1870 and edited a radical periodical. Translated several English works into French. (*Dictionnaire des parlementaires Français*, Royal Military Academy Archives.)

Talleyrand-Périgord, Charles Maurice de (1754–1838). French statesman. (*NBU.*)

Taylor, Alfred Swaine (1806–80). Medical jurist. Professor of medical jurisprudence, Guy's Hospital, London, 1831–77. Appointed joint lecturer on chemistry at Guy's, 1832; held the post alone from 1851 to 1870. Often a witness in trials involving suspected murder by poisoning; engaged in the case of William Palmer, the Rugeley poisoner. (*DNB.*)

10 December 1860

Tegetmeier, William Bernhard (1816–1912). Editor, journalist, lecturer, and naturalist. Pigeon fancier and an expert on fowls. Secretary of the Apiarian Society of London. (Richardson 1916.)
 20 January [1860]; 24 [February? 1860]; 17 April [1860]; 8 May [1860]; 20 July [1860]; 30 July [1860]; 1 September [1860]

Temple, Frederick (1821–1902). Fellow and tutor of Balliol College, Oxford, 1842–8. Headmaster of Rugby School, 1857–69. Bishop of Exeter, 1869–85; of London, 1885–96. Archbishop of Canterbury, 1896–1902. (*DNB*.)

Tennent, James Emerson (1804–69). Traveller, politician, and author. Assumed the name Tennent on his marriage in 1832. MP for Belfast, 1832–45. Civil secretary to the colonial government of Ceylon, 1845–50. Appointed governor of St Helena in 1850, but never assumed office. FRS 1862. (*DNB*.)

Thomson, Charles Wyville (1830–82). Naturalist and oceanographer. Professor of mineralogy and geology, Belfast, 1854–60; of natural science, 1860–8. Professor of botany, Royal College of Science, Dublin, 1868–70. Appointed Regius professor of natural history, Edinburgh, 1870. Interested since 1860 in deep-sea researches, appointed chief of the civilian scientific staff of the *Challenger* expedition, 1872–6. (*DNB*, *DSB*.)

Thomson, Thomas (1817–78). East India Company surgeon and botanist in India. Collaborated with Joseph Dalton Hooker on various publications and joined him on his Himalayan expedition in 1849. Curator of the Asiatic Society's museum, Calcutta, 1840. Superintendent of the Calcutta botanic garden and professor of botany at the Calcutta Medical College, 1854–61. FRS 1855. (R. Desmond 1977, *DNB*.)

Thwaites, George Henry Kendrick (1811–82). Botanist and entomologist. Superintendent of the Peradeniya botanic gardens, Ceylon, 1849; director, 1857–80. FRS 1865. (R. Desmond 1977, *DNB*.)
 [14 February 1860]; 21 March [1860]; 20 October [1860]

Ticknor and Fields. Boston publishing company, founded in 1832 by William Davis Ticknor (1810–64). Soon joined by James Thomas Fields (1817–81), the company took the name of Ticknor and Fields in 1854. Published the *Atlantic Monthly* as well as the works of many leading writers of England and America. (*DAB*.)

Tollet, Ellen Harriet (d. 1890). Daughter of George Tollet (1767–1855). A close friend of the Wedgwood and Darwin families. (*Burke's landed gentry* 1846.)

Tollet, Georgina. Daughter of George Tollet (1767–1855). A close friend of the Wedgwood and Darwin families. Edited the manuscript of *Origin*. (Freeman 1978.)

Tomes, Robert Fisher (b. 1823). An authority on bats. Fellow of the Zoological Society of London. Lived in Welford, Warwickshire. (CD's Address book, *Post Office directory of Birmingham, Warwickshire (Staffordshire, and Worcestershire)* 1864.)

Tookey, Charles. Fellow of the Chemical Society. Worked at the metallurgical laboratory, Royal School of Mines, before joining the Royal Mint, Hong Kong. Later became assayer in the Japanese Imperial Mint.

Torrey, John (1796–1873). American botanist and chemist. Collected and described plants from various government explorations. Professor of chemistry at the College of Physicians and Surgeons, New York City, 1827–55. (*DAB*, *DSB*.)

Town, Daniel. Carpenter in Down. (*Post Office directory of the six home counties* 1859.)

Traill, Thomas Stewart (1781–1862). Practised medicine in Liverpool, 1803–32. Appointed professor of medical jurisprudence, Edinburgh University, 1832. Fellow, Royal College of Physicians, Edinburgh, 1833; president, 1852. (*DNB*.)

Treadwell, Daniel (1791–1871). American inventor. Trained as a silversmith but preferred experimenting with machinery. Patented a power printing press in 1826; by 1835, had secured several patents for his machines for spinning hemp. President, Boston Mechanics Institute, 1829. Appointed Rumford professor, Harvard University, 1834. (*DAB*.)

Trécul, Auguste Adolphe Lucien (1818–96). French hospital doctor who wrote many articles on botany. (*NUC*.)

Trimmer, Joshua (1795–1857). Geologist. Employed on the Geological Survey of England, 1846–54. Studied the relationship between geology and agriculture. (*DNB*, Flett 1937, Sarjeant 1980.)

Tristram, Henry Baker (1822–1906). Divine and naturalist. Went to Bermuda for health reasons in 1846; became secretary to the governor, 1847–9. Developed an interest in studying birds and shells. Rector of Castle Eden, Durham, 1849–60. Appointed canon residentiary of Durham, 1874. Travelled extensively, visiting Algeria, Palestine, Egypt, and Japan. Representative of the Church Missionary Society for many years. FRS 1868. (*DNB*.)

Trübner, Johann Nicolaus (1817–84). German-born publisher. Worked for Longman's, 1843–51. Entered into partnership with Thomas Delf and, later, David Nutt and established a successful publishing business. His main interests were philology, philosophy, religion, and oriental studies. Studied Sanskrit and Hebrew. (*ADB*, *DNB*.)
14 October [1860]

Tyndall, John (1820–93). Irish physicist, lecturer, and populariser of science. Professor of natural philosophy at the Royal Institution, London, 1853; superintendent, 1867–87. FRS 1852. (*DNB*, *DSB*.)

Unger, Franz (1800–70). Austrian botanist and palaeobotanist. Professor of plant anatomy and physiology, University of Vienna, 1849–66. (*ADB*, *DSB*.)

Venables, Edmund (1819–95). Antiquary and divine. Curate at Bonchurch, Isle of Wight, 1853–5, remaining on the Isle of Wight until the mid-1860s. Compiled a guide to the island, published in 1860. Moved to Lincoln, and was appointed to a prebendal stall in the cathedral, 1865; precentor and canon-residentiary, 1867. Wrote several guides to Lincoln. (*DNB*.)

Victoria, Queen of Great Britain and Empress of India (1819–1901). Succeeded to the throne on the death of her uncle, William IV, on 20 June 1837. (*DNB*.)

Vilmorin, Pierre Louis François Léveque de (1816–60). French botanist. President of the Paris seed firm Vilmorin-Andrieux et Cie from 1843. Published papers on the breeding of cereal grains, potatoes, sugar-beet, and flowers. (*DSB*, *NUC*.)

Virgil (Publius Vergilius Maro) (70–19 BC). Roman poet. (*EB*.)

Wagner, Rudolph (1805–64). German anatomist, physiologist, and anthropologist. Appointed ordinary professor of physiology, comparative anatomy, and natural history at Göttingen, 1840. (*ADB*, *DSB*.)

Wallace, Alfred Russel (1823–1913). Collector in the Amazon, 1848–52; in the Malay Archipelago, 1854–62. Independently formulated a theory of natural selection in 1858. Lecturer and author of works on protective coloration, mimicry, and zoogeography. Became a spiritualist in 1865 and was considered controversial because of his views on spiritualism, vaccination, land nationalisation, women's rights, and socialism. FRS 1893. (*DNB*, *DSB*.)

18 May 1860; *[December? 1860]*

Wallich, George Charles (1815–99). Surgeon and zoologist. Served with the Indian army, 1838–57. Joined HMS *Bulldog* as naturalist to the North Atlantic telegraphic expedition in 1860; discovered evidence of animal life at great oceanic depths and published his findings. Also published on the Protozoa. (*DNB*, *DSB*.)

12 December 1860; *14 December [1860]*

Wallis, William. Surgeon and registrar of births and deaths for the parishes of Hartfield and Withyam in Sussex. Collector of orchids. (*Post Office directory of the six home counties* 1859.)

Washington, John (1800–63). Naval officer and hydrographer. Secretary of the Royal Geographical Society, 1836–41. Involved in surveys of the east coast of England, 1841–7. Hydrographer to the admiralty, 1855–63. FRS 1845. (*DNB*.)

Waterhouse, George Robert (1810–88). Naturalist. A founder of the Entomological Society, 1833. Curator of the Zoological Society, 1836–43. Joined the mineralogical and geological branch of the natural history department of the British Museum in 1843; keeper, geology department, 1857–80. Described CD's mammalian and entomological specimens from the *Beagle* voyage. (*DNB*, Gilbert 1977.)

[February 1860]; 1 April [1860]

Watkins, Frederick (1808–88). Clergyman. Attended Shrewsbury School, 1823–6. A friend of CD's at Cambridge University. HM Inspector of Schools, 1841–73. Rector of Long Marston, Yorkshire, 1873–88. Archdeacon of York, 1874–88. (*Alum. Cantab.*)

30 July [1860]

Watson, Hewett Cottrell (1804–81). Botanist, phytogeographer, and phrenologist. Published various guides to the distribution of British plants. Edited the *Phrenological Journal*, 1837–40. Collected plants in the Azores, 1842. An early supporter of the idea of the progressive development of plant species. (*DNB*, *DSB*.)

[3? January 1860]; [5–11 January 1860]; *10 May 1860*

Watt, James (1736–1819). Engineer and chemist. Trained as a scientific instrument maker. Mathematical instrument maker to the University of Glasgow, 1757–66. Practised as a civil engineer in Glasgow, 1766–74. Moved to Birmingham in 1774 and, in partnership with Matthew Boulton, successfully commercialised his improved steam engine design. Retired from the business in 1800. (*DNB*, *DSB*.)

Way, Albert (1805–74). Antiquary and traveller. A friend of CD's at Cambridge: the two went on entomological outings together. Fellow, Society of Antiquaries, 1839; director, 1842–6. A founder of the Archaeological Institute, 1845. (*Alum. Cantab.*, *DNB*.)
7 April [1860]

Wedgwood, Amy (1835–1910). Daughter of Francis and Frances Mosley Wedgwood. Emma Darwin's niece. (Freeman 1978.)

Wedgwood, Caroline Sarah (1800–88). CD's sister. Married Josiah Wedgwood III, her cousin, in 1837. (*Darwin pedigree*.)

Wedgwood, Frances Mackintosh (Fanny Hensleigh) (1800–89). Married Hensleigh Wedgwood in 1832. (*Burke's peerage* 1980, Freeman 1978.)

Wedgwood, Frances Mosley (Fanny Frank) (1808–74). Married Francis Wedgwood in 1832. (*Burke's peerage* 1980.)

Wedgwood, Francis (Frank) (1800–88). Master-potter and partner in the Wedgwood pottery works at Etruria, Staffordshire, until 1876. Emma Darwin's brother. Married Frances Mosley in 1832. (*Alum. Cantab.*)

Wedgwood, Hensleigh (1803–91). Philologist and barrister. Fellow, Christ's College, Cambridge, 1829–30. Metropolitan police magistrate at Lambeth, 1832–7. Registrar of metropolitan carriages, 1838–49. An original member of the Philological Society, 1842. Published the first volume of his *Dictionary of English etymology* in 1859. Emma Darwin's brother. Married Frances Mackintosh (1800–89) in 1832. (*DNB*.)
[January? 1860]

Wedgwood, Sarah Elizabeth (Elizabeth) (1793–1880). Emma Darwin's sister. Resided at Maer Hall, Staffordshire, until 1847, then at The Ridge, Hartfield, Sussex, 1847–62. She moved to London before settling in Down in 1868. (*Burke's peerage* 1980, *Emma Darwin*.)
25 December [1860?]

Weir, Harrison William (1824–1906). Animal painter and illustrator. (*DNB*.)

Westwood, John Obadiah (1805–93). Entomologist and palaeographer. An active member of the Entomological Society. Hope professor of zoology, Oxford University, 1861–93. Entomological referee for the *Gardeners' Chronicle*. Awarded Royal Medal of the Royal Society, 1855. (*DNB*, Gilbert 1977.)
25 June [1860]; 9 [July 1860]

Whately, Richard (1787–1863). Principal of St Alban Hall, Oxford, 1825–9. Professor of political economy, Oxford, 1829–31. Archbishop of Dublin, 1831–63. (*DNB*.)

Whewell, William (1794–1866). Mathematician and historian and philosopher of science. Master of Trinity College, Cambridge, 1841–66. Professor of moral philosophy, Cambridge University, 1838–55. FRS 1820. (*DNB*, *DSB*.)
2 January 1860

White, Gilbert (1720–93). Naturalist and clergyman. Author of *The natural history and antiquities of Selborne* (1789). (*DNB*, *DSB*.)

White, Joseph Blanco (1775–1841). Spanish-born theological writer. Ordained priest, 1800. Left for England in 1810, as the French advanced into Spain. Published a monthly periodical called the *Español* for a time before qualifying as an English clergyman in 1814. Formed a close friendship with Richard Whately at Oxford, joining him in Dublin in 1832. (*DNB*.)

Wight, Robert (1796–1872). Botanist. Assistant surgeon, then naturalist, to the East India Company, 1829–31 and 1834–53. Published numerous works on the botany of India. FRS 1855. (R. Desmond 1977, *DNB*.)

Wilberforce, Samuel (1805–73). Clergyman. Rector of Brighstone, Isle of Wight, 1830–40; of Alverstoke, Hampshire, 1840–3. Appointed dean of Westminster, 1845. Chaplain to Prince Albert, 1841. Bishop of Oxford, 1845–69. Opposed CD's theory of evolution. (*DNB*, Sarjeant 1980.)

Williams, Edward Augustus. Surgeon of Bromley High Street. Member of the Royal College of Surgeons of England, 1824. (*London and provincial medical directory* 1860, *Post Office directory of the six home counties* 1859.)

Williams and Norgate. London booksellers and publishers who specialised in foreign scientific literature. The business was a partnership between Edmund Sydney Williams (1817–91) and Frederick Norgate. Publishers of the *Natural History Review*. (*Modern English biography*.)
16 January [1860]; 29 [January 1860]; 6 March [1860]; *29 March 1860*; 1 April [1860]

Wilson, William (1799–1871). Botanist. Articled to a firm of solicitors; intense work led to serious illness. Taking much outdoor exercise to recuperate, became interested in botany. From 1824, devoted himself entirely to botany. Became well known as a bryologist. (R. Desmond 1977, *DNB*.)

Wollaston, Thomas Vernon (1822–78). Entomologist and conchologist. Collected insects and shells on Madeira during visits to the country made for health reasons. (*DNB*, Gilbert 1977.)
[16 September 1860]

Wolstenholme, Joseph (1829–91). Mathematician. Fellow of Christ's College, Cambridge, 1852–69; tutor, 1869. Professor of mathematics at the Royal Indian Engineering College, Cooper's Hill, 1871–89. (*Alum. Cantab.*, *DNB*.)

Wood, Searles Valentine (1798–1880). Geologist. Practised as a solicitor in London, 1851–65. Retired from business after his partner's death. Studied the fossils of the East Anglian Crag, specialising in Eocene fossil Mollusca. (*DNB*, Sarjeant 1980.)

Woodward, Bernard Bolingbroke (1816–69). Older brother of Samuel Pickworth Woodward. As a young man became interested in the congregational church. Pastor of the independent church of Worwell-with-Harleston in Norfolk, 1843–8. Resigned his pastorate in order to devote himself solely to literature. Appointed librarian in ordinary to Queen Victoria, 1860. (*DNB*.)
6 March [1860]; 9 [July 1860]

Woodward, Samuel Pickworth (1821–65). Naturalist. Sub-curator, Geological Society, 1839–45. Appointed professor of geology and natural history at the Royal Agricultural College, Cirencester, 1845. First-class assistant in the department of geology and mineralogy, British Museum, 1848–65. (*DNB*, Sarjeant 1980.)
6 March [1860]; 9 [July 1869]

Wrangel, Ferdinand Petrovich (1796–1870). Russian navigator who explored northern waters. Director of the Russian American Company, 1840–9. Naval minister, 1855–7. (*GSE*.)

Wright, Charles (1811–85). American botanical collector. Botanist to the United States North Pacific Expedition, 1853–6. Investigated the botany of Cuba, 1856–67. (*DAB*.)

Wright, Chauncey (1830–75). American mathematician and philosopher. Recording secretary, American Academy of Arts and Sciences, Boston, 1863–70. Published the first of a series of philosophical essays in the *North American Review* in 1864. (*DAB*.)

Wright, Edward Perceval (1834–1910). Irish naturalist. Edited the *Natural History Review*, 1854–66. Appointed director of the university museum in Dublin, 1857. Lecturer in zoology, Trinity College, Dublin, 1858–68; professor of botany and keeper of the herbarium, 1869. Resigned the chair in 1904 due to ill health, but continued to superintend the herbarium. (*DNB*.)

Wright, Joseph (1734–97). Painter known as Wright of Derby. Known particularly for his use of strong effects of light and shade, his most famous paintings being those of figures illuminated by artificial (chiefly candle) light. A friend of Josiah Wedgwood I and Erasmus Darwin. (*DNB*.)

Wyman, Jeffries (1814–74). American anatomist and ethnologist. Curator of the Lowell Institute, Boston, 1840; lecturer, 1840–1 and 1848–9. Travelled in Europe, 1841–3. Professor of anatomy and physiology, Hampden-Sydney Medical College, Virginia, 1843–8. Hersey professor of anatomy, Harvard University, 1847–74. Collected in South America, 1857–8. Curator of the Peabody Museum of American Archaeology and Ethnology, Harvard, 1866–74. (*DAB*, *DSB*.)
[c. *15*] *September 1860*; 3 October [1860]; 3 December [1860]

Unidentified.
4 March [1860]; 25 [April 1860]; 13 [May 1860]

Bibliography of biographical sources

ADB: *Allgemeine deutsche Biographie*. Under the auspices of the Historical Commission of the Royal Academy of Sciences. 56 vols. Leipzig: Duncker and Humblot. 1875–1912.

Allan, Mea. 1967. *The Hookers of Kew, 1785–1911*. London: Michael Joseph.

Alum. Cantab.: *Alumni Cantabrigienses. A biographical list of all known students, graduates and holders of office at the University of Cambridge, from the earliest times to 1900*. Compiled by J. A. Venn. Part 2. From 1752 to 1900. 6 vols. Cambridge: Cambridge University Press. 1940–54.

Alum. Oxon.: *Alumni Oxonienses: the members of the University of Oxford, 1715–1886*. By Joseph Foster. 4 vols. Oxford. 1888.

Army list: *The army list*. London. 1815–81.

Aust. dict. biog.: *Australian dictionary of biography: 1788–1850; 1851–1890; 1891–1939*. Edited by Douglas Pike, Bede Nairn, Geoffrey Serle, and John Ritchie. 12 vols. and index. Melbourne: Melbourne University Press; London and New York: Cambridge University Press. 1966–91.

Barrett, Charles Raymond Booth. 1893. *The Trinity House of Deptford Strond*. London.

BHGW: *Biographisch-literarisches Handwörterbuch zur Geschichte der exacten Wissenschaften*. By Johann Christian Poggendorff. Vols. 1–4, Leipzig: Johann Ambrosius Barth. 1863–1904. Vol. 5, Leipzig and Berlin: Verlag Chemie. 1926.

BLC: *The British Library general catalogue of printed books to 1975*. 360 vols. and supplement (6 vols.). London. 1979–88.

BLKO: *Biographisches Lexicon des Kaiserthums Oesterreich, enthaltend die Lebensskizzen der denkwürdigen Personen, welche seit 1750 in den österreichischen Kronländern geboren wurden oder darin gelebt und gewirkt haben*. By Constant von Wurzbach. 60 vols. Vienna. 1856–90.

Blyakher, L. Y. 1982. *History of embryology in Russia from the middle of the eighteenth century to the middle of the nineteenth century*. Translated from the Russian by H. I. Youssef and B. A. Malek, with an introduction by Jane Maienschein. Washington, D.C.: Smithsonian Institution and the National Science Foundation.

BNB: *Biographie nationale publiée par l'Académie royale des sciences, des lettres et des beaux-arts de Belgique*. 28 vols. Brussels. 1866–1944.

Bridges, E. Lucas. 1948. *Uttermost part of the earth*. London.

British authors: *British authors of the nineteenth century*. Edited by Stanley J. Kunitz. New York. 1964.

Burke's landed gentry: *Burke's genealogical and heraldic history of the landed gentry*. By John Burke. 1–18 editions. London: Burke's Peerage Ltd. 1836–1972.

Burke's peerage: *Burke's peerage and baronetage*. 1–105 editions. London: Burke's Peerage (Genealogical Books). 1826–1980.

BWN: *Biographisch Woordenboek der Nederlanden, bevattende Levensbeschrijvingen van zoodanige Personen, die zich op eenigerlei wijze in ons Vaderland hebben vermaard gemaakt*. Edited by K. J. R. van Harderwijk and G. D. J. Schotel. 20 vols. Haarlem. 1852–77.

Catalogue of graduates, University of Dublin: *A catalogue of graduates who have proceeded to degrees in the University of Dublin, from the earliest recorded commencements to July, 1866, with supplement to December 16, 1868*. Dublin: Hodges, Smith, and Foster. 1869.

CDEL: *A critical dictionary of English literature and British and American authors, living and deceased.* By S. Austin Allibone. 3 vols. Philadelphia and London. 1859–71. Supplement by John Foster Kirk. 2 vols. Philadelphia and London. 1891.

Complete peerage: *The complete peerage of England Scotland Ireland Great Britain and the United Kingdom extant extinct or dormant.* By George Edward Cokayne. New edition, revised and much enlarged. Edited by the Hon. Vicary Gibbs and others. 12 vols. London: The St Catherine Press. 1910–59.

Crockford's: *The clerical directory, a biographical and statistical book of reference for facts relating to the clergy and the Church.* London. 1858. Continued as *Crockford's clerical directory etc.* London. 1860–.

Cundall, Frank. 1920. Richard Hill. *Journal of Negro History* 5: 37–44.

DAB: *Dictionary of American biography.* Under the auspices of the American Council of Learned Societies. 20 vols., index, and 7 supplements. New York: Charles Scribner's Sons; London: Oxford University Press. 1928–81.

Darwin pedigree: *Pedigree of the family of Darwin.* Compiled by H. Farnham Burke. Privately printed. 1888. Reprinted in Freeman, Richard Broke, *Darwin pedigrees.* London: printed for the author. 1984.

DBF: *Dictionnaire de biographie Française.* Under the direction of J. Balteau, M. Barroux, M. Prevost, and others. 17 vols. (A–Humann). Paris: Librairie Letouzey et Ané. 1933–89.

Desmond, Ray. 1977. *Dictionary of British and Irish botanists and horticulturists, including plant collectors and botanical artists.* 3d ed. London: Taylor and Francis.

Dictionnaire des parlementaires Français. Edited by Adolphe Robert, Edgar Bourloton, and Gaston Cougny. 5 vols. Paris. 1889–91.

Dictionnaire universel des contemporains: *Dictionnaire universel des contemporains contenant toutes les personnes notables de la France et des pays étrangers*. Edited by G. Vapereau. Paris. 1858. 3d ed., 1865. 4th ed., 1870. 5th ed., 1880.

DNB: *Dictionary of national biography.* Edited by Leslie Stephen and Sidney Lee. 63 vols. and 2 supplements (6 vols.). London: Smith, Elder, and Co. 1885–1912. *Dictionary of national biography 1912–85.* Edited by H. W. C. Davis, J. R. H. Weaver, and others. 8 vols. London: Oxford University Press. 1927–90.

DSAB: *Dictionary of South African biography.* Edited by W. J. de Kock, D. W. Krüger, and C. J. Beyers. Vol. 1, Cape Town: Nasionale Boekhandel Beperk; vols. 2–3, Cape Town: Tafelberg Publishers; vol. 4, Durban and Pretoria: Butterworth & Co. 1968–81.

DSB: *Dictionary of scientific biography.* Edited by Charles Coulston Gillispie. 14 vols., supplement, and index. New York: Charles Scribner's Sons. 1970–80.

Dupree, A. Hunter. 1959. *Asa Gray, 1810–1888.* Cambridge, Mass.: The Belknap Press of Harvard University.

East-India register and army list: *The East-India register and army list.* Compiled from the official returns, by permission of the Honourable the East-India Company. 16 vols. London. 1845–60.

EB: *Encyclopædia Britannica.* 11th ed. 29 vols. Cambridge: Cambridge University Press. 1910–11.

Elliot, Clark A. 1979. *Biographical dictionary of American science: the seventeenth through the nineteenth centuries.* Westport, Conn. and London: Greenwood Press.

Emma Darwin (1904): *Emma Darwin, wife of Charles Darwin: a century of family letters.* Edited by Henrietta Litchfield. 2 vols. Cambridge: privately printed by Cambridge University Press. 1904.

Emma Darwin: *Emma Darwin: a century of family letters, 1792–1896.* Edited by Henrietta Litchfield. Revised edition. 2 vols. London: John Murray. 1915.

Flett, John Smith. 1937. *The first hundred years of the Geological Survey of Great Britain.* London: His Majesty's Stationery Office.

Foskett, Daphne. 1972. *A dictionary of British miniature painters.* 2 vols. London: Faber and Faber.

Freeman, Richard Broke. 1978. *Charles Darwin: a companion.* Folkestone: W. Dawson & Sons; Hamden, Conn.: Archon Books, Shoe String Press.

Fulton, Robert. [1874–6]. *The illustrated book of pigeons. With standards for judging.* London, Paris & New York: Cassell Petter & Galpin.

Gascoigne, Robert Mortimer. 1984. *A historical catalogue of scientists and scientific books: from the earliest times to the close of the nineteenth century.* New York and London: Garland Publishing.

Gilbert, Pamela. 1977. *A compendium of the biographical literature on deceased entomologists.* London: British Museum (Natural History).

Graves, Algernon. 1901. *A dictionary of artists who have exhibited works in the principal London exhibitions from 1760 to 1893.* 3d ed., with additions and corrections. London: Henry Graves and Co.

GSE: *Great Soviet encyclopedia* (translation of 3d ed. of *Bol'shaia Sovetskaia entsiklopediia.* Moscow. 1970–9). 31 vols. New York: Macmillan; London: Collier Macmillan. 1973–82.

Harvard University catalogue: *Harvard University quinquennial catalogue of the officers and graduates 1636–1925.* Cambridge, Mass: Harvard University Press. 1925.

Huxley, Leonard, ed. 1900. *Life and letters of Thomas Henry Huxley.* 2 vols. London: Macmillan and Co.

IBN: *Index bio-bibliographicus notorum hominum.* Part C: *Corpus alphabeticum sectio generalis.* Edited by Jean-Pierre Lobies and Otto and Wolfram Zeller. 52 vols. (A–Demópoulos, Demetrios). Osnabrück: Biblio Verlag. 1974–91.

James, Frank A. J. L., ed. 1987. *Chemistry and theology in mid-Victorian London: the diary of Herbert McLeod, 1860–1870.* London and New York: Mansell.

Larousse XX: *Larousse du XX*[e] siècle. Under the direction of Paul Augé. 6 vols. Paris: Librairie Larousse. 1928–33.

London and provincial medical directory: *London and provincial medical directory, and general medical register*. London. 1848–60.

MDCB: *The Macmillan dictionary of Canadian biography*. Edited by W. Stewart Wallace. 4th ed., revised, enlarged, and updated by W. A. McKay. Toronto: Macmillan of Canada. 1978.

Medical directory: *The Medical directory . . ., including the London and provincial medical directory, the medical directory for Scotland, the medical directory for Ireland, etc*. London. 1870–.

Medical directory for Ireland: *The medical directory and general medical register for Ireland*. London. 1852–60.

Modern English biography: *Modern English biography containing many thousand concise memoirs of persons who have died since the year 1850*. By Frederic Boase. 3 vols. and supplement (3 vols.). Truro: printed for the author. 1892–1921.

Moore, James R. 1985. Darwin of Down: the evolutionist as squarson-naturalist. In Kohn, David, ed., *The Darwinian heritage*. Princeton: Princeton University Press in association with Nova Pacifica.

Nägele, Erwin, ed. 1926. *Jubiläums-Katalog der E. Schweizerbart'schen Verlagshandlung*. Stuttgart: E. Schweizerbart.

Navy list: *The navy list*. London. 1815–70.

NBU: *Nouvelle biographie universelle*. Edited by Jean Chrétien Ferdinand Hoefer. 46 vols. Paris. 1852–66.

NDB: *Neue deutsche Biographie*. Under the auspices of the Historical Commission of the Bavarian Academy of Sciences. 16 vols. (A–Melanchthon). Berlin: Duncker and Humblot. 1953–90.

NDBA: *Nuevo Diccionario Biografico Argentino (1750–1930)*. Edited by Vicente Osvaldo Cutolo. 7 vols. Buenos Aires: Editorial Elche. 1968–85.

NUC: *The national union catalog*. Pre-1956 imprints. 685 vols. and supplement (vols. 686–754). London and Chicago: Mansell. 1968–81.

Physicians: *The roll of the Royal College of Physicians of London*. Vol. 3: 1801–25. By William Munk. 2d ed., revised and enlarged. London. 1878. Vol. 4: *Munk's roll, vol. 4: Lives of the fellows of the Royal College of Physicians of London 1826–1925*. Compiled by G. H. Brown. London: Royal College of Physicians. 1955.

Plarr, Victor Gustave. 1930. *Plarr's lives of the fellows of the Royal College of Surgeons of England*. Revised by Sir D'Arcy Power. 2 vols. London: Simpkin Marshall

Post Office directory of Birmingham, Warwickshire (Staffordshire, and Worcestershire). London. 1845–.

Post Office directory of Cheshire. London. 1857–.

Post Office directory of Gloucestershire, with Bath and Bristol, Herefordshire, and Shropshire. London. 1856–.

Post Office directory of Monmouthshire and . . . South Wales. London. 1871–.

Post Office directory of the six home counties, viz., Essex, Herts, Kent, Middlesex, Surrey and Sussex. London. 1845–.

Post Office London directory: *Post Office London directory*. London. 1802–.

Pritchard, Michael. 1986. *A directory of London photographers, 1841–1908.* Watford: ALLM Books.

Richardson, Edmund William. 1916. *A veteran naturalist; being the life and work of W. B. Tegetmeier.* London: Witherby & Co.

Roll of graduates of the University of Glasgow: *A roll of the graduates of the University of Glasgow from 31st December, 1727 to 31st December, 1897.* Compiled by W. Innes Addison. Glasgow: James MacLehose & Sons. 1898.

Rugby School register: *Rugby School register.* Vol. 1 from April 1675 to October 1857. Revised and annotated for the Old Rugbeian Society by Godfrey A. Solly. Rugby: George Over. 1933.

Sarjeant, William A. S. 1980. *Geologists and the history of geology: an international bibliography from the origins to 1978.* 5 vols. Supplement (2 vols). 1987. London: Macmillan.

SBL: *Svenskt biografiskt lexikon.* Edited by Bertil Boëthius, Erik Grill, Birgitta Lager-Kromnow, and Göran Nilzén. 26 vols. (A–Nilsson). Stockholm: Albert Bonnier and P. A. Norstedt. 1918–89.

Scherren, Henry. 1905. *The Zoological Society of London. A sketch of its foundation and development and the story of its farm, museum, gardens, menagerie and library.* London, Paris, New York, and Melbourne: Cassell & Co.

Steinberg, S. H. ed. 1953. *Cassell's Encyclopaedia of Literature.* 2 vols. London: Cassell and Company.

Taxonomic literature: *Taxonomic literature. A selective guide to botanical publications and collections with dates, commentaries and types.* 2d ed. By Frans A. Stafleu and Richard S. Cowan. 7 vols. Utrecht: Bohn, Scheltema & Holkema. The Hague: dr. W. Junk b.v., Publishers. 1976–8.

Wedgwood, Barbara and Wedgwood, Hensleigh. 1980. *The Wedgwood circle, 1730–1897: four generations of a family and their friends.* London: Studio Vista.

Wellesley index: *The Wellesley index to Victorian periodicals 1824–1900.* Edited by Walter E. Houghton and Jean Harris Slingerland. 4 vols. and epitome and index (1989). Toronto: University of Toronto Press; London: Routledge and Kegan Paul. 1966–87.

Whitaker's Almanack: *An Almanack.* By Joseph Whitaker. London. 1869–.

WWW: *Who was who.* A companion to *Who's who.* 7 vols. and cumulated index, 1897–1980. London: Adam & Charles Black. 1920–81.

INDEX

The dates of letters to and from Darwin's correspondents are listed in the Biographical register and index to correspondents and are not repeated here. Darwin's works are indexed under the short titles used throughout this volume and listed in the bibliography.

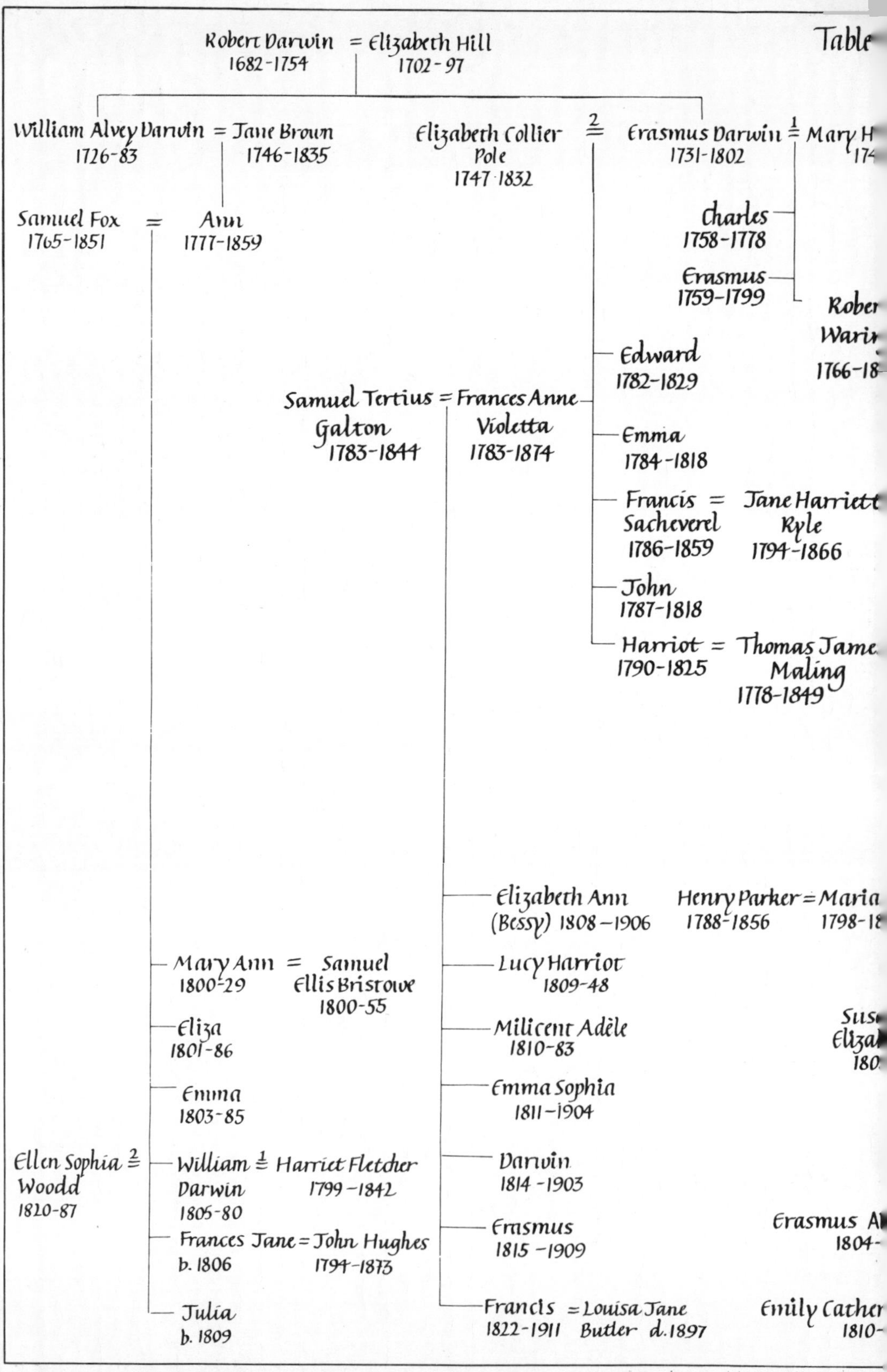
Table
Robert Darwin = Elizabeth Hill
1682-1754
1702-97
William Alvey Darwin = Jane Brown
1726-83
1746-1835
Elizabeth Collier Pole
1747-1832
2
Erasmus Darwin
1731-1802
1
Mary H
Samuel Fox = Ann
1765-1851
1777-1859
Charles
1758-1778
Erasmus
1759-1799
Rober
Warir
1766-18
Edward
1782-1829
Samuel Tertius Galton = Frances Anne Violetta
1783-1844
1783-1874
Emma
1784-1818
Francis Sacheverel = Jane Harriett Ryle
1786-1859
1794-1866
John
1787-1818
Harriot = Thomas James Maling
1790-1825
1778-1849
Elizabeth Ann (Bessy) 1808-1906
Henry Parker = Maria
1788-1856
1798-18
Mary Ann = Samuel Ellis Bristowe
1800-29
1800-55
Lucy Harriot
1809-48
Eliza
1801-86
Milicent Adèle
1810-83
Emma
1803-85
Emma Sophia
1811-1904
Ellen Sophia Woodd
1820-87
2
William Darwin
1805-80
1
Harriet Fletcher
1799-1842
Darwin
1814-1903
Frances Jane = John Hughes
b. 1806
1794-1873
Erasmus
1815-1909
Erasmus A
1804-
Julia
b. 1809
Francis = Louisa Jane Butler d. 1897
1822-1911
Emily Cather
1810-